HANDBOOK OF COMPUTING AND STATISTICS
WITH APPLICATIONS
VOLUME 1

Handbook of Computing and Statistics with Applications

VOLUME 1

Series Editor
E.J. Kontoghiorghes

Handbook of Latent Variable and Related Models

Edited by

Sik-Yum Lee
Department of Statistics
The Chinese University of Hong Kong
Shatin, N.T., Hong Kong

ELSEVIER
Amsterdam • Boston • Heidelberg • London • New York • Oxford
Paris • San Diego • San Francisco • Singapore • Sydney • Tokyo
N·H

North-Holland is an imprint of Elsevier
Radarweg 29, PO Box 211, 1000 AE Amsterdam, The Netherlands
The Boulevard, Langford Lane, Kidlington, Oxford OX5 1GB, UK

First edition 2007

Library of Congress Cataloging-in-Publication Data
A catalog record for this book is available from the Library of Congress

British Library Cataloguing in Publication Data
A catalogue record for this book is available from the British Library

ISBN-13: 978-0-444-52044-9
ISBN-10: 0-444-52044-9
Series ISSN: 1871-0301

For information on all North-Holland publications
visit our website at books.elsevier.com

Transferred to Digital Printing in 2010

Handbook Series on Computing and Statistics with Applications

The series will publish high quality volumes at the interface of computing and statistics. Of particular interest are handbooks in important statistical applications areas where both computing techniques and numerical methods have a major impact. The aim is twofold: first, to bring together research results in computational statistics that are scattered throughout publications in specialized areas; second, to provide scientists with reference books and unrivaled sources of information about the most recent developments in computational statistics and applications. Emphasis will be given to computational methods with computational statisticians being the primary target readership.

The series focuses on all computational aspects of statistics. The scope is broad enough to include handbooks in areas of computing which have a major impact on statistical techniques and methods of data analysis. All aspects of statistics which make use, directly or indirectly, of computing are considered. Applications of computational statistics in diverse disciplines will be strongly represented. These areas include, but are not limited to, economics, medicine and epidemiology, biology, finance, physics, chemistry, climatology and communication.

Erricos John Kontoghiorghes

Preface

Latent variable models (LVMs), including but not limited to the factor analysis model and structural equation models (SEMs), are very useful for investigating the relationships among observed and latent variables. Historically, the factor analysis model is the most basic LVM which was developed by psychometricians to test hypotheses about organization of mental ability. Nowadays, this model still represents a powerful multivariate method, and has very wide applications in substantive research. An important development of SEM is due to Karl Jöreskog who integrated the confirmatory factor analysis and the simultaneous equation model within the LISREL model. Mainly due to his LISREL program, and other user friendly software such as EQS 6.0 and Mx, SEMs have been extensively applied not only to behavioral, educational, social and psychological research, but also to environmental, biological, and medical sciences in recent years.

The exciting field of LVMs provides plenty of interesting research topics for psychometricians, statisticians, and quantitative social scientists from a variety of disciplines. The field has had a very rapid and healthy growth in recent years, both in methodological developments and in practical applications. To achieve new results researchers have used various approaches to establish new models, theories, and computing methods. This Handbook is intended to provide a comprehensive overview of some recent developments of latent variable and related models to researchers from a wide variety of disciplines including biology, business, economics, education, medicine, psychology, public health, and sociology.

From a model perspective, this Handbook includes a class important models, ranging from the most basic but general covariance and/or moment structures; to models with specific formulations that include the factor analysis model, linear SEMs, nonlinear SEM, multisample SEM, multilevel SEM, mixture of SEMs, as well as the normal mixed effect models, generalized linear mixed effect models, and spatial mixed models. For the above mentioned models, many interesting and important topics are discussed. These not only include the fundamental issues of estimation of parameters, goodness-of-fit assessments of the hypothesized models, hypothesis testing, and model comparison; but also cover factor rotation, reliability, and some basic properties in relation to the factor analysis model (and/or covariance structural models); robustness of inferences for covariance and moment structures in relation to the normal theory; selection of manifest variables; meta-analysis; and local influence analysis. To address these problems, the authors used new derivations or new theoretical and computational methodologies, and/or novel applications of the existing tools.

Essentially, most of the chapters used two major approaches. One approach focused on the sample covariance matrix (or its related statistics), and can be regarded as the commonly known covariance structural analysis. The other approach focused on specific formulations of the models with raw observations, and used maximum likelihood and/or its related methods, as well as some Bayesian methods, for statistical analysis. Results given in the chapters were obtained through a wide variety of techniques, including matrix methods; asymptotic and large sampling theories; geometrical concepts such as the conformal normal curvature; commonly used numerical methods such as adaptive quadrature; powerful tools in statistical computing, including the Expectation-Maximization (EM) algorithm and its related algorithms such as the Monte Carlo EM algorithm and the stochastic approximation EM algorithm, and Markov chain Monte Carlo (MCMC) methods such as the Gibbs sampler and the Metropolis–Hastings algorithm, bridge sampling, and path sampling; and finally the user friendly software EQS 6.0, gllamn, Mx, and WinBUGS.

As expected, the rich class of models and statistical methods described in this Handbook provide efficient and powerful tools for analyzing a wide spectrum of different kinds of complex data. In addition to the common continuous normal data, the nonstandard data that can be analyzed by these tools include arbitrary non-normal data, binary data, dichotomous and ordered categorical (ordinal) data, ranking data, missing data with ignorable and non-ignorable missing mechanisms, hierarchical data, and heterogeneous data. The analyses of these complex data are illustrated by examples with real data sets from business, education, medicine, public health, and sociology.

This Handbook owes much to many people. I am most thankful to Professor Erricos Kontoghiorghes, who initiated the idea to produce a Handbook of Latent Variable and Related Models, and gave valuable suggestions. I owe a great debt to all the contributors for their generous support and hard work in contributing the excellent chapters. These chapters will greatly enhance knowledge of the theoretical and computational techniques in factor analysis, SEMs, and LVMs. Without their contributions, this Handbook would not exist. All chapters have been reviewed, either by the authors of the other chapters, or by some experts in the field, including Asim Ansari, Douglas G. Bonett, Michael M. Browne, Chib-ping Chou, Masansori Ichikawa, Man-lai Tang, Bo-cheng Wei, and Qiwei Yao. I wish to express my deepest thanks to all reviewers who read the chapters and made constructive comments for revision. Finally, I am grateful to all the wonderful people on the editorial staff, particularly Andy Deelen and Sweitze Roffel of Elsevier Science B.V. for their continued assistance, encouragement, and support of our work.

Sik-Yum Lee

About the Authors

Peter M. Bentler is Distinguished Professor of Psychology and Statistics at the University of California, Los Angeles. A developer of the EQS Structural Equations Program, he has been an elected president of the Society of Multivariate Experimental Psychology (SMEP), the Psychometric Society, and the Division of Evaluation, Measurement, and Statistics of the American Psychological Association. In 1996, he received the Distinguished Scientific Contributions Award from the latter society, and in 2005, the Sells Award for Outstanding Career Contributions to Multivariate Experimental Psychology from SMEP.

Mortaza (Mori) Jamshidian is Professor of Statistics and Mathematics in the Department of Mathematics at the California State University, Fullerton. He received his Ph.D. in Applied Mathematics under Robert I. Jennrich from the University of California, Los Angeles in 1988. His previous positions include Senior Statistician at BMDP Statistical Software, and Assistant and Associate Professor at University of Central Florida, Orlando. His research interests include computational statistics, psychometrics, biometrics, analysis of incomplete data, simultaneous inference, EM algorithm, and linear and nonlinear regression modeling.

Matthew Mata is currently enrolled in the Masters program for Applied Mathematics at California State University, Fullerton and will graduate in the summer of 2006. He graduated summa cum laude with a Bachelor of Arts degree in Applied Mathematics from California State University, Fullerton in 2004. He plans to begin the Ph.D. program for Mathematics at University of California, Los Angeles in the fall of 2006. He has spent the last year working as a research assistant for Dr. Mortaza Jamshidian in the field of missing data.

Robert I. Jennrich is Professor Emeritus in the Department of Mathematics at the University of California at Los Angeles. He designed many of the components of the original BMDP statistical software system, is a former chair of the Statistical Computing Section of the American Statistical Association, is a Fellow of the Institute of Mathematical Statistics and a Fellow of the American Statistical Association. His research interests are in statistical computing, nonlinear regression, and psychometric methods. He has published a book entitled *An Introduction to Computational Statistics: Regression Analysis.*

Yutaka Kano is Professor of Statistics in the Graduate School of Engineering Science in Osaka University in Japan. He is an elected member of the International Statistical Institute, Past Editor of *Behaviormetrika*, an official journal of the Behaviormetric Society of Japan, and currently a Member of Editorial Board of *Psychometrika* and *Journal of Multivariate Analysis* among others. His research interests include mathematical statistics and multivariate analysis, particularly statistical methodology in psychometrics such as factor analysis and structural equation modeling.

Sik-Yum Lee is a Professor of Statistics in the Department of Statistics at the Chinese University of Hong Kong. He received a distinguished service award from the International Chinese Statistical Association in 1993, is a Former President of the Hong Kong Statistical Society, is an elected member of the International Statistical Institute and a Fellow of the American Statistical Association. His research interests are in structural equation models, latent variable models, and statistical diagnostics. He has published a book entitled *Structural Equation Modelling: A Bayesian Approach.*

Lu Bin is an Associate Professor in the School of Finance at the Nanjing University of Economics and Finance. His research interests are in structural equation models, latent variable models, statistical diagnostics, statistical computing, and semiparametric Bayesian method.

Fernand Mac-Moune Lai, MD (France), FRCPA (Australia), FHKCPath (Hong Kong), FHKAM (Hong Kong), MScFI (United Kingdom). He is Professor in the Department of Anatomical & Cellular Pathology, of the Faculty of Medicine at the Chinese University of Hong Kong. His field of interest covers the investigation of many diseases of the kidney, focusing on those with clinical and epidemiological relevance to the region of South-East Asia and China, with most publications in the area of renal pathology and investigation.

Xin-Yuan Song is an Assistant Professor in the Department of Statistics at the Chinese University of Hong Kong. Her research interests are in structural equation models, latent variable models, statistical computing, and statistical diagnostics.

Martin Knott is a Senior Lecturer in Statistics at the London School of Economics and Political Science. His research interests are in latent variable models, multivariate distributions and optimization. He is co-author with David Bartholomew of a book entitled *Latent Variable Models and Factor Analysis.*

Dimitris Mavridis is a Ph.D. student in the Department of Statistics at the Athens University of Economics and Business, Greece. His research interests are on goodness-of-fit measures and on identifying extreme observations/response patterns using latent variable models for categorical and metric manifest variables.

Irini Moustaki is an Associate Professor in Statistics at the Athens University of Economics and Business, Greece. Her research interest are latent variable models, missing values, categorical data and issues related to goodness-of-fit tests and contaminated data. She is co-author of a book titled *The Analysis and Interpretation of Multivariate Data for Social Scientists* and co-Editor of a book titled *Latent Variable and Latent Structure Models*. She is Associate Editor for *Structural Equation Modelling, Computational Statistics and Data Analysis and Methodology: European Journal of Research Methods for the Behavioral and Social Sciences*.

Jesus Palomo is currently Assistant Professor at the Department of Statistics and Operations Research, Rey Juan Carlos University, Spain. Previously, he was Postdoctoral Fellow at the National Institute of Environmental Health and the Statistical and Applied Mathematical Sciences Institute. Dr. Palomo's primary areas of research are Bayesian statistics applied to decision sciences and multivariate analysis. He is the Year 2005 recipient of the Young Statistician Award given by the European Network for Business and Industrial Statistics. The International Society for Bayesian Analysis awarded him the Honorable mention for the 2005 Savage Award for Bayesian dissertations in the Applications section. He participated as a postdoc in the Latent Variable Models in the Social Sciences program and worked under Dr. Bollen and Dr. Dunson, where this work was developed.

Kenneth A. Bollen is Director of the Odum Institute for Research in Social Science and the H.R. Immerwahr Distinguished Professor of Sociology and Statistics at the University of North Carolina at Chapel Hill. Bollen's primary areas of statistical research are structural equation models and latent curve models. He is the Year 2000 recipient of the Paul F. Lazarsfeld Memorial Award for Distinguished Contributions in the Field of Sociological Methodology by the American Sociological Association (Methodology Section). The ISI named him among the World's Most Cited Authors in the Social Sciences. He is coauthor of *Latent Curve Models: A Structural Equations Approach* (with P. Curran) and author of *Structural Equation Models with Latent Variables* and of over 100 papers. Bollen has continuously taught courses and workshops on structural equation models at ICPSR, UNC, and elsewhere since 1980.

David B. Dunson is Senior Investigator in the Biostatistics Branch of the National Institute of Environmental Health Sciences of the U.S. National Institutes of Health. Dr. Dunson is also Adjunct Associate Professor of Statistics in the Institute of Statistics and Decision Sciences at Duke University.

His primary areas of statistical research focus on Bayesian methods for correlated and multivariate data, with a particular emphasis on latent variables and nonparametric methods. He is also actively involved in collaborations in reproductive, environmental and genetic epidemiology. He is on the editorial boards of *Biometrics* and the *Journal of the American Statistical Association*, and has published widely in the statistics and epidemiologic literature, with over 80 papers published or in press.

Wai-Yin Poon is a Professor in the Department of Statistics at The Chinese University of Hong Kong. She has been the Associate Dean of Science (Education) since 2004, and is dedicated to promote science education. Her research interests mainly focus on structural equation modeling, categorical data, ranking data, and influence analysis.

Sophia Rabe-Hesketh is Professor at the Graduate School of Education and the Graduate Group in Biostatistics at the University of California, Berkeley. She is also Chair of Social Statistics at the Institute of Education, University of London. Her research interests are in applied statistics and in developing the Generalized Linear Latent and Mixed (GLLAMM) modeling framework that unifies and extends multilevel and latent variable models. She has written the gllamm software to estimate models within the GLLAMM framework. Rabe-Hesketh is co-Editor of *Statistical Methods in Medical Research*, Associate Editor for *Psychometrika*, *Statistical Modelling*, and *The Stata Journal*, and a member of the editorial board of *Structural Equation Modeling*. She has co-authored five books, including *Generalized Latent Variable Modeling: Multilevel, Longitudinal and Structural Equation Models* published by Chapman and Hall/CRC in 2004.

Anders Skrondal is Professor of Statistics and Chair in Social Statistics in the Department of Statistics and Professor of Research Methodology at the Methodology Institute at the London School of Economics. He is also Senior Biostatistician in the Division of Epidemiology, Norwegian Institute of Public Health. His research interests include topics in biostatistics, social statistics, econometrics, and psychometrics. Recently, Skrondal has concentrated on the development of the Generalized Linear Latent and Mixed Model (GLLAMM) framework. Outcomes of this project include papers published in (bio)statistical, psychometric and econometric journals as well as two bestselling books: *Generalized Latent Variable Modeling: Multilevel, Longitudinal and Structural Equation Models* published by Chapman and Hall/CRC in 2004 and *Multilevel and Longitudinal Modeling using Stata* published by Stata Press in 2005. Skrondal is also involved in numerous collaborative projects within medical, social and behavioural research. He is currently co-Editor of *Statistical Methods in Medical Research*.

Xiaohui Zheng is a Ph.D. student in the Graduate School of Education at the University of California, Berkeley. She has a master's degree in Statistics from Michigan State University. Her research interests include latent variable models and multilevel measurement models. She is currently a graduate student researcher at the Berkeley Evaluation and Assessment Research (BEAR) Center.

Alexander Shapiro is a Professor in the School of Industrial and Systems Engineering at Georgia Institute of Technology. His areas of interest include: stochastic programming, multivariate statistical analysis, simulation-based optimization of stochastic systems, nondifferentiable optimization. He has more than 100 refereed journal publications and is co-author of books: *Discrete Event Systems: Sensitivity Analysis and Stochastic Optimization by the Score Function Method*, John Wiley and Sons, New York, 1993, and *Perturbation Analysis of Optimization Problems*, Springer, New York, 2000. He is co-Editor in Chief of the journal *Mathematical Methods of Operations Research*, and on editorial board of journals: *Mathematics of Operations Research, Mathematical*

Programming, *ESAIM: Control, Optimization and Calculus of Variations*, *Encyclopedia of Statistical Sciences*, *Journal of Multivariate Analysis* (1998–2002).

Jian Qing Shi is Senior Lecturer in Statistics in the School of Mathematics & Statistics at the University of Newcastle Upon Tyne in the United Kingdom. He is a Fellow of the Royal Statistical Society. His research interests are in nonparametric curve fitting and prediction, meta-analysis, covariance structural analysis and nonlinear system control.

Jos ten Berge is Professor of Differential Psychology and Psychometrics at the University of Groningen, and is Director of the Heymans Institute of Psychological Research. His research topics include factor analysis, reliability theory, and theory and methods of three-way component analysis. He published a monograph on *Least Squares Optimization in Multivariate Analysis*.

Gregor Sočan studied psychology at the University of Ljubljana and received his doctorate from the University of Groningen. He is an Assistant Professor at the Department of Psychology of the University of Ljubljana. His research interests include factor analysis, reliability assessment and speed of information-processing.

Melanie M. Wall is Associate Professor in the Division of Biostatistics within the School of Public Health at the University of Minnesota. Her research interests are in statistical methods for latent variable modeling and extending traditional latent variable models (e.g., to include nonlinearities, to include spatial structure, to include both categorical and continuous latent variables) making them more attractive to a variety of researchers. In particular, she works on applying latent variable models to answer research questions relevant for behavioral public health.

Yasuo Amemiya is Manager, Statistical Analysis & Forecasting at the IBM Thomas J. Watson Research Center. He is an Elected Fellow of the American Statistical Association, and has served on the editorial boards of various statistical journals. Prior to joining IBM, he was Professor of Statistics at Iowa State University. His research areas include multivariate statistical analysis. For latent variable modeling, he was a pioneer in developing asymptotic robustness theory and in initiating research on modern nonlinear model analysis.

Haruo Yanai is currently Professor of Statistics at St. Luke's College of Nursing, Tokyo, Japan. He received a Ph.D. in Psychological Statistics (1972) and a Ph.D. in Medical Statistics (1982), both from the University of Tokyo. From 1986 to 2006, he was Professor at the Research Division of the National Center for University Entrance Examinations in Tokyo. He has published over twenty books in Japanese and two in English, and a number of papers in the areas of multivariate analysis, linear algebra with special emphasis on generalized inverse matrices and projection matrices, data analysis in psychology, entrance examinations, epidemiology, and other behavioral sciences. Dr. Yanai was one of the founders of the Behaviormetric Society of Japan, and has been an Associate Editor of *Psychometrika* since 1992.

Yoshio Takane is Professor of Psychology, McGill University, Montreal, Canada, which he has held sine 1977. He received a D.L. (Doctor of Letters) in Psychology in 1976 from the University of Tokyo, and a Ph.D. in Quantitative Psychology in 1977 from the University of North Carolina at Chapel Hill. He has published extensively in multivariate analysis and related areas (http://takane.brinkster.net/Yoshio/). He is a Past President of the Psychometric Society. His recent interests lie in the development of methods for structured analysis of multivariate data, and artificial neural network simulations.

Ke-Hai Yuan is William J. and Dorothy K. O'Neill III Associate Professor at the University of Notre Dame. His research interests are in the areas of psychometric theory and applied multivariate statistics. He has developed many reliable procedures for structural equation models and multilevel models with non-normal data, missing data and data containing outliers. He received the Raymond B. Cattell award from the Society of Multivariate Experimental Psychology in 2002 and is a recipient of the 2005–2006 James McKeen Cattell Fund Fellowship.

Hongtu Zhu is an Associate Professor in the Department of Biostatistics at the University of North Carolina at Chapel Hill. His research interests include structural equation models, statistics computing, longitudinal data analysis, and neuroimaging statistics as they are applied to the study of neuropsychiatric disorders.

Bradley S. Peterson is the Suzanne Crosby Murphy Professor in Pediatric Neuropsychiatry and Director of MRI Research in the Department of Psychiatry at Columbia University and the New York State Psychiatric Institute. His research interests include brain imaging statistics and the pathophysiology of childhood neuropsychiatric disorders.

Minggao Gu is a Professor in the Department of Statistics, The Chinese University of Hong Kong. His research interests include survival analysis, Monte Carlo method, semiparametric method, biostatistics, and ranking data and risk management.

Faming Liang is an Associate Professor in the Department of Statistics, Texas A&M University. His research focus is on Markov chain Monte Carlo, machine learning, bioinformatics, and computational physics. Dr. Liang has developed several advanced Monte Carlo algorithms, including dynamic importance sampling, evolutionary Monte Carlo, and stochastic approximation Monte Carlo, which have wide applications in statistics and bioinformatics.

Contributors

Amemiya, Yasuo, *IBM T.J. Watson Research Center, Room 33-222, 1101 Kitchawan Road, Route 134, Yorktown Heights, NY 10598, USA*; *e-mail: yasuo@us.ibm.com* (Ch. 15).

Bentler, Peter M., *University of California, UCLA Psychology Department, P.O. Box 951563, Los Angeles, CA 90095-1563, USA*; *e-mail: bentler@ucla.edu* (Chs. 1, 17).

Bollen, Ken, *University of North Carolina at Chapel Hill, Department of Sociology, 220 Hamilton Hall, Chapel Hill, NC 27599, USA*; *e-mail: bollen@unc.edu* (Ch. 8).

Dunson, David B., *National Institute of Environmental Health Sciences, Biostatistics Branch, MD A3-03, U.S. National Institutes of Health, P.O. Box 12233, Research Triangle Park, NC 27709, USA*; *e-mail: dunson1@niehs.nih.gov* (Ch. 8).

Gu, Minggao, *The Chinese University of Hong Kong, Department of Statistics, Shatin, Hong Kong, NT, China*; *e-mail: minggao@cuhk.edu.hk* (Ch. 18).

Jamshidian, Mortaza, *California State University, Fullerton, Department of Mathematics, 800 N. State College Blvd., Fullerton, CA 92834, USA*; *e-mail: mori@fullerton.edu* (Ch. 2).

Jennrich, R.I., *UCLA, Department of Mathematics, 405 Hilgard Avenue, Los Angeles, CA 90024-1554, USA*; *e-mail: rij@math.ucla.edu* (Ch. 3).

Kano, Yutaka, *Osaka University, Graduate School of Engineering Science, Toyonaka, Osaka 560-8531, Japan*; *e-mail: kano@sigmath.es.osaka-u.ac.jp* (Ch. 4).

Knott, Martin, *London School of Economics & Political Science, Department of Statistics, London, UK*; *e-mail: m.knott@lse.ac.uk* (Ch. 7).

Lai, Fernand Mac-Moune, *Department of Anatomical and Cellular Pathology, The Chinese University of Hong Kong, Shatin, NT, Hong Kong, China* (Ch. 6).

Lee, Sik-Yum, *The Chinese University of Hong Kong, Department of Statistics, Shatin, NT, Hong Kong, China*; *e-mail: sylee@sparc2.sta.cuhk.hk* (Preface, Chs. 5, 6).

Liang, Faming, *Texas A&M University, Department of Statistics, 3143 TAMU, College Station, TX 77843-3143, USA*; *email: fliang@stat.tamu.edu* (Ch. 18).

Lu, Bin, *School of Finance, Nanjing University of Finance and Economics, Nanjing, Jiangsu 210003, P.R. China* (Ch. 6).

Mata, Matthew, *California State University, Fullerton, Department of Mathematics, 800 N. State College Blvd., Fullerton, CA 92834, USA*; *e-mail: matthew.mata@verizon.net* (Ch. 2).

Mavridis, Dimitris, *Athens University of Economics and Business, Department of Statistics, 76 Patission Street, 104 34 Athens, Greece*; *e-mail: maurid@aueb.gr* (Ch. 7).

Moustaki, I., *Athens University of Economics and Business, Department of Statistics, 76 Patission Street, 104 34 Athens, Greece*; *e-mail: moustaki@aueb.gr* (Ch. 7).

Palomo, Jesus, *Statistics and Applied Mathematical, Sciences Institute (SAMSI), P.O. Box 14006, RTP, NC 27709-4006, USA*; *e-mail: jpalomo@escet.urjc.es* (Ch. 8).

Peterson, Bradley S., *Columbia University, Department of Psychiatry, and the New York State Psychiatric Institute, Department of Child Psychiatry, 1051 Riverside Drive, Unit 74, New York, NY 10032, USA*; *email: petersob@childpsych.columbia.edu* (Ch. 18).

Poon, Wai-Yin, *The Chinese University of Hong Kong, Department of Statistics, Shatin, NT, Hong Kong, China*; *e-mail: wypoon@sta.cuhk.edu.hk* (Ch. 9).

Rabe-Hesketh, Sophia, *University of California, Graduate School of Education, 3659 Tolman Hall, Berkeley, CA 94720-1670, USA*; *e-mail: sophiarh@berkeley.edu* (Ch. 10).

Shapiro, Alexander, *Georgia Institute of Technology, School of Industrial and Systems Engineering, Atlanta, GA 30332, USA*; *e-mail: ashapiro@isye.gatech.edu* (Ch. 11)

Shi, Jian Qing, *University of Newcastle, School of Mathematics and Statistics, Newcastle, UK*; *e-mail: j.q.shi@ncl.ac.uk* (Ch. 12).

Skrondal, Anders, *Department of Statistics, London School of Economics, Houghton Street, London WC2A 2AE, UK*; *e-mail: a.skrondal@lse.ac.uk* (Ch. 10).

Sočan, Gregor, *University of Ljubljana, Slovenia*; *e-mail: gregor.socan@ff.uni-lj.si* (Ch. 14).

Song, Xin-Yuan, *The Chinese University of Hong Kong, Department of Statistics, Shatin, NT, Hong Kong, China*; *e-mail: xysong@u4000.sta.cuhk.edu.hk* (Chs. 6, 13).

Takane, Yoshio, *McGill University, Department of Psychology, Stewart Biological Sciences Bldg., Room N7/3, Montreal, Canada*; *e-mail: takane@takane2.psych.mcgill.ca* (Ch. 16).

Ten Berge, Jos M.F., *University of Groningen, Department of Psychology, Grote Kruisstraat 2/1, 9712 TS Groningen, The Netherlands*; *e-mail: j.m.f.ten.berge@rug.nl* (Ch. 14).

Wall, Melanie M., *University of Minnesota, School of Public Health, A460 Mayo Building, 420 Delaware Street S.E., Minneapolis, MN 55455, USA*; *e-mail: melanie@biostat.umn.edu* (Ch. 15).

Yanai, Haruo, *St. Luke's College of Nursing, 10-1, Akashi-cho, Chuo-ku, Tokyo 104-0044, Japan*; *e-mail: kyanai@slcn.ac.jp* (Ch. 16).

Yuan, Ke-Hai, *105 Haggar Hall, Notre Dame, IN 46556, USA*; *e-mail: kyuan@nd.edu* (Ch. 17).

Zheng, Xiaohui, *Tolman Hall, University of California, Graduate School of Education, Berkeley, CA 94720-1670, USA*; *zhengx@berkeley.edu* (Ch. 10).

Zhu, Hongtu, *University of North Carolina at Chapel Hill, Department of Biostatistics, Chapel Hill, NC 27599-7420, USA*; *e-mail: hzhu@bios.unc.edu* (Ch. 18).

Table of contents

Handbook of Computing and Statistics with Applications, Vol. 1
ISSN: 1871-0301

DOI: 10.1016/S1871-0301(06)01001-8

Covariance Structure Models for Maximal Reliability of Unit-Weighted Composites

Peter M. Bentler

Abstract

When developing or evaluating scales, the internal consistency reliability of the scale based on its items or parts is always an important issue. The growth of structural modeling has allowed the easy computation of model-based estimates of reliability. These are typically touted as replacements for coefficient alpha, which remains the most widely used measure of internal consistency. Among model-based estimates, coefficients based on a 1-factor model have been most widely recommended. However, when the 1-factor model does not fit the data, the meaning of such a coefficient is unclear. A new identification condition for factor analytic models is proposed that assures the composite can be modeled with only one common factor even if the components are multidimensional. This common factor is maximally correlated with the composite, and the reliability of the composite is the maximal internal consistency coefficient for a unit-weighted composite. The coefficient also describes k-factor reliability, the greatest lower bound to reliability, and reliability for any composite from a latent variable model with additive errors. Reliability coefficients for differentially-weighted composites are also described, and differentially-weighted maximal reliability is contrasted with unit-weighted maximal reliability. Computational methods for these coefficients are described.

Structural equation models, especially factor analytic covariance structure models, provide a new way to think about a very old problem, the assessment of the internal consistency of a composite. Although Spearman (1904, 1907) had invented factor analysis and reliability as separate methodologies, internal consistency reliability can be viewed as a special case of covariance structure analysis with an added concern for composite scores. Composite scores or scale scores are frequently used in psychology and related social and behavioral sciences. A composite variable is a sum of other variables. In the typical case, a composite X is a simple sum of p unit-weighted components such as $X = X_1 + X_2 + \cdots + X_p$. Our primary discussion in this chapter emphasizes unit-weighted composites, but in concluding sections we also discuss differentially-weighted composites such as $Y = w_1X_1 + w_2X_2 + \cdots + w_pX_p$. Examples of composites include the total score on a test composed of items, an attitude score based on summed responses to a survey, and so on. An internal consistency reliability coefficient describes the quality of the composite or scale in terms of hypothesized constituents of the components X_i. These might represent true and error parts based on classical test theory ($X_i = T_i + E_i$), common and unique parts based on common factor analysis ($X_i = C_i + U_i$), or the loading of the component on its factor plus residual error ($X_i = \lambda_i F + E_i$). Covariance structure analysis provides models for the decomposition of the variables used in defining the reliability of the composite.

By far the most widely used measure of internal consistency is Cronbach's (1951) coefficient α (Hogan et al., 2000). In the population, it is defined as

$$\alpha = \frac{p}{p-1}\left(1 - \frac{\mathbf{1}'D\mathbf{1}}{\mathbf{1}'\Sigma\mathbf{1}}\right),$$

where D is the diagonal of the covariance matrix Σ of the components X_i, and $\mathbf{1}$ is a column vector of unit elements which serves as a summing vector. Thus $\mathbf{1}'D\mathbf{1}$ is the sum of the variances of the p component variables, and $\mathbf{1}'\Sigma\mathbf{1}$, the sum of all the elements of the p by p covariance matrix, is the variance of the total score X. In practice, α is estimated by substituting the sample covariance matrix S in place of Σ, yielding what we might call $\hat{\alpha}$. The popularity of this coefficient stems from several facts: it can easily be computed, it is available in many program packages as a default, it can be applied without fitting or validating any specific model to the components X_i, and, importantly, under appropriate conditions it is a lower bound to reliability $\alpha \leqslant \rho_{xx}$ (see, e.g., Novick and Lewis, 1967). The latter property arises if the variables have a decomposition $X_i = T_i + E_i$, where T_i and E_i are uncorrelated with covariance matrices Σ_T and diagonal Ψ_E, so that the covariance matrix is decomposed into two orthogonal parts $\Sigma = \Sigma_T + \Psi_E$. Then the composite has a similar decomposition $X = T + E$ where $T = \sum_1^p T_i$, $E = \sum_1^p E_i$, and the reliability of the composite is defined as the ratio of $\mathrm{var}(T)/\mathrm{var}(X)$, or

$$\rho_{xx} = \frac{\sigma_T^2}{\sigma_X^2} = \frac{\mathbf{1}'\Sigma_T\mathbf{1}}{\mathbf{1}'\Sigma\mathbf{1}} = 1 - \frac{\mathbf{1}'\Psi_E\mathbf{1}}{\mathbf{1}'\Sigma\mathbf{1}}.$$

There are many good recent discussions of α, its problems, and its alternatives (e.g., Barchard and Hakstian, 1997; Becker, 2000; Bonett, 2003; Enders, 2003; Enders and Bandalos, 1999; Feldt and Charter, 2003; Green, 2003; Green and Hershberger, 2000; Hakstian and Barchard, 2000; Komaroff, 1997; Miller, 1995; Osburn, 2000; Raykov, 1997, 1998, 2001, 2004a; Raykov and Shrout, 2002; Schmidt et al., 2003; Schmitt, 1996; Shevlin et al., 2000; Vautier and Jmel, 2003; Zinbarg et al., 2005). For the purposes of this chapter, two issues are important. First, the lower-bound property $\alpha \leqslant \rho_{xx}$ has been questioned. When correlated errors are present so that Ψ_E is not diagonal, α can exceed ρ_{xx}. Second, the size of α provides no information on the degree of unidimensional reliability, sometimes called homogeneity, that is, on the proportion of total variance due only to the main or only underlying common true score factor. In order to deal with both of these problems, the recent theoretical literature has suggested abandoning coefficient α and using a coefficient based on a 1-factor covariance structure model matrix of the parts that make up the composite.

In this approach, the covariance matrix of the true scores is presumed to be unidimensional, that is $\Sigma_T = \lambda\lambda'$, where $\lambda(p \times 1)$ is the factor loading vector of the p variables on a single common factor. Hence the covariance matrix of the observed scores is decomposed as

$$\Sigma = \lambda\lambda' + \Psi_u,$$

where Ψ_u is the covariance matrix of the unique variables or residual errors. Then

$$\rho_{11} = \frac{\sigma_T^2}{\sigma_X^2} = \frac{\mathbf{1}'\lambda\lambda'\mathbf{1}}{\mathbf{1}'\Sigma\mathbf{1}} = \frac{(\mathbf{1}'\lambda)^2}{\mathbf{1}'\Sigma\mathbf{1}} = \frac{\left(\sum_1^p \lambda_i\right)^2}{\mathbf{1}'\Sigma\mathbf{1}} = 1 - \frac{\mathbf{1}'\Psi_u\mathbf{1}}{\mathbf{1}'\Sigma\mathbf{1}}$$

defines reliability ρ_{11} ($\leqslant \rho_{xx}$) based on the hypothesis of a unidimensional latent variable (see, e.g., Jöreskog, 1971, p. 112). Zinbarg et al. (2005) equate ρ_{11} with McDonald's (1985) ω_H. Typically, Ψ_u is taken to be a diagonal matrix representing the hypothesis of uncorrelated error components, but in some circumstances correlated errors may be hypothesized (see Bollen, 1980). In practice, of course, ρ_{11} is not operational. In order to estimate ρ_{11}, the model $\Sigma = \lambda\lambda' + \Psi_u$ is fit to a sample covariance matrix S, and estimates $\hat{\lambda}$ and $\widehat{\Psi}_u$ are obtained. These are plugged into the defining formula, yielding $\hat{\rho}_{11}$. The approach also provides important information about the contribution of a given component variable to reliability via the factor loading $\hat{\lambda}_i$. Recent discussions of this approach are given by Kano and Azuma (2003) and Raykov (2004a).

Although ρ_{11} is certainly an improvement over α, it has a serious and fundamental flaw that has been largely overlooked. In realistic applications of covariance structure analysis, especially with a large number of variables X_i (e.g., $p = 40$) such as might be used in a reliability study, the null hypothesis $\Sigma = \lambda\lambda' + \Psi_u$ of a single common factor may hardly be tenable. ten Berge and Sočan (2004, p. 613) made the stronger point that the unidimensional hypothesis "will invariably be rejected when there are more than three test parts". If this null hypothesis is rejected, it is hard to know what ρ_{11} describes. We would argue that if $\Sigma \neq \lambda\lambda' + \Psi_u$, estimating $\hat{\rho}_{11}$ based on an incorrect 1-factor model is inappropriate. A similar point of view was espoused by McDonald (1999, p. 89) who indicated that if the 1-factor fit is poor "... we should not be using the coefficient anyway". In this chapter, we propose an extension of factor-based reliability so that it yields an appropriate coefficient of unidimensional internal consistency for all covariance matrices that can be fit by a general covariance structure model with additive errors. Among these, the exploratory factor analytic (EFA) model is the prototype. We show that among coefficients based on unit-weighted composites, our coefficient gives the largest reliability. This result is then applied to the case of differentially-weighted composites.

1. Proposed identification condition for factor models

Suppose that an exploratory factor analytic model for a random set of scores

$$x = \mu + \Lambda\xi + \varepsilon \tag{1}$$

holds in the population. Under the usual assumptions that the means μ are unstructured, ξ and ε are uncorrelated, the covariance matrix of ξ is I, and the covariance matrix of the ε is given by a diagonal matrix Ψ, the covariance matrix of $(x - \mu)$ has the structure of the standard covariance structure model

$$\Sigma = \Lambda\Lambda' + \Psi. \tag{2}$$

Here we allow the factor loading matrix Λ to be $(p \times k)$, where the number of factors $k \leqslant (p - 1)$ can be any appropriate number. When standard approaches to estimation of factor models are used, k has to be small enough so that there are positive degrees of freedom when fitting the model to a sample covariance matrix S. We will call this the "small-k" situation. In such a case, it is well known that without further restrictions this model is not identified (e.g., Jöreskog, 1967). Based on the partition of the factor loading matrix into $\Lambda = [\lambda|\bar{\Lambda}]$, where λ is $(p \times 1)$ and $\bar{\Lambda}$ is $(p \times (k - 1))$, we propose the following identification conditions:

(1) λ contains unrestricted free parameters.
(2) $\mathbf{1}'\bar{\Lambda} = 0$, that is, the $k - 1$ columns of $\bar{\Lambda}$ sum to zero.
(3) $\bar{\Lambda}$ contains free parameters subject to $(k - 1)(k - 2)/2$ restrictions. Some examples of such restrictions are:
    (a) $\bar{\Lambda}'\Psi^{-1}\bar{\Lambda}$ is diagonal. This is similar to the standard identification condition in exploratory maximum likelihood factor analysis, where it is based on all k factors.
    (b) $\bar{\Lambda}$ contains free parameters except for a triangle of fixed zero elements. A simple way is to fix $\bar{\Lambda}_{ij} = 0$ if $j > i$. To illustrate, if $k = 5$, $\bar{\Lambda}$ is $p \times 4$, and the first 4 rows (out of p) are given as $\begin{bmatrix} * & 0 & 0 & 0 \\ * & * & 0 & 0 \\ * & * & * & 0 \\ * & * & * & * \end{bmatrix}$, where "*" represents a free parameter and "0" is a fixed zero. An advantage of this approach is that any structural equation modeling program can be used to estimate the free parameters of the model.
    (c) Rotational criteria are imposed on $\bar{\Lambda}$ so that it is in some simple structure form. If oblique transformations are considered, the defining model may contain correlated factors, that is, $\Sigma = \Lambda\Phi\Lambda' + \Psi$, where Φ is the covariance matrix of the factors. In this approach, the factor corresponding to λ remains uncorrelated with the remaining factors.

It follows from the above that the number of identification conditions imposed on the model is $(k-1)+(k-1)(k-2)/2$, which equals $k(k-1)/2$, the precise number imposed on the standard exploratory factor analysis model. Thus the proposed representation is simply an alternative form of the exploratory factor model.

However, the identification condition is defined more generally. It also holds under conditions where the number of factors generates negative degrees of freedom, i.e., exceeds the Ledermann (1937) bound of $0.5(2p + 1 - \sqrt{8p + 1})$ in the standard exploratory factor model. We will call this the "large-k" situation, which may require a number of factors near p. Such a large number of factors does not occur in ordinary exploratory factor analysis, but it occurs in such contexts as minimum trace factor analysis (e.g., Bentler, 1972; Shapiro, 1982a; Shapiro and ten Berge, 2000), constrained minimum trace factor analysis (e.g., Bentler and Woodward, 1980, 1983; ten Berge et al., 1981; Shapiro, 1982a), or minimum rank factor analysis (e.g., della Riccia and Shapiro, 1982; ten Berge and Kiers, 1991; Shapiro and ten Berge, 2002). See also ten Berge (2000). In the large-k situation, identification condition three is not necessary since its primary purpose is to enable standard exploratory factor analytic estimation with positive degrees of freedom.

2. Reliability based on proposed parameterization

Under our model, the total score X is obtained as

$$X = \mathbf{1}'x = \mathbf{1}'\mu + \mathbf{1}'\Lambda\xi + \mathbf{1}'\varepsilon.$$

But identification condition two has the consequence that

$$\mathbf{1}'\Lambda = \mathbf{1}'[\lambda|\bar{\Lambda}] = [\mathbf{1}'\lambda|\mathbf{1}'\bar{\Lambda}] = [\mathbf{1}'\lambda|0] = \left[\sum_{1}^{p}\lambda_i|0\right] = [\lambda_X|0].$$

This, in turn, has the consequence that the composite score has a factor analytic decomposition such that it is solely a function of the first factor ξ_1

$$X = \mu_X + \lambda_X\xi_1 + \varepsilon_X, \tag{3}$$

where $\mu_X = \mathbf{1}'\mu$ and $\varepsilon_X = \mathbf{1}'\varepsilon$. Thus, with $T = \lambda_X\xi_1$ being unidimensional, and $E = \varepsilon_X$, the true and error scores are uncorrelated, and we obtain the unidimensional internal consistency coefficient

$$\rho_{kk} = \frac{\sigma_T^2}{\sigma_X^2} = \frac{\lambda_X^2}{\lambda_X^2 + \sigma_{\varepsilon X}^2}. \tag{4}$$

Some algebra verifies that the reliability coefficient ρ_{kk} represents the squared correlation between X and ξ_1, or, stated differently, ξ_1 is the factor that has the highest correlation with the composite X.[1] This correlation is reduced if there are negative factor loadings since, for a fixed total error variance $\sigma_{\varepsilon X}^2$, the sum of factor loadings on our special first factor determines λ_X^2 and hence the size of ρ_{kk}. Negative loadings in λ reduce reliability.[2]

A key feature of our coefficient is that it equally well represents the internal consistency of all k dimensions. But in our model, the remaining $k-1$ dimensions contribute nothing to the reliability of the composite. That is, reliability based on all k dimensions is identical to reliability based only on our first dimension. Since $(\mathbf{1}'\lambda)^2 = \mathbf{1}'\Lambda\Lambda'\mathbf{1}$, the

[1] This correlation is given as $\rho_{X\xi_1} = \mathrm{Cov}\{(X-\mu_X)/\sigma_X, \xi_1\} = \lambda_X/\sigma_X = \sqrt{\rho_{kk}}$.

[2] This is with respect to a unit-weighted composite. Existence of a negative factor loading implies that a weight of "−1" should be used instead of a weight of "+1" for that variable to obtain the composite having maximal reliability. Alternatively, the variable should be reverse-keyed prior to computing the covariance matrix.

proportion of common variance across all dimensions to total variance is

$$\rho_{kk} = \frac{\mathbf{1}'\Lambda\Lambda'\mathbf{1}}{\mathbf{1}'(\Lambda\Lambda' + \Psi)\mathbf{1}} = \frac{(\mathbf{1}'\lambda)^2}{(\mathbf{1}'\lambda)^2 + \mathbf{1}'\Psi\mathbf{1}} = \frac{\lambda_X^2}{\lambda_X^2 + \sigma_{\varepsilon X}^2}, \tag{5}$$

which just takes us back to our coefficient (4). In this setup, the actual number of factors k is irrelevant.

In practice, the model under the proposed parameterization has to be estimated from the data. In a saturated means model, $\hat{\mu} = \overline{X}$, the sample mean of X, and a number k has to be chosen so that the model-reproduced covariance matrix $\widehat{\Sigma}$ under the given identification conditions approximates the sample covariance matrix S closely enough from a statistical point of view. The above formula yields the estimator

$$\hat{\rho}_{kk} = \frac{\mathbf{1}'\hat{\Lambda}\hat{\Lambda}'\mathbf{1}}{\mathbf{1}'\widehat{\Sigma}\mathbf{1}} = 1 - \frac{\mathbf{1}'\widehat{\Psi}\mathbf{1}}{\mathbf{1}'\widehat{\Sigma}\mathbf{1}} = \frac{\hat{\lambda}_X^2}{\hat{\lambda}_X^2 + \hat{\sigma}_{\varepsilon X}^2}. \tag{6}$$

As noted above, in the standard situation of exploratory factor analysis, k will be a relatively small number. Also, then $\widehat{\Sigma} \neq S$. In contrast, in minimum trace or minimum rank modeling situations, k will be quite large and while $\Sigma = \Lambda\Lambda' + \Psi$ as before, also $\widehat{\Sigma} = S$. As a result, $\hat{\lambda}$ as well as the estimated total variance $\mathbf{1}'\widehat{\Sigma}\mathbf{1}$ being explained under the types of models (small-k vs. large-k) are liable to be different, and hence these diverse approaches no doubt will yield different sample estimates $\hat{\rho}_{kk}$.

3. Properties of the coefficient

The coefficient ρ_{kk} can be computed without imposing our proposed identification conditions. That is, $\rho_{kk} = \mathbf{1}'\Lambda\Lambda'\mathbf{1}/\mathbf{1}'\Sigma\mathbf{1}$ is invariant to any particular rotation or transformation of the matrix Λ. As noted in (6), only the product $\Lambda\Lambda'$ is required, and this product is invariant to orthogonal or oblique transformations. Any factor solution is good enough. Representation of the latent factors using the proposed set of identification conditions is not needed for defining or computing ρ_{kk}. The interpretation remains the same: the reliability of a unidimensional composite (3), and also, the proportion of common variance attributable to all k factors. Furthermore, as is discussed further below, if Λ is based on minimum trace factor analysis, it is Bentler's (1972) dimension-free coefficient, and when based on constrained minimum trace factor analysis, it will be the greatest lower bound to reliability (Jackson and Agunwamba, 1977).

Nonetheless, if interest centers on *unidimensional* reliability, and especially, *maximal* unidimensional reliability, (4) must be the largest among possible choices of the given factor. Specifically, the factor ξ_1 in (3) must be chosen so that the proportion of total score variance due to this factor, hence the reliability (4), is maximum. The coefficient defined by our identification conditions accomplishes this maximum. To see this, we start with some arbitrary factor loading matrix $\tilde{\Lambda}$ and finding a maximizing rotation.

THEOREM. *Let $\Sigma = \tilde{\Lambda}\tilde{\Lambda}' + \Psi$, and let t be a normal vector ($t't = 1$). Then the factor loading vector $\lambda = \tilde{\Lambda}t$ that maximizes $(\mathbf{1}'\lambda)^2$ is given by $\lambda = (\mathbf{1}'\tilde{\Lambda}\tilde{\Lambda}'\mathbf{1})^{-1/2}\tilde{\Lambda}\tilde{\Lambda}'\mathbf{1}$, and the residual factors $\bar{\Lambda}$, where $\bar{\Lambda}\bar{\Lambda}' = \tilde{\Lambda}\tilde{\Lambda}' - \lambda\lambda'$, have zero column sums ($\mathbf{1}'\bar{\Lambda} = 0$).*

For proof, see Appendix A.

The proposed identification conditions one and two are thus not arbitrary. They are necessary to defining the factor ξ_1 that yields the maximal unidimensional internal consistency reliability (4). The theorem also shows that the maximal reliability factor can be obtained as a rotation from any starting factor solution, and in fact a rotation is not needed, since the factor loading vector can be computed as $\lambda = \{\mathbf{1}'(\Sigma - \Psi)\mathbf{1}\}^{-1/2}(\Sigma - \Psi)\mathbf{1}$ if desired. Furthermore, if one has no interest in the individual factor loadings λ_i themselves, coefficient (4) can be computed directly from any initial factor solution, such as an unrotated maximum likelihood solution. Even though coefficient (4) is based on $(\mathbf{1}'\lambda)^2$ in the numerator, under the theorem it can be computed as

$$(\mathbf{1}'\lambda)^2 = \mathbf{1}'\tilde{\Lambda}\tilde{\Lambda}'\mathbf{1} = \mathbf{1}'(\Sigma - \Psi)\mathbf{1} = \mathbf{1}'\Sigma\mathbf{1} - \mathbf{1}'\Psi\mathbf{1}.$$

The maximal total common variance based on one factor is just the total variance of the composite minus the total residual variance. There is no need to compute the optimal factor loadings λ, although these are informative in their own right. Of course, in practice, the estimators $\widehat{\Sigma}$ and $\widehat{\Psi}$ are used in such computations.

It might be noticed that the factor loading matrix $\Lambda = [\lambda|\bar{\Lambda}]$ is in a form very similar to that obtained with centroid factor analysis. For a recent discussion of certain aspects of this very old method and relevant references, see Choulakian (2003). However, centroid factor analysis was developed as a method of factor extraction. In the above defining formulae, nothing has been stated about what method of estimation is used to provide estimates of the factor loadings. The theoretical coefficient ρ_{kk} in (4) is independent of any estimation method, while the estimator $\hat{\rho}_{kk}$ in (6) can be obtained from a factor solution obtained by any appropriate method of factor analysis. In the small-k situation, it often will be based on maximum likelihood, generalized least squares, or least squares estimation, while in the atypical large-k situation, it will be based on minimum trace, minimum rank, or a similar methodology.

Although the maximal reliability coefficient was developed under the usual factor analytic assumption that Ψ is a diagonal matrix, all the key results – including the theorem – also hold when Ψ is a more general covariance matrix of residuals. That is, "correlated errors" are allowed.

4. Illustration with exploratory factor analysis

Table 1 gives the correlation matrix for the widely known nine psychological variables based on 101 cases (Harman, 1976, p. 244), which we take as a covariance matrix for purposes of illustration. A one-factor maximum likelihood solution is presented in the left part of Table 2. This solution does not fit the data. It has a likelihood ratio $\chi^2_{27} = 190.6$. Nonetheless, ρ_{11} was estimated, yielding $\hat{\rho}_{11} = 0.880$. Actually, this coefficient – for a factor that does not explain the data – is not an improvement over $\hat{\alpha} = 0.886$. A three-factor model fits these data extremely well, with $\chi^2_{12} = 1.6$. The loadings $\hat{\lambda}$ for this maximal reliability factor are given in the right part of Table 2. It will be seen that the sum of factor loadings is larger than was the case for the 1-factor model. The corresponding $\hat{\rho}_{kk}$ is 0.939.

Table 1
Correlation matrix of nine psychological variables

1.00								
0.75	1.00							
0.78	0.72	1.00						
0.44	0.52	0.47	1.00					
0.45	0.53	0.48	0.82	1.00				
0.51	0.58	0.54	0.82	0.74	1.00			
0.21	0.23	0.28	0.33	0.37	0.35	1.00		
0.30	0.32	0.37	0.33	0.36	0.38	0.45	1.00	
0.31	0.30	0.37	0.31	0.36	0.38	0.52	0.67	1.00

Table 2
Factor loadings $\hat{\lambda}$ in two exploratory factor analysis solutions

Variable number	1-Factor model	3-Factor model
1	0.636	0.727
2	0.697	0.738
3	0.667	0.754
4	0.867	0.789
5	0.844	0.767
6	0.879	0.803
7	0.424	0.492
8	0.466	0.597
9	0.462	0.635
Sum	5.942	6.302

The maximal reliability factor loadings given in the right part of Table 2 can be computed with any covariance structure modeling program that allows linear constraints. Such a setup using EQS (Bentler, In press) is shown in Table 3, conforming to our proposed identification conditions. The equation setup contains a nearly completely full matrix of free factor loadings, except that F3 does not influence V1. Any other choice of fixed zero loading to prevent F2 and F3 from being rotated would do as well. The /CONSTRAINTS section forces the sum of factor loadings on F2 to be zero, and also forces the sum of loadings on F3 to be zero. Other than this, the model setup is completely standard. The EQS output provides the factor loadings $\hat{\lambda}$ as already presented in the right part of Table 2. In addition, in the output section labeled "reliability coefficients", EQS provides the estimated coefficient (6) as

```
Reliability coefficient RHO = .939
```

5. Reliability with general latent variable models

Our proposed conceptualization of maximal unit-weighted reliability does not require that the structure of the variables involved be based on an exploratory factor analy-

Table 3
EQS setup for maximal unit-weighted reliability with exploratory factor analysis model

```
/TITLE
Maximum Reliability Model Setup
/SPECIFICATIONS
VARIABLES= 9; CASES=101;
MATRIX=COVARIANCE; METHOD=ML;
/EQUATIONS
V1=*F1+*F2+0F3+E1;
V2=*F1+*F2+*F3+E2;
V3=*F1+*F2+*F3+E3;
V4=*F1+*F2+*F3+E4;
V5=*F1+*F2+*F3+E5;
V6=*F1+*F2+*F3+E6;
V7=*F1+*F2+*F3+E7;
V8=*F1+*F2+*F3+E8;
V9=*F1+*F2+*F3+E9;
/VARIANCES
F1 TO F3 = 1.0;
E1 TO E9 =  .5*;
/CONSTRAINTS
(V1,F2)+(V2,F2)+(V3,F2)+(V4,F2)+(V5,F2)+(V6,F2)+(V7,F2)
+(V8,F2)+(V9,F2)=0;
(V2,F3)+(V3,F3)+(V4,F3)+(V5,F3)+(V6,F3)+(V7,F3)+(V8,F3)+(V9,F3)=0;
/MATRIX
1.00
 .75 1.00
 .78  .72 1.00
 .44  .52  .47 1.00
 .45  .53  .48  .82 1.00
 .51  .58  .54  .82  .74 1.00
 .21  .23  .28  .33  .37  .35 1.00
 .30  .32  .37  .33  .36  .38  .45 1.00
 .31  .30  .37  .31  .36  .38  .52  .67 1.00
/END
```

sis model. To see this, we consider a structural equation model in which the observed variables can be decomposed into common scores with additive errors. Specifically, consider the measurement model and latent variable regressions

$$x = \mu + \Lambda\xi + \varepsilon, \tag{7a}$$

$$\xi = B\xi + \zeta. \tag{7b}$$

The first Eq. (7a) is identical to Eq. (1), but here we conceive of it as representing any confirmatory factor analysis (CFA) model. The second Eq. (7b) allows any factor ξ_i to be regressed on any other factor ξ_j. Assuming no correlations between ξ, ζ, ε, and a full rank$(I - B)$, it follows that we can rewrite

$$\xi = (I - B)^{-1}\zeta,$$
$$x = \mu + \Lambda(I - B)^{-1}\zeta + \varepsilon. \tag{8}$$

Table 4
Partial EQS setup for maximal unit-weighted reliability with confirmatory factor analysis model

```
/EQUATIONS
V1=*F1          +E1;
V2=*F1          +E2;
V3=*F1          +E3;
V4=      *F2    +E4;
V5=      *F2    +E5;
V6=      *F2    +E6;
V7=         *F3+E7;
V8=         *F3+E8;
V9=         *F3+E9;
/VARIANCES
F1 TO F3 = 1.0;
E1 TO E9 =  .5*;
/COVARIANCES
F1 TO F3 =*;
```

With the covariance matrix of the ζ given as Φ, the covariance structure of the model is given by

$$\Sigma = \Lambda(I - B)^{-1}\Phi(I - B)^{-1'}\Lambda' + \Psi. \tag{9}$$

We now obtain an expression for reliability of the composite based on (9). Our theorem utilized the decomposition $\Sigma = \tilde{\Lambda}\tilde{\Lambda}' + \Psi$. With the definition of $\tilde{\Lambda} = \Lambda(I - B)^{-1}\Phi^{1/2}$, for any square root $\Phi^{1/2}$ such that $\Phi^{1/2}\Phi^{1/2'} = \Phi$, the theorem can be directly applied. We have

COROLLARY. *Let* $\Sigma = \Lambda(I - B)^{-1}\Phi(I - B)^{-1'}\Lambda' + \Psi$ *and* $\tilde{\Lambda} = \Lambda(I - B)^{-1}\Phi^{1/2}$. *Then the factor loading vector that maximizes* $(\mathbf{1}'\lambda)^2$ *is given by* $\lambda = (\mathbf{1}'\tilde{\Lambda}\tilde{\Lambda}'\mathbf{1})^{-1/2}\tilde{\Lambda}\tilde{\Lambda}'\mathbf{1}$, *and the residual factors* $\bar{\Lambda}$, *where* $\bar{\Lambda}\bar{\Lambda}' = \tilde{\Lambda}\tilde{\Lambda}' - \lambda\lambda'$, *have zero column loading sums* $(\mathbf{1}'\bar{\Lambda} = 0)$.

As a result, we can obtain the unit-weighted maximal reliability (4) in the same way whether the structural model is an exploratory factor analysis model (2) or a more general covariance structure model (9) based on a fairly general structural equation model of the form (7). Even though the model structure may be complex, the composite will be able to be represented as unidimensional via (3).

Again, any covariance structure modeling program can be used to obtain the reliability estimator. When a general model is specified in EQS, the program checks whether the covariance structure of the model can be translated into a form such as (9). If so, maximal unit-weighted reliability is computed using Eq. (5). This is done whether Ψ is diagonal or not; that is, correlated errors are allowed. An illustration of (9) with B being the null matrix is the 3-factor confirmatory factor analysis model with correlated factors given in Table 4, again based on the data of Table 1. This model fits the data well. The goodness of fit is $\chi^2_{24} = 19.1$ $(p > 0.05)$. The theorem again provides the maximal unit weighted internal consistency coefficient. This is printed as

```
Reliability coefficient RHO = .936
```

It is seen that the coefficient is somewhat smaller than the one obtained from the exploratory factor analysis model. But this model has twice as many degrees of freedom.

6. Dimension-free and greatest lower bound reliability

In the large-k exploratory factor analysis model, the matrix expression of the model is as before in (2), with $\Sigma = \Lambda\Lambda' + \Psi$ and Ψ diagonal. Instead of fitting this model to S, the sample covariance matrix, in an *approximate* way with a small k number of factors, Bentler's (1972) dimension-free reliability coefficient is defined to be that coefficient (5) with factor loading matrix Λ (of arbitrary dimension k) chosen so that $trace(\Lambda\Lambda')$ is minimized while the model $\Sigma = \Lambda\Lambda' + \Psi$ holds *precisely*.[3] But

$$\min trace(\Lambda\Lambda') = \max trace(\Psi) = \max \mathbf{1}'\Psi\mathbf{1}$$

so that Bentler's dimension-free lower bound to reliability is

$$\rho_{blb} = 1 - \max \frac{\mathbf{1}'\Psi\mathbf{1}}{\mathbf{1}'\Sigma\mathbf{1}} = \min\left(1 - \frac{\mathbf{1}'\Psi\mathbf{1}}{\mathbf{1}'\Sigma\mathbf{1}}\right), \quad \text{with } (\Sigma - \Psi)_{psd}. \tag{10}$$

There is no smaller reliability coefficient for which $(\Sigma - \Psi)$ is positive semidefinite, i.e., for which the factors are real, and not imaginary. The dimension-free lower bound is generally associated with a number of factors k larger than the Ledermann bound. If, at the solution, the unique variance matrix Ψ has non-negative variances, the dimension free lower bound is the greatest lower bound to reliability.

If in minimizing (10) we also constrain $\psi_{ii} \geqslant 0$ (i.e., we disallow Heywood cases), we obtain the greatest lower bound to internal consistency reliability. That is,

$$\rho_{glb} = 1 - \max \frac{\mathbf{1}'\Psi\mathbf{1}}{\mathbf{1}'\Sigma\mathbf{1}} = \min\left(1 - \frac{\mathbf{1}'\Psi\mathbf{1}}{\mathbf{1}'\Sigma\mathbf{1}}\right),$$
$$\text{with } (\Sigma - \Psi) \text{ and } \Psi_{psd}. \tag{11}$$

Since (11) is an optimization problem with an additional constraint as compared to (10), $\rho_{blb} \leqslant \rho_{glb}$, with equality when there are no negative unique variances in ρ_{blb}.[4]

Again the theorem holds directly. As a result, although ρ_{blb} and ρ_{glb} are dimension-free coefficients of internal consistency with k-dimensional factor spaces, their composite scores can be represented to be based on a single latent variable via (3). Among the large-k factors is a single dimension that has maximal unit weighted internal consistency.

---

[3] This criterion makes clear the more recent terminology for this method as "minimum trace factor analysis" (e.g., della Riccia and Shapiro, 1982; Shapiro, 1982a).

[4] Hence, this constrained optimization problem is sometimes called constrained minimum trace factor analysis (e.g., ten Berge et al., 1981).

Table 5
Factor loading vector $\hat{\lambda}$ for 3-factor CFA and greatest lower bound solutions

Variable number	CFA loadings	*blb* and *glb* loadings
1	0.752	0.727
2	0.721	0.743
3	0.743	0.756
4	0.806	0.790
5	0.754	0.772
6	0.762	0.801
7	0.461	0.499
8	0.596	0.598
9	0.643	0.635
Sum	6.240	6.321

In practice, the population covariance matrix is not available for either Bentler's dimension-free lower bound or the greatest lower bound. Hence in these procedures, we take $\widehat{\Sigma} = S$, and the sample decomposition is given by

$$S = \hat{\Lambda}\hat{\Lambda}' + \widehat{\Psi}.$$

Thus the factors precisely reproduce the covariances among all variables, with zero residuals. This leads to a small-sample bias. Bias corrections are available for Bentler's coefficient (Shapiro and ten Berge, 2000; Li and Bentler, 2004). At this writing, most structural modeling programs do not compute blb and glb solutions, but EQS does so. EQS uses the computational algorithms of Bentler (1972), Bentler and Woodward (1980, 1983); see also ten Berge et al. (1981) and Jamshidian and Bentler (1998) to obtain these coefficients without any prespecification of the number of factors k, which is obtainable at the solution. The results are printed out as follows for the example of Table 1.

```
BENTLER'S DIMENSION-FREE LOWER BOUND RELIABILITY = .945
GREATEST LOWER BOUND RELIABILITY                 = .945
```

These coefficients are the same since there are no negative unique variances. The 0.945 value exceeds both the 3-factor EFA coefficient of 0.939, and the 3-factor CFA coefficient of 0.936, previously reported. The factor loadings for the maximal reliability factor from the solution for the 3-factor CFA model (see Table 4) are given in the left column of Table 5. The right column of Table 5 gives the loadings for the comparable maximal reliability factor from the dimension-free solution. The estimates are quite similar.

7. Reliability of weighted composites

The maximal reliability composite discussed above is based on simple summation of parts, since unweighted sums are typically used to obtain a total score. However, sometimes known or unknown weights w_i may be used to give some variables more influence than others, resulting in the weighted composite $Y = w_1X_1 + w_2X_2 + \cdots + w_pX_p$. Since a unit-weighted composite is a special case with $w_i = 1$, it may be desirable to place the earlier discussion into the context of the more general case of weighted composites. First we discuss composites with known weights.

It is easy to show that all of the previous results apply directly to the case of weighted composites. Suppose that w is the p-length column vector of weights, and that D_w is the diagonal matrix with w in its diagonal. Then rescaling the x variables in (1) with these weights yields the new variables $y = D_w x$. Their unit-weighted sum $Y = \mathbf{1}'y = \mathbf{1}'D_w x$ is the weighted composite. The rescaled variables possess a comparable factor analytic decomposition

$$y = D_w x = D_w \mu + D_w \Lambda\xi + D_w \varepsilon = \mu_y + \Lambda_y \xi + \varepsilon_y, \tag{12}$$

and, with obvious notation, the covariance structure of these rescaled variables is given by

$$\Sigma_y = \Lambda_y \Lambda_y' + \Psi_y. \tag{13}$$

As a result, the latent structure of the rescaled variables parallels that of the original variables in the previous sections. Hence, the preceding results, including the theorem, apply directly to the situation of weighted composites.

In the first part of this paper, we used the decomposition $\Sigma = \Sigma_T + \Psi_E$ to define reliability as $\rho_{xx} = \sigma_T^2/\sigma_X^2 = \mathbf{1}'\Sigma_T\mathbf{1}/\mathbf{1}'\Sigma\mathbf{1} = 1 - \mathbf{1}'\Psi_E\mathbf{1}/\mathbf{1}'\Sigma\mathbf{1}$. Using the same decomposition, we can also define reliability of a weighted composite Y as

$$\rho_{yy} = \frac{\sigma_{Ty}^2}{\sigma_Y^2} = \frac{w'\Sigma_T w}{w'\Sigma w} = 1 - \frac{w'\Psi_E w}{w'\Sigma w}. \tag{14}$$

The unit-weighted composite is a special case. When the error variances are not known, it is hard to estimate (14) directly. Bentler (1968) noted that factor analytic concepts can be fruitfully employed. Let the covariance matrix of the original unweighted variables X_i have the composition $\Sigma = \Sigma_c + \Psi$, where Σ_c is a non-negative definite covariance matrix of the common variables, and Ψ is the covariance matrix of the unique variables. Then Bentler's (1968, Eq. (12)) general formula for internal consistency reliability of a weighted composite is

$$\rho_{w'x} = \frac{w'\Sigma_c w}{w'\Sigma w} = \left(1 - \frac{w'\Psi w}{w'\Sigma w}\right). \tag{15}$$

See also Heise and Bohrnstedt (1970, Eq. (32)). Eq. (15) is a lower bound to reliability (14) under the usual factor analytic assumption $\Psi = \Psi_E + \Psi_S$, where unique covariance matrix Ψ is a sum of the non-negative definite covariance matrices of random errors and specific but reliable variables. Then $w'\Psi w = w'\Psi_E w + w'\Psi_S w$, and some algebra shows that

$$\rho_{yy} = \rho_{w'x} + (w'\Psi_S w)/(w'\Sigma w). \tag{16}$$

Thus $\rho_{w'x} \leqslant \rho_{yy}$, i.e., (15) is a lower bound to (14). The coefficients will be equal when there is no specificity.

Actually, several of the key reliability coefficients defined in earlier parts of this chapter are just special cases of (15). Obviously, with equal weights and $\Sigma_c = \lambda\lambda'$, a single common factor model, and $\Psi = \Psi_u$, (15) simplifies to ρ_{11} as discussed previously, namely reliability under a 1-factor hypothesis as popularized by McDonald (1999, p. 89) with coefficient ω. With equal weights and a general k-factor decomposition $\Sigma_c = \Lambda\Lambda'$, (15) becomes ρ_{kk} as given in (5) and elsewhere. With unit weights and optimization criteria to define Ψ, this is the dimension-free and greatest lower bound given in (10) and (11).

8. Selection of weights for maximal reliability

With our given definitions of reliability of weighted composites, we now consider the case where the weight vector w is not known, but must be estimated. Following the earlier work of Green (1950), Bentler (1968) reviewed a method for finding w to maximize the reliability coefficient (14) when the error variances are known. He then developed a parallel method to maximize (15) for the case where the factor model parameters are not known, and related this method to Rao's (1955) canonical factor analysis, a variant of maximum likelihood factor analysis. In this approach, the weights and reliability coefficients are derived from the eigenvalue-eigenvector decomposition of $\Psi^{-1/2}(\Sigma - \Psi)\Psi^{-1/2}$, with S replacing Σ when the parameters need to be estimated. Consider the special case of one common factor $\Sigma - \Psi = \lambda\lambda'$. Then the weights by Bentler's method are given by

$$w = (\lambda'\Psi^{-1}\lambda)^{-1/2}\Psi^{-1}\lambda \tag{17}$$

and the maximal reliability is given by

$$\rho_{w'x(\max)} = \left(\frac{\lambda'\Psi^{-1}\lambda}{\lambda'\Psi^{-1}\lambda + 1}\right). \tag{18}$$

This coefficient has been rediscovered in the recent literature, where it has been referred to as "maximal reliability" (e.g., Drewes, 2000; Li, 1997; Li et al., 1996; Raykov, 2004b) or "construct reliability" (Hancock and Mueller, 2001). It is not necessarily written in the form (18), e.g., Hancock and Mueller write it as $\lambda'\Sigma^{-1}\lambda$ for standardized variables. EQS computes weighted composite reliability (18) while allowing various constraints on parameters, thus permitting modifications to the basic coefficient.

If weights (17) are used to define (12) and (13), the internal consistency coefficient (15) is defined with these weights, and Λ_y has our identification structure, it can be shown that parallel to (3) the weighted total score is based on a single common factor

$$Y = \mu_Y + \lambda_Y\xi_1 + \varepsilon_Y, \tag{19}$$

where μ_Y, λ_Y and ε_Y are unit-weighted sums of μ_y, the first column of Λ_y, and ε_y, respectively. While we can write

$$\rho_{w'x} = \frac{\lambda_Y^2}{\lambda_Y^2 + \sigma_{\varepsilon_Y}^2}, \tag{20}$$

the numerator of (20) is just $w'\Lambda\Lambda'w = w'\Sigma_c w$, and the denominator is $w'\Sigma w$. Thus the coefficient (20) is the same as (15).

An additional type of optimum weighted reliability coefficient based on (15) was proposed by Shapiro (1982b). Consider the set of possible $\rho_{w'x}$ coefficients as Ψ is varied, and consider its minimal value $\min_\Psi(\rho_{w'x})$. Shapiro's method is to find that w (excluding the null vector) so that the value of $\{\min_\Psi(\rho_{w'x})\}$ is maximized, i.e., is as large as possible. This is done without any assumption of unidimensionality. Thus Shapiro's min–max weighted greatest lower bound reliability is less restricted conceptually than that given by (18), which is based on unidimensionality. EQS computes Shapiro's coefficient using an algorithm of Jamshidian and Bentler (1998).

9. Conclusions

Like the 1-factor based internal consistency reliability coefficients, the proposed approach to maximal unit-weighted reliability requires modeling the sample covariance matrix. This must be a successful enterprise, as estimation of reliability only makes sense when the model does an acceptable job of explaining the sample data. Of course, since a k-factor model rather than a 1-factor model would typically be used in the proposed approach, the odds of adequately modeling the sample covariance matrix are greatly improved. The selected model can be a member of a much wider class of models, including exploratory or confirmatory factor models or any arbitrary structural model with additive errors. Whatever the resulting dimensionality k, and whether the typical "small-k" or theoretical alternative "large-k" approach is used, or whether a general structural model with additive errors is used, the proposed identification conditions assure that the resulting internal consistency coefficient $\hat{\rho}_{kk}$ represents the proportion of variance in the unit-weighted composite score that is attributable to the common factor generating maximal internal consistency. The proposed coefficient can be interpreted as representing unidimensional reliability even when the instrument under study is multifactorial, since, as was seen in (3), the composite score can be modeled by a single factor as $X = \mu_X + \lambda_X\xi_1 + \varepsilon_X$. Nonetheless, it equally well has an interpretation as summarizing the internal consistency of the k-dimensional composite. Although generally there is a conflict between unidimensionality and reliability, as noted by ten Berge and Sočan (2004), our approach reconciles this conflict.

When the large-k approach is used to model the covariance matrix, the theory of dimension-free and greatest lower bound coefficients can be applied. This is based on a tautological model that defines factors that precisely reproduce the covariance matrix. As these theories have been developed in the past, reliability is defined on scores that are explicitly multidimensional. However, we have shown here that the dimension-free and greatest lower-bound coefficients can equivalently be defined for the single most reliable dimension among the many that are extracted. Based on this observation, new approaches to these coefficients may be possible.

The optimal unit-weighted coefficient developed here also applies to composites obtained from any latent variable covariance structure model with p-dimensional additive errors. Although we used a specific type of model as given in (7) to develop this approach, we are in no way limited to that specific linear model structure. Our approach holds equally for any completely arbitrary structural model with additive errors that we can write in the form $\Sigma = \Sigma(\theta) + \Psi$, where $\Sigma(\theta)$ and Ψ are non-negative definite matrices. Then we can decompose $\Sigma(\theta) = \Lambda\Lambda'$ and proceed as described previously, e.g., we can compute the factor loadings for the maximally reliable factor via $\lambda = \{\mathbf{1}'(\Sigma - \Psi)\mathbf{1}\}^{-1/2}(\Sigma - \Psi)\mathbf{1}$. The maximal reliability coefficient for a unit weighted composite based on any model that can be specified as a Bentler and Weeks (1980) model has been available in EQS 6 for several years.

With regard to reliability of differentially weighted composites, we extended our maximal reliability for a unit-weighted composite (4) to maximal reliability for a weighted composite (18). While both coefficients have a similar sounding name, they represent quite different types of "maximal" reliability, and, in spite of the recent enthusiasm for (18), in our opinion this coefficient ought to be used only rarely. Weighted maximal reliability does not describe the reliability of a typical total score or scale. Such a scale score, in common practice, is a unit-weighted composite of a set of items or components. Our unit-weighted maximal reliability (4) or (5) describes this reliability. In contrast, maximal reliability (18) gives the reliability of a differentially weighted composite, and if one is not using such a composite, to report it would be misleading. Of course, if a researcher actually might consider differentially weighting of items or parts in computing a total score, then a comparison of (4) to (18) can be very instructive. If (18) is only a marginal improvement over (4), there would be no point to differential weighting. On the other hand, if (4) is not too large while (18) is substantially larger, differential weighting might make sense.

Finally, all of the theoretical coefficients described in this chapter were specified in terms of population parameters, and their estimation was assumed to be associated with the usual and simple case of independent and identically distributed observations. When data have special features, such as containing missing data, estimators of the population parameters will be somewhat different in obvious ways. For example, maximum likelihood estimators based on an EM algorithm may be used to obtain structured or unstructured estimates of Σ (e.g., Jamshidian and Bentler, 1999) for use in the defining formulae. In more complicated situations, such as multilevel modeling (e.g., Liang and Bentler, 2004), it may be desirable to define and evaluate internal consistency reliability separately for within-cluster and for between-cluster variation. The defining formulae described here can be generalized to such situations in the obvious ways and are already available in EQS 6.

Acknowledgements

This research, based on Bentler (2003, 2004), was supported in part by grants DA00017 and DA01070 from the National Institute on Drug Abuse.

Appendix A

PROOF OF THEOREM. Let $\phi = (\mathbf{1}'\lambda)^2 - \gamma(t't - 1)$. Taking derivatives $\partial\phi/\partial\gamma$ and setting to zero establishes $t't = 1$. Then $\partial\phi/\partial t$ yields the eigenequation $(\tilde{\Lambda}'\mathbf{1}\mathbf{1}'\tilde{\Lambda} -$

$\gamma I)t = 0$. Solving this yields $\gamma = (\mathbf{1}'\tilde{\Lambda}\tilde{\Lambda}'\mathbf{1})$ and $t = (\mathbf{1}'\tilde{\Lambda}\tilde{\Lambda}'\mathbf{1})^{-1/2}\tilde{\Lambda}'\mathbf{1}$. Substituting into $\lambda = \tilde{\Lambda}t$ and simplifying gives $\lambda = (\mathbf{1}'\tilde{\Lambda}\tilde{\Lambda}'\mathbf{1})^{-1/2}\tilde{\Lambda}\tilde{\Lambda}'\mathbf{1}$. It follows that $\bar{\Lambda}\bar{\Lambda}'\mathbf{1} = (\tilde{\Lambda}\tilde{\Lambda}' - \lambda\lambda')\mathbf{1} = 0$, which means that $\mathbf{1}'\bar{\Lambda} = 0$. Finally, $(\mathbf{1}'\lambda)^2$ is maximized rather than minimized since the minimum ϕ occurs with $\lambda = 0$. □

References

Barchard, K.A., Hakstian, A.R. (1997). The effects of sampling model on inference with coefficient alpha. *Educational & Psychological Measurement* **57**, 893–905.

Becker, G. (2000). Coefficient alpha: Some terminological ambiguities and related misconceptions. *Psychological Reports* **86**, 365–372.

Bentler, P.M. (1968). Alpha-maximized factor analysis (Alphamax): Its relation to alpha and canonical factor analysis. *Psychometrika* **33**, 335–345.

Bentler, P.M. (1972). A lower-bound method for the dimension-free measurement of internal consistency. *Social Science Research* **1**, 343–357.

Bentler, P.M. (2003). Should coefficient alpha be replaced by model-based reliability coefficients? Invited address. Psychometric Society IMPS-2003, Cagliari, Italy.

Bentler, P.M. (2004). Maximal reliability for unit-weighted composites. UCLA Statistics Preprint No. 405. Department of Statistics, UCLA.

Bentler, P.M. (In press). EQS 6 Structural Equations Program Manual. Multivariate Software. Encino, CA. http://www.mvsoft.com.

Bentler, P.M., Weeks, D.G. (1980). Linear structural equations with latent variables. *Psychometrika* **45**, 289–308.

Bentler, P.M., Woodward, J.A. (1980). Inequalities among lower bounds to reliability: With applications to test construction and factor analysis. *Psychometrika* **45**, 249–267.

Bentler, P.M., Woodward, J.A. (1983). The greatest lower bound to reliability. In: Wainer, H., Messick, S. (Eds.), *Principals of Modern Psychological Measurement: A Festschrift for Frederic M. Lord*. Erlbaum, Hillsdale, NJ, pp. 237–253.

Bollen, K.A. (1980). Issues in the comparative measurement of political democracy. *American Sociological Review* **45**, 370–390.

Bonett, D.G. (2003). Sample size requirement for testing and estimating coefficient alpha. *Journal of Educational and Behavioral Statistics* **27**, 335–340.

Choulakian, V. (2003). The optimality of the centroid method. *Psychometrika* **68**, 473–475.

Cronbach, L.J. (1951). Coefficient alpha and the internal structure of tests. *Psychometrika* **16**, 297–334.

della Riccia, G., Shapiro, A. (1982). Minimum rank and minimum trace of covariance matrices. *Psychometrika* **47**, 443–448.

Drewes, D.W. (2000). Beyond the Spearman–Brown: A structural approach to maximal reliability. *Psychological Methods* **5**, 214–227.

Enders, C.K. (2003). Using the expectation maximization algorithm to estimate coefficient alpha for scales with item-level missing data. *Psychological Methods* **8**, 322–337.

Enders, C.K., Bandalos, D.L. (1999). The effects of heterogeneous item distributions on reliability. *Applied Measurement in Education* **12**, 133–150.

Feldt, L.S., Charter, R.A. (2003). Estimating the reliability of a test split into two parts of equal or unequal length. *Psychological Methods* **8**, 102–109.

Green Jr., B.F. (1950). A note on the calculation of weights for maximum battery reliability. *Psychometrika* **15**, 57–61.

Green, S.B. (2003). A coefficient alpha for test-retest data. *Psychological Methods* **8**, 88–101.

Green, S.B., Hershberger, S.L. (2000). Correlated errors in true score models and their effect on coefficient alpha. *Structural Equation Modeling* **7**, 251–270.

Hakstian, A.R., Barchard, K.A. (2000). Toward more robust inferential procedures for coefficient alpha under sampling of both subjects and conditions. *Multivariate Behavioral Research* **35**, 427–456.

Hancock, G.R., Mueller, R.O. (2001). Rethinking construct reliability. In: Cudeck, R., du Toit, S., Sörbom, D. (Eds.), *Structural Equation Modeling: Present and Future*. Scientific Software International, Lincolnwood, IL, pp. 195–216.

Harman, H. (1976). *Modern Factor Analysis*, third ed. University of Chicago Press, Chicago.

Heise, D.R., Bohrnstedt, G.W. (1970). Validity, invalidity, and reliability. In: Borgatta, E.F. (Ed.), *Sociological Methodology 1970*. Jossey-Bass, San Francisco, pp. 104–129.

Hogan, T.P., Benjamin, A., Brezinsky, K.L. (2000). Reliability methods: A note on the frequency of use of various types. *Educational and Psychological Measurement* **60**, 523–531.

Jackson, P.H., Agunwamba, C.C. (1977). Lower bounds for the reliability of total scores on a test composed of nonhomogeneous items: I. Algebraic lower bounds. *Psychometrika* **42**, 567–578.

Jamshidian, M., Bentler, P.M. (1998). A quasi-Newton method for minimum trace factor analysis. *Journal of Statistical Computation and Simulation* **62**, 73–89.

Jamshidian, M., Bentler, P.M. (1999). ML estimation of mean and covariance structures with missing data using complete data routines. *Journal of Educational and Behavioral Statistics* **24**, 21–41.

Jöreskog, K.G. (1967). Some contributions to maximum likelihood factor analysis. *Psychometrika* **32**, 443–482.

Jöreskog, K.G. (1971). Statistical analysis of sets of congeneric tests. *Psychometrika* **36**, 109–133.

Kano, Y., Azuma, Y. (2003). Use of SEM programs to precisely measure scale reliability. In: Yanai, H., Okada, A., Shigemasu, Y., Kano, J.J., Meulman, K. (Eds.), *New Developments in Psychometrics*. Springer-Verlag, Tokyo, pp. 141–148.

Komaroff, E. (1997). Effect of simultaneous violations of essential tau-equivalence and uncorrelated error on coefficient alpha. *Applied Psychological Measurement* **21**, 337–348.

Ledermann, W. (1937). On the rank of the reduced correlation matrix in multiple-factor analysis. *Psychometrika* **2**, 85–93.

Li, H. (1997). A unifying expression for the maximal reliability of a linear composite. *Psychometrika* **62**, 245–249.

Li, H., Rosenthal, R., Rubin, D.B. (1996). Reliability of measurement in psychology: From Spearman–Brown to maximal reliability. *Psychological Methods* **1**, 98–107.

Li, L., Bentler, P.M. (2004). The greatest lower bound to reliability: Corrected and resampling estimators. Unpublished manuscript.

Liang, J.-J., Bentler, P.M. (2004). An EM algorithm for fitting two-level structural equation models. *Psychometrika* **69**, 101–122.

McDonald, R.P. (1985). *Factor Analysis and Related Methods*. Erlbaum, Hillsdale, NJ.

McDonald, R.P. (1999). *Test Theory: A Unified Treatment*. Erlbaum, Mahwah, NJ.

Miller, M.B. (1995). Coefficient alpha: A basic introduction from the perspectives of classical test theory and structural equation modeling. *Structural Equation Modeling* **2**, 255–273.

Novick, M.R., Lewis, C. (1967). Coefficient alpha and the reliability of composite measurements. *Psychometrika* **32**, 1–13.

Osburn, H.G. (2000). Coefficient alpha and related internal consistency reliability coefficients. *Psychological Methods* **5**, 343–355.

Rao, C.R. (1955). Estimation and tests of significance in factor analysis. *Psychometrika* **20**, 93–111.

Raykov, T. (1997). Scale reliability, Cronbach's coefficient alpha, and violations of essential tau-equivalence with fixed congeneric components. *Multivariate Behavioral Research* **32**, 329–353.

Raykov, T. (1998). Coefficient alpha and composite reliability with interrelated nonhomogeneous items. *Applied Psychological Measurement* **22**, 375–385.

Raykov, T. (2001). Bias of coefficient α for fixed congeneric measures with correlated errors. *Applied Psychological Measurement* **25**, 69–76.

Raykov, T. (2004a). Point and interval estimation of reliability for multiple-component measuring instruments via linear constraint covariance structure modeling. *Structural Equation Modeling* **11**, 342–356.

Raykov, T. (2004b). Estimation of maximal reliability: A note on a covariance structure modelling approach. *British Journal of Mathematical and Statistical Psychology* **57**, 21–27.

Raykov, T., Shrout, P.E. (2002). Reliability of scales with general structure: Point and interval estimation using a structural equation modeling approach. *Structural Equation Modeling* **9**, 195–212.

Schmidt, F.L., Le, H., Ileus, R. (2003). Beyond alpha: An empirical examination of the effects of different sources of measurement error on reliability estimates for measures of individual-differences constructs. *Psychological Methods* **8**, 206–224.

Schmitt, N. (1996). Uses and abuses of coefficient alpha. *Psychological Assessment* **8**, 350–353.

Shapiro, A. (1982a). Rank reducibility of a symmetric matrix and sampling theory of minimum trace factor analysis. *Psychometrika* **47**, 187–199.

Shapiro, A. (1982b). Weighted minimum trace factor analysis. *Psychometrika* **47**, 243–264.

Shapiro, A., ten Berge, J.M.F. (2000). The asymptotic bias of minimum trace factor analysis, with applications to the greatest lower bound to reliability. *Psychometrika* **65**, 413–425.

Shapiro, A., ten Berge, J.M.F. (2002). Statistical inference of minimum rank factor analysis. *Psychometrika* **67**, 79–94.

Shevlin, M., Miles, J.N.V., Davies, M.N.O., Walker, S. (2000). Coefficient alpha: A useful indicator of reliability?. *Personality Individual Differences* **28**, 229–237.

Spearman, C. (1904). "General intelligence", objectively determined and measured. *American Journal of Psychology* **15**, 201–293.

Spearman, C. (1907). Demonstration of formulae for true measurement of correlation. *American Journal of Psychology* **18**, 161–169.

ten Berge, J.M.F. (2000). Linking reliability and factor analysis: Recent developments in some classical psychometric problems. In: Hampson, S.E. (Ed.), *Advances in Personality Psychology*, vol. 1. Psychology Press/Taylor & Francis, Philadelphia, pp. 138–156.

ten Berge, J.M.F., Kiers, H.A.L. (1991). A numerical approach to the exact and the approximate minimum rank of a covariance matrix. *Psychometrika* **56**, 309–315.

ten Berge, J.M.F., Sočan, G. (2004). The greatest lower bound to the reliability of a test and the hypothesis of unidimensionality. *Psychometrika* **69**, 613–625.

ten Berge, J.M.F., Snijders, T.A.B., Zegers, F.E. (1981). Computational aspects of the greatest lower bound to reliability and constrained minimum trace factor analysis. *Psychometrika* **46**, 201–213.

Vautier, S., Jmel, S. (2003). Transient error or specificity? An alternative to the staggered equivalent split-half procedure. *Psychological Methods* **8**, 225–238.

Zinbarg, R.E., Revelle, W., Yovel, I., Li, W. (2005). Cronbach's α, Revelle's β, and McDonald's ω_H: Their relations with each other and two alternate conceptualizations of reliability. *Psychometrika* **70**, 1–11.

Handbook of Computing and Statistics with Applications, Vol. 1
ISSN: 1871-0301

DOI: 10.1016/S1871-0301(06)01002-X

2

Advances in Analysis of Mean and Covariance Structure when Data are Incomplete*

Mortaza Jamshidian and Matthew Mata

Abstract

Missing data arise in many areas of empirical research. One such area is in the context of structural equation models (SEM). A review is presented of the methodological advances in fitting data to SEM and, more generally, to mean and covariance structure models when there is missing data. This encompasses common missing data mechanisms and some widely used methods for handling missing data. The methods fall under the classifications of ad-hoc, likelihood-based, and simulation-based. Also included are the results of some of the published simulation studies. In order to encourage further research, a method is proposed for performing sensitivity analysis, which up to now has been seemingly lacking. A simulation study was done to demonstrate the method using a three-factor factor analysis model, focusing on MCAR and MNAR data. Parameter estimates from samples of all available data, in the form of box plots, are compared with parameter estimates from only the complete data. The results indicate a possible distinction for determining missing data mechanisms.

1. Introduction

Missing data are broadly experienced in almost all areas of empirical research. A few examples of well-known situations where missing data arise are: omitted or mistakenly recorded information during data entry, non-response to some sensitive questions (e.g., age, income, drug use), and variables being too expensive to measure (e.g., measurement may require destroying expensive parts, or interviewer needs to travel a long distance). These examples and, more broadly, situations in which the respondents provide partial responses are referred to as *item non-response*, in line with situations where responses to some items in a questionnaire are missing. Missing data can also occur as a result of *drop-outs*, for example, when an experiment is run on a group of individuals

*This research was supported in part by the National Science Foundation Grant DMS-0437258.

Table 1
Number of indexed articles in ISI Web of Science by given keywords. ISI Web of Science consists of five high-quality databases containing information gathered from thousands of scholarly journals in all areas of research including Science Citation Index Expanded, Social Sciences Citation Index, Arts & Humanities Citation Index, Index Chemicus, and Current Chemical Reactions

Keywords	1966–79	1980–84	1985–89	1990–94	1995–99	2000–05
Missing data OR incomplete data	104	65	107	554	1191	2341
(Missing data OR incomplete data) AND (covariance structure OR structural equation model OR factor analysis)	3	1	0	16	28	66

over a period of time as in clinical studies. Yet another form of missing data is *unit non-response* in which case no responses are available for a subject that was to be included in the sample. In one of its broadest definitions, Efron (1994) defines missing data as a class of problems made difficult by the absence of some part of a familiar data structure. In the examples mentioned above, the missing structure is an observable covariate that is not recorded or observed. Efron's definition of missing data covers a broader class of problems. For instance, the latent variables in factor analysis would be considered missing data by his definition. The main task in analyzing missing data seems to be development of methodology to realize "the familiar structure". This, for example, is the main theme in development of the famous EM algorithm (Dempster et al., 1977) which is frequently used in analysis of incomplete data.

There is a rich body of statistics literature related to analysis of incomplete data, going back to the 1960s on survey methodology, and even going further back to 1930s on experiments. In this chapter, we focus our attention on a portion of this literature that is related to the structural equation models (SEM) with continuous responses which we define shortly. Table 1 lists the number of published articles, related to missing data and SEM, cited in the ISI Web of Science database since 1966. We chose the starting date of 1966 because the first missing data note on SEM, indexed in the ISI Web of Science, appeared in Woodbury and Siler (1966). As evident from Table 1, research in analysis of missing data, both in general and in SEM, has had a notable increase in activity since the early 1990s, about the beginning of the availability of inexpensive computing resources.

Structural equation modeling mainly consists of placing structures on the population covariance matrix Σ and sometimes the population mean $\boldsymbol{\mu}$. These structures arise from a combination of measurement models and latent variable models. Generally, a random sample of observations, say $\mathbf{x}_1, \ldots, \mathbf{x}_n$, from the population is obtained where each $\mathbf{x}_i$ is a $p \times 1$ vector consisting of observations on p variables. Then a plausible model of the form $\boldsymbol{\mu}(\boldsymbol{\theta})$ and $\Sigma(\boldsymbol{\theta})$, where $\boldsymbol{\theta}$ denotes the parameters of the model, is fitted to the data. Obviously, the least restricted model is the *saturated model* in which no structure is imposed, namely $\boldsymbol{\mu}$ has p free parameters and Σ is an unrestricted symmetric $p \times p$ matrix. A popular example of a covariance structure model, also known as the LISREL model (Jöreskog and Sörbom, 2004), has the measurement model

$$\mathbf{x}_i = \boldsymbol{\mu} + \Lambda \boldsymbol{\zeta}_i + \boldsymbol{\varepsilon}_i, \quad i = 1, \ldots, n, \tag{1}$$

where $\boldsymbol{\mu}$ is a $p \times 1$ intercept vector, Λ is a $p \times k$ ($< p$) matrix of loadings, $\boldsymbol{\zeta}_i$ is a random vector of latent variables, and $\boldsymbol{\varepsilon}_i$ is a random vector of measurement errors, independent of $\boldsymbol{\zeta}_i$, with $\mathrm{E}(\boldsymbol{\varepsilon}_i) = 0$ and $\mathrm{Cov}(\boldsymbol{\varepsilon}_i) = \Psi$.

A special case is the basic *confirmatory factor analysis* model, where $\boldsymbol{\mu} = 0$ and $\mathrm{Cov}(\boldsymbol{\zeta}_i) = \Phi$, a $k \times k$ matrix. In this model, usually some of the parameters in Λ, Φ, and/or Ψ are fixed to one or more constants and thus the parameter vector $\boldsymbol{\theta}$ consists of the free parameters in (Λ, Φ, Ψ). The implied covariance matrix for this model is

$$\Sigma(\boldsymbol{\theta}) = \Lambda \Phi \Lambda^{\mathrm{T}} + \Psi. \tag{2}$$

A more general model arises when the measurements $\mathbf{x}_i$ are split into two groups of endogenous and exogenous variables and the other components of the measurement model (1) are partitioned accordingly. More specifically, consider the partition $\boldsymbol{\zeta}_i = (\boldsymbol{\eta}_i^{\mathrm{T}}, \boldsymbol{\xi}_i^{\mathrm{T}})^{\mathrm{T}}$, where $\boldsymbol{\eta}_i$ and $\boldsymbol{\xi}_i$ are a $k_1 \times 1$ and $k_2 \times 1$ endogenous and exogenous latent vectors with $k_1 + k_2 = k$. Furthermore consider the latent variable model

$$\boldsymbol{\eta}_i = B\boldsymbol{\eta}_i + \Gamma\boldsymbol{\xi}_i + \boldsymbol{\delta}_i, \tag{3}$$

where B and Γ are $k_1 \times k_1$ and $k_1 \times k_2$ matrices of unknown parameters such that $B_0 = I - B$ is nonsingular. It is assumed that the random vectors $\boldsymbol{\xi}_i$ and $\boldsymbol{\delta}_i$ are independent with $\mathrm{Cov}(\boldsymbol{\xi}_i) = \Phi_\xi$ and $\mathrm{Cov}(\boldsymbol{\delta}_i) = \Psi_\delta$, with Ψ_δ diagonal. The parameters $\boldsymbol{\theta}$ in this model consist of the free elements in Λ, B, Γ, Ψ, Φ_ξ, and Ψ_δ, and the implied covariance matrix for this model is given by

$$\Sigma(\boldsymbol{\theta}) = \Lambda \begin{pmatrix} B_0^{-1}(\Gamma\Phi_\xi\Gamma^{\mathrm{T}} + \Psi_\delta)(B_0^{-1})^{\mathrm{T}} & B_0^{-1}\Gamma\Phi_\xi \\ \Phi_\xi\Gamma^{\mathrm{T}}(B_0^{-1})^{\mathrm{T}} & \Phi_\xi \end{pmatrix} \Lambda^{\mathrm{T}} + \Psi. \tag{4}$$

Our aim in this chapter is to give an overview of methodological advances in fitting data to the models described above, and other more general mean and covariance structures, when data are incomplete (i.e., $\mathbf{x}_i$'s are not all completely observed). It is well known that the use of inappropriate methods for handling missing data can lead to bias in parameter estimates, bias in standard errors and test statistics, and inefficient use of data (see, e.g., Little and Rubin (2002) and references therein).

It turns out that one of the most relevant aspects in selecting an appropriate method is related to the process by which the data are missing, often referred to as the *missing data mechanism*. In Section 2, we will review the often cited missing data mechanisms of missing completely at random (MCAR), missing at random (MAR), and missing not at random (MNAR). Most of the published methodologies aim at coping with various missing data mechanisms as well as various distributional and model assumptions. In Section 3, we will give a review of some of these methods under the three classifications of *ad hoc, likelihood-based*, and *simulation-based* methods. A body of missing data SEM literature is devoted to simulation studies and comparison of the methodology in this area. In Section 4, we will point out the results of some of these simulation studies. Finally, as we will see, most of the work done in missing data are based on the assumption of MCAR or MAR. In general it is difficult, if not impossible, to test whether data are MNAR. It may, however, be plausible to perform some sort of sensitivity analysis in this case. This has been mentioned by a few authors, but to our knowledge no specific proposals for doing sensitivity analysis has been given. In Section 5, we will give a specific method for sensitivity analysis. While our proposed methodology leaves a lot to be desired, it is aimed at getting the ball rolling in related research in this area. Tests have been proposed for MCAR, and we will mention those in Section 5. Section 6 contains a discussion of the available missing data methods in SEM statistical software.

2. Missing data mechanism

To begin analyzing a set of data containing missing values, we must have a general understanding of why the data are missing. There is usually some underlying reason for why the missing data. A failure to acknowledge this reason and address it in the analysis may result in biased inference. *Missing data mechanism* is the term commonly used to describe the method by which data are missing. Rubin (1976) seems to be the first to formally introduce the missing data mechanisms of *missing completely at random*, and *missing at random*. These mechanisms are by far the most popular ones used in the missing data literature.

To elaborate on each missing data mechanism, we begin with missing completely at random. Let Y denote the matrix of complete data. Following the definition by Rubin (1987), data are missing completely at random if the missingness does not depend on either the observed or missing values in Y. An example of this would be when questions are accidently skipped over when answering a survey. The reason the questions were skipped has no dependence on the answers to the previous or subsequent questions or the actual answer that would have been given had the question not been skipped, thus making it MCAR.

The next missing data mechanism is missing at random. Let Y_{obs} denote the observed parts of the matrix Y and Y_{mis} the missing parts of Y. We then classify the missing data mechanism as missing at random if the missingness depends only on Y_{obs}. Returning to the example above, if age is a completely observed variable and older respondents are more likely to skip questions, then the data would be missing at random because the missingness depends on the age of the respondent, which is observed.

When data are not MCAR or MAR, then sometimes they are referred to as missing not at random (MNAR). In particular, when the missingness depends on the missing values (i.e., Y_{mis}), data would be MNAR. Returning to the survey example, data missing not at random would include questions that are left unanswered because they relate to a sensitive topic to the respondent such as domestic life, income, or substance abuse. Since there are not any clues as to why the data are missing, the analysis of MNAR data is challenging in the sense that one may never know the validity of the final inferences made based on the data.

Values are not always missing from a collection of data because of a failure by the respondent or omissions in the recording of the data. Some data values are missing because the researcher chooses for them to be missing. Missing data such as this are referred to as *missing by design*. There are many different ways to design a study so that a portion of the data are missing by design. A basic example of this would be collecting data that are easy or inexpensive on all the respondents in a study and gathering data that are more difficult or more expensive on only a few of the respondents. Kogovsek et al. (2002) states that a few advantages to using a missing data design are reduction of respondent burden, reduction of time and cost of the survey, shortening the elapsed time between waves of questioning, and increasing the quality of the data gathered. For a more detailed discussion of these issues and examples of studies with data missing by design, for example, see Arbuckle (1996), Kamakura and Wedel (2000), and Kogovsek et al. (2002).

In general, to determine the *type* of missing data mechanism, one must acquire information about the missing data. This information may be obtained by developing and relying on some reasonable theory about the missing data, or by collecting additional data; for example, by follow-ups, to make the mechanism accessible (Graham and Donaldson, 1993; Little and Rubin, 1987, Section 12.6). A few statistical tests have been developed to check MCAR. For example, MCAR can be checked by testing the equality of the distribution of observed variables across the missing patterns using a t test for location (BMDP8D, Dixon, 1988; Little, 1988). In this procedure, for each variable with missing values, the sample is split into two groups consisting of cases with that variable observed and cases with that variable missing. Then for each of the other variables, we can compare the means of their observed values in the two groups with the two sample t tests. If any of the t tests show significant differences between the means, then there is evidence that the data are *not* MCAR. Little (1988) and more recently Krishnamoorthy and Pannala (1998) have proposed other procedures that are more computationally efficient than the procedure just described.

Kim and Bentler (2002) have extended the work of Little (1988) to propose a few tests for MCAR via testing homogeneity of mean and covariances across groups comprised of cases with similar missing data patterns. They rationalize this test by stating that "the various patterns of missing data come from a single population, that is, that the data exhibit homogeneity of means and covariances (HMC). Rejection of homogeneity implies rejection of MCAR. On the other hand, acceptance of HMC does not prove that a data set is MCAR-though it is unclear when, if ever, data could be HMC but not also be MCAR." So Kim and Bentler's (2002) test is informative when the null hypothesis (i.e., the hypothesis of HMC) is rejected. In a simulation study, however, we have found that the Kim and Bentler's test is very conservative, and it especially has a very low power when the number of missing data patterns (groups) is large.

It seems to us that, in general, tests of MCAR should not be confined to test differences between various data patterns. Two subjects may have the same missing data pattern, but have different reasons (mechanisms) for missingness. To accommodate such situations, the data should be grouped in a meaningful manner, and then the HMC test between the resulting group be performed (for an example of testing homogeneity of covariances among groups in the context of missing data see Jamshidian and Schott (2005)). Additional work in this area is indeed warranted and is deemed valuable. We are not aware of tests for MAR or MNAR. It is clear that the data itself may not be used to test for MNAR, because the missingness depends on the missing value itself. As we will see in Section 6, however, some sensitivity analysis may be useful in assessing MNAR.

3. Methods for handling missing data

3.1. Ad hoc methods

Among the various ways in which missing data are handled, the ad hoc methods are relatively easy to understand and execute, but their simplicity, as has been pointed out in various SEM papers, comes at the price of often generating biased or poor estimates. Here we describe a few of such methods along with their possible defects.

3.1.1. Complete case analysis

Complete case analysis, also known as *listwise deletion* (LD), utilizes only the cases in a data set for which there are no missing values on any of the variables. This can result in loss of significant amount of information even in data sets that contain a modest number of variables. For example, when ten variables are independently measured with a 90% chance of observing a single case of each variable, the probability that a case contains no missing values is only 35%. When data are MCAR, the complete cases form a random subsample from the population, thus the estimates obtained will not be biased. But obviously, there can be a significant loss of efficiency in parameter estimates when a large amount of data are discarded. Of course, if the data are MNAR the results of LD will most likely be biased.

3.1.2. Available case analysis

Available case analysis, also known as *pairwise deletion* (PD), uses all the available data rather than just cases which have no missing values. This avoids throwing away possibly useful information, especially if there are few or no complete cases. In almost all SEM packages, the sample mean and sample covariance are sufficient input to fit a model. In the available case analysis, the sample mean and the sample variance for each variable are computed based on all the observed cases for the corresponding variable, and the covariance between a pair of variables is computed based on all the observed cases for that pair. Brown (1983) investigated this method in the context of factor analysis. This method leads to unbiased estimates if data are MCAR, and it leads to biased estimates for MAR data. One major shortcoming of the available case method is that it can result in a covariance matrix that is not positive definite. Notwithstanding this problem, because it uses more data, the available case analysis is expected to be more efficient than the complete case analysis. If data are MCAR and the correlations between variables are modest, a simulation by Kim and Curry (1977) supports this expected conclusion. Other simulations, however, indicate superiority of complete case analysis in presence of large correlations (Azen and Van Guilder, 1981). Marsh (1998) performed a simulation that indicates this method can lead to substantially biased test statistics, depending on the percentage of missing data and the sample size.

3.1.3. Single imputation methods

In a *single imputation method* the missing data are filled by some means and the resulting completed data set is used for inference. *Mean imputation* (MI) is one such method in which the mean of the observed values for each variable is computed and the missing values for that variable are imputed by this mean. This method can lead into severely biased estimates even if data are MCAR (see, e.g., Jamshidian and Bentler, 1999). Clearly, if the number of missing values in a variable is large, and these values are imputed by the observed sample mean, then the resulting variance estimate for that variable can be severely underestimated.

To take advantage of the correlations that may exist between variables in a data set, Buck (1960) proposed imputing the missing values by predictions from regression models that are fitted using the mean and covariance matrix estimated by complete case analysis. The mean estimates from this method are consistent estimates of the population mean (Buck, 1960) for MCAR data and, under some mild regularity conditions, for MAR data. The variances and covariances are, however, underestimated by this method but the extent of underestimation is usually less than that of the unconditional mean imputation. Sometimes a random noise is added to the imputation values obtained based on Buck's method. Such imputations are referred to as *stochastic regression imputation*.

Other methods of imputation impute the missing data based on the observed cases for subjects that agree, or approximately agree, on some observed covariates, for example, age, gender, etc. An example of such a method is the *similar response pattern imputation* (SRPI) in which missing values are replaced by observed values from a case that scored similarly, where the similarity is determined by a set of user-specified matching variables. Rubin (1987) has discussed a number of other such methods.

3.1.4. A common problem with ad hoc methods in SEM

To fit a structural equation model when using the above methods, with the exception of the complete case analysis, a two stage method is followed. In the first stage, the missing data are imputed and the resulting completed data are used to obtain a sample mean and sample covariance matrix. In the second stage, these values are used in an SEM program to fit a model. It should be noted that even if the parameter estimates are unbiased, the standard errors produced by the SEM programs obviously do not take into account the variability inherent in the imputed values and thus, most likely, the resulting standard errors are underestimates. Thus one has to be cautious in taking the resulting standard errors at their face values when making inference. If the resulting mean and covariance estimates are consistent, as we will discuss in Section 3.2, adjustments to the standard errors are possible to make them valid.

3.2. Likelihood-based approaches for some standard and non-standard SEMs

A major portion of the SEM missing data literature is devoted to normal-theory maximum likelihood methods for model fitting and inference. This literature explores computational methods for parameter estimation of various structural equation models, discusses standard error estimation and tests of hypotheses, and performs simulation studies to compare the methods.

As in Section 1, let $\mathbf{x}_1, \ldots, \mathbf{x}_n$ denote a random sample from a population with mean $\boldsymbol{\mu}(\boldsymbol{\theta})$ and covariance $\Sigma(\boldsymbol{\theta})$, where $\boldsymbol{\theta}$ is the parameter vector to be estimated. Furthermore, assume that some of the $\mathbf{x}_i$'s are not observed completely and denote the observed part of $\mathbf{x}_i$ by $\mathbf{y}_i$. Assuming that $\mathbf{x}_i$ are from a p-variate normal distribution with mean

$\boldsymbol{\mu}(\boldsymbol{\theta})$ and covariance $\Sigma(\boldsymbol{\theta})$, then $\mathbf{y}_i$ has a p_i ($\leqslant p$)-variate normal distribution with mean $\boldsymbol{\mu}_i(\boldsymbol{\theta})$ and covariance $\Sigma_i(\boldsymbol{\theta})$, where p_i is the number of components of $\mathbf{y}_i$, and $\boldsymbol{\mu}_i(\boldsymbol{\theta})$ and $\Sigma_i(\boldsymbol{\theta})$ are, respectively, the subvector and submatrix of $\boldsymbol{\mu}(\boldsymbol{\theta})$ and $\Sigma(\boldsymbol{\theta})$ corresponding to the observed components of $\mathbf{x}_i$. Assuming that $\mathbf{y}_1, \ldots, \mathbf{y}_n$ are independent, then the contribution of a case $\mathbf{y}_i$ to the log-likelihood is

$$l_i(\boldsymbol{\theta}|\mathbf{y}_i) = \frac{p_i}{2}\log(2\pi) - \frac{1}{2}\bigl\{\log\bigl|\Sigma_i(\boldsymbol{\theta})\bigr| + \bigl(\mathbf{y}_i - \boldsymbol{\mu}_i(\boldsymbol{\theta})\bigr)^{\mathrm{T}}\Sigma_i^{-1}(\boldsymbol{\theta})\bigl(\mathbf{y}_i - \boldsymbol{\mu}_i(\boldsymbol{\theta})\bigr)\bigr\}. \quad (5)$$

Thus, the maximum likelihood of $\boldsymbol{\theta}$, is the value that maximizes the log-likelihood function

$$l(\boldsymbol{\theta}|\mathbf{y}_i) = \sum_{i=1}^{n} l_i(\boldsymbol{\theta}). \quad (6)$$

For the special case of the saturated model and when the data are observed completely, the maximum likelihood of $\boldsymbol{\mu}(\boldsymbol{\theta})$ and $\Sigma(\boldsymbol{\theta})$ are the usual sample mean and sample covariance (with divisor of n rather than $n-1$). When the data are incomplete, then maximization of (6) requires iterative methods. The most common method to maximize (6) is the EM algorithm of Dempster et al. (1977). An EM algorithm is defined by a "complete" data with a mapping to the "observed" data. In our context, we let $\mathbf{x}_1, \ldots, \mathbf{x}_n$ denote the complete data. Then the EM algorithm consists of two steps: an expectation step (E-step) and a maximization step (M-step). At a point $\boldsymbol{\theta}$, the E-step computes

$$Q(\boldsymbol{\theta}', \boldsymbol{\theta}) = E^*\bigl[l(\boldsymbol{\theta}'|\mathbf{x}_i)\bigr], \quad (7)$$

where $E^*(\cdot) = E(\cdot|\mathbf{y}_i, \boldsymbol{\theta})$. The M-step consists of maximizing $Q(\boldsymbol{\theta}', \boldsymbol{\theta})$ with respect to $\boldsymbol{\theta}'$ to obtain a new point, $\tilde{\boldsymbol{\theta}}$. The iteration process is then as follows: At each step replace $\boldsymbol{\theta}$ by $\tilde{\boldsymbol{\theta}}$ and repeat the E-step and M-step until the values of $\boldsymbol{\theta}$ converge.

It turns out that using the EM algorithm for obtaining $\bar{\boldsymbol{\theta}}$, the ML estimate of $\boldsymbol{\theta}$ under the saturated model is fairly straightforward. For this method, EM is commonly used for this purpose and the resulting $\boldsymbol{\mu}(\bar{\boldsymbol{\theta}})$ and $\Sigma(\bar{\boldsymbol{\theta}})$ in the SEM literature are often referred to as the EM mean and covariance. This brings us to the first type of likelihood-based estimation method.

3.2.1. *EM mean and covariance*

Because, in the complete data case, the sufficient statistics to obtain ML estimates of $\boldsymbol{\theta}$ in SEM are the sample mean and sample covariance, most SEM software allow these quantities as input, rather than the raw data. Obviously, these sufficient statistics cannot be computed based on a set of incomplete data. However, the EM mean $\boldsymbol{\mu}(\bar{\boldsymbol{\theta}})$ and covariance $\Sigma(\bar{\boldsymbol{\theta}})$ obtained under the saturated model seem to be a natural surrogate for the complete data sample mean and covariance. This suggests a method to fit an SEM to incomplete data; namely, to utilize $\boldsymbol{\mu}(\bar{\boldsymbol{\theta}})$ and $\Sigma(\bar{\boldsymbol{\theta}})$ in place of the complete data sample mean and covariance and proceed with model fitting. We denote the estimate of $\boldsymbol{\theta}$ obtained based on this two-stage procedure by $\tilde{\boldsymbol{\theta}}$. Finkbeiner (1979) and Brown (1983) investigated this procedure in the context of factor analysis. Based on a simulation study, Brown (1983) favored this method over both listwise and pairwise deletion. Arminger and Sobel (1990) pointed out the difficulty of obtaining standard errors for $\tilde{\boldsymbol{\theta}}$, and that this method is generally not as efficient as the full information maximum likelihood, which we will discuss shortly.

When one of the two assumptions of (I) data are MAR and normally distributed, or (II) data are MCAR and non-normal, is satisfied, Yuan and Bentler (2000) showed that $\tilde{\boldsymbol{\theta}}$ is strongly consistent and asymptotically normally distributed (see also, Laird, 1988). In their Eq. (6b), they give a *sandwich type* estimate of the asymptotic covariance of $\tilde{\boldsymbol{\theta}}$. The sandwich type estimator is a function of the observed information matrix and the empirical information matrix. Because the EM mean and covariance are not based on n complete cases, if the sample size of n is input in the SEM software along with these quantities, the standard errors output by the SEM will be underestimates of the true standard errors. Enders and Peugh (2004) have investigated input of a smaller sample size to adjust the standard errors in this circumstance. As expected, and as they point out, “there is no single value of n that is appropriate when using an EM covariance matrix as input into an SEM analysis”.

3.2.2. *Full information maximum likelihood*

A natural method of obtaining ML estimate of $\boldsymbol{\theta}$ is to maximize (6) with respect to $\boldsymbol{\theta}$, using the $\boldsymbol{\mu}(\boldsymbol{\theta})$ and $\Sigma(\boldsymbol{\theta})$ induced by the model in (3). We denote this estimate by $\hat{\boldsymbol{\theta}}$. In factor analysis, one would use $\boldsymbol{\mu}(\boldsymbol{\theta}) = 0$ and $\Sigma(\boldsymbol{\theta}) = \Lambda\Phi\Lambda^{\mathrm{T}} + \Psi$ and maximize with respect to elements of $\boldsymbol{\theta}$ which in this case consist of free elements in Λ, Φ, and Ψ. In the SEM literature, this method is often referred to as the *full information maximum likelihood* (FIML). Under the assumption that the data are normal and data are either MCAR or MAR, the FIML estimates are asymptotically normal, unbiased, and fully efficient with standard errors that can be obtained from the observed information matrix. Note that the Fisher information standard errors are valid for MCAR data, but are not valid for MAR data (Kenward and Molenberghs, 1998), and it is recommended that standard errors be computed based on the observed information matrix.

Computation of FIML estimator depends on the type of SEM model being fit. Finkbeiner (1979) and Lee (1986) respectively gave a quasi-Newton method and a Fisher-scoring algorithm to directly maximize (6). Jamshidian (1997) proposed an EM algorithm to obtain FIML estimates for the confirmatory factor analysis in which the expectation in the E-step and the maximization in the M-step have closed from. Extension of this method to more general covariance structure models, such as the LISREL model with implied covariance (4), seem not to be straightforward.

For a more general SEM than the factor model, Muthén et al. (1987) proposed using a multiple group option of available packages to obtain $\hat{\boldsymbol{\theta}}$ (also see Allison, 1987). In their procedure, a group consists of all the data with similar missing data patterns. Thus the usual sample mean and covariance are computed based on the observed data for each group and the multiple group model is fit under the restriction that $\boldsymbol{\mu}(\boldsymbol{\theta})$ and $\Sigma(\boldsymbol{\theta})$ are equal across all groups. This method works fine, except for the limitation that there has to be sufficient number of cases in each group to obtain a positive definite sample covariance for the groups. Jamshidian and Bentler (1999) gave a general framework for implementing EM to obtain FIML estimates. Their algorithm takes advantage of the machinery used in obtaining FIML when data are complete. Because, for a general structural equation model, the M-step of their algorithm is iterative, they also proposed a generalized EM algorithm which is very simple to implement, given a complete data routine.

Tang and Bentler (1998) developed the statistical theory and proposed an EM algorithm for FIML estimates under equality constraints on parameters. This work can be extended to inequality constraints on parameters using the methods discussed in Jamshidian (2004a). In another extension, Graham (2003) has discussed the problem of adding missing-data-relevant (auxiliary) variables to FIML-based SEM in order to improve efficiency and bias. He has given Amos and LISREL 8.5 code for implementation of his methods.

3.2.3. Nonlinear SEM

A nonlinear SEM allows modeling a nonlinear relationship between the latent variables, for example, quadratic and interaction effects amongst the latent variables. To give a specific model, the linear latent variable model (3) can be replaced by

$$\eta_i = B\eta_i + \Gamma F(\boldsymbol{\xi}_i) + \delta_i, \tag{8}$$

where $F(\boldsymbol{\xi}_i)$ is a $k_2 \times 1$ vector valued function. Historically, nonlinear SEM goes as far back as McDonald (1962), where he considered nonlinear factor analysis. Since then, a number of papers have appeared on nonlinear SEM, proposing various approaches to this problem. Lee and Zhu (2002) give a review of this literature and develop the maximum likelihood approach for nonlinear SEM of continuous and complete data.

Given that a method for complete data nonlinear SEM is available, a natural approach to estimate the parameters in the incomplete data case is to utilize the EM algorithm. It turns out, however, that because of the nonlinearity, neither the E-step nor the M-step of the EM algorithm have closed from solutions. Lee et al. (2003) have utilized the method of Monte Carlo EM given by Wei and Tanner (1990) to approximate the E-step, and they use a sequence of conditional maximization, as in the ECM algorithm of Meng and Rubin (1993) to perform the maximization step. Lee et al. (2003) utilize the method from Louis (1982) to obtain standard errors. In a simpler approach to obtaining standard errors, one may utilize one of the methods proposed by Jamshidian and Jennrich (2000). Lee et al. (2003) have illustrated their method using a numerical example, and have suggested methodologies for assessing some of the distributional assumptions made. More recently, Lee and Tang (2006) have developed a Bayesian approach for analyzing nonlinear structural equation models with non-ignorable missing data. In general, nonlinear SEM is fairly complicated even for complete data, from both computational and modeling perspectives. Computations are further exasperated by missing data. It will be useful to see some real applications of nonlinear SEM and simulation studies that would reveal the advantages of the this method to the linear SEM method. Finally, we would like to mention that as in FIML, the models proposed are valid only if data are MCAR or MAR.

3.2.4. Mixture SEM

When a data set is comprised of several groups, it would be sensible to fit a mixture model to the data. For example, if it is assumed that the data are comprised of g groups, then a g-component mixture model is fit in which each $\mathbf{x}_i$ is assumed to come from $\mathcal{N}(\boldsymbol{\mu}^{(h)}(\boldsymbol{\theta}), \Sigma^{(h)}(\boldsymbol{\theta}))$, for $h = 1, \ldots, g$ with some probability π_h, where $\sum_{h=1}^{h} \pi_h = 1$. Accordingly, if $\mathbf{y}_i$ is the observed part of $\mathbf{x}_i$, then its contribution to the likelihood will be

$$l_i(\boldsymbol{\theta}|\mathbf{y}_i) = \sum_{h=1}^{g} \pi_h l_i^{(h)}(\boldsymbol{\theta}|\mathbf{y}_i), \tag{9}$$

where

$$\begin{aligned} l_i^{(h)}(\boldsymbol{\theta}|\mathbf{y}_i) = \frac{p_i}{2}\log(2\pi) - \frac{1}{2}\big\{\log\big|\Sigma_i^{(h)}(\boldsymbol{\theta})\big| \\ + \big(\mathbf{y}_i - \boldsymbol{\mu}_i^{(h)}(\boldsymbol{\theta})\big)^{\mathrm{T}}\big(\Sigma_i^{(h)}\big)^{-1}(\boldsymbol{\theta})\big(\mathbf{y}_i - \boldsymbol{\mu}_i^{(h)}(\boldsymbol{\theta})\big)\big\}, \end{aligned}$$

where, as before, p_i is the number of components of $\mathbf{y}_i$, and $\boldsymbol{\mu}_i^{(h)}$ and $\Sigma_i^{(h)}$ denote the subvector and submatrix of $\boldsymbol{\mu}^{(h)}$ and $\Sigma^{(h)}$ corresponding to the observed components of $\mathbf{y}_i$. In this case, the overall number of parameters is the aggregate of the parameters in the mean and implied covariance for each group plus the admixture parameters $\pi_1, \ldots, \pi_g$.

By far the most popular algorithm to handle mixture models, even in the complete data case, is the EM algorithm where the admixture parameters π_h are considered as incomplete data. Recently, Lee and Song (2003) formulated the EM algorithm for the mixture SEM with the implied covariance structure (4) for each group for the case when data are incomplete. Their algorithm maximizes (4) with $l_i(\boldsymbol{\theta}|\mathbf{y}_i)$ defined by (9). The stochastic version of EM (Wei and Tanner, 1990), mentioned above, is used as the E-step does not have a closed form. The M-step is performed via a sequence of conditional maximization as in ECM. Lee and Song (2003) performed a set of simulation studies that showed the superiority of the ML method over the listwise deletion method. They have also applied this model to real data set collected in the project World Values Survey 1981–1984 and 1990–1993.

3.2.5. Generalized least squares and minimum chi-square

Lee (1986) has proposed a generalized least squares method for estimating the parameters $\boldsymbol{\theta}$ in SEM in an effort to do without the normality assumption. Suppose that there are m missing data patterns and for each pattern j, there exist n_j cases, sufficiently large, based on which a positive definite sample covariance S_i is obtained. Lee (1986) proposed estimating $\boldsymbol{\theta}$ by minimizing

$$G(\boldsymbol{\theta}) = \sum_{j=1}^{m} \frac{n_j}{n} \operatorname{trace}\big\{\big(S_j - \Sigma_j(\boldsymbol{\theta})\big)W_j\big\}, \tag{10}$$

where $\Sigma_j(\boldsymbol{\theta})$ is the subset of $\Sigma(\boldsymbol{\theta})$ related to the observed components in the pattern j, and W_j is a positive definite weight matrix which converges in probability to the true

$\Sigma^{-1}(\boldsymbol{\theta})$. He gave an iterative algorithm to accomplish this and gave formulas for the standard error of estimates.

With the same intention of moving away from assumption of normality, recently Yuan and Bentler (2000) gave an estimation method that utilizes the minimum chi-squared method of Ferguson (1996, Chapter 23). Let vech(.) be an operator that transforms a symmetric matrix into a vector by stacking the column of the matrix, leaving out the elements above the diagonal. Let $\boldsymbol{\beta}(\boldsymbol{\theta}) = (\mathrm{vech}(\Sigma(\boldsymbol{\theta}))^{\mathrm{T}}, \boldsymbol{\mu}(\boldsymbol{\theta})^{\mathrm{T}})^{\mathrm{T}}$. Let $\hat{\boldsymbol{\beta}} = (\mathrm{vech}(\Sigma(\bar{\mathbf{b}}))^{\mathrm{T}}, \boldsymbol{\mu}(\bar{\mathbf{b}})^{\mathrm{T}})^{\mathrm{T}}$ be the estimate of $\boldsymbol{\beta}$ obtained from what we called the EM estimates of $\boldsymbol{\mu}$ and Σ from the saturated model. Furthermore, let $\widehat{\Omega}$ denote the sandwich type estimate of the asymptotic covariance of $\hat{\boldsymbol{\beta}}$. Then the minimum chi-square estimate of $\boldsymbol{\theta}$ is obtained by minimizing

$$Q(\boldsymbol{\theta}) = \big(\hat{\beta} - \beta(\boldsymbol{\theta})\big)^{\mathrm{T}} \Omega^{-1} \big(\hat{\beta} - \beta(\boldsymbol{\theta})\big) \tag{11}$$

with respect to $\boldsymbol{\theta}$. Yuan and Bentler (2000) gave the asymptotic standard error formulas for this estimator and stated that it is asymptotically normal. They state that when data are not normal, the minimum chi-square estimator is asymptotically at least as efficient as the FIML and the ML estimate $\tilde{\boldsymbol{\theta}}$ that uses the EM mean and covariance.

3.2.6. *Tests of goodness of fit*

Tests of goodness of fit are usually performed in order to evaluate the validity of the structural model $\boldsymbol{\mu}(\boldsymbol{\theta})$ and $\Sigma(\boldsymbol{\theta})$. If $\hat{\boldsymbol{\theta}}$ is the FIML estimator and $\bar{\boldsymbol{\theta}}$ is the estimate of the parameters under the saturated model, then the test statistics $T_1 = 2[l(\bar{\boldsymbol{\theta}}) - l(\hat{\boldsymbol{\theta}})]$ is the usual likelihood ratio test and under the null hypothesis and with the assumptions of MCAR and normality has an asymptotic χ^2 distribution (see Jamshidian and Bentler, 1999). For the estimator $\tilde{\boldsymbol{\theta}}$, a surrogate test to the complete data case is

$$\begin{aligned} T_2 = n\Big\{&\mathrm{trace}\big[\Sigma^{-1}(\tilde{\boldsymbol{\theta}})\big(\Sigma(\bar{\boldsymbol{\theta}}) + \big(\boldsymbol{\mu}(\bar{\boldsymbol{\theta}}) - \boldsymbol{\mu}(\tilde{\boldsymbol{\theta}})\big)\big(\boldsymbol{\mu}(\bar{\boldsymbol{\theta}}) - \boldsymbol{\mu}(\tilde{\boldsymbol{\theta}})\big)^{\mathrm{T}}\big)\big] \\ &- \log\big|\Sigma(\bar{\boldsymbol{\theta}})\Sigma^{-1}(\tilde{\boldsymbol{\theta}})\big| - p\Big\}. \end{aligned}$$

While T_1 and T_2 are equal in the complete data case, it is not clear whether the asymptotic distribution of T_2 is χ^2 with incomplete data. When data are non-normal and MCAR, Yuan and Bentler (2000) have shown that both T_1 and T_2 follow a mixture of χ^2 distributions, and give rescaled versions of T_1 and T_2 as well. Since neither T_1 nor T_2 or their rescaled versions asymptotically follow a chi-square distribution, Yuan and Bentler (2000) have proposed a test statistics, that can be evaluated at either $\hat{\boldsymbol{\theta}}$ or $\tilde{\boldsymbol{\theta}}$, that under some regularity conditions has an asymptotically χ^2 distribution. Finally, Yuan and Bentler (2000) have proposed yet another test $T_3 = nQ(\boldsymbol{\theta}^*)$, where $\boldsymbol{\theta}^*$ is the minimizer of (11). This statistics is asymptotically χ^2.

3.2.7. *SEM with polytomous data*

In all of the procedures described above, it is assumed that data are continuous. In social and behavioral sciences, sometimes data are dichotomous or, more generally, polytomous. Routine application of the continuous data methodology to polytomous data can result in biased estimates. Lee and Tang (1992) provided a two-stage procedure for parameter estimation in SEM of polytomous incomplete data. The first stage of their method uses the partition maximum likelihood approach of Lee and Chiu (1990) to estimate the correlation matrix. Then, in the second stage, this correlation matrix is utilized to estimate parameters via the generalized least squares approach. Recently, Song and Lee (2002) used a Bayesian approach to fitting SEM to MCAR or MAR data that consists of mixed continuous and polytomous data. They provide standard error of estimates and related test statistics for their procedure. Lee and Song (2003) have extended this methodology to nonlinear SEM. Rabe-Hesketh et al. (2004a) have introduced a set of models called generalized linear latent and mixed models (GLLAMM), which combine features of generalized linear mixed models (GLMM) and SEM and consist of a response model and a structural model for the latent variables. GLLAMMs can handle responses of mixed type including continuous responses, counts, dutarion/survival data, dichotomous, and ordered and unordered categorical responses.

3.3. Simulation-based methods

Multiple imputation and bootstrap are two main simulation-based methods that can be used in analysis of incomplete data with the latter receiving noticeably more attention in this context. The methods based on the data augmentation of Wei and Tanner (1990) and the Gibbs sampler, mentioned in the previous section, may also be considered simulation-based methods, but we would like to think of them more as machineries to accomplish the E-step of the EM algorithm. Thus in this view, they can be considered computational algorithms to obtain maximum likelihood estimates. In the context of SEM of incomplete data, a few articles have recently discussed multiple imputation and much fewer have been devoted to bootstrap.

3.3.1. Multiple imputation

Multiple imputation consists of producing, say m, complete data sets from the incomplete data by imputing the missing data m times by some reasonable method. Then each completed data set is analyzed using a complete data method and the resulting methods are combined to achieve inference. Multiple imputation is motivated by the Bayesian framework and as such, the general methodology suggested for imputation is to impute using the posterior predictive distribution of the missing data given the observed data and some estimate of the parameters. For generating imputations from this distribution, one, for example, can use the ML estimates for the parameters or use the Bayesian framework to generate m sets of parameter estimates from the posterior distribution of $\boldsymbol{\theta}$ in some Bayesian analysis of the data. More detailed imputation methods and methods of combining parameters can be found in Rubin (1987). A short overview can be found in Schafer and Graham (2002) or Jamshidian (2004b).

Schafer and Graham (2002) and Allison (2003) seem to be the two major articles that discuss employment of multiple imputation in SEM. Both articles discuss various available software for multiple imputation and their utility for SEM. Clearly the method of imputation plays a key role in success of the multiple imputation methods. To our knowledge, to date, the imputation methods that are employed in SEM do not utilize the

SEM being fit. In a different context than SEM, Song and Belin (2004) discuss the use of the factor analysis model for imputing incomplete high-dimensional data. We think that there is much work to be done with regard to the utility of multiple imputation in the area of SEM. The main job is the method of imputation. Comparison between proper and improper imputations, employing SEM in place of saturated models for data imputation, and the effect of data mechanism on the results are future valuable work in SEM.

3.3.2. Bootstrap

One of the major motivations in using bootstrap is to do away with distributional assumptions and produce nonparametric inference. This methodology can be used for standard error estimation, construction of confidence intervals, and test of hypotheses. The bootstrap method has been discussed by several authors in the context of SEM and complete data. There is much less work in the context of incomplete data. Enders (2001) is the only work that we could find that discusses the method of bootstrap for testing goodness of fit in the incomplete data setting in SEM. We will discuss his simulation results in the next section. More work in this area might include examining various bootstrap methodologies for incomplete data. Following Efron (1994), Jamshidian (2004b) gives an overview of three methods of nonparametric bootstrap, full mechanism bootstrap, and multiple imputation bootstrap. He has examined, via simulation, these methodologies in the context of a simple problem, and in some cases the methods are promising. It would be useful to investigate applications of these methods in the context of structural equations models. The utility of the methods in analysis of non-normal data and again their robustness to missing data mechanism is of interest.

4. Simulation studies

Most of the methodological papers mentioned above perform some kind of a simulation study to empirically examine their theoretical results. In this section, we do not attempt to report all of these simulation studies, but rather we focus on studies that have made comparison between various methods. Specifically, in the following two subsections we summarize simulation studies done with regard to normal and non-normal data.

4.1. Comparison of methods under the normality assumption

Arbuckle (1996) conducted a simulation to demonstrate the efficiency of ML estimates relative to pairwise and listwise deletion for a typical estimation problem. In his setting, he used an SEM consisting of two latent variables, each with three observed variables. He generated data with sample sizes 145 and 500 and considered missing data mechanisms of MCAR and MAR with missing rates of 0, 5, 10, 20, and 30%. He used the parameter estimates from the data with no missing values as a benchmark and compared the estimates obtained based on incomplete data to the benchmark estimates. His study found that for MCAR as well as MAR data, ML estimation was superior to both the pairwise and listwise deletion, with the superiority being more pronounced for MAR data. He noted that increasing the sample size could not compensate for the bias of the listwise and pairwise deletion methods and that ML did not appear to fully compensate for the bias created by the missing data process.

Bernaards and Sijtsma (1999, 2000) performed an extensive simulation study to compare FIML to several imputation methods for factor analysis of incomplete data. They implemented each imputation method with and without adding a random residual and studied the effects of sample size, percentage of missingness, and missing data mechanism. They compared how well each method recovered the factor loadings and concluded that FIML performed best as compared to the considered imputation methods. Among the imputation methods, however, they recommended methods that impute mean per person across the available scores for that person.

4.2. Comparison of methods for non-normal data

Enders (2001) explored the impact of non-normality on FIML estimation for SEM with missing data. Three studies were conducted with the first two concerning MCAR and MAR data. The methods of LD, PD, MI and SRPI were examined and compared to ML. These five methods were compared in the four contexts of bias, mean square errors, standard errors (confidence interval coverage), and model rejection rates. The simulations for Studies 1 and 2 included a full structural equation model with three latent variables, each with three observed variables. Samples sizes of 250, 500, and 750; missing data rates of 0, 5, 10, 15, and 25%; and seven distributional conditions of different levels of non-normality, skewness, and kurtosis were used. For each of the 105 possible between-subject designs, 250 raw data matrices were generated.

Enders defined parameter estimate bias as

$$\%\text{Bias} = \left(\frac{\hat{\bar{\theta}}_j - \theta_j}{\theta_j}\right) \times 100,$$

where θ_j was the true population value for parameter j and $\hat{\bar{\theta}}_j$ was the mean of the corresponding parameter estimate from the 250 replications. He considered percent bias values of less than 10–15% as non-problematic. For the MCAR data, Enders reported little or no bias in the structural model parameters for all the methods with non-normality having no noticeable impact on the bias observed. MI did yield biased factor loadings, with the bias increasing as the missing data rate increased, but the bias did not exceed problematic levels. For the MAR data, Enders reported that all methods, except FIML, yielded biased estimates at problematic levels with an increase in bias as the missing data rates increased. Surprisingly, his study showed that as non-normality, and particularly skewness, increased the bias from the ad hoc methods decreased. Enders cautioned that a different result would most likely have occurred in this case had the data been negatively, rather than positively, skewed or a different MAR technique was used.

Enders compared the relative efficiency of FIML to the other ad hoc methods by using the ratio of the mean squares error (MSE) for these methods. The main message here was that for the MCAR data the relative efficiency of FIML increased as the missing data rates increased, but the distribution shape had little or no impact on the relative efficiencies. For the MAR data, the results were more sensitive to the shape and as with the bias, the ad hoc methods yielded more efficient estimates under extreme non-normality than FIML. Again, one has to be cautioned in concluding that general ad hoc methods work better under non-normality.

In comparing coverage probabilities, Enders' results for both MCAR and MAR were nearly the same with the coverage rates dropping below their nominal levels as the non-normality increased for all the methods. The rejection rates across all methods were also above their nominal level especially when the non-normality increased, with the level of deviation depending on the missing data mechanism. Enders' third study concluded much more reasonable results in terms of coverage and rejection rates for the Bollen–Stein bootstrap method under non-normal data. An overall conclusion of Enders was that the non-normal data had the same negative effect on ML estimation for both missing and complete data and the presence or amount of missing data did not increase the problems due to non-normality.

Yuan and Bentler (2000) performed simulation studies to assess their theoretical contributions to the three methods under their study. Their simulation aimed at studying the two main assumptions of (I) normal data and MAR mechanism, and (II) non-normal data and MCAR mechanism. They concluded that under the normality assumption, the estimates under MCAR are generally less biased than those under MAR for the three methods. They point out that a similar inaccuracy is observed when the distributional assumptions are incorrect, and emphasize that using a normal distribution and a MAR missing data mechanism leads to the same parameter estimates as using an unknown distribution and an MCAR missing data mechanism. Their simulation studies, however, do not indicate noticeable biases for non-normal data that are MAR. Taking into account their analytical and simulation results, they recommend use of the minimum chi-squared method when sample size is large, use of the two-stage method with the sandwich-type covariance matrices for standard error for medium sample size, and they recognize that the problem of small samples is still an open one.

5. Sensitivity analysis for missing data mechanism

It is evident that the quality of inference made when data are incomplete critically depends on the missing data mechanism. Testing for the type of missing data mechanisms in absence of auxiliary information is fairly difficult. MCAR may be the easiest mechanism to test, and tests for MCAR have been proposed (see, e.g., Little, 1988; Kim and Bentler, 2002). The proposed tests, however, work in some special cases and the problem of testing MCAR for more general cases seems to be still open. To test whether data are MNAR is virtually impossible because the missingness in this case would depend on the missing data itself, thus in the absence of information no tests can be developed. In the SEM literature (e.g., Allison, 2003; Schafer and Graham, 2002) and elsewhere, performing sensitivity analysis has been encouraged as a possible way to detect deviations from an assumed missing data mechanism. To our knowledge, however, there does not seem to be any specific proposals about how one should go about performing such sensitivity analyses. As we have found out through some effort, the answer to this question may not be straightforward and indeed deserves some attention. In this section, we will give one specific method for sensitivity analysis and report our simulation studies. In the outset we would like to caution the reader that the methods that we discuss here are not well developed and have been tested under specific assumptions that we will mention. Our hope is that this work will encourage further research in this topic and indeed we have such research underway ourselves.

The method that we propose aims at determining whether data are MCAR. It differs from the previously proposed methodology (e.g., Kim and Bentler, 2002; Muthén et al., 1987) in that we do not group the data by their missing data pattern; you can have two cases with the same missing data pattern that do not follow the same missing data mechanism. Suppose that a data set consists of n cases, n_c of which are completely observed. We assume that the n_c cases are observed at random, and therefore they constitute a random sample from the population. Note that one has to examine this assumption carefully, as it may not hold in some cases. If data are MCAR, then any subsample of the n cases would be a random sample of the population. Thus, if an unbiased estimator of model parameters is applied to a random subsample, the result would be an unbiased estimate of the parameters. Indeed the deviation of the estimates from the true parameter would depend on the size of the sample. On the other hand, if data are MNAR, and a subsample is taken, then that subsample may not be considered a "random sample" from the population in the sense that it may satisfy a different model as compared to the population as a whole. Thus, if we apply our estimator on this subsample, the resulting estimate may be different as compared to the estimates obtained from complete cases only. This difference motivates the sensitivity analysis that we describe next. Because we will use ML estimates, the estimates for both MCAR and MAR data would be consistent, and in that sense we would say that our methodology is sensitive to MNAR data.

The sensitivity analysis usually should follow once we have decided on a model and an estimation method. Having the model and an estimation method at hand, we propose to perform a sensitivity analysis as follows:

(1) Obtain an estimate of the parameters, using the method and based on the n_c complete cases.
(2) Choose a random subsample of size n_c from all the n cases and obtain the parameter estimates based on this random subsample.
(3) Repeat step 2, r times to capture the variability of the estimates.
(4) Compare the estimate obtained in step 1 to the r parameters obtained in step 3. Significant differences between the parameter estimate in step 1 and those in step 3 can be an indication that data might not be MCAR.

Of course, one can use variations of the above procedure. Some of the variations include resampling the complete cases to capture the variability of the estimate based on the complete data, use a sample size different from n_c in step 2, and if the number of complete cases is sufficiently large, we may choose a portion in step 1 and do not use that portion in the remaining steps. Indeed, success of this method would also depend on the sample sizes and the number of complete cases. Moreover, statistical methods should be developed to do the comparison in 4. These are subject of a current research underway by the authors. For this chapter, however, we will employ the steps mentioned above in a simulation study to assess sensitivity to missing data mechanism. For step 4, we use a simple method of comparison, as we explain.

For our simulation study, we use a three-factor factor analysis model with population parameters

$$\Lambda^{\mathrm{T}} = \begin{pmatrix} \lambda & \lambda & \lambda & 0 & 0 & 0 & 0 & 0 & 0 \\ 0 & 0 & 0 & \lambda & \lambda & \lambda & 0 & 0 & 0 \\ 0 & 0 & 0 & 0 & 0 & 0 & \lambda & \lambda & \lambda \end{pmatrix}, \qquad \Phi = \begin{pmatrix} 1 & 0.5 & 0.5 \\ 0.5 & 1 & 0.5 \\ 5 & 0.5 & 1 \end{pmatrix},$$

and Ψ a diagonal matrix with diagonal elements $1 - \lambda$. Note that the zero values and the ones on the diagonal of Φ are fixed. We generated data from a multivariate normal distribution with mean zero and covariance $\Sigma = \Lambda\Phi\Lambda^{\mathrm{T}} + \Psi$. The parameters that we estimate are the factor loadings λ, factor covariances, unique variances, as well as the mean of the nine variables.

As a *benchmark* setting we use $\lambda = 0.8$, and a sample of size $n = 1000$ with the $n_c = 200$ complete cases. The remaining 800 cases have roughly 25% of their data missing. We also use $r = 25$ in step 3. The data was generated by first generating 200 complete cases and setting them aside. Then an additional 800 cases were created ensuring that each case had at least one incomplete datum. We did not keep the cases that "survived" the missing data mechanism, because when generating data that were MNAR these complete cases would be observed not at random and thus contaminate our data set. Note that one assumption of our sensitivity analysis is that the complete data cases are observed at random.

To create data that were MCAR, each datum was assigned a random number from 0 to 1. If that number exceeded 0.75, the datum was deleted and otherwise, it was left alone. To create the MNAR data, all nine variables were normalized to have mean 0 and variance 1, creating a set of v_{ij} corresponding to the (i, j) component of the data matrix. Then we deleted the (i, j) datum in our data matrix, if $v_{ij} > 0.67$. This created about 25% missing data.

In our simulation, we have used the benchmark or varied one of the components of the benchmark. For example, the boxplots shown in Figures 1 and 2, respectively, correspond to two cases of data that are MCAR and MNAR and we changed the percentage of missing from the benchmark of 25% to the value of 15%. The boxplots on the figures are a summary of the $r = 25$ estimates of each of the parameters in μ, Λ, and Ψ (step 3) and the circle indicates the parameter estimate obtained for the complete data (step 1). For simplicity, parameters in Φ are not shown. What stands out from these figures is that the circles are outside of the boxplots' whiskers in many of the instances for the MNAR data, but this is not the case for MCAR data. This motivated us to examine the percentage of times that the complete-data estimate falls outside of the whiskers as a gauge for step 4 of our procedure. It is interesting to note that in Figure 2, each of the first boxplots for $\boldsymbol{\mu}$, Λ, and Ψ has a different location than others. This is due to the fact that we did not install any missing data in the first variable.

To account for variability, we have repeated each of the steps 1–4 ten times, and observed the number of times the circles (estimates based on the complete data) fall

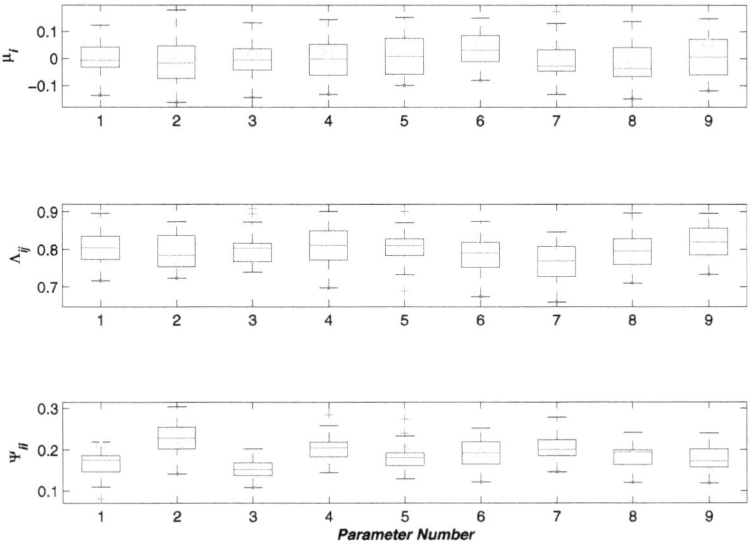

Fig. 1. Sensitivity analysis boxplots. Benchmark setting with data MCAR.

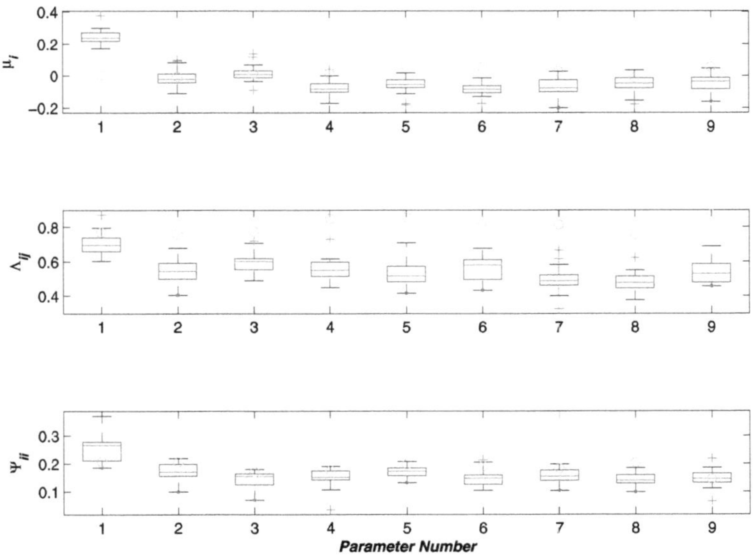

Fig. 2. Sensitivity analysis boxplots. Benchmark setting with data MNAR.

outside of the whiskers for each parameter under conditions of MCAR and MNAR data. Table 2 gives a summary of our observations. Overall, as we had hoped, the average number of circles outside the boxplots for the MNAR data is significantly larger than that for MCAR data. An interesting observation is that the parameters in Ψ are

Table 2
Average number of complete data parameter estimates outside the whiskers when the experiment was repeated 10 times

		μ	Λ	Φ	Φ	Total
Benchmark	MCAR	0.4	0.2	0.1	0.5	1.2
	MNAR	5.3	8.1	3	2	18.4
$r = 50$	MCAR	0.4	0	0.1	0.1	0.6
	MNAR	4.8	8	3	1.3	17.1
15% missing data	MCAR	0.6	0.4	0.2	0.7	1.9
	MNAR	7.3	8	3	1.3	19.5
35% missing data	MCAR	0.2	0.4	0.1	0.5	1.2
	MNAR	8.9	8	1.8	2.4	21.1
$n = 500$, $n_c = 100$	MCAR	0.2	0.8	0.2	0.2	1.4
	MNAR	2.8	8.5	1.5	0.6	13.4
$n = 1000$, $n_c = 100$	MCAR	0	1.1	0.1	0.3	1.5
	MNAR	4	8.1	2.4	1.6	16.1
$\lambda = 0.4$	MCAR	0.7	1.1	0	1	2.3
	MNAR	8	3.3	0	7	18.3

less sensitive to the missing data mechanism. This, of course may be an artifact of the way we have generated our missing data. Ignoring the sensitivities on parameter Ψ, our observations are summarized as follows: For the benchmark problem, the average number of complete data estimates falling outside the boxplots ($16.4/21 = 78\%$) is significantly higher for MNAR data than that for MCAR data ($0.7/21 = 3\%$).

To ensure that the number of replications $r = 25$ was sufficient, we run the benchmark using $r = 50$. From the row labeled "$r = 50$" in Table 2, it is clear that increasing the replications from 25 to 50 does not significantly change the results. In an attempt to see the performance of this type of sensitivity analysis with other rates of missing, we run the benchmark with rates of missing of 15 and 35%. The results for these cases are very much in line with that of the benchmark, with slightly more sensitivity to missing data mechanism when the percentage of missing data increases. Experiments with sample sizes of $n = 1000$ and $n_c = 100$ and $n = 500$ and $n_c = 100$ continue to show a similar pattern to the benchmark case. Finally, to see the effect of factor loading values, we changed the benchmark value of $\lambda = 0.8$ to $\lambda = 0.4$. Interestingly, in this case the sensitivity of parameter estimates in Λ to missing data mechanism decreased, but that for Ψ increased significantly. Even for this case, however, overall there is a significant difference between the MCAR and the MNAR case.

While our simulation study is far from complete, it has all the indications that one may be successful in developing sensitivity analyses to detect missing data mechanisms. We are in the process of doing further experiments with these types of methods, which, for example, includes comparison to data that are MAR. What seems to be the case, however, is that such sensitivity analysis may produce useful results and detect deviations from model assumptions in practice.

6. SEM software for incomplete data

Schafer and Graham (2002) provide some discussion of the available software for structural equation modeling of missing data. This section provides some updates and additions to their note.

We start with EQS 6.1 (Bentler and Wu, In press). Of the ad hoc methods mentioned, EQS has implemented LD, PD, mean imputation, regression imputation (Buck's method), and stochastic regression imputation. The default in the software is LD. Although classified under imputation methods, EQS has also implemented what we called EM mean and covariance in Section 3.2. The FIML estimator is also available in EQS, but a word of caution is that by default the standard errors are based on the Fisher information matrix. If data are MAR, one should use standard errors based on the observed information which are also available in EQS. This software includes tests for multivariate normality when data are incomplete (Yuan et al., 2004) as well as tests for MCAR (Kim and Bentler, 2002). We have pointed out some of the limitations of the MCAR tests, and some limitations also have been mentioned in the EQS manual. Methodologies for obtaining appropriate standard errors and test statistics for non-normal data are available in EQS. These are mainly based on the published methodology in Yuan and Bentler (2000).

Mplus (Muthén and Muthén, 2006) provides FIML and least squares estimation for continuous, censored, binary, ordinal, nominal, counts, or combinations of these variable types. A nice feature of Mplus is that the default FIML standard errors are computed based on the observed information, rather than the Fisher information. Additionally, bootstrap standard errors and confidence intervals are also available. Mplus also has facilities for obtaining parameter and standard error estimates based on the multiple imputation methodology. Rabe-Hesketh et al. (2004b) also have a Stata program for the GLLAMMs, mentioned in Section 3.2, which handles incomplete data of various forms mentioned in Section 3.2.

The currently available Release 5.0.1 of Amos, distributed by SPSS, offers FIML estimation. The upcoming Release 6.0, will add three imputation methods: regression imputation (Buck's method), stochastic regression imputation, and Bayesian imputation using a Markov chain Monte Carlo (MCMC) algorithm (Metropolis). The latter two methods can be used in Amos to create multiple data sets with imputations for multiple imputation analyses. Amos won't automatically conduct the multiple imputation analyses (i.e., automatically combine analyses from the multiple data sets), but a calculator is being made available in SPSS that will combine the results from the multiple analyses, whether done in Amos or elsewhere.

LISREL 8.7 (Jöreskog and Sörbom, 2004), distributed by Scientific Software International, includes FIML, multiple imputation, and imputation by matching (SRPI). The default method for this program is FIML. The multiple imputation procedure uses the Monte Carlo Markov Chain, and the EM procedure (Schafer, 1997).

There are other software, free and commercial, for SEM that we have not covered here. A list of some of this software can be found at http://www.smallwaters.com/weblinks/ and http://www.gsm.uci.edu/~joelwest/SEM/Software.html.

References

Allison, P.D. (1987). Estimation of linear models with incomplete data. In: Clogg, C. (Ed.), *Sociological Methodology*. American Sociological Association, Washington, DC, pp. 71–103.

Allison, P.D. (2003). Missing data techniques for structural equation modeling. *Journal of Abnormal Psychology* **112**, 545–557.

Arbuckle, J.L. (1996). Full information estimation in the presence of incomplete data. In: Marcoulides, G.E., Schumacker, R.E. (Eds.), *Advanced Structural Equation Modeling*. Lawrence Erlbaum, New Jersey.

Arminger, G., Sobel, M.E. (1990). Pseudo maximum likelihood estimation of mean and covariance structures with missing data. *Journal of the American Statistical Association* **85**, 195–203.

Azen, S., Van Guilder, M. (1981). Conclusions regarding algorithms for handling incomplete data. In: *Proceedings of the Statistical Computing Section*. American Statistical Association, pp. 53–56.

Bentler, P.M., Wu, E.J.C. (In press). *EQS for Windows User's Guide*.

Bernaards, C.A., Sijtsma, K. (1999). Factor analysis of multidimensional polytomous item response data suffering from ignorable item nonresponse. *Multivariate Behavioral Research* **34**, 277–313.

Bernaards, C.A., Sijtsma, K. (2000). Influence of imputation and EM methods on factor analysis when item nonresponse in questionnaire data is nonignorable. *Multivariate Behavioral Research* **35**, 321–364.

Brown, C.H. (1983). Asymptotic comparison of missing data procedures for estimating factor loadings. *Psychometrika* **48**, 269–291.

Buck, S.F. (1960). A method for estimation of missing values in multivariate data suitable for use with an electronic computer. *Journal of the Royal Statistical Society B* **22**, 302–306.

Dempster, A.P., Laird, N.M., Rubin, D.B. (1977). Maximum likelihood from incomplete data via the EM algorithm (with discussion). *Journal of the Royal Statistical Society B* **39**, 1–38.

Dixon, W.J. (Ed.) (1988). *BMDP Statistical Software*. University of California Press, Berkeley.

Efron, B. (1994). Missing data, imputation, and the bootstrap (with discussion). *Journal of the American Statistical Association* **89**, 463–479.

Enders, C.K. (2001). The impact of non-normality on full information maximum-likelihood estimation for structural equation models with missing data. *Psychological Methods* **6**, 352–370.

Enders, C.K., Peugh, J.L. (2004). Using an EM covariance matrix to estimate structural equation models with missing data: Choosing an adjusted sample size to improve the accuracy of inferences. *Structural Equations Modeling* **11**, 1–19.

Ferguson, T.S. (1996). *A Course in Large Sample Theory*. Chapman and Hall, London.

Finkbeiner, C. (1979). Estimation for the multiple factor model when data are missing. *Psychometrika* **44**, 409–420.

Graham, J.W. (2003). Adding missing-data relevant variables to FIML-based structural equations models. *Structural Equations Modeling* **10**, 80–100.

Graham, J.W., Donaldson, S.I. (1993). Evaluating interventions with differential attrition: The importance of non-response mechanism and use of follow-up data. *Journal of Applied Psychology* **78**, 119–128.

Jamshidian, M. (1997). An EM algorithm for ML factor analysis with missing data. In: Berkane, M. (Ed.), *Latent Variable Modeling and Applications to Causality*. Springer-Verlag, pp. 247–258.

Jamshidian, M. (2004a). On algorithms for restricted maximum likelihood estimation. *Computational Statistics and Data Analysis* **45**, 137–157.

Jamshidian, M. (2004b). Strategies for analysis of missing data. In: Hardy, M., Bryman, A. (Eds.), *Handbook of Data Analysis*. Sage Publication, pp. 113–130.

Jamshidian, M., Bentler, P.M. (1999). Using complete data routines for ML estimation of mean and covariance structures with missing data. *Journal of Educational and Behavioral Statistics* **24**, 21–41.

Jamshidian, M., Jennrich, R.I. (2000). Standard errors for EM estimation. *Journal of the Royal Statistical Society B* **62**, 257–270.

Jamshidian, M., Schott, J. (2005). Testing equality of covariance matrices when data are incomplete. *Computational Statistics and Data Analysis*. In press.

Jöreskog, K.G., Sörbom, D. (2004). *LISREL 8.7 [Computer Software]*. Scientific Software International, Inc., Licolnwood, IL.

Kamakura, W.A., Wedel, M. (2000). Factor analysis and missing data. *Journal of Marketing Research* **37**, 490–498.

Kenward, M.G., Molenberghs, G. (1998). Likelihood based frequentist inference when data are missing at random. *Statistical Science* **13**, 236–247.

Kim, J.O., Curry, J. (1977). The treatment of missing data in multivariate analysis. *Sociological Methods Research* **6**, 215–240.

Kim, K.H., Bentler, P.M. (2002). Tests of homogeneity of means and covariance matrices for multivariate incomplete data. *Psychometrika* **67**, 609–623.

Kogovsek, T., Ferligoj, A., Coenders, G., Saris, W.E. (2002). Estimating the reliability and validity of personal support measures: full information ML estimation with planned incomplete data. *Social Networks* **24**, 1–20.

Krishnamoorthy, K., Pannala, M.K. (1998). Some simple test procedures for normal mean vector with incomplete data. *Annals of Institute of Statistical Mathematics* **50**, 531–542.

Laird, N.M. (1988). Missing data in longitudinal-studies. *Statistics in Medicine* **7**, 305–315.

Lee, S.Y. (1986). Estimation for structural equation models with missing data. *Psychometrika* **51**, 93–99.

Lee, S.Y., Chiu, Y.M. (1990). Analysis of multivariate polychoric correlation models with incomplete data. *British Journal of Mathematical and Statistical Psychology* **43**, 145–154.

Lee, S.-Y., Song, X.-Y. (2003). Maximum likelihood estimation and model comparison for mixtures of structural equation models with ignorable missing data. *Journal of Classification* **20**, 221–225.

Lee, S.Y., Tang, M.L. (1992). Analysis of structural equations models with incomplete polytomous data. *Communications in Statistics – Theory and Methods* **21**, 213–232.

Lee, S.Y., Tang, N.S. (2006). Bayesian analysis of nonlinear structural equation models with nonignorable missing outcome data. *Psychometrika*. In press.

Lee, S.Y., Zhu, H.T. (2002). Maximum likelihood estimation of nonlinear structural equation models. *Psychometrika* **67**, 189–210.

Lee, S.-Y., Song, X.-Y., Lee, J.C.K. (2003). Maximum likelihood estimation of nonlinear structural equation models with ignorable missing data. *Journal of Educational and Behavioral Statistics* **28**, 111–134.

Little, R.J.A. (1988). A test of missing completely at random for multivariate data with missing values. *Journal of the American Statistical Association* **83**, 1198–1202.

Little, R.J.A., Rubin, D.B. (1987). *Statistical Analysis with Missing Data*. Wiley, New York.

Little, R.J.A., Rubin, D.B. (2002). *Statistical Analysis with Missing Data*, second ed. Wiley, New York.

Louis, T.A. (1982). Finding the observed information matrix when using EM algorithm. *Journal of the Royal Statistical Society B* **44**, 226–233.

Marsh, H.W. (1998). Pairwise deletion for missing data in structural equations model: nonpositive definite matrices, parameter estimates, goodness of fit, and adjusted sample sizes. *Structural Equations Modeling* **5**, 22–36.

McDonald, R.P. (1962). A general approach to nonlinear factor analysis. *Psychometrika* **27**, 123–157.

Meng, X.L., Rubin, D.B. (1993). Maximum likelihood via the ECM algorithm: a general framework. *Biometrika* **80**, 267–278.

Muthén, L.K., Muthén, B.O. (2006). *Mplus User's Guide*, fourth ed.. Muthén & Muthén, Los Angeles, CA.

Muthén, B., Kaplan, D., Hollis, M. (1987). On structural equation modeling with data that are not missing completely at random. *Psychometrika* **52**, 431–462.

Rabe-Hesketh, S., Skrondal, A., Pickles, A. (2004a). Generalized multilevel structural equation modeling. *Psychometrika* **69**, 167–190.

Rabe-Hesketh, S., Skrondal, A., Pickles, A. (2004b). GLLAMM Manual. Working Paper 160. U.C. Berkeley Division of Biostatistics Working paper Series. http://www.bepress.com/ucbbiostat/paper160/.

Rubin, D.B. (1976). Inference and missing data. *Biometrika* **63**, 581–592.

Rubin, D.B. (1987). *Multiple Imputation for Nonresponse in Surveys*. Wiley, New York.

Schafer, J.L. (1997). *Analysis of Incomplete Multivariate Data. Monographs on Statistics and Applied Probability*, vol. 72. Chapman and Hall.

Schafer, J.L., Graham, J.W. (2002). Missing data: Our view of the state of the art. *Psychological Methods* **7**, 147–177.

Song, J.W., Belin, T.R. (2004). Imputation for incomplete high-dimensional multivariate normal data using a common factor model. *Statistics in Medicine* **23**, 2827–2843.

Song, X.-Y., Lee, S.Y. (2002). Analysis of structural equation model with ignorable missing continuous and polytomous data. *Psychometrika* **67**, 261–288.

Tang, M., Bentler, P.M. (1998). Theory and methods for constrained estimation in structural equation models with incomplete data. *Computational Statistics and Data Analysis* **27**, 257–270.

Wei, G.C.G., Tanner, M.A. (1990). A Monte Carlo implementation of the EM algorithm and the poor man's data augmentation algorithms. *Journal of the American Statistical Association* **82**, 805–811.

Woodbury, M.A., Siler, W. (1966). Factor analysis with missing data. *Annals of the New York Academy of Sciences* **128** (A3), 746.

Yuan, K.-H., Bentler, P.M. (2000). Three likelihood-based methods for mean and covariance structure analysis with nonnormal missing data. *Sociological Methodology* **30**, 167–202.

Yuan, K.-H., Lambert, P.L., Fouladi, R.T. (2004). Mardia's multivariate kurtosis with missing data. *Multivariate Behavioral Research* **39**, 413–437.

Handbook of Computing and Statistics with Applications, Vol. 1
ISSN: 1871-0301

DOI: 10.1016/S1871-0301(06)01003-1

Rotation Algorithms: From Beginning to End

R.I. Jennrich

Abstract
Rotation algorithms used in exploratory factor analysis are discussed. A mostly historical overview beginning with the graphical method of Thurstone and proceeding to the present is given. Early methods, graphical and analytic, were indirect in that they attempted to produce simple reference structures rather than simple loadings. The first methods designed to produce simple loadings were the orthogonal methods. Later less restrictive oblique methods for simple loadings were introduced. These early methods were problem specific. More recently simple general orthogonal and oblique methods have been developed. The title is a bit presumptuous. It will be argued that in some sense the rotation algorithm problem in exploratory factor analysis has been solved. There are now very simple, very general, and reliable algorithms for orthogonal and oblique rotation. But one can always do more so the "End" in the title probably represents only a plateau.

1. Introduction

Rotation algorithms began with the graphical methods of Thurstone (1947) for producing simple structure in factor analysis. Beginning with an initial reference structure he produced a sequence of simpler reference structures. Each was constructed from a graphical analysis of plots produced from a current reference structure. These rather labor intensive methods actually worked quite well. Simple reference structures tend to correspond to simple loadings so simplifying reference structures may be viewed as an "indirect" method of producing simple loadings in the terminology of Harman (1976).

A number of factor analysts Carroll (1953), Neuhaus and Wrigley (1954), and Saunders (1953) independently proposed the first analytic rotation method. This was based on maximizing a criterion designed to measure the simplicity of a factor loading matrix. Their algorithm used a sequence of two factor rotations found in each case by analytically optimizing the criterion rather than by a graphical analysis of plots. The common criterion used by these authors is called the quartimax criterion. Unlike Thurstone's method, these methods required the factors to be orthogonal and because of this restriction often failed to produced results as nice as those obtained from graphical methods.

A number of alternative rotation criteria were proposed and optimized by sequences of analytic two factor rotations. Some of these, in particular, varimax (Kaiser, 1958), worked better than quartimax, but as with quartimax these were restricted to orthogonal rotation.

One difficulty with these early methods was that algorithms for each criterion were specific to that criterion. A new criterion required a new algorithm. The first step to remove this difficulty was a pairwise orthogonal rotation algorithm proposed by Jennrich (1970) for optimizing arbitrary quartic criteria which included most of the orthogonal rotation criteria in use at that time. The only criterion specific code required was a formula to define the criterion.

Carroll (1953) was the first to propose an analytic oblique method. He used a criterion appropriate for oblique rotation called the quartimin criterion and applied it to the reference structure. He showed how to make a sequence of one factor at a time rotations to optimize the criterion. Two problems with this approach were that it was restricted to the quartimin criterion and some modest generalizations of it and like Thurstone's method it was indirect.

Jennrich and Sampson (1966) were the first to provide a direct analytic method for oblique rotation. They showed how to optimize the quartimin criterion applied directly to the factor loadings using a sequence of one factor rotations. Unlike Carroll's, their method was direct and generalized easily to other rotation criteria.

Today there are many nonquartic criteria of interest for both orthogonal and oblique rotation. A breakthrough came when Browne and Cudeck (see Section 8.3 below) proposed a very simple approach to optimizing arbitrary criteria using pairwise rotation and a line search algorithm. This can be used for either orthogonal or oblique rotation. The only criterion specific code required is a formula to define the criterion.

Along the same line Jennrich (2001, 2002) proposed orthogonal and oblique gradient projection (GP) algorithms for optimizing arbitrary criteria. These methods used gradients to optimize the criteria directly without requiring pairwise rotations. They require a formula for the criterion and its gradient. When used with numerical gradients, they require only a formula for the criterion. With analytic gradients they can be considerably faster than Browne and Cudeck's pairwise line search method.

What follows provides details for the overview just given.

2. Factor analysis

The factor analysis model we consider (see, e.g., Harman, 1976) has the form

$$x = \mu + \Lambda f + u, \tag{1}$$

where x is a vector of observed responses, f is a vector of common factors, and u is a vector of unique factors defined on a population. The matrix Λ is a p by k matrix of factor loadings. It is assumed that the vectors f and u have mean zero and are uncorrelated, that the components of f have variance one, and that the components of u are uncorrelated. The vector μ is the mean of x.

Table 1
Examples of perfect and Thurstone simple structure

Perfect			Thurstone		
1	0	0	1	0	0
1	0	0	0	1	0
1	0	0	0	0	1
0	1	0	0.89	0.45	0
0	1	0	0.89	0	0.45
0	0	1	0	0.71	0.71
0	0	1	0.71	0	0.71

Under these assumptions, the covariance matrix Σ of x has the structure

$$\Sigma = \Lambda\Phi\Lambda' + \Psi,$$

where $\Phi = \operatorname{cov} f$, $\Psi = \operatorname{cov} u$, and Ψ is diagonal.

If there are no further constraints, (1) is called an exploratory factor analysis model. If there are enough constraints to uniquely identify Λ and Φ it is called a confirmatory model. Models that are neither exploratory nor confirmatory do not seem to have a name and are seldom considered. The two named models represent a major division in the study and application of factor analysis. Often an exploratory analysis is used to help formulate a confirmatory analysis. Here only exploratory analysis is considered.

For an exploratory analysis there are two steps. The first is estimating

$$\Omega = \Lambda\Phi\Lambda' \tag{2}$$

and Ψ from a sample of values of x. This is called extraction. The second is estimating Λ and the correlation matrix Φ from the estimate of Ω. This is called rotation for reasons that will become clear shortly. The rotation problem is the major component of exploratory factor analysis and is the problem considered here.

Given Ω there are many Λ and Φ that satisfy (2). The usual approach to estimating Λ and Φ, that is to the rotation problem, is to find a Λ that looks nice or slightly more specifically has a simple form or structure. The main problem is what does this vague statement mean? One case is clear. If each row of Λ has at most one nonzero element, Λ is said to have perfect simple structure an example of which is displayed in Table 1. Thurstone (1935) proposed a less demanding definition of simple structure. The second loading matrix in Table 1 has Thurstone simple structure. Thurstone simple structure requires a fair number of zeros, but far fewer than perfect simple structure. The difficulty is that among all factorizations (2) of Ω there may not be a Λ with perfect simple structure or with Thurstone simple structure and that is the usual case. It may, however, be possible to find a Λ that approximates Thurstone simple structure or even perfect simple structure.

Today the usual approach to the rotation problem is to choose a rotation criterion Q that assigns a numerical complexity $Q(\Lambda)$ to Λ. The Λ that satisfies (2) for some correlation matrix Φ and minimizes Q is the rotated value of Λ corresponding to Q. Unfortunately there are many choices for Q. Most rotation research is centered on finding useful choices. Fortunately for us, our problem is not to find a desirable criterion Q, but rather to find general and easy to use algorithms to optimize a given Q. A very nice overview of rotation criteria can be found in Browne (2001). A common special version of the exploratory factor analysis model assumes the factors are uncorrelated. These are called the orthogonal factor analysis models. For them Φ is an identity matrix and (2) becomes

$$\Omega = \Lambda\Lambda'.$$

When the factors may be correlated, the model is called an oblique factor analysis model. Because of its greater generality, the oblique model can produce significantly simpler loading matrices.

3. A parameterization for Λ and Φ

It is helpful to parameterize Λ and Φ in (2) appropriately. Choose an A so

$$\Omega = AA'.$$

This may be done by using a principal components factorization of Ω. Actually the extraction step usually presents Ω in this form. The matrix A is called an initial loading matrix which we assume has full column rank. When this is the case Λ and Φ satisfy (2) if and only if

$$\Lambda = AT^{-1} \quad \text{and} \quad \Phi = TT' \tag{3}$$

for some nonsingular matrix T with rows of length one. Thus T provides a parameterization for Λ and Φ. The rows of T correspond to the factors f. Indeed with some abuse of notation one might write $f = T$. We will call T the factor matrix.

In the case of orthogonal factor analysis $\Phi = I$, T is an orthogonal matrix, $T^{-1} = T'$, and the rows of Λ are orthogonal transformations of the rows of A. This motivates calling the Λ that satisfy (3) with T orthogonal the orthogonal rotations of A. While it makes less sense, this terminology is also used in the oblique case. Thus the Λ that satisfy (3) when T is not orthogonal are called the oblique rotations of A when in fact the rows of Λ are not actually rotations of the rows of A.

Finding an oblique rotation Λ of A to minimize a criterion $Q(\Lambda)$ reduces to finding a nonsingular T with rows of length one to minimize

$$f(T) = Q(AT^{-1}). \tag{4}$$

Finding an orthogonal rotation Λ of A to minimize $Q(\Lambda)$ reduces to finding an orthogonal matrix T to minimize

$$f(T) = Q(AT'). \tag{5}$$

4. Reference structures

To avoid the T^{-1} in the definition (3) of the rotated loadings Λ and the difficulty of working with it, the first approaches to the rotation problem rotated reference structures rather than loadings. Following Thurstone (1947), let the rows of a nonsingular matrix U be bi-orthogonal to the rows of T and have length one. Bi-orthogonal means the rth row of U is orthogonal to the sth row of T whenever $r \neq s$. Let

$$R = AU'.$$

This is called the reference structure rotation of A. It is of interest because

$$R = AT^{-1}TU' = \Lambda\Delta,$$

where Δ is diagonal because the rows of T and U are bi-orthogonal. Because Δ is diagonal the columns of R are rescaled versions of the columns of Λ and this suggests that R is simple when Λ is simple and conversely. Rather than finding a loading matrix Λ with simple structure a reference matrix R with simple structure is sought. This means finding a nonsingular matrix U with rows of length one to make R as simple as possible. Harman (1976) calls making R simple an indirect method and making Λ simple a direct method. Given R and U it is easy to find the corresponding Λ and T. The rows of U correspond to what are called reference factors. We will call U the reference factor matrix.

In the orthogonal case direct and indirect methods are the same because $R = \Lambda$ and $U = T$. To see this note that when T is orthogonal it is bi-orthogonal to itself so $U = T$ and

$$\Lambda = AT^{-1} = AT' = AU' = R.$$

5. Thurstone's graphical rotation method

Let R be the current value of a reference structure. Assume first that R has two columns. Then each row of R is an ordered pair of numbers and may be viewed as a point in a two-dimensional space. Let Figure 1 be a plot of these points, one for each row of R.

Note that none of the points are close to the horizontal or vertical axes and hence none have a small first or second component. As a consequence R is not simple because it has no components that are small in magnitude. The slanted line in the plot passes through a cluster of points and the vector n_1 is perpendicular to it. Assuming the points corresponding to the rows of R are ordered from left to right,

$$Rn_1' = \begin{pmatrix} * \\ * \\ * \\ \varepsilon \\ \varepsilon \\ \varepsilon \end{pmatrix},$$

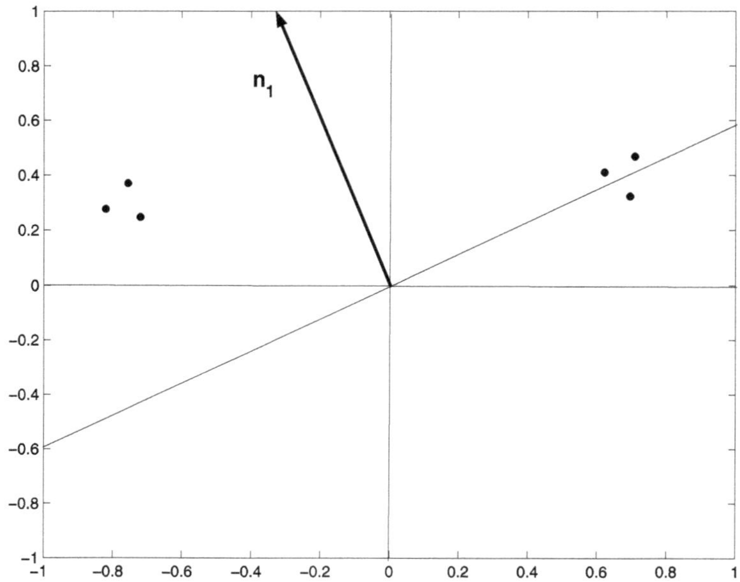

Fig. 1. A plot of the rows of R denoted by the dots, a line through the cluster on the right, and a vector n_1 perpendicular to the line.

where "$*$" denotes a fairly large value and "ε" denotes a rather small value. Because of its three small values this column looks simpler than either of the columns of R. This is the basic idea behind graphical rotation. One can play the same game with the cluster of points on the left to produce a normal vector n_2. Let

$$N = \begin{pmatrix} n_1 \\ n_2 \end{pmatrix}.$$

Then

$$R^* = RN' = \begin{pmatrix} * & \varepsilon \\ * & \varepsilon \\ * & \varepsilon \\ \varepsilon & * \\ \varepsilon & * \\ \varepsilon & * \end{pmatrix}$$

which has rather simple structure.

When R has more than two columns Thurstone recommends plotting every pair of columns and selecting several normal vectors from these. A given plot may produce zero, one, or two normal vectors. The number of normal vectors selected must equal the number of columns of R. Let N be the matrix containing the normal vectors with zeros inserted so their components match the appropriate columns of R and let

$$\widetilde{R} = RN'.$$

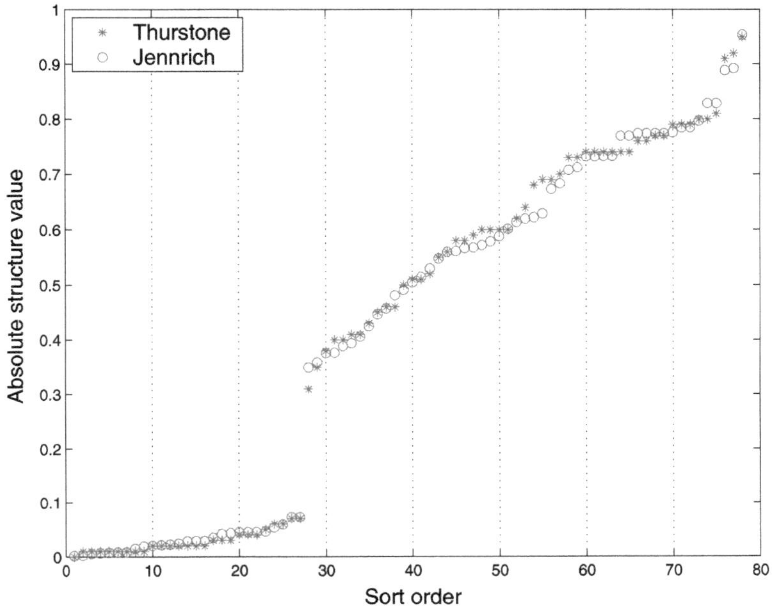

Fig. 2. Sorted absolute loading plots for the Thurstone and Jennrich solutions to the 26 variable box problem.

Hopefully $\widetilde{R}$ will be simpler than R. A difficulty is that $\widetilde{R}$ may not be a reference structure. This can be fixed by an appropriate scaling of the rows of N. Note that

$$\widetilde{R} = AU'N' = A(NU)'$$

and that $\widetilde{R}$ will be a reference structure if the rows of NU have length one. Re-scale the rows of N so this is the case. Then $\widetilde{R}$ is a reference structure and may be viewed as an update to R. The algorithm proceeds by repeating this process until it converges.

Because it is better to actually do it than simply talk about doing it, the author attempted to graphically rotate the well-known Thurstone 26 variable box data. Figure 2 is a sorted absolute loading plot (Jennrich, 2004) applied to the reference structure values. It shows the result of the author's effort and that of Thurstone (1947).

A proper solution is known to have 27 small values. Thurstone and the author both got 27 small values. It is also known that a proper solution has three pure indicators and these should produce three distinct large values. Thurstone found three clearly distinct large values. The author's three largest values are not quite as distinct and not quite as large. Overall, however, the two plots are very similar and the author would like to claim he did almost as well as Thurstone. This suggests graphical methods can work even in the hands of a novice.

6. Early analytic oblique rotation methods

Rather than using graphical methods to produce simple reference structures early analytic methods minimized the value $Q(R)$ of a complexity function Q applied to the reference structure R.

6.1. Carroll's one column at a time method

Carroll (1953) provided a rather ingenious way to minimize the quartimin criterion

$$Q(R) = \sum_{a \neq b} \sum \sum_{i} r_{ia}^2 r_{ib}^2. \tag{6}$$

He proposed optimizing $Q(R)$ by viewing it as a function of a single column r_a of R, choosing r_a to minimize this function, and cycling through columns of R. For the quartimin criterion this is surprisingly easy because while $Q(R)$ is a quartic function of R, as a function of r_a it is quadratic.

To see this note that (6) may be expressed in the form

$$Q(R) = \sum_{a \neq b} \sum (r_a^2, r_b^2), \tag{7}$$

where r_a^2 is the element-wise square of r_a and (x, y) denotes the inner-product between two vectors x and y. Note that $Q(R)$ may also be expressed in the form

$$Q(R) = \sum_{b \neq a} (r_a^2, r_b^2) + \text{other terms},$$

where the r_b^2 and "other terms" do not involve r_a. Thus $Q(R)$ is a quadratic function of r_a and the problem reduces to minimizing

$$q = \sum_{b \neq a} (r_a^2, r_b^2)$$

with respect to r_a. It follows from the definition of R that

$$r_a = A u_a', \tag{8}$$

where u_a is the ath row of the reference factor matrix U. Let $d = \sum_{b \neq a} r_b^2$, D be the diagonal matrix whose diagonal components are the components of d, and $M = A'DA$. Then

$$q = (r_a^2, d) = r_a' D r_a = u_a A' D A u_a' = u_a M u_a'.$$

Thus q and hence $Q(R)$ is a quadratic function of u_a. Moreover, it is minimized when u_a, which must have length one, is the transpose of the eigenvector of M that has the smallest eigenvalue. This is easily found using standard computer software. It provides an updated value for u_a and using (8) gives the corresponding updated value for r_a.

Actually this approach can be generalized to the entire orthomin family of criteria in which the quartimin criterion is by far the most popular. It does not, however, seem to generalize to other criteria. For some time Carroll's indirect method was the standard for oblique analytic rotation.

6.2. Procrustes rotation to a target

In indirect target rotation one attempts to find a reference structure R that is as close as possible to a target H in the least squares sense. This is called procrustes rotation. One of the earliest methods for doing this was that of Mosier (1939). It proceeds as follows. Given an initial loading matrix A let AB be a least squares approximation to the target H. This may not be a reference structure because the columns of B may not have unit length. Mosier re-scaled the columns of B so they have unit length. Then $R = AB$ is a reference structure that approximates H, but in general it will not be a least squares reference structure approximation.

Browne (1967) proposed an algorithm for finding an exact least squares reference structure approximation for the procrustes rotation problem. For this he assumed A had full column rank and that for every column h of H, $A'h$ was not orthogonal to the eigenspace corresponding to the smallest eigenvalue of $A'A$. These assumptions are almost always satisfied in practice. Cramer (1974) and Ten Berge and Nevels (1977) have shown how to remove Browne's assumptions.

While Mosier's method fails to solve the least squares reference structure problem it is remarkably simple and this has turned out to be important. Korth and Tucker (1976) have shown that Mosier's approximation solves a different problem. It maximizes the sum of the Tucker congruence coefficients between the columns of H and R. Of far greater importance Mosier's method has led to the very popular promax method to be discussed next.

6.3. Promax rotation

Hendrickson and White (1964) proposed the following method for oblique rotation. Begin with a varimax or some other orthogonal rotation Λ of an initial loading matrix A – orthogonal rotation is discussed in Section 8. Let H be the element-wise cube of Λ. Some other power may be used, but a minor modification is required for even powers. The next step is to find a Mosier reference structure approximation to H. The result is the promax rotation of A.

The idea behind the promax method is that cubing the components of Λ will make components close to zero even closer and increase the ratio between large components and small components. This tends to make H appear simpler than Λ and a Mosier approximation to H will hopefully produce a reference structure simpler than that provided by Λ.

Promax may be viewed as orthogonal rotation with oblique polish. Assuming one has an orthogonal rotation method, promax provides a very simple way to extend it to oblique rotation. It is still very popular probably because of its early introduction into statistical software. In some cases it is the only form of oblique rotation provided.

7. Pairwise algorithms

To this point our discussion has focused on indirect methods. Almost all direct rotation algorithms proceed by modifying the loading matrix Λ two columns at a time and cycling through pairs of columns until convergence is obtained. These are called pairwise methods. To see how they proceed note that from (3)

$$A = \Lambda T.$$

Let Λ_1 be a pair of columns of Λ, and Λ_2 be the other columns. Let T_1 be the rows of T corresponding to the selected columns of Λ and let T_2 be the other rows of T. Then

$$A = \Lambda_1 T_1 + \Lambda_2 T_2.$$

To modify Λ_1 let

$$\tilde{\Lambda}_1 = \Lambda_1 M^{-1},$$

where M is a nonsingular two by two matrix such that

$$\widetilde{T}_1 = M T_1$$

has rows of length one in the oblique case and orthonormal rows in the orthogonal case. The aim is to choose M so $\tilde{\Lambda}_1$ is simpler than Λ_1. However this is done, replace the Λ_1 columns of Λ by $\tilde{\Lambda}_1$ and the T_1 rows of T by $\widetilde{T}_1$ and call the results $\tilde{\Lambda}$ and $\widetilde{T}$. If simplicity is measured by a complexity function Q, one can choose M to minimize $Q(\tilde{\Lambda})$. How to do this will be discussed below. The pairwise algorithm proceeds by cycling through pairs of columns of Λ until convergence is obtained.

8. Analytic rotation methods: Orthogonal

We have discussed indirect graphical and analytic methods for oblique rotation. The first direct methods were the orthogonal methods. For these T is an orthogonal matrix. The rotation criterion Q is applied directly to the loading matrix Λ and optimized over all orthogonal rotations Λ of an initial loading matrix A. More precisely the problem is to find an orthogonal matrix T to minimize (5).

8.1. Early pairwise algorithms

Almost all orthogonal rotation algorithms are pairwise algorithms and have the form outlined in Section 7. Using the notation of Section 7, let M be a two by two matrix whose rows are chosen so MT_1 has orthonormal rows. Since in the orthogonal case the rows of T_1 are orthonormal, M must be an orthogonal matrix and assuming $Q(\Lambda)$ is invariant with respect to sign changes in the columns of Λ one may assume M is a rotation of the form

$$M = \begin{pmatrix} \cos\theta & \sin\theta \\ -\sin\theta & \cos\theta \end{pmatrix}. \tag{9}$$

The invariance assumption is necessary but almost never mentioned. Fortunately all rotation criteria known to the author have the required invariance property and hence requiring M to have the form given in (9) is a harmless restriction. Since M is orthogonal $M^{-1} = M'$ and in the notation of Section 7, $\tilde{\Lambda}_1 = \Lambda_1 M'$. Note that M is a function of

θ and as a consequence so is $\tilde{\Lambda}$. Let

$$q(\theta) = Q(\tilde{\Lambda}).$$

The next step is to optimize $q(\theta)$ with respect to θ. Use the optimizing value of θ and (9) to define the optimizing value of M and proceed as in Section 7. To do this one must have a way to optimize $q(\theta)$ with respect to θ. The remainder of this section will be devoted to this problem.

By far the most popular criterion for orthogonal rotation is Kaiser's (1958) varimax criterion. This is a simplicity criterion so the object is to maximize rather than minimize it. The criterion is given by

$$Q(\Lambda) = \sum\sum \lambda_{ir}^4 - \frac{1}{p}\sum_r\left(\sum_i \lambda_{ir}^2\right)^2.$$

The corresponding function $q(\theta)$ is a quartic function of $\sin\theta$ and $\cos\theta$ that depends on the loadings λ_{ir} in Λ_1 and is rather complex. Nevertheless Kaiser (1958) has shown that with a sufficient amount of algebraic manipulation and a sufficient number of trigonometric identities the optimizing value of θ satisfies

$$\tan(4\theta) = \frac{D - AB/p}{C - (A^2 - B^2)/p}, \tag{10}$$

where

$$\begin{aligned} u_i &= \lambda_{i1}^4 - \lambda_{i2}^4, \\ v_i &= 2\lambda_{i1}^2\lambda_{i2}^2, \\ A &= \sum u_i, \\ B &= \sum v_i, \\ C &= \sum\left(u_i^2 - v_i^2\right), \\ D &= 2\sum u_i v_i. \end{aligned}$$

Unfortunately both the maximizing and minimizing value of θ satisfy (10). The following table chooses the proper value:

Numerator	Denominator	Quadrant of $4\hat{\theta}$
+	+	$0° \leqslant 4\hat{\theta} \leqslant 90°$
+	−	$90° \leqslant 4\hat{\theta} \leqslant 180°$
−	−	$-180° \leqslant 4\hat{\theta} \leqslant -90°$
−	+	$-90° \leqslant 4\hat{\theta} \leqslant 0°$

where "numerator" is the sign of the numerator in (10) and "denominator" is the sign of the denominator. This is a rather complex algorithm. More recently Nevels (1986)

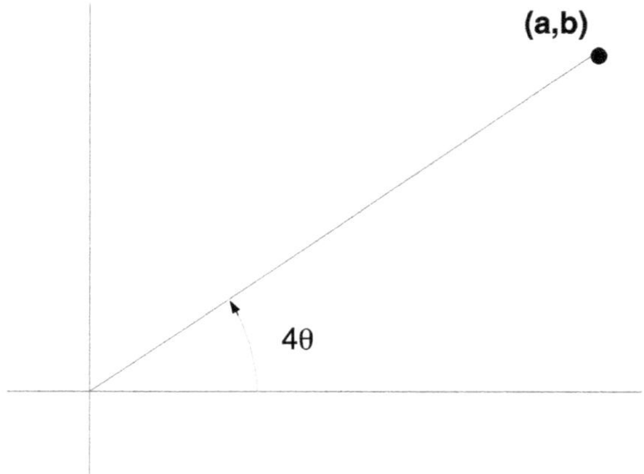

Fig. 3. The optimizing value of 4θ.

developed a pairwise algorithm for varimax rotation that does not require a table to find an optimal rotation.

A fair number of other quartic criteria have been proposed and used including the quartimax criterion of Carroll (1953), Neuhaus and Wrigley (1954), and Saunders (1953), and the orthomax family (Harman, 1960) and the Crawford and Ferguson (1970) family of criteria. In each case the authors derived formulas similar to Kaiser's to optimize their criteria.

8.2. *A general pairwise algorithm for quartic criteria*

Rather than have a special algorithm for each criterion it would be nice to have a single algorithm for all quartic criteria that requires no more than the definition of the criterion. Let $Q(\Lambda)$ be an arbitrary quartic simplicity criterion. Jennrich (1970) noted that because $q(\theta)$ in the previous section is a quartic function of $\cos\theta$ and $\sin\theta$ it must have the form

$$\begin{aligned} q(\theta) = {} & \gamma_0 + \alpha_1 \cos\theta + \beta_1 \sin\theta + \alpha_2 \cos 2\theta + \beta_2 \sin 2\theta \\ & + \alpha_3 \cos 3\theta + \beta_3 \sin 3\theta + \alpha_4 \cos 4\theta + \beta_4 \sin 4\theta \end{aligned}$$

and he showed that if $Q(\Lambda)$ is invariant under permutation and sign changes in the columns of Λ, $q(\theta)$ has period $\pi/2$ and as a consequence must have the simpler form

$$q(\theta) = c + a\cos 4\theta + b\sin 4\theta. \tag{11}$$

This is maximized by choosing θ so the vector $(\cos 4\theta \; \sin 4\theta)$ has the same direction as the vector (a, b) as displayed in Figure 3.

Thus to produce the optimizing value of θ, it is sufficient to find a and b in (11). This can be done for a general quartic criteria by evaluating $q(\theta)$ at three values of θ. This leads to the equations

$$\begin{aligned} Q(\theta_1) &= c + a\cos 4\theta_1 + b\sin 4\theta_1, \\ Q(\theta_2) &= c + a\cos 4\theta_2 + b\sin 4\theta_2, \end{aligned}$$

$$Q(\theta_3) = c + a\cos 4\theta_3 + b\sin 4\theta_3$$

which may be solved for a, b, and c. If

$$\theta_1 = 0, \quad \theta_2 = \pi/8, \quad \theta_3 = -\pi/8$$

solving the equations gives

$$a = Q(0) - Q(\pi/8)/2 - Q(-\pi/8)/2,$$
$$b = Q(\pi/8)/2 - Q(-\pi/8)/2.$$

Using these values and Figure 3 gives the optimizing value of θ. While other ranges might be used, choosing θ so $\pi/4 < \theta \leqslant \pi/4$ has the advantage that rotations near the identity are represented by θ near zero.

This defines a pairwise algorithm for optimizing any quartic criterion that is invariant under permutation and sign changes in the columns of Λ. All current quartic criteria have this property, at least all known to the author.

8.3. A pairwise algorithm for general criteria

Some of the newer rotation criteria are not quartic, for example, the Yates' (1987) geomin criterion and Jennrich's (2004) entropy criterion. A dramatic breakthrough came when Browne and Cudeck showed how to construct an incredibly simple pairwise algorithm that worked for arbitrary criteria. Rather than analytically finding the value of θ that optimized $q(\theta)$ they simply plotted $q(\theta)$ and read the optimizing value of θ from the plot. More precisely they used a standard line search algorithm to optimize $q(\theta)$. It is important to note that the only user supplied input required is a definition for $Q(\Lambda)$.

Using line searching increases computation time, but not enough to cause concern. The Browne and Cudeck method has been used successfully with many different criteria. No paper on this algorithm has been published. The algorithm, however, is part of the CEFA (Comprehensive Exploratory Factor Analysis) software (Browne et al., 2002). This free software deals with almost every aspect of exploratory factor analysis including a broad variety of methods for extraction and rotation, factoring correlation matrices, and providing standard errors for the estimates produced. It has a graphical user interface and a nice manual. The software and manual may be downloaded from http://quantrm2.psy.ohio-state.edu/browne/.

8.4. A gradient projection algorithm for general criteria

To this point in our discussion all orthogonal rotation algorithms have been pairwise algorithms. We consider here an alternate approach based on gradients that does not require pairwise steps. This algorithm proposed by Jennrich (2001) is called a gradient projection (GP) algorithm.

As noted all orthogonal rotations Λ of an initial loading matrix A are of the form $\Lambda = AT'$ for some orthogonal matrix T. And finding a Λ to minimize $Q(\Lambda)$ means finding an orthogonal matrix T to minimize the function f defined by (5). Let G be the gradient of f at the current value of T. One might consider moving T in the negative gradient direction say to $X = T - \alpha G$ to decrease the value of $f(T)$. The problem with

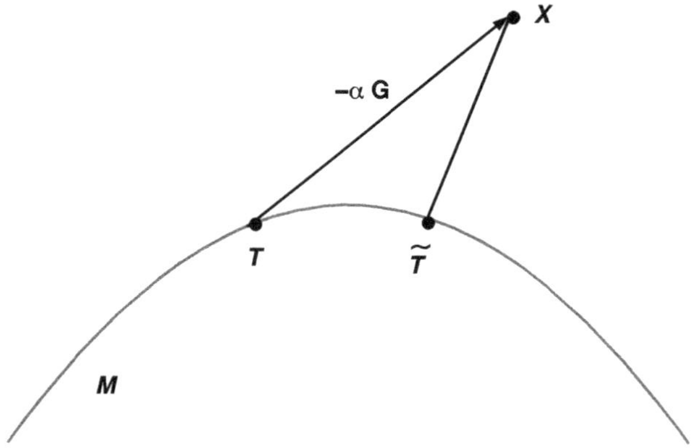

Fig. 4. Graphical representation of the gradient projection algorithm.

this is that X may not be an orthogonal matrix. That is X may not be on the manifold $\mathcal{M}$ of orthogonal matrices. See Figure 4. Jennrich suggested dealing with this by projecting X onto $\mathcal{M}$ to produce the $\widetilde{T}$ in Figure 4. In general projecting onto a nonlinear manifold is a difficult problem, but $\mathcal{M}$ is a special manifold and projection is easy. The projection of X on $\mathcal{M}$ is simply

$$\widetilde{T} = UV', \tag{12}$$

where $X = UDV'$ is a singular value decomposition of X.

Jennrich (2001) showed:

THEOREM 1. *If T is not a stationary point of f restricted to $\mathcal{M}$, then*

$$f(\widetilde{T}) < f(T) \tag{13}$$

for all $\alpha > 0$ and sufficiently small.

A strictly decreasing algorithm is obtained by halving α, if necessary, until (13) is satisfied, replacing T by $\widetilde{T}$, and repeating this process until it converges. Strictly decreasing algorithms are important because under mild assumptions they must converge to a stationary point and since local minimizers are the only points of attraction of a decreasing algorithm, strictly decreasing algorithms almost always converge to at least a local minimizer.

This algorithm can be considerably faster than the Browne and Cudeck pairwise algorithm, but has the disadvantage of requiring both the value and gradient of f at the current value of T. One way to deal with this is to use numerical gradients. Then only values of f are required. Using numerical gradients slows the algorithm, but very little is lost in numerical precision. When the output is printed to a reasonable number of decimal places the results of using analytic and numerical gradients are essentially identical. Many psychometricians and statisticians fear numerical derivatives. This is a big mistake. What they should fear are the consequences of failing to use them.

One can find free SAS, SPSS, R/S, and Matlab code for GP rotation at http://www.stat.ucla.edu/research/gpa. Thus for almost any computing environment one is working in, one can find code written specifically for that environment and hence code that may be used immediately without any need for translation. There is code using analytic or numerical gradients. Also given is code for a variety of criteria and their gradients. A discussion of this software and the gradients provided may be found in Bernaards and Jennrich (2005).

9. Direct analytic methods: Oblique

Indirect oblique analytic methods were discussed in Section 6. Here we consider direct oblique methods. Most of these have been pairwise methods.

9.1. A parameterization for pairwise oblique methods

In pairwise methods one chooses a pair of factors or more precisely rows of T and rotates these to improve the resulting loading matrix Λ. In the orthogonal case both of the selected rows of T were modified. While this could also be done in the oblique case it is much simpler to modify just one row or more precisely replace one row of the selected pair by a linear combination of the two rows. For this we need to replace pairs of rows of T by ordered pairs of rows. In the notation of Section 7, let T_1 be an arbitrary ordered pair of rows of T and denote these by t_1 and t_2. The rotated rows will have the form

$$\tilde{t}_1 = \alpha_1 t_1 + \alpha_2 t_2,$$
$$\tilde{t}_2 = t_2.$$

In the notation of Section 7,

$$M = \begin{pmatrix} \alpha_1 & \alpha_2 \\ 0 & 1 \end{pmatrix}.$$

The values α_1 and α_2 are not arbitrary. They must be chosen so $\tilde{t}_1$ is of length one. This means that

$$\alpha_1^2 + 2\alpha_1\alpha_2\phi + \alpha_2^2 = 1, \tag{14}$$

where ϕ is the inner-product of t_1 and t_2. It is called ϕ because it is also the current estimate of the correlation between the factors corresponding to t_1 and t_2. Note that

$$M^{-1} = \begin{pmatrix} 1/\alpha_1 & -\alpha_2/\alpha_1 \\ 0 & 1 \end{pmatrix} = \begin{pmatrix} \gamma & -\delta \\ 0 & 1 \end{pmatrix},$$

where $\gamma = 1/\alpha_1$ and $\delta = \alpha_2/\alpha_1$. The values γ and δ may be viewed as a re-parameterization for M^{-1}. In terms of these parameters (14) becomes

$$\gamma^2 = 1 + 2\delta\phi + \delta^2.$$

In the notation of Section 7,

$$\tilde{\Lambda}_1 = \Lambda_1 \begin{pmatrix} \gamma & -\delta \\ 0 & 1 \end{pmatrix}. \tag{15}$$

If the criterion $Q(\Lambda)$ is invariant under sign changes in the columns of Λ we may assume γ is nonnegative. Then $\tilde{\Lambda}_1$ may be viewed as a function of δ by letting

$$\gamma = (1 + 2\delta\phi + \delta^2)^{1/2}. \tag{16}$$

Thus $\tilde{\Lambda}$ is a function of δ as is $Q(\tilde{\Lambda})$. This may be summarized by a theorem proved by Jennrich and Sampson (1966).

THEOREM 2. *If $Q(\Lambda)$ is invariant under sign changes in the columns of Λ and $\tilde{\Lambda}$ is the result of the pairwise rotation defined by* (15) *and* (16)*, then $Q(\tilde{\Lambda})$ is a function of the parameter δ.*

Let

$$q(\delta) = Q(\tilde{\Lambda}) \tag{17}$$

denote the function identified in Theorem 2. Most oblique rotation algorithms are pairwise algorithms using the parameterization in this section.

9.2. Pairwise methods for quartic criteria

Most quartic criteria are of the form $Q(\Lambda) = F(\Lambda^2)$ where Λ^2 is the element wise square of Λ and F is a quadratic function. When this is the case $q(\delta)$ in the previous subsection is a quartic function of δ. Most pairwise algorithms proceed by finding formulas for the coefficients of $q(\delta)$ in terms of the current values of Λ and T. Then $q(\delta)$ is optimized and the optimizing value of δ used to complete the pairwise step.

Unfortunately new formulas are required for each criterion making this is a rather labor intensive approach similar to that used in early orthogonal pairwise algorithms. A much more efficient approach would be to simply evaluate $q(\delta)$ at five values of δ and solve the resulting linear equations for the five coefficients of the quartic $q(\delta)$. This method will work without change for any quartic criterion. To the author's knowledge, however, it has not been used.

Whatever method is used to find $q(\delta)$, once it is found, the optimizing value can be found in closed form by equating its derivative, which is a cubic, to zero and solving for δ.

9.3. A pairwise algorithm for general criteria

The methods of the previous subsection are restricted to quartic criteria and as in the orthogonal case there are criteria of interest that are not quartic. Browne and Cudeck use a minor modification of their orthogonal pairwise algorithm in Section 8.3 to obtain an oblique pairwise algorithm for general criteria.

For this they used the δ parameterization and function $q(\delta)$ in Section 9.1. As in the orthogonal case, $q(\delta)$ is optimized using a line search algorithm and as in that case the only criterion specific requirement is a formula to define the criterion. This method has been used successfully with many criteria and is used in the free CEFA software that may be downloaded as described in Section 8.3.

9.4. *A gradient projection algorithm for general criteria*

An alternative to pairwise algorithms for oblique rotation is provided by an oblique GP algorithm (Jennrich, 2002). This uses a minor modification to the algorithm given in Section 8.4 for orthogonal rotation. For oblique rotation we seek a nonsingular matrix T with rows of length one to minimize the function f defined by (4). As in Section 8.4, let G be the gradient of f at the current value of T and let $X = T - \alpha G$. In the oblique case this is projected onto the manifold $\mathcal{M}$ of nonsingular matrices T with rows of length one. Let $\widetilde{T}$ denote the projection. See Figure 4. As in the orthogonal case projection is easy, in fact easier than in the orthogonal case. It is shown by Jennrich that $\widetilde{T}$ is simply X with its rows scaled to have length one. That is

$$\widetilde{T} = \left(\mathrm{dg}(XX')\right)^{-1/2} X. \tag{18}$$

The remainder of the algorithm is as described in Section 8.4. The only algorithm change required is to replace Eq. (12) for the orthogonal algorithm by (18).

Jennrich has shown that Theorem 1 also holds for the oblique GP algorithm which means that as in the orthogonal case the oblique GP algorithm is a strictly decreasing.

The oblique GP algorithm may be used with numerical gradients and in this case like the Browne and Cudeck algorithm the only criterion specific requirement is a formula for the criterion.

One can find free SAS, SPSS, R/S, and Matlab code for this oblique GP algorithm at the web site identified in Section 8.4 including code for a variety of specific criteria and their gradients.

10. Discussion

We have given an historical review of the development of rotation algorithms for exploratory factor analysis. Our discussion began with the indirect graphical methods of Thurstone for oblique rotation. This was followed by a discussion of indirect analytic methods for oblique rotation including the one factor at a time method of Carroll (1953), the procrustes method of Mosier (1939), and the promax method of Hendrickson and White (1964).

Next analytic methods for orthogonal rotation were reviewed. The first direct analytic methods of any kind were the pairwise orthogonal methods for quartic criteria. These included the methods of Carroll (1953), Neuhaus and Wrigley (1954), and Saunders (1953) for the quartimax criterion, Kaiser's (1958) method for the varimax criterion, and others. Each criterion had its own algorithm. Jennrich (1970) introduced a general pairwise algorithm for arbitrary quartic criteria. A breakthrough came with the line search pairwise algorithm of Browne and Cudeck that can be used with essentially any criterion. Jennrich (2001) introduced a GP algorithm that is not pairwise and can be used with arbitrary criteria.

Finally direct, as opposed to indirect, analytic methods for oblique rotation were reviewed. The first were pairwise algorithms using ordered pairs and a variety of quartic criteria. These were generalized to arbitrary criteria by using the pairwise line search algorithm of Browne and Cudeck and the GP algorithm of Jennrich (2002).

By way of comparison the pairwise quartic algorithms have the advantage that no line search is required and they are probably the fastest of the algorithms discussed. Their main disadvantage is that they are restricted to quartic criteria.

The main advantage of the general pairwise line search algorithms is that they apply to arbitrary criteria and the only problem specific code required is that to evaluate the criterion. Also they are very simple algorithms. Minor disadvantages are that they require cycling through pairs and require a line search sub-algorithm.

The main advantage of the GP algorithms is that they apply to arbitrary criteria, do not require stepping through pairs of factors, and when using numerical gradients are very simple to use. When using analytic gradients they appear to be significantly faster than the general pairwise algorithms at least in the limited experience of the author. Their main disadvantage is that when used with analytic gradients they require problem specific code to produce the gradients. While these can be avoided with almost no loss of precision by using numerical gradients, the use of numerical gradients sacrifices the speed advantage of these algorithms.

A number of parameterizations have been important along the way. One is the general parameterization for Λ and Φ by the nonsingular matrices T with rows of length one given in Section 3 and the parameterization for the reference structure R by the nonsingular matrices U with rows of length one given in Section 4. The two by two matrix M for pairwise rotation was parameterized by the parameter θ given in Section 8.1 for orthogonal rotation and by a parameter δ given in Section 9.1 for oblique rotation.

In a sense the Browne and Cudeck line search and the Jennrich gradient projection algorithms solve the rotation algorithm problem because they provide simple, reliable, and reasonably efficient algorithms for arbitrary criteria. It is this that motivated the title. We of course are not really done with rotation algorithm development.

A number of basic optimization methods have essentially not been tried on the rotation problem. These include derivative free optimization, quasi-Newton, and Newton–Raphson methods. Actually Bentler (1977) did use Newton–Raphson to optimize his invariant factor simplicity criterion, but there has been no follow up, no generalization to other criteria, and no comparison with other methods. Because of the complexity of the second derivatives required for rotation problems, Newton–Raphson methods may not be too promising. Because rotation criteria tend to be fairly smooth functions, derivative free optimization methods may be slow compared to methods that use derivatives. Quasi-Newton methods, especially those that use numerical derivatives, seem more promising. For these, the only problem specific code required is that to evaluate the criterion. In this regard they are similar to the line search and GP algorithms using numerical derivatives. As noted, however, quasi-Newton algorithms don't seem to have been formulated and evaluated in the literature.

Some other areas that need further investigation include:

- More testing of the algorithms we have to evaluate precision, reliability, and speed.
- A general comparison of pairwise line search, gradient projection and quasi-Newton algorithms to possibly recommend the use of one over the others.
- Some applications require more speed. These include simulation studies that require many executions of a rotation algorithm and local minima problems that are dealt

with by using many random starts. For these faster algorithms than we presently have may be required.

- There are applications of rotation in other areas that can be considered. Signal processing, for example, uses independent components analysis to recover signals from noisy composites of them. This proceeds by extracting principal components from the composites and orthogonally rotating these to independence rather than to simplicity. Algorithms similar to those used in factor analysis may be useful here.

References

Bernaards, C.A., Jennrich, R.I. (2005). Gradient projection algorithms and software for arbitrary rotation criteria in factor analysis. *Educational and Psychological Measurement* **65**, 676–696.

Bentler, P.M. (1977). Factor simplicity index and transformations. *Psychometrika* **42**, 277–295.

Browne, M.W. (1967). On oblique procrustes rotation. *Psychometrika* **32**, 125–132.

Browne, M.W. (2001). An overview of analytic rotation in exploratory factor analysis. *Multivariate Behavioral Research* **36**, 111–150.

Browne, M.W., Cudeck, R., Tateneni, K., Mels, G. (2002). CEFA: Comprehensive Exploratory Factor Analysis, Version 1.10 [Computer software and manual]. Retrieved from http://quantrm2.psy.ohio-state.edu/browne/.

Cramer, E.M. (1974). On Browne's solution for oblique procrustes rotation. *Psychometrika* **39**, 159–263.

Carroll, J.B. (1953). An analytical solution for approximating simple structure in factor analysis. *Psychometrika* **18**, 23–28.

Crawford, C.B., Ferguson, G.A. (1970). A general rotation criterion and its use in orthogonal rotation. *Psychometrika* **35**, 321–332.

Harman, H.H. (1960). Factor analysis. In: Wilf, H.S., Ralston, A. (Eds.), *Mathematical Methods for Digital Computers*. John Wiley & Sons, New York, pp. 204–212.

Harman, H.H. (1976). *Modern Factor Analysis*, third ed. University of Chicago Press, Chicago.

Hendrickson, A.E., White, P.O. (1964). A quick method for rotation to oblique simple structure. *British Journal of Statistical Psychology* **17**, 65–70.

Jennrich, R.I. (1970). Orthogonal rotation algorithms. *Psychometrika* **35**, 229–235.

Jennrich, R.I. (2001). A simple general procedure for orthogonal rotation. *Psychometrika* **66**, 289–306.

Jennrich, R.I. (2002). A simple general procedure for oblique rotation. *Psychometrika* **67**, 7–19.

Jennrich, R.I. (2004). Rotation to simple loadings using component loss functions: The orthogonal case. *Psychometrika* **69**, 257–273.

Jennrich, R.I., Sampson, P.F. (1966). Rotation for simple loadings. *Psychometrika* **31**, 313–323.

Kaiser, H.F. (1958). The varimax criterion for analytic rotation in factor analysis. *Psychometrika* **23**, 187–200.

Korth, B., Tucker, L.R. (1976). Procrustes matching by congruence coefficients. *Psychometrika* **41**, 531–535.

Mosier, C.I. (1939). Determining a simple structure when loadings for certain tests are known. *Psychometrika* **4**, 149–162.

Neuhaus, J.O., Wrigley, C. (1954). The quartimax method: An analytical approach to orthogonal simple structure. *British Journal of Mathematical and Statistical Psychology* **7**, 81–91.

Nevels, K. (1986). A direct solution for pairwise rotations in Kaiser's varimax method. *Psychometrika* **51**, 327–329.

Saunders, D.R. (1953). An analytic method for rotation to orthogonal simple structure. Research Bulletin 53-10. Educational Testing Service, Princeton, NJ.

Ten Berge, J.M.F., Nevels, K. (1977). A general solution to Mosier's oblique procrustes problem. *Psychometrika* **42**, 593–600.

Thurstone, L.L. (1935). *Vectors of the Mind*. University of Chicago Press, Chicago.

Thurstone, L.L. (1947). *Multiple Factor Analysis*. University of Chicago Press, Chicago.

Yates, A. (1987). *Multivariate Exploratory Data Analysis: A Perspective on Exploratory Factor Analysis*. State University of New York Press, Albany.

Handbook of Computing and Statistics with Applications, Vol. 1
ISSN: 1871-0301

DOI: 10.1016/S1871-0301(06)01004-3

4

Selection of Manifest Variables*

Yutaka Kano

Abstract
Manifest variable selection in factor analysis and structural equation modeling (SEM) is an important process because a set of manifest variables defines a construct that researchers study and use to give scores to respondents. Emphasis on a model fit is placed in selection of variables. It is shown that the model fit criterion is as important as traditional psychometric properties, including variable content, communality or reliability and the number of variables. It is also shown how serious bias can be created by analysis with a model containing inconsistent variables, particularly, in reliability analysis. A web-based variable selection program in factor analysis is introduced. The program is called SEFA. Two examples are provided with empirical data to illustrate usefulness of the SEFA for manifest variable selection in scale construction.

Keywords: Coefficient alpha; Error covariances; Indicators; Latent construct; Measurement model; Model fit; Reliability analysis; Scale construction; SEFA

1. Introduction

Variable selection is an important issue in statistics and has been discussed extensively in the literature. Many important fruitful consequences have been implemented in statistical programs. Almost all the discussion on variable selection, however, focuses on the selection of independent variables in models with clear dependent (criteria) variables, e.g., regression analysis, discriminant analysis and time series analysis. This does not mean that variable selection in such models as factor analysis and principal component analysis is not important. In analyses with those models, variable selection is a very important step. According to Fabrigar et al. (1999), manifest variable selection is one of the five major methodological issues for proper use of factor analysis. Little et

*This work is prepared based on many invited lectures and talks on variable selections in factor analysis and structural equation modeling. Those include the IMPS2001 (Osaka), International Symposium on Structural Equation Modeling (Chicago) and Japanese meetings on statistics, behaviormetrics and psychology. The work is partially supported by a Grant-in-Aid #15500185 for Scientific Research from the Japan Society for the Promotion of Science.

al. (1999), mention that selecting indicators is as important for the generalizability of research designs as selecting persons or occasions of measurement. Indeed, when one makes an analysis of data from a questionnaire, there are usually many items (variables) in it from which appropriate items are selected and analyzed.

Scale construction is recognized as a major research topic that provides a measure of a latent construct in social sciences (e.g., Bartholomew, 1998). Scale construction using observational questionnaire data normally conducts factor analysis to select a set of appropriate items or manifest variables. It is nothing but variable selection in factor analysis. A measurement model in structural equation modeling (e.g., Bollen, 1989) is a mechanism that defines latent constructs or factors through manifest variables using a factor analysis model. Manifest variables with a common factor or a latent construct are therefore called indicators. Selection of manifest variables is related to the definition of the latent construct, and thus determination of a measurement model is a core of structural equation modeling. Consequently, manifest variable selection in factor analysis and related methods is an important process that methodologists have to study to develop proper and easy-to-use procedures.

Even though the importance of variable selection is recognized in factor analysis and related multivariate models, research methodologists have paid very little attention to variable selection in those models (e.g., Hogarty et al., 2004). There is no well-established procedure, and no option for variable selection is supplied in statistical programs. As a result, the selection of indicators (or variables) has typically relied on informal or intuitive reasoning or historical precedent (Little et al., 1999).

There are several psychometric properties that applied researchers have used in manifest variable selection. Cattell proposed defining a content domain or a domain of interest in a study and considered that manifest variables are sampled from the domain (e.g., Cattell, 1952, 1978; Nunnally and Bernstein, 1994). In the domain, the common factor scores are assumed to be unique, i.e., factor score indeterminacy vanishes in the domain (Guttman, 1955; Williams, 1978; Mulaik and McDonald, 1978; Krijnen, 2002). Ideally, the manifest variables selected are those sampled suitably from the domain. The theoretical idea is related to more practical thinking that gives practitioners useful guidelines on variable selection. The criteria used routinely include (i) meanings or contents of variables, (ii) the size of communality and reliability of variables, and (iii) the number of manifest variables. Reliability includes item reliability and scale reliability. A large factor loading of a variable usually ensures that the variable has a strong connection with the latent construct and will result in high reliability. There is a rule-of-thumb on the number of manifest variables. The rule is that 3 to 5 manifest variables for each factor should be chosen (e.g., Fabrigar et al., 1999, p. 273). The fourth criterion that might be mentioned here is to remove a manifest variable loaded on two or more factors. The criterion is often used in scale construction. Usually it is preferable to have subscales each of which has a distinct set of items.

In addition to the consideration of the psychometric properties, it is also important to take statistical aspects into account. There are two statistical tests related to the selection of manifest variables when factor analysis is conducted, namely to test significance of factor loadings or communalities and to test a model fit. Testing significance of factor loadings is a statistical examination of the size of the factor loadings. We first focus on the latter here since a model fit examination is of primary importance in practical research, particularly when applying structural equation modeling. The test of significance of communalities will be used to cope with a certain difficulty in selecting variables with a model fit.

Inclusion of inconsistent variables in a model under consideration may influence other manifest variables. For instance, inconsistency of X_i could cause unduly low factor-loading estimates and communality estimates for some variables other than X_i. Thus, deletion of manifest variables with low communalities does not work for those cases. Without examination of a model fit, it would be difficult to see whether the low communality and factor loading estimate for X_i are caused by the inconsistency of X_i itself or by that of other variables. See Section 4 for detailed discussions. As a result, a model fit should be verified before examining the psychometric properties.

In Section 2, we make a literature review on variable selection in factor analysis and pay attention to selection with a model fit. Some technical details are given. In Section 3, we introduce a web-based variable selection program SEFA, and give two analyses of empirical data to demonstrate backward elimination and forward selection procedures in factor analysis by SEFA. Section 4 discusses the meaning of variable selection with a model fit in relation to reliability analysis.

2. Manifest variable selection in factor analysis

We shall continue further the discussion on manifest variable selection based on the psychometric properties (i) to (iii) described in the introduction, which properties, although psychometric, do have statistical aspects.

The issues of communality and the number of variables per factor are strongly related to required sample sizes in estimation. Ihara and Okamoto (1985) among others confirmed that a model with smaller communalities makes the optimization problem in estimation more difficult, via a simulation study. Kano et al. (1993) and Yuan et al. (1997) have shown that the asymptotic variance of a factor loading estimator becomes smaller if a latent factor has more indicators. Fabrigar et al. (1999, p. 274) summarized those results to claim that adequate sample size is influenced by the extent to which factors are overdetermined and the level of the communalities of the manifest variables. A larger sample size is required for the smaller number of indicators for each factor and/or for lower communalities. In addition, it is known that increase of indicators in number reduces factor score indeterminacy (e.g., Williams, 1978). In theory, applied researchers are suggested to have more indicators.

Most important is the issue of content or meaning of manifest variables, i.e., whether the variables selected can define a latent factor that accurately corresponds to the (psychological) construct the researcher defines in his or her research objective. Although the issue is considered purely psychological, there is a statistical aspect that gives important insights to this problem. We examine criterion-related validity for choosing an appropriate set of manifest variables in the context of scale construction. Suppose that a construct under development is expected to be moderately positively correlated with established Construct X, moderately negatively correlated with established Construct Y, and uncorrelated with established Construct Z. Prepare a broad set of candidate indicators of the construct and use them to conduct a (strict) confirmatory factor analysis (CFA) of the four constructs. Then select indicators so that the expected correlations are achieved. The procedure owes entirely to Little et al. (1999, p. 208). Here a fit of the CFA model proves useful. In fact, once a CFA model with a set of indicators is well fitted, CFA models obtained by removing some of the indicators from it are also well fitted and the correlation structure among the four constructs remains the same, ignoring small fluctuations. As a result, researchers can reduce indicators from the largest well-fitted CFA model by applying other selection criteria, i.e., reliability and internal consistency, without any consideration of criterion-related validity.

Practical researchers are likely to remove manifest variables loaded on more than one factor, particularly when factor analysis is conducted for scale construction. It is done because items that can be an indicator for more than one latent construct can invalidate discriminant validity of a (sub)scale. Besides, in our limited experience, such items or manifest variables are often inconsistent with the factor analysis model, and removal of the items can improve a fit of the model for the case.

Yanai (1980) and Tanaka (1983) have suggested an alternative approach, where manifest variables are selected so that factor score configuration is maintained.

Kano and his collaborators have suggested selecting manifest variables so as to well fit a model considered to a data set in factor analysis (Kano and Ihara, 1994; Kano and Harada, 2000a, 2000b). The following section will explain it in some details. Kano and Harada (2000a) developed a computer program named SEFA to implement variable selection on a WWW server. We shall explain SEFA in Section 3.

2.1. *Basic idea on variable selection with a model fit*

We shall discuss here a basic idea of the selection of manifest variables with a model fit in factor analysis. Some mathematical background will be described in Section 2.2.

A factor analysis model with k common (latent) factors for a p-vector $\boldsymbol{X}$ of manifest variables is defined as

$$\boldsymbol{X} = \boldsymbol{\mu} + \Lambda \boldsymbol{f} + \boldsymbol{e},$$

where $\boldsymbol{\mu}$ $(= E(\boldsymbol{X}))$ is a general mean vector, $\boldsymbol{f}$ is a random k-vector of common factors with $E(\boldsymbol{f}) = \boldsymbol{0}$ and $V(\boldsymbol{f}) = \Phi$, $\boldsymbol{e}$ being a random p-vector of unique factors with $E(\boldsymbol{e}) = \boldsymbol{0}$ and $V(\boldsymbol{e}) = \Psi$ a diagonal matrix, and $\mathrm{Cov}(\boldsymbol{f}, \boldsymbol{e}) = 0$. The matrix Λ of $p \times k$ consists of factor loadings. In factor analysis, the general mean vector $\boldsymbol{\mu}$ is typically a nuisance and can be estimated by a sample mean vector. The parameters of interest to be estimated are Λ, Φ and Ψ. The covariance structure of the factor analysis model is then expressed as

$$V(\boldsymbol{X}) = \Lambda \Phi \Lambda' + \Psi.$$

Partition

$$\begin{bmatrix} X_1 \\ \boldsymbol{X}_2 \end{bmatrix} = \begin{bmatrix} \mu_1 \\ \boldsymbol{\mu}_2 \end{bmatrix} + \begin{bmatrix} \boldsymbol{\lambda}_1 \\ \Lambda_2 \end{bmatrix} \boldsymbol{f} + \begin{bmatrix} e_1 \\ \boldsymbol{e}_2 \end{bmatrix} \quad \text{and} \quad \Psi = \begin{bmatrix} \psi_{11} & \boldsymbol{0} \\ \boldsymbol{0} & \Psi_{22} \end{bmatrix}. \tag{1}$$

Here we do not discuss the typical issues in factor analysis such as parameter identification and factor rotation. For simplicity, we consider the orthogonal factor case, i.e., $\Phi = I_k$.

Consider the following hypothesis testing and associate test statistics T_0 and $T_{2'}$:

$$T_0 \quad \cdots \quad H_0\colon V(\boldsymbol{X}) = \Lambda\Lambda' + \Psi \quad \text{versus} \quad A_0\colon \text{not } H_0$$

and

$$T_{2'} \quad \cdots \quad H_{2'}\colon V(\boldsymbol{X}) = \begin{bmatrix} \sigma_{11} & \boldsymbol{\sigma}_{12} \\ \boldsymbol{\sigma}_{21} & \Lambda_2\Lambda_2' + \Psi_{22} \end{bmatrix} \quad \text{versus} \quad A_{2'}\colon \text{not } H_{2'}, \tag{2}$$

where σ_{11} and $\boldsymbol{\sigma}_{21}$ are all free parameters.

Kano and Ihara (1994) considered that X_1 is inconsistent when T_0 suggests rejection of H_0 and $T_{2'}$ suggests acceptance of $H_{2'}$. Kano and Harada (2000a) used the idea of Lagrange multiplier (LM) tests to develop a new statistic, calculated with minimal computational effort, that can test goodness-of-fit of all p marginal models with $p-1$ manifest variables. The statistic is useful in implementing a backward elimination procedure. They also derived an approximate test statistic for goodness-of-fit for models obtained by adding an external variable. The statistic is used to make forward selection.

In practice, one makes backward eliminations and forward selections many times to identify a well-fitted model. One may ask the following question: Does the procedure tell us to remove a consistent variable when there are inconsistent variates? Browne (1998) anticipated that the newly developed LM statistic could not work if X_1 were inconsistent with the model. Regarding this question, Kano (2002) proved that a small misspecification for the X_1 does not fatally influence the performance of the LM test.

Hogarty et al. (2004) experimentally compared variable selection by a model fit and that reached by the traditional way with the size of factor loadings. They pointed out that variable selection by a model fit cannot identify any variables that are not correlated with the latent factors. It is obvious that any factor analysis model with a zero row vector in Λ is also a factor analysis model. Such manifest variables may not be useful for analysis, however. SEFA advises against using variables with small communalities. Harada and Kano (2001) developed a methodology that can statistically test whether communality is small (or zero) based on the result by Ichikawa (1992), who gave the asymptotic distribution of a communality estimator. The new methodology has been incorporated into SEFA.

2.2. *Some mathematical details*

In this section, we shall describe some mathematical derivations for the consequences stated in the previous section within the context of covariance structure analysis. The covariance structure model is written as $\{V(\boldsymbol{X}) = \Sigma(\boldsymbol{\theta}) \mid \boldsymbol{\theta} \in \Theta\}$, where Θ $(\subset R^q)$ is a parameter space of $\boldsymbol{\theta}$ and $\Sigma(\boldsymbol{\theta})$ is a twice continuously differentiable matrix-valued function on Θ. Let $\boldsymbol{X} = [X_1, \boldsymbol{X}_2']'$ with $\boldsymbol{X}_2$ a $(p-1)$-vector. The covariance matrix is also partitioned correspondingly. In particular, $V(\boldsymbol{X}_2) = \Sigma_{22}(\boldsymbol{\theta}_2)$, where $\boldsymbol{\theta}_2$ is part of $\boldsymbol{\theta}$. Consider the following hypothesis testing and associate test statistics T_0, T_2, $T_{2'}$

and $T_{02'}$ as in (2):

$$\begin{aligned} &T_0 \quad \cdots \quad H_0\colon V(\boldsymbol{X}) = \Sigma(\boldsymbol{\theta}) \quad \text{versus} \quad A_0\colon \text{not } H_0, \\ &T_2 \quad \cdots \quad H_2\colon V(\boldsymbol{X}_2) = \Sigma_{22}(\boldsymbol{\theta}_2) \quad \text{versus} \quad A_2\colon \text{not } H_2, \\ &T_{2'} \quad \cdots \quad H_{2'}\colon V(\boldsymbol{X}) = \begin{bmatrix} \sigma_{11} & \boldsymbol{\sigma}_{12} \\ \boldsymbol{\sigma}_{21} & \Sigma_{22}(\boldsymbol{\theta}_2) \end{bmatrix} \quad \text{versus} \quad A_{2'}\colon \text{not } H_{2'} \end{aligned} \tag{3}$$

and

$$T_{02'} \quad \cdots \quad H_0\colon V(\boldsymbol{X}) = \Sigma(\boldsymbol{\theta}) \quad \text{versus} \quad H_{2'}\colon V(\boldsymbol{X}) = \begin{bmatrix} \sigma_{11} & \boldsymbol{\sigma}_{12} \\ \boldsymbol{\sigma}_{21} & \Sigma_{22}(\boldsymbol{\theta}_2) \end{bmatrix}.$$

Suppose that $\boldsymbol{X}_1, \ldots, \boldsymbol{X}_N$ be a random sample from a multivariate normal population $N(\boldsymbol{\mu}, \Sigma)$. Let $n = N - 1$ and let S be an unbiased sample covariance matrix. The likelihood ratio test statistic T_0 can be expressed as

$$T_0 = n\left(\ln\left|\Sigma(\hat{\boldsymbol{\theta}})\right| - \ln|S| + \operatorname{tr}\left[\Sigma(\hat{\boldsymbol{\theta}})^{-1}\left(S - \Sigma(\hat{\boldsymbol{\theta}})\right)\right]\right),$$

where $\hat{\boldsymbol{\theta}}$ is the MLE under the model in H_0. The statistic T_2 can be represented in the same way. The other statistics can be expressed similarly. Notice that the MLE $\tilde{\boldsymbol{\theta}}_2$ in the model in $H_{2'}$ is the solution to

$$\operatorname{tr}\left[\Sigma_{22}(\boldsymbol{\theta}_2)^{-1}\left(\Sigma_{22}(\boldsymbol{\theta}_2) - S_{22}\right)\Sigma_{22}(\boldsymbol{\theta}_2)^{-1}\frac{\partial}{\partial\theta_i}\Sigma_{22}(\boldsymbol{\theta}_2)\right] = 0$$
$$(i = 1, \ldots, q),$$

which are thus the same equations as those in H_2. The MLEs for σ_{11} and $\boldsymbol{\sigma}_{21}$ are obtained using the relations:

$$\begin{aligned} \sigma_{11} &= s_{11} - \boldsymbol{s}_{12}S_{22}^{-1}\boldsymbol{s}_{21} + \boldsymbol{s}_{12}S_{22}^{-1}\Sigma_{22}(\boldsymbol{\theta}_2)S_{22}^{-1}\boldsymbol{s}_{21}, \\ \boldsymbol{\sigma}_{21} &= \Sigma_{22}(\boldsymbol{\theta}_2)S_{22}^{-1}\boldsymbol{s}_{21}. \end{aligned}$$

See Kano and Ihara (1994) for a proof. Notice that T_2 and $T_{2'}$ are statistics for testing a fit of the model for $\boldsymbol{X}_2$, and so they are asymptotically equivalent.

In order to implement a backward elimination procedure, it is necessary to compute T_2 for all p models obtained by deleting manifest variables, one by one. The computational effort for it is not ignorable if p is large, because optimization problems have to be solved p times. In addition, some computational difficulties have been reported for estimation in factor analysis, e.g., nonconvergence of iterative processes and/or occurrence of improper solutions. Thus, it is particularly useful if the p test statistics T_2's can be obtained as explicit functions of $\hat{\boldsymbol{\theta}}$ in H_0 and S. For this purpose, Kano and Harada (2000a) derived

$$T_2 = T_{2'} + o_p(1) = T_0 - (T_0 - T_{2'}) + o_p(1) = T_0 - T_{02'} + o_p(1).$$

Let $T_{02'}$ be defined as an LM test statistic, and then $T_{02'}$ and therefore $T_2 = T_0 - T_{02'}$ are functions of S and $\hat{\boldsymbol{\theta}}$ under H_0. A similar derivation holds for any model deleting one variable other than X_1. Thus, one can obtain p test statistics for testing p marginal models by solving an optimization problem *once* to obtain the MLE $\hat{\boldsymbol{\theta}}$ and maximum likelihood under H_0. Of course, a similar result holds when several variables are removed simultaneously. Note that in factor analysis the procedure is valid only when the number of factors is held constant.

We shall now derive the LM statistic $T_{02'}$. Let $v(A)$ and $\mathrm{vec}(A)$ for a matrix A of order p denote a $p(p+1)/2$-vector of elements of lower triangular part of A and a p^2-vector obtained by stacking all column vectors of A in order. Let D_p be the duplication matrix of order $p^2 \times p(p+1)/2$ which is defined by the relation $\mathrm{vec}(A) = D_p v(A)$ for any symmetric matrix A of order p. Let $D_p^+ = (D_p' D_p)^{-1} D_p'$. See Magnus and Neudecker (1999) for the notation and their properties. Let $\Sigma_{2'}(\underline{\theta})$ denote the covariance structure in $H_{2'}$ with $\underline{\theta} = [\underline{\theta}_1', \theta_2']' = [\sigma_{11}, \boldsymbol{\sigma}_{12}, \theta_2']'$. Define

$$\underline{\Delta}(\underline{\theta}) = \frac{\partial v(\Sigma_{2'}(\underline{\theta}))}{\partial \underline{\theta}} = \begin{bmatrix} I_p & O \\ O & \frac{\partial v(\Sigma_{22}(\theta_2))}{\partial \theta_2'} \end{bmatrix} \left(= \begin{bmatrix} I_p & O \\ O & \underline{\Delta}_2(\theta_2) \end{bmatrix}, \text{ say} \right),$$
$$\underline{\Gamma}_N(\underline{\theta}) = 2D_p^+ \big(\Sigma_{2'}(\underline{\theta}) \otimes \Sigma_{2'}(\underline{\theta})\big) D_p^{+'}.$$

The score vector $s(\underline{\theta})$ and the Fisher information matrix $I(\underline{\theta})$ can be expressed as

$$s(\underline{\theta}) = n \cdot \underline{\Delta}(\underline{\theta})' \underline{\Gamma}_N(\underline{\theta})^{-1} v\big(S - \Sigma_{2'}(\underline{\theta})\big),$$
$$I(\underline{\theta}) = n \cdot \underline{\Delta}(\underline{\theta})' \underline{\Gamma}_N(\underline{\theta})^{-1} \underline{\Delta}(\underline{\theta}).$$

The statistic $T_{2'}$ can then be expressed as

$$T_{2'} = s(\hat{\underline{\theta}})' I(\hat{\underline{\theta}})^{-1} s(\hat{\underline{\theta}}).$$

Here $\hat{\underline{\theta}}$ is formed by the estimator $\hat{\theta}$ in the model of H_0, and hence $T_{2'}$ is an explicit function of $\hat{\theta}$ and S.

A question arises here. Does the newly developed statistic $T_0 - T_{02'}$ produce an appropriate statistic for a fit of the model for $\boldsymbol{X}_2$ even if X_1 is inconsistent? The answer is yes. To study this, Kano (2002) introduced Pitman's local alternative:

$$V(\boldsymbol{X}) = \Sigma(\boldsymbol{\theta}) + \frac{1}{\sqrt{n}} D = \Sigma(\boldsymbol{\theta}) + \frac{1}{\sqrt{n}} \begin{bmatrix} d_{11} & \boldsymbol{d}_{12} \\ \boldsymbol{d}_{21} & D_{22} \end{bmatrix}. \tag{4}$$

The inconsistency of X_1 creates a bias of MLE. Let

$$\Delta = \frac{\partial v(\Sigma(\boldsymbol{\theta}))}{\partial \boldsymbol{\theta}'} \quad \text{and} \quad \Gamma_N = 2D_p^+ \big(\Sigma(\boldsymbol{\theta}) \otimes \Sigma(\boldsymbol{\theta})\big) D_p^{+'}.$$

The asymptotic bias of $\sqrt{n}(\hat{\boldsymbol{\theta}} - \boldsymbol{\theta})$ is then given as

$$\big(\Delta' \Gamma_N^{-1} \Delta\big)^{-1} \Delta \Gamma_N^{-1} \,\mathrm{vec}(D), \tag{5}$$

and the bias may not be zero even if $D_{22} = O$. We could make the same assertion for $\hat{\boldsymbol{\theta}}_2$ because it is part of $\hat{\boldsymbol{\theta}}$.

Let

$$\Delta_2 = \frac{\partial v(\Sigma_{22}(\boldsymbol{\theta}_2))}{\partial \boldsymbol{\theta}_2'} \quad \text{and} \quad \Gamma_{N2} = 2D_{p-1}^+ \big(\Sigma_{22}(\boldsymbol{\theta}_2) \otimes \Sigma_{22}(\boldsymbol{\theta}_2)\big) D_{p-1}^{+'}.$$

The test statistic $T_0 - T_{02'}$ has the stochastic expansion as

$$T_0 - T_{02'} = nv\big(S_{22} - \Sigma_{22}(\hat{\boldsymbol{\theta}}_2)\big)'\big(\Gamma_{N2}^{-1} - \Gamma_{N2}^{-1}\Delta_2\big(\Delta_2'\Gamma_{N2}^{-1}\Delta_2\big)^{-1}\Delta_2'\Gamma_{N2}^{-1}\big) \\ \times v\big(S_{22} - \Sigma_{22}(\hat{\boldsymbol{\theta}}_2)\big) + o_p(1), \tag{6}$$

and we note that

$$\sqrt{n}v\big(S_{22} - \Sigma_{22}(\hat{\boldsymbol{\theta}}_2)\big) = \sqrt{n}v\left(S_{22} - \Sigma_{22}(\boldsymbol{\theta}_2) - \frac{D_{22}}{\sqrt{n}}\right) \\ + v(D_{22}) + \Delta_2\sqrt{n}(\hat{\boldsymbol{\theta}}_2 - \boldsymbol{\theta}_2) + o_p(1).$$

Since the term involving Δ_2 in the above disappears in the quadratic form in (6), the asymptotic distribution of the statistic $T_0 - T_{02'}$ is free from the bias of $\hat{\boldsymbol{\theta}}_2$, and hence it converges in distribution to the *central* chi-square distribution if $D_{22} = O$, even when $[d_{11}, \boldsymbol{d}_{12}']' \neq \mathbf{0}$ in (4). The statistic can surely detect deviation of $\boldsymbol{X}_2$ from the model $\Sigma_{22}(\boldsymbol{\theta}_2)$, which has been expressed as D_{22}, because the statistic converges in law to the noncentral chi-square distribution if $D_{22} \neq O$. In other words, whether D_{22} is null or not, the effect of $[d_{11}, \boldsymbol{d}_{12}']'$ asymptotically drops in the distribution of $T_0 - T_{02'}$ if the deviation $[d_{11}, \boldsymbol{d}_{12}']'/\sqrt{n}$ is not so large.

Finally we shall refer to testing a hypothesis concerning a communality. Using the notation in (1), the communality of X_1 is expressible as $\boldsymbol{\lambda}_1\boldsymbol{\lambda}_1^{\mathrm{T}}$. Harada and Kano (2001) used a Wald type test statistic to test

$$H\colon \boldsymbol{\lambda}_1\boldsymbol{\lambda}_1^{\mathrm{T}} = c_0 \quad \text{versus} \quad A\colon \boldsymbol{\lambda}_1\boldsymbol{\lambda}_1^{\mathrm{T}} \geqslant c_0$$

for some known small constant $c_0 \geqslant 0$, where we have assumed that $\Phi = I_k$. We used the formula of asymptotic variance of a communality estimator derived by Ichikawa (1992). Let $A = (a_{ij}) = \Psi^{-1} - \Psi^{-1}\Lambda(\Lambda'\Psi^{-1}\Lambda)^{-1}\Lambda'\Psi^{-1}$ and $\Xi = (a_{ij}^2)$. Ichikawa derived the asymptotic variance of the MLE $\widehat{\boldsymbol{\lambda}_1\boldsymbol{\lambda}_1^{\mathrm{T}}}$ in the form:

$$2\xi^{11} + 2\sigma_{11}^2 - 4\psi_1^2,$$

where ξ^{11} is the (1, 1)th element of ξ^{-1}.

It should be noted that the theoretical derivation is valid not only for (exploratory) factor analysis models but also for any other covariance structure models, except for the communality testing. Thus, it is easily applicable to confirmatory factor analysis (CFA) models, for instance, and we are now developing a new program for variable selection in CFA models.

3. SEFA and examples with empirical data

Kano and Harada (2000a) developed the program SEFA (Stepwise Variable Selection in Exploratory Factor Analysis) based on the theory described in the previous sections. The SEFA is delivered in a virtually platform-independent manner with only minimal requirements on a user's hardware or software. Anyone with a WWW (World Wide Web) browser can use the program. The URL of the SEFA is as follows:

```
http://koko16.hus.osaka-u.ac.jp/~harada/sefa2002/stepwise/
```

SEFA can print a list of several goodness-of-fit measures for models that are obtained by deletion or addition of one variable and is extremely useful in backward elimination and forward selection of manifest variables. SEFA provides an excellent circumstance for variable selection even though practical users are not interested in fit measures.

On the top page of the SEFA program, a user is asked to input a sample correlation matrix, the number of variables, sample size, and the number of factors. Click a submit button, and then (s)he will obtain a table of fit measures for every one-variable-deleted model and every one-variable-added model, as well as those for the current model. When the required information is entered, PROC factor in SAS runs once, and fit measures of those models are computed, using $T_0 - T_{02'}$, from the sample correlation matrix and communality estimates imported from the SAS output. When a user wants to drop manifest variables, (s)he just de-selects the check boxes of the variables, and submits the job again.

There are some options in SEFA. Any user can choose his (or her) favorite fit measures from χ^2, GFI, AGFI, CFI, IFI and RMSEA (see, e.g., Tanaka, 1993; Hu and Bentler, 1999, for details of these measures); (s)he can choose significance levels for the chi-square goodness-of-fit test and the test of communalities. For the latter, (s)he can specify null hypotheses, i.e., how small a communality can be. After submission of a job to SEFA, users can obtain an output webpage as in Figure 1. The webpage shows a table that includes, from the flush-left: (i) selecting box, (ii) variable name, (iii) factor loading estimate, (iv) communality estimate with p-value of testing communality in parenthesis, (v) chi-square statistic and other fit measures.

There are several facets on manifest variable selection in factor analysis. The first facet is backward elimination or forward selection, as noted. The second is conducting variable selection starting at a model with all variables, as will be demonstrated in Section 3.1, or to focus on part of the model, that is, a part of the manifest variables, which includes only indicators of a single or a few factor(s), is selected and examined. Practitioners can add or delete all manifest variables loaded on a factor simultaneously, and then the number of factors increases or decreases by one.

We demonstrate a backward elimination procedure in manifest variable selection by SEFA in Section 3.1; and a forward selection procedure is described in Section 3.2. We shall use the traditional psychometric properties as well as a model fit.

3.1. Analysis of perception on physical exercise

Using SEFA, we analyzed a data set of a questionnaire of perception on physical exercise, which was collected and analyzed by Oka et al. (2002). The data set has 15 variables and the sample size is 653. These applied researchers expected the data set to have a unidimensional structure. A one-factor analysis model receives a rather poor fit because the chi-square goodness-of-fit statistic gives a value of 460.11 with 90 degrees

Variable	F1	Communality (P-value)	df	Chi-square	Prob > chi**2	GFI	AGFI	CFI	IFI	RMSEA
Current Model (15 in 15)			90	460.1068	0.0000	0.9196	0.8928	0.8918	0.8936	0.0799
☑ X1	0.60	0.364(0.00)	77	386.0540	0.0000	0.9257	0.8987	0.9006	0.9022	0.0789
☑ X2	0.67	0.452(0.00)	77	348.0551	0.0000	0.9328	0.9084	0.9094	0.9109	0.0739
☑ X3	0.54	0.295(0.00)	77	433.0019	0.0000	0.9205	0.8916	0.8891	0.8909	0.0847
☑ X4	0.63	0.402(0.00)	77	376.3264	0.0000	0.9303	0.9050	0.9024	0.9040	0.0777
☑ X5	0.65	0.417(0.00)	77	427.3836	0.0000	0.9207	0.8919	0.8871	0.8889	0.0840
☑ X6	0.64	0.404(0.00)	77	413.0557	0.0000	0.9206	0.8917	0.8916	0.8933	0.0823
☑ X7	0.47	0.224(0.00)	77	437.3357	0.0000	0.9190	0.8896	0.8898	0.8915	0.0852
☑ X8	0.69	0.477(0.00)	77	380.6088	0.0000	0.9271	0.9006	0.8988	0.9005	0.0782
☑ X9	0.67	0.451(0.00)	77	375.3770	0.0000	0.9278	0.9015	0.9012	0.9028	0.0776
☑ X10	0.64	0.411(0.00)	77	441.5062	0.0000	0.9174	0.8874	0.8833	0.8851	0.0857
☑ X11	0.59	0.343(0.00)	77	417.7270	0.0000	0.9217	0.8933	0.8921	0.8938	0.0829
☑ X12	0.62	0.381(0.00)	77	385.7568	0.0000	0.9287	0.9028	0.9002	0.9018	0.0789
☑ X13	0.58	0.338(0.00)	77	377.4551	0.0000	0.9279	0.9016	0.9037	0.9053	0.0778
☑ X14	0.68	0.457(0.00)	77	343.2054	0.0000	0.9337	0.9096	0.9108	0.9123	0.0733
☑ X15	0.38	0.147(0.06)	77	404.1958	0.0000	0.9217	0.8932	0.9005	0.9021	0.0812

Fig. 1. SEFA output for the initial model with 15 manifest variables.

Variable	F1	Communality (P-value)	df	Chi-square	Prob > chi**2	GFI	AGFI	CFI	IFI	RMSEA
Current Model (11 in 15)			44	107.3664	0.0000	0.9710	0.9565	0.9666	0.9672	0.0473
☑ X1	0.60	0.363(0.00)	35	78.0372	0.0000	0.9759	0.9622	0.9737	0.9742	0.0437
☐ X2			54	185.9781	0.0000	0.9552	0.9354	0.9432	0.9442	0.0616
☑ X3	0.55	0.304(0.00)	35	93.3829	0.0000	0.9724	0.9567	0.9657	0.9663	0.0509
☑ X4	0.64	0.415(0.00)	35	81.0675	0.0000	0.9750	0.9607	0.9712	0.9718	0.0452
☑ X5	0.66	0.442(0.00)	35	95.0681	0.0000	0.9715	0.9553	0.9623	0.9630	0.0516
☑ X6	0.65	0.417(0.00)	35	85.9517	0.0000	0.9746	0.9600	0.9683	0.9689	0.0475
☑ X7	0.49	0.239(0.00)	35	93.0315	0.0000	0.9724	0.9567	0.9668	0.9673	0.0507
☑ X8	0.68	0.465(0.00)	35	82.3526	0.0000	0.9753	0.9612	0.9697	0.9703	0.0458
☐ X9			54	175.2335	0.0000	0.9584	0.9399	0.9470	0.9479	0.0590
☑ X10	0.64	0.412(0.00)	35	89.9277	0.0000	0.9727	0.9570	0.9660	0.9666	0.0494
☑ X11	0.56	0.311(0.00)	35	90.9249	0.0000	0.9730	0.9576	0.9670	0.9675	0.0498
☑ X12	0.58	0.332(0.00)	35	78.6887	0.0000	0.9768	0.9635	0.9737	0.9742	0.0440
☐ X13			54	173.0444	0.0000	0.9590	0.9408	0.9448	0.9458	0.0585
☐ X14			54	200.8494	0.0000	0.9527	0.9317	0.9361	0.9372	0.0650
☑ X15	0.39	0.152(0.05)	35	77.2362	0.0001	0.9773	0.9643	0.9763	0.9768	0.0433

Fig. 2. SEFA output for the final model with 11 manifest variables.

of freedom, and GFI = 0.920, CFI = 0.892 and RMSEA = 0.080. The stepwise variable selection program SEFA gives an output in Figure 1, suggesting deletion of the variable X_{14}.

Recall that the size of communality has been often used as a criterion for variable selection in factor analysis. Results of the communality estimates in Figure 1 show that X_{15} and X_7 have small communalities $(0.147, 0.224)$. The communality criterion makes a totally different choice of variables.

We dropped X_{14} and ran the SEFA again. The SEFA has suggested dropping X_2, X_9 and X_{13} in order. We followed SEFA's suggestion until we reached a final model with 11 variables (Figure 2), where $\chi^2_{44} = 107.37$, GFI $= 0.971$, CFI $= 0.967$ and RMSEA $= 0.047$. Removing the four variables, the chi-square statistic improves from 460.11 to 107.37. Their difference follows according to a chi-square distribution with 46 degrees of freedom.

3.2. Analysis of low self-control data

We analyzed a date set in criminal psychology by SEFA. Gottfredson and Hirschi (1990) considered low self-control and criminal opportunity as principal causes of criminal behaviors. Grasmick et al. (1993) developed a scale of low self-control which consists of the following six subscales: impulsivity (F_1), simple tasks (F_2), risk seeking (F_3), physical activities (F_4), self-centered (F_5) and temper (F_6). Each subscale has four items so that the scale has a total of 24 items. Kono and Okamoto (1999, 2001) translated the low self-control scale to Japanese. Murakami (2000) distributed to university students questionnaires including the Japanese version of the low self-control scale items, where 220 students returned the questionnaires. Here we shall use her data set removing observations with missing values, where the sample size is 213. See Grasmick et al. (1993) for the original English questionnaire items.

Table 1 (left) shows results of exploratory factor analysis for the data set with all variables, where the factor pattern matrix is rotated by the promax method.

According to Grasmick et al. (1993), manifest variables X_1 to X_4 are indicators of the latent construct F_1, X_5 to X_8 are indicators of the latent construct F_2, and so on. The results for the data set are not very consistent with the theory by Grasmick. Indeed, discriminant validity between F_1 and F_2 does not hold, and there are several substantial estimates that destroy simple structure in the factor loading matrix.

The low self-control scale was developed in U.S. and will be valid within the Western culture. It is not sure if its direct translation is applicable to Japanese as well. The results could be improved by appropriately selecting manifest variables, as will be tried next.

Because the most serious problem appears in the factor loading estimates for variables X_1 to X_8, we first analyzed these variables with two common factors. A fit of the estimated model is fairly good. We, however, found that X_6 is loaded on both factors, impulsivity and simple tasks, which destroys simple structure. Besides, psychologists mention that the meaning of the item X_6 could relate to the both constructs. As a result, we decided to delete X_6. Next, we chose variables X_{17} to X_{24} to examine their suitableness to a two-factor analysis model. An output of SEFA is shown in Figure 3 (left), indicating that a fit of the model is rather poor ($\chi^2_{13} = 39.2723$). As far as a model fit is concerned, SEFA suggests deletion of X_{17} and X_{22}. Notice that X_{22} is more loaded on `F1`, which is against the theory. Psychologists mentioned that X_{17} is an important item that should stay. Accordingly, we decided to delete X_{22}.

Table 1
Manifest variable selection for the low self-control data by SEFA

	Initial model with all 24 variables						Final model with X_6, X_{11}, X_{22} deleted					
	F_1	F_2	F_3	F_4	F_5	F_6	F_1	F_2	F_3	F_4	F_5	F_6
X_1	**0.46**	0.18	0.12	0.04	−0.05	0.16	0.39	−0.07	0.11	0.11	0.12	0.21
X_2	0.21	**0.42**	0.17	−0.08	−0.04	−0.15	**0.62**	0.02	0.05	−0.04	−0.01	−0.08
X_3	0.17	**0.42**	0.22	−0.01	−0.12	−0.01	**0.44**	0.12	0.14	0.04	−0.08	0.01
X_4	−0.12	**0.53**	0.08	0.28	0.02	0.04	−0.15	**0.72**	0.18	0.23	−0.04	0.02
X_5	0.01	**0.66**	−0.08	−0.12	0.15	−0.09	0.16	**0.56**	−0.07	−0.16	0.12	−0.06
X_6	0.16	**0.75**	0.00	−0.06	−0.10	0.08						
X_7	−0.05	**0.56**	−0.11	0.07	0.13	0.04	0.09	**0.55**	−0.08	0.01	0.06	0.06
X_8	0.09	**0.54**	−0.33	0.07	0.02	0.02	0.15	**0.45**	−0.31	0.06	0.05	0.05
X_9	0.01	0.00	**0.83**	0.02	0.02	0.07	0.09	0.00	**0.83**	0.00	−0.02	0.08
X_{10}	0.04	0.03	**0.76**	−0.05	0.12	0.03	0.06	0.07	**0.81**	−0.11	0.06	0.07
X_{11}	0.13	0.02	0.25	−0.13	−0.01	0.39						
X_{12}	0.04	−0.15	**0.57**	0.28	0.01	−0.14	0.09	−0.18	**0.51**	0.31	0.02	−0.16
X_{13}	0.11	0.07	0.14	**0.59**	−0.16	−0.01	0.05	0.04	0.11	**0.66**	−0.08	−0.04
X_{14}	0.09	−0.01	−0.06	**0.78**	0.00	0.02	0.07	−0.01	−0.11	**0.79**	0.02	0.06
X_{15}	−0.09	0.18	−0.01	**0.76**	−0.01	−0.09	0.06	0.20	−0.06	**0.73**	−0.08	−0.08
X_{16}	0.00	−0.33	0.04	**0.56**	0.09	0.10	−0.35	−0.09	0.08	**0.58**	0.14	0.06
X_{17}	**0.40**	−0.07	−0.18	0.02	**0.53**	0.02	0.00	−0.06	−0.16	0.04	**0.68**	0.03
X_{18}	0.07	0.07	0.08	−0.07	**0.67**	−0.09	−0.07	0.13	0.08	−0.13	**0.61**	−0.10
X_{19}	0.01	−0.03	0.11	−0.03	**0.64**	−0.03	−0.11	0.11	0.14	−0.11	**0.54**	−0.04
X_{20}	**0.52**	0.05	0.05	0.09	**0.40**	−0.03	0.24	−0.07	0.05	0.14	**0.56**	0.00
X_{21}	0.09	−0.01	−0.08	0.09	0.02	**0.73**	0.00	0.05	−0.01	0.03	0.02	**0.79**
X_{22}	−0.16	0.18	0.15	−0.03	0.38	0.31						
X_{23}	0.02	−0.11	0.02	−0.10	−0.06	**0.54**	−0.01	−0.06	0.07	−0.13	−0.06	**0.52**
X_{24}	−0.04	0.10	−0.01	0.08	−0.04	0.37	−0.02	0.13	0.02	0.06	−0.06	0.30

Figures in bold face and with a underline denote factor loading estimates greater than or equal to 0.4.

We added variables X_{13} to X_{16} and added one factor; and then added X_9 to X_{12} and added one factor again. Results are shown in Figure 3 (right). While there is no manifest variable that can improve a model fit much by its deletion, X_{11} has a problem which is loaded more on `F4` than on `F2`. When we ran SEFA for variables X_9 to X_{24} without X_{22} to study the problem in more detail, we found the same tendency. Psychologists mentioned that X_{11} should not have any relation to the factor 'temper'. Thus, we decided to delete X_{11}.

We obtained an improper solution at X_{23}, i.e., its communality estimate is greater than or equal to one, when we ran SEFA for variables X_9 to X_{12} and X_{21} to X_{24} without X_{11} and X_{22}. See, e.g., Van Driel (1978), Kano (1998) and Chen et al. (2001) for improper solutions. SEFA can deal with improper solutions, where we have used S^{-1} for $\hat{\Psi}^{-1}$ in the program. SEFA alerts users by coloring the corresponding cell yellow for an improper solution. The improper solution occurs because the three-indicator rule

Variable	F1	F2	Communality (P-value)	df	Chi-square
Current Model (8 in 24)				13	39.2723
X1				19	68.7385
X2				19	43.5343
X3				19	54.6309
X4				19	43.8614
X5				19	41.2852
X6				19	48.5997
X7				19	42.3224
X8				19	50.0101
X9				19	47.5988
X10				19	52.6698
X11				19	44.9928
X12				19	45.9801
X13				19	46.4331
X14				19	52.9666
X15				19	47.4222
X16				19	43.1654
X17	0.52	0.03	0.280(0.01)	8	15.1684
X18	0.75	-0.10	0.546(0.00)	8	34.5089
X19	0.63	-0.05	0.384(0.00)	8	34.4320
X20	0.49	0.05	0.249(0.02)	8	22.6799
X21	0.03	0.74	0.557(0.00)	8	27.5565
X22	0.43	0.29	0.321(0.00)	8	18.4655
X23	-0.06	0.55	0.289(0.03)	8	24.9717
X24	0.01	0.33	0.112(0.42)	8	25.0833

Variable	F1	F2	F3	F4	Communality (P-value)	df	Chi-square
Current Model (15 in 24)						51	72.8899
X1						62	109.6295
X2						62	104.4953
X3						62	94.7262
X4						62	93.6485
X5						62	101.9556
X6						62	106.1904
X7						62	96.5319
X8						62	97.8036
X9	0.02	0.87	-0.02	0.07	0.786(0.00)	41	59.7189
X10	-0.05	0.74	0.10	0.04	0.582(0.00)	41	55.6600
X11	-0.13	0.29	0.03	0.37	0.254(0.02)	41	53.3307
X12	0.28	0.59	-0.04	-0.14	0.497(0.00)	41	60.9181
X13	0.60	0.16	-0.09	-0.02	0.448(0.00)	41	50.9611
X14	0.83	-0.07	0.07	0.07	0.659(0.00)	41	56.0625
X15	0.72	-0.06	0.01	-0.04	0.499(0.00)	41	67.0580
X16	0.52	0.11	-0.02	0.02	0.316(0.00)	41	59.8150
X17	0.03	-0.10	0.62	0.03	0.372(0.00)	41	51.3680
X18	-0.10	0.02	0.65	-0.08	0.450(0.00)	41	59.9781
X19	-0.07	0.06	0.58	-0.04	0.361(0.00)	41	56.5341
X20	0.12	0.11	0.57	0.03	0.360(0.00)	41	52.9115
X21	0.11	-0.06	0.05	0.83	0.720(0.00)	41	49.2247
X22						62	104.5376
X23	-0.10	0.06	-0.09	0.52	0.265(0.03)	41	56.2707
X24	0.07	-0.02	-0.03	0.28	0.083(0.64)	41	68.0097

Fig. 3. SEFA outputs for low self-control data.

for identification seems to be violated empirically, that is, there are only two indicators with enough large loading estimates 0.83 and 0.52 for the common factor and the factor loading estimate 0.28 for X_{24} is not large. For the situation, the estimates are rather unstable, and can be improper or proper, depending on variables jointly analyzed.

At a final stage, we added variables X_1 to X_8 without X_6, which results in estimates in the final model in Table 1 (right). In the final result, one variable X_4 is grouped into the subscale for F_2 although X_4 is an item for F_1 originally. Psychologists mentioned that the content of X_4 is closer to F_2 than F_1 (in Japan). Looking at the final result in Table 1, the result after the manifest variable selection represents the Grasmick theory more accurately.

4. Variable selection with a model fit and reliability analysis

The first version of SEFA came out in 1999. Since then, we have received many questions from practical users. Some major questions are as follows:

(a) What is the meaning of choosing variables with a model fit?

(b) Why do they need to choose variables such that their model receives a good fit? What happens if inconsistent variables remain in their model?
(c) What should they do if SEFA indicates that a marker variable is inconsistent with their model?
(d) They claim that SEFA is more likely to suggest deleting a variable with a larger communality estimate, e.g., 0.6 or more. Why is it so? How should they cope with it?

An all-too-common answer to the questions would be that examination of model adequacy is the first step of statistical analysis and that a model with a poor fit is often misleading. In this section we consider a factor analysis model with correlated errors and a reliability analysis to provide more specific and persuasive examples in reply to these questions.

Recall that one purpose of factor analysis is to construct a scale of a psychological construct and to assess reliability of the scale constructed.

4.1. Model with correlated errors

Probably it was Bollen (1980) who first emphasized importance of introducing error covariances in researches using factor analysis. We shall reanalyze the data set in Section 3.1 to achieve a good fit not by removing manifest variables but by allowing error covariances. We analyzed it with the help of LM tests offered by EQS (Bentler, 2004). The path diagram of a final model is shown in Figure 4. Fit indices of the final model are $\chi^2 = 250.375$ ($df = 83, n = 653$), GFI $= 0.950$, CFI $= 0.952$, RMSEA $= 0.056$. Those statistics indicate a fairly good fit.

Recall that SEFA suggests deletion of the variables X_2, X_9, X_{13} and X_{14}. We see that SEFA chooses such variables to be removed that those error covariances are eliminated. As a result, the final model with 11 manifest variables is a factor analysis model with uncorrelated errors. This is an interpretation of how SEFA suggests manifest variables to be removed. *This gives an answer to question (a).* Note that there is some freedom in eliminating error covariances. For example, if either X_8 or X_9 is removed, the covariance between E_8 and E_9 can be eliminated. If the model in Figure 4 is true, the choice between X_8 and X_9 is completely arbitrary from the viewpoint of a model fit. The situation is slightly more complicated for the variables X_{11} to X_{14}. It is possible to remove three of the four variables. Oka et al. (2002) dropped a total of five variables slightly different from ours by taking into account the meaning of the variables.

There are many cases where it is relevant to introduce error covariances, one of which is the case where there exist confounding third unmeasured variables that connect two or more manifest variables. A method factor is a typical example of confounding variables in studies of social sciences. While common factors cannot usually explain covariances between manifest variables caused by method factors, error covariances can explain them. Method factors are often discussed in the analysis of MTMM matrices (e.g., Campbell and Fiske, 1959; Campbell and O'Connell, 1967).

Green and Hershberger (2000) suggested several kinds of measurement models for longitudinal data. One of their models is a true score model with a moving average

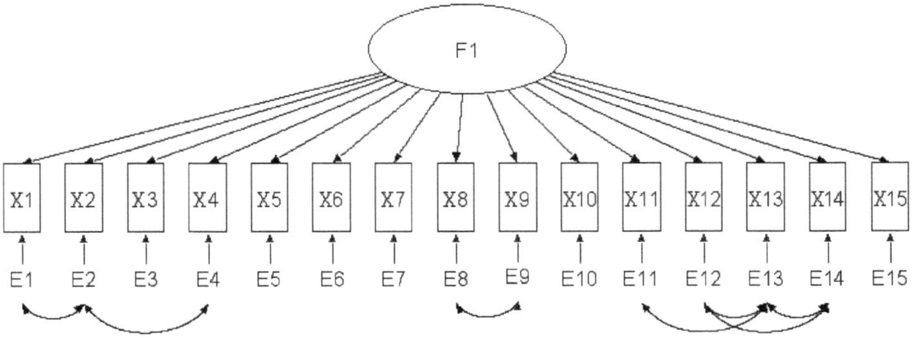

Fig. 4. The final model with error covariances.

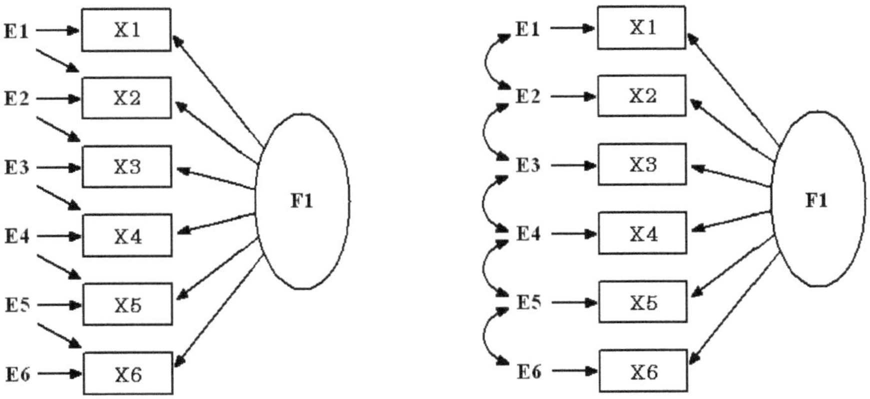

Fig. 5. A model with a moving average component of order one and its equivalent model.

component of order one given in Figure 5 (left). The model is equivalent to a one-factor analysis model with errors correlated next to each other (Figure 5 (right)).

A third one is nonlinearity. Suppose that the following nonlinear factor analysis model holds:

$$\begin{aligned} X_1 &= \mu_1 + \lambda_{11} f + \lambda_{12} f^2 + e_1 \\ X_2 &= \mu_2 + \lambda_{21} f + \lambda_{22} f^2 + e_2 \\ X_3 &= \mu_3 + \lambda_{31} f \qquad\qquad + e_3 \\ &\cdots \\ X_p &= \mu_p + \lambda_{p1} f \qquad\qquad + e_p, \end{aligned}$$

where the usual assumptions of a factor analysis model are made. Assuming that the distribution of f_1 is symmetric about the origin, we have $\mathrm{Cov}(f, f^2) = 0$, and $\mathrm{Cov}(\lambda_{12} f^2 + e_1, \lambda_{22} f^2 + e_2) = \lambda_{12}\lambda_{22}\,\mathrm{Var}(f^2)$ ($\neq 0$) results in an error covariance. When the effect of quadratic terms is not large and does not interest the researcher, one can approximate the effect of f by the linear terms, and the quadratic terms can be regarded as errors. See Lee and Zhu (2002) for general theory of nonlinear structural equation models.

What happens if there exist error covariances in the model and one fits a model without error covariances? A fit of the model will be poor. In addition, factor loadings and communality estimates can be biased. This problem will be discussed in the next section. Which model to employ becomes an issue: a model removing some manifest variables or a model introducing error covariances. We shall discuss this issue in the context of reliability analysis in the next section.

4.2. *Model fit and reliability analysis*

We shall study what happens in reliability analysis or traditional test theory when error covariances are ignored. See McDonald (1999) for test theory and reliability analysis.

In test theory, suppose first that an observable test score (or scale score) X can be decomposed into a true score T and a measurement error E, that is,

$$X = T + E,$$

where T and E are independently distributed. Define

$$\rho = \frac{V(T)}{V(X)} = \frac{V(T)}{V(T) + V(E)}.$$

The coefficient ρ denotes proportion of true score variation to the test score variation, and is called reliability of X. While a single observation of X cannot determine the reliability of X, one can evaluate it if multiple observations or indicators for T are available. Traditional test theory is highly related with factor analysis, as will be stated below.

Consider a one-factor analysis model:

$$X_i = \mu_i + \lambda_i f + e_i \quad (i = 1, \ldots, p), \tag{7}$$

where $E(f) = E(e_i) = 0$, $V(f) = 1$, $V(e_i) = \psi_i$, $\mathrm{Cov}(f, e_i) = 0$ and $\mathrm{Cov}(e_i, e_j) = 0$ $(i \neq j)$. The common factor f is considered a true score for a construct. Precisely speaking, e_i is the sum of a specific factor and an error factor. There is a structural equation model that can separate these factors if multiple measures for each X_i are observed. In this chapter, the e_i itself is regarded as an error since multiple measures are hardly observed. Reliability of an item X_i is then given as

$$\rho_i = \frac{V(\lambda_i f)}{V(X_i)} = \frac{\lambda_i^2}{\lambda_i^2 + \psi_i} \quad (i = 1, \ldots, p).$$

For a single-factor analysis model, the reliability above is identical with a communality of X_i if X_i is standardized.

A scale score of X is defined as the total sum of X_i, i.e., $X = \sum_{i=1}^p X_i$. Reliability of the scale X is expressible in the form:

$$\rho = \frac{V(\sum_{i=1}^p \lambda_i f)}{V(X)} = \frac{(\sum_{i=1}^p \lambda_i)^2}{(\sum_{i=1}^p \lambda_i)^2 + \sum_{i=1}^p \psi_i}. \tag{8}$$

Cronbach's coefficient α (Cronbach, 1951) is a most frequently used measure of reliability, which is defined as

$$\alpha = \frac{p}{p-1}\left(\frac{\sum_{i\neq j}^{p} \mathrm{Cov}(X_i, X_j)}{V(X)}\right).$$

When a one-factor analysis model in (7) holds true, it is known that $\rho \geqslant \alpha$. In fact,

$$\rho - \alpha = \frac{p\sum_{i=1}^{p}(\lambda_i - \bar{\lambda})^2}{(p-1)V(X)} \geqslant 0.$$

See McDonald (1999, p. 93) for a proof. The coefficient α is merely a lower bound of the true reliability, and the equality holds if and only if λ_i's are equal to each other. The equality condition is said to be essentially τ-equivalent.

What happens if the factor analysis model fails to fit to the data? As an example of unfitted models, we consider the case where errors are correlated. Let $\psi_{ij} = \mathrm{Cov}(e_i, e_j)$. Suppose that some of ψ_{ij}'s $(i \neq j)$ are not zero. Then the correct reliability is no longer given as in (8), but given as follows:

$$\rho' = \frac{(\sum_{i=1}^{p}\lambda_i)^2}{(\sum_{i=1}^{p}\lambda_i)^2 + \sum_{i=1}^{p}\psi_i + \sum_{i\neq j}^{p}\psi_{ij}}. \tag{9}$$

There might be a case where ψ_{ij} should be counted in true score variance rather than in error variance as above (Bentler, 2001). The formula in (9) should be used when the error covariances are caused by method factors because the method factors are usually not related to the latent construct f and should be regarded as 'errors'.

Raykov (2001) and Green and Hershberger (2000), among others, have discussed the bias of reliability coefficients or Cronbach's α (Cronbach, 1951) caused by error covariances. The examination of a fit of a (strict) one-factor analysis model gives a useful criterion to the problem whether the ρ in (8) can be used, as noted by Bentler (2003).

Here we shall introduce an artificial example of Kano and Azuma (2003) to see how error covariances influence reliability coefficients. They considered three models (Models 1 to 3) in Figure 6 as a data generation process, and computed α and ρ' as reliability measures based on the three true models and the correlation matrices implied by the models. In addition, for the correlation matrices from Models 2 and 3, they estimated a one-factor analysis model without error covariances, i.e., Model 1, which is a misspecified model. Results are shown in Models 2$'$ and 3$'$ also in Figure 6 (estimates of Models 2$'$ and 3$'$ were not reported in Kano and Azuma (2003)). We applied the formula (8) to figure out the coefficient ρ, where the estimates in Models 2$'$ and 3$'$ were used. The coefficient ρ' is always true. The example shows that error covariances invalidate use of ρ and α. Models 2$'$ and 3$'$ should receive very poor fits. If practical users ignored the caution of fit measures, biased results of the analysis including reliability could mislead them. Thus, a model-fit examination is important. *This is an answer to question (b).*

Another important implication to be made here is on the relation between the magnitude of a factor loading estimate and inconsistency of the variable. Clearly a variable with a low factor loading estimate or a low communality is not necessarily inconsistent.

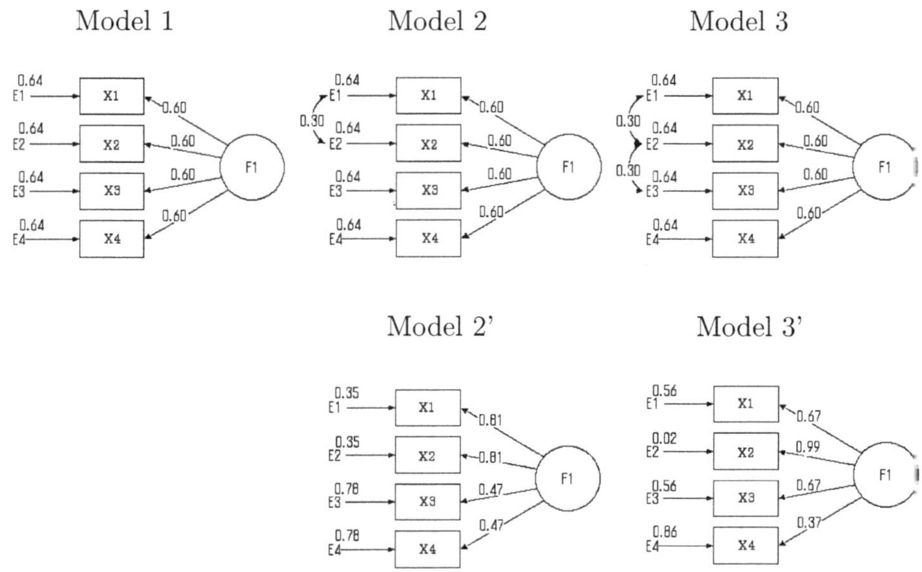

	Model 1	Model 2 / 2'	Model 3 / 3'
α :	0.69	0.74	0.77
ρ :	0.69	0.74	0.78
ρ' :	0.69	0.65	0.61

Fig. 6. Correlated errors and reliability coefficients (Kano and Azuma, 2003).

In an analysis using a model without any error covariances, factor loading estimates are boosted for positively correlated errors and reduced for negatively correlated errors. These biases are created because the common factor is forced to explain the error covariances. In our limited experience, we find that errors are more likely to be positively correlated. This fact explains why SEFA often suggests deletion of variables with large factor loading estimates, provided that our experience can be generalized. *This gives an answer to question (d).*

As discussed thus far, there are two options to cope with the problem of error covariances, namely, deleting manifest variables and modeling error covariances appropriately. Let us study which to choose in terms of reliability. Clearly we obtain a well-fitted model for Model 2 if either X_1 or X_2 is removed, and we can delete X_2 to obtain a well-fitted model for Model 3. Resultant models are then the same factor analysis model with three manifest variables and one factor, and reliability coefficient of the scale consisting of the three variables is 0.63. As a result, it is seen that in the situation of Model 2, the factor analysis model with an error covariance has a slightly better reliability (0.65) than the model removing X_1 or X_2. On the other hand, in Model 3, the model removing X_2 is slightly better than the model with the two error covariances. Therefore which option produces a scale with better reliability depends. It is unknown until the reliability is figured out.

As a matter of fact, the program SEFA does not necessarily suggest deletion of X_1 or X_2 for Model 2 nor deletion of X_2 for Model 3, because any one-factor analysis model for three manifest variables is saturate and can be fitted perfectly to data basically, so that deletion of any one variable in Models 2 or 3 creates a perfectly fitted model. However, the estimates are seriously biased when X_3 or X_4 is omitted for Model 2 or when any variable other then X_2 is omitted for Model 3. More importantly, there is no caution about model examination for the case. One cannot notice that the variable not to be deleted has been deleted. SEFA does not work when deletion of a manifest variable creates a saturate model. For a five-or-more-variable model with a similar structure, SEFA can make a proper suggestion on the variable to be deleted, as expected.

In addition to reliability, construct validity is an important factor for consideration in scale construction. Because a measurement model and a scale measure a latent psychological construct that practical researchers have defined to achieve their research purpose, it is meaningless unless a set of variables defines the target construct legitimately. Feasibility is also important. It is tough for respondents to answer too many items in a questionnaire. Any questionnaire with fewer items will be better if reliability and validity of a scale or a measurement model maintain. Researchers need to make a comprehensive decision on selecting manifest variables, taking into account reliability, validity and feasibility. Analyses with SEFA and models with error covariances will provide useful information on final decision making on the choice of manifest variables.

Related to the discussion above is the question of whether variables with large communalities but inconsistent with the model should be dropped. In general, removal of a variable with a large communality causes reduction of reliability. On the other hand, a variable unfitted to a one-factor analysis model can reduce reliability as shown above. Thus, one cannot mention anything about whether such a variable should be included, without examination of reliability by formula (9), as far as reliability analysis is concerned. In general, any marker variables that practical researchers consider important in light of their research purpose should not be deleted even though their deletion improves a model fit. Use of a model with error covariances will then be recommended. *This is an answer to question (c).*

A similar problem is whether a variable should be included which is consistent with the model but whose communality is small. Harada and Kano (2001) offered an option in the SEFA to print results of testing whether the communality is small (or zero), as mentioned in Section 3. Examination of the size of communality estimates and their statistical testing by SEFA enable practical users to easily identify manifest variables that are not sufficiently correlated to constructs under consideration. It is also possible to decide whether to include manifest variables with small communalities according to whether or not their inclusion contributes to improvement of reliability.

5. Conclusion and final remarks

In this chapter, we have emphasized the importance of manifest variable selection via a model fit and made an introduction of the useful web-based program SEFA. Two examples with empirical data were provided to illustrate the usefulness of SEFA. The statistical theory behind it was summarized.

We should emphasize here that we do not claim that practical users should select manifest variables based solely on the criterion of a model fit. We would say that a model fit is as important as the traditional psychometric properties (i) to (iii) described in the introduction. The model fit criterion is not an option but has become the one that practical users have to cope with in order to make variable selection or scale construction in factor analysis appropriately.

We pointed out some problems in factor analysis and reliability analysis if a model fit is ignored. The specific problems are a bias of the reliability coefficients α and ρ which can result in unduly boosted reliability and a bias of factor loading estimates which can mislead researchers to identify inconsistent variables. Two choices are given to deal with the problem of badly fitted models: one is to delete some manifest variables and the other is to introduce error covariances. Which to employ depends on reliability, and programs of structural equation modeling are useful in decision on the two options. In particular, EQS 6 (Bentler, 2004) is highly recommended because EQS 6 offers efficient LM tests to find pairs of error variables to be covariated and can print many kinds of reliability measures.

In this chapter we have emphasized the importance of error covariances as a cause of violation of the assumptions of a standard factor analysis model. Practical users must keep in mind that error covariances can be introduced only if a sound theoretical reason can be found for why they are covariated (e.g., Browne, 1982, p. 101).

We have discussed model unfitness due to inconsistent variables. Of course, there are other possibilities for unfitted models. Inclusion of outliers into a data set can cause terrible values of fit measures. Indeed, Bollen (1987) reported an improper solution due to outliers. Modification by error covariances may not work when there are unknown latent variables that substantively influence manifest variables. In the case, exploratory analysis has to be conducted. In our discussion, we have assumed that the model employed is approximately true while small modification may be needed.

Currently we are working on an extension of the program SEFA to cover a model with correlated errors along with reliability analysis.

Acknowledgements

The author would like to express his sincere thanks to Professor Sik-Yum Lee who gave him an opportunity of writing this chapter. He would like to thank Mr. Akira Harada, Mr. Yusuke Miyamoto, Dr. Asako Miura and Dr. Kei Hirai for their collaboration and assistance particularly on the program SEFA. Dr. Oka's kindness permitting him use of the data set in Section 3 is much appreciated. Finally the author is obliged to an anonymous reviewer for his comments.

References

Bartholomew, D.J. (1998). Scaling unobservable constructs in social science. *Applied Statistics* **47**, 1–13.

Bentler, P.M. (2001). Personal communication.

Bentler, P.M. (2003). Should coefficient alpha be replaced by model-based reliability coefficients? Invited paper presented at the IMPS2003, Cagliari, Italy.

Bentler, P.M. (2004). *EQS 6 Structural Equations Program Manual*. Multivariate Software, Inc., Encino, CA.

Bollen, K.A. (1980). Issues in the comparative measurement of political democracy. *American Sociological Review* **45**, 370–390.

Bollen, K.A. (1987). Outliers and improper solutions: A confirmatory factor analysis example. *Sociological Methods and Research* **15**, 375–384.

Bollen, K.A. (1989). *Structural Equations with Latent Variables*. Wiley-Interscience Publication, New York.

Browne, M.W. (1982). Covariance structures. In: Hawkins, D.M. (Ed.), *Topics in Applied Multivariate Analysis*. Cambridge University Press, Cambridge, England, pp. 72–141.

Browne, M.W. (1998). Personal communication.

Campbell, D.T., Fiske, D.W. (1959). Convergent and discriminant validation by the multitrait-multimethod matrix. *Psychological Bulletin* **56**, 81–105.

Campbell, D.T., O'Connell, E.J. (1967). Method factors in multitrait-multimethod matrices: Multiplicative rather than additive?. *Multivariate Behavioral Research* **2**, 409–426.

Cattell, R.B. (1952). *Factor Analysis*. Harper, New York.

Cattell, R.B. (1978). *The Scientific Use of Factor Analysis in Behavioral and Life Sciences*. Plenum, New York.

Chen, F., Bollen, K.A., Paxton, P., Curran, P.J., Kirby, J.B. (2001). Improper solutions in structural equation models: Causes, consequences, and strategies. *Sociological Methods and Research* **29**, 468–508.

Cronbach, L.J. (1951). Coefficient alpha and the internal structure of a test. *Psychometrika* **16**, 297–334.

Fabrigar, L.R., Wegener, D.T., MacCallum, R.C., Strahan, E.J. (1999). Evaluating the use of exploratory factor analysis in psychological research. *Psychological Methods* **4**, 272–299.

Gottfredson, M.R., Hirschi, T. (1990). *A General Theory of Crime*. Stanford University Press, Stanford, CA.

Grasmick, H.G., Tittle, C.R., Bursik, R.J., Arneklev, B.J. (1993). Testing the core empirical implications of Gottfredson and Hirschi's general theory of crime. *Journal of Research in Crime and Delenquency* **30**, 5–29.

Green, S.B., Hershberger, S.L. (2000). Correlated errors in true score models and their effect on coefficient alpha. *Structural Equation Modeling* **7**, 251–270.

Guttman, L. (1955). The determinacy of factor score matrices with applications for five other problems of common factor theory. *British Journal of Statistical Psychology* **8**, 65–82.

Harada, A., Kano, Y. (2001). Variable selection and test of communality in exploratory factor analysis. Paper presented at the IMPS2001 (IMPS2001 abstracts, p. 16).

Hogarty, K.Y., Kromrey, J.D., Ferron, J.M., Hines, C.V. (2004). Selection of variables in exploratory factor analysis: An empirical comparison of a stepwise and traditional approach. *Psychometrika* **69**, 593–612.

Hu, L., Bentler, P.M. (1999). Cutoff criteria for fit indexes in covariance structure analysis: Conventional criteria versus new alternatives. *Structural Equation Modeling* **6**, 1–55.

Ichikawa, M. (1992). Asymptotic distributions of the estimators of communalities in factor analysis. *Psychometrika* **57**, 399–404.

Ihara, M., Okamoto, M. (1985). Experimental comparison of least-squares and maximum likelihood methods in factor analysis. *Statistics and Probability Letters* **3**, 287–293.

Kano, Y. (1998). Improper solutions in exploratory factor analysis: Causes and treatments. In: Rizzi, A., Vichi, M., Bock, H. (Eds.), *Advances in Data Sciences and Classification*. Springer, Berlin, pp. 375–382.

Kano, Y. (2002). Variable selection for structural models. *Journal of Statistical Planning and Inference* **108**, 173–187.

Kano, Y., Azuma, Y. (2003). Use of SEM programs to precisely measure scale reliability. In: Yanai, H., et al. (Eds.), *New Developments in Psychometrics*. Springer-Verlag, Tokyo, pp. 141–148.

Kano, Y., Harada, A. (2000a). Stepwise variable selection in factor analysis. *Psychometrika* **65**, 7–22.

Kano, Y., Harada, A. (2000b). Variable selection for factor analysis and structural equation modeling. Paper presented at the International Symposium on Structural Equation Modeling, St. Charles, IL.

Kano, Y., Ihara, M. (1994). Identification of inconsistent variates in factor analysis. *Psychometrika* **59**, 5–20.

Kano, Y., Bentler, P.M., Mooijaart, A. (1993). Additional information and precision of estimators in multivariate structural models. In: Matusita, K., Puri, M.L., Hayakawa, T. (Eds.), *Statistical Sciences and Data Analysis*. VSP International Science Publisher, Zeist, The Netherlands, pp. 187–196.

Kono, S., Okamoto, H. (1999). Self-control of prisoners. *Japanese Journal of Criminal Psychology* **37** (Special number), 122–123 (in Japanese).
Kono, S., Okamoto, H. (2001). A study of relations among factors of self-control, degree of criminal progress and domestic conditions in offenders. *Japanese Journal of Criminal Psychology* **39**, 1–13 (in Japanese).
Krijnen, W.P. (2002). On the construction of all factors of the model for factor analysis. *Psychometrika* **67**, 161–172.
Lee, S.Y., Zhu, H.T. (2002). Maximum likelihood estimation of nonlinear structural equation models. *Psychometrika* **67**, 189–210.
Little, T.D., Lindenberger, U., Nesselroade, J.R. (1999). On selecting indicators for multivariate measurement and modeling with latent variables: When "good" indicators are bad and "bad" indicators are good. *Psychological Methods* **4**, 192–211.
Magnus, J.R., Neudecker, H. (1999). *Matrix Differential Calculus with Applications in Statistics and Econometrics*, revised ed. John Wiley and Sons, Ltd., Chichester.
McDonald, R.P. (1999). *Test Theory: A Unified Approach*. LEA, Mahwah, NJ.
Mulaik, S.A., McDonald, R.P. (1978). The effect of additional variables on factor indeterminacy in models with a single common factor. *Psychometrika* **43**, 177–192.
Murakami, N. (2000). A study on criminal behavior via low self-control. Unpublished.
Nunnally, J.C., Bernstein, I.H. (1994). *Psychometric Theory*, third ed. McGraw-Hill, New York.
Oka, K., Hirai, K., Tsutsumi, T. (2002). Psychological factors associated with physical inactivity among middle-aged adults: Decisional balance for exercise. *Japanese Journal of Behavioral Medicine* **9**, 23–30 (in Japanese).
Raykov, T. (2001). Bias of Cronbach's coefficient alpha for fixed congeneric measures with correlated errors. *Applied Psychological Measurement* **26**, 69–76.
Tanaka, J.S. (1993). Multifaceted conceptions of fit in structural equation models. In: Bollen, K.A., Long, J.S. (Eds.), *Testing Structural Equation Models*. Sage, Newbury Park, CA, pp. 1–39.
Tanaka, Y. (1983). Some criteria for variable selection in factor analysis. *Behaviormetrika* **13**, 31–45.
Van Driel, O.P. (1978). On various causes of improper solutions in maximum likelihood factor analysis. *Psychometrika* **43**, 225–243.
Williams, J.S. (1978). A definition for the common factor model and the elimination of problems of factor score indeterminacy. *Psychometrika* **43**, 293–306.
Yanai, H. (1980). A proposition of generalized method for forward selection of variables. *Behaviormetrika* **7**, 95–107.
Yuan, K.-H., Bentler, P.M., Kano, Y. (1997). On averaging variables in a confirmatory factor analysis model. *Behaviormetrika* **24**, 71–83.

Handbook of Computing and Statistics with Applications, Vol. 1
ISSN: 1871-0301

DOI: 10.1016/S1871-0301(06)01005-5

5

Bayesian Analysis of Mixtures Structural Equation Models with Missing Data

Sik-Yum Lee

Abstract
The main objective is to develop an algorithm with a permutation sampler for a Bayesian approach for analyzing mixture of structural equation models with an unknown number of components and ignorable missing data that are missing at random. The permutation sampler is implemented in the posterior simulation for selecting an appropriate identifiability constraint in order to cope with the important label switching problem. The Bayes factor, which is computed via a path sampling procedure, is used for model selection. It is shown by means of simulation studies that (i) the Bayesian estimates are accurate for models with poorly separated components, and (ii) an inappropriate identifiability constraint may give incorrect results. Sensitivity analysis of the results with respect to different prior inputs in the conjugate prior distributions is also conducted via simulation studies. Bayesian classification is discussed. An illustrative real example is also presented.

Keywords: MAR missing data; Permutation sampler; Identifiability constraint; Gibbs sampler; Bayes factor; Path sampling; Bayesian classification

1. Introduction

Mixture models have been found to be very useful in behavioral, medical, and psychological research. For example, they have been used for modeling heterogeneity, handling outliers, and density estimation. Analysis of mixture models has received a lot of past and recent attention in the field of statistics (see, Titterington et al., 1985; Richardson and Green, 1997; Stephens, 2000). Recently, analysis of mixture structural equation models (SEMs) has received a lot of attention in psychometrics. For example, see Jedidi et al. (1997), Yung (1997), Dolan and van der Maas (1998), and Arminger et al. (1999) for two-stage method and maximum likelihood based methods with various algorithms. Zhu and Lee (2001) proposed a Bayesian analysis coupled with MCMC methods to analyze mixture SEMs; and more recently, Lee and Song (2003a) developed a procedure to compute the Bayes factor for model comparison.

In practice, missing data are very common. It is well-recognized that the impact of the incomplete data should be taken into account to achieve correct results. The approach of replacing the missing entries by estimates obtained from the sample means or the predicted values by regression on the basis of the fully observed data creates dependent observations which are very difficult to handle. Moreover, for mixture models, the component memberships of the observations are not identified, hence one does not know which part of the data should be used to compute the mean estimates or the predicted values from regression. Lee and Song (2003b) developed a maximum likelihood (ML) approach for analyzing mixture SEMs with missing data that are missing at random (MAR, Little and Rubin, 1987). They pointed out that existing methods proposed by Finkbeiner (1979), Lee (1986), Jamshidian and Bentler (1999), and Song and Lee (2002) in analyzing various SEMs with missing data cannot be applied to mixture SEMs with an unknown number of components. In this chapter, a Bayesian approach for analyzing mixtures of SEMs with missing data and an unknown number of components is proposed. The justifications for proposing the Bayesian approach as an alternative are:

(i) It directly incorporates prior knowledge in the analysis. More precise estimates of the parameters can be obtained under situations in which good prior information is available.
(ii) As claimed by many important articles on Bayesian analysis of SEMs (Schines et al., 1999; Lee and Song, 2003b), the sampling-based Bayesian methods give reliable statistical inference even with small sample sizes.
(iii) The posterior distributions of parameters and latent variables can be estimated, and means as well as quantiles of posterior distributions can be obtained.
(iv) For model comparison, the Bayesian information criterion (BIC) in the ML approach is only a rough approximation of the Bayes factor in the Bayesian approach.

For a mixture SEM with K components, it is well known that the model is invariant with respect to permutation of the labels $k = 1, \dots, K$. Hence, the model is not identified, and adoption of an unique labeling for identifiability is important. In the literature, an identifiability constraint on some entries of the components' mean vectors is used to force a unique labeling. However, it is important that the identifiability constraint in a Bayesian analysis has to be selected more carefully. Arbitrary constraints may not be able to solve the important labeling switching problem, and may lead to incorrect results. We will apply the permutation sampler (Frühwirth-Schnatter, 2001) to solve the labeling switching problem in the Bayesian analysis of mixture of SEMs. By means of a simulation study (see Section 4.1), we show that accurate Bayesian estimates can be obtained by using the permutation sampler even for mixture models with rather poor separation.

The comparison of two mixture SEMs with different numbers of components is based on the Bayes factor (Berger, 1985), which is an important statistic in Bayesian model selection and has been applied widely. Inspired by Lee and Song (2003a), an algorithm on the basis of path sampling (Gelman and Meng, 1998) for computing the logarithm of the Bayes factor is developed with MAR data. Conjugate prior distributions with given hyper-parameter values are used. The important issue on the sensitivity of results with respect to prior inputs is addressed. According to their natures and characteristics, we group the hyper-parameters into five categories, then we conduct a symmetric sensitivity analysis with respect to prior inputs in each category. Moreover, we also study the sensitivity analysis with respect to the assumption of MAR.

The chapter is organized as follows. Section 2 defines the mixture SEMs with an unknown number of components and missing data. Here, the issue on the unique labeling for identifiability is discussed. Section 3 presents Bayesian estimation of the mixture SEMs. A permutation sampler is implemented to select an appropriate identifiability constraint for solving the important label switching problem. An algorithm based on path sampling for computing the log Bayes factor for model selection is also developed. Moreover, Bayesian classification is discussed. Results of simulation studies for investigating the empirical performance of the Bayesian estimates with the selected identifiability constraints, and the sensitivity analyses with respect to prior inputs, are reported in Section 4. A real example is presented in Section 5. The freely available software WinBUGS to get the Bayesian estimates is introduced in Section 6, followed by a discussion in Section 7. Technical details about the implementation of the permutation sampler are presented in Appendices A–C.

2. Model description

A mixture SEMs for a $p \times 1$ random vector $\boldsymbol{y}_i$ is defined as follows:

$$f(\boldsymbol{y}_i) = \sum_{k=1}^{K} \pi_k f_k(\boldsymbol{y}_i | \boldsymbol{\mu}_k, \boldsymbol{\Sigma}_k), \quad i = 1, \ldots, n, \tag{1}$$

where K is the number of components which can be unknown, π_k's are component probabilities which are nonnegative and sum to 1.0, $f_k(\boldsymbol{y}|\boldsymbol{\mu}_k, \boldsymbol{\Sigma}_k)$ is a multivariate normal density function with an unknown mean vector $\boldsymbol{\mu}_k$ and a covariance matrix $\boldsymbol{\Sigma}_k$. Conditional on the kth component, suppose that $\boldsymbol{y}$ satisfies the following measurement model:

$$\boldsymbol{y} = \boldsymbol{\mu}_k + \boldsymbol{\Lambda}_k \boldsymbol{\omega}_k + \boldsymbol{\varepsilon}_k, \tag{2}$$

where $\boldsymbol{\mu}_k$ is an $p \times 1$ intercept vector, $\boldsymbol{\Lambda}_k$ is a $p \times q$ factor loading matrix, $\boldsymbol{\omega}_k$ is a $q \times 1$ random vector of latent variables, and $\boldsymbol{\varepsilon}_k$ is a $p \times 1$ random vector of error measurements with distribution $N(\boldsymbol{0}, \boldsymbol{\Psi}_k)$, which is independent of $\boldsymbol{\omega}_k$, and $\boldsymbol{\Psi}_k$ is a diagonal matrix. Let $\boldsymbol{\omega}_k$ be partitioned into $(\boldsymbol{\eta}_k^{\mathrm{T}}, \boldsymbol{\xi}_k^{\mathrm{T}})^{\mathrm{T}}$, where $\boldsymbol{\eta}_k$ is a $q_1 \times 1$ vector, $\boldsymbol{\xi}_k$ is a $q_2 \times 1$ vector, and $q_1 + q_2 = q$. The structural equation is defined as

$$\boldsymbol{\eta}_k = \boldsymbol{B}_k \boldsymbol{\eta}_k + \boldsymbol{\Gamma}_k \boldsymbol{\xi}_k + \boldsymbol{\delta}_k, \tag{3}$$

where $\boldsymbol{B}_k$ and $\boldsymbol{\Gamma}_k$ are $q_1 \times q_1$ and $q_1 \times q_2$ matrices of unknown parameters; and random vectors $\boldsymbol{\xi}_k$ and $\boldsymbol{\delta}_k$ are independently distributed as $N(\boldsymbol{0}, \boldsymbol{\Phi}_k)$ and $N(0, \boldsymbol{\Psi}_{\delta k})$, respectively; and $\boldsymbol{\Psi}_{\delta k}$ is a diagonal matrix. We assume that $\boldsymbol{B}_{0k} = (\boldsymbol{I}_{q_1} - \boldsymbol{B}_k)$ is nonsingular and $|\boldsymbol{I}_{q_1} - \boldsymbol{B}_k|$ is independent of any elements in $\boldsymbol{B}_k$. One specific form of $\boldsymbol{B}_k$ that satisfies this assumption is the lower or upper triangular matrix.

As the mixture model defined in (1) is invariant with respect to permutation of labels $k = 1, \ldots, K$, adoption of an unique labeling for identifiability is important. Roeder and Wasserman (1997), and Zhu and Lee (2001) proposed to impose the ordering $\mu_{1,1} < \cdots < \mu_{K,1}$ for eliminating the label switching (jumping between the various labeling subspace), where $\mu_{k,1}$ is the first element of the mean vector $\boldsymbol{\mu}_k$. This method works fine if $\mu_{1,1}, \ldots, \mu_{K,1}$ are well separated. However, if $\mu_{1,1}, \ldots, \mu_{K,1}$ are close to each other, it may not be able to eliminate the label switching, and may introduce incorrect results. Hence, it is necessary to find a sensible identifiability constraint. In this chapter, the random permutation sampler developed by Frühwirth-Schnatter (2001) will be applied for finding the suitable identifiability constraints. See the following sections for more details.

Moreover, for each $k = 1, \ldots, K$, structural parameters in the covariance matrix $\boldsymbol{\Sigma}_k$ corresponding to the model defined by (2) and (3) are not identified. A common method in structural equation modeling for identifying the model is to fix appropriate elements in $\boldsymbol{\Lambda}_k$, $\boldsymbol{B}_k$, and/or $\boldsymbol{\Gamma}_k$ at preassigned values. The positions of the preassigned values of the fixed elements in these matrices of regression coefficients can be chosen on a problem-by-problem basis, as long as each $\boldsymbol{\Sigma}_k$ is identified. In practice, most manifest variables are usually clear indicators of their corresponding latent variables. This give rather clear prior information to specify the zero values to appropriate elements in these parameter matrices. See the illustrative example in Section 5 for a more concrete example. For clear discussion of the proposed method, we let $\boldsymbol{\pi} = (\pi_1, \ldots, \pi_K)$, and $\boldsymbol{\vartheta}$ be the vector which contains all unknown parameters in the covariance matrices that defines an identified model.

3. Bayesian analysis of the models

3.1. Bayesian estimation and the permutation sampler

To deal with the missing data problem, let $\boldsymbol{y}_i = \{\boldsymbol{y}_{i,\text{obs}}, \boldsymbol{y}_{i,\text{mis}}\}$, where $\boldsymbol{y}_{i,\text{obs}}$ represents the observed estimates of $\boldsymbol{y}_i$, whereas $\boldsymbol{y}_{i,\text{mis}}$ represents the missing entries. We assume that missing data are MAR with an ignorable mechanism (Little and Rubin, 1987). For a fully observed data point $\boldsymbol{y}_i$, $\boldsymbol{y}_{i,\text{mis}}$ does not exist. Bayesian analysis of the current mixture of SEMs will be studied on the basis of the observed data set $\{\boldsymbol{y}_{i,\text{obs}};\ i = 1, \ldots, n\}$. Let $\boldsymbol{Y}_{\text{obs}} = \{\boldsymbol{y}_{i,\text{obs}};\ i = 1, \ldots, n\}$, $\boldsymbol{Y}_{\text{mis}} = \{\boldsymbol{y}_{i,\text{mis}};\ i = 1, \ldots, n\}$ be the collection of missing data, $\boldsymbol{Y} = (\boldsymbol{Y}_{\text{obs}}, \boldsymbol{Y}_{\text{mis}})$, and $\boldsymbol{Z} = \{\boldsymbol{\omega}_1, \ldots, \boldsymbol{\omega}_n\}$ be the matrix of latent variables. Using the data augmentation idea for overcoming the computational difficulties, the observed data $\boldsymbol{Y}_{\text{obs}}$ is augmented with $\boldsymbol{Y}_{\text{mis}}$ and $\boldsymbol{Z}$ in the posterior analysis. Moreover, inspired by the existing work on mixture SEMs (see, Zhu and Lee, 2001; Lee and Song, 2003a), we further introduce a grouping variable w_i for $\boldsymbol{y}_i$ as a latent allocation variable in the analysis. We assume that w_i is independently drawn from the following distribution

$$p(w_i = k) = \pi_k, \quad k = 1, \ldots, K, \tag{4}$$

and given w_i, observations are drawn independently from the respective subpopulation. Let $\boldsymbol{W} = (w_1, \ldots, w_n)$ be the collection of latent allocation variables, then the Bayesian estimates of $\boldsymbol{\theta}$ and $\boldsymbol{\pi}$ are obtained from a sequence of observations simulated by some MCMC methods from the joint posterior distribution $[\boldsymbol{Y}_{\text{mis}}, \boldsymbol{Z}, \boldsymbol{W}, \boldsymbol{\theta}, \boldsymbol{\pi} | \boldsymbol{Y}_{\text{obs}}]$.

It is well known that for general mixture models with K-components, the unconstrained parameter space contains $K!$ subspaces, each one corresponding to a different way to label the states. In the current mixture of SEM, the likelihood is invariant to relabeling the states. If the priors of $\boldsymbol{\theta}$ and $\boldsymbol{\pi}$ are also invariant, the unconstrained posterior is invariant to relabeling the states and identical on all labeling subspaces. This induces a multimodel posterior and has a serious impact on Bayesian estimation.

The MCMC approach proposed by Frühwirth-Schnatter (2001) will be used in this paper to deal with the above label switching problem. In this approach, an unidentified model is first estimated by sampling from the unconstrained posterior using the so-called random permutation sampler, where each sweep is concluded by a random permutation of the current labeling of the components. The random permutation sampler delivers a sample that explores the whole unconstrained parameter space and jump between the various labeling subspace in a balanced fashion. As emphasized in Frühwirth-Schnatter (2001), although the model is unidentified, the output of the random permutation sampler can be used to estimate unknown parameters that are invariant to relabeling the states and can be explored to find a suitable identifiability constraints. Then, the model is re-estimated by sampling from the posterior distribution under the imposed identifiability constraints, again using the permutation sampler. The implementation of the permutation sampler in relation to the mixtures of SEMs and the method of selecting the identifiability constraint are briefly described in Appendices A and B, respectively.

The MCMC sampling scheme under the selected identifiability constraints is implemented as below:

Step (a): Generate $(\boldsymbol{Y}_{\text{mis}}, \boldsymbol{Z}, \boldsymbol{W}, \boldsymbol{\theta}, \boldsymbol{\pi})$ from the unconstrained posterior $p(\boldsymbol{Y}_{\text{mis}}, \boldsymbol{Z}, \boldsymbol{W}, \boldsymbol{\theta}, \boldsymbol{\pi} | \boldsymbol{Y}_{\text{obs}})$ according to the following steps via the Gibbs sampler (Geman and Geman, 1984). At the $(r+1)$th iteration with a current $\boldsymbol{Y}_{\text{mis}}^{(r)}, \boldsymbol{Z}^{(r)}, \boldsymbol{W}^{(r)}, \boldsymbol{\theta}^{(r)}, \boldsymbol{\pi}^{(r)}$: iteratively generate $\boldsymbol{W}^{(r+1)}$ from $p(\boldsymbol{W}|\boldsymbol{\theta}^{(r)}, \boldsymbol{\pi}^{(r)}, \boldsymbol{Y}_{\text{mis}}^{(r)}, \boldsymbol{Y}_{\text{obs}})$; $\boldsymbol{Z}^{(r+1)}$ from $p(\boldsymbol{Z}|\boldsymbol{\theta}^{(r)}, \boldsymbol{\pi}^{(r)}, \boldsymbol{Y}_{\text{mis}}^{(r)}, \boldsymbol{W}^{(r+1)}, \boldsymbol{Y}_{\text{obs}})$; $\boldsymbol{Y}_{\text{mis}}^{(r+1)}$ from $p(\boldsymbol{Y}_{\text{mis}}|\boldsymbol{\theta}^{(r)}, \boldsymbol{\pi}^{(r)}, \boldsymbol{Z}^{(r+1)}, \boldsymbol{W}^{(r+1)}, \boldsymbol{Y}_{\text{obs}})$; and $(\boldsymbol{\theta}^{(r+1)}, \boldsymbol{\pi}^{(r+1)})$ from $p(\boldsymbol{\theta}, \boldsymbol{\pi}|\boldsymbol{Z}^{(r+1)}, \boldsymbol{W}^{(r+1)}, \boldsymbol{Y}_{\text{mis}}^{(r+1)}, \boldsymbol{Y}_{\text{obs}})$. Note that $p(\boldsymbol{W}|\boldsymbol{\theta}, \boldsymbol{\pi}, \boldsymbol{Y}_{\text{obs}}, \boldsymbol{Y}_{\text{mis}})$ is simpler than $p(\boldsymbol{W}|\boldsymbol{\theta}, \boldsymbol{\pi}, \boldsymbol{Z}, \boldsymbol{Y}_{\text{obs}}, \boldsymbol{Y}_{\text{mis}})$ and does not involve $\boldsymbol{Z}$.

Step (b): Reordering the labeling through the permutation which fulfills the identifiability constraints.

The conditional distributions associated with the Gibbs sampler involve the prior distributions of $\boldsymbol{\theta}$ and $\boldsymbol{\pi}$. In this paper, we follow the suggestion of Zhu and Lee (2001), and Lee and Song (2003a) to use the conjugate prior distributions for $\boldsymbol{\theta}$. More specifically, for $m = 1, \ldots, p$; $l = 1, \ldots, q_1$, we take

$$
\begin{aligned}
&p(\boldsymbol{\Lambda}_{km}|\psi_{km}) \stackrel{D}{=} N[\boldsymbol{\Lambda}_{0km}, \psi_{km}\boldsymbol{H}_{0ykm}], \\
&p(\psi_{km}^{-1}) \stackrel{D}{=} Gamma[\alpha_{0\varepsilon k}, \beta_{0\varepsilon k}], \qquad p(\boldsymbol{\mu}_k) \sim N[\boldsymbol{\mu}_0, \boldsymbol{\Sigma}_0], \\
&p(\boldsymbol{\Pi}_{kl}|\psi_{\delta kl}) \stackrel{D}{=} N[\boldsymbol{\Pi}_{0kl}, \psi_{\delta kl}\boldsymbol{H}_{0\omega kl}], \\
&p(\psi_{\delta kl}^{-1}) \stackrel{D}{=} Gamma[\alpha_{0\delta k}, \beta_{0\delta k}], \qquad p(\boldsymbol{\Phi}_k) \stackrel{D}{=} IW_{q_2}[\boldsymbol{R}_0, \rho_0],
\end{aligned} \tag{5}
$$

where ψ_{km} and $\psi_{\delta kl}$ are the mth diagonal element of $\boldsymbol{\Psi}_k$ and the lth diagonal element of $\boldsymbol{\Psi}_{\delta k}$, respectively; $\boldsymbol{\Lambda}_{km}^{\mathrm{T}}$ and $\boldsymbol{\Pi}_{kl}^{\mathrm{T}}$ are vectors that contain unknown parameters in the mth row of $\boldsymbol{\Lambda}_k$ and the lth row of $\boldsymbol{\Pi}_k$, respectively; $\alpha_{0\varepsilon k}$, $\beta_{0\varepsilon k}$, $\boldsymbol{\Lambda}_{0km}$, $\alpha_{0\delta k}$, $\beta_{0\delta k}$, $\boldsymbol{\Pi}_{0kl}$, ρ_0, and positive definite matrices $\boldsymbol{H}_{0ykm}$, $\boldsymbol{H}_{0\omega kl}$, and $\boldsymbol{R}_0$ are hyper-parameters whose values are assumed to be given, and $IW_{q_2}[\cdot,\cdot]$ denotes the inverted Wishart distribution of dimension q_2. The given values of the hyper-parameters represent the prior information of the corresponding parameters. Subjective hyper-parameter values can be taken for situations with accurate prior information, either from analysis of closed related data or knowledge of experts. For other situations, data-dependent prior inputs are rather common in analysis of mixture models (see Raftery, 1996; Richardson and Green, 1997; Song and Lee, 2001, among others). The prior distribution of $\boldsymbol{\pi}$ is taken to be the following symmetric Dirichlet distribution: $p(\boldsymbol{\pi}) \propto D(\alpha_1, \ldots, \alpha_K)$, with hyper-parameters $\alpha_1, \ldots, \alpha_K$. The mild conditions on the independence of these prior distributions as specified in Zhu and Lee (2001) are also assumed.

As the conditional distributions corresponding to $\boldsymbol{W}$, $\boldsymbol{Z}$ and $\boldsymbol{\theta}$ are conditional on $\boldsymbol{Y}$, they are equal to those obtained on the basis of mixture SEMs without missing data. As the prior distributions involved are the same, these conditional distributions can be directly obtained from Lee and Song (2003a).

Conditional distribution $p(\boldsymbol{Y}_{\mathrm{mis}}|\boldsymbol{\theta}, \boldsymbol{\pi}, \boldsymbol{Z}, \boldsymbol{W}, \boldsymbol{Y}_{\mathrm{obs}})$ is derived as below. For $i = 1, \ldots, n$, as $\boldsymbol{y}_i$ are mutually independent, $\boldsymbol{y}_{i,\mathrm{mis}}$ are also mutually independent. As $\boldsymbol{\Psi}_\varepsilon$ is diagonal, $\boldsymbol{y}_{i,\mathrm{mis}}$ is independent of $\boldsymbol{y}_{i,\mathrm{obs}}$. Also, with given $\boldsymbol{\omega}_i$, we know the component membership of $\boldsymbol{y}_{i,\mathrm{mis}}$, hence $\boldsymbol{\pi}$ is irrelevant. Let p_i be the dimension of $\boldsymbol{y}_{i,\mathrm{mis}}$, we have

$$p(\boldsymbol{Y}_{\mathrm{mis}}|\boldsymbol{\theta}, \boldsymbol{W}, \boldsymbol{Y}_{\mathrm{obs}}, \boldsymbol{Z}) = \prod_{i=1}^{n} p(\boldsymbol{y}_{i,\mathrm{mis}}|\boldsymbol{\theta}, \boldsymbol{\omega}_i, w_i), \quad \text{and}$$

$$[\boldsymbol{y}_{i,\mathrm{mis}}|\boldsymbol{\theta}, \boldsymbol{\omega}_i, w_i = k] \stackrel{D}{=} N[\boldsymbol{\mu}_{i,\mathrm{mis},k} + \boldsymbol{\Lambda}_{i,\mathrm{mis},k}\boldsymbol{\omega}_i, \boldsymbol{\Psi}_{\varepsilon i,\mathrm{mis},k}], \tag{6}$$

where $\boldsymbol{\mu}_{i,\mathrm{mis},k}$ is a $p_i \times 1$ subvector of $\boldsymbol{\mu}$ with elements corresponding to observed components deleted, $\boldsymbol{\Lambda}_{i,\mathrm{mis},k}$ is the corresponding $p_i \times q$ submatrix of $\boldsymbol{\Lambda}_k$, and $\boldsymbol{\Psi}_{\varepsilon i,\mathrm{mis},k}$ is the corresponding $p_i \times p_i$ submatrix of $\boldsymbol{\Psi}_{\varepsilon k}$ with the appropriate rows and columns deleted. Hence, this conditional distribution only involves a product of simple normal distributions. The computational burden involved is light. Convergence of the Gibbs sampler is monitored by the 'estimated potential scale reduction (EPSR)' values as proposed by Gelman (1996). Bayesian estimates of $\boldsymbol{\theta}$, $\boldsymbol{\pi}$ and $\boldsymbol{\omega}_i$ are obtained from the sample means of the simulated observations after convergence. Standard error estimates are computed via the corresponding sample covariance matrices. The marginal posterior distribution can be summarized by tabulating $100(1-\alpha)\%$ highest probability density (HPD) interval (see Chen and Shao, 1999) for the parameters of interest. The HPD intervals presented in the real example are estimated via the algorithm given in Chen and Shao (1999).

3.2. *A path sampling procedure for Bayes factor computation*

Let M_0 and M_1 be the two competing mixtures of SEMs with different number of components. For $v = 0, 1$, let $p(\boldsymbol{Y}_{\mathrm{obs}}|M_v)$ be the probability density of $\boldsymbol{Y}_{\mathrm{obs}}$ given M_v. The choice between M_0 and M_1 is based on the following Bayes factor:

$$B_{10} = \frac{p(\boldsymbol{Y}_{\text{obs}}|M_1)}{p(\boldsymbol{Y}_{\text{obs}}|M_0)}. \tag{7}$$

B_{10} is a summary of the evidence provided by the data in favor of M_1 as opposed to M_0. It measures how well M_1 predicts the data relative to M_0. Usually, the natural logarithm of B_{10} is considered and interpreted via the criterion in Kass and Raftery (1995).

Following similar reasoning as in Gelman and Meng (1998), the procedure proposed by Lee and Song (2003a) for computing the Bayes factor can be extended to mixture SEMs with MAR missing as below. Consider the following class of densities with a continuous parameter $t \in [0, 1]$: $p(\boldsymbol{Y}_{\text{mis}}, \boldsymbol{Z}, \boldsymbol{\theta}, \boldsymbol{\pi}|\boldsymbol{Y}_{\text{obs}}, t) = p(\boldsymbol{Y}_{\text{mis}}, \boldsymbol{Z}, \boldsymbol{\theta}, \boldsymbol{\pi}|t)/z(t)$, where

$$z(t) = p(\boldsymbol{Y}_{\text{obs}}|t) = \int p(\boldsymbol{Y}_{\text{obs}}, \boldsymbol{Y}_{\text{mis}}, \boldsymbol{Z}|\boldsymbol{\theta}, \boldsymbol{\pi}, t)p(\boldsymbol{\theta}, \boldsymbol{\pi})\, \mathrm{d}\boldsymbol{Y}_{\text{mis}}\, \mathrm{d}\boldsymbol{Z}\, \mathrm{d}\boldsymbol{\theta}\, \mathrm{d}\boldsymbol{\pi}, \tag{8}$$

and $p(\boldsymbol{Y}_{\text{obs}}, \boldsymbol{Y}_{\text{mis}}, \boldsymbol{Z}|\boldsymbol{\theta}, \boldsymbol{\pi}, t)$ is the complete-data likelihood function of $\boldsymbol{\theta}$ and $\boldsymbol{\pi}$ for a given t, and $p(\boldsymbol{\theta}, \boldsymbol{\pi})$ be the prior density of $(\boldsymbol{\theta}, \boldsymbol{\pi})$ which is independent of t. We need to construct a path using $t \in [0, 1]$ to link M_1 and M_0, so that $B_{10} = z(1)/z(0)$. Taking logarithm and differentiating (8) with respect to t, and let $U(\boldsymbol{Y}_{\text{obs}}, \boldsymbol{Y}_{\text{mis}}, \boldsymbol{Z}, \boldsymbol{\theta}, \boldsymbol{\pi}, t) = \mathrm{d}\log p(\boldsymbol{Y}_{\text{obs}}, \boldsymbol{Y}_{\text{mis}}, \boldsymbol{Z}|\boldsymbol{\theta}, \boldsymbol{\pi}, t)/\mathrm{d}t$, it can be shown similarly as in Lee and Song (2003a) that

$$\widehat{\log B_{10}} = \frac{1}{2}\sum_{s=0}^{S}(t_{(s+1)} - t_{(s)})(\overline{U}_{(s+1)} + \overline{U}_{(s)}), \tag{9}$$

where $t_{(0)} = 0 < t_{(1)} < t_{(2)} < \cdots < t_{(S)} < t_{(S+1)} = 1$ are fixed grids in [0, 1], and

$$\overline{U}_{(s)} = J^{-1}\sum_{j=1}^{J} U\big(\boldsymbol{Y}_{\text{obs}}, \boldsymbol{Y}_{\text{mis}}^{(j)}, \boldsymbol{Z}^{(j)}, \boldsymbol{\theta}^{(j)}, \boldsymbol{\pi}^{(j)}, t_{(s)}\big), \tag{10}$$

in which $\{\boldsymbol{Y}_{\text{mis}}^{(j)}, \boldsymbol{Z}^{(j)}, \boldsymbol{\theta}^{(j)}, \boldsymbol{\pi}^{(j)}, j = 1, \ldots, J\}$ be simulated observations drawn from $p(\boldsymbol{Y}_{\text{mis}}, \boldsymbol{Z}, \boldsymbol{\theta}, \boldsymbol{\pi}|\boldsymbol{Y}_{\text{obs}}, t_{(s)})$. See Lee and Song (2003a) for more details in finding a path to link two competing mixture SEMs with different components.

Note that in computing Bayes factors, the inclusion of an identifiability constraint in simulating $\{\boldsymbol{Y}_{\text{mis}}^{(j)}, \boldsymbol{Z}^{(j)}, \boldsymbol{\theta}^{(j)}, \boldsymbol{\pi}^{(j)}, j = 1, \ldots, J\}$ for evaluating $\overline{U}_{(s)}$ in (10) is not necessary. As the likelihood is invariant to relabeling the states, the inclusion of such a constraint will not change the values of $U(\boldsymbol{Y}_{\text{obs}}, \boldsymbol{Y}_{\text{mis}}, \boldsymbol{Z}, \boldsymbol{\theta}, \boldsymbol{\pi}, t)$. As a result, the log Bayes factors estimated by (9) and (10) will not be changed.

3.3. Bayesian classification

In addition to their role in facilitating computation for estimation and model comparison, the allocation variables in $\boldsymbol{W}$ are also important for Bayesian classification of observations with and without missing entries. Using the "percentage correctly classified" loss function (see Richardson and Green, 1997), the Bayesian classification of an existing observation $\boldsymbol{y}_{i,\text{obs}}$, $i = 1, \ldots, n$, is given by

$$\widehat{w}_i = \operatorname{argmax}_k \Pr\bigl\{(w_i = k | \boldsymbol{Y}_{\text{obs}})\bigr\},$$

where w_i is the allocation variable corresponding to $\boldsymbol{y}_{i,\text{obs}}$. As $\boldsymbol{y}_{i,\text{obs}}$, $i = 1, \ldots, n$, are independent, $\widehat{w}_i = \operatorname{argmax}_k \Pr\{(w_i = k | \boldsymbol{y}_{i,\text{obs}})\}$. This posterior probability can be directly estimated via the sample mean of the corresponding observations in $\{w_i^{(j)}, j = 1, \ldots, J\}$ that are generated by the Gibbs sampler:

$$\Pr\{w_i = k | \boldsymbol{y}_{i,\text{obs}}\} \approx J^{-1} \sum_{j=1}^{J} I\bigl(w_i^{(j)} = k\bigr), \quad k = 1, \ldots, K.$$

For classifying a new incomplete observation $\boldsymbol{y}_{\text{obs}}^*$, let w^* be the corresponding allocation variable. The Bayesian classification requires to evaluate $\Pr\{w^* = k | \boldsymbol{Y}_{\text{obs}}, \boldsymbol{y}_{\text{obs}}^*\}$. Theoretically, the addition of $\boldsymbol{y}_{\text{obs}}^*$ in the original sample slightly changes the posterior distributions, and it seems that the simulation process should be rerun for each new $\boldsymbol{y}_{\text{obs}}^*$. However, this is clearly impractical. Hence, we follow the idea of Zhu and Lee (2001) to employ the following approximation:

$$\Pr\bigl(w^* = k | \boldsymbol{Y}_{\text{obs}}, \boldsymbol{y}_{\text{obs}}^*\bigr) \approx J^{-1} \sum_{j=1}^{J} \frac{\pi_k^{(j)} f_k(\boldsymbol{y}_{\text{obs}}^* | \boldsymbol{\theta}^{(j)})}{\{\sum_{k=1}^{K} \pi_k^{(j)} f_k(\boldsymbol{y}_{\text{obs}}^* | \boldsymbol{\theta}^{(j)})\}},$$

where $f_k(\boldsymbol{y}_{\text{obs}}^* | \boldsymbol{\theta})$ denotes the marginal density function, and $(\pi_k^{(j)}, \boldsymbol{\theta}^{(j)})$ are observations that are simulated by the Gibbs sampler.

4. Simulation studies

4.1. Simulation study 1: Identifiability constraints and separation of components

Yung (1997), and Dolan and van der Maas (1998) pointed out that some statistical results they achieved cannot be trusted when the separation of the component is poor. Yung (1997) considered $d_{kh} = \max_{l \in \{k,h\}} \{(\boldsymbol{\mu}_k - \boldsymbol{\mu}_h)^{\mathrm{T}} \boldsymbol{\Sigma}_l^{-1} (\boldsymbol{\mu}_k - \boldsymbol{\mu}_h)\}^{1/2}$ as a measure of separation, and recommended that d_{hk} should be about 3.8 or over. The main objective of this simulation study is to demonstrate the random permutation sampler for finding a suitable identifiability constraints. Another objective is to investigate the performance of the proposed Bayesian approach in analyzing a mixture of SEMs with two poorly separated components. Random observations are simulated from a mixture SEMs with two components defined by (1), (2), and (3). The SEM for each $k = 1, 2$ involves nine manifest variables which are indicators for three latent variables η, ξ_1 and ξ_2. The loading matrix in each component takes the following common non-overlapping structure:

$$\boldsymbol{\Lambda}_k^{\mathrm{T}} = \begin{bmatrix} 1.0 & \lambda_{k,21} & \lambda_{k,31} & 0 & 0 & 0 & 0 & 0 & 0 \\ 0 & 0 & 0 & 1.0 & \lambda_{k,52} & \lambda_{k,62} & 0 & 0 & 0 \\ 0 & 0 & 0 & 0 & 0 & 0 & 1.0 & \lambda_{k,83} & \lambda_{k,93} \end{bmatrix},$$

where the one's and zero's are fixed parameters for achieving an identified covariance structure, whilst the others are distinct unknown parameters. In the kth component, the structural equation is given by: $\eta = \gamma_{k,1}\xi_1 + \gamma_{k,2}\xi_2 + \delta$, where $\gamma_{k,1}$ and $\gamma_{k,2}$ are unknown parameters. The true population values are given by $\pi_1 = \pi_2 = 0.5$, $\boldsymbol{\mu}_1 = (0.0, 0.0, 0.0, 0.0, 0.0, 1.0, 1.0, 1.0, 1.0)^T$, $\boldsymbol{\mu}_2 = (0.0, 0.0, 0.0, 0.5, 1.5, 0.0, 1.0, 1.0, 1.0)^T$; $\lambda_{1,21} = \lambda_{1,31} = \lambda_{1,83} = \lambda_{1,93} = 0.4$, $\lambda_{1,52} = \lambda_{1,62} = 0.8$, $\lambda_{2,21} = \lambda_{2,31} = \lambda_{2,83} = \lambda_{2,93} = 0.8$, $\lambda_{2,52} = \lambda_{2,62} = 0.4$, $\gamma_{1,1} = 0.2$, $\gamma_{1,2} = 0.7$, $\gamma_{2,1} = 0.7$, $\gamma_{2,2} = 0.2$, $\phi_{1,11} = \phi_{1,22} = \phi_{2,11} = \phi_{2,22} = 1.0$, $\phi_{1,12} = \phi_{2,12} = 0.3$, $\psi_{1,11} = \cdots = \psi_{1,99} = \psi_{2,11} = \cdots = \psi_{2,99} = \psi_{1,\delta} = \psi_{2,\delta} = 0.5$. In this 2-component mixture SEM, the total number of unknown parameters is 62. The separation d_{12} is equal to 2.56, which is less than the suggested value by Yung (1997).

Based on the above settings, we simulate 400 observations from each component, hence the total sample size is 800. The MAR missing data are created via the following steps: (i) 200 fully observed data points are randomly selected among the 800 observations, and sample means $\bar{y}_{(1)}$, $\bar{y}_{(4)}$, and $\bar{y}_{(7)}$ are computed on the basis of these data points. (ii) In each and every of the remaining 600 observations, we decide whether its elements $y_{(1)}$, $y_{(4)}$ and/or $y_{(7)}$ are missing or not by randomly generating an observation v from $N[0, 1]$. More specifically, we randomly generate independent observations $v_{(1)}$, $v_{(2)}$ and $v_{(3)}$ from $N[0, 1]$, then $y_{(1)}$ is deleted only if $v_{(1)} > \bar{y}_{(1)}$, $y_{(4)}$ is deleted only if $v_{(2)} < \bar{y}_{(4)} - 1$, and $y_{(7)}$ is deleted only if $v_{(3)} > \bar{y}_{(7)} - 1.5$. In this missing data set, $y_{(2)}$, $y_{(3)}$, $y_{(5)}$, $y_{(6)}$, $y_{(8)}$ and $y_{(9)}$ are retained and a number of $y_{(1)}$, $y_{(4)}$ and/or $y_{(7)}$ are missing at random.

We focus on $\boldsymbol{\mu}_1$ (or $\boldsymbol{\mu}_2$) in finding a suitable identifiability constraint. The first step is to apply the random permutation sampler to produce a MCMC sample from the unconstrained posterior with size 5000 after a burn-in phase of 500 simulations. This random permutation sampler delivers a sample that explores a whole unconstrained parameter space and jumps between the various labeling subspaces in a balanced fashion. For a mixture of two SEMs, we just have 2! labeling subspaces. In the random permutation sampler, after each sweep the 1s and 2s are permuted randomly; that is, with probability 0.5, the 1s stay as 1s and with probability 0.5 they become 2s. The output can be explored to find a suitable identifiability constraint. Based on the reasoning given in Appendix B, it suffices to consider the parameters in $\boldsymbol{\mu}_1$. To search for an appropriate identifiability constraint, we look at scatterplots of $\mu_{1,1}$ versus $\mu_{1,l}$, $l = 2, \ldots, 9$, for getting information on aspects of the states that are most different. These scatterplots are presented in Figure 1. They clearly indicate that the most two significant differences between the two components are sampled values corresponding to $\mu_{1,5}$ and $\mu_{1,6}$. If permutation sampling is based on the constraint $\mu_{1,5} < \mu_{2,5}$ or $\mu_{1,6} > \mu_{2,6}$, label switching will not appear. We can see from Figure 1 that if permutation sampling is combined with the constraint $\mu_{1,1} < \mu_{1,2}$ (or $\mu_{1,1} < \mu_{1,3}$, etc.), label switching may still present in the MCMC outputs.

Bayesian estimates are obtained by using permutation sampler with the identifiability constraint $\mu_{1,5} < \mu_{2,5}$, and utilizing the incomplete data with missing entries. Values of hyper-parameters in the conjugate prior distributions (see (6)) are taken as: For $h = 1, \ldots, 9$, $\mu_{0,h}$ equals to the sample mean $\bar{y}_h$, $\boldsymbol{\Sigma}_0 = 10^2 \boldsymbol{I}$, elements in $\boldsymbol{\Lambda}_{0km}$, and $\boldsymbol{\Pi}_{0kl}$ (which only involves the γ's) are taken to be true parameter values, $\boldsymbol{H}_{0ykm}$ and $\boldsymbol{H}_{0\omega kl}$

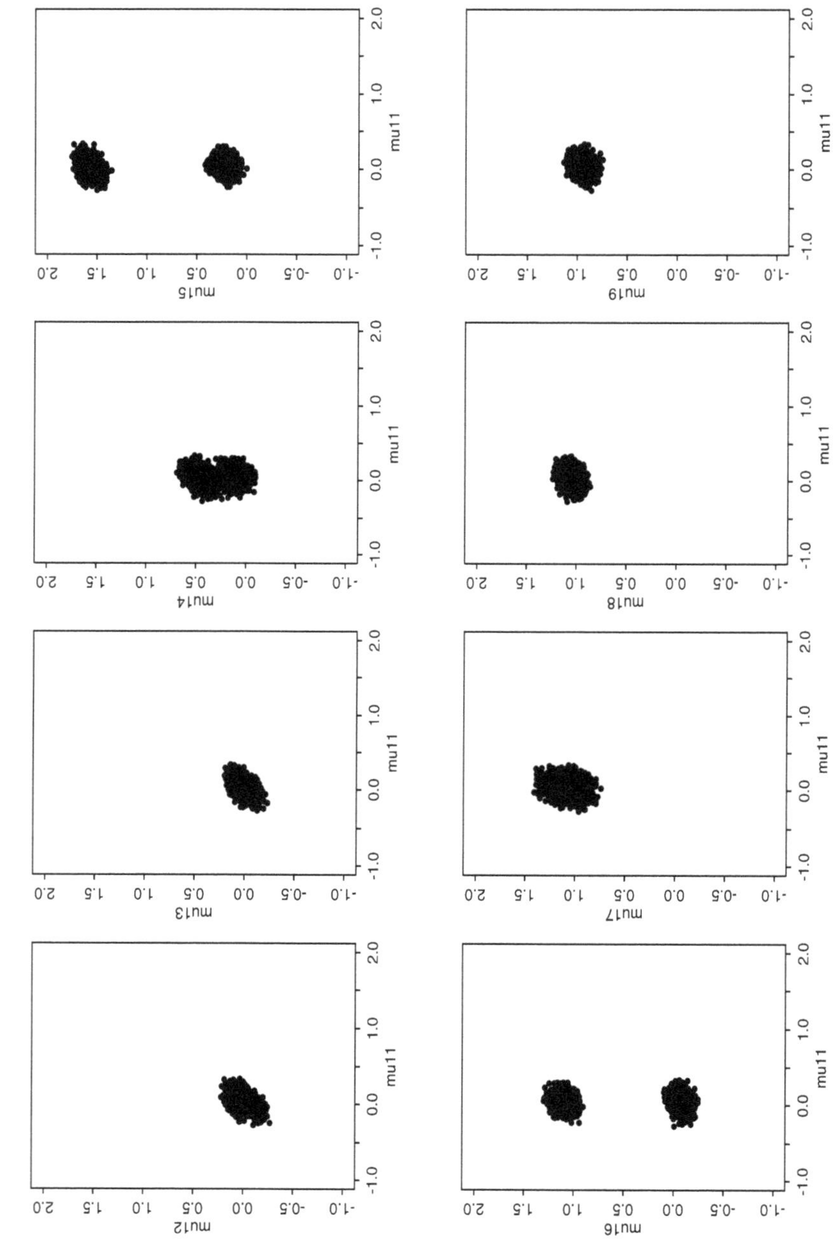

Fig. 1. Scatterplots of MCMC output for components of $\boldsymbol{\mu}_1$.

Table 1
Bayesian estimates of parameters, their standard deviations and root mean squares obtained in Simulation 1

Component 1				Component 2			
Par	Mean	SD	RMS	Par	Mean	SD	RMS
$\pi_1 = 0.5$	0.510	0.030	0.031	$\pi_2 = 0.5$	0.490	0.030	0.031
$\mu_{1,1} = 0.0$	−0.010	0.089	0.101	$\mu_{2,1} = 0.0$	0.010	0.089	0.087
$\mu_{1,2} = 0.0$	−0.007	0.048	0.057	$\mu_{2,2} = 0.0$	0.005	0.061	0.061
$\mu_{1,3} = 0.0$	−0.003	0.047	0.055	$\mu_{2,3} = 0.0$	0.006	0.061	0.061
$\mu_{1,4} = 0.0$	−0.010	0.079	0.080	$\mu_{2,4} = 0.5$	0.524	0.079	0.078
$\mu_{1,5} = 0.0$	0.028	0.071	0.084	$\mu_{2,5} = 1.5$	1.499	0.058	0.077
$\mu_{1,6} = 1.0$	0.982	0.066	0.109	$\mu_{2,6} = 0.0$	0.012	0.059	0.090
$\mu_{1,7} = 1.0$	0.994	0.094	0.100	$\mu_{2,7} = 1.0$	1.005	0.094	0.097
$\mu_{1,8} = 1.0$	0.994	0.045	0.042	$\mu_{2,8} = 1.0$	1.002	0.057	0.056
$\mu_{1,9} = 1.0$	1.005	0.046	0.045	$\mu_{2,9} = 1.0$	0.999	0.058	0.057
$\lambda_{1,21} = 0.4$	0.414	0.055	0.053	$\lambda_{2,21} = 0.8$	0.783	0.060	0.056
$\lambda_{1,31} = 0.4$	0.402	0.056	0.054	$\lambda_{2,31} = 0.8$	0.778	0.060	0.063
$\lambda_{1,52} = 0.8$	0.786	0.070	0.084	$\lambda_{2,52} = 0.4$	0.406	0.062	0.073
$\lambda_{1,62} = 0.8$	0.784	0.074	0.089	$\lambda_{2,62} = 0.4$	0.412	0.062	0.065
$\lambda_{1,83} = 0.4$	0.440	0.071	0.076	$\lambda_{2,83} = 0.8$	0.803	0.077	0.059
$\lambda_{1,93} = 0.4$	0.444	0.070	0.081	$\lambda_{2,93} = 0.8$	0.809	0.078	0.069
$\gamma_{1,1} = 0.2$	0.227	0.085	0.076	$\gamma_{2,1} = 0.7$	0.722	0.099	0.088
$\gamma_{1,2} = 0.7$	0.705	0.117	0.089	$\gamma_{2,2} = 0.2$	0.222	0.076	0.061
$\phi_{1,11} = 1.0$	1.008	0.127	0.122	$\phi_{2,11} = 1.0$	0.961	0.137	0.142
$\phi_{1,12} = 0.3$	0.272	0.080	0.078	$\phi_{2,12} = 0.3$	0.302	0.079	0.073
$\phi_{1,22} = 1.0$	0.865	0.152	0.181	$\phi_{2,22} = 1.0$	0.985	0.154	0.128
$\psi_{1,11} = 0.5$	0.567	0.094	0.088	$\psi_{2,11} = 0.5$	0.553	0.078	0.081
$\psi_{1,22} = 0.5$	0.514	0.046	0.046	$\psi_{2,22} = 0.5$	0.522	0.055	0.049
$\psi_{1,33} = 0.5$	0.520	0.046	0.049	$\psi_{2,33} = 0.5$	0.523	0.055	0.052
$\psi_{1,44} = 0.5$	0.530	0.071	0.062	$\psi_{2,44} = 0.5$	0.568	0.087	0.092
$\psi_{1,55} = 0.5$	0.563	0.069	0.091	$\psi_{2,55} = 0.5$	0.527	0.053	0.050
$\psi_{1,66} = 0.5$	0.540	0.065	0.066	$\psi_{2,66} = 0.5$	0.522	0.053	0.061
$\psi_{1,77} = 0.5$	0.608	0.107	0.128	$\psi_{2,77} = 0.5$	0.588	0.092	0.108
$\psi_{1,88} = 0.5$	0.517	0.047	0.045	$\psi_{2,88} = 0.5$	0.530	0.059	0.056
$\psi_{1,99} = 0.5$	0.519	0.047	0.047	$\psi_{2,99} = 0.5$	0.528	0.060	0.054
$\psi_{1,\delta} = 0.5$	0.550	0.092	0.071	$\psi_{2,\delta} = 0.5$	0.561	0.084	0.086

are the identity matrices, $\alpha_{0\varepsilon k} = \alpha_{0\delta k} = 10$, $\beta_{0\varepsilon k} = \beta_{0\delta k} = 8$, $\rho_0 = 6$ and $\boldsymbol{R}_0^{-1} = 5\boldsymbol{I}$. The α in the Dirichlet distribution of $\boldsymbol{\pi}$ is taken as 1. We conduct a few test runs, and find that the algorithm converged in less than 500 iterations. Hence, Bayesian estimates in 100 replications are obtained using a burn-in phase of 500 iterations, and a total of $J = 2000$ observations collected after the burn-in phase. Based on 100 replications, the mean (Mean), standard deviations (SD), and root mean squares (RMS) between estimates and true values are computed. Results are reported in Table 1. We observe that the means of the Bayesian estimates are pretty close to their true parameter values, and the RMS values are reasonably small. These results indicate that the permutation sampler under a suitable identifiability constraint produces rather accurate Bayesian estimates for a mixture of SEMs with poorly separated components.

To see the importance of choosing a good identifiability constraint, we have conducted a simulation on the basis of the same settings of the two-component SEMs, but using the inappropriate constraint $\mu_{1,1} < \mu_{2,1}$. We observe that the means of Bayesian estimates for the parameters $\{\mu_{1,5}, \mu_{1,6}, \boldsymbol{\Lambda}_1, \boldsymbol{\Gamma}_1\}$ and $\{\mu_{2,5}, \mu_{2,6}, \boldsymbol{\Lambda}_2, \boldsymbol{\Gamma}_2\}$ in 100 replications are quite close. Hence, the method fails to identify a two-component SEM, and the means of these parameters in 100 replications are not close to their corresponding true values. For example, the means of the estimates for $(\mu_{k,5}, \mu_{k,6})$ are (0.711, 0.514) when $k = 1$, and (0.809, 0.487) when $k = 2$; the means of the estimates of $(\lambda_{k,21}, \lambda_{k,31}, \lambda_{k,52}, \lambda_{k,62}, \lambda_{k,83}, \lambda_{k,93})$ for $k = 1, 2$ are (0.589, 0.583, 0.605, 0.615, 0.613, 0.616), and (0.606, 0.595, 0.575, 0.596, 0.628, 0.634); and the means of the estimates of $(\gamma_{k,1}, \gamma_{k,2})$ for $k = 1, 2$ are (0.459, 0.480), and (0.488, 0.449); respectively. The reason for these findings can be revealed from the first scatterplot in Figure 1, in relation to $\mu_{1,1}$ and $\mu_{1,2}$.

4.2. Simulation study 2: Sensitivity analysis with respect to prior inputs

The objectives of this simulation study is to investigate the sensitivity of model selection results with respect to different types of prior hyper-parameter inputs. The data set is generated from a mixture SEMs with two components defined by (1), (2), and (3). For each $k = 1, 2$, the model involves six manifest variables which are indicators for three latent variables η, ξ_1 and ξ_2. The structure of the loading matrix in each component is

$$\boldsymbol{\Lambda}^{\mathrm{T}} = \begin{bmatrix} 1.0 & \lambda_{k,21} & 0 & 0 & 0 & 0 \\ 0 & 0 & 1.0 & \lambda_{k,42} & 0 & 0 \\ 0 & 0 & 0 & 0 & 1.0 & \lambda_{k,63} \end{bmatrix},$$

where the one's and zero's are fixed parameters for achieving an identified covariance structure, whilst $\lambda_{k,21}$, $\lambda_{k,42}$ and $\lambda_{k,63}$ are unknown parameters. In the kth component, the structural equation is of the following form $\eta = \gamma_{k,1}\xi_1 + \gamma_{k,2}\xi_2 + \delta$, where $\gamma_{k,1}$ and $\gamma_{k,2}$ are unknown parameters. The true population values are given by $\pi_1 = \pi_2 = 0.5$, $\boldsymbol{\mu}_1 = (0, \ldots, 0)^{\mathrm{T}}$, $\boldsymbol{\mu}_2 = (0, 0, 0, 2, 2, 2)^{\mathrm{T}}$; $\lambda_{1,21} = \lambda_{1,42} = \lambda_{1,63} = 0.4$, $\lambda_{2,21} = \lambda_{2,42} = \lambda_{2,63} = 0.8$, $\gamma_{1,1} = \gamma_{1,2} = 0.6$, $\gamma_{2,1} = 0.6$, $\gamma_{2,2} = -0.6$, $\phi_{1,11} = \phi_{1,22} = \phi_{2,11} = \phi_{2,22} = 1.0$, $\phi_{1,12} = \phi_{2,12} = 0.3$, $\psi_{1,11} = \cdots = \psi_{1,66} = 0.5$, $\psi_{2,11} = \cdots = \psi_{2,66} = 0.64$, $\psi_{1,\delta} = 0.36$ and $\psi_{2,\delta} = 0.80$. In this 2-component model, the total number of unknown parameters is 40. The MAR missing data are created via similar steps as in Section 4.1. We compute the logarithm of the Bayes factor by the proposed path sampling procedure with 20 grids in [0, 1]. Based on the convergence behaviors of a few test runs, $J = 1000$ observations collected after 500 burn-in iterations are used in the computation of $\overline{U}_{(s)}$ at each grid, see (9) and (10).

The sensitivity analysis is conducted to reveal the impact of different prior hyper-parameters inputs in the conjugate prior distributions given in (6). As there are various kinds of distributions, we consider five different groups of hyper-parameters according to the characteristics of the corresponding parameters. These groups are:

(I) α's in the Dirichlet distribution associated with $\boldsymbol{\pi}$;
(II) $(\boldsymbol{\mu}_0, \boldsymbol{\Sigma}_0)$ in the normal distribution associated with $\boldsymbol{\mu}_k$;
(III) $(\alpha_{0\varepsilon k}, \beta_{0\varepsilon k})$, and $(\alpha_{0\delta k}, \beta_{0\delta k})$ respectively associated with ψ_{km}^{-1} and $\psi_{\delta kl}^{-1}$;

Table 2
Different prior inputs

Type	[Hyper-parameters]	Prior inputs I	II	III
A	$[\alpha_1, \alpha_2]$	(1, 1)	(1, 2)	(2, 1)
B	$[\boldsymbol{\mu}_{0h}, \boldsymbol{\Sigma}_0]$	$(\bar{y}_{(h)}, 4^2\boldsymbol{I})$	$(\bar{y}_{(h)}, 2^2\boldsymbol{I})$	$(\bar{y}_{(h)}, 8^2\boldsymbol{I})$
C	$[(\alpha_{0\varepsilon k}, \beta_{0\varepsilon k}) = (\alpha_{0\delta k}, \beta_{0\delta k})]$	(3, 8)	(3, 4)	(3, 12)
D	$[\boldsymbol{\Lambda}_{0km}, \boldsymbol{H}_{0ykm}]$	(T.V., $\boldsymbol{I}$)	(T.V./2, $2\boldsymbol{I}$)	(2 × T.V., $3\boldsymbol{I}$)
	$[\boldsymbol{\Pi}_{0km}, \boldsymbol{H}_{0\omega kl}]$	(T.V., $\boldsymbol{I}$)	(T.V./2, $2\boldsymbol{I}$)	(2 × T.V., $3\boldsymbol{I}$)
E	$[\boldsymbol{R}_0^{-1}, \rho_0]$	$(3\boldsymbol{I}, 8)$	$(2\boldsymbol{I}, 4)$	$(4\boldsymbol{I}, 12)$

Note: T.V. means true values. Type A means the other prior inputs in (ii), (iii), (iv), and (v) are fixed at the *basic prior inputs*. Types B, C, D and E are similarly defined.

(IV) $(\boldsymbol{\Lambda}_{0km}, \boldsymbol{H}_{0ykm})$ and $(\boldsymbol{\Pi}_{0kl}, \boldsymbol{H}_{0\omega kl})$ respectively associated with $\boldsymbol{\Lambda}_{km}$ and $\boldsymbol{\Pi}_{kl}$,
(V) $(\boldsymbol{R}_0^{-1}, \rho_0)$ associated with $\boldsymbol{\Phi}$.

We study each group separately by holding the hyper-parameter values in the other groups fixed. We use the following values as the *basic prior inputs*:

(i) For $D(\alpha_1, \alpha_2)$: $\alpha_1 = \alpha_2 = 1$.
(ii) For $N[\boldsymbol{\mu}_0, \boldsymbol{\Sigma}_0]$: $\mu_{0h} = \bar{\boldsymbol{y}}_h$, $h = 1, \ldots, 6$, $\boldsymbol{\Sigma}_0 = 4^2\boldsymbol{I}$.
(iii) For $Gamma(\alpha_{0\varepsilon k}, \beta_{0\varepsilon k})$ and $Gamma(\alpha_{0\delta k}, \beta_{0\delta k})$: $(\alpha_{0\varepsilon k}, \beta_{0\varepsilon k}) = (\alpha_{0\delta k}, \beta_{0\delta k}) = (3, 8)$.
(iv) For $N[\boldsymbol{\Lambda}_{0km}, \boldsymbol{H}_{0ykm}]$ and $N[\boldsymbol{\Pi}_{0kl}, \boldsymbol{H}_{0\omega kl}]$: $\boldsymbol{\Lambda}_{0km}$ and $\boldsymbol{\Pi}_{0kl}$ are the true parameter matrices, $\boldsymbol{H}_{0ykm}$ and $\boldsymbol{H}_{0\omega kl}$ are the appropriate identity matrices.
(v) For $IW(\boldsymbol{R}_0, \rho_0)$: $\rho_0 = 8$ and $\boldsymbol{R}_0^{-1} = 3\boldsymbol{I}$.

According to the suggestion of Kass and Raftery (1995), we perturb the prior inputs of the hyper-parameter values within the particular group. For example, to study the sensitivity of the results with respect to $\boldsymbol{\mu}_0$ and $\boldsymbol{\Sigma}_0$, we additionally consider $(\boldsymbol{\mu}_0, \boldsymbol{\Sigma}_0) = (\bar{y}_{(h)}/2, 2^2\boldsymbol{I})$, and $(\boldsymbol{\mu}_0, \boldsymbol{\Sigma}_0) = (2\bar{y}_{(h)}, 8^2\boldsymbol{I})$, but holding the prior inputs in (i), (iii), (iv), and (v) fixed at the basic prior inputs. Different types (A, B, C, D, E) of prior inputs are summarized in Table 2. The total number of different settings is 15.

We consider the performance of path sampling for computing the Bayes factor in comparing an one-component mixture SEM M_1 against a two-component mixture SEM M_2. Due to the heavy computation burden and the fact that it is not absolutely necessary to use many replications for revealing the performance and the sensitivity of the Bayes factor, we only taken ten replications. The means and standard deviations that are obtained from the ten replications are presented in Table 3. For the sensitivity analysis, we compare the means of the log B_{21} estimates that are obtained under prior inputs I, II, and III associated with Types A, ..., E. We observe that the log B_{21} estimates that are associated with Types A, E and probably D are not significantly different, whilst the estimates that are associated with Types B and C are quite different. These results indicate that the log Bayes factor is more sensitive to the prior inputs of Types B and C. Comparing the log B_{21} estimates under priors I, II, and III associated with Type B, we observe

Table 3
Mean and standard deviation (SD) of estimated log B_{21}

Type	Prior inputs					
	I		II		III	
	Mean	SD	Mean	SD	Mean	SD
A	14.57	8.00	16.92	5.65	19.45	6.34
B	14.34	8.05	19.13	5.03	8.99	6.18
C	14.75	9.74	29.85	5.26	9.45	5.19
D	16.46	6.69	12.14	6.96	11.55	5.94
E	16.70	7.71	13.60	7.79	17.68	4.95

that the estimates under prior III with $\boldsymbol{\Sigma}_0 = 8^2\boldsymbol{I}$ are generally smaller. According to its definition and interpretation, the log Bayes factors estimated with prior input III tend to favor the simpler (one-component) model. This is an natural phenomenon in model selection of mixture models, because a large $\boldsymbol{\Sigma}_0$ gives more freedom for $\boldsymbol{\mu}_k$ to vary and hence has less power to separate the components. However, the Bayes factor still has enough power to select the correct model. Similarly, from results in relation to Type C (note that the prior distributions are corresponding to ψ_{km}^{-1} and $\psi_{\delta kl}^{-1}$), we observe that the log Bayes factors estimated on the basis of prior distributions with larger variances (e.g., prior input III), so that the corresponding parameters have more freedom to vary, tend to favor the simpler one-component model. Note also that the Bayes factor still has the power to select the correct model.

We consider the performance of the procedure for comparing a two-component and a three-component model. The log B_{32} estimates are computed under the different prior inputs as before. We observe that the log Bayes factors are within the range -0.3 to -2.0. According to the criterion for interpreting the log Bayes factors, the correct two-component mixture SEMs is selected by the log Bayes factor under all different choices of prior inputs.

The above simulation study has been conducted with the listwise deletion approach that only uses the fully observed data. We found that an incorrect one-component model is selected under all the different cases.

5. An illustrative example

The illustrative example is based on a small portion of ICPSR data set collected in the project WORLD VALUES SURVEY 1981–1984 AND 1990–1993 (World Values Study Group, ICPSR version, 1994). Although the whole data set was collected in 45 societies around the world, only the data obtained from the United Kingdom are used. Six variables that are related to respondents' job and homelife are taken as manifest variables in $\boldsymbol{y} = (y_{(1)}, \ldots, y_{(6)})^{\mathrm{T}}$. For completeness, the corresponding questions are presented in Appendix C. These variables are measured via a 10-point scale and hence are treated as continuous. There are 1483 random observations, many of them are with missing entries and there are only 196 fully observed data. From the questions associated with the manifest variables, it is natural to consider a measurement model with three latent variables: η, ξ_1 and ξ_2, such that the first two variables are indicators for η, the 3rd and 4th variables are indicators for ξ_1, and the remaining manifest variables are indicators for ξ_2. We choose the structure that gives non-overlapping latent variables for clear interpretation and for identifying the model. The following specifications on the parameter matrices of component are used: $\boldsymbol{B} = \boldsymbol{0}$, $\boldsymbol{\Gamma} = (\gamma_1, \gamma_2)$, $\boldsymbol{\Psi}_\delta = \psi_\delta$,

$$\boldsymbol{\Lambda}^{\mathrm{T}} = \begin{pmatrix} 1.0^* & \lambda_{21} & 0^* & 0^* & 0^* & 0^* \\ 0^* & 0^* & 1.0^* & \lambda_{42} & 0^* & 0^* \\ 0^* & 0^* & 0^* & 0^* & 1.0^* & \lambda_{63} \end{pmatrix}, \qquad \boldsymbol{\Phi} = \begin{pmatrix} \phi_{11} & \phi_{12} \\ \phi_{21} & \phi_{22} \end{pmatrix}, \tag{11}$$

and $\boldsymbol{\Psi} = \mathrm{diag}(\psi_1, \ldots, \psi_6)$. The latent variables can be roughly interpreted as 'job satisfaction, η', 'homelife, ξ_1', and 'job attitude, ξ_2'.

We first conduct the model selection analysis to choose an appropriate number of components for the mixture of SEM, then estimate the unknown parameters in the selected model via the MCMC procedure with a permutation sampler as described in Section 3.1. Here, the formulation of the model in every component is taken to be the same as in (11). The Bayes factor computed via path sampling is used to select a model with the most appropriate number of components. The following hyper-parameters values are used: $\boldsymbol{\mu}_0 = \bar{\boldsymbol{y}}$, where $\bar{\boldsymbol{y}}$ is obtained on the basis of the 196 fully observed data, $\boldsymbol{\Sigma}_0 = 9^2\boldsymbol{I}$; $\alpha_1 = \alpha_2 = 1$; $\boldsymbol{\Lambda}_{0km} = \boldsymbol{0}$, $\boldsymbol{\Pi}_{0kl} = \boldsymbol{0}$, $\boldsymbol{H}_{0ykm} = \boldsymbol{I}$, $\boldsymbol{H}_{0\omega kl} = \boldsymbol{I}$; $(\alpha_{0\varepsilon k}, \beta_{0\varepsilon k}) = (\alpha_{0\delta k}, \beta_{0\delta k}) = (2, 30)$; and $\rho_0 = 20$, $\boldsymbol{R}_0^{-1} = 5\boldsymbol{I}$. Again, for each $t_{(s)}$, we observe that the Gibbs sampler algorithm converged quickly within 500 iterations. A total of $J = 1000$ additional observations are simulated by the Gibbs sampler after convergence for computing $\overline{U}_{(s)}$ in (10), and then log Bayes factors are estimated via (9), using 20 fixed grids in [0, 1]. Let M_k denotes the mixture model with k components, the log Bayes factor estimates are equal to $\widehat{\log B_{21}} = 61.42$, and $\widehat{\log B_{32}} = -0.96$. According to the criterion given in Kass and Raftery (1995), a two-component model is selected. The selection results that are obtained via the log Bayes factors have been cross-validated with different prior inputs. The same conclusion of selecting a two-component model is obtained.

In estimation, based on MCMC samples simulated by the random permutation sampler, we find that $\mu_{1,1} < \mu_{2,1}$ is a suitable identifiability constraint. Bayesian estimates of the selected mixture model with two components are obtained by using permutation sampler combined with this identifiability constraint. Results are presented in Table 4, together with the HPD intervals. We observe that parameter estimates of μ_1, μ_2, μ_3, μ_5, μ_6, λ_{21}, λ_{63}, γ_2, ϕ_{12}, $\psi_{11}, \ldots, \psi_{66}$, and ψ_δ under components 1 and 2 are quite different. This finding roughly cross-validates the model selection result that a two-component model is better than an one-component model. Hence, it is concluded that there are two heterogeneous groups in the data set, which have some significantly different parameters in their mean vectors and covariance structures. For other structural models that produce reasonably close $\boldsymbol{\Sigma}_k$, we expect to get a similar result of selecting a two-component model. As we have not included all possible models in the comparison, we cannot conclude that the selecting model is the globally best fitting model.

Table 4
Bayesian solution under the 2-component model in analyzing the ICPSR data set

Par	Component 1		Component 2	
	Est	HPD	Est	HPD
π	0.336	[0.279, 0.387]	0.664	[0.613, 0.721]
μ_1	6.955	[6.616, 7.245]	8.862	[8.762, 8.967]
μ_2	6.069	[5.720, 6.391]	8.200	[8.081, 8.315]
μ_3	1.864	[1.283, 2.410]	2.543	[2.299, 2.789]
μ_4	5.125	[4.772, 5.471]	5.481	[5.243, 5.714]
μ_5	5.843	[5.448, 6.248]	8.186	[8.025, 8.342]
μ_6	5.024	[4.393, 5.715]	7.839	[7.539, 8.130]
λ_{21}	0.564	[0.309, 0.813]	0.999	[0.857, 1.132]
λ_{42}	2.266	[1.604, 2.942]	2.339	[1.978, 2.683]
λ_{63}	2.714	[0.975, 4.542]	1.069	[0.682, 1.504]
γ_1	0.214	[−0.161, 0.582]	0.180	[0.088, 0.277]
γ_2	−0.940	[−1.674, −0.248]	0.370	[0.201, 0.545]
ϕ_{11}	1.111	[0.542, 1.764]	1.310	[0.957, 1.748]
ϕ_{12}	−0.114	[−0.340, 0.108]	0.180	[0.012, 0.340]
ϕ_{22}	0.508	[0.211, 0.932]	0.766	[0.467, 1.076]
ψ_{11}	2.645	[1.677, 3.491]	0.615	[0.509, 0.732]
ψ_{22}	3.810	[3.137, 4.610]	0.954	[0.790, 1.136]
ψ_{33}	2.210	[1.310, 3.136]	1.346	[1.043, 1.725]
ψ_{44}	4.396	[2.501, 6.630]	2.453	[1.651, 3.283]
ψ_{55}	5.842	[4.592, 7.072]	1.324	[1.014, 1.642]
ψ_{66}	5.826	[3.736, 8.078]	2.884	[2.030, 3.925]
ψ_{δ}	2.576	[1.643, 3.452]	0.752	[0.615, 0.897]

6. Analysis via WinBUGS

In practice, the freely available software WinBUGS (Windows Version of Bayesian Inference Using Gibbs Sampling, Spiegelhalter et al., 2003) is very useful for the production of reliable Bayesian statistics for a very wide range of statistical models (see Congdon, 2003), including most SEMs (Lee, 2007). WinBUGS was mainly developed using MCMC techniques, such as the Gibbs sampler (Geman and Geman, 1984) and the Metropolis–Hastings (MH) algorithm. Under broad conditions, this software can provide simulated samples from the joint posterior distribution of the unknown quantities, such as the parameters and latent variables in the proposed model. As mentioned, empirical summary statistics can be obtained from these samples to conduct statistical inferences, such as obtaining Bayesian estimates and their standard error estimates.

The most advanced version of BUGS is WinBUGS 1.4, which iteratively runs with Windows and was developed by the Medical Research Council (MRC) Biostatistics Unit (Cambridge, UK) and the Department of Epidemiology and Public Health of the Imperial College School of Medicine at St Mary's Hospital (London). It can be freely downloaded from the website http://www.mrc-bsu.cam.ac.uk/bugs/. The free version of WinBUGS is a restricted version, and it is necessary to email the BUGS project for a key that will allow the user to use the full version. The WinBUGS manual (Spiegelhalter

Table 5
Results obtained via WinBUGS in analyzing the artificial data set

Component 1			Component 2		
Par	Est	HPD	Par	Est	HPD
$\pi_1 = 0.5$	0.503	[0.449, 0.547]	$\pi_2 = 0.5$	0.497	[0.442, 0.551]
$\mu_{1,1} = 0.0$	−0.076	[−0.224, 0.077]	$\mu_{2,1} = 0.0$	0.116	[−0.038, 0.264]
$\mu_{1,2} = 0.0$	−0.004	[−0.096, 0.091]	$\mu_{2,2} = 0.0$	0.076	[−0.057, 0.203]
$\mu_{1,3} = 0.0$	−0.010	[−0.095, 0.075]	$\mu_{2,3} = 0.0$	0.028	[−0.057, 0.203]
$\mu_{1,4} = 0.0$	−0.078	[−0.229, 0.087]	$\mu_{2,4} = 0.5$	0.560	[0.418, 0.701]
$\mu_{1,5} = 0.0$	0.033	[−0.111, 0.181]	$\mu_{2,5} = 1.5$	1.580	[1.471, 1.690]
$\mu_{1,6} = 1.0$	1.012	[0.903, 1.121]	$\mu_{2,6} = 0.0$	0.049	[−0.063, 0.150]
$\mu_{1,7} = 1.0$	0.962	[0.803, 1.119]	$\mu_{2,7} = 1.0$	1.059	[0.887, 1.235]
$\mu_{1,8} = 1.0$	1.011	[0.924, 1.092]	$\mu_{2,8} = 1.0$	1.084	[0.969, 1.200]
$\mu_{1,9} = 1.0$	1.002	[0.909, 1.093]	$\mu_{2,9} = 1.0$	1.110	[1.002, 1.230]
$\lambda_{1,21} = 0.4$	0.313	[0.214, 0.418]	$\lambda_{2,21} = 0.8$	0.861	[0.744, 0.995]
$\lambda_{1,31} = 0.4$	0.402	[0.303, 0.506]	$\lambda_{2,31} = 0.8$	0.816	[0.699, 0.951]
$\lambda_{1,52} = 0.8$	0.724	[0.610, 0.847]	$\lambda_{2,52} = 0.4$	0.516	[0.388, 0.665]
$\lambda_{1,62} = 0.8$	0.709	[0.589, 0.837]	$\lambda_{2,62} = 0.4$	0.496	[0.365, 0.633]
$\lambda_{1,83} = 0.4$	0.367	[0.235, 0.524]	$\lambda_{2,83} = 0.8$	0.768	[0.613, 0.963]
$\lambda_{1,93} = 0.4$	0.418	[0.276, 0.601]	$\lambda_{2,93} = 0.8$	0.865	[0.693, 1.067]
$\gamma_{1,1} = 0.2$	0.212	[0.021, 0.394]	$\gamma_{2,1} = 0.7$	0.856	[0.648, 1.080]
$\gamma_{1,2} = 0.7$	0.787	[0.549, 1.103]	$\gamma_{2,2} = 0.2$	0.194	[0.044, 0.346]
$\phi_{1,11} = 1.0$	0.992	[0.776, 1.239]	$\phi_{2,11} = 1.0$	0.777	[0.560, 1.027]
$\phi_{1,21} = 0.3$	0.263	[0.095, 0.459]	$\phi_{2,21} = 0.3$	0.270	[0.143, 0.411]
$\phi_{1,22} = 1.0$	0.934	[0.549, 1.372]	$\phi_{2,22} = 1.0$	0.925	[0.624, 1.293]
$\psi_{1,11} = 0.5$	0.526	[0.376, 0.711]	$\psi_{2,11} = 0.5$	0.572	[0.437, 0.729]
$\psi_{1,22} = 0.5$	0.557	[0.470, 0.656]	$\psi_{2,22} = 0.5$	0.485	[0.384, 0.600]
$\psi_{1,33} = 0.5$	0.507	[0.425, 0.598]	$\psi_{2,33} = 0.5$	0.569	[0.465, 0.688]
$\psi_{1,44} = 0.5$	0.492	[0.368, 0.626]	$\psi_{2,44} = 0.5$	0.639	[0.484, 0.811]
$\psi_{1,55} = 0.5$	0.486	[0.384, 0.600]	$\psi_{2,55} = 0.5$	0.551	[0.447, 0.665]
$\psi_{1,66} = 0.5$	0.548	[0.433, 0.680]	$\psi_{2,66} = 0.5$	0.505	[0.412, 0.613]
$\psi_{1,77} = 0.5$	0.680	[0.452, 0.966]	$\psi_{2,77} = 0.5$	0.555	[0.387, 0.761]
$\psi_{1,88} = 0.5$	0.573	[0.474, 0.681]	$\psi_{2,88} = 0.5$	0.568	[0.445, 0.701]
$\psi_{1,99} = 0.5$	0.569	[0.471, 0.679]	$\psi_{2,99} = 0.5$	0.494	[0.370, 0.632]
$\psi_{1,\delta} = 0.5$	0.527	[0.356, 0.736]	$\psi_{2,\delta} = 0.5$	0.520	[0.379, 0.691]

et al., 2003) is available online, and gives brief instructions on WinBUGS. See also Lawson et al. (2003, Chapter 4) for supplementary descriptions.

In analyzing mixtures of SEMs with missing data, WinBUGS can be applied to obtain Bayesian estimates of the latent variables, the unknown parameters and their HPD intervals. However, for model comparison, it does not directly produce the Bayes factor. Moreover, Spiegelhalter et al. (2003) pointed out that the Deviance Information Criteria (DIC) may not be appropriate for model comparison of mixture models. Thus, WinBUGS also does not give the DIC value for mixture model.

To illustrate the application of WinBUGS in estimation for mixtures of SEMs with missing data, an artificial data set is simulated on the basis of exactly the same settings presented in the Simulation study I of Section 4. This simulated data set was reanalyzed by WinBUGS. Again, we first conducted an initial estimation to identify the appropriate identifiability constraint $\mu_{1,5} < \mu_{2,5}$ from the output as before. The model is then reanalyzed with this constraint, and three starting values of the parameters that are obtained from the sample mean, the 5th, and the 95th percentile of the corresponding simulated samples. In analyzing this data set under the same hyperparameter values as before, WinBUGS converged in less than 2000 iterations. Bayesian estimates are obtained from 2000 observations collected after convergence. These estimates are presented in Table 5, together with the HPD intervals. We observe that the results produced by WinBUGS are satisfactory.

7. Discussion

Missing data are particularly important in the analysis of mixture models, because ignoring these data may lead to an incorrect decision in selecting the number of components, and hence may give very misleading conclusion. We develop a Bayesian approach for analyzing mixture SEMs with MAR data and an unknown number of components. The main novel contributions are:

(i) A Bayesian estimation procedure that is coupled with a permutation sampler for selecting an identifiability constraint to solve the label switching problem. It is shown that the proposed procedure gives accurate estimates even for mixture SEMs with poorly separated components, and that inappropriate constraints may give incorrect results.
(ii) Extensions of the procedures for estimation and model comparison in Lee and Song (2003a) are developed for analysis of mixture SEMs with MAR data.
(iii) Sensitivity analyses of the statistical results with respect to the assumption of MAR and to prior inputs of the hyper-parameters.
(iv) Bayesian classification of incomplete observation.

In general, it is well known that (see Kass and Raftery, 1995; Richardson and Green, 1997; among others) Bayesian model selection tends to be more sensitive to priors than estimation. This is indeed a characteristic of model selection problems that is reflected by reasonable statistical methods such as the Bayes factor. On the basis of our sensitivity results on prior inputs, more attention should be paid on the prior inputs of $\boldsymbol{\Sigma}_0$ in the prior distribution of $\boldsymbol{\mu}_k$, and $[(\alpha_{0\varepsilon k}, \beta_{0\varepsilon k}), (\alpha_{0\delta k}, \beta_{0\delta k})]$ in the prior distributions corresponding to the inverses of the error measurement variances. In practice, it is desirable to compute several log Bayes factors with different prior inputs to cross-validate the decision. Of course knowledge of experts should not be ignored.

Clearly, there are several useful extensions of the current model; for example, models involve missing data with an non-ignorable missing mechanism, and/or SEMs with more complex structures. Developments of good statistical methods for handling these extensions represent interesting topics for future research.

Acknowledgements

The research is partially supported by a direct grant from the Chinese University of Hong Kong and by a grant from the Research Grant Council of the Hong Kong Special

Administrative Region, CUHK 4243/03H. The authors are grateful to ICPSR and the relevant funding agency for allowing the use of the data.

Appendix A. The permutation sampler

Let $\boldsymbol{\psi} = (\boldsymbol{Y}_{\text{mis}}, \boldsymbol{Z}, \boldsymbol{W}, \boldsymbol{\theta}, \boldsymbol{\pi})$, the permutation sampler for simulating $\boldsymbol{\psi}$ from the posterior $p(\boldsymbol{\psi}|\boldsymbol{Y}_{\text{obs}})$ is implemented as below:

1. First generate $\hat{\boldsymbol{\psi}}$ from the unconstrained posterior $p(\boldsymbol{\psi}|\boldsymbol{Y}_{\text{obs}})$ using standard Gibbs sampling steps;
2. Select some permutation $\rho(1), \ldots, \rho(K)$ of the current labeling of the states and define $\boldsymbol{\psi} = \rho(\hat{\boldsymbol{\psi}})$ from $\hat{\boldsymbol{\psi}}$ by reordering the labeling through this permutation, $(\boldsymbol{\theta}_1, \ldots, \boldsymbol{\theta}_K) := (\boldsymbol{\theta}_{\rho(1)}, \ldots, \boldsymbol{\theta}_{\rho(K)})$, $\boldsymbol{\pi} = (\pi_1, \ldots, \pi_K) := (\pi_{\rho(1)}, \ldots, \pi_{\rho(K)})$ and $\boldsymbol{W} = (w_1, \ldots, w_n) := (\rho(w_1), \ldots, \rho(w_n))$.

One application of permutation sampling is the random permutation sampler, where each sweep of the MCMC chain is concluded by relabeling the states through a random permutation of $\{1, \ldots, K\}$. This procedure delivers a sample that explores the whole unconstrained parameter space and jumps between the various labeling subspaces in a balanced fashion. Another application is the permutation sampling under identifiability constraints. A way to include an identifiability constraint is to use a permutation sampler, where the permutation is selected in such a way that the identifiability constraint is fulfilled.

Appendix B. Searching for identifiability constraints

For $k = 1, \ldots, K$, let $\boldsymbol{\theta}_k$ denote parameter vector corresponding to the kth component. The MCMC output of the random permutation sampler can be explored to find a suitable identifiability constraint, see Frühwirth-Schnatter (2001). It is sufficient to consider the parameters in $\boldsymbol{\theta}_1$, because a balanced sample from the unconstrained posterior will contain the same information for all parameters in $\boldsymbol{\theta}_k$ with $k \neq 1$. As the random permutation sampler jumps between the various labeling subspaces, part of the values simulated for $\boldsymbol{\theta}_1$ will belong to the first state, part will belong to the second state, and so on. To differ for various states, it is useful to consider bivariate scatterplots of $\boldsymbol{\theta}_{1,i}$ versus $\boldsymbol{\theta}_{1,a}$ for possible combinations of i and a. Jumping between the labeling subspaces produces groups in these scatterplots that correspond to different states. By describing the difference between the various groups geometrically, identification of a unique labeling subspace through conditions on the state-specific parameters is attempted. If the values simulated for a certain component of $\boldsymbol{\theta}$ differ significantly between the groups when jumping between the labeling subspaces, then an order condition on this component could be used to separate the labeling subspaces, whilst if the values sampled for a certain component of $\boldsymbol{\theta}$ hardly differ between the states when jumping between the labeling subspaces, then this component will be a poor candidate for separating the labeling subspaces.

Appendix C. Manifest variables in the ICPSR example

The number of the variable corresponding to the original data set is given in parenthesis at the end of each statement.

$y_{(1)}$: Overall, how satisfied are you with your home life? (V180)
$y_{(2)}$: All things considered, how satisfied are you with your life as a whole in these day? (V96)
$y_{(3)}$: Thinking about your reasons for doing voluntary work, how important the religious beliefs in your own case? (V62)
$y_{(4)}$: How important is God in your life? (V176)
$y_{(5)}$: Overall, how satisfied or dissatisfied are you with your job? (V116)
$y_{(6)}$: How free are you to make decisions in your job? (V117)

References

Arminger, G., Stein, P., Witterberg, J. (1999). Mixtures of conditional mean- and covariance-structure models. *Psychometrika* **64**, 475–494.

Berger, J.O. (1985). *Statistical Decision Theory and Bayesian Analysis*. Springer-Verlag, New York.

Chen, M.H., Shao, Q.H. (1999). Monte Carlo estimation of Bayesian credible and HPD intervals. *Journal of Computational and Graphical Statistics* **8**, 69–92.

Congdon, P. (2003). *Applied Bayesian Modeling*. John Wiley, New York.

Dolan, C.V., van der Maas, J.J.L. (1998). Fitting multivariate normal mixtures subject to structural equation modeling. *Psychometrika* **63**, 227–253.

Finkbeiner, C. (1979). Estimation for the multiple factor models when data are missing. *Psychometrika* **44**, 409–420.

Frühwirth-Schnatter, S. (2001). Markov chain Monte Carlo estimation of classical and dynamic switching and mixture models. *Journal of the American Statistical Association* **96**, 194–208.

Gelman, A. (1996). Inference and monitoring convergence. In: Gilk, W.R., Richardson, S., Spiegelhalter, D.J. (Eds.), *Markov Chain Monte Carlo in Practice*. Chapman, Hall, London, pp. 131–144.

Gelman, A., Meng, X.L. (1998). Simulating normalizing constants: From importance sampling to bridge sampling to path sampling. *Statistical Science* **13**, 163–185.

Geman, S., Geman, D. (1984). Stochastic relaxation, Gibbs distributions, and the Bayesian restoration of images. *IEEE Transactions on Pattern Analysis and Machine Intelligence* **6**, 721–741.

Jamshidian, M., Bentler, P.M. (1999). Maximum likelihood of mean and covariance structure with missing data using complete data routines. *Journal of Educational and Behavioral Statistics* **24**, 21–41.

Jedidi, K., Jagpal, H.S., DeSarbo, W.S. (1997). STEMM: A general finite mixture structural equation model. *Journal of Classification* **14**, 23–50.

Kass, R.E., Raftery, A.E. (1995). Bayes factors. *Journal of the American Statistical Association* **90**, 773–795.

Lawson, A.B., Browne, W.T., Vidal Rodeiro, C.L. (2003). *Disease Mapping with WinBUGS and MLWiN*. John Wiley, England.

Lee, S.Y. (1986). Estimation for structural equation models with missing data. *Psychometrika* **51**, 93–99.

Lee, S.Y. (2007). *Structural Equation Modeling: A Bayesian Approach*. John Wiley, England.

Lee, S.Y., Song, X.Y. (2003a). Bayesian model selection for mixtures of structural equation models with an unknown number of components. *British Journal of Mathematical and Statistical Psychology* **56**, 145–165.

Lee, S.Y., Song, X.Y. (2003b). Maximum likelihood estimation and model comparison for mixtures of structural equation model with ignorable missing data. *Journal of Classification* **20**, 221–255.

Little, R.J.A., Rubin, D.B. (1987). *Statistical Analysis with Missing Data*. Wiley, New York.

Raftery, A.E. (1996). Approximate Bayes factors and accounting for model uncertainty in generalized linear models. *Biometrika* **83**, 251–266.

Richardson, S., Green, P.J. (1997). On Bayesian analysis of mixtures with unknown number of components, with discussion. *Journal of the Royal Statistical Society, Series B* **59**, 731–792.

Roeder, K., Wasserman, L. (1997). Practical Bayesian density estimation using mixtures of normals. *Journal of the American Statistical Association* **92**, 894–902.

Schines, R., Hoijtink, H., Boomsma, A. (1999). Bayesian estimation and testing of structural equation models. *Psychometrika* **64**, 37–52.

Song, X.Y., Lee, S.Y. (2001). Bayesian estimation and test for factor analysis model with continuous and polytomous data in several populations. *British Journal of Mathematical and Statistical Psychology* **54**, 237–263.

Song, X.Y., Lee, S.Y. (2002). Analysis of structural equation model with ignorable missing continuous and polytomous data. *Psychometrika* **67**, 261–288.

Spiegelhalter, D.J., Thomas, A., Best, N.G., Lunn, D. (2003). *WinBUGS User Manual. Version 1.4.* MRC Biostatistics Unit, Cambridge, England.

Stephens, M. (2000). Dealing with label switching in mixture models. *Journal of Royal Statistical Society, Series B* **62**, 795–809.

Titterington, D.M., Smith, A.F.M., Makov, U.E. (1985). *Statistical Analysis of Finite Mixture Distributions.* John Wiley, New York.

WORLD VALUES SURVEY, 1981–1984 AND 1990–1993, ICPSR version (1994). Institute for Social Research, Ann Arbor, MI [producer]. Interuniversity Consortium for Political and Social Research, Ann Arbon, MI, 1994 [distribution].

Yung, Y.F. (1997). Finite mixtures in confirmatory factor-analysis models. *Psychometrika* **62**, 297–330.

Zhu, H.T., Lee, S.Y. (2001). A Bayesian analysis of finite mixtures in the LISREL model. *Psychometrika* **66**, 133–152.

Handbook of Computing and Statistics with Applications, Vol. 1
ISSN: 1871-0301

DOI: 10.1016/S1871-0301(06)01006-7

6

Local Influence Analysis for Latent Variable Models with Non-Ignorable Missing Responses

Bin Lu, Xin-Yuan Song, Sik-Yum Lee and Fernand Mac-Moune Lai

Abstract
A general procedure for local influence analysis of a generic latent variable model with non-ignorable missing data is provided. The local influence measures are based on the conditional expectation of the complete-data log-likelihood function in the corresponding EM algorithm. It is shown that the proposed methodology is feasible for a wide variety of perturbation schemes. Especially, a minor perturbation to the missing model is introduced to investigate the effect of how minor perturbations on the missing mechanism model can lead a large impact on key features of the whole model. To illustrate the methodology, two models are considered here. For the normal mixed effects model, results that are obtained from analyses of a simulation study, and a real example on a longitudinal study of a kidney disease are presented. And for the generalized linear mixed model, an artificial example is used to show the performance of our method.

Keywords: Conformal normal curvature; Generalized linear mixed model; Local influence; Non-ignorable missing mechanism; Normal mixed effects model

1. Introduction

Local influence analysis has been regarded as a crucial component in a thorough statistical analysis. It is an important statistical technique to assess the stability of the estimation output with respect to the model inputs. Model inputs may include data, parameters to be estimated, errors and model specifications, assumptions or other characteristics. Output may include the parameters estimates, final objective function values, estimates of residuals and standard errors, etc. One main objective of local influence is to assess whether the model output is particularly sensitive to minor perturbations of the model input. The perturbations may take the various forms including deletion of part of the data, changes in model specifications, etc. Cook (1986) proposed a unified approach for the assessment of local influence in minor perturbations of a statistical model. In the past years, this approach dominated the local influence analysis in biomedical statistics. Typical examples are the application to proportional hazards models (Cain and Lange, 1984; Weissfeld, 1990), regression models with censored data (Escobar and Meeker, 1992), nonlinear regression (Laurent and Cook, 1993), generalized linear models (Thomas and Cook, 1989), linear mixed models (Lesaffre and Verbeke, 1998; Demidenko and Stukel, 2005), and generalized linear mixed models (Ouwens et al., 2001; Zhu and Lee, 2003), among others.

It is well recognized that latent variable models, such as random effect models (Laird and Ware, 1982), and generalized linear mixed models (GLMMs) (Zeger and Karim, 1991; Breslow and Clayton, 1993; Diggle et al., 1994) are particularly useful for analyzing longitudinal data in biomedical research. In longitudinal studies, missing response data are very common due to treatment dropout, study dropout, mistimed measurements, subjects' inability to participate due to sickness, and so forth. A subject's response can be missing at one follow-up time and then can be measured at the next follow-up time, resulting in arbitrary nonmonotone missing data patterns. Often, missing response data in these studies are non-ignorable in the sense that the reason for the missing data often depends on the missing values themselves (Little and Rubin, 1987). For example, the side effects of the treatment may make the patients worse and thereby affect their participation. Hence, it is important to develop a statistical method for analyzing latent variable models with missing data that are missing with a non-ignorable missing mechanism. Recently, Ibrahim et al. (2001) proposed a maximum likelihood (ML) method for estimating parameters in the GLMMs with non-ignorable missing data that are also missing with nonmonotone patterns. However, very limited work has been done so far on developing local influence methods for latent variable models with nonmonotone and non-ignorable missing data.

Recently, based on Cook's approach (Cook, 1986), Verbeke et al. (2001) developed local influence measures for linear mixed models with dropouts. van Steen et al. (2001) applied their method to the multivariate model of Dale (1986). The principal idea of their method is to explore how small perturbations around a missing at random (MAR) dropout model in the direction of a missing not at random (MNAR) mechanism can have a large impact. However, their work is limited to monotone missing data; hence, missing data with arbitrary missing patterns cannot be assessed. Moreover, they applied Cook's approach which cannot be applied to some complex models because the building blocks in the associated diagnostic measures involve intractable integrals. Hence, their method cannot be applied to analyze the general nonmonotone and non-ignorable missing data in more complex models like GLMMs.

The development of local influence measures for latent variable models with non-ignorable missing data on the basis of Cook's approach is rather difficult (see Davidian and Giltinan, 1995). The reason for this is that the observed data likelihood function involves intractable integrals, and hence the objective function and second derivatives in the basic building blocks of Cook's local influence measures are difficult to evaluate. In this article, we will develop a general procedure for local influence analysis of a generic latent variable model with non-ignorable missing data based on the approach given in Zhu and Lee (2001). As this procedure is tied up with the powerful EM algorithm (Dempster et al., 1977) that is effective in handling missing data, and can take the advantages of the Markov chain Monte Carlo (MCMC) methods in creating the diagnostic measures, we expect that this procedure has wide application to the local influence of models in biomedical research.

This paper is organized as follows. Section 2 deals with local influence measures for a generic latent variable model (LVM) with non-ignorable missing data. The application of the newly developed methodologies to the normal mixed effects models, and to the GLMMs are presented in Sections 3 and 4, respectively. Finally, a discussion is given in Section 5. Some technical details are given in Appendices A–C.

2. Local influence of latent variable models with non-ignorable missing data

2.1. *A generic latent variable model with non-ignorable missing data*

Consider a data set with observations that are composed of a response, y_{ij}, and covariate vectors $\mathbf{x}_{ij}(s_1 \times 1)$ and $\mathbf{z}_{ij}(s_2 \times 1)$, where $j = 1, \ldots, n_i$ within clusters $i = 1, \ldots, N$. For example, a subject can be considered as a cluster, and repeated measurements for this subject i can be obtained at n_i different time points. Let M_0 be a latent variable model to fit $\mathbf{y}_i = (y_{i1}, \ldots, y_{in_i})'$ which involves latent variables $\mathbf{b}_i(s_2 \times 1)$, $i = 1, \ldots, N$. Let $\boldsymbol{\theta}$ be the vector of unknown parameters in M_0 which may involve regression coefficients and/or variances and covariances of some random vectors and let $f(\mathbf{y}_i; \boldsymbol{\theta})$ be the probability density function of $\mathbf{y}_i$. Responses y_{ij} and y_{ik} are correlated, but $\mathbf{y}_i$, $i = 1, \ldots, N$, are independent.

To accommodate the missing data, we define a missing indicator $\mathbf{r}_i = (r_{i1}, \ldots, r_{in_i})'$ for $\mathbf{y}_i$ such that $r_{ij} = 1$ if y_{ij} is missing, and $r_{ij} = 0$ if y_{ij} is observed. Let $\mathbf{R} = (\mathbf{r}_1, \ldots, \mathbf{r}_N)$, and $\mathbf{Y}_m$ and $\mathbf{Y}_o$ be the missing data and the observed data, respectively. If the distribution of $\mathbf{r}$ is independent of $\mathbf{Y}_m$, the missing mechanism is defined to be MAR; otherwise, the missing mechanism is non-ignorable (Little and Rubin, 1987). For analyzing missing data with a non-ignorable and unknown mechanism, the basic issues include specifying specify a reasonable model for $\mathbf{r}$ given $\mathbf{Y}_m$ and $\mathbf{Y}_o$, and then developing statistical methods for analyzing the posited latent variable model together with the missing model that accounts for the non-ignorable missing mechanism. Let $\mathbf{y}_i = (\mathbf{y}'_{oi}, \mathbf{y}'_{mi})'$, where $\mathbf{y}_{oi}$ is a vector of observed manifest variables and $\mathbf{y}_{mi}$ is a vector of missing components of the random vector $\mathbf{y}_i$. Here, we assume an arbitrary pattern of missing data in $\mathbf{y}_i$; thus, $\mathbf{y}_i = (\mathbf{y}'_{oi}, \mathbf{y}'_{mi})'$ may represent some permutation of the indices of the original $\mathbf{y}_i$. Hence, the missing pattern is nonmonotone. Let $[\mathbf{r}_i|\mathbf{y}_i, \mathbf{X}_i, \mathbf{Z}_i, \mathbf{b}_i, \boldsymbol{\gamma}]$ be the conditional distribution of $\mathbf{r}_i$ given $\mathbf{y}_i, \mathbf{X}_i, \mathbf{Z}_i$ and $\mathbf{b}_i$ with parameter $\boldsymbol{\gamma}$, where $\mathbf{X}_i = (\mathbf{x}_{i1}, \ldots, \mathbf{x}_{in_i})'$ and $\mathbf{Z}_i = (\mathbf{z}_{i1}, \ldots, \mathbf{z}_{in_i})'$. The observed-data likelihood of $\boldsymbol{\theta}$ and $\boldsymbol{\gamma}$ based on $\mathbf{Y}_o$ and $\mathbf{R}$ is given by:

$$L_o(\boldsymbol{\theta}, \boldsymbol{\gamma}; \mathbf{Y}_o, \mathbf{R}) \propto \prod_{i=1}^{N} \int_{\mathbf{b}_i, \mathbf{y}_{mi}} f(\mathbf{y}_i|\mathbf{b}_i, \boldsymbol{\theta}) f(\mathbf{r}_i|\mathbf{y}_i, \mathbf{X}_i, \mathbf{Z}_i, \mathbf{b}_i, \boldsymbol{\gamma}) f(\mathbf{b}_i|\boldsymbol{\theta})\, d\mathbf{b}_i\, d\mathbf{y}_{mi}. \tag{1}$$

In general, the integral in (1) does not have a closed form and its dimension is equal to the sum of the dimensions of $\mathbf{b}_i$ and $\mathbf{y}_{mi}$. Here, $f(\mathbf{r}_i|\mathbf{y}_i, \mathbf{X}_i, \mathbf{Z}_i, \mathbf{b}_i, \boldsymbol{\gamma})$ is related to the non-ignorable missingness mechanism. We now consider the selection of a model for the non-ignorable missingness mechanism. Theoretically, any general model can be taken. However, one must be careful not to use a too complicated or large model, since it can easily become unidentifiable. Moreover, a too complex model will also induce difficulty in deriving the corresponding conditional distributions of the missing responses given the observed data, and inefficient sampling from those conditional distributions. Based on the suggestion in Ibrahim et al. (2001), we propose the following model for the non-ignorable missingness mechanism:

$$f(\mathbf{r}_i|\mathbf{y}_i, \mathbf{X}_i, \mathbf{Z}_i, \mathbf{b}_i, \boldsymbol{\gamma}) = \prod_{j=1}^{n_i} \pi_{ij}^{r_{ij}} (1 - \pi_{ij})^{1-r_{ij}}, \tag{2}$$

where $\pi_{ij} = \Pr(r_{ij} = 1|\mathbf{y}_i, \mathbf{x}_{ij}, \mathbf{z}_{ij}, \mathbf{b}_i, \boldsymbol{\gamma})$. Ibrahim et al. (2001) pointed out that since r_{ij} is binary, one can use a sequence of logistic regressions for modeling $\Pr(r_{ij} = 1|\mathbf{y}_i, \mathbf{x}_{ij}, \mathbf{z}_{ij}, \mathbf{b}_i, \boldsymbol{\gamma})$ in (2). They also pointed out that this model has the potential for reducing the number of parameters in the missing data mechanism. Furthermore it yields correlation structures between the r_{ij}'s, allows more flexibility in specifying the missing data model, and facilities efficient sampling from the conditional distribution of the missing response given the observed data. Hence, the following logistic regression model is used:

$$m(\mathbf{y}_i, \mathbf{x}_{ij}, \mathbf{z}_{ij}, \mathbf{b}_i, \boldsymbol{\gamma}) = \text{logit}\{\Pr(r_{ij} = 1|\mathbf{y}_i, \mathbf{x}_{ij}, \mathbf{z}_{ij}, \mathbf{b}_i, \boldsymbol{\gamma})\} = \boldsymbol{\gamma}'\mathbf{F}_{ij}, \tag{3}$$

where $\mathbf{F}_{ij} = (1, \mathbf{y}_i', \mathbf{x}_{ij}', \mathbf{z}_{ij}')'$ and $\boldsymbol{\gamma}$ is the corresponding regression coefficient. Any parameter in the $\boldsymbol{\gamma}$ can be fixed to zero. Hence, the non-ignorable missing mechanism defined in the above is rather flexible. It can be handled in special cases in which $\mathbf{r}_i$ just depends on a subset of entries in $\mathbf{y}_i$, or a subset of entries in $\mathbf{x}_{ij}$ and $\mathbf{z}_{ij}$, or both. Due to the complexity of the latent variable model, the complicated pattern of the missing data, and the presence of the random effects $\mathbf{b}_i$, the integral in (1) does not usually have an analytic form. Therefore, the observed data likelihood is complicated. Hence, it is difficult to calculate local influence measures based on Cook's approach that depends heavily on the observed data likelihood.

2.2. *Local influence analysis*

Consider a perturbation vector $\boldsymbol{\omega} = (\omega_1, \ldots, \omega_m)'$. Let $\boldsymbol{\psi} = (\boldsymbol{\theta}', \boldsymbol{\gamma}')'$, and the observed data log-likelihood $\ell_o(\boldsymbol{\psi}; \mathbf{Y}_o, \mathbf{R}) = \log L_o(\boldsymbol{\psi}; \mathbf{Y}_o, \mathbf{R})$. In Cook's approach (Cook, 1986), the following likelihood-displacement function was considered:

$$LD(\boldsymbol{\omega}) = 2\{\ell_o(\hat{\boldsymbol{\psi}}; \mathbf{Y}_o, \mathbf{R}) - \ell_o(\hat{\boldsymbol{\psi}}_{\omega}^*; \mathbf{Y}_o, \mathbf{R}, \boldsymbol{\omega})\},$$

where $\hat{\boldsymbol{\psi}}_{\omega}^*$ is the vector that maximizes $\ell_o(\boldsymbol{\psi}; \mathbf{Y}_o, \mathbf{R}, \boldsymbol{\omega})$, and the local behavior of $LD(\boldsymbol{\omega})$ is studied by examining the normal curvature of the influence graph $\mathcal{G}_L(\boldsymbol{\omega}) = (\boldsymbol{\omega}', LD(\boldsymbol{\omega}))'$ for developing the local influence measures. As $\ell_o(\hat{\boldsymbol{\psi}}; \mathbf{Y}_o, \mathbf{R})$ is usually very complicated, the Cook approach will encounter serious difficulties.

As pointed out by Ibrahim et al. (2001), the ML estimate $\hat{\boldsymbol{\psi}}$ of $\boldsymbol{\psi}$ can be obtained by applying a Monte Carlo EM (MCEM) type of algorithm on the complete data log-likelihood, $\ell_c(\boldsymbol{\psi}; \mathbf{Y}_m, \boldsymbol{\Omega}, \mathbf{Y}_o, \mathbf{R})$, where $\boldsymbol{\Omega} = (\mathbf{b}_1, \ldots, \mathbf{b}_N)$ is treated as hypothetical missing data. The E-step at the rth iteration of the MCEM algorithm evaluates

$$Q(\boldsymbol{\psi}|\boldsymbol{\psi}^{(r)}) = E\big[\ell_c(\boldsymbol{\psi}; \mathbf{Y}_m, \boldsymbol{\Omega}, \mathbf{Y}_o, \mathbf{R})|\mathbf{Y}_o, \mathbf{R}, \boldsymbol{\psi}^{(r)}\big], \tag{4}$$

where the expectation is taken with respect to the conditional distribution of $(\mathbf{Y}_m, \boldsymbol{\Omega})$ given $\mathbf{Y}_o$ and $\mathbf{R}$ at $\boldsymbol{\psi}^{(r)}$. We derive the local influence measure based on an objective function that is defined by (4). As usual, we assume that there is a null point $\boldsymbol{\omega}^0$ such that $\ell_\omega(\boldsymbol{\psi}; \mathbf{Y}_m, \boldsymbol{\Omega}, \mathbf{Y}_o, \mathbf{R}, \boldsymbol{\omega}^0) = \ell_c(\boldsymbol{\psi}; \mathbf{Y}_m, \boldsymbol{\Omega}, \mathbf{Y}_o, \mathbf{R})$ for all $\boldsymbol{\psi}$. Let $\hat{\boldsymbol{\psi}}(\boldsymbol{\omega})$ be the ML estimate of $\boldsymbol{\psi}$ for the perturbed model which maximizes

$$Q(\boldsymbol{\psi}, \boldsymbol{\omega}|\hat{\boldsymbol{\psi}}) = E\big[\ell_\omega(\boldsymbol{\psi}; \mathbf{Y}_m, \boldsymbol{\Omega}, \mathbf{Y}_o, \mathbf{R}, \boldsymbol{\omega})|\mathbf{Y}_o, \mathbf{R}, \hat{\boldsymbol{\psi}}\big]. \tag{5}$$

Obviously, $\hat{\boldsymbol{\psi}}(\boldsymbol{\omega}^0) = \hat{\boldsymbol{\psi}}$. Inspired by Zhu and Lee (2001), we consider the following Q-displacement function:

$$f_Q(\boldsymbol{\omega}) = 2\big\{Q(\hat{\boldsymbol{\psi}}|\hat{\boldsymbol{\psi}}) - Q\big(\hat{\boldsymbol{\psi}}(\boldsymbol{\omega})|\hat{\boldsymbol{\psi}}\big)\big\}. \tag{6}$$

Similar to Cook (1986), and Zhu and Lee (2001), we measure the influence of an observation on the local behavior of the Q-displacement function, which is defined in terms of its normal or conformal normal curvatures for the small perturbation of $\boldsymbol{\omega}$ from $\boldsymbol{\omega}^0$. We define the following notations: $\ddot{\mathbf{Q}}_{\omega^0} = \partial^2 Q(\hat{\boldsymbol{\psi}}(\boldsymbol{\omega})|\hat{\boldsymbol{\psi}})/\partial\boldsymbol{\omega}\partial\boldsymbol{\omega}'|_{\omega=\omega^0}$, $\ddot{\mathbf{Q}}_{\hat{\psi}} = \partial^2 Q(\boldsymbol{\psi}|\hat{\boldsymbol{\psi}})/\partial\boldsymbol{\psi}\partial\boldsymbol{\psi}'|_{\psi=\hat{\psi}}$ and $\boldsymbol{\Delta}_{\omega^0} = \partial^2 Q(\boldsymbol{\psi}, \boldsymbol{\omega}|\hat{\boldsymbol{\psi}})/\partial\boldsymbol{\psi}\partial\boldsymbol{\omega}'|_{\psi=\hat{\psi},\omega=\omega^0}$. Under some mild regularity conditions, $-\ddot{\mathbf{Q}}_{\hat{\psi}}$ and $-\ddot{\mathbf{Q}}_{\omega^0}$ are semi-positive definite. Based on the reasoning given in Zhu and Lee (2001), the conformal normal curvature $\mathcal{B}_{f_Q,h}$ at $\boldsymbol{\omega}^0$ in the direction of a unit vector $\mathbf{h}$ is given as follows:

$$\mathcal{B}_{f_Q,h} = \frac{-2\mathbf{h}'\ddot{\mathbf{Q}}_{\omega^0}\mathbf{h}}{\mathrm{tr}\{-2\ddot{\mathbf{Q}}_{\omega^0}\}}, \tag{7}$$

where

$$\ddot{\mathbf{Q}}_{\omega^0} = \boldsymbol{\Delta}'_{\omega^0}\ddot{\mathbf{Q}}_{\hat{\psi}}^{-1}\boldsymbol{\Delta}_{\omega^0}. \tag{8}$$

Let $\mathbf{Q} = -2\ddot{\mathbf{Q}}_{\omega^0}/\,\mathrm{tr}\{-2\ddot{\mathbf{Q}}_{\omega^0}\}$, and $\lambda_1 \geqslant \cdots \geqslant \lambda_r > 0$ be the r nonzero eigenvalues of $\mathbf{Q}$, and $\mathbf{e}_1, \ldots, \mathbf{e}_r$ be the corresponding orthogonal eigenvectors. The following aggregate contribution vector (Lesaffre and Verbeke, 1998; Poon and Poon, 1999; Zhu and Lee, 2001) of all eigenvectors that are associated with all nonzero eigenvalues

$$\mathbf{M}(0) = \sum_{i=1}^{r} \lambda_i \mathbf{e}_i^2,$$

where $\mathbf{e}_i^2 = (e_{i1}^2, \ldots, e_{im}^2)'$, is used for assessing local influence. For $j = 1, \ldots, m$, it follows from Zhu and Lee (2001) that the jth component of $\mathbf{M}(0)$, $M(0)_j = q_j$ for $j = 1, \ldots, m$, where q_j is the jth diagonal element of the matrix $\mathbf{Q}$. Therefore, it is very simple to compute q_j, because no eigenfunctions and eigenvalues are involved. The unusual aspects of model input can be detected from the relatively large elements in $\{M(0)_j,\ j = 1, \ldots, m\}$. Let $\overline{M}(0)$ and $SM(0)$ be the mean and standard deviation of $\{M(0)_j,\ j = 1, \ldots, m\}$. We have $\overline{M}(0) = 1/m$ (see Zhu and Lee, 2001), which just depends on m. Therefore, it is reasonable to regard observations whose $M(0)_j$ is significantly larger than $\overline{M}(0) = 1/m$ as influential. Therefore, a relatively large $M(0)_j$ play important roles in determining the influence of single observations. Taking into account the variation of $M(0)_j$, $c_1\overline{M}(0) + c_2 SM(0)$ may be used as a benchmark, where c_1 and c_2 are selected constants. Hence, the jth observation may be regarded as influential if $M(0)_j > c_1\overline{M}(0) + c_2 SM(0)$.

In order to get the conformal normal curvature $\mathcal{B}_{f_Q,h}$, we need to calculate $\boldsymbol{\Delta}_{\omega^0}$ and $\ddot{\mathbf{Q}}_{\hat{\psi}}$ (see (7) and (8)). Due to the existence of the non-ignorable missing data and the random effects, these expectations are evaluated by Monte Carlo integrations. The technical details are given in Appendices A and B.

3. Normal mixed effects model

3.1. Model and local influence analysis

Suppose we use the following normal mixed effects model (Laird and Ware, 1982) to fit the data described in Section 2. For the jth response for the ith subject, it can be modelled as:

$$\begin{aligned} y_{ij} &= \mathbf{x}'_{ij}\boldsymbol{\beta} + \mathbf{z}'_{ij}\mathbf{b}_i + \varepsilon_{ij}, \quad i = 1, \dots, N;\ j = 1, \dots, n_i, \\ \mathbf{b}_i &\sim \mathcal{N}_{s_2}(\mathbf{0}, \boldsymbol{\Sigma}), \\ \varepsilon_{ij} &\sim \mathcal{N}(0, \sigma_f^2), \end{aligned} \tag{9}$$

where $\boldsymbol{\beta}(s_1 \times 1)$ is a vector of the fixed effects associated with covariate vector $\mathbf{x}_{ij}$; $\mathbf{b}_i$ are mutually independent $s_2 \times 1$ subject-specific random effects associated with covariate vector $\mathbf{z}_{ij}$; $\boldsymbol{\Sigma}$ is the $s_2 \times s_2$ covariance matrix, and $\boldsymbol{\Sigma} = \boldsymbol{\Sigma}(\boldsymbol{\sigma})$ depends on $\boldsymbol{\sigma}(s_3 \times 1)$, a vector of unknown variance components. ε_{ij} are mutually independent errors, and $\mathbf{b}_i$ and ε_{ij} are independent. As for the missing data, we use the model given in (2) and (3) to describe the non-ignorable missing mechanism. Let $\boldsymbol{\psi} = (\boldsymbol{\beta}', \sigma_f^2, \boldsymbol{\sigma}', \boldsymbol{\gamma}')'$. Then the complete data log-likelihood apart from a constant is given by:

$$\begin{aligned} &\ell_c(\boldsymbol{\psi}; \mathbf{Y}_m, \boldsymbol{\Omega}, \mathbf{Y}_o, \mathbf{R}) \\ &\quad = \ell_1\big(\boldsymbol{\beta}, \sigma_f^2; \mathbf{Y}_m, \boldsymbol{\Omega}, \mathbf{Y}_o, \mathbf{R}\big) + \ell_2(\boldsymbol{\sigma}; \boldsymbol{\Omega}) + \ell_3(\boldsymbol{\gamma}; \mathbf{Y}_m, \mathbf{Y}_o, \mathbf{R}), \end{aligned} \tag{10}$$

where

$$\begin{aligned} &\ell_1\big(\boldsymbol{\beta}, \sigma_f^2; \mathbf{Y}_m, \boldsymbol{\Omega}, \mathbf{Y}_o, \mathbf{R}\big) \\ &\quad = -\frac{1}{2}\sum_{i=1}^{N}\sum_{j=1}^{n_i}\left\{\log\sigma_f^2 + \frac{1}{\sigma_f^2}\big(y_{ij} - \mathbf{x}'_{ij}\boldsymbol{\beta} - \mathbf{z}'_{ij}\mathbf{b}_i\big)^2\right\}, \end{aligned} \tag{11}$$

$$\ell_2(\boldsymbol{\sigma}; \boldsymbol{\Omega}) = -\frac{1}{2}\sum_{i=1}^{N}\big\{\log|\boldsymbol{\Sigma}| + \mathbf{b}_i^{\mathrm{T}}\boldsymbol{\Sigma}^{-1}\mathbf{b}_i\big\}, \tag{12}$$

$$\ell_3(\boldsymbol{\gamma}; \mathbf{Y}_m, \mathbf{Y}_o, \mathbf{R}) = \sum_{i=1}^{N}\sum_{j=1}^{n_i}\big\{r_{ij}\log\pi_{ij} + (1 - r_{ij})\log(1 - \pi_{ij})\big\}. \tag{13}$$

To study the sensitivity of uncertainties in the data or model, we can proceed by specifying a perturbation scheme via the vector $\boldsymbol{\omega}$. The following are the four useful perturbation schemes which we considered for the normal mixed effects model. To calculate the building blocks for $\ddot{\mathbf{Q}}_{\omega^0}$, we need to calculate $\partial^2 \ell_\omega(\boldsymbol{\psi}; \mathbf{Y}_m, \boldsymbol{\Omega}, \mathbf{Y}_o, \mathbf{R}, \boldsymbol{\omega})/\partial\boldsymbol{\psi}\partial\boldsymbol{\omega}'$ for each perturbation scheme and $\partial^2 \ell_c(\boldsymbol{\psi}; \mathbf{Y}_m, \boldsymbol{\Omega}, \mathbf{Y}_o, \mathbf{R})/\partial\boldsymbol{\psi}\partial\boldsymbol{\psi}'$; see (8). The formulae for these two second derivatives are listed in Appendix C.

(1) *Case weights within clusters*

Without considering the data structure, we are only interested in finding out influential data points in all observations. A useful strategy is to add a weight to every data point. Let $\boldsymbol{\omega}$ be an $I \times 1$ perturbation vector, where $I = \sum_{i=1}^N n_i$. The perturbed complete data log-likelihood $\ell_\omega(\boldsymbol{\psi}; \mathbf{Y}_m, \boldsymbol{\Omega}, \mathbf{Y}_o, \mathbf{R}, \boldsymbol{\omega})$ apart from a constant is given by:

$$\ell_{1\omega}\big(\boldsymbol{\beta}, \sigma_f^2; \mathbf{Y}_m, \boldsymbol{\Omega}, \mathbf{Y}_o, \mathbf{R}, \boldsymbol{\omega}\big) + \ell_2(\boldsymbol{\sigma}; \boldsymbol{\Omega}) + \ell_3(\boldsymbol{\gamma}; \mathbf{Y}_m, \mathbf{Y}_o, \mathbf{R}), \tag{14}$$

where

$$\begin{aligned}&\ell_{1\omega}\big(\boldsymbol{\beta}, \sigma_f^2; \mathbf{Y}_m, \boldsymbol{\Omega}, \mathbf{Y}_o, \mathbf{R}, \boldsymbol{\omega}\big)\\ &\quad= -\frac{1}{2}\sum_{i=1}^N\sum_{j=1}^{n_i}\omega_{ij}\left\{\log\sigma_f^2 + \frac{1}{\sigma_f^2}\big(y_{ij} - \mathbf{x}_{ij}'\boldsymbol{\beta} - \mathbf{z}_{ij}'\mathbf{b}_i\big)^2\right\},\end{aligned}$$

and the other two parts are the same as in (12) and (13). Then $\boldsymbol{\omega}^0 = \mathbf{1}_I$, where $\mathbf{1}_I$ is an $I \times 1$ vector with all elements equal to 1. This is the most popular perturbation scheme in the diagnostic literature.

(2) *Case weights among clusters*

Suppose that we are interested in identifying the clusters which are outlying among other clusters. Now, we consider simultaneous changes in the weights of all clusters via $\boldsymbol{\omega} = (\omega_1, \ldots, \omega_N)'$, an $N \times 1$ vector of weights. The perturbed complete data log-likelihood apart from a constant is similar to (14), but here,

$$\begin{aligned}&\ell_{1\omega}\big(\boldsymbol{\beta}, \sigma_f^2; \mathbf{Y}_m, \boldsymbol{\Omega}, \mathbf{Y}_o, \mathbf{R}, \boldsymbol{\omega}\big)\\ &\quad= -\frac{1}{2}\sum_{i=1}^N\omega_i\left(\sum_{j=1}^{n_i}\left\{\log\sigma_f^2 + \frac{1}{\sigma_f^2}\big(y_{ij} - \mathbf{x}_{ij}'\boldsymbol{\beta} - \mathbf{z}_{ij}'\mathbf{b}_i\big)^2\right\}\right).\end{aligned}$$

In this perturbation scheme, the null point $\boldsymbol{\omega}^0 = \mathbf{1}_N$.

(3) *Multiplicative perturbation on random effects*

Consider a perturbation scheme via an $N \times 1$ vector $\boldsymbol{\omega}$ such that $\mathbf{b}_i(\boldsymbol{\omega}) = \omega_i \otimes \mathbf{b}_i$. In this case, $\boldsymbol{\omega}^0 = (1, \ldots, 1)'$. Ignoring a constant, the perturbed complete data log-likelihood is given by:

$$\ell_{1\omega}\big(\boldsymbol{\beta}, \sigma_f^2; \mathbf{Y}_m, \boldsymbol{\Omega}, \mathbf{Y}_o, \mathbf{R}, \boldsymbol{\omega}\big) + \ell_{2\omega}(\boldsymbol{\sigma}; \boldsymbol{\Omega}, \boldsymbol{\omega}) + \ell_3(\boldsymbol{\gamma}; \mathbf{Y}_m, \mathbf{Y}_o, \mathbf{R}), \tag{15}$$

where

$$\ell_{1\omega}\big(\boldsymbol{\beta}, \sigma_f^2; \mathbf{Y}_m, \boldsymbol{\Omega}, \mathbf{Y}_o, \mathbf{R}\big)$$
$$= -\frac{1}{2}\sum_{i=1}^{N}\sum_{j=1}^{n_i}\left\{\log\sigma_f^2 + \frac{1}{\sigma_f^2}\big(y_{ij} - \mathbf{x}_{ij}'\boldsymbol{\beta} - \mathbf{z}_{ij}'\mathbf{b}_i\omega_i\big)^2\right\},$$

$$\ell_{2\omega}(\boldsymbol{\sigma}; \boldsymbol{\Omega}, \boldsymbol{\omega}) = -\frac{1}{2}\sum_{i=1}^{N}\big\{\log|\boldsymbol{\Sigma}| + \omega_i^2\mathbf{b}_i^{\mathrm{T}}\boldsymbol{\Sigma}^{-1}\mathbf{b}_i\big\},$$

and the other part is the same as in (13). The null point for this case is $\boldsymbol{\omega}^0 = \mathbf{1}_N$, where $\mathbf{1}_N$ is an $N \times 1$ vector with all elements equal to 1.

(4) *Case weights perturbation on missing mechanism*

As we use the logistic regression model to represent the missing mechanism, it is necessary for us to investigate the effect of how minor perturbations on the missing mechanism model can effect a large impact on the key features of the whole model. Let $\boldsymbol{\omega}$ be an $I \times 1$ perturbation vector. The perturbed complete data log-likelihood apart from a constant is given by:

$$\ell_1\big(\boldsymbol{\beta}, \sigma_f^2; \mathbf{Y}_m, \boldsymbol{\Omega}, \mathbf{Y}_o, \mathbf{R}\big) + \ell_2(\boldsymbol{\sigma}; \boldsymbol{\Omega}) + \ell_{3\omega}(\boldsymbol{\gamma}; \mathbf{Y}_m, \mathbf{Y}_o, \mathbf{R}, \boldsymbol{\omega}), \tag{16}$$

where

$$\ell_{3\omega}(\boldsymbol{\gamma}; \mathbf{Y}_m, \mathbf{Y}_o, \mathbf{R}, \boldsymbol{\omega}) = \sum_{i=1}^{N}\sum_{j=1}^{n_i}\omega_{ij}\big[r_{ij}\boldsymbol{\gamma}'\mathbf{F}_{ij} - \log\big(1 + \exp(\boldsymbol{\gamma}'\mathbf{F}_{ij})\big)\big],$$

and the other two parts are the same as in (11) and (12). Then $\boldsymbol{\omega}^0 = \mathbf{1}_I$, where $\mathbf{1}_I$ is an $I \times 1$ vector with all elements equal to 1.

3.2. Simulation study

In the simulation study, we consider the following normal random effects model:

$$y_{ij} = \beta_1 x_{ij1} + \beta_2 x_{ij2} + b_i + \varepsilon_{ij}, \quad i = 1, \ldots, 50, \tag{17}$$

where $x_{ij1} = j - 3$ and $x_{ij2} = 1$ if $i \leqslant 25$ and 0 if otherwise. The data set involves 50 clusters of size $n_i = 7$. To create the non-ignorable missing data, we consider the following logistic regression model:

$$\operatorname{logit}\Pr(r_{ij} = 1|\mathbf{y}_i, \boldsymbol{\gamma}) = \gamma_0 + \gamma_1 y_{i,j-1} + \gamma_2 y_{ij}, \tag{18}$$

for $i = 1, \ldots, 50$; $j = 2, \ldots, n_i$.

The parameters involved in (17) are set as follows: $\boldsymbol{\beta}' = (2.0, 1.0)$, $\varepsilon_{ij} \sim \mathcal{N}(0, \sigma_f^2)$ with $\sigma_f^2 = 0.36$, and the random effects $b_i \sim \mathcal{N}(0, \sigma_b^2)$ with $\sigma_b^2 = 0.25$. The parameters involved in the missing mechanism model (18) are set at $\boldsymbol{\gamma}' = (1.0, 1.0, -1.0)$. Under the above settings, the average rate of missingness is about 30%.

To meet the needs of this paper, we are ready to generate three data sets with outliers. For data set I, we set $b_{50} = 12$ and regenerate $\{y_{ij}\colon i = 50; j = 1, \ldots, 7\}$ according to

Table 1
ML estimates in the simulation for the normal mixed effects model

Para.	MLE		
	Data I	Data II	Data III
β_1	2.008	2.016	2.007
β_2	0.994	0.453	0.985
γ_0	0.516	2.791	0.306
γ_1	0.823	2.983	0.772
γ_2	−0.861	−3.218	−0.761
σ_f^2	0.365	4.301	0.365
σ_b^2	8.280	0.222	0.347

the above normal random effects model. Therefore, the last cluster is an outlying cluster. For data set II, we use $\{b_i = 12\colon i = 46, \ldots, 50\}$ to generate $\{y_{i,n_i}\colon i = 46, \ldots, 50\}$ from the same model. There are five outliers in the second data set. Data set III is generated as follows. To create outliers, for $i = 46, 47, 48, 49, 50$; $j = 7$, we set the missing probabilities of y_{ij} to follow the following model:

$$\operatorname{logit} \Pr(r_{ij} = 1 | \mathbf{y}_i, \boldsymbol{\gamma}) = \gamma_0.$$

That is, for these five responses, their missing probabilities are the constants, which are different from the other responses.

For each of the above three data sets, the ML estimates of unknown parameters were obtained via the MCEM algorithm (Ibrahim et al., 2001), and they are reported in Table 1. To estimate the local influence measures, we ran the Gibbs sampler to collect 52 000 random observations from the joint conditional distribution $[\mathbf{Y}_m, \boldsymbol{\Omega} | \mathbf{Y}_o, \mathbf{R}, \hat{\boldsymbol{\psi}}]$ based on the ML estimates were derived. The first 2000 observations were discarded as the burn-in phase, the last 50 000 random observations were used to calculate $\boldsymbol{\Delta}_{\omega^0}$ and $\ddot{\mathbf{Q}}_\psi(\hat{\boldsymbol{\psi}})$ via the formulae (A.3) and (A.4) in Appendix A.

For data set I, we consider the perturbation schemes 2 and 3. The plots of $M(0)_j$ for these four perturbation schemes are shown in Figures 1 and 2. As we expected, only the last cluster is identified as an outlying cluster in Figures 1 and 2.

For data set II, we consider perturbation schemes 1 and 2. The plots of $M(0)_j$ for these three perturbation schemes are shown in Figures 3 and 4. From Figure 3, we see that five artificial outlying observations are detected and from Figure 4, we also see that the clusters with outlying observations are identified.

For data set III, we consider the perturbation on the missing mechanism (perturbation scheme 4). The plots of $M(0)_j$ are shown in Figure 5. As expected, only the five responses with different missing mechanisms are identified as outliers.

3.3. *Real example: the renal data*

The data set in relation to this longitudinal study is obtained from 131 patients with IgA nephropathy. The resulting variable is the patients' serum levels of creatinine

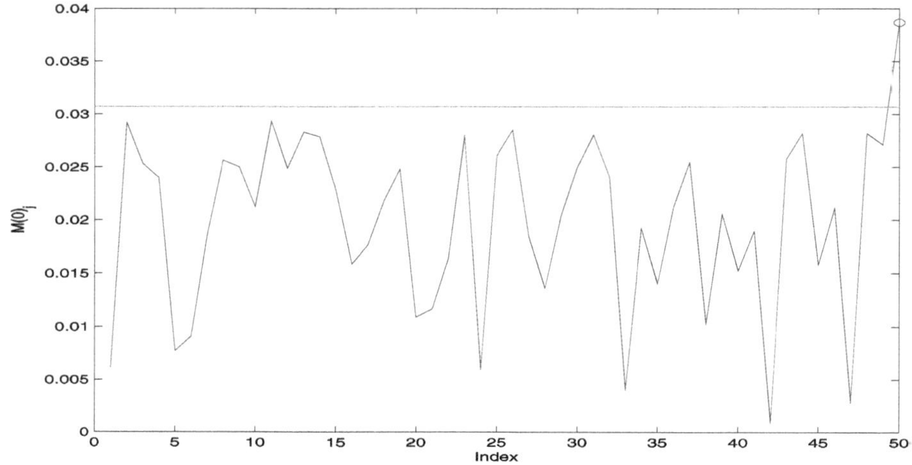

Fig. 1. Index plots of $M(0)_j$ and benchmark (—) for case weights among clusters: Artificial data I for normal mixed effects model.

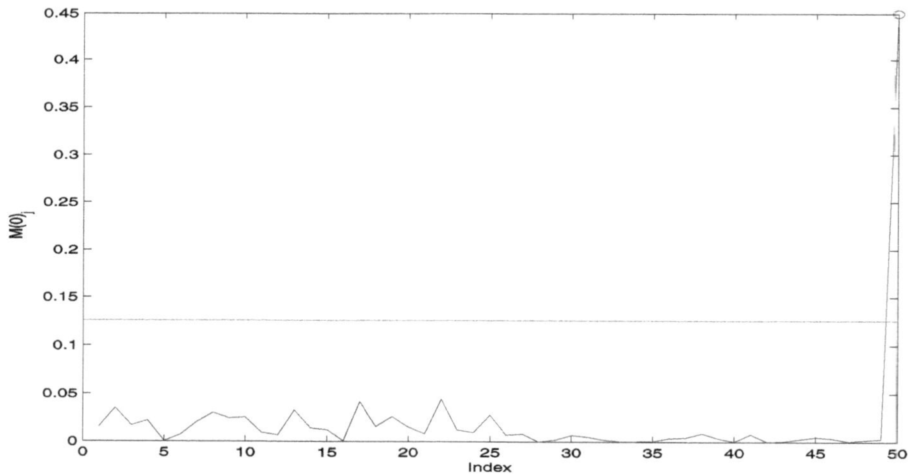

Fig. 2. Index plots of $M(0)_j$ and benchmark (—) for multiplicative perturbation on random effects: Artificial data I for normal mixed effects model.

(creat) which best reflect the renal kidney function. The covariates include cortex (co), glomerular grade (gg), tubulointerstitial grade (tig), sex (= 0 for 'male', and = 1 for 'female'), total protein in 24-hour urine (24utp), and serum calcium (ca). Let $y_{ij} = \log \mathtt{creat}_{ij}$ be the jth observation for the ith patient. For this data set, we consider the following normal random effects model:

$$
\begin{aligned}
y_{ij} &= \beta_0 + \beta_1 \mathtt{co}_i + \beta_2 \mathtt{gg}_i + \beta_3 \mathtt{tig}_i + \beta_4 \mathtt{sex}_i + \beta_5 \mathtt{24utp}_{ij} \\
&\quad + \beta_6 \mathtt{ca}_{ij} + b_i + \varepsilon_{ij}, \quad i = 1, \ldots, 131,
\end{aligned}
$$

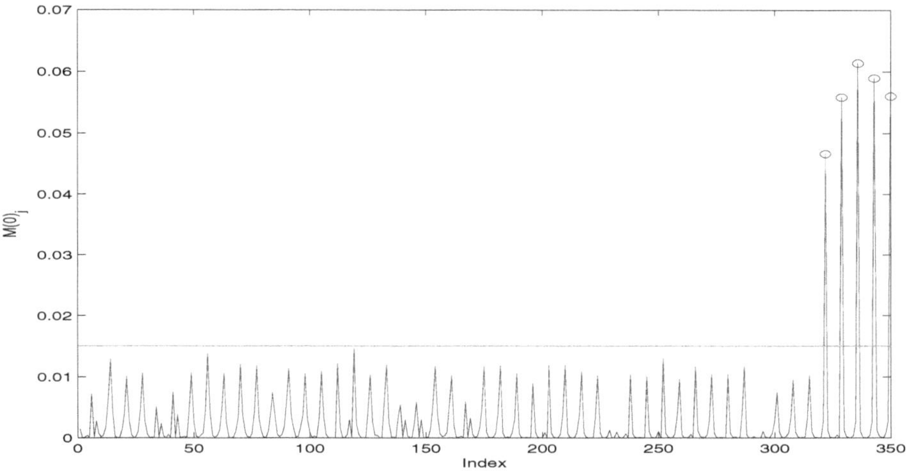

Fig. 3. Index plots of $M(0)_j$ and benchmark (—) for case weights within clusters: Artificial data II for normal mixed effects model.

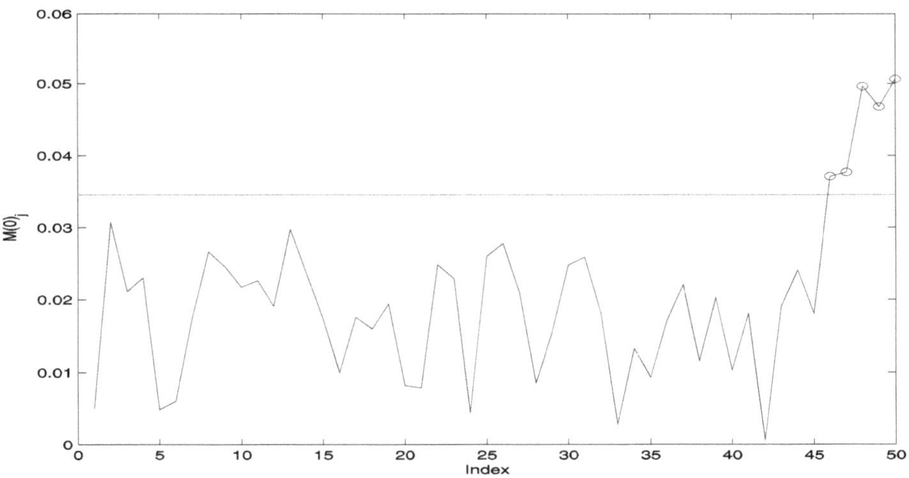

Fig. 4. Index plots of $M(0)_j$ and benchmark (—) for case weights among clusters: Artificial data II for normal mixed effects model.

where $b_i \sim \mathcal{N}(0, \sigma_b^2)$ and $\varepsilon_{ij} \sim \mathcal{N}(0, \sigma_f^2)$. To cope with the missing responses in this data set, we use the following logistic regression to model the missing mechanism:

$$\begin{aligned}\text{logit}\,\Pr(r_{ij} = 1|\mathbf{y}_i, \boldsymbol{\gamma}) = \gamma_0 + \gamma_1 \mathtt{y}_{i,j-1} + \gamma_2 \mathtt{y}_{i,j} + \gamma_3 \mathtt{co}_i + \gamma_4 \mathtt{gg}_i + \gamma_5 \mathtt{tig}_i \\ + \gamma_6 \mathtt{sex}_i + \gamma_7 \mathtt{24utp}_{ij} + \gamma_8 \mathtt{ca}_{ij}.\end{aligned} \quad (19)$$

The ML estimates of the model parameters obtained via the MCEM algorithm (Ibrahim et al., 2001) are reported in Table 2. To estimate the local influence measures, we used

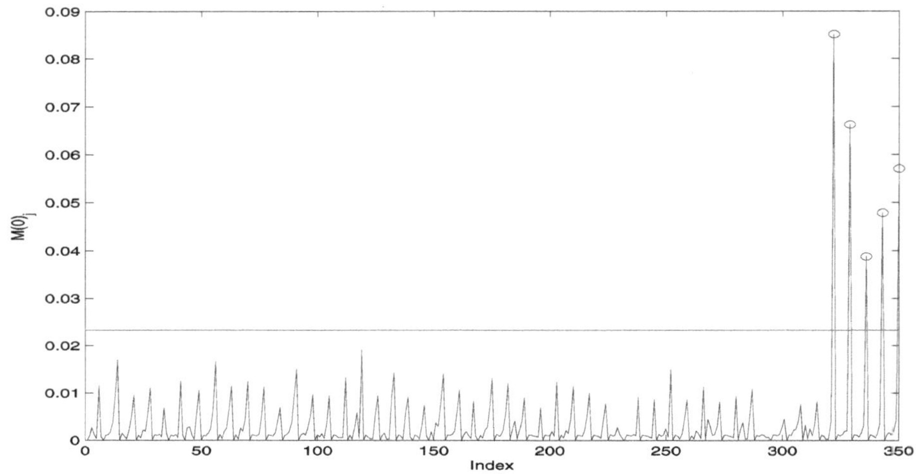

Fig. 5. Index plots of $M(0)_j$ and benchmark (—) for case weights perturbation on missing mechanism: Artificial data III for normal mixed effects model.

Table 2
Results for the renal data

Para.	MLE	SE	Para.	MLE	SE
β_0	3.799	0.037	γ_0	1.136	0.341
β_1	−0.017	0.003	γ_1	−0.188	0.031
β_2	0.134	0.012	γ_2	−0.142	0.072
β_3	0.344	0.012	γ_3	−0.091	0.023
β_4	0.242	0.016	γ_4	0.401	0.026
β_5	−0.032	0.009	γ_5	−0.601	0.036
β_6	0.034	0.015	γ_6	−0.643	0.039
			γ_7	−1.438	0.313
σ_f^2	0.122	0.005			
σ_b^2	0.124	0.012			

the Gibbs sampler to collect 52 000 random observations from the joint conditional distribution $[\mathbf{Y}_m, \boldsymbol{\Omega}|\mathbf{Y}_o, \mathbf{R}, \hat{\boldsymbol{\psi}}]$ based on the ML estimates were derived. After discarding the first 2000 observations as burn-in phase, the last 50 000 random observations were used to calculate $\boldsymbol{\Delta}_{\omega^0}$ and $\ddot{\mathbf{Q}}_{\psi}(\hat{\boldsymbol{\psi}})$ via the formulae (A.3) and (A.4) given in Appendix A.

The four perturbation schemes given in Section 3.1 are considered. Plots of $M(0)_j$ for case weights perturbation within patients (perturbation scheme 1) are shown in Figure 6. From this figure, we know that 14 observations stand out as the influential or potential outliers, and they are (2, 7), (9, 8), (49, 9), (62, 5), (66, 3), (88, 2), (88, 3), (88, 4), (99, 2), (108, 9), (112, 2), (112, 3), (113, 2), and (127, 9), in which the first entry in the bracket represents the patient's number, and the second entry denotes the ob-

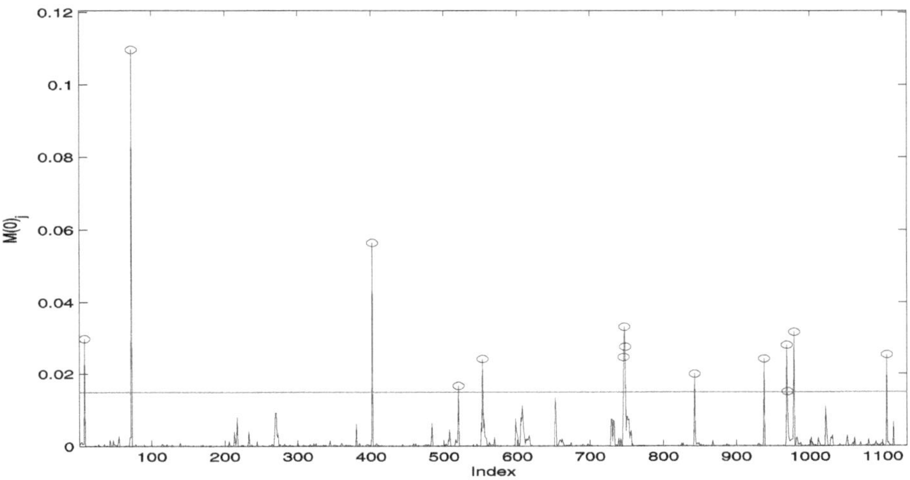

Fig. 6. Index plots of $M(0)_j$ and benchmark (—) for case weights within clusters: Renal data.

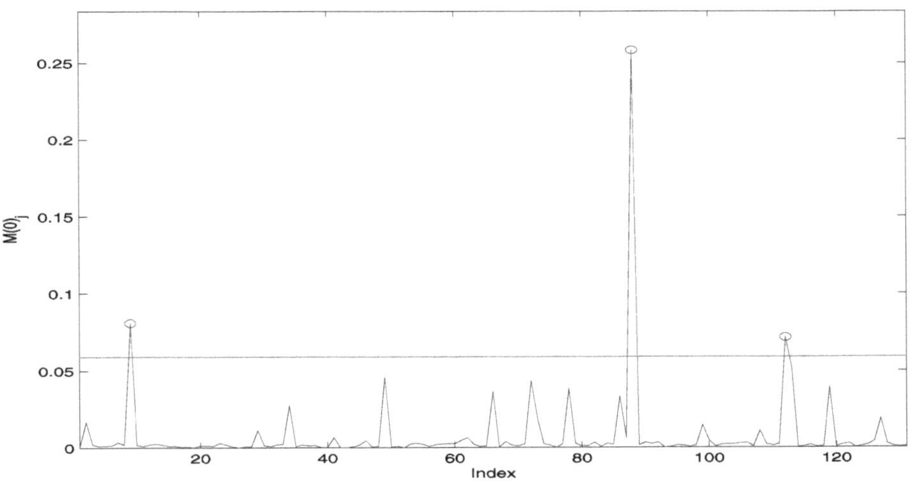

Fig. 7. Index plots of $M(0)_j$ and benchmark (—) for case weights among clusters: Renal data.

servation's number. We find out that these data points are significantly larger or smaller than the other observations in the data set. For example, the seventh observation of the second patient (2, 7) is 146, but the other observations of this patient are from 998 to 1358. Therefore, we should pay more attention to this patient; specifically, why the patient's serum levels of creatinine change so sharply should be studied.

To identify the influential clusters, we considered the case weights perturbation among clusters (perturbation scheme 2). Plots of $M(0)_j$ associated with this perturbation scheme are presented in Figure 7. There are three patients who were detected as influential. They are the 9th, 88th, and 112th patients. From the above results, we can

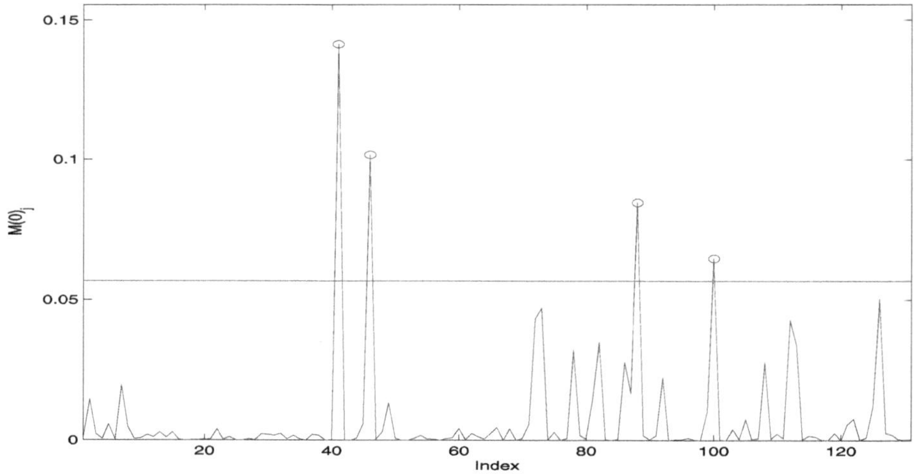

Fig. 8. Index plots of $M(0)_j$ and benchmark (—) for multiplicative perturbation on random effects: Renal data.

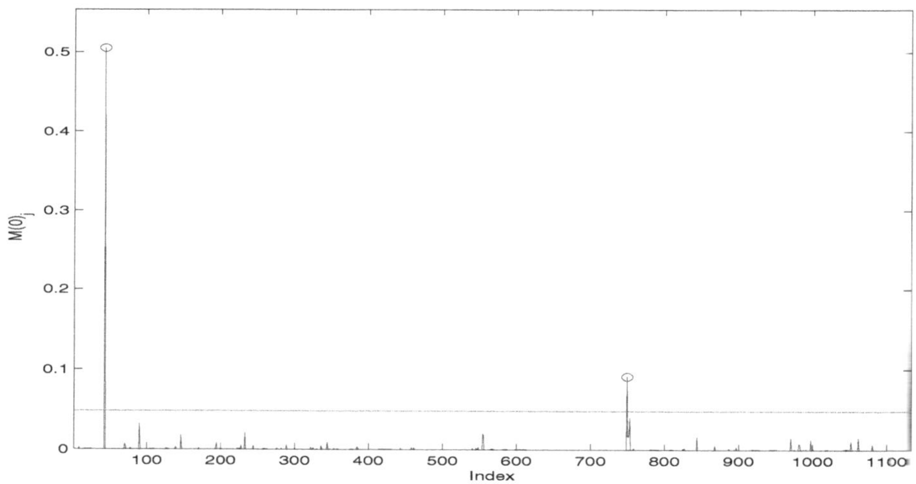

Fig. 9. Index plots of $M(0)_j$ and benchmark (—) for case weights perturbation on missing mechanism: Renal data.

see that some patients with influential observations can also be detected as influential clusters by the perturbation scheme 2. The 9th patient has the most influential observation (9, 8) which has the largest $M(0)_j$ in the perturbation scheme 1. Meanwhile, the 88th and 112th patients who have more than one influential observations.

To study the effects of departure from the assumption that $b_i \sim \mathcal{N}(0, \sigma_b^2)$, we considered the multiplicative perturbation on random effects (perturbation scheme 3). The diagnostic measures are presented in Figure 8 which also shows us that the 41st, 46th,

88th, and 100th patients are influential clusters. That is to say that about 3% of the observations (i.e., four among 131 patients) are more sensitive to the variance σ_b^2. Therefore, it seems that the assumption of homogeneity of the random effects is reasonable.

To investigate the sensitivity of the missing mechanism, we also considered perturbation scheme 4 for this data set. The diagnostic measures of this perturbation scheme are plotted in Figure 9. From this figure, we see that there are only two responses which are more influential. They are the 9th observation of the 6th patient, and the 5th observation of the 88th patient. Therefore, almost all responses are robust to missing model (19).

4. Generalized linear mixed model

4.1. Model and local influence analysis

In this section, we use another specific latent variable model to fit the data which are described in Section 2. It is assumed that conditional on a latent vector $\mathbf{b}_i\,(s_2 \times 1)$, y_{ij} follows an exponential family distribution (see, McCullagh and Nelder, 1989) of the following form:

$$f(y_{ij}|\boldsymbol{\beta}, \mathbf{b}_i, \phi) = \exp\bigl[\phi\{y_{ij}\theta_{ij} - a(\theta_{ij})\} + c(y_{ij}, \phi)\bigr], \tag{20}$$

where ϕ is a scalar dispersion parameter, and $\theta(\cdot)$ is a link function. The conditional mean and conditional variance of y_{ij} given $\mathbf{b}_i$ are respectively $E(y_{ij}|\mathbf{b}_i) = \mu_{ij} = \dot{a}(\theta_{ij})$ and $\operatorname{var}(y_{ij}|\mathbf{b}_i) = \ddot{a}(\theta_{ij})/\phi$, where $\dot{a}(u) = \mathrm{d}a/\mathrm{d}u$ and $\ddot{a}(u) = \mathrm{d}^2a/\mathrm{d}u^2$. The GLMMs are defined by (20) and the systematic component

$$g(\mu_{ij}) = \eta_{ij} = \mathbf{x}_{ij}'\boldsymbol{\beta} + \mathbf{z}_{ij}'\mathbf{b}_i, \quad \text{or} \quad \theta_{ij} = k\bigl(\mathbf{x}_{ij}'\boldsymbol{\beta} + \mathbf{z}_{ij}'\mathbf{b}_i\bigr), \tag{21}$$

where $\boldsymbol{\beta} = (\beta_1, \ldots, \beta_{s_1})'$ is a vector of regression coefficients, and $k(\cdot)$ and $g(\cdot)$ are known continuous differentiable functions. Moreover, $k(\cdot)$ and $g(\cdot)$ satisfy $k(u) = \dot{a}^{-1}(g^{-1}(u))$, where $\dot{a}^{-1}(\cdot)$ and $g^{-1}(\cdot)$ are the inverse functions of $\dot{a}(\cdot)$ and $g(\cdot)$, respectively. The distribution of $\mathbf{b}_i$ is assumed to be normal $\mathcal{N}_{s_2}(\mathbf{0}, \boldsymbol{\Sigma})$, where $\boldsymbol{\Sigma} = \boldsymbol{\Sigma}(\boldsymbol{\sigma})$ depends on $\boldsymbol{\sigma}\,(s_3 \times 1)$, which is a vector of unknown variance components. To deal with the missing data problem, we also use the model given in (2) and (3) to describe the non-ignorable missing mechanism. Let $\boldsymbol{\psi} = (\boldsymbol{\beta}', \phi, \boldsymbol{\sigma}', \boldsymbol{\gamma}')'$. Then the complete data log-likelihood apart from a constant is given by:

$$\begin{aligned}
&\ell_c(\boldsymbol{\psi}; \mathbf{Y}_m, \boldsymbol{\Omega}, \mathbf{Y}_o, \mathbf{R}) \\
&\quad = \ell_1(\boldsymbol{\beta}, \phi; \mathbf{Y}_m, \boldsymbol{\Omega}, \mathbf{Y}_o, \mathbf{R}) + \ell_2(\boldsymbol{\sigma}; \boldsymbol{\Omega}) + \ell_3(\boldsymbol{\gamma}; \mathbf{Y}_m, \mathbf{Y}_o, \mathbf{R}),
\end{aligned} \tag{22}$$

where

$$\ell_1(\boldsymbol{\beta}, \phi; \mathbf{Y}_m, \boldsymbol{\Omega}, \mathbf{Y}_o, \mathbf{R}) = \sum_{i=1}^{N}\sum_{j=1}^{n_i}\bigl[\phi\{y_{ij}\theta_{ij} - a(\theta_{ij})\} + c(y_{ij}, \phi)\bigr], \tag{23}$$

$$\ell_2(\boldsymbol{\sigma}; \boldsymbol{\Omega}) = -\frac{1}{2}\sum_{i=1}^{N}\bigl\{\log|\boldsymbol{\Sigma}| + \mathbf{b}_i'\boldsymbol{\Sigma}^{-1}\mathbf{b}_i\bigr\}, \tag{24}$$

$$\ell_3(\boldsymbol{\gamma}; \mathbf{Y}_m, \mathbf{Y}_o, \mathbf{R}) = \sum_{i=1}^{N}\sum_{j=1}^{n_i}\{r_{ij}\log\pi_{ij} + (1-r_{ij})\log(1-\pi_{ij})\}. \tag{25}$$

Similar to Section 3.1, we also considered the following four perturbation schemes for this model:

(1) *Case weights within clusters*

Let $\boldsymbol{\omega}$ be an $I \times 1$ perturbation vector, where $I = \sum_{i=1}^{N} n_i$. The perturbed complete data log-likelihood, $\ell_\omega(\boldsymbol{\psi}; \mathbf{Y}_m, \boldsymbol{\Omega}, \mathbf{Y}_o, \mathbf{R}, \boldsymbol{\omega})$, ignoring a constant is given by:

$$\ell_{1\omega}(\boldsymbol{\beta}, \phi; \mathbf{Y}_m, \boldsymbol{\Omega}, \mathbf{Y}_o, \mathbf{R}, \boldsymbol{\omega}) + \ell_2(\boldsymbol{\sigma}; \boldsymbol{\Omega}) + \ell_3(\boldsymbol{\gamma}; \mathbf{Y}_m, \mathbf{Y}_o, \mathbf{R}), \tag{26}$$

where

$$\ell_{1\omega}(\boldsymbol{\beta}, \phi; \mathbf{Y}_m, \boldsymbol{\Omega}, \mathbf{Y}_o, \mathbf{R}, \boldsymbol{\omega}) = \sum_{i=1}^{N}\sum_{j=1}^{n_i}\omega_{ij}\big[\phi\{y_{ij}\theta_{ij} - a(\theta_{ij})\} + c(y_{ij}, \phi)\big],$$

and the other two parts are the same as in (24) and (25).

(2) *Case weights among clusters*

Let $\boldsymbol{\omega} = (\omega_1, \ldots, \omega_N)'$, then the perturbed complete data log-likelihood apart from a constant is similar to (26), but here:

$$\ell_{1\omega}(\boldsymbol{\beta}, \phi; \mathbf{Y}_m, \boldsymbol{\Omega}, \mathbf{Y}_o, \mathbf{R}, \boldsymbol{\omega}) = \sum_{i=1}^{N}\omega_i\left\{\sum_{j=1}^{n_i}\big[\phi\{y_{ij}\theta_{ij} - a(\theta_{ij})\} + c(y_{ij}, \phi)\big]\right\}.$$

(3) *Multiplicative perturbation on random effects*

Consider a perturbation scheme via an $N \times 1$ vector $\boldsymbol{\omega}$ such that $\mathbf{b}_i(\boldsymbol{\omega}) = \omega_i \otimes \mathbf{b}_i$. The perturbed complete data log-likelihood apart from a constant is given by:

$$\ell_{1\omega}(\boldsymbol{\beta}, \phi; \mathbf{Y}_m, \boldsymbol{\Omega}, \mathbf{Y}_o, \mathbf{R}, \boldsymbol{\omega}) + \ell_{2\omega}(\boldsymbol{\sigma}; \boldsymbol{\Omega}, \boldsymbol{\omega}) + \ell_3(\boldsymbol{\gamma}; \mathbf{Y}_m, \mathbf{Y}_o, \mathbf{R}), \tag{27}$$

where

$$\ell_{1\omega}(\boldsymbol{\beta}, \phi; \mathbf{Y}_m, \boldsymbol{\Omega}, \mathbf{Y}_o, \mathbf{R}, \boldsymbol{\omega}) = \sum_{i=1}^{N}\sum_{j=1}^{n_i}\big[\phi\{y_{ij}k(\mathbf{x}_{ij}'\boldsymbol{\beta} + \mathbf{z}_{ij}'\mathbf{b}_i\omega_i) - a\big(k(\mathbf{x}_{ij}'\boldsymbol{\beta} + \mathbf{z}_{ij}'\mathbf{b}_i\omega_i)\big)\} + c(y_{ij}, \phi)\big].$$

$$\ell_{2\omega}(\boldsymbol{\sigma}; \boldsymbol{\Omega}, \boldsymbol{\omega}) = -\frac{1}{2}\sum_{i=1}^{N}\{\log|\boldsymbol{\Sigma}| + \omega_i^2\mathbf{b}_i'\boldsymbol{\Sigma}^{-1}\mathbf{b}_i\},$$

and the other part is the same as in (25).

(4) *Case weights perturbation on missing mechanism*

Let $\boldsymbol{\omega}$ be an $I \times 1$ perturbation vector. The perturbed complete data log-likelihood apart from a constant is given by:

$$\ell_1(\boldsymbol{\beta}, \sigma^2; \mathbf{Y}_m, \boldsymbol{\Omega}, \mathbf{Y}_o, \mathbf{R}) + \ell_2(\boldsymbol{\sigma}; \boldsymbol{\Omega}) + \ell_{3\omega}(\boldsymbol{\gamma}; \mathbf{Y}_m, \mathbf{Y}_o, \mathbf{R}, \boldsymbol{\omega}), \tag{28}$$

where

$$\ell_{3\omega}(\boldsymbol{\gamma}; \mathbf{Y}_m, \mathbf{Y}_o, \mathbf{R}, \boldsymbol{\omega}) = \sum_{i=1}^{N} \sum_{j=1}^{n_i} \omega_{ij} \big[r_{ij} \boldsymbol{\gamma}' \mathbf{F}_{ij} - \log\big(1 + \exp(\boldsymbol{\gamma}' \mathbf{F}_{ij})\big)\big],$$

and the other two parts are the same as in (23) and (24).

To get the conformal normal curvature $\mathcal{B}_{f_Q,\mathrm{h}}$, we need to calculate the second derivatives $\partial^2 \ell_\omega(\boldsymbol{\psi}; \mathbf{Y}_m, \boldsymbol{\Omega}, \mathbf{Y}_o, \mathbf{R}, \boldsymbol{\omega})/\partial \boldsymbol{\psi} \partial \boldsymbol{\omega}'$ and $\partial^2 \ell_c(\boldsymbol{\psi}; \mathbf{Y}_m, \boldsymbol{\Omega}, \mathbf{Y}_o, \mathbf{R})/\partial \boldsymbol{\psi} \partial \boldsymbol{\psi}'$; see (7) and (8). The formulae for these two second derivatives are given in Appendix C.

4.2. *Artificial example*

In this artificial example, we follow the design of Zeger and Karim (1991) and we consider the following model:

$$\begin{aligned} &\operatorname{logit} \Pr(\mathrm{y}_{ij} = 1 | \mathbf{b}_i) \\ &\quad = \beta_0 + \beta_1 t_j + \beta_2 x_i + \beta_3 x_i t_j + b_{i1} + b_{i1} t_j, \quad i = 1, \ldots, 100, \end{aligned} \tag{29}$$

where y_{ij} is a conditionally independent binary observation, $x_i = 1$ for half of the samples $x_i = 0$ for the other half, and $t_j = j - 4$ for $j = 1, \ldots, 7$. The data set involves 100 clusters of size $n_i = 7$. To create the non-ignorable missing data, the logistic regression model is used to model the missing probability of y_{ij}:

$$\operatorname{logit} \Pr(r_{ij} = 1 | \mathbf{y}_i, \boldsymbol{\gamma}) = \gamma_0 + \gamma_1 \mathrm{y}_{i,j-1} + \gamma_2 \mathrm{y}_{ij}, \tag{30}$$

for $i = 1, \ldots, 100$; $j = 2, \ldots, n_i$.

The fixed effects coefficients are set at $\boldsymbol{\beta}' = (-2.5, 1.0, -1.0, -0.5)$ while the random effects (b_{i1}, b_{i2}) are generated as a series of 100 independent and identically distributed normal variables with mean $\mathbf{0}$ and covariance matrix $\boldsymbol{\Sigma} = \operatorname{diag}(0.5, 0.25)$, a 2×2 diagonal matrix. The parameters involved in the missing model (30) are set at $\boldsymbol{\gamma}' = (-1.0, 1.0, 1.0)$. Under the above settings, the average rate of missingness is about 30%.

Three data sets with artificial outliers are created to illustrate the performance of the method proposed in this paper. In data set I, we use $x_i t_7 + 10, i = 1, \ldots, 5$, as the covariate for β_3 to generate $\{\mathrm{y}_{i7}: i = 1, \ldots, 5\}$ from the same model. There are five outliers in this data set. For data set II, we use $\{b_{ij} = 10: i = 1, 2; j = 1, 2\}$ to generate $\{\mathbf{y}_i: i = 1, 2\}$. Hence, there are two artificial outlying clusters. Data set III is generated as follows. To create outliers, for $i = 96, \ldots, 100$; $j = 7$, we set the missing probabilities of $\mathbf{y}_{ij}$ to follow the following model:

$$\operatorname{logit} \Pr(r_{ij} = 1 | \mathbf{y}_i, \boldsymbol{\gamma}) = \gamma_0.$$

Table 3
ML estimates in the simulation for the generalized linear mixed model

Para.	MLE		
	Data I	Data II	Data III
β_0	−2.726	−2.866	−2.801
β_1	1.090	1.116	1.113
β_2	−0.920	−0.588	−0.917
β_3	−0.724	−0.720	−0.768
γ_0	−0.935	−0.92	−0.924
γ_1	0.150	0.326	0.123
γ_2	0.956	0.804	0.862
σ_{11}	0.331	0.445	0.365
σ_{22}	0.117	0.116	0.118

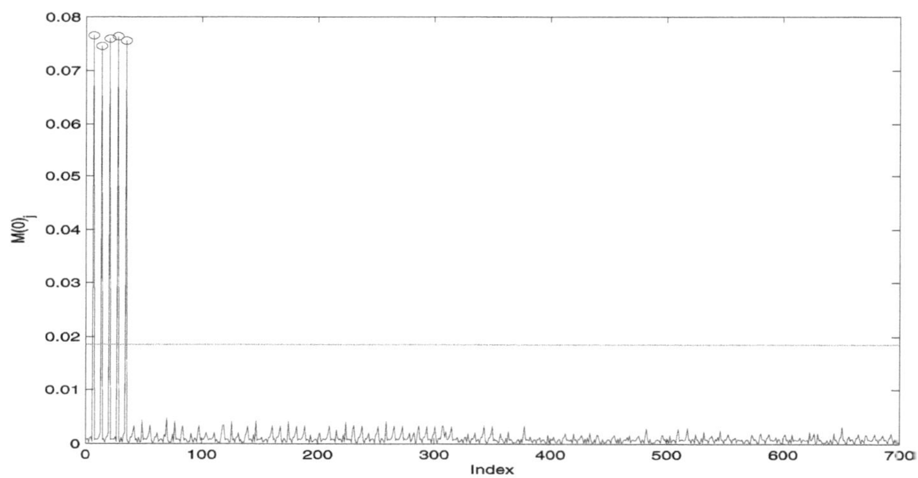

Fig. 10. Index plots of $M(0)_j$ and benchmark (—) for case weights within clusters: Artificial data I for GLMMs.

Therefore, the missing mechanism for these five responses is missing completely at random, which are different from the other responses.

The ML estimates of unknown parameters were obtained via the MCEM algorithm proposed by Ibrahim et al. (2001). These estimates are reported in Table 3 for each of the above-simulated data sets. Based on the ML estimates, 52 000 random observations were simulated from the joint conditional distribution $[\mathbf{Y}_m, \boldsymbol{\Omega}|\mathbf{Y}_o, \mathbf{R}, \hat{\boldsymbol{\psi}}]$ via the Gibbs sampler (see Appendices A and B). The first 2000 observations were discarded as burn-in phase, while the last 50 000 random observations were used to calculate the building blocks, $\boldsymbol{\Delta}_{\omega 0}$ and $\ddot{\mathbf{Q}}_{\psi}(\hat{\boldsymbol{\psi}})$, via the formulae (A.3) and (A.4) in Appendix A.

For data set I, we consider perturbation schemes 1 and 2. The plots of $M(0)_j$ for these four perturbation schemes are shown in Figures 10 and 11. As we expected, only the five created outliers stand out with significantly large local influence measures in

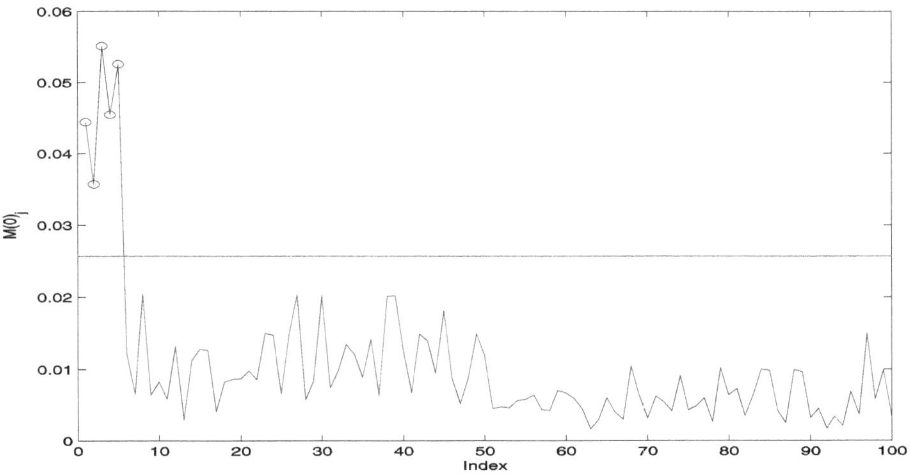

Fig. 11. Index plots of $M(0)_j$ and benchmark (—) for case weights among clusters: Artificial data I for GLMMs.

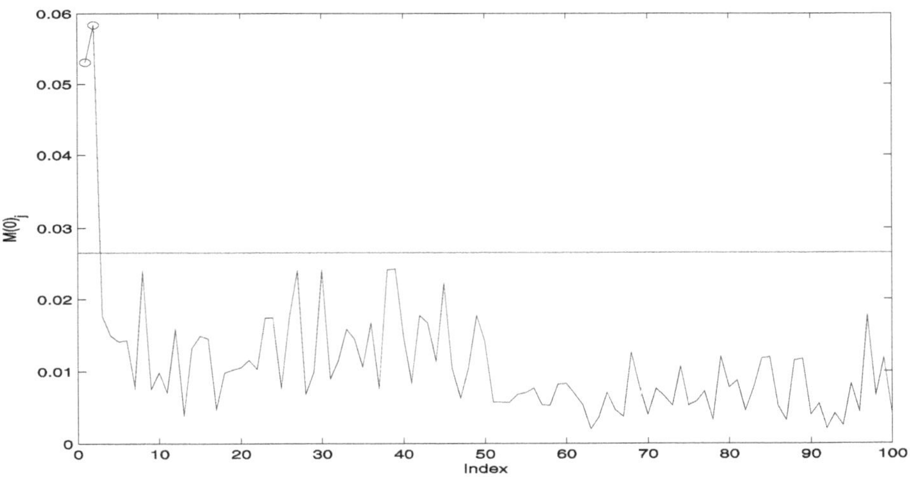

Fig. 12. Index plots of $M(0)_j$ and benchmark (—) for case weights among clusters: Artificial data II for GLMMs.

Figure 10. From Figure 11, we also see that the clusters with outlying observations are identified.

For data set II, we consider perturbation schemes 2 and 3. The plots of $M(0)_j$ for these two perturbation schemes are shown in Figures 12 and 13. From these two figures, we see that only two artificial outlying clusters are detected.

For data set III, we consider the perturbation on the missing mechanism (perturbation scheme 4). The plots of $M(0)_j$ are shown in Figure 14. As expected, only the five responses with different missing mechanisms are identified as outliers.

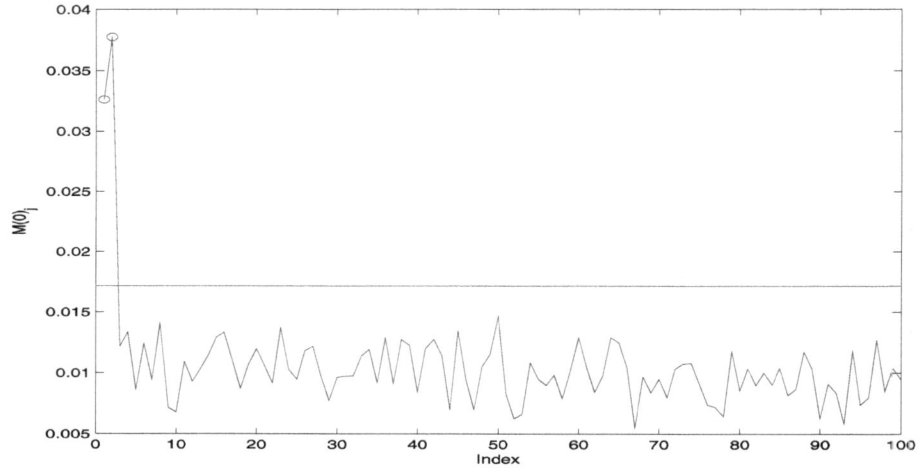

Fig. 13. Index plots of $M(0)_j$ and benchmark (—) for multiplicative perturbation on random effects: Artificial data II for GLMMs.

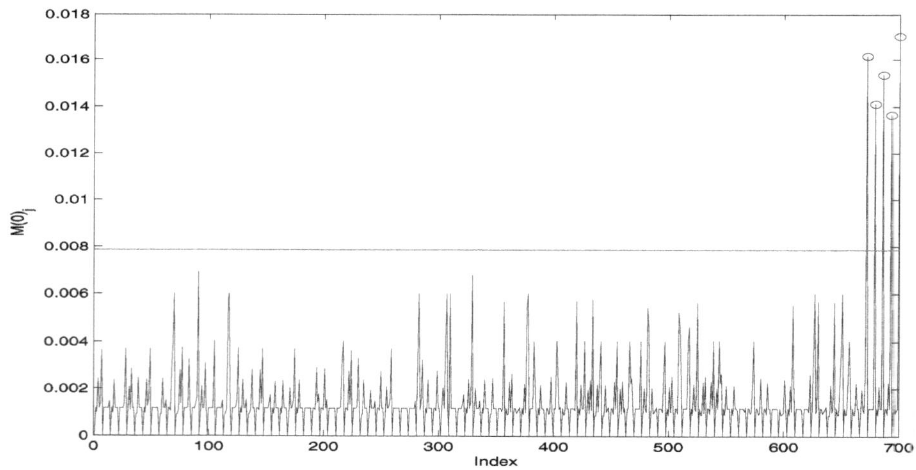

Fig. 14. Index plots of $M(0)_j$ and benchmark (—) for case weights perturbation on missing mechanism: Artificial data III for GLMMs.

5. Conclusion

Due to the complexity of the observed data log-likelihood functions of the model considered in this paper, it is very difficult to obtain influence measures based on Cook's approach (Cook, 1986). To overcome this difficulty, we developed the influence measures on the basis of the Q-displacement function instead of the troublesome observed data log-likelihood function. The results obtained from our simulation studies indicate that this method works very well, and it can detect influential observations efficiently.

Appendix A

To get the basic building blocks of local influence measures, we need to derive expressions for $\boldsymbol{\Delta}_{\omega^0}$ and $\ddot{\mathbf{Q}}_{\hat{\psi}}$ (see (8)). Assuming the legitimacy of interchange of integration and differentiation, we have

$$\ddot{\mathbf{Q}}_{\hat{\psi}} = E\left[\frac{\partial^2 \ell_c(\boldsymbol{\psi}; \mathbf{Y}_m, \boldsymbol{\Omega}, \mathbf{Y}_o, \mathbf{R})}{\partial\boldsymbol{\psi}\partial\boldsymbol{\psi}'}\bigg|\mathbf{Y}_o, \mathbf{R}, \hat{\boldsymbol{\psi}}\right]\bigg|_{\boldsymbol{\psi}=\hat{\boldsymbol{\psi}}}, \tag{A.1}$$

$$\boldsymbol{\Delta}_{\omega^0} = E\left[\frac{\partial^2 \ell_\omega(\boldsymbol{\psi}; \mathbf{Y}_m, \boldsymbol{\Omega}, \mathbf{Y}_o, \mathbf{R}, \boldsymbol{\omega})}{\partial\boldsymbol{\psi}\partial\boldsymbol{\omega}'}\bigg|\mathbf{Y}_o, \mathbf{R}, \hat{\boldsymbol{\psi}}\right]\bigg|_{\boldsymbol{\psi}=\hat{\boldsymbol{\psi}}, \boldsymbol{\omega}=\boldsymbol{\omega}^0}. \tag{A.2}$$

Due to the existence of the real missing data, it is difficult to evaluate directly the conditional expectations of the second derivatives in (A.1) and (A.2). Inspired by the idea given in (Wei and Tanner, 1990), we can overcome this difficulty via the Monte Carlo approximation. Therefore, a sufficiently large number of observations need to be generated from the conditional distribution of $(\mathbf{Y}_m, \boldsymbol{\Omega})$ given $\mathbf{Y}_o$, $\mathbf{R}$ and $\hat{\boldsymbol{\psi}}$. The Gibbs sampler (Geman and Geman, 1984) can be used for this aim. The basic algorithm of the Gibbs sampler is briefly given as below: At the jth iteration with current values $\mathbf{Y}_m^j$ and $\boldsymbol{\Omega}^j$:

Step (a): Generate $\mathbf{Y}_m^{j+1}$ from $[\mathbf{Y}_m|\boldsymbol{\Omega}^j, \mathbf{Y}_o, \mathbf{R}, \hat{\boldsymbol{\psi}}]$;
Step (b): Generate $\boldsymbol{\Omega}^{j+1}$ from $[\boldsymbol{\Omega}|\mathbf{Y}_m^{j+1}, \mathbf{Y}_o, \mathbf{R}, \hat{\boldsymbol{\psi}}]$.

The details of how to simulate the values from the above two full conditional distributions are given in Appendix B.

Let $\{(\mathbf{Y}_m^j, \boldsymbol{\Omega}^j); j = 1, \ldots, J\}$ be a sample randomly drawn from the joint conditional distribution $[\mathbf{Y}_m, \boldsymbol{\Omega}|\mathbf{Y}_o, \mathbf{R}, \hat{\boldsymbol{\psi}}]$, the building blocks can be approximated by

$$\ddot{\mathbf{Q}}_{\hat{\psi}} \approx \frac{1}{J}\sum_{j=1}^{J}\frac{\partial^2 \ell_c(\boldsymbol{\psi}; \mathbf{Y}_o, \mathbf{Y}_m^j, \boldsymbol{\Omega}^j, \mathbf{R})}{\partial\boldsymbol{\psi}\partial\boldsymbol{\psi}'}\bigg|_{\boldsymbol{\psi}=\hat{\boldsymbol{\psi}}}, \tag{A.3}$$

$$\boldsymbol{\Delta}_{\omega^0} \approx \frac{1}{J}\sum_{j=1}^{J}\frac{\partial^2 \ell_\omega(\boldsymbol{\psi}; \mathbf{Y}_o, \mathbf{Y}_m^j, \boldsymbol{\Omega}^j, \mathbf{R}, \boldsymbol{\omega})}{\partial\boldsymbol{\psi}\partial\boldsymbol{\omega}'}\bigg|_{\boldsymbol{\psi}=\hat{\boldsymbol{\psi}}, \boldsymbol{\omega}=\boldsymbol{\omega}^0}. \tag{A.4}$$

This random sample can usually be obtained by the sampling-based procedure that is developed for simulating observations in the MCEM procedure for maximum likelihood estimation. Hence, the additional programming effort is light.

Appendix B

Sampling from $f(y_{ij}|r_{ij} = 1, \boldsymbol{\psi})$:

Based on suggestion given in (Roberts, 1996), we choose $\mathcal{N}[\cdot, \alpha_1\tau^2]$ as the proposal distribution, where $\tau^2 = (\sigma^{-2} + \gamma_{ij}^2 \exp(\mathbf{F}_{-ij}'\boldsymbol{\gamma}_{-ij})(1+\exp(\mathbf{F}_{-ij}'\boldsymbol{\gamma}_{-ij}))^{-2})^{-1}$, γ_{ij} is the coefficient of y_{ij} in logistic regression (3) and $\boldsymbol{\gamma}_{-ij}$ is remaining part with γ_{ij} removed. The MH algorithm (Metropolis et al., 1953; Hastings, 1970) is implemented as follows:

At the kth iteration with a current value $u_{ij}^{(k)}$, a new candidate u_{ij}^{*} is generated from $\mathcal{N}[u_{ij}^{(k)}, \alpha_1\tau^2]$, and accepting this new candidate u_{ij}^{*} as $u_{ij}^{(k+1)}$ with probability

$$\min\left\{1, \frac{f(u_{ij}^{*}|r_{ij}=1, \boldsymbol{\psi})}{f(u_{ij}^{(k)}|r_{ij}=1, \boldsymbol{\psi})}\right\}.$$

Sampling from $f(\mathbf{b}_i|\mathbf{y}_i, \mathbf{r}_i, \boldsymbol{\psi})$:

Following the arguments in (Quintana et al., 1999), a reasonable proposal distribution for $\mathbf{b}_i$ is the multivariate normal distribution with mean $\mathbf{C}(\hat{\mathbf{b}}_i)\mathbf{Z}_i'\mathbf{W}_i(\hat{\mathbf{b}}_i)\mathbf{Z}_i\hat{\mathbf{b}}_i$ and covariance matrix $\mathbf{C}(\hat{\mathbf{b}}_i)$, where For the convenience of expression, we define the notations: $\mathbf{Z}_i' = (\mathbf{z}_{i1}, \ldots, \mathbf{z}_{in_i})$, $W_{ij}^{-1} = \ddot{a}(\theta_{ij})\{\dot{g}(\mu_{ij})\}^2$, $\mathbf{W}_i(\mathbf{b}_i)$ is the $n_i \times n_i$ diagonal matrix with diagonal elements W_{ij} for $j = 1, \ldots, n_i$, $\hat{\mathbf{b}}_i$ is the maximizer of $\prod_{j=1}^{n_i} f(\mathbf{y}_i|\mathbf{b}_i, \boldsymbol{\psi})$ and $\mathbf{C}(\mathbf{b}_i) = \{\boldsymbol{\Sigma}^{-1} + \mathbf{Z}_i'\mathbf{W}_i(\mathbf{b}_i)\mathbf{Z}_i\}^{-1}$. However, evaluation of $\hat{\mathbf{b}}_i$ wastes a lot of computing time and resulting an inefficient algorithm. Hence, the following algorithm is implemented to generate observations from the target density $f(\mathbf{b}_i|\mathbf{y}_i, \mathbf{r}_i, \boldsymbol{\psi})$: At the kth iteration of the MH algorithm with current value $\mathbf{b}_i^{(k)}$ is generated from $\mathcal{N}[\mathbf{b}_i^{(k)}, \alpha_2\mathbf{C}(\mathbf{0})]$, and accepting this new candidate $\mathbf{b}_i^{*}$ as $\mathbf{b}_i^{(k+1)}$ with probability

$$\min\left\{1, \frac{f(\mathbf{b}_i^{*}|\mathbf{y}_i, \mathbf{r}_i, \boldsymbol{\psi})}{f(\mathbf{b}_i^{(k)}|\mathbf{y}_i, \mathbf{r}_i, \boldsymbol{\psi})}\right\}.$$

Appendix C

C.1. The derivatives for normal mixed effects model

Let σ_{t_1} be the t_1th component of $\boldsymbol{\sigma}$, $\dot{\boldsymbol{\Sigma}}(t_1) = (\partial\boldsymbol{\Sigma}/\partial\sigma_{t_1})$, $\ddot{\boldsymbol{\Sigma}} = (\partial^2\boldsymbol{\Sigma}/\partial\sigma_{t_1}\partial\sigma_{t_2})$, and $\mathbf{S}_b = \sum_{i=1}^{N}\mathbf{b}_i\mathbf{b}_i'/N$.

$$\frac{\partial^2\ell_\omega(\boldsymbol{\psi}; \mathbf{Y}_m, \boldsymbol{\Omega}, \mathbf{Y}_o, \mathbf{R})}{\partial\sigma_f^4} = \frac{1}{\sigma_f^4}\sum_{i=1}^{N}\sum_{j=1}^{n_i}\left\{\frac{1}{2} - \frac{1}{\sigma_f^2}\left(y_{ij} - \mathbf{x}_{ij}'\boldsymbol{\beta} - \mathbf{z}_{ij}'\mathbf{b}_i\right)^2\right\},$$

$$\frac{\partial^2\ell_\omega(\boldsymbol{\psi}; \mathbf{Y}_m, \boldsymbol{\Omega}, \mathbf{Y}_o, \mathbf{R})}{\partial\sigma_f^2\partial\boldsymbol{\beta}'} = -\frac{1}{\sigma_f^4}\sum_{i=1}^{N}\sum_{j=1}^{n_i}\left(y_{ij} - \mathbf{x}_{ij}'\boldsymbol{\beta} - \mathbf{z}_{ij}'\mathbf{b}_i\right)\mathbf{x}_{ij}',$$

$$\frac{\partial^2\ell_\omega(\boldsymbol{\psi}; \mathbf{Y}_m, \boldsymbol{\Omega}, \mathbf{Y}_o, \mathbf{R})}{\partial\boldsymbol{\beta}\partial\boldsymbol{\beta}'} = -\frac{1}{\sigma_f^2}\sum_{i=1}^{N}\sum_{j=1}^{n_i}\mathbf{x}_{ij}\mathbf{x}_{ij}',$$

$$\begin{aligned}\frac{\partial^2\ell_\omega(\boldsymbol{\psi}; \mathbf{Y}_m, \boldsymbol{\Omega}, \mathbf{Y}_o, \mathbf{R})}{\partial\sigma_{t_1}\partial\sigma_{t_2}} = &-\frac{N}{2}\operatorname{tr}\left\{\boldsymbol{\Sigma}^{-1}\ddot{\boldsymbol{\Sigma}}(t_1, t_2)\boldsymbol{\Sigma}^{-1}(\boldsymbol{\Sigma} - \mathbf{S}_b)\right.\\ &\left.+\boldsymbol{\Sigma}^{-1}\dot{\boldsymbol{\Sigma}}(t_1)\boldsymbol{\Sigma}^{-1}\dot{\boldsymbol{\Sigma}}(t_2)\boldsymbol{\Sigma}^{-1}(2\mathbf{S}_b - \boldsymbol{\Sigma})\right\},\end{aligned}$$

$$\frac{\partial^2 \ell_\omega(\boldsymbol{\psi}; \mathbf{Y}_m, \boldsymbol{\Omega}, \mathbf{Y}_o, \mathbf{R})}{\partial \boldsymbol{\gamma} \partial \boldsymbol{\gamma}'} = \sum_{i=1}^{N} \sum_{j=1}^{n_i} \left\{ \left[\frac{r_{ij}}{\pi_{ij}} - \frac{1 - r_{ij}}{1 - \pi_{ij}} \right] \frac{\partial^2 \pi_{ij}}{\partial \boldsymbol{\gamma} \partial \boldsymbol{\gamma}'} - \left[\frac{r_{ij}}{\pi_{ij}^2} + \frac{1 - r_{ij}}{(1 - \pi_{ij})^2} \right] \frac{\partial \pi_{ij}}{\partial \boldsymbol{\gamma}} \frac{\partial \pi_{ij}}{\partial \boldsymbol{\gamma}'} \right\},$$

whereas $\partial^2 \ell_c(\boldsymbol{\psi}; \mathbf{Y}_m, \boldsymbol{\Omega}, \mathbf{Y}_o, \mathbf{R})/\partial \boldsymbol{\psi} \partial \boldsymbol{\psi}'$ other terms are equal to zero.

C.1.1. Case weights within clusters

$$\frac{\partial^2 \ell_\omega(\boldsymbol{\psi}; \mathbf{Y}_m, \boldsymbol{\Omega}, \mathbf{Y}_o, \mathbf{R}, \boldsymbol{\omega})}{\partial \omega_{ij} \partial \boldsymbol{\beta}'} = \frac{1}{\sigma_f^2} (\mathrm{y}_{ij} - \mathbf{x}_{ij}' \boldsymbol{\beta} - \mathbf{z}_{ij}' \mathbf{b}_i) \mathbf{x}_{ij}',$$

$$\frac{\partial^2 \ell_\omega(\boldsymbol{\psi}; \mathbf{Y}_m, \boldsymbol{\Omega}, \mathbf{Y}_o, \mathbf{R}, \boldsymbol{\omega})}{\partial \omega_{ij} \partial \sigma_f^2} = -\frac{1}{2\sigma_f^2} \left\{ 1 + \frac{1}{\sigma_f^2} (\mathrm{y}_{ij} - \mathbf{x}_{ij}' \boldsymbol{\beta} - \mathbf{z}_{ij}' \mathbf{b}_i)^2 \right\}.$$

C.1.2. Case weights among clusters

$$\frac{\partial^2 \ell_\omega(\boldsymbol{\psi}; \mathbf{Y}_m, \boldsymbol{\Omega}, \mathbf{Y}_o, \mathbf{R}, \boldsymbol{\omega})}{\partial \omega_i \partial \boldsymbol{\beta}'} = \frac{1}{\sigma_f^2} \sum_{j=1}^{n_i} (\mathrm{y}_{ij} - \mathbf{x}_{ij}' \boldsymbol{\beta} - \mathbf{z}_{ij}' \mathbf{b}_i) \mathbf{x}_{ij}',$$

$$\frac{\partial^2 \ell_\omega(\boldsymbol{\psi}; \mathbf{Y}_m, \boldsymbol{\Omega}, \mathbf{Y}_o, \mathbf{R}, \boldsymbol{\omega})}{\partial \omega_i \partial \sigma_f^2} = -\frac{1}{2\sigma_f^2} \sum_{j=1}^{n_i} \left\{ 1 + \frac{1}{\sigma_f^2} (\mathrm{y}_{ij} - \mathbf{x}_{ij}' \boldsymbol{\beta} - \mathbf{z}_{ij}' \mathbf{b}_i)^2 \right\}.$$

C.1.3. Multiplicative perturbation on random effects

$$\frac{\partial^2 \ell_\omega(\boldsymbol{\psi}; \mathbf{Y}_m, \boldsymbol{\Omega}, \mathbf{Y}_o, \mathbf{R}, \boldsymbol{\omega})}{\partial \omega_i \partial \boldsymbol{\beta}'} = -\frac{1}{\sigma_f^2} \sum_{j=1}^{n_i} \mathbf{z}_{ij}' \mathbf{b}_i \mathbf{x}_{ij}',$$

$$\frac{\partial^2 \ell_\omega(\boldsymbol{\psi}; \mathbf{Y}_m, \boldsymbol{\Omega}, \mathbf{Y}_o, \mathbf{R}, \boldsymbol{\omega})}{\partial \omega_i \partial \boldsymbol{\sigma}_f^2} = -\frac{1}{\sigma_f^2} \sum_{j=1}^{n_i} (\mathrm{y}_{ij} - \mathbf{x}' \boldsymbol{\beta} - \mathbf{z}_{ij}' \mathbf{b}_i) \mathbf{z}_{ij}' \mathbf{b}_i,$$

$$\frac{\partial^2 \ell_\omega(\boldsymbol{\psi}; \mathbf{Y}_m, \boldsymbol{\Omega}, \mathbf{Y}_o, \mathbf{R}, \boldsymbol{\omega})}{\partial \omega_i \partial \sigma_{t_1}} = \mathbf{b}_i' \boldsymbol{\Sigma}^{-1} \dot{\boldsymbol{\Sigma}}(t_1) \boldsymbol{\Sigma}^{-1} \mathbf{b}_i.$$

C.1.4. Case weights perturbation on missing mechanism

$$\frac{\partial^2 \ell_\omega(\boldsymbol{\psi}; \mathbf{Y}_m, \boldsymbol{\Omega}, \mathbf{Y}_o, \mathbf{R}, \boldsymbol{\omega})}{\partial \omega_{ij} \partial \boldsymbol{\gamma}'} = (r_{ij} - \pi_{ij}) \mathbf{F}_{ij}'.$$

C.2. The derivatives for GLMMs

For the convenience of expression, we define the following notations: $\dot{d}_{ij} = (\mathrm{y}_{ij} - \mu_{ij}) \dot{k}(\mathbf{x}_{ij}' \boldsymbol{\beta} + \mathbf{z}_{ij}' \mathbf{b}_i)$ and $\ddot{d}_{ij} = \ddot{a}(\theta_{ij}) \{\dot{k}(\mathbf{x}_{ij}' \boldsymbol{\beta} + \mathbf{z}_{ij}' \mathbf{b}_i)\}^2 - (\mathrm{y}_{ij} - \mu_{ij}) \ddot{k}(\mathbf{x}_{ij}' \boldsymbol{\beta} + \mathbf{z}_{ij}' \mathbf{b}_i)$.

$$\frac{\partial^2 \ell_\omega(\boldsymbol{\psi}; \mathbf{Y}_m, \boldsymbol{\Omega}, \mathbf{Y}_o, \mathbf{R})}{\partial \phi^2} = \sum_{i=1}^{N} \sum_{j=1}^{n_i} \frac{\partial^2 c(\mathrm{y}_{ij}, \phi)}{\partial \phi^2},$$

$$\frac{\partial^2 \ell_\omega(\boldsymbol{\psi}; \mathbf{Y}_m, \boldsymbol{\Omega}, \mathbf{Y}_o, \mathbf{R})}{\partial\phi\partial\boldsymbol{\beta}'} = \sum_{i=1}^{N}\sum_{j=1}^{n_i} \dot{d}_{ij}\mathbf{x}'_{ij},$$

$$\frac{\partial^2 \ell_\omega(\boldsymbol{\psi}; \mathbf{Y}_m, \boldsymbol{\Omega}, \mathbf{Y}_o, \mathbf{R})}{\partial\boldsymbol{\beta}\partial\boldsymbol{\beta}'} = -\phi\sum_{i=1}^{N}\sum_{j=1}^{n_i} \ddot{d}_{ij}\mathbf{x}_{ij}\mathbf{x}'_{ij},$$

$$\begin{aligned}\frac{\partial^2 \ell_\omega(\boldsymbol{\psi}; \mathbf{Y}_m, \boldsymbol{\Omega}, \mathbf{Y}_o, \mathbf{R})}{\partial\sigma_{t_1}\partial\sigma_{t_2}} &= -\frac{N}{2}\operatorname{tr}\big\{\boldsymbol{\Sigma}^{-1}\ddot{\boldsymbol{\Sigma}}(t_1,t_2)\boldsymbol{\Sigma}^{-1}(\boldsymbol{\Sigma}-\mathbf{S}_b)\\ &\quad + \boldsymbol{\Sigma}^{-1}\dot{\boldsymbol{\Sigma}}(t_1)\boldsymbol{\Sigma}^{-1}\dot{\boldsymbol{\Sigma}}(t_2)\boldsymbol{\Sigma}^{-1}(2\mathbf{S}_b-\boldsymbol{\Sigma})\big\},\end{aligned}$$

$$\begin{aligned}\frac{\partial^2 \ell_\omega(\boldsymbol{\psi}; \mathbf{Y}_m, \boldsymbol{\Omega}, \mathbf{Y}_o, \mathbf{R})}{\partial\boldsymbol{\gamma}\partial\boldsymbol{\gamma}'} &= \sum_{i=1}^{N}\sum_{j=1}^{n_i}\left\{\left[\frac{r_{ij}}{\pi_{ij}} - \frac{1-r_{ij}}{1-\pi_{ij}}\right]\frac{\partial^2\pi_{ij}}{\partial\boldsymbol{\gamma}\partial\boldsymbol{\gamma}'}\right.\\ &\quad \left. - \left[\frac{r_{ij}}{\pi_{ij}^2} + \frac{1-r_{ij}}{(1-\pi_{ij})^2}\right]\frac{\partial\pi_{ij}}{\partial\boldsymbol{\gamma}}\frac{\partial\pi_{ij}}{\partial\boldsymbol{\gamma}'}\right\},\end{aligned}$$

whereas $\partial^2\ell_c(\boldsymbol{\psi}; \mathbf{Y}_m, \boldsymbol{\Omega}, \mathbf{Y}_o, \mathbf{R})/\partial\boldsymbol{\psi}\partial\boldsymbol{\psi}'$ other terms are equal to zero.

C.2.1. Case weights within clusters

$$\frac{\partial^2 \ell_\omega(\boldsymbol{\psi}; \mathbf{Y}_m, \boldsymbol{\Omega}, \mathbf{Y}_o, \mathbf{R}, \boldsymbol{\omega})}{\partial\omega_{ij}\partial\phi} = \mathrm{y}_{ij}\theta_{ij} - a(\theta_{ij}) + \frac{\partial c(\mathrm{y}_{ij},\phi)}{\partial\phi},$$

$$\frac{\partial^2 \ell_\omega(\boldsymbol{\psi}; \mathbf{Y}_m, \boldsymbol{\Omega}, \mathbf{Y}_o, \mathbf{R}, \boldsymbol{\omega})}{\partial\omega_{ij}\partial\boldsymbol{\beta}'} = \phi\dot{d}_{ij}\mathbf{x}'_{ij}.$$

C.2.2. Case weights among clusters

$$\frac{\partial^2 \ell_\omega(\boldsymbol{\psi}; \mathbf{Y}_m, \boldsymbol{\Omega}, \mathbf{Y}_o, \mathbf{R}, \boldsymbol{\omega})}{\partial\omega_i\partial\phi} = \sum_{j=1}^{n_i}\left[\mathrm{y}_{ij}\theta_{ij} - a(\theta_{ij}) + \frac{\partial c(\mathrm{y}_{ij},\phi)}{\partial\phi}\right],$$

$$\frac{\partial^2 \ell_\omega(\boldsymbol{\psi}; \mathbf{Y}_m, \boldsymbol{\Omega}, \mathbf{Y}_o, \mathbf{R}, \boldsymbol{\omega})}{\partial\omega_i\partial\boldsymbol{\beta}'} = \phi\sum_{j=1}^{n_i}\dot{d}_{ij}\mathbf{x}'_{ij}.$$

C.2.3. Multiplicative perturbation on random effects

$$\frac{\partial^2 \ell_\omega(\boldsymbol{\psi}; \mathbf{Y}_m, \boldsymbol{\Omega}, \mathbf{Y}_o, \mathbf{R}, \boldsymbol{\omega})}{\partial\omega_i\partial\phi} = \sum_{j=1}^{n_i}\dot{d}_{ij}\mathbf{z}'_{ij}\mathbf{b}_i,$$

$$\frac{\partial^2 \ell_\omega(\boldsymbol{\psi}; \mathbf{Y}_m, \boldsymbol{\Omega}, \mathbf{Y}_o, \mathbf{R}, \boldsymbol{\omega})}{\partial\omega_i\partial\boldsymbol{\beta}'} = -\phi\sum_{j=1}^{n_i}\ddot{d}_{ij}\mathbf{z}'_{ij}\mathbf{b}_i\mathbf{x}'_{ij},$$

$$\frac{\partial^2 \ell_\omega(\boldsymbol{\psi}; \mathbf{Y}_m, \boldsymbol{\Omega}, \mathbf{Y}_o, \mathbf{R}, \boldsymbol{\omega})}{\partial\omega_i\partial\sigma_{t_1}} = \mathbf{b}'_i\boldsymbol{\Sigma}^{-1}\dot{\boldsymbol{\Sigma}}(t_1)\boldsymbol{\Sigma}^{-1}\mathbf{b}_i.$$

C.2.4. Case weights perturbation on missing mechanism

$$\frac{\partial^2 \ell_\omega(\boldsymbol{\psi}; \mathbf{Y}_m, \boldsymbol{\Omega}, \mathbf{Y}_o, \mathbf{R}, \boldsymbol{\omega})}{\partial \omega_{ij} \partial \boldsymbol{\gamma}'} = (r_{ij} - \pi_{ij})\mathbf{F}'_{ij}.$$

References

Breslow, N.E., Clayton, D.G. (1993). Approximate inference in generalized linear mixed models. *Journal of the American Statistical Association* **88**, 9–25.

Cain, K.C., Lange, N.T. (1984). Approximate case influence for the proportional hazards regression model with censored data. *Biometrics* **40**, 493–499.

Cook, D.R. (1986). Assessment of local influence (with discussion). *Journal of the Royal Statistical Society, Series B* **48**, 133–169.

Dale, J.R. (1986). Global cross-ratio models for bivariate, discrete, ordered responses. *Biometrics* **42**, 909–917.

Davidian, M., Giltinan, D.M. (1995). *Nonlinear Models for Repeated Measurement Data*. Chapman, Hall, London.

Demidenko, E., Stukel, T.A. (2005). Influence analysis for linear mixed-effects models. *Statistics in Medicine* **24**, 893–909.

Dempster, A.P., Laird, N.M., Rubin, D.B. (1977). Maximum likelihood from incomplete data via EM algorithm (with discussion). *Journal of the Royal Statistical Society, Series B* **39**, 1–38.

Diggle, P., Liang, K.Y., Zeger, S.L. (1994). *Analysis of Longitudinal Data*. Oxford University Press, New York.

Escobar, L.A., Meeker, W.Q. (1992). Assessing influence in regression analysis with censored data. *Biometrics* **48**, 507–528.

Geman, S., Geman, D. (1984). Stochastic relaxation, Gibbs distribution, and the Bayesian restoration of images. *IEEE Transactions on Pattern Analysis and Machine Intelligence* **6**, 721–741.

Hastings, W.K. (1970). Monte Carlo sampling methods using Markov chains and their application. *Biometrika* **57**, 97–109.

Ibrahim, J.G., Chen, M.H., Lipsitz, S.R. (2001). Missing responses in generalised linear mixed models when the missing data mechanism is nonignorable. *Biometrika* **88**, 551–564.

Laird, N.M., Ware, J.H. (1982). Random effects models for longitudinal data. *Biometrics* **38**, 963–974.

St. Laurent, R.T., Cook, R.D. (1993). Leverage, local influence and curvature in nonlinear regression. *Biometrika* **80**, 99–106.

Lesaffre, E., Verbeke, G. (1998). Local influence in linear mixed models. *Biometrics* **54**, 570–582.

Little, R.J.A., Rubin, D.B. (1987). *Statistical Analysis with Missing Data*. Wiley, New York.

McCullagh, P., Nelder, J.A. (1989). *Generalized Linear Models*, second ed. Chapman, Hall, London.

Metropolis, N., Rosenbluth, A.W., Rosenbluth, M.N., Teller, A.H., Teller, E. (1953). Equations of state calculations by fast computing machine. *Journal of Chemical Physics* **21**, 1087–1091.

Ouwens, M.J.N.M., Tan, F.E.S., Berger, M.P.F. (2001). Local influence to detect influential data structures for generalized linear mixed models. *Biometrics* **57**, 1166–1172.

Poon, W.Y., Poon, Y.S. (1999). Conformal normal curvature and assessment of local influence. *Journal of the Royal Statistical Society, Series B* **61**, 51–61.

Quintana, F.A., Liu, J.S., del Pino, G.E. (1999). Monte Carlo EM with importance reweighting and its applications in random effect models. *Computational Statistics and Data Analysis* **29**, 429–444.

Roberts, G.O. (1996). Markov Chain concepts related to sampling algorithms. In: Gilks, W.R., Richardson, S., Spiegelhalter, D.J. (Eds.), *Markov Chain Monte Carlo in Practice*. Chapman, Hall, London, pp. 45–57.

Thomas, W., Cook, R.D. (1989). Assessing influence on regression coefficients in generalized linear models. *Biometrika* **76**, 741–749.

van Steen, K., Molenberghs, G., Verbeke, G., Thijs, H. (2001). A local influence approach to sensitivity analysis of incomplete longitudinal ordinal data. *Statistical Modelling: An International Journal* **1**, 125–142.

Verbeke, G., Molenberghs, G., Thijs, H., Lesaffre, E., Kenward, M.G. (2001). Sensitivity analysis for nonrandom dropout: A local influence approach. *Biometrics* **57**, 7–14.

Wei, G.C.G., Tanner, M.A. (1990). A Monte Carlo implementation of the EM algorithm and the Poor man's data augmentation algorithm. *Journal of the American Statistical Association* **85**, 699–704.

Weissfeld, L.A. (1990). Influence diagnostics for the proportional hazards models. *Statistics and Probability Letters* **10**, 411–417.

Zeger, S.L., Karim, M.R. (1991). Generalized linear models with random effects: A Gibbs sampling approach. *Journal of the American Statistical Association* **86**, 79–86.

Zhu, H.T., Lee, S.Y. (2001). Local influence for incomplete data models. *Journal of the Royal Statistical Society, Series B* **63**, 111–126.

Zhu, H.T., Lee, S.Y. (2003). Local influence for generalized linear mixed models. *The Canadian Journal of Statistics* **31**, 293–309.

Handbook of Computing and Statistics with Applications, Vol. 1
ISSN: 1871-0301

DOI: 10.1016/S1871-0301(06)01007-9

7

Goodness-of-Fit Measures for Latent Variable Models for Binary Data

D. Mavridis, I. Moustaki and M. Knott

Abstract
Goodness-of-fit measures for latent variable models for binary responses are discussed. Overall goodness-of-fit statistics such as the Pearson chi-squared test and the likelihood ratio test can only be used when the observed and expected frequencies under the model are large enough. When sparseness is present, limited information statistics that use information from the lower-order margins have been proposed in the literature. Those statistics and a new one that is based on the odds ratio are presented and compared in terms of Type I error and their statistical power. Simulated results and a real example are used for exploring the performance of the overall and limited information goodness-of-fit tests under different number of items, sample size and degree of sparseness. Standardized and adjusted residuals are also studied.

1. Introduction

Goodness-of-fit tests are used to evaluate how well a proposed model fits or predicts a particular data set. Usually, test statistics compute deviations between the observed data and predictions from the model. The value of a test statistic is said to be statistically significant if it is found to be within the rejection area of the distribution of the test statistic under the assumption that the model is true. The rejection area is often the upper 5 or 1% of the distribution's tail.

This chapter discusses goodness-of-fit tests and goodness-of-fit measures for latent variable models for binary data. Binary manifest variables are very common in the social sciences where, for example, a yes/no or correct/incorrect response is required. Latent variable models have been greatly extended in the recent years in many different directions. More specifically, latent variable models can handle different types of observed variables (categorical, metric, survival, mixed variables), variables measured across time (dynamic models) as well as sampling units nested within levels (multilevel analysis). Different estimation methods such as the E-M algorithm (Bock and

Aitkin, 1981; Bartholomew and Knott, 1999), the Newton–Raphson algorithm (Rabe-Hesketh et al., 2001), as well as purely Bayesian approaches (Patz and Junker, 1999; Dunson, 2000) have been developed for handling the complexity of those multivariate models. However, less has been done on testing the appropriateness of the fitted latent variable models.

For k binary variables, the data can be described by a contingency table consisting of 2^k cells. The Pearson's statistic X^2 and the likelihood ratio statistic G^2 are probably the most well-known goodness-of-fit statistics for models that are based on the multinomial distribution. These goodness-of-fit statistics require a large number of observations in each cell for their asymptotic distributions to hold. However, for a moderate sample size and large number of variables, sparseness in the 2^k table is inevitable. To overcome the problem of sparseness, limited information statistics similar to Pearson's X^2 but using only information from the lower order margins are available (Christoffersson, 1975; Muthén, 1978; Reiser, 1996; Bartholomew and Leung, 2002; Maydeu-Olivares and Joe, 2005).

The chapter is organized as follows: in Section 2, we give a brief description of the latent variable model for binary variables and the notation that will be used for the rest of the chapter. In Section 3, we discuss overall goodness-of-fit test statistics and we define the problem of sparseness and ways to tackle it. Limited information statistics that are based on lower order margins are presented in Section 4. In Section 5 we propose the odds ratio for testing model fit. Finally, in Section 6 a large scale simulation study is performed that compares the available goodness-of-fit statistics in terms of Type I error and statistical power.

2. Latent variable models for binary responses

Suppose that we have k observed or manifest variables also known as items, and q latent or unobserved variables. The responses given to a set of k items are called a response pattern. For a data set with k binary items there are 2^k possible response patterns. The vector of observed variables will be denoted by $\mathbf{y}$ where $\mathbf{y}' = (y_1, y_2, \dots, y_k)$ and the vector of latent variables will be denoted by $\mathbf{z}$ where $\mathbf{z}' = (z_1, z_2, \dots, z_q)$.

The responses given to k binary items by n individuals can be presented in two different ways. One way is to define a data matrix, $\mathbf{Y}$, of dimension $n \times k$, where n is the number of individuals and k the number of items. The element y_{mi} denotes the response of the mth individual to the ith item, where $m = 1, \dots, n$. Another way is to define a matrix $\mathbf{Y}^r$ of dimensions $2^k \times k$ that contains all possible response patterns in its rows. The element y^r_{si} of that matrix denotes the value of the response pattern s to the ith item, where $s = 1, \dots, 2^k$.

The joint distribution of the k observed variables for an individual m is:

$$f(\mathbf{y}_m) = \int_{R_{z_1}} \cdots \int_{R_{z_q}} g(\mathbf{y}_m|\mathbf{z})h(\mathbf{z})\,\mathrm{d}\mathbf{z}, \tag{1}$$

where $\mathbf{y}_m$ is the mth row of matrix $\mathbf{Y}$, $h(\mathbf{z})$ is the prior distribution of the latent variables $\mathbf{z}$, $g(\mathbf{y}_m|\mathbf{z})$ is the conditional distribution of the observed variables $\mathbf{y}$ given the latent variables $\mathbf{z}$ and R_{z_j} denotes the range of values of the jth latent variable. The latent variables are assumed to be independent with standard normal distributions ($z_j \sim N(0, 1)$ for all j).

We make the assumption of conditional independence, that is, observed variables are assumed to be independent conditional on the latent variables. In other words, the associations among the observed variables are adequately explained by the latent variables $\mathbf{z}$ giving:

$$g(\mathbf{y}_m|\mathbf{z}) = \prod_{i=1}^{k} g(y_{mi}|\mathbf{z}). \tag{2}$$

Eq. (1) becomes:

$$f(\mathbf{y}_m) = \int_{-\infty}^{+\infty} \cdots \int_{-\infty}^{+\infty} \prod_{i=1}^{k} g(y_{mi}|\mathbf{z}) h(\mathbf{z})\, d\mathbf{z}. \tag{3}$$

For the case of binary variables the distribution of $g(y_{mi}|\mathbf{z})$ is the Bernoulli distribution:

$$g(y_{mi}|\mathbf{z}) = \left[P(y_{mi} = 1|\mathbf{z}, \boldsymbol{\beta})\right]^{y_{mi}}\left[1 - P(y_{mi} = 1|\mathbf{z}, \boldsymbol{\beta})\right]^{1-y_{mi}},$$
$$i = 1, \ldots, k;\ m = 1, \ldots, n,$$

where $P(y_{mi} = 1|\mathbf{z}, \boldsymbol{\beta})$ is the probability of individual m answering 'positively' to item i ($y_{mi} = 1$) conditional on the latent variables ($\mathbf{z}$). The probability $P(y_{mi} = 1|\mathbf{z}, \boldsymbol{\beta})$ is modelled with a logistic link written as:

$$\operatorname{logit}\left[P(y_{mi} = 1|\mathbf{z}, \boldsymbol{\beta})\right] = \beta_{0i} + \sum_{j=1}^{q} \beta_{ij} z_j,$$
$$i = 1, \ldots, k;\ m = 1, \ldots, n, \tag{4}$$

where the parameter β_{0i} is an intercept or a location parameter. In psychometrics and educational testing, the intercept is also known as 'difficulty' parameter. The parameters β_{ij} are the factor loadings also known as 'discrimination' parameters that measure the discriminating power of an item. Their size determines the effect that a change in $\mathbf{z}$ has on logit $P(y_{mi} = 1|\mathbf{z}, \boldsymbol{\beta})$.

It follows that the conditional probability $P(y_{mi} = 1|\mathbf{z}, \boldsymbol{\beta})$ of answering positively to item i is:

$$P(y_{mi} = 1|\mathbf{z}, \boldsymbol{\beta}) = \frac{\exp\left(\beta_{0i} + \sum_{j=1}^{q} \beta_{ij} z_j\right)}{1 + \exp\left(\beta_{0i} + \sum_{j=1}^{q} \beta_{ij} z_j\right)}. \tag{5}$$

We denote the vector of all model parameters by $\boldsymbol{\beta}' = (\beta_{01}, \ldots, \beta_{0k}, \beta_{11}, \ldots, \beta_{1q}, \ldots, \beta_{k1}, \ldots, \beta_{kq})$. In this paper, the parameters are estimated using maximum likelihood with the E-M algorithm, treating the latent variables as missing data (Bock and Aitkin, 1981; Bartholomew and Knott, 1999). The estimated vector of parameters is denoted by $\hat{\boldsymbol{\beta}}$.

3. Goodness-of-fit tests for latent variable models for binary data

The goodness-of-fit of a model can be checked in different ways. One way is to compare the observed and estimated frequencies under a model across the response patterns. That will be considered as an overall goodness-of-fit test. Another way is to compare the observed with the estimated frequencies for lower order margins. Finally, models can be compared using model selection criteria such as the Akaike (AIC) or the Bayesian Information Criterion (BIC). In this chapter, we discuss and compare the performance of tests for overall fit and for fit on the lower order margins.

Each response pattern s occurs in the sample with a frequency n_s. The vector $\mathbf{n}' = (n_1, n_2, \ldots, n_{2^k})$ contains the observed frequencies of the 2^k cells. The 2^k vector of sample (observed) proportions for the response patterns is denoted by $\hat{\boldsymbol{p}}' = (\hat{p}_1, \ldots, \hat{p}_{2^k})$. The corresponding vector of true probabilities under the model is denoted by $\boldsymbol{\pi}(\boldsymbol{\beta})' = (\pi_1(\boldsymbol{\beta}), \ldots, \pi_{2^k}(\boldsymbol{\beta}))$. Each element of $\boldsymbol{\pi}(\boldsymbol{\beta})'$ is computed from:

$$P(\mathbf{y} = \mathbf{y}_s^r) = \pi_s(\boldsymbol{\beta}) = \int_{-\infty}^{+\infty} \cdots \int_{-\infty}^{+\infty} \prod_{i=1}^{k} g(y_{si}^r|\mathbf{z}) h(\mathbf{z})\, \mathrm{d}\mathbf{z}, \quad s = 1, \ldots, 2^k, \tag{6}$$

where $\mathbf{y}_s^r$ denotes the sth row of the data matrix $\mathbf{Y}^r$. The estimated probabilities under the fitted model are denoted by $\boldsymbol{\pi}(\hat{\boldsymbol{\beta}})' = (\pi_1(\hat{\boldsymbol{\beta}}), \ldots, \pi_{2^k}(\hat{\boldsymbol{\beta}}))$ and they are computed by replacing in (6) the vector $\boldsymbol{\beta}$ with the estimated one $\hat{\boldsymbol{\beta}}$.

In general, the fit of a model is judged by how close the estimated proportions $\boldsymbol{\pi}(\hat{\boldsymbol{\beta}})$ are to the sample proportions $\hat{\boldsymbol{p}}$.

3.1. Overall goodness-of-fit tests

The frequencies of the response patterns are considered to follow the multinomial distribution with parameters the total sample size n and the true probabilities estimated for each of the 2^k response patterns. The multivariate central limit theorem states that, for large sample size n, the multinomial distribution can be approximated by the multivariate normal. This result has been used to obtain approximate distributions for many goodness-of-fit statistics, such as, the Pearson statistic

$$X^2_{\text{Pearson}} = \sum_{s=1}^{2^k} \frac{(n\hat{p}_s - n\pi_s(\hat{\boldsymbol{\beta}}))^2}{n\pi_s(\hat{\boldsymbol{\beta}})}, \tag{7}$$

where n is the sample size, $\hat{p}_s$ and $\pi_s(\hat{\boldsymbol{\beta}})$ are the observed and estimated proportions respectively for the response pattern s. The estimated probabilities $\pi_s(\hat{\boldsymbol{\beta}})$ are computed from (6) by replacing the true parameters with their corresponding estimates. Cressie and Read (1984) define a power-divergence family of test statistics given by:

$$CR(\lambda) = \sum_{s=1}^{2^k} \frac{2}{\lambda(\lambda+1)} n\hat{p}_s \left[\left(\frac{\hat{p}_s}{\pi_s(\hat{\boldsymbol{\beta}})} \right)^{\lambda} - 1 \right]. \tag{8}$$

This family contains the Pearson ($\lambda = 1$) and the likelihood ratio test ($\lambda = 0$). Under a correct model, all the statistics defined in (8) are asymptotically distributed as χ^2 with (2^k-number of parameters estimated-1) degrees of freedom, where k denotes the number of items.

3.2. Sparseness

When the sample size n is small and the number of items big then the table of frequencies for the response patterns becomes so sparse that the asymptotic results do not give a good approximation to the distribution of the test statistics.

There is no universally accepted definition of sparseness. Sparseness occurs when the expected frequency for many response patterns is small. Most researchers claim that the expected frequency for every response pattern should be at least five (Cochran, 1954) while others claim that the expected frequency should be at least ten (Cramer, 1946) or even twenty (Kendall, 1952; Tate and Hyer, 1973).

Many solutions to the problems of sparseness have been proposed in the literature. One suggestion is to combine cells (Hosmer and Lemeshow, 1980), so that small expected frequencies vanish. Combining cells is more successful when the extent of sparseness is not severe. We should note that combining cells eventually means that the model is fitted on fewer cells which might lead to a non-identified model. In general, combining cells by a method based on the data leads to an asymptotic distribution of the statistics of the power-divergence family with an unknown distribution.

Another solution to the problem of sparseness is adding a small constant to the frequency of every response pattern. That leads inevitably to an increase in the sample size with most response patterns having the same probability. A third and more recent solution is to fit other distributions for the goodness-of-fit statistics. Morris (1975) and Koehler and Larntz (1980) claim that if the number of items increases as the sample size increases, then the goodness-of-fit statistics follow the normal distribution. Simonoff (1985) proposes a test of goodness-of-fit for this situation. However, the normal distribution is not realistic for most applications.

3.3. Goodness-of-fit tests on the lower margins

Tests similar to the Pearson X^2, have been developed for latent variable models using only the information from the lower-order margins (Christoffersson, 1975; Muthén, 1978; Reiser and VandenBerg, 1994; Bartholomew and Tzamourani, 1999; Bartholomew and Leung, 2002; Maydeu-Olivares and Joe, 2005). Expected frequencies for lower-order margins are rarely small.

More specifically, the frequency distribution for a pair of binary items (i, j) is expressed through a (2×2) contingency table. The cells of this table give the observed frequency of the pair responses $((1,1), (1,0), (0,1), (0,0))$, where $(1,1)$ indicates a 'positive' response to items i and j. If the frequency for one pair of responses is known, the others are determined from the one-way margins that are regarded as fixed. In total $k(k-1)/2$ contingency tables exist for all pair of items. In addition to the observed cell frequencies, expected frequencies under the model are computed for the cells of the $k(k-1)/2$ two-way tables. A Pearson X^2 statistic can be computed for each cell of the

$k(k-1)/2$ two-way tables. Assuming that the distribution of that statistic in each cell follows a chi-square distribution with one degree of freedom then any value greater than four would be an indication of poor fit. We should also note that the chi-square values obtained from the $k(k-1)/2$ cells are not independent and therefore if one sums to produce an aggregate measure of fit its distribution will be unknown.

In the same way that estimated probability is computed for a response pattern $\mathbf{y}_s$ using (6) one can compute estimated lower-order probabilities for the items. More specifically, the estimated univariate probabilities for item i are

$$\mathrm{P}(y_i = 1|\hat{\boldsymbol{\beta}}) = \sum_{s=1}^{2^k} y_{si}^r \pi_s(\hat{\boldsymbol{\beta}}), \quad i = 1, \ldots, k, \tag{9}$$

where y_{si}^r is the ith element of response pattern s. The estimated bivariate probabilities for items i and j are

$$\mathrm{P}(y_i = 1, y_j = 1|\hat{\boldsymbol{\beta}}) = \sum_{s=1}^{2^k} y_{si}^r y_{sj}^r \pi_s(\hat{\boldsymbol{\beta}}), \quad i, j = 1, \ldots, k, \tag{10}$$

where y_{si}^r and y_{sj}^r are the ith and jth element, respectively, of response pattern s. By replacing $\pi_s(\hat{\boldsymbol{\beta}})$ in (9) and (10) with $\hat{p}_s$ we get the univariate and bivariate sample proportions, respectively. Higher-order probabilities can be defined in the same way as above.

Lower-order probabilities (see, e.g., Eqs. (9) and (10)) can be easily obtained using a matrix notation. This is achieved by multiplying the $(2^k \times 1)$ vector of observed or expected proportions with an indicator matrix M consisting of zeros and ones. In the case of univariate and bivariate margins, matrix M has $k(k+1)/2$ rows (number of univariate and bivariate margins) and 2^k columns (number of all possible response patterns). Hence, every univariate and bivariate frequency is linked with a response pattern through an element of matrix M. Consider a row of matrix M which refers to the bivariate margin of positive responses to items (i, j), then if a response pattern has 'positive' answers to items i and j then the cell that links this bivariate margin with this response pattern will take the value 1 and 0 otherwise. Therefore, the univariate and bivariate observed and estimated proportions denoted by the vectors $\hat{\boldsymbol{p}}^l$ and $\boldsymbol{\pi}^l(\hat{\boldsymbol{\beta}})$ respectively are obtained from:

$$\hat{\boldsymbol{p}}^l = M\hat{\boldsymbol{p}} \tag{11}$$

and

$$\boldsymbol{\pi}^l(\hat{\boldsymbol{\beta}}) = M\boldsymbol{\pi}(\hat{\boldsymbol{\beta}}), \tag{12}$$

where $\hat{\boldsymbol{p}}$ and $\boldsymbol{\pi}(\hat{\boldsymbol{\beta}})$ are the vectors of the observed and estimated proportions for the whole response pattern.

The dimension of matrix M is defined according to which lower-order margins are included in the test statistics. For example, when only bivariate proportions are used, the dimension of matrix M is reduced to $(\frac{k(k-1)}{2} \times 2^k)$ and the vectors $\hat{\boldsymbol{p}}^l$ and $\boldsymbol{\pi}^l(\hat{\boldsymbol{\beta}})$ from Eqs. (11) and (12), respectively, refer to bivariate margins only.

Model deviations can be expressed through unstandardized residuals of the whole response pattern denoted by:

$$\boldsymbol{e} = \hat{\boldsymbol{p}} - \boldsymbol{\pi}(\hat{\boldsymbol{\beta}}), \tag{13}$$

where $\hat{\boldsymbol{p}}$ and $\boldsymbol{\pi}(\hat{\boldsymbol{\beta}})$ are the observed and estimated proportions, respectively.

However, tests discussed in this chapter concentrate on deviations on the lower-order margins denoted by:

$$\boldsymbol{e}^l = \hat{\boldsymbol{p}}^l - \boldsymbol{\pi}^l(\hat{\boldsymbol{\beta}}), \tag{14}$$

where the index l indicates residuals obtained from lower-order margins such as univariate ($l = 1$), bivariate ($l = 2$), etc. The notation $l = 2$ refers to bivariate margins only. All limited information tests presented here study deviations between positive observed and expected responses to items. The hypothesis of interest is:

$$H_0\colon \boldsymbol{\varepsilon}^l = \boldsymbol{p}^l - \boldsymbol{\pi}^l(\boldsymbol{\beta}) = 0 \tag{15}$$

against the alternative hypothesis that

$$H_1\colon \boldsymbol{\varepsilon}^l = \boldsymbol{p}^l - \boldsymbol{\pi}^l(\boldsymbol{\beta}) \neq 0, \tag{16}$$

where $\boldsymbol{p}$ and $\boldsymbol{\pi}(\boldsymbol{\beta})$ are the true proportions and the proportions under the model for the lth order lower margins. Since $(\hat{\boldsymbol{p}} - \boldsymbol{\pi}(\hat{\boldsymbol{\beta}})) \sim N_{2^k}(0, \Sigma)$, see (Rao, 1973), and $\boldsymbol{e}^l$ can be computed from the whole response pattern using the M matrix, it follows:

$$\boldsymbol{e}^l = M\big(\hat{\boldsymbol{p}} - \boldsymbol{\pi}(\hat{\boldsymbol{\beta}})\big) \sim N_{k(k+1)/2}(0, \Omega), \tag{17}$$

where $\Omega = M\Sigma M'$ is the variance–covariance matrix of $\boldsymbol{e}^l$ and Σ is the covariance matrix of $\boldsymbol{e}$.

All the tests on the lower-order margins in this paper are of the form:

$$W = \boldsymbol{e}^{l'}(M\Sigma M')^{-1}\boldsymbol{e}^l. \tag{18}$$

Using known results from multivariate theory, W follows a χ^2 with degrees of freedom equal to the rank of the asymptotic variance–covariance matrix given by $M\Sigma M'$. The differences among the various tests presented in Section 4 are in the definition of the covariance matrix used in place of $M\Sigma M'$ in (18).

4. Limited information statistics

As it has been shown in Section 3.3, limited information statistics based on residuals are given in the form of (18). For the computation of (18), the asymptotic variance–covariance matrix of the residuals $\boldsymbol{e}^l$ must be computed. In Section 4.1, the asymptotic variance–covariance matrix of $\boldsymbol{e}^l$ is computed based on the hypothesis that the parameter values are known in advance. In Section 4.2, the parameter values are not considered known but they are estimated from the data. The limited information statistics that are presented in Section 4.1 claim that the estimation of the model parameters incur an non-negligible impact on the estimation of the variance–covariance matrix of $\boldsymbol{e}^l$.

4.1. Parameter values are considered to be known

In this section, we review the two limited information test statistics that assume known parameter values mainly the Bartholomew and Leung (2002) and Christoffersson (1975) tests. When parameters are considered to be known the lower-order residuals $\boldsymbol{e}^l$ have a covariance matrix given by:

$$\Omega = M\big(\mathrm{diag}\big(\boldsymbol{\pi}(\boldsymbol{\beta})\big) - \boldsymbol{\pi}(\boldsymbol{\beta})\boldsymbol{\pi}(\boldsymbol{\beta})'\big)M'. \tag{19}$$

4.1.1. Christoffersson's test

Christoffersson's test (X^2_{C}) is the test defined in (18) where the covariance matrix Ω defined in (19) of the bivariate residuals $\boldsymbol{e}^l$ is estimated by:

$$\widehat{\Omega}_{\mathrm{C}} = M\big(\mathrm{diag}(\hat{\boldsymbol{p}}) - \hat{\boldsymbol{p}}\hat{\boldsymbol{p}}'\big)M', \tag{20}$$

where the vector $\hat{\boldsymbol{p}}$ contains the observed proportions. Christoffersson (1975) claims that sample proportions $\hat{\mathbf{p}}$ provide a good approximation to the true probabilities.

4.1.2. Bartholomew and Leung test

The Bartholomew and Leung statistic (X^2_{BL}) is:

$$X^2_{\mathrm{BL}} = \sum_{i=1}^{k}\sum_{j=i+1}^{k} \frac{(n_{ij} - n\hat{\pi}_{ij})^2}{n\hat{\pi}_{ij}(1-\hat{\pi}_{ij})}, \tag{21}$$

where n_{ij} is the number of individuals that answered positively to items i and j and $\hat{\pi}_{ij}$ is the estimated from the fitted model probability. The test accounts only for bivariate margins and it uses the sum of all $k(k-1)/2$ chi-squared values. It is shown in their paper that the test is also a sum of squares of standardized residuals.

Their simulation studies have shown that the distribution of the X^2_{BL} statistic resembles the distribution of χ^2, especially in the upper tail. They give the exact moments of the statistic and they also derive a linear function of χ^2 that has the same moments as the X^2_{BL} statistic. The statistic in (21) can be also written as:

$$X^2_{\mathrm{BL}} = (\hat{p}_{ij} - \hat{\pi}_{ij})'\big(\mathrm{diag}(\widehat{\Omega}_{\mathrm{BL}})\big)^{-1}(\hat{p}_{ij} - \hat{\pi}_{ij}), \tag{22}$$

where $\widehat{\Omega}_{\mathrm{BL}}$ denotes the asymptotic covariance matrix of $\boldsymbol{e}^l$ that is defined by

$$\widehat{\Omega}_{\mathrm{BL}} = M\big(\mathrm{diag}\big(\boldsymbol{\pi}(\hat{\boldsymbol{\beta}})\big) - \boldsymbol{\pi}(\hat{\boldsymbol{\beta}})\boldsymbol{\pi}(\hat{\boldsymbol{\beta}})'\big)M'. \tag{23}$$

They report that no significant differences exist between the use of their test when parameters are known and when they are estimated. They claim that even with a sample size as small as 100, their procedure will be adequate. However, Cai et al. (2004) found that the X^2_{BL} has low power. They increased the power of the test by applying a correction that accounts for the parameters being estimated from the data. Their test is denoted as X^2_{L}.

Note that there are two main differences between χ^2_{BL} and the X^2_{C} statistic. First only the diagonal elements of the matrix $\widehat{\Omega}_{\mathrm{BL}}$ are used in the X^2_{BL} where X^2_{C} uses information both from the three way and the four way margins. Second, estimated bivariate probabilities are used instead of observed.

4.2. Composite null hypothesis

Both X^2_{BL} and X^2_{C} test statistics assess the fit with Σ defined as if the parameters are known in advance. In most applications, one is interested in assessing the fit of a model when the parameters are estimated from the data. When the parameters are replaced by estimators, there will be an effect on both the test statistic and its sampling distribution. In computing a test, it is necessary to compensate for the fact that the parameters have been fitted using the same data.

Reiser (1996) proposed a limited-information test of fit that uses univariate and bivariate margins (denoted here as X^2_{R}). When the three-way margins are also included the test is denoted as $X^2_{\text{R(3)}}$. Maydeu-Olivares and Joe (2005) proposed a class of quadratic form statistics based on the residuals of margins up to order r (denoted here as X^2_{OJ}).

To account for parameter estimation, the multivariate delta method can be used to find the asymptotic covariance matrix of $\boldsymbol{e}^l$. Theoretical results can be found in Birch (1964), Agresti (1990) and Bishop et al. (1975). The asymptotic covariance matrix of $\boldsymbol{e}^l$ is given in Reiser (1996) and Maydeu-Olivares and Joe (2005) and is defined as

$$\Omega = M\big(\text{diag}\big(\boldsymbol{\pi}(\boldsymbol{\beta})\big) - \boldsymbol{\pi}(\boldsymbol{\beta})\boldsymbol{\pi}(\boldsymbol{\beta})' - G(A'A)^{-1}G'\big)M', \tag{24}$$

where the M matrix is defined in Section 3.3, $A = \text{diag}(\boldsymbol{\pi}(\boldsymbol{\beta}))^{-1/2}\partial\boldsymbol{\pi}(\boldsymbol{\beta})/\partial\boldsymbol{\beta}$ and $G = \partial\boldsymbol{\pi}(\boldsymbol{\beta})/\partial\boldsymbol{\beta}$. The $\boldsymbol{\pi}(\boldsymbol{\beta})$ are replaced by the estimated ones. Agresti (1990) has shown that the estimator of the asymptotic variance–covariance matrix is more efficient when the maximum likelihood estimates rather than the sample proportions are used. The product $A'A$ is the Fisher information matrix; its inverse, divided by the sample size n, gives the asymptotic variance–covariance matrix of the parameter vector $\hat{\boldsymbol{\beta}}$.

A common problem encountered in the computation of the asymptotic variance–covariance matrix of $\boldsymbol{e}^l$ given in (24) is that it may be ill-conditioned due to very small eigenvalues that stem from the high multicollinearity in matrix M. The Reiser (1996) test considers the Moore–Penrose inverse of Ω whereas Maydeu-Olivares and Joe (2005) replaces the generalized inverse by a matrix that they claim to be more stable. For the Moore–Penrose inverse a tolerance level needs to be set. In this paper the eigenvalues of the matrix and the corresponding percentage of the explained variance for each eigenvalue were computed and eigenvalues that explained just a small percentage of the variance, say 1%, were treated as zero. The tolerance level can be adjusted to the nature of the data set and the estimated parameters.

4.3. Residuals

When one of the limited information tests presented in Section 4 shows a poor fit, our interest focus on the source of the misfit. The X^2_{Pearson} in Eq. (7) as well as all the tests discussed in Section 4 can be decomposed to individual terms.

Two different types of residuals can be computed as in Rao (1973), mainly, standardized and adjusted residuals. Standardized residuals for a response pattern s are given by:

$$e_{\text{st}} = \sqrt{n}\frac{e_s}{\sqrt{\hat{\pi}_s(1-\hat{\pi}_s)}}, \tag{25}$$

where $e_s = \hat{p}_s - \hat{\pi}_s$ and $\hat{p}_s$ and $\hat{\pi}_s$ are the observed and expected proportions, respectively, for a response pattern s.

Adjusted residuals are defined as:

$$e_{\text{adj}} = \sqrt{n}\frac{e_s}{\sigma_s}, \tag{26}$$

where σ_s is the standard deviation for e_s. Both standardized and adjusted residuals follow approximately the standard normal distribution.

Similarly, one can compute standardized and adjusted residuals for the lower-order margins. Standardized residuals for the two-way margins are:

$$e_{\text{st}}^{l} = \sqrt{n}\frac{e_{ij}^{l}}{\sqrt{\hat{\pi}_{ij}(1-\hat{\pi}_{ij})}}, \tag{27}$$

where $e_{ij}^{l} = \hat{p}_{ij} - \hat{\pi}_{ij}$ and $\hat{p}_{ij}$ and $\hat{\pi}_{ij}$ are the observed and expected proportions respectively for a pair of items (i, j).

The adjusted residuals are:

$$e_{\text{adj}}^{l} = \sqrt{n}\frac{e_{ij}}{\sigma_{ij}}, \tag{28}$$

where σ_{ij} is the standard deviation for e_{ij} obtained from the diagonal element of (24).

5. Test based on the log-odds ratio

A measure of association in 2×2 contingency tables is the odds ratio. Consider the 2×2 contingency table for items i and j given in Table 1. Cell entry n_{00} denotes the number of individuals who responded negatively to items i and j. The odds ratio for Table 1 is defined as:

$$\hat{\theta} = \frac{n_{00}n_{11}}{n_{01}n_{10}}, \tag{29}$$

where a value close to one shows independence between the items. The estimated odds ratio under the fitted model based on the expected frequencies is:

$$\theta(\hat{\beta}) = \frac{\hat{n}_{00}\hat{n}_{11}}{\hat{n}_{01}\hat{n}_{10}}, \tag{30}$$

Table 1
Two-way contingency table for items i and j

Items		y_j 0	y_j 1
y_i	0	n_{00}	n_{01}
	1	n_{10}	n_{11}

where the quantities $\hat{n}_{00}$, $\hat{n}_{01}$, $\hat{n}_{10}$ and $\hat{n}_{11}$ are expected frequencies under the fitted model. One would expect the odds ratio computed from the observed and the expected frequencies to be close under the assumption that the fitted model is correct. The logarithm of the odds ratio approximates its asymptotic normal distribution (Agresti, 1990).

In a data set with k binary items, there are $k(k-1)/2$ possible pair of items and corresponding log-odds ratios. We denote with $\log\hat{\boldsymbol{\theta}}$ the $(k(k-1)/2 \times 1)$ vector that contains the log-odds ratios of all pair of items.

Clearly $\log\hat{\boldsymbol{\theta}}$ can be written as a linear function of $\hat{\mathbf{p}}$ and $\log\boldsymbol{\theta}(\hat{\boldsymbol{\beta}})$ can be written as a linear function of $\boldsymbol{\pi}(\hat{\boldsymbol{\beta}})$. Hence $\log\hat{\boldsymbol{\theta}} = g(\hat{\mathbf{p}})$ and $\log\boldsymbol{\theta}(\hat{\boldsymbol{\beta}}) = g(\boldsymbol{\pi}(\hat{\boldsymbol{\beta}}))$. Using the delta method, the asymptotic distribution of $\log\hat{\theta}$ is:

$$\log\hat{\boldsymbol{\theta}} \sim N\big(\log\boldsymbol{\theta}\big(\boldsymbol{\pi}(\hat{\boldsymbol{\beta}})\big), D\Omega D'\big), \tag{31}$$

where $D = \frac{\partial\log\boldsymbol{\theta}(\boldsymbol{\pi})}{\partial\boldsymbol{\pi}}\big|_{\boldsymbol{\pi}(\hat{\boldsymbol{\beta}})}$.

Let us define the difference between the observed and the expected log-odds ratio:

$$g(\boldsymbol{e}) = \log\hat{\boldsymbol{\theta}} - \log\boldsymbol{\theta}\big(\boldsymbol{\pi}(\hat{\boldsymbol{\beta}})\big). \tag{32}$$

To test the hypothesis that $g(\boldsymbol{\varepsilon})$ is zero in the population we use the test statistic:

$$X^2_{\text{lort}} = g(\boldsymbol{e})(D\Omega D')^{-1}g(\boldsymbol{e})' \sim \chi^2. \tag{33}$$

The degrees of freedom in (33) are equal to the rank of the matrix $D\Omega D'$. In our examples, we use the estimated covariance matrix given in Christoffersson (1975) for the case of known parameters and the estimated covariance matrix given in Reiser (1996) for the case of estimated parameters. In the application section, the first test is denoted with X^2_{lor} and the latter with X^2_{lorc}.

5.1. *Log-odds ratio residuals for pair of items*

Similar to the residuals defined in Section 4.3, we give here residuals based on the log-odds ratio. For every pair of items, residuals are denoted as:

$$e_{\text{lor}(i)} = \log\hat{\theta}_i - \log\theta_i(\hat{\boldsymbol{\beta}}), \quad i = 1, \ldots, \frac{k(k-1)}{2}. \tag{34}$$

The standardized log-odds ratio residuals are given as

$$\frac{\log\hat{\theta}_i - \log\theta_i(\hat{\boldsymbol{\beta}})}{\sqrt{\omega_i^2}}, \tag{35}$$

where ω_i^2 is the ith diagonal element of $D\Omega D'$. We give two different types of residuals based on the covariance matrix used. When parameters are considered known residuals will be denoted by $\boldsymbol{e}_{\text{lor}}$ and when parameters are estimated from the data residuals will be denoted by $\boldsymbol{e}_{\text{lorc}}$.

5.2. An extension of Bartholomew and Leung test using the log-odds ratio

A similar test to that of Bartholomew and Leung (2002) that approximates the distribution of the sum of squares of the standardized log-odds ratio residuals $\boldsymbol{e}_{\mathrm{lor}}$ by a linear function of a χ^2 distribution is developed. The statistic is $\chi^2_{\mathrm{lor\,BL}} = \boldsymbol{e}'_{\mathrm{lor}}\boldsymbol{e}_{\mathrm{lor}} \sim a + b\chi^2_c$. For the formulation of the test, the hypothesis that the parameter estimation has a non-negligible impact on the asymptotic variance–covariance matrix of $g(\boldsymbol{e})$ is made. Cai et al. (2004) have focused on the same hypothesis for approximating the distribution of the sum of squares of the standardized residuals. Consider that $\Phi_{\mathrm{st}} = D\Omega D'$ where Ω is taken from (19) and $\Phi_{\mathrm{adj}} = D\Omega D'$ where Ω is taken from (24). The $\boldsymbol{\pi}(\boldsymbol{\beta})$ are replaced by the estimated ones to get the estimated asymptotic matrices $\widehat{\Phi}_{\mathrm{st}}$ and $\widehat{\Phi}_{\mathrm{adj}}$. The first three asymptotic moments of the sum of squares of the standardized log-odds ratio residuals are

$$\begin{aligned}
\mu_1\big(\chi^2_{\mathrm{lor\,BL}}\big) &= \mathrm{tr}\big(\widehat{\Phi}_{\mathrm{st}}^{-1}\widehat{\Phi}_{\mathrm{adj}}\big),\\
\mu_2\big(\chi^2_{\mathrm{lor\,BL}}\big) &= 2\,\mathrm{tr}\big((\widehat{\Phi}_{\mathrm{st}}^{-1}\widehat{\Phi}_{\mathrm{adj}})^2\big),\\
\mu_3\big(\chi^2_{\mathrm{lor\,BL}}\big) &= 8\,\mathrm{tr}\big((\widehat{\Phi}_{\mathrm{st}}^{-1}\widehat{\Phi}_{\mathrm{adj}})^3\big).
\end{aligned} \tag{36}$$

By equating these moments to the theoretical moments of a linear function of χ^2 distribution the parameters of interest (a, b, c) are estimated (see Bartholomew and Leung (2002) for details).

6. Simulations

The distribution of overall goodness-of-fit test statistics such as the Pearson and the likelihood ratio statistic for a latent variable model for binary responses has been investigated using parametric bootstrapping in many studies (Reiser and VandenBerg, 1994; Langeheine et al., 1996; Von Davier, 1997; Bartholomew and Tzamourani, 1999; Tollenaar and Mooijaart, 2003).

We performed simulations using the method of parametric bootstrapping for judging the goodness-of-fit of the model as well as for comparing the performance of the tests presented in the chapter in terms of Type I error and power. The steps of parametric bootstrap are:

(1) Estimate the hypothesized model using the data and compute the test statistics of interest.
(2) Treat the estimated parameters as true and generate from the hypothesized model a large number of random samples of same size as the original one.
(3) In each generated data set, estimate the model and compute the goodness-of-fit test statistics.
(4) Compare the actual value of the test statistic from step 1 with reference to the bootstrap sampling distribution of the test statistic obtained in step 3.

By this procedure, we assess both the fit of the model and we also study the asymptotic behavior of the test statistics. More specifically, Type I error is computed for the test statistic of interest and an assessment is made on how satisfactorily it resembles its asymptotic distribution by comparing the empirical moments to the asymptotic ones and by plotting histograms and qqplots.

Von Davier (1997) claims that for the power-divergence family, given in (8), the parametric bootstrapping method fails for the likelihood ratio statistic and the Freeman–Tukey statistic whereas it works for the Pearson and the Cressie–Read statistic.

The power of a test is the probability of rejecting the null hypothesis when the alternative hypothesis is correct. In latent variable models, goodness-of-fit tests have no clear alternatives and these tests are usually used as omnibus tests against all alternatives. For example, when we compute the power of a t-test for a population mean, we choose values under the alternative hypothesis for the population mean but in latent variable modelling there are many parameters involved. In order to assess the power of the tests we assume the one-factor model under the null hypothesis and the two-factor model under the alternative. A large number of data sets is generated from a two-factor model and a one-factor model is fitted in each data set. The data have been generated assuming independent and distinct latent variables. For all the simulated examples, the loadings of the first factor for all items are one and for the second factor half of the items have loadings 1 and half -1. The power of a test statistic, for a given nominal statistical level α_0, is defined, as the proportion of times the test statistic obtained from each simulation yields a p-value lower than α_0. The p-value is computed from the theoretical distribution of the assumed test statistic.

6.1. Examples

We compare the performance of overall and limited information test statistics using real and simulated data. The comparisons are performed under different levels of sparseness. We investigate how satisfactorily the empirical distributions approximate the asymptotic distributions of the tests and we also evaluate them in terms of Type I error and power.

The first example is a real data set that consists of $n = 257$ individuals and four items. The second and third examples are based on simulated data with $n = 200$ individuals and six items and $n = 200$ individuals and eight items, respectively. All simulations are based on 1000 bootstraps.

6.1.1. Example 1: A real data set

This data set is taken from Duncan (1979) and it is on sex role expectations. In the 1953 Detroit Area Study, a sample of 257 mothers were asked the following question regarding sex role expectations: 'Here are some things that might be done by a boy or a girl. Suppose the person were 13 years old. As I read each of these to you, I would like you to tell me if it should be done as a regular task by a boy, by a girl, or by both' (1) Shovelling walks, (2) Washing the car, (3) Dusting furniture, (4) Making beds. Responses of 'boy' to items (1) and (2) and 'girl' to items (3) and (4) were coded as '0' and are referred as the traditional answers. Responses of 'both', which Duncan refers to as 'egalitarian' answers are coded as '1'.

For testing the overall goodness-of-fit of the one-factor model we computed the Pearson chi-square and the likelihood ratio test statistics. These are found to be 23.59 and

Table 2
Example 1: Asymptotic p-values for the limited information statistics, one-factor model

Statistic	X^2_{BL}	X^2_{C}	X^2_{L}	X^2_{OJ}	$X^2_{\mathrm{OJ(3)}}$	X^2_{R}	$X^2_{\mathrm{R(3)}}$	X^2_{lor}	X^2_{lorc}	$X^2_{\mathrm{lor\,BL}}$
	1.95	7.61	1.95	8.60	17.80	13.99	16.99	8.94	9.20	5.86
	0.811	0.022	0.007	0.014	0.007	0.003	0.018	0.022	0.027	0.01

Table 3
Example 1: Bivariate standardized and adjusted residuals, 'log-odds' residuals, one-factor model

Pairs	e^l_{st}	e^l_{adj}	e_{lor}	e_{lorc}
(1, 2)	1.10	2.73	1.94	2.79
(1, 3)	−0.48	−2.51	−0.86	−2.55
(1, 4)	−0.22	−1.75	−0.39	−1.83
(2, 3)	−0.17	−0.54	−0.34	−0.58
(2, 4)	−0.55	−2.38	1.029	−2.36
(3, 4)	−0.03	−0.41	−0.12	−0.59

24.08, respectively, with a corresponding p-value 0.001 for both tests. Both tests reject the one-factor model. The absence of sparseness in the data is judged by obtaining a reasonably large value for the ratio $n/2^k = 16.063$ and by having only four response patterns with an observed frequency lower than five. Therefore, we expect the overall goodness-of-fit tests to approximate satisfactorily their asymptotic distributions.

Table 2 gives the asymptotic p-values for the limited information statistics. All limited information tests except from the log-odds ratio investigate the fit of the model on pairs and triples of positive responses only. From Table 2, we see that all tests but the X^2_{BL} reject the one-factor model. Note that the only difference between the X^2_{BL} and X^2_{L} test is that the later accounts for the estimated parameters. The tests X^2_{C}, X^2_{OJ} and X^2_{lor} have two degrees of freedom whereas X^2_{R} and X^2_{lorc} have three. Tests that use information up to the three-way margins contain more information and have more degrees of freedom with $X^2_{\mathrm{R(3)}}$ having an extra degree of freedom than $X^2_{\mathrm{OJ(3)}}$.

Table 3 gives the standardized and adjusted residuals computed from (27) and (28), respectively, for positive responses only. Assuming that standardized residuals follow the standard normal distribution, absolute values smaller than 2 indicate a good fit. In Table 3, all standardized residuals are smaller that 1.2. However, some of the adjusted residuals indicate a poor fit. It should be noted that adjusted residuals are computed under the hypothesis that the model parameters are not known in advance but they are estimated from the model whereas standardized residuals are based on the assumption that the model parameters are known in advance. Note that the adjusted residuals e^l_{adj} are very close to the adjusted log-odds ratio residuals e_{lorc}. The results of the parametric bootstrapping presented in Table 6 aim to investigate the asymptotic distribution of the overall and the limited goodness-of-fit tests as well as the asymptotic behavior of the standardized and the adjusted residuals. Tables 4 and 5 give the p-values obtained from

Table 4
Example 1: Empirical p-values based on 1000 bootstrap samples, one-factor model

Statistic	X^2_{BL}	X^2_{C}	X^2_{L}	X^2_{OJ}	$X^2_{\mathrm{OJ}(3)}$	X^2_{R}	$X^2_{\mathrm{R}(3)}$	X^2_{lor}	X^2_{lorc}	$X^2_{\mathrm{lor\,BL}}$
p-value	0.006	0.014	0.006	0.009	0.006	0.013	0.006	0.008	0.024	0.004

Table 5
Example 1: Empirical p-values for the bivariate residuals based on 1000 bootstrap values, one-factor model

Pairs	Empirical p-values			
	e^l_{st}	e^l_{adj}	e_{lor}	e_{lorc}
(1, 2)	0.008	0.004	0.003	0.004
(1, 3)	0.025	0.018	0.023	0.013
(1, 4)	0.441	0.44	0.428	0.416
(2, 3)	0.247	0.264	0.219	0.225
(2, 4)	0.021	0.018	0.022	0.019
(3, 4)	0.701	0.701	0.598	0.577

the empirical distributions using the value of the test statistic of the original data set. All tests reject the one-factor model. It is clear that there is a big difference between the asymptotic and the empirical p-value of the X^2_{BL}. The empirical p-values of the residuals in Table 5 show a misfit in three bivariate margins. This result is in accordance with the large adjusted residuals in Table 3.

Figure 1 shows the empirical distribution of the standardized bivariate residuals that were derived from 1000 bootstrap samples ($k(k-1)/2 = 6$ in total since only positive responses to pair of items are examined). It is clear that the mean value is 0 but the variance seems much smaller. We hardly observe any value larger than 1 in any of the standardized residuals. It is known that for a standard normal distribution the first four theoretical moments are 0, 1, 0 and 3. From Table 6, we see that the empirical second and fourth moments for the standardized residuals are not close to the theoretical ones. As far as the adjusted residuals are concerned, the empirical moments are close to the theoretical ones with the exception of the third adjusted residual.

This is also seen from Figure 2 where the empirical qqplots of the quantiles of the adjusted residuals e^l_{adj} versus the quantiles of a standard Normal distribution are plotted. The qqplot for the adjusted residuals for pair (1, 4) shows some deviation from the tails. The results are similar for the log-odds ratio residuals but they are not given here.

Table 7 gives Type I error rates for the limited information statistics based on 1000 bootstrap samples. With the exception of X^2_{BL}, all statistics give a Type I error rate close to the nominal level.[1] Table 8 gives the corresponding Type I errors for the overall statistics.

[1] The 95% confidence interval for $\alpha_0 = 0.01$ is [0.038, 0.0162], for $\alpha_0 = 0.05$ is [0.0365, 0.0635] and for $\alpha_0 = 0.1$ is [0.0814, 0.1186].

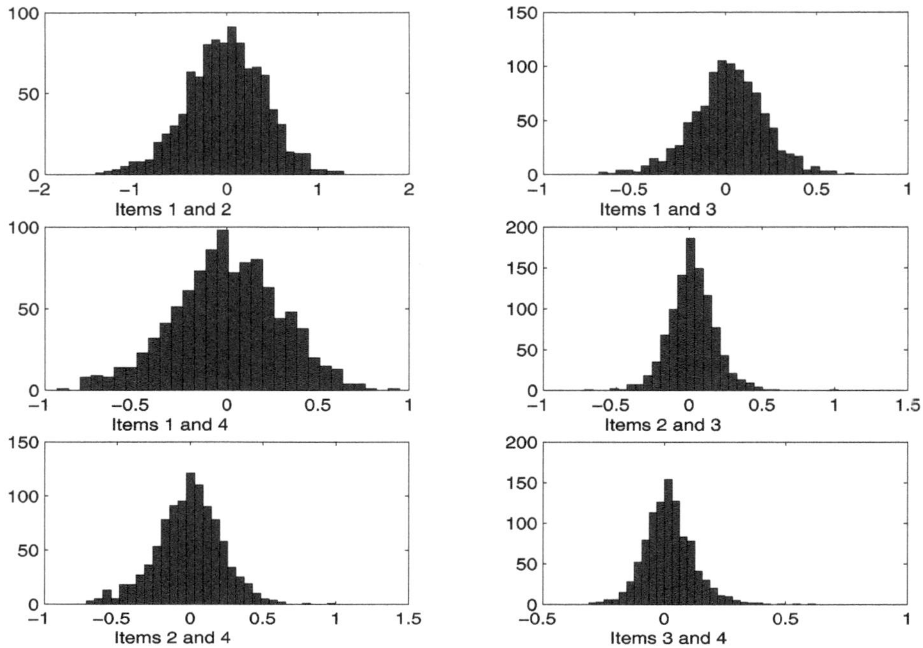

Fig. 1. Empirical distributions for the bivariate standardized residuals.

Table 6
Example 1: Moments for the standardized and the adjusted bivariate residuals from 1000 bootstrap samples, one-factor model

	Empirical moments								
e^l_{st}	1st	2nd	3rd	4th	e^l_{adj}	1st	2nd	3rd	4th
(1, 2)	−0.017	0.425	−0.012	0.103	(1, 2)	−0.014	0.997	0.007	2.831
(1, 3)	0.016	0.199	−0.001	0.006	(1, 3)	0.114	1.021	0.05	3.574
(1, 4)	−0.002	0.297	−0.003	0.023	(1, 4)	0.001	2.696	3.794	267.369
(2, 3)	0.022	0.156	0.001	0.003	(2, 3)	0.09	0.607	0.435	3.504
(2, 4)	−0.008	0.226	−0.001	0.01	(2, 4)	−0.019	0.994	−0.044	2.97
(3, 4)	0.021	0.104	0.001	0.001	(3, 4)	0.228	1.068	0.549	4.685

Table 7
Example 1: Type I error rates from 1000 bootstrap samples for the limited information statistics

α_0	X^2_{BL}	X^2_{C}	X^2_{L}	X^2_{OJ}	$X^2_{OJ(3)}$	X^2_{R}	$X^2_{R(3)}$	X^2_{lor}	X^2_{lorc}	$X^2_{lor\,BL}$
0.01	0	0.01	0.006	0.007	0.007	0.006	0.008	0.01	0.01	0.008
0.05	0	0.055	0.04	0.037	0.044	0.044	0.045	0.053	0.05	0.037
0.1	0	0.103	0.1	0.098	0.11	0.096	0.092	0.101	0.083	0.084

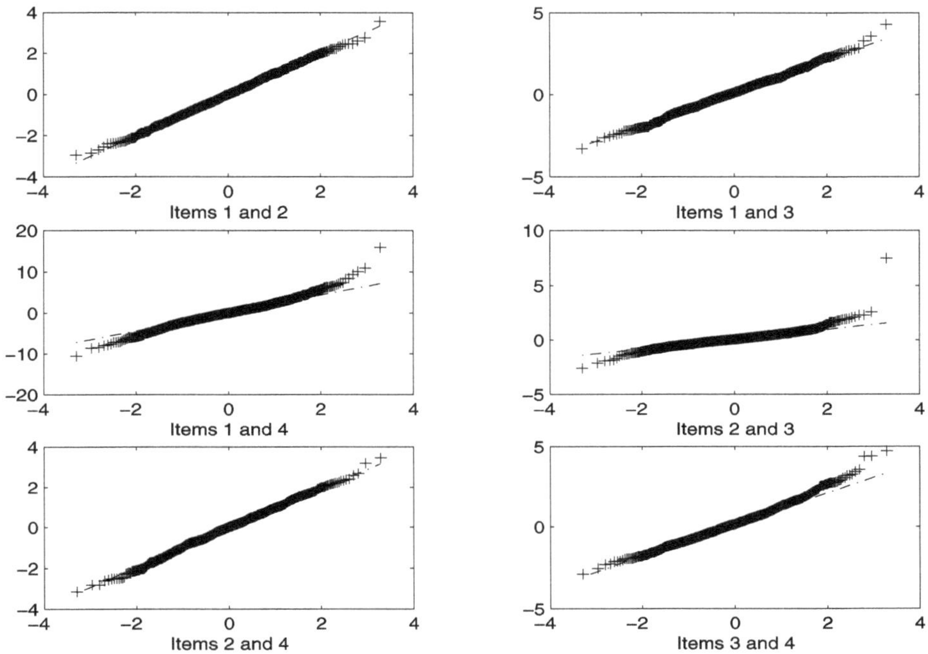

Fig. 2. qqplots for the adjusted bivariate residuals.

Table 8
Example 1: Type I error rates from 1000 bootstrap samples for the overall goodness-of-fit statistics

Nominal level α_0	X^2_{Pearson}	X^2_{CR}
0.01	0.011	0.015
0.05	0.058	0.076
0.1	0.122	0.147

We investigate below how close the empirical distributions of the limited information statistics are to the corresponding asymptotic ones. All statistics presented in Sections 4 and 5 claim to follow a chi-square distribution. The theoretical moments of a χ^2 distribution are

$$\begin{aligned}
\mu_1 &= df,\\
\mu_2 &= 2df,\\
\mu_3 &= 8df,\\
\mu_4 &= 48df + 12df^2. \qquad (37)
\end{aligned}$$

Table 9 gives the empirical moments of the limited information statistics from the 1000 bootstrap samples. The theoretical moments that are derived from the asymptotic

Table 9
Example 1: Empirical moments for the limited information statistics, one-factor model

Moments	X^2_{C}	X^2_{OJ}	X^2_{lor}	X^2_{lorc}	X^2_{R}	$X^2_{\mathrm{OJ(3)}}$	$X^2_{\mathrm{R(3)}}$
1st	2.01(2)	1.97	1.99	2.06	2.75(3)	6.13(6)	6.76(7)
2nd	4.05(4)	3.62	4.03	4.43	6.01(6)	11.33(12)	13.85(14)
3rd	16.84(16)	12.62	16.727	17.01	21.45(24)	35.99(48)	55.45(56)
4th	155.18(144)	104.55	153.15	133.75	208.129(252)	**500**(720)	892.78(924)

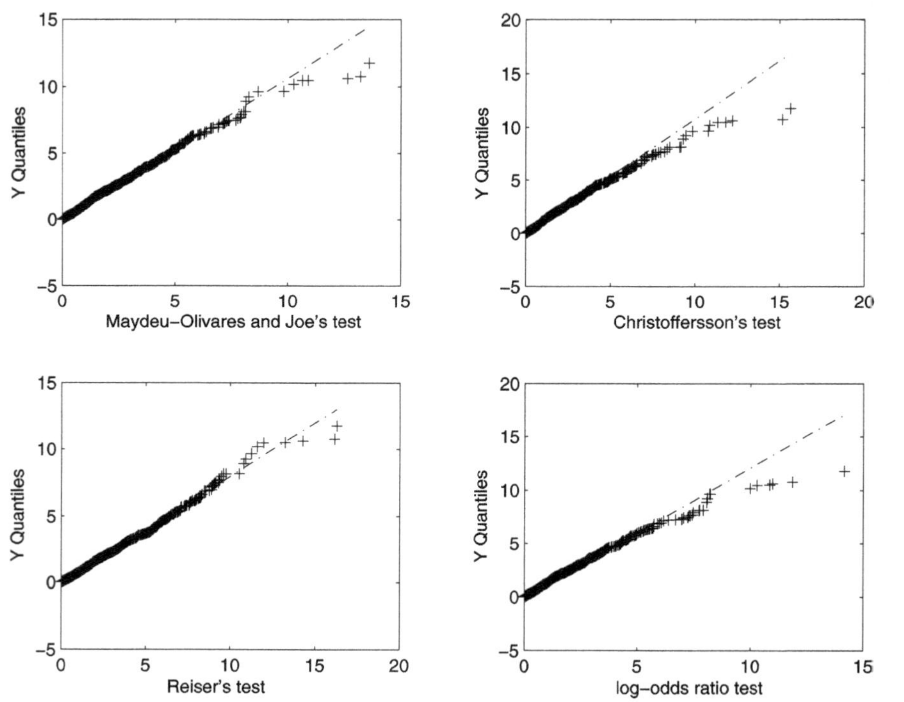

Fig. 3. qqplots for the limited information statistics that use information up to the two-way margins.

distribution of the statistics are given in parenthesis. The tests given in the first four columns of Table 9 have two degrees of freedom. Only in X^2_{OJ} and $X^2_{\mathrm{OJ(3)}}$, there seems to be some deviation between the third and fourth empirical moments from the corresponding asymptotic ones but the difference is found to be non-significant for X^2_{OJ} whereas it is found to be significant for the fourth moment of $X^2_{\mathrm{OJ(3)}}$. Cases where the empirical moment deviates from the asymptotic ones are printed in bold.

Figure 3 gives the empirical qqplots of the quantiles of the limited information statistics versus the quantiles of the corresponding χ^2 distribution. Figure 4 gives the empirical qqplots of the quantiles of the limited information statistics that use the information up to the three-way margins versus the quantiles of the corresponding χ^2

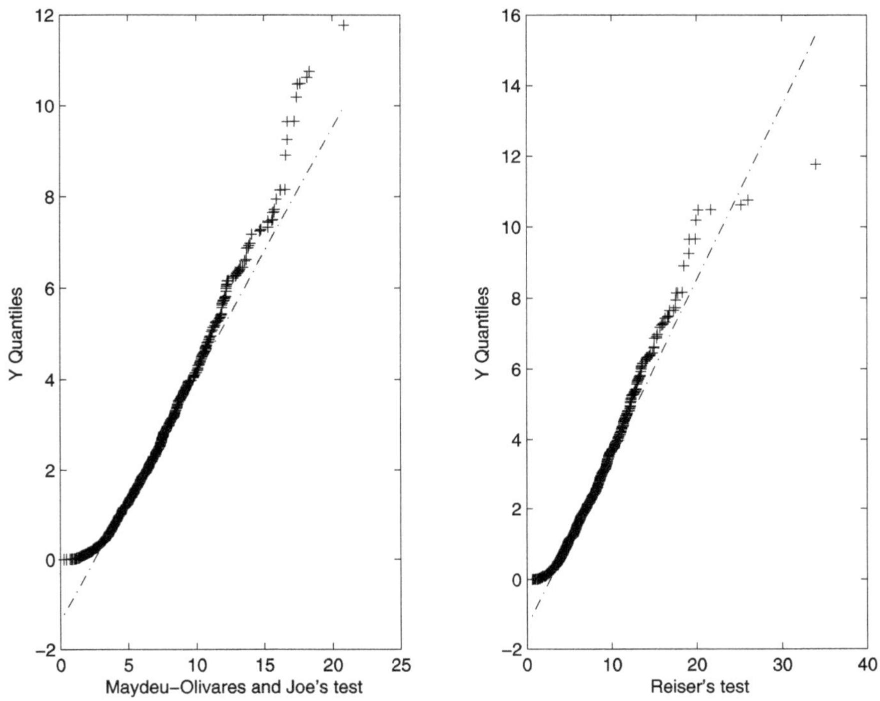

Fig. 4. qqplots for the limited information statistics that use information up to the three-way margins.

distribution. In Figure 3, the log-odds ratio test for the case where the parameters are estimated from the data is plotted. Most limited information statistics seem to follow their asymptotic distribution though there seem to be some deviations in the upper tail especially for the test-statistics that use information up to the three-way margins. With an example of only 4 items, a test that is based on a comparison between the observed and the expected, under the fitted model, three-way margins is very close to an overall goodness-of-fit test.

Table 10 gives the power of the limited information statistics and the overall goodness-of-fit measures. Tests that consider that the parameters are known in advance are less powerful with X^2_{BL} appearing to have no power. Also tests that use information from higher margins than the two-way margins are less powerful.

6.1.2. Example 2: A simulated data set with six items

A more sparse, with reference to the previous example, data set is constructed by increasing the number of items to six and reducing the sample size to 200. The parameters that were used for the generation of the data set were zeros for the thresholds ($\boldsymbol{\beta}_0$) and ones for the factor loadings ($\boldsymbol{\beta}_1$) of all items. The ratio $n/2^k = 3.125$ is very small and there are 16 response patterns that were not observed. Type I error rates obtained from 1000 bootstrap samples for the overall goodness-fit-tests are given in Table 11. They are all very different from the nominal levels of significance.

Table 10
Example 1: Power for the limited information statistics when the data are generated from a two-factor model $n = 257$, $k = 4$

Statistic	Nominal level α_0		
	0.01	0.05	0.1
X^2_{BL}	0	0	0
X^2_{C}	0.29	0.57	0.8
X^2_{L}	0.82	0.97	1
X^2_{OJ}	0.77	0.89	0.97
$X^2_{\mathrm{OJ(3)}}$	0.52	0.74	0.86
X^2_{R}	0.69	0.88	0.94
$X^2_{\mathrm{R(3)}}$	0.61	0.81	0.91
X^2_{lor}	0.32	0.59	0.81
X^2_{lorc}	0.77	0.91	0.96
$X^2_{\mathrm{lor\,BL}}$	0.9	0.95	0.98
X^2_{Pearson}	0.5	0.77	0.86
X^2_{CR}	0.51	0.78	0.87

Table 11
Example 2: Type I error rates from 1000 bootstrap samples for the overall goodness-of-fit statistics, one-factor model

Nominal p-value	X^2_{Pearson}	X^2_{CR}
0.01	0.233	0.215
0.05	0.434	0.422
0.1	0.576	0.575

Table 12
Example 2: Type I error rates from 1000 bootstrap samples for the limited information statistics, one-factor model

α_0	X^2_{BL}	X^2_{C}	X^2_{L}	X^2_{OJ}	$X^2_{\mathrm{OJ(3)}}$	X^2_{R}	$X^2_{\mathrm{R(3)}}$	X^2_{lor}	X^2_{lorc}	$X^2_{\mathrm{lor\,BL}}$
0.01	**0**	0.005	0.007	0.009	0.008	0.01	0.01	0.005	0.011	**0.041**
0.05	**0**	**0.012**	0.041	0.047	0.046	0.046	0.052	**0.013**	0.051	**0.104**
0.1	**0**	**0.029**	**0.08**	0.097	0.095	0.097	0.098	**0.029**	0.09	**0.177**

Table 12 gives Type I error rates for the limited information statistics. Type I error rates for some statistics are worse from the corresponding rates computed in the less sparse example given in Table 7. Type I error rates that lie outside their 95% confidence

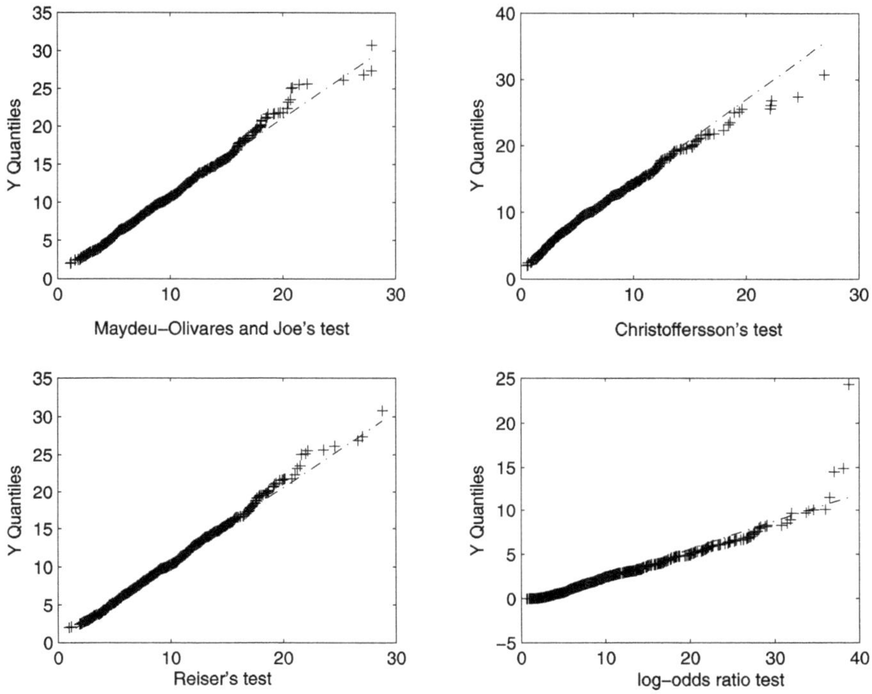

Fig. 5. qqplots for the limited information statistics that use information up to the two-way margins.

intervals are printed in bold. Only the empirical Type I error rates of X^2_{OJ}, X^2_{R} and X^2_{lorc} lie within the 95% confidence interval of the theoretical ones. Those three tests assume that parameters are estimated and are not known in advance.

Figures 5 and 6 give the qqplots for the limited information statistics that use information up to the two-way and three-way margins, respectively. There is only some concern for the upper tail of the X^2_{C}. It should be noted that in Figure 5 the log-odds ratio under the composite hypothesis, X^2_{lorc}, is plotted.

Table 13 gives the empirical moments of the limited-information test statistics where the theoretical moments are given in parenthesis. The theoretical moments for X^2_{OJ} and X^2_{lor} are not given since they have the same limiting distribution with X^2_{C}. Significant differences between empirical and theoretical moments have been found for X^2_{C} and X^2_{lor} and they are printed in bold. These are the test statistics that assume that the model parameters are known in advance and this discrepancy is a further evidence that those test statistics do not follow their asymptotic distribution.

Table 14 gives the power of the tests. The high power of the overall goodness-of-fit measures is an indication of its high tendency to reject too often even a correct model.

6.1.3. *Example 3*

The third example consists of $n = 300$ subjects and $k = 8$ items generated from a one-factor model. The parameter values that were used for generating the data are zeros

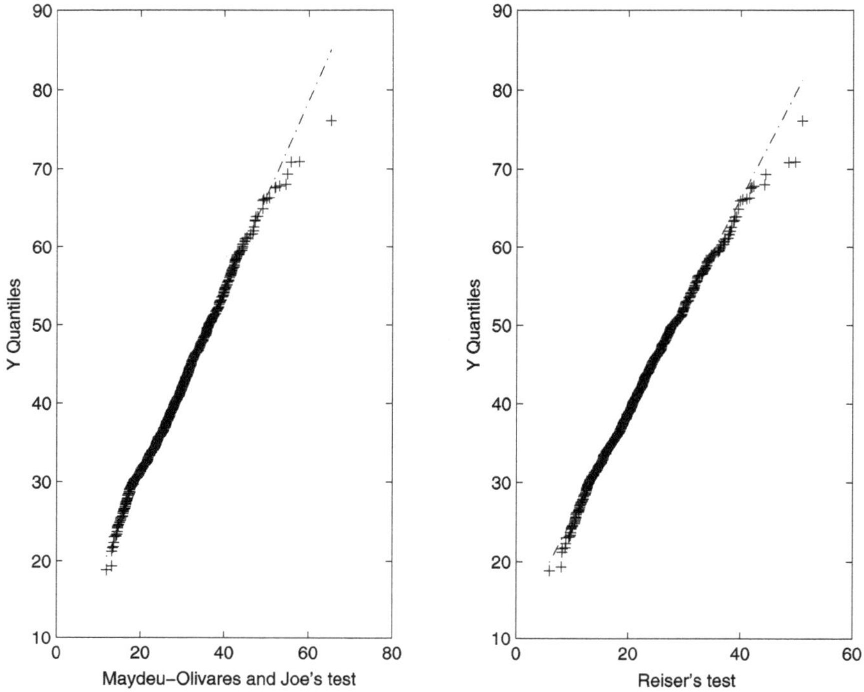

Fig. 6. qqplots for the limited information statistics that use information up to the three-way margins.

Table 13
Example 2: Empirical moments for the limited information statistics, one-factor model

Moments	X^2_{C}	X^2_{OJ}	X^2_{lor}	X^2_{lorc}	X^2_{R}	$X^2_{\mathrm{OJ(3)}}$	$X^2_{\mathrm{R(3)}}$
1st	**6.02**(9)	8.78	**6.11**	7.12(7)	9.07(9)	28.74(29)	21.38(21)
2nd	12.64(18)	15.67	13.13	16.49(14)	16.52(18)	52.22(58)	38.55(42)
3rd	65.51(72)	59.75	69.94	59.15(56)	62.64(72)	178.39(232)	123.06(168)
4th	1023.8(1404)	1118.9	1119.9	866.33(924)	1179.3(1404)	8953.9(69513)	4969.8(36729)

for the thresholds and ones for the factor loadings for all items. Table 15 gives Type I error rates from 1000 bootstrap samples.The test statistics, X^2_{C}, X^2_{lor} and $X^2_{\mathrm{lor\,BL}}$ seem to be too liberal while X^2_{BL} is found once again to be a very conservative test. Limited information statistics that are computed under the simple hypothesis fail to produce Type I error rates that will match their asymptotic Type I error rates.

Note that X^2_{lorc} includes deviations between observed and expected under the model responses not only for positive responses but for all combinations. More specifically, for one of the simulated data sets the big discrepancies found among X^2_{OJ}, X^2_{lor} and X^2_{lorc} were explained by the fact that for items 4 and 7 the standardized residuals for positive responses was found to be 0.535 while the corresponding standardized residual for a

Table 14
Example 2: Power for the limited information statistics when the data are generated from a two-factor model $n = 200$, $k = 6$

Statistic	Nominal level α_0		
	0.01	0.05	0.1
X^2_{BL}	0	0	0
X^2_{C}	0.28	0.41	0.48
X^2_{L}	0.78	0.85	0.95
X^2_{OJ}	0.74	0.84	0.91
$X^2_{\mathrm{OJ(3)}}$	0.52	0.74	0.8
X^2_{R}	0.69	0.79	0.89
$X^2_{\mathrm{R(3)}}$	0.55	0.75	0.83
X^2_{lor}	0.32	0.48	0.56
X^2_{lorc}	0.48	0.59	0.68
$X^2_{\mathrm{lor\,BL}}$	0.94	0.98	1
X^2_{Pearson}	0.85	0.95	0.97
X^2_{CR}	0.85	0.95	0.97

Table 15
Example 3: Type I error rates from 1000 bootstrap samples for the limited information statistics, one-factor model

α_0	X^2_{BL}	X^2_{C}	X^2_{L}	X^2_{OJ}	$X^2_{\mathrm{OJ(3)}}$	X^2_{R}	$X^2_{\mathrm{R(3)}}$	X^2_{lor}	X^2_{lorc}	$X^2_{\mathrm{lor\,BL}}$
0.01	**0**	**0.038**	0.004	0.006	0.004	0.075	0.004	**0.049**	0.016	**0.03**
0.05	**0**	**0.096**	0.06	0.05	0.039	0.039	0.039	**0.108**	0.048	**0.068**
0.1	**0**	**0.159**	0.103	0.109	**0.075**	0.096	0.081	**0.179**	0.088	**0.116**

negative response to item 4 and positive to item 7 was -1.761. Having good fit on the lower-order margins for positive responses and bad fit on the remaining combinations is more common as the number of items increases.

Table 16 gives the power of the limited and overall goodness-of-fit tests. It is clear that power has increased considerably in reference with the previous examples (Tables 10 and 14). Power results should always be interpreted with some caution since power comparisons require equal Type I error rates. Liberal tests, such as overall goodness-of-fit tests have high power because they tend to overestimate the significance. These tests often reject a correct model, let alone a false one. Surprisingly, although the X^2_{C} and X^2_{lor} are liberal for the case of eight items are not as powerful as the rest of the tests.

Table 16
Example 3: Power for the limited information statistics when the data are generated from a two-factor model $n = 300$, $k = 8$

Statistic	Nominal level α_0		
	0.01	0.05	0.1
X^2_{BL}	0.01	0.04	0.19
X^2_{C}	0.54	0.66	0.71
X^2_{L}	1	1	1
X^2_{OJ}	1	1	1
$X^2_{\text{OJ(3)}}$	0.98	1	1
X^2_{R}	1	1	1
$X^2_{\text{R(3)}}$	1	1	1
X^2_{lor}	0.54	0.68	0.75
X^2_{lorc}	0.98	0.99	0.99
$X^2_{\text{lor BL}}$	1	1	1
X^2_{Pearson}	1	1	1
X^2_{CR}	1	1	1

7. Conclusion

The chapter reviews overall and limited information goodness-of-fit statistics for latent variable models with binary items. The fit of the model is examined on a contingency table consisting of 2^k cells where k is the number of items. The usual practices for investigating the fit of such a model involve the Pearson X^2 and the likelihood ratio statistic. Overall goodness-of-fit tests are proven to be liberal when data are sparse. These statistics do not hold their asymptotic distributions under the presence of sparseness and they tend to overestimate the significance by a large amount. This was obvious in the second and third example where the empirical Type I error probabilities for the overall measures of fit obtained from the simulation study are very different from the theoretical nominal levels. Even with four items the overall goodness-of-fit tests tend to overestimate the level of significance when $\alpha_0 = 0.1$. Sparseness is not easily defined and is affected by the sample size, the number of items as well as model parameters. To overcome the problem of sparseness, limited information statistics that evaluate the fit of the model in the lower order margins have been proposed in the literature.

There does not seem to be a panacea for judging the fit of a latent variable model. Each of the limited information statistics has its pros and cons. Along with the limited information statistics the estimation of different types of residuals is proposed. The standardized residuals do not follow the standard normal distribution whereas the adjusted residuals seem to approximate it more satisfactorily. The same holds for the log-odds ratio residuals.

The limited information statistics can be divided into two categories those that assume known parameters and those that assume that parameters are being estimated from the data. Tests that are based on the latter assumption are more powerful and their empirical Type I error rates are closer to the nominal ones. On the other hand, tests that consider the parameters known are easily computed. Finally, the asymptotic variance–covariance matrix of $\boldsymbol{e}$, when the parameters are estimated from the data, is usually singular and its inversion might be based on subjective criteria.

Another division of the tests is between those that use only one cell (out of four) from the pairwise associations and tests that use all four cells of a contingency table of any two items. Although it is not a usual phenomenon, there might be some examples where the model fits satisfactorily one cell of a two-way contingency table but not some or all of the rest. This occurs more often as the number of items increases. Tests that use the whole information from a two-way contingency table are based on the log-odds ratio and are discussed here.

The X^2_{BL} test statistic is very conservative and it rarely rejects even a 'false' model. However, the correction to this test suggested by Cai et al. (2004) has improved significantly the performance of the X^2_{BL} test.

The test-statistic X^2_{lor} is the analogue of X^2_{C} that uses all the information from a two-way contingency table. Similarly, X^2_{lorc} is the analogue for X^2_{R}. The statistics X^2_{C} and X^2_{lor} have similar behavior, similar Type I error rates, similar power and empirical moments. They are less powerful than X^2_{R} and X_{lorc}. In the first example X^2_{C} and X^2_{lor} hold their asymptotic distribution but this is not the case in the rest of the examples where the number of items has increased to six and eight. The statistics X^2_{OJ}, X^2_{R} and X^2_{lorc} seem to have a valid asymptotic distribution. The statistic X^2_{OJ} is a variant of X^2_{R} that uses less information but at the same time proposes a more stable method for the inversion of the asymptotic variance–covariance matrix of $\sqrt{n}\boldsymbol{e}^l$ than that of Reiser's test. Maydell-Olivares and Joe's test, as well as Reiser's test can easily be adjusted to include higher order margins. In our simulations these tests have been computed up to the three-way margins. The statistics $X_{\mathrm{OJ}(3)}$ and $X^2_{\mathrm{R}(3)}$ are less powerful than their counterparts that use only the univariate and bivariate margins. All tests increase their power as the number of items increases. However, we should bear in mind that power comparisons between tests are more meaningful when tests have the same Type I error rates. Finally we expanded X^2_{L} so as to incorporate all cells of the two-way contingency table. This test, $X^2_{\mathrm{lor\,BL}}$, though powerful, does not behave satisfactorily in terms of Type I error rates.

A potential problem with those tests is that sparsity might still occur in the three-way margins especially as the number of items increases and the sample size remains moderate.

From the simulations reported here and others conducted, we have come to the conclusion that the use of limited information criteria looks promising and useful and more research is needed in order to draw safe conclusions about whether or not their asymptotic distributions hold. Many combinations of different sample sizes and number of items should be explored. In results not shown here, it is found that the bootstrap test does not perform well with respect to Type I error. Tollenaar and Mooijaart (2003) and Reiser (2004) have come to the same conclusion as well. This fact, along with the invalid asymptotic distribution of overall goodness-of-fit measures when data are sparse, constitutes the use of limited information test statistics of great importance.

Acknowledgements

D. Mavridis is supported by the research program 'Iraklitos: Fellowships for research of Athens University of Economics and Business', co-funded by the European Union.

References

Agresti, A. (1990). *Categorical Data Analysis*. Wiley, New York.

Bartholomew, D.J., Knott, M. (1999). *Latent Variable Models and Factor Analysis*, second ed. *Kendall Library of Statistics*, vol. 7. Arnold, London.

Bartholomew, D.J., Leung, S.O. (2002). A goodness of fit test for sparse 2^p contingency tables. *British Journal of Mathematical and Statistical Psychology* **55**, 1–15.

Bartholomew, D.J., Tzamourani, P. (1999). The goodness-of-fit of latent trait models in attitude measurement. *Sociological Methods and Research* **27**, 525–546.

Birch, M. (1964). A new proof of the Pearson–Fischer theorem. *Annals of Mathematical Statistics* **35**, 818–824.

Bishop, Y.M., Fienberg, S., Holland, P. (1975). *Discrete Multivariate Analysis*. MIT Press, Cambridge, MA.

Bock, R.D., Aitkin, M. (1981). Marginal maximum likelihood estimation of item parameters: application of an EM algorithm. *Psychometrika* **46** (4), 443–459.

Cai, L., Maydeu-Olivares, A., Coffman, L., Thissen, D. (2004). Limited information goodness of fit testing of item response theory models for sparse 2^p tables. *British Journal of Mathematical and Statistical Psychology*.

Christoffersson, A. (1975). Factor analysis of dichotomized variables. *Psychometrika* **40** (1), 5–31.

Cochran, W.G. (1954). Some methods for strengthening the common χ^2 tests. *Biometrics* **10**, 417–451.

Cramer, H. (1946). *Mathematical Methods of Statistics*. Princeton University Press, Princeton, NJ.

Cressie, N., Read, T. (1984). Multinomial goodness-of-fit tests. *Journal of the Royal Statistical Society, Series B* **46** (3), 440–464.

Duncan, O.D. (1979). Indicators of sex typing: Traditional and egalitarian, situational and ideological responses. *American Journal of Sociology* **85**, 251–260.

Dunson, D.B. (2000). Bayesian latent variable models for clustered mixed outcomes. *Journal of the Royal Statistical Society, Series B* **62**, 355–366.

Hosmer, D., Lemeshow, S. (1980). A goodness-of-fit test for the multiple logistic regression model. *Communications in Statistics A* **10**, 1043–1069.

Kendall, M.G. (1952). *The Advanced Theory of Statistics*. Griffin, London.

Koehler, K., Larntz, K. (1980). An empirical investigation of goodness-of-fit statistics for sparse multinomials. *Journal of the American Statistical Association* **75**, 336–344.

Langeheine, R., Pannekoek, J., van de Pol, F. (1996). Bootstrapping goodness-of-fit measures in categorical data analysis. *Sociological Methods and Research* **24** (4), 492–516.

Maydeu-Olivares, A., Joe, H. (2005). Limited and full-information estimation and goodness-of-fit testing in 2^n contingency tables: A unified framework. *Journal of the American Statistical Association* (6), 1009–1020.

Morris, C. (1975). Central limit theorems for multinomial sums. *Annals of Statistics* **3**, 365–384.

Muthén, B. (1978). Contributions to factor analysis of dichotomous variables. *Psychometrika* **43** (4), 551–560.

Patz, R.J., Junker, B.W. (1999). A straightforward approach to Markov Chain Monte Carlo methods for item response models. *Journal of Educational and Behavioral Statistics* **24**, 146–178.

Rabe-Hesketh, S., Pickles, A., Skrondal, A. (2001). GLLAMM: Stata program to fit generalised linear latent and mixed models. Manual Technical Report. Department of Biostatistics and Computing.

Rao, C.R. (1973). *Linear Statistical Inference and its Applications*, third ed. John Wiley & Sons, New York.

Reiser, M. (1996). Analysis of residuals for the multinomial item response model. *Psychometrika* **61**, 509–528.

Reiser, M. (2004). A comparison of full and limited-information methods for testing goodness of fit with applications to latent structure models. Technical Report. Arizona State University.

Reiser, M., VandenBerg, M. (1994). Validity of the chi-square test in dichotomous variable factor analysis when expected frequencies are small. *British Journal of Mathematical and Statistical Psychology* **47**, 85–107.

Simonoff, J. (1985). An improved goodness-of-fit statistic for sparse multinomials. *Journal of the American Statistical Association* **80** (391), 671–677.

Tate, M.W., Hyer, L. (1973). Inaccuracy of the chi-squared test of goodness of fit when expected frequencies are small. *Journal of the American Statistical Association* **68**, 836–841.

Tollenaar, N., Mooijaart, A. (2003). Type I errors and power of the parametric bootstrap goodness-of-fit test: Full and limited information. *British Journal of Mathematical and Statistical Psychology* **56**, 271–288.

Von Davier, M. (1997). Bootstrapping goodness-of-fit statistics for sparse categorical data: Results of a Monte-Carlo study. *Methods of Psychological Research Online* **2** (2).

Handbook of Computing and Statistics with Applications, Vol. 1
ISSN: 1871-0301

DOI: 10.1016/S1871-0301(06)01008-0

8

Bayesian Structural Equation Modeling

Jesus Palomo, David B. Dunson and Ken Bollen

Abstract

Structural equation models (SEMs) with latent variables are routinely used in social science research, and are of increasing importance in biomedical applications. Standard practice in implementing SEMs relies on frequentist methods. A simple and concise description of an alternative Bayesian approach is developed. Furthermore, a brief overview of the literature, a description of Bayesian specification of SEMs, and an outline of a Gibbs sampling strategy for model fitting is provided. Bayesian inferences are illustrated through an industrialization and democratization case study from the literature. The Bayesian approach has some distinct advantages, due to the availability of samples from the joint posterior distribution of the model parameters and latent variables, that are highlighted. These posterior samples provide important information not contained in the measurement and structural parameters. As is illustrated using the case study, this information can often provide valuable insight into structural relationships.

1. Introduction

Structural equation models (SEMs) with latent variables provide a very general framework for modeling of relationships in multivariate data (Bollen, 1989). Although SEMs are most commonly used in studies involving intrinsically latent variables, such as happiness, quality of life, or stress, they also provide a parsimonious framework for covariance structure modeling. For this reason, they have become increasingly used outside of the traditional social science applications.

Software available for routine fitting of SEMs, including LISREL (Jöreskog and Sörbom, 1996), MPLUS (Muthén and Muthén, 1998, 2003) and BMDP (Bentler, 1992), rely on frequentist methods. Most commonly, SEMs are fitted using either full information maximum likelihood estimation (Jöreskog and Sörbom, 1985) or generalized least squares procedures (Browne, 1974). Such methods can easily allow mixtures of continuous and ordered categorical observed variables by using an underlying variable structure (Muthén, 1984; Arminger and Küsters, 1988). Recent research has developed extensions to allow interactions and nonlinear structures (Jöreskog and Yang, 1996;

Bollen and Paxton, 1998; Wall and Amemiya, 2000). Frequentist inferences are typically based on point estimates and hypothesis tests for the measurement and latent variable parameters, marginalizing out the latent variables.

Although the overwhelming majority of the literature on SEMs is frequentist in nature, Bayesian approaches have been proposed by a number of authors. For factor models, which are a special case of SEMs, there is a long history of Bayesian methods (see, for example, Martin and McDonald, 1975; Lee, 1981; Ansari and Jedidi, 2000; Lopes and West, 2003). For more general SEMs, early work was done by Bauwens (1984) and Lee (1992). Recent articles have focused on the use of Markov chain Monte Carlo (MCMC) methods to implement Bayesian analysis in complex cases, involving nonlinear structures (Arminger and Muthén, 1998; Lee and Song, 2004), heterogeneity (Ansari et al., 2000; Lee and Song, 2003a, 2003b), and multilevel data (Dunson, 2000; Jedidi and Ansari, 2001; Song and Lee 2004a, 2004b; Jackman, 2004). In addition, Raftery (1993) considers the important problem of model selection in SEMs from a Bayesian perspective. Additional articles on Bayesian SEMs have been published by Scheines et al. (1999) and Lee and Shi (2000a, 2000b).

The goal of this chapter is not to review all of these approaches, but instead to provide an easily accessible overview of a Bayesian approach to SEMs, illustrating some of the advantages over standard frequentist practice. The flexibility of the Bayesian approach allows to apply the method in a very broad class of SEM-type modeling frameworks, such as nonlinear interactions, missing data, mixed categorical, count, and continuous observed variables, etc. The WinBUGS software package,[1] which is freely available, can be used to implement Bayesian SEM analysis, see, e.g., Clinton et al. (2004) for a Bayesian spatial voting model using WinBUGS.

There are several important differences between the Bayesian and frequentist approaches, which will be highlighted. First, the Bayesian approach requires the specification of prior distributions for each of the model unknowns, including the latent variables and the parameters from the measurement and structural models. Frequentists typically assume Gaussian distributions for the latent variables, but do not specify priors for mean or covariance parameters.[2] Because the posterior distributions upon which Bayesian inferences are based depend both on the prior distribution and the likelihood of the data, the prior plays an important role. In particular, specification of the prior allows for the incorporation of substantive information about structural relationships, which may be available from previous studies or social science theory. In the absence of such information, vague priors can be chosen. As the sample size increases, the posterior distribution will be driven less by the prior, and frequentist and Bayesian estimates will tend to agree closely.

A second difference is computational. Bayesian model fitting typically relies on MCMC, which involves simulating draws from the joint posterior distribution of the model unknowns (parameters and latent variables) through a computationally intensive procedure. The advantage of MCMC is that there is no need to rely on large sample

[1] http://www.mrc=bsu.cam.ac.uk/bugs/.

[2] Researchers can incorporate observed variables that come from distributions with excess kurtosis by using corrected likelihood ratio tests, bootstrapping methods, or sandwich estimators for asymptotic standard errors (Satorra and Bentler, 1988; Bollen and Stine, 1990, 1993).

assumptions (e.g., asymptotic normality), because exact posterior distributions can be estimated for any functional of the model unknowns. In small to moderate samples, these exact posteriors can provide a more realistic measure of model uncertainty, reflecting asymmetry and not requiring the use of a delta method or other approximations. The downside is that it may take a long time (e.g., several hours) to obtain enough samples from the posterior so that Monte Carlo (MC) error in posterior summaries is negligible. This is particularly true in SEMs, because there can be problems with slow mixing producing high autocorrelation in the MCMC samples. This autocorrelation, which can be reduced greatly through careful parametrization or computation tricks (e.g., blocking and parameter expansion), makes it necessary to collect more samples to produce an acceptable level of MC error.

An additional benefit that is gained by paying this computational price is that samples are available from the joint posterior distribution of the latent variables. Often, these samples can be used to obtain important insights into structural relationships, which may not be apparent from estimates (Bayesian or frequentist) of the structural parameters. This is certainly the case in the industrialization and democratization application (Bollen, 1989), which we will use to illustrate the concepts starting in Section 3.

Section 2 reviews the basic SEM modeling framework and introduces the notation. Section 3 describes the Bayesian approach, focusing on normal linear SEMs for simplicity in exposition, introduces the conditionally-conjugate priors for the parameters from the measurement and latent variable models, and outlines a simple Gibbs sampling algorithm for posterior computation. Section 4 applies the approach to the industrialization and democratization case study. Section 5 contains a discussion, including recommendations for important areas for future research.

2. Structural equation models

SEMs provide a broad framework for modeling of means and covariance relationships in multivariate data. Although the Bayesian approach is flexible enough to allow several extensions, our focus here is on the usual normal linear SEM, which is often referred to as a linear structural relations or LISREL model. LISREL models generalize many commonly-used statistical models, including ANOVA, MANOVA, multiple linear regression, path analysis, and confirmatory factor analysis. Because SEMs are setup to model relationships among endogenous and exogenous latent variables, accounting for measurement error, they are routinely used in social science applications. Social scientists have embraced latent variable models, realizing that it is typically not possible to obtain one perfect measure of a trait of interest. In contrast, biomedical researchers and epidemiologists tend to collapse multiple items related to a latent variable, such as stress, into a single arbitrarily-defined score prior to analysis (Herring and Dunson, 2004).

In factor models, a vector of observed variables $\boldsymbol{Y}_i$ is considered to arise through random sampling from a multivariate normal distribution denoted by $N(\boldsymbol{\nu} + \boldsymbol{\Lambda} \boldsymbol{f}_i, \boldsymbol{\Sigma})$, where $\boldsymbol{f}_i$ is the vector of latent variables; $\boldsymbol{\Lambda}$ is the factor loadings matrix describing the effects of the latent variables on the observed variables; $\boldsymbol{\nu}$ is the vector of intercepts and

$\boldsymbol{\Sigma}$ is the covariance matrix. However, in SEMs the focus is also on studying relationships among factors. For this purpose, the distinction between the *measurement* model and *structural* (latent) model is common. The former specifies the relationships of the latent to the observed variables, whereas the latter specifies the relationships among the latent variables. Following the standard LISREL notation, as in Bollen (1989) and Jöreskog and Sörbom (1996), the measurement model is, for $i = 1, \dots, N$ observations,

$$\boldsymbol{y}_i = \boldsymbol{\nu}_y + \boldsymbol{\Lambda}_y \boldsymbol{\eta}_i + \boldsymbol{\delta}_i^y, \tag{1a}$$

$$\boldsymbol{x}_i = \boldsymbol{\nu}_x + \boldsymbol{\Lambda}_x \boldsymbol{\xi}_i + \boldsymbol{\delta}_i^x, \tag{1b}$$

where model (1a) relates the vector of indicators $\boldsymbol{y}_i = (y_{i1}, \dots, y_{ip})'$ to an underlying m-vector of latent variables $\boldsymbol{\eta}_i = (\eta_{i1}, \dots, \eta_{im})'$, $m \leqslant p$, through the $p \times m$ factor loadings matrix $\boldsymbol{\Lambda}_y$. Similarly, (1b) relates $\boldsymbol{x}_i = (x_{i1}, \dots, x_{iq})'$ to an n-vector of latent variables $\boldsymbol{\xi}_i = (\xi_{i1}, \dots, \xi_{in})'$, $n \leqslant q$, through the $q \times n$ matrix $\boldsymbol{\Lambda}_x$. The vectors $\boldsymbol{\delta}_i^y$ and $\boldsymbol{\delta}_i^x$ are the measurement error terms, with dimensions $p \times 1$ and $q \times 1$, respectively. The vectors $\boldsymbol{\nu}_y$, $p \times 1$, and $\boldsymbol{\nu}_x$, $q \times 1$, are the intercept terms of the measurement models.

In Eqs. (1a) and (1b), it is assumed that the observed variables are continuous. However, as in Muthén (1984), the model remains valid for categorical or censored observed variables $(\boldsymbol{y}_i, \boldsymbol{x}_i)$ since they can be linked to their underlying continuous counterparts $(\boldsymbol{y}_i^*, \boldsymbol{x}_i^*)$ through a threshold model. Potentially, one can also define separate generalized linear models for each of the observed variables in the measurement model (as in Sammuel et al., 1997; Moustaki and Knott, 2000; Dunson, 2000, 2003; Dunson et al., 2003) to allow a broader class of measurement models.

On the other hand, the *structural (latent variable) model* is focused on studying the relationships among latent variables, $\boldsymbol{\eta}$ and $\boldsymbol{\xi}$. This is performed by regressing the dependent vector, $\boldsymbol{\eta}$, on the explanatory vector $\boldsymbol{\xi}$ as follows, $i = 1, \dots, N$,

$$\boldsymbol{\eta}_i = \boldsymbol{\alpha} + \boldsymbol{B}\boldsymbol{\eta}_i + \boldsymbol{\Gamma}\boldsymbol{\xi}_i + \boldsymbol{\zeta}_i, \tag{2}$$

where the $m \times m$ matrix $\boldsymbol{B}$ describes the relationships among latent variables in $\boldsymbol{\eta}_i$. Clearly, the elements of the diagonal of $\boldsymbol{B}$ are all zero. The $m \times n$ matrix $\boldsymbol{\Gamma}$ quantifies the influence of $\boldsymbol{\xi}_i$ on $\boldsymbol{\eta}_i$. The $m \times 1$ vectors $\boldsymbol{\alpha}$ and $\boldsymbol{\zeta}_i$ represent the intercept and the unexplained parts of $\boldsymbol{\eta}_i$, respectively.

Under this parametrization, common assumptions in SEMs are:

(i) the elements of $\boldsymbol{\xi}_i$ and $\boldsymbol{\zeta}_i$ are independent and normally distributed, $\boldsymbol{\xi}_i \sim N_n(\boldsymbol{\mu}_\xi, \boldsymbol{\Omega}_\xi)$, $\boldsymbol{\Omega}_\xi = \mathrm{diag}(\omega_{\xi 1}^2, \dots, \omega_{\xi n}^2)$, and $\boldsymbol{\zeta}_i \sim N_m(\boldsymbol{0}, \Omega_\zeta)$, $\boldsymbol{\Omega}_\zeta = \mathrm{diag}(\omega_{\zeta 1}^2, \dots, \omega_{\zeta n}^2)$;
(ii) the measurement error vectors $\boldsymbol{\delta}_i^y \sim N_p(\boldsymbol{0}, \boldsymbol{\Sigma}_y)$, $\boldsymbol{\Sigma}_y = \mathrm{diag}(\sigma_{1y}^2, \dots, \sigma_{py}^2)$, and $\boldsymbol{\delta}_i^x \sim N_q(\boldsymbol{0}, \boldsymbol{\Sigma}_x)$, $\boldsymbol{\Sigma}_x = \mathrm{diag}(\sigma_{1x}^2, \dots, \sigma_{qx}^2)$ are assumed independent; and
(iii) $\boldsymbol{\delta}' = (\boldsymbol{\delta}^{y\prime}, \boldsymbol{\delta}^{x\prime})$, $\mathrm{Cov}(\boldsymbol{\zeta}, \boldsymbol{\delta}') = \boldsymbol{0}$, $\mathrm{Cov}(\boldsymbol{\xi}, \boldsymbol{\delta}') = \boldsymbol{0}$, $\mathrm{Cov}(\boldsymbol{\xi}, \boldsymbol{\zeta}') = \boldsymbol{0}$, and $(\boldsymbol{I} - \boldsymbol{B})$ is nonsingular.

In addition, some constraints need to be placed on $\boldsymbol{\Lambda}_x$ and $\boldsymbol{\Lambda}_y$ for identifiability. The standard LISREL formulation considers correlations among $\boldsymbol{\xi}$. Our proposed parameterization allows for such correlations by including additional *pseudo*-latent variables, as is illustrated in detail through the case study later in the chapter.

3. Bayesian approach

Instead of relying on point estimates (MLEs, least squares, etc.) and asymptotically-justified confidence bounds and test statistics, the Bayesian approach we describe bases inferences on exact posterior distributions for the parameters and latent variables estimated by Markov chain Monte Carlo. As sample sizes increase, Bayesian and frequentist estimators of the parameters should converge. However, an appealing feature of the Bayesian approach is that posterior distributions are obtained not only for the parameters, but also for the latent variables. Although the posterior distribution for the latent variables is shrunk back towards the normal prior, lack of fit can be captured, including non-normality, non-linearity, and relationships that are not immediately apparent from the parameter estimates. Although frequentist two-stage approaches that fit the measurement model first and, then, compute factor scores can similarly be used to capture lack of fit, estimates are biased and measures of uncertainty in the factors scores are difficult to obtain (Croon and Bolck, 1997). For goodness-of-fit, a number of authors have proposed the use of posterior predictive (PP) p-values (Scheines et al., 1999; Lee and Shi, 2000a, 2000b; Zhu and Lee, 2001), though PP p-values can be conservative due to the double use of the data (Bayarri and Berger, 2000). An alternative is to use the BIC-criterion advocated by Raftery (1993), though some authors have questioned the use of the BIC in hierarchical models. Recent work has focused instead on computational methods for estimating Bayes factors using path sampling, including Lee and Song (2002), in the context of a nonlinear SEM with covariates; Lee and Song (2003a, 2003b) for mixture SEMs; Lee and Song (2004) for nonlinear SEM with missing categorical data; and Song and Lee (2004a, 2004b) for two-level SEMs, among others.

Since the Bayesian approach yields estimates of the exact joint posterior distribution of the latent variables, it can be used flexibly to, for example,

(1) Obtain point and interval estimates for the factor scores of each individual.
(2) Formally compare the factor scores for different subjects (e.g., through a posterior probability that the score is higher for a particular subject).
(3) Assess whether a particular subject's factor score has changed over time.
(4) Identify outlying subjects in the tails of the latent variable distribution.
(5) Assess relationships that may not be fully captured by the basic modeling structure (e.g., is the association between latent traits linear and apparent across the range of factor scores or predominantly due to the more extreme individuals?).

Potentially, one could use a richer model that allows nonlinear and more complex relationships among the latent variables. However, it is often not apparent *a priori* how such relationships should be specified, and important insights can be obtained through careful examination of posterior distributions of the latent variables obtained under a simple LISREL model.

3.1. Specification

The Bayesian model requires the specification of a full likelihood and prior distributions for the parameters. The complete data likelihood, including the latent variables, has the following form:

$$\mathcal{L}(\boldsymbol{y}, \boldsymbol{x}, \boldsymbol{\eta}, \boldsymbol{\xi}; \boldsymbol{\Theta}) = \prod_{i=1}^{N} \{N_p(\boldsymbol{y}_i; \boldsymbol{\nu}_y + \boldsymbol{\Lambda}_y \boldsymbol{\eta}_i, \boldsymbol{\Sigma}_y) N_q(\boldsymbol{x}_i; \boldsymbol{\nu}_x + \boldsymbol{\Lambda}_x \boldsymbol{\xi}_i, \boldsymbol{\Sigma}_x) \times N_m(\boldsymbol{\eta}_i; \boldsymbol{\alpha} + \boldsymbol{B}\boldsymbol{\eta}_i + \boldsymbol{\Gamma}\boldsymbol{\xi}_i, \boldsymbol{\Omega}_\zeta) N_n(\boldsymbol{\xi}_i; \boldsymbol{\mu}_\xi, \boldsymbol{\Omega}_\xi)\},$$

where $\boldsymbol{\Theta} = (\boldsymbol{\alpha}, \boldsymbol{b}, \boldsymbol{\gamma}, \boldsymbol{\nu}_y, \boldsymbol{\nu}_x, \boldsymbol{\lambda}_y, \boldsymbol{\lambda}_x, \boldsymbol{\sigma}_y^2, \boldsymbol{\sigma}_x^2, \boldsymbol{\omega}_\zeta^2, \boldsymbol{\mu}_\xi, \boldsymbol{\omega}_\xi^2)$ is the vector of model parameters. Here, the lower case bold letters denote that only the free elements are included in the parameter vector $\boldsymbol{\Theta}$, with the remaining elements being fixed in advance in the model specification process.

To complete a Bayesian specification of the model, we choose priors for each of the parameters in $\boldsymbol{\Theta}$. For convenience in elicitation and computation, we choose normal or truncated normal priors for the free elements of the intercept vectors, $\boldsymbol{\nu}_y$, $\boldsymbol{\nu}_x$ and $\boldsymbol{\alpha}$, the factor loadings, $\boldsymbol{\lambda}_y$ and $\boldsymbol{\lambda}_x$, and the structural parameters $\boldsymbol{b}$ and $\boldsymbol{\gamma}$. For the variance component parameters, including the diagonal elements of $\boldsymbol{\Sigma}_y$, $\boldsymbol{\Sigma}_x$, $\boldsymbol{\Omega}_\zeta$ and $\boldsymbol{\Omega}_\xi$, we choose independent inverse-gamma priors (avoiding high variance priors for the latent variable variances, which have well known problems). The bounds on the truncated normal are chosen to restrict parameters that are known in advance to fall within a certain range. For example, positivity constraints are often appropriate and may be necessary for identifiability based on the data. It is important to distinguish between frequentist identifiability, which implies that all the model parameters can be estimated based on the data given sufficient sample size, and Bayesian identifiability, which implies Bayesian learning. In particular, Bayesian learning occurs when the posterior distributions can differ from the prior distributions, reflecting that we have updated our beliefs based on the current data. Potentially, one can choose informative prior distributions for the parameters in a model that is underidentified from a frequentist perspective, and still obtain Bayesian identifiability for unknowns of interest. However, we prefer to focus on models which are identified in a frequentist sense to avoid relying so strongly on the prior specification.

The joint posterior distribution for the parameters and latent variables is computed, following Bayes' rule, as

$$\pi(\boldsymbol{\Theta}, \boldsymbol{\xi}, \boldsymbol{\eta} | \boldsymbol{y}, \boldsymbol{x}) = \frac{\mathcal{L}(\boldsymbol{y}, \boldsymbol{x}, \boldsymbol{\eta}, \boldsymbol{\xi}; \boldsymbol{\Theta}) \pi(\boldsymbol{\Theta})}{\int \mathcal{L}(\boldsymbol{y}, \boldsymbol{x}, \boldsymbol{\eta}, \boldsymbol{\xi}; \boldsymbol{\Theta}) \pi(\boldsymbol{\Theta})\, \mathrm{d}\boldsymbol{\eta}\, \mathrm{d}\boldsymbol{\xi}\, \mathrm{d}\boldsymbol{\Theta}}, \tag{3}$$

which is simply the complete data likelihood multiplied by the prior and divided by a normalizing constant referred to as the marginal likelihood. Clearly, calculation of the marginal likelihood (the term in the denominator) is very challenging, because it typically involves a high-dimensional integration of the likelihood over the prior distribution. Fortunately, MCMC techniques can be used to generate draws from the joint posterior distribution without need to calculate the marginal likelihood. For an overview of MCMC algorithms, refer to the recent books by Robert and Casella (2004), Gilks et al. (1996), Gamerman (1997) and Chen et al. (2000). Due to the conditionally normal linear structure of the SEM and to the choice of conditionally conjugate truncated normal and inverse-gamma priors for the parameters, MCMC computation can proceed through a straightforward Gibbs sampling algorithm, see Geman and Geman (1984) or Gelfand and Smith (1990) for more details.

3.2. Gibbs sampler

The Gibbs sampler is an MCMC technique that alternately samples from the full conditional posterior distributions of each unknown, or blocks of unknowns, including the parameters and latent variables. Before proceeding to the next step, the sampled parameter or group of parameters value is updated. Under mild regularity conditions, these samples converge to a stationary distribution, which is the joint posterior distribution. Hence, we can run the Gibbs sampler, discard a burn-in to allow convergence (diagnosed by trace plots and standard tests), and then calculate posterior summaries based on the collected samples. For illustration, we focus here on full conditional posterior distributions for the latent variables and structural parameters. Derivation of the conditional distribution for the remaining parameters follows simpler algebraic results and, in general, is not necessary since black-box sampling algorithms exist. For example, packages such as WinBUGS, Spiegelhalter et al. (2003), can automatically run Gibbs sampler algorithms based only on model and prior specifications.

We focus now on deriving the conditional posterior distributions for the latent variables, $\boldsymbol{\eta}_i$, $\boldsymbol{\xi}_i$, and the structural parameters $\boldsymbol{\alpha}$, $\boldsymbol{b}$, $\boldsymbol{\gamma}$. As introduced previously, MCMC methods use the joint posterior distribution (3) in terms of $\pi(\boldsymbol{\Theta}, \boldsymbol{\xi}, \boldsymbol{\eta} | \boldsymbol{y}, \boldsymbol{x}) \propto \mathcal{L}(\boldsymbol{x}, \boldsymbol{y}, \boldsymbol{\eta}, \boldsymbol{\xi}; \boldsymbol{\Theta})\pi(\boldsymbol{\Theta})$. Based on this property and factoring the joint posterior, we compute the conditional posterior for the endogenous latent variable as follows

$$\pi\left(\boldsymbol{\eta}_i | \boldsymbol{\nu}_y, \boldsymbol{\Lambda}_y, \boldsymbol{\Sigma}_y, \tilde{\boldsymbol{\mu}}_{\eta i}, \widetilde{\boldsymbol{\Omega}}_\eta, \boldsymbol{y}_i\right) \propto \pi(\boldsymbol{y}_i; \boldsymbol{\nu}_y + \boldsymbol{\Lambda}_y \boldsymbol{\eta}_i, \boldsymbol{\Sigma}_y) \cdot \pi\left(\boldsymbol{\eta}_i; \tilde{\boldsymbol{\mu}}_{\eta i}, \widetilde{\boldsymbol{\Omega}}_\eta\right)$$

with $\tilde{\boldsymbol{\mu}}_{\eta i} = \boldsymbol{A}[\boldsymbol{\alpha} + \boldsymbol{\Gamma}\boldsymbol{\xi}_i]$, $\widetilde{\boldsymbol{\Omega}}_\eta = \boldsymbol{A}\,\boldsymbol{\Omega}_\zeta \boldsymbol{A}'$ and $\boldsymbol{A} = [\boldsymbol{I}_{m\times m} - \boldsymbol{B}]^{-1}$. After straightforward computations it is distributed as $N_m(\hat{\boldsymbol{\eta}}_i, \widehat{\boldsymbol{\Omega}}_\eta)$ with

$$\hat{\boldsymbol{\eta}}_i = \widehat{\boldsymbol{\Omega}}_\eta\left[\boldsymbol{\Lambda}_y' \boldsymbol{\Sigma}_y^{-1}(\boldsymbol{y}_i - \boldsymbol{\nu}_y) + \widetilde{\boldsymbol{\Omega}}_\eta^{-1} \tilde{\boldsymbol{\mu}}_{\eta i}\right],$$
$$\widehat{\boldsymbol{\Omega}}_\eta^{-1} = \boldsymbol{\Lambda}_y' \boldsymbol{\Sigma}_y^{-1} \boldsymbol{\Lambda}_y + \widetilde{\boldsymbol{\Omega}}_\eta^{-1}.$$

The conditional posterior for the exogenous latent variable is obtained as follows

$$\begin{aligned}&\pi(\boldsymbol{\xi}_i | \boldsymbol{\eta}_i, \boldsymbol{\Omega}_\zeta, \boldsymbol{\nu}_x, \boldsymbol{\lambda}_x, \boldsymbol{\Sigma}_x, \boldsymbol{\alpha}, \boldsymbol{B}, \boldsymbol{\Gamma}, \boldsymbol{\mu}_\xi, \boldsymbol{\Omega}_\xi, \boldsymbol{x}_i)\\ &\quad\propto \pi(\boldsymbol{x}_i; \boldsymbol{\nu}_x + \boldsymbol{\lambda}_x \boldsymbol{\xi}_i, \boldsymbol{\Sigma}_x)\pi(\boldsymbol{\eta}_i; \boldsymbol{\alpha} + \boldsymbol{B}\boldsymbol{\eta}_i - \boldsymbol{\Gamma}\boldsymbol{\xi}_i, \boldsymbol{\Omega}_\zeta)\pi(\boldsymbol{\xi}_i; \boldsymbol{\mu}_\xi, \boldsymbol{\Omega}_\xi)\end{aligned}$$

which, after computations, is distributed as $N_n(\hat{\boldsymbol{\xi}}_i, \widehat{\boldsymbol{\Omega}}_\xi)$ with

$$\hat{\boldsymbol{\xi}}_i = \widehat{\boldsymbol{\Omega}}_\xi\left[\boldsymbol{\Lambda}_x' \boldsymbol{\Sigma}_x^{-1}(\boldsymbol{x}_i - \boldsymbol{\nu}_x) + \boldsymbol{\Gamma}' \boldsymbol{\Omega}_\zeta^{-1}(\boldsymbol{\eta}_i - \boldsymbol{\alpha} - \boldsymbol{B}\boldsymbol{\eta}_i) + \boldsymbol{\Omega}_\xi^{-1} \boldsymbol{\mu}_\xi\right],$$
$$\widehat{\boldsymbol{\Omega}}_\xi^{-1} = \boldsymbol{\Lambda}_x' \boldsymbol{\Sigma}_x^{-1} \boldsymbol{\Lambda}_x + \boldsymbol{\Gamma}' \boldsymbol{\Omega}_\zeta^{-1} \boldsymbol{\Gamma} + \boldsymbol{\Omega}_\xi^{-1}.$$

Since the estimates of $\boldsymbol{\eta}_i$ and $\boldsymbol{\xi}_i$ are based on one observation, $\boldsymbol{y}_i$ and $\boldsymbol{x}_i$, respectively, their accuracy will not increase with the sample size. However, a point estimate obtained by the proposed MCMC approach is more accurate than the classical regression estimate methods. See Lee and Shi (2000a, 2000b) for a comparison between Bayesian and classical methods.

The structural parameters have the following conditional posteriors:

- For the vector of intercepts,

$$\pi(\boldsymbol{\alpha}|\boldsymbol{B},\boldsymbol{\eta}_i,\boldsymbol{\Gamma},\boldsymbol{\xi}_i,\boldsymbol{\Omega}_\zeta,\boldsymbol{\mu}_\alpha,\boldsymbol{\Omega}_\alpha) \propto \prod_{i=1}^{N}\pi(\boldsymbol{\eta}_i;\boldsymbol{\alpha}+\boldsymbol{B}\boldsymbol{\eta}_i+\boldsymbol{\Gamma}\boldsymbol{\xi}_i,\boldsymbol{\Omega}_\zeta)\pi(\boldsymbol{\alpha};\boldsymbol{\mu}_\alpha,\boldsymbol{\Omega}_\alpha)$$

which is distributed as $N_m(\hat{\boldsymbol{\alpha}},\widehat{\boldsymbol{\Omega}}_\alpha)$ with

$$\hat{\boldsymbol{\alpha}} = \widehat{\boldsymbol{\Omega}}_\alpha\left[\boldsymbol{\mu}'_\alpha\boldsymbol{\Omega}_\alpha^{-1}+\sum_{i=1}^{N}(\boldsymbol{\eta}_i-\boldsymbol{B}\boldsymbol{\eta}_i-\boldsymbol{\Gamma}\boldsymbol{\xi}_i)'\boldsymbol{\Omega}_\zeta^{-1}\right],$$

$$\widehat{\boldsymbol{\Omega}}_\alpha^{-1} = N\boldsymbol{\Omega}_\zeta^{-1}+\boldsymbol{\Omega}_\alpha^{-1}.$$

- For the coefficient b_{rj}, which measures the impact of $\boldsymbol{\eta}^j$ on $\boldsymbol{\eta}^r$, with $r, j = 1,\ldots,m$ and $b_{rr}=0$,

$$\pi\left(b_{rj}|\boldsymbol{b}_{-rj},\boldsymbol{\alpha},\boldsymbol{\eta}_i,\boldsymbol{\Gamma},\boldsymbol{\xi}_i,\boldsymbol{\Omega}_\zeta,\mu_b,\omega_b^2\right) \propto \prod_{i=1}^{N}\pi(\boldsymbol{\eta}_i;\boldsymbol{\alpha}+\boldsymbol{B}\boldsymbol{\eta}_i+\boldsymbol{\Gamma}\boldsymbol{\xi}_i,\boldsymbol{\Omega}_\zeta)\pi\left(b_{rj};\mu_b,\omega_b^2\right)$$

which is distributed as $N(\hat{b}_{rj},\widehat{\omega}_b)$ with

$$\hat{b}_{rj} = \widehat{\omega}_b\left[\frac{\mu_b}{\omega_b^2}+\sum_{i=1}^{N}\frac{\eta_i^j}{\omega_{\zeta^r}^2}\left(\eta_i^r-\alpha^r-\sum_{s=1}^{n}(\gamma_{rs}\cdot\xi_i^s)-\sum_{\substack{t=1\\t\neq j}}^{m}b_{rt}\cdot\eta_i^t\right)\right],$$

$$\widehat{\omega}_b^{-1} = \frac{\sum_{i=1}^{N}(\eta_i^j)^2}{\omega_{\zeta^r}^2}+\frac{1}{\omega_b^2}.$$

- For the coefficient γ_{rj}, which measures the effect of $\boldsymbol{\xi}^j$ on $\boldsymbol{\eta}^r$, with $r = 1,\ldots,m$, $j=1,\ldots,n$,

$$\pi\left(\gamma_{rj}|\boldsymbol{b},\boldsymbol{\alpha},\boldsymbol{\gamma}_{-rj},\boldsymbol{\eta}_i,\boldsymbol{\xi}_i,\boldsymbol{\Omega}_\zeta,\mu_\gamma,\omega_\gamma^2\right) \propto \prod_{i=1}^{N}\pi(\boldsymbol{\eta}_i;\boldsymbol{\alpha}+\boldsymbol{B}\boldsymbol{\eta}_i+\boldsymbol{\Gamma}\boldsymbol{\xi}_i,\boldsymbol{\Omega}_\zeta)\pi\left(\gamma_{rj};\mu_\gamma,\omega_\gamma^2\right)$$

which is distributed as $N(\hat{\gamma}_{rj},\widehat{\omega}_\gamma)$ with

$$\hat{\gamma}_{rj} = \widehat{\omega}_\gamma\left[\frac{\mu_\gamma}{\omega_\gamma^2}+\sum_{i=1}^{N}\frac{\xi_i^j}{\omega_{\zeta^r}^2}\left(\eta_i^r-\alpha^r-\sum_{s=1}^{m}(b_{rs}\cdot\eta_i^s)-\sum_{\substack{t=1\\t\neq j}}^{n}\gamma_{rt}\cdot\xi_i^t\right)\right],$$

$$\widehat{\omega}_\gamma^{-1} = \frac{\sum_{i=1}^{N}(\xi_i^j)^2}{\omega_{\zeta^r}^2}+\frac{1}{\omega_\gamma^2}.$$

Once all the full conditional posteriors are computed, the following Gibbs sampling algorithm can be implemented:

```
Given Θ^0, ξ^0, η^0
  for (k = 1,...,# iterations)
    for (i = 1,...,N)
      Sample η_i^k ~ π(η_i | μ̃_ηi^{k-1}, Ω̃_η^{k-1}, ξ_i^{k-1}, Θ^{k-1}, y_i)
      Sample ξ_i^k ~ π(ξ_i | η_i^k, Θ^{k-1}, x_i)
  Sample (ν_y^k, λ_y^k) ~ π(ν_y, λ_y | Θ^{k-1}, η^k, y)
  Sample (ν_x^k, λ_x^k) ~ π(ν_x, λ_x | Θ^{k-1}, ξ^k, x)
  Sample (α^k, b^k, γ^k) ~ π(α, b, γ | η^k, ξ^k, Θ^{k-1}, x, y)
  Sample (σ_y^2)^k ~ π(σ_y^2 | η^k, ν_y^k, λ_y^k, Θ^{k-1}, y)
  Sample (σ_x^2)^k ~ π(σ_x^2 | ξ^k, ν_x^k, λ_x^y, Θ^{k-1}, x)
  Sample μ_ξ^k ~ π(μ_ξ | ξ^k, m, M)
  Sample Ω_ξ^k ~ π(Ω_ξ | ξ^k, a_ξ, β_ξ)
  Sample Ω_ζ^k ~ π(Ω_ζ | η^k, a_η, β_η)
  Output= {η^k, ξ^k, Θ^k}
```

where $\boldsymbol{m}$, $\boldsymbol{M}$, $\boldsymbol{a}_\xi$, $\boldsymbol{\beta}_\xi$, $\boldsymbol{a}_\eta$ and $\boldsymbol{\beta}_\eta$ are the hyper-parameters for $\boldsymbol{\mu}_\xi$, $\boldsymbol{\Omega}_\xi$ and $\boldsymbol{\Omega}_\zeta$, respectively.

Along with the benefits of Bayesian SEMs come the need to carefully consider certain computational issues. A particular concern is *slow mixing* of the MCMC algorithm, which can lead to very high autocorrelation in the samples and slow convergence rates. Parametrization has a large impact on computation in hierarchical models, including SEMs. For a given implied multivariate normal model, there is an equivalence class of SEMs having identical MLEs, but with different constraints made to ensure identifiability. The level of slow mixing can vary dramatically across SEMs in such an equivalence class, ranging from autocorrelation values near 1 to values near 0. Fortunately, it is easy to preselect an SEM in each equivalence class to limit slow mixing by choosing a *centered parametrization*. This simply involves incorporating free mean and variance parameters for each of the latent variables, with constraints instead incorporated in the intercepts and factor loadings in the measurement model. Following such a rule of thumb has a dramatic impact on computational efficiency without limiting inferences – one can always obtain posterior samples under a different parametrization by appropriately transforming draws obtained under the centered parametrization. In addition to centering, techniques that can be used to improve mixing include data augmentation or parameter expansion (Hills and Smith, 1992) updating parameters in blocks instead of one by one, and randomizing the order of updating (Liu et al., 1994; Roberts and Sahu, 1997). Techniques to determine the effective number of Gibbs samples necessary to produce a given level of precision in a posterior quantile of interest are available (Raftery and Lewis, 1992). In addition, there are many tests to diagnose convergence of the Markov chain (cf., Brooks and Gelman, 1998; Brooks and Giudici, 2000).

Table 1
Summary of the industrialization and democratization data

Indicator	Min	1st qu.	Median	Mean	3rd qu.	Max	Sd
y_1	1.250	2.900	5.400	5.465	7.500	10.000	2.623
y_2	0	0	3.333	4.256	8.283	10.000	3.947
y_3	0	3.767	6.667	6.563	10.000	10.000	3.281
y_4	0	1.581	3.333	4.453	6.667	10.000	3.349
y_5	0	3.692	5.000	5.136	7.500	10.000	2.613
y_6	0	0	2.233	2.978	4.207	10.000	3.373
y_7	0	3.478	6.667	6.196	10.000	10.000	3.286
y_8	0	1.301	3.333	4.043	6.667	10.000	3.246
x_1	3.784	4.477	5.075	5.054	5.515	6.737	0.733
x_2	1.386	3.663	4.963	4.792	5.830	7.872	1.511
x_3	1.002	2.300	3.568	3.558	4.523	6.425	1.406

4. Democratization and industrialization application

We will illustrate the Bayesian approach and highlight differences with frequentist methods using a democratization and industrialization example from the literature (Bollen, 1980, 1989). There has long been interest in studying relationships between industrialization in developing countries and democratization. To obtain insight into this relationship, our focus is on assessing whether industrialization level (IL) in Third World countries is positively associated with current and future political democracy level (PDL). The common political instabilities make these associations unclear. In the proposed model, it is assumed that some of the consequences of industrialization, for example societal wealth, an educated population, advances in living standards, etc., enhance the chances of democracy. To test this theory, measures of PDL (in 1960 and 1965) and IL indicators (in 1960) were collected in 75 developing countries. These include all developing countries, excluding micro-states, for which complete data were available.

Since political democracy refers to the extent of political rights and political liberties, we define a vector $\boldsymbol{y}$ of measures based on expert ratings: freedom of the press ($\boldsymbol{y}_1^{1960}$, $\boldsymbol{y}_5^{1965}$), freedom of group opposition ($\boldsymbol{y}_2^{1960}$, $\boldsymbol{y}_6^{1965}$), fairness of elections ($\boldsymbol{y}_3^{1960}$, $\boldsymbol{y}_7^{1965}$), and elective nature of the legislative body ($\boldsymbol{y}_4^{1960}$, $\boldsymbol{y}_8^{1965}$). Each of the rates were arbitrarily linearly transformed to the scale [0, 10]. See Treier and Jackman (2005) for a Bayesian latent factor model applied to Polity indicators of democracy.

Industrialization is defined as the degree to which a society's economy is characterized by mechanized manufacturing processes, and the following vector of indicators $\boldsymbol{x}$ is compiled for consideration: gross national product per capita ($\boldsymbol{x}_1^{1960}$), inanimate energy consumption per capita ($\boldsymbol{x}_2^{1960}$) and the percentage of the labor force in industry ($\boldsymbol{x}_3^{1960}$). For simplicity in the notation, we will hereafter remove the superscripts indicating the year.

The data collected are summarized in Table 1 and plotted in Figure 1.

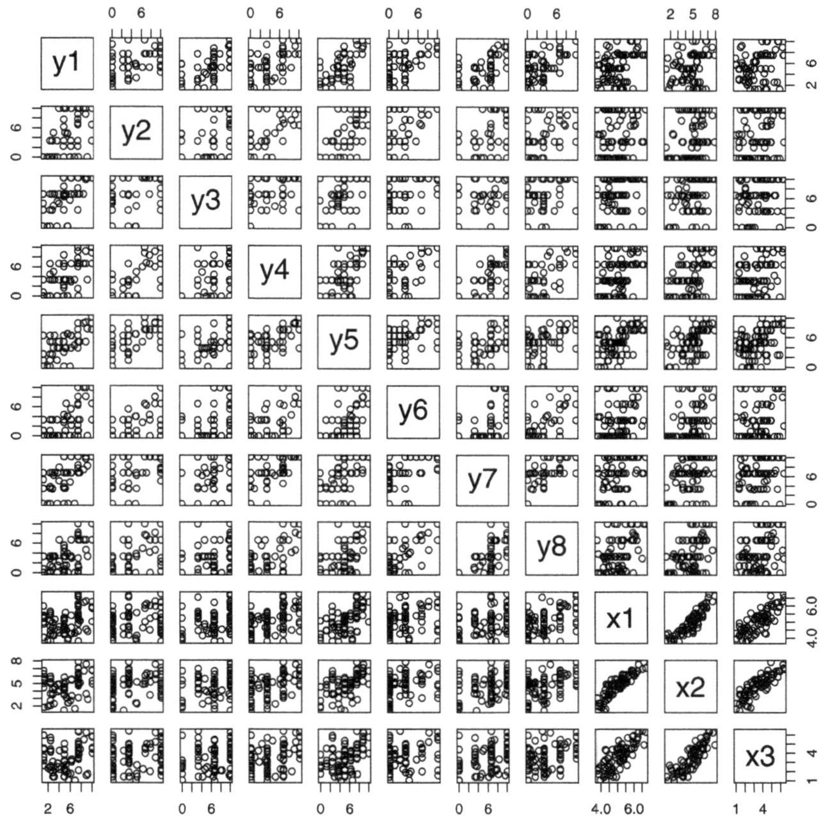

Fig. 1. Industrialization and democratization data.

4.1. Model structure

We show in the path diagram of Figure 2 the assumed model, where, for the countries under study, the PDL in 1960 and 1965 is represented by $\boldsymbol{\eta}^{60}$ and $\boldsymbol{\eta}^{65}$, respectively, and the IL in 1960 is symbolized by $\boldsymbol{\xi}$. Following the convention, circles represent latent variables, the squares the observed variables and the arrows linear relations. The relations assumed imply that the IL in 1960, $\boldsymbol{\xi}$, affects the PDL both in 1960, $\boldsymbol{\eta}^{60}$, and 1965, $\boldsymbol{\eta}^{65}$, through the regression coefficients γ^{60} and γ^{65}, respectively. The impact of the PDL in 1960 on the level in 1965 is represented by the arrow b_{21}. The pseudo-latent variables, $\boldsymbol{D}^{15}$, $\boldsymbol{D}^{24}$, $\boldsymbol{D}^{26}$, $\boldsymbol{D}^{37}$, $\boldsymbol{D}^{48}$, $\boldsymbol{D}^{68}$, are used to represent the correlation among the errors in the ratings that were elicited by the same expert in two points of time.

For the ith country, the latent variable model, as introduced in (2), is now formulated in matrix form as follows,

$$\begin{pmatrix} \eta_i^{60} \\ \eta_i^{65} \end{pmatrix} = \begin{pmatrix} \alpha^{60} \\ \alpha^{65} \end{pmatrix} + \begin{pmatrix} 0 & 0 \\ b_{21} & 0 \end{pmatrix} \begin{pmatrix} \eta_i^{60} \\ \eta_i^{65} \end{pmatrix} + \begin{pmatrix} \gamma^{60} \\ \gamma^{65} \end{pmatrix} \xi_i + \begin{pmatrix} \zeta_i^{60} \\ \zeta_i^{65} \end{pmatrix}, \tag{4}$$

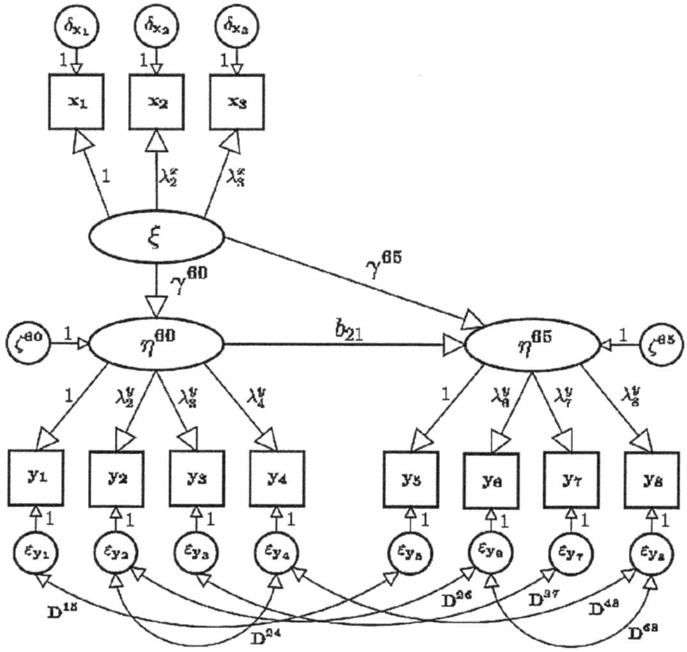

Fig. 2. Path diagram for the democratization study.

where the disturbances $\boldsymbol{\zeta}_i = (\zeta_i^{60}, \zeta_i^{65})$ are assumed to be independent normally distributed with mean zero and precision parameters $\omega_{\zeta^{60}}^{-1}$ and $\omega_{\zeta^{65}}^{-1}$, respectively. The measurement model, as introduced in (1), is now formulated as follows

$$\begin{pmatrix} y_{1i} \\ y_{2i} \\ y_{3i} \\ y_{4i} \\ y_{5i} \\ y_{6i} \\ y_{7i} \\ y_{8i} \end{pmatrix} = \begin{pmatrix} 0 \\ \nu_2^y \\ \nu_3^y \\ \nu_4^y \\ 0 \\ \nu_6^y \\ \nu_7^y \\ \nu_8^y \end{pmatrix} + \begin{pmatrix} 1 & 0 \\ \lambda_2^y & 0 \\ \lambda_3^y & 0 \\ \lambda_4^y & 0 \\ 0 & 1 \\ 0 & \lambda_6^y \\ 0 & \lambda_7^y \\ 0 & \lambda_8^y \end{pmatrix} \begin{pmatrix} \eta_i^{60} \\ \eta_i^{65} \end{pmatrix}$$

$$+ \begin{pmatrix} 1 & 0 & 0 & 0 & 0 & 0 \\ 0 & 1 & 1 & 0 & 0 & 0 \\ 0 & 0 & 0 & 1 & 0 & 0 \\ 0 & 1 & 0 & 0 & 1 & 0 \\ 1 & 0 & 0 & 0 & 0 & 0 \\ 0 & 0 & 1 & 0 & 0 & 1 \\ 0 & 0 & 0 & 1 & 0 & 0 \\ 0 & 0 & 0 & 0 & 1 & 1 \end{pmatrix} \begin{pmatrix} D_i^{15} \\ D_i^{24} \\ D_i^{26} \\ D_i^{37} \\ D_i^{48} \\ D_i^{68} \end{pmatrix} + \begin{pmatrix} \delta_{1i}^y \\ \delta_{2i}^y \\ \delta_{3i}^y \\ \delta_{4i}^y \\ \delta_{5i}^y \\ \delta_{6i}^y \\ \delta_{7i}^y \\ \delta_{8i}^y \end{pmatrix}, \tag{5a}$$

$$\begin{pmatrix} x_{1i} \\ x_{2i} \\ x_{3i} \end{pmatrix} = \begin{pmatrix} 0 \\ \nu_2^x \\ \nu_3^x \end{pmatrix} + \begin{pmatrix} 1 \\ \lambda_2^x \\ \lambda_3^x \end{pmatrix} \xi_i + \begin{pmatrix} \delta_{1i}^x \\ \delta_{2i}^x \\ \delta_{3i}^x \end{pmatrix}, \tag{5b}$$

where λ_j^y is the influence of PDL on the indicator $\boldsymbol{y}_j$, $j = 1, \ldots, 8$, $\boldsymbol{D}^{rs}$ is a pseudo-latent variable to model the correlations among the measurement errors $\boldsymbol{\delta}_r^y$ and $\boldsymbol{\delta}_s^y$. We fix the intercepts, $\nu_1^y = \nu_5^y = \nu_1^x = 0$, and factor loadings, $\lambda_1^y = \lambda_5^y = \lambda_1^x = 1$, for identifiability of the model and to scale the latent variables. Therefore, PDL will be scaled in terms of freedom in press and IL in terms of gross national product per capita. Furthermore, this approach results in a centered parametrization, which has appealing computational properties as discussed in Section 3.2.

Under expressions (4) and (5), and for $i = 1, \ldots, 75$ developing countries, the complete data likelihood including the latent variables $\boldsymbol{\eta}$ and ξ is as follows

$$\begin{aligned} \mathcal{L}(\boldsymbol{y}, \boldsymbol{x}, \boldsymbol{\eta}, \xi, \boldsymbol{D}; \boldsymbol{\Theta}) = \prod_{i=1}^{75} \big\{ & N_8(\boldsymbol{y}_i; \boldsymbol{\nu}_y + \boldsymbol{\Lambda}_y \boldsymbol{\eta}_i, \boldsymbol{\Sigma}_y) \\ & \times N_3(\boldsymbol{x}_i; \boldsymbol{\nu}_x + \boldsymbol{\Lambda}_x \xi_i, \boldsymbol{\Sigma}_x) N(\xi_i; \mu_\xi, \omega_\xi^2) \\ & \times N_2(\boldsymbol{\eta}_i; \boldsymbol{\alpha} + \boldsymbol{B}\boldsymbol{\eta}_i + \boldsymbol{\Gamma}\xi_i, \boldsymbol{\Omega}_\zeta) N_6(\boldsymbol{D}_i; \boldsymbol{0}, \boldsymbol{\Omega}_D) \big\} \end{aligned}$$

with $\boldsymbol{\Sigma}_y = \mathrm{diag}(\sigma_{y1}^2, \ldots, \sigma_{y8}^2)$, $\boldsymbol{\Sigma}_x = \mathrm{diag}(\sigma_{x1}^2, \sigma_{x2}^2, \sigma_{x3}^2)$, $\boldsymbol{\Omega}_\zeta = \sigma_{y1}^2 \cdot \mathrm{diag}(\omega_{\zeta 60}^{-1}, \omega_{\zeta 65}^{-1})$, $\boldsymbol{\Omega}_D = \mathrm{diag}(\omega_{D15}^2, \omega_{D24}^2, \omega_{D26}^2, \omega_{D37}^2, \omega_{D48}^2, \omega_{D68}^2)$, and $\boldsymbol{\Theta}$ includes the free elements of $(\boldsymbol{\nu}_y, \boldsymbol{\Lambda}_y, \boldsymbol{\nu}_x, \boldsymbol{\Lambda}_x, \boldsymbol{B})$ and the parameters $(\boldsymbol{\sigma}_y^2, \boldsymbol{\sigma}_x^2, \mu_\xi, \omega_\xi^2, \boldsymbol{\alpha}, \boldsymbol{\gamma}, \boldsymbol{\Omega}_\zeta, \boldsymbol{\omega}_D^2)$.

In the Bayesian analysis, the prior specification involves quantifying expert's uncertainty in the model parameters $\boldsymbol{\Theta}$. In the cases where not much information is available beyond the observed data, non-informative or objective priors are the usual selection (Berger, 1985; Bernardo and Smith, 1994). Here, we consider a variety of alternative priors, with the primary choice based on expert elicitation, choosing a specification that assigns high probability to a plausible range for the parameter values based on Ken Bollen's experience in this area. We also consider priors centered on the MLEs, but with inflated variance, for sake of comparison. Refer to Appendix A for more details on the hyperparameters used.

In this case, the joint posterior is computed, following Bayes' rule, as

$$\pi(\boldsymbol{\Theta}, \xi_i, \boldsymbol{\eta}_i, \boldsymbol{D}_i | \boldsymbol{x}, \boldsymbol{y}) \propto \mathcal{L}(\boldsymbol{y}, \boldsymbol{x}, \boldsymbol{\eta}, \boldsymbol{\xi}, \boldsymbol{D}; \boldsymbol{\Theta}) \cdot \pi(\boldsymbol{\Theta}).$$

Although this joint posterior distribution is complex, all the corresponding full conditional posterior distributions have simple conjugate forms due to the model assumed. A Gibbs sampling algorithm based on the general scheme introduced before is used to obtain samples from the posterior distributions of the parameters of interest, for example, PDL and IL for every single country in the study in both periods (1960 and 1965), or the impact of the PDL in 1960 on the PDL in 1965. The implementation of the algorithm was written in *R*, and run 50 000 iterations, discarding the first 10 000 for burn-in, and keeping one every 400 iterations to reduce the correlation among the posterior samples. WinBUGS could also be used, but our *R* implementation gave us

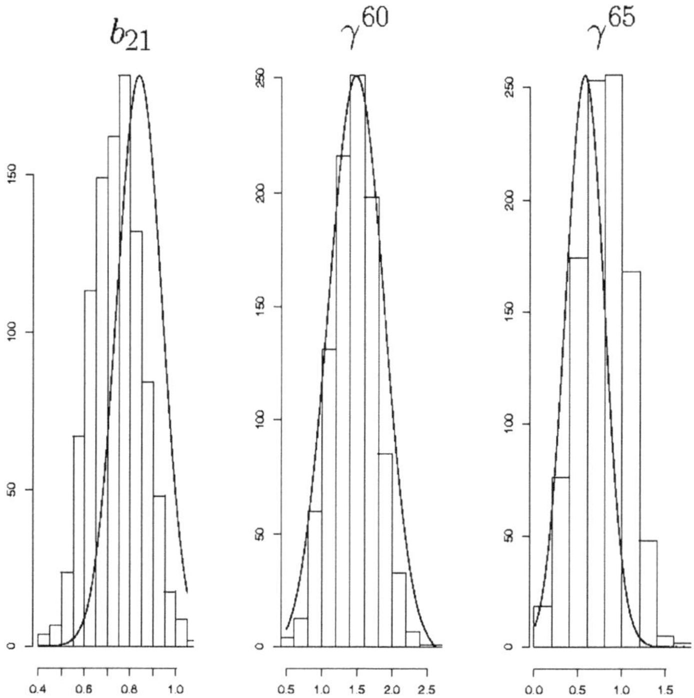

Fig. 3. Histograms for the posterior samples under the subjective priors scheme. From left to right: b_{21}, γ^{60} and γ^{65}. The confidence intervals for the MLEs are represented with straight lines.

greater flexibility regarding the computational algorithms and priors we could consider.

4.2. Results

We start by comparing the frequentist (Maximum Likelihood) and Bayesian estimates for the aforementioned parameters of interest, see Appendix B for a full list of parameters estimates. In Figures 3 and 4 we show graphically the histograms of the posterior samples for the parameters b_{21}, γ^{60}, γ^{65}, μ_ξ and σ_ξ^2 along with a confidence interval for the MLEs. The learning process experimented in updating the prior to the posterior beliefs based on the observed data are presented in Table 2. For example, a prior 95% probability interval for the influence of PDL in 1960 on the level in 1965 is: $[-4.062, 5.736]$, under the centered MLE priors scheme, and $[-1.828, 3.828]$ under the subjective priors scheme. These probability intervals, a posteriori, are narrowed to $[0.523, 0.959]$ and $[0.522, 1.1]$, respectively. This shows a convergence, after observing the data, regardless of the starting prior knowledge.

As measures of the goodness-of-fit of the frequentist model, we report the R-square indicators in Table 3. In Bayesian SEMs, we shall use some loss function $L(y_i, \hat{y}_i)$ to measure the goodness-of-fit based on the predictive distribution. For example, the Root Mean Squared Error (RMSE) and the Mean Absolute Error (MAE) are common

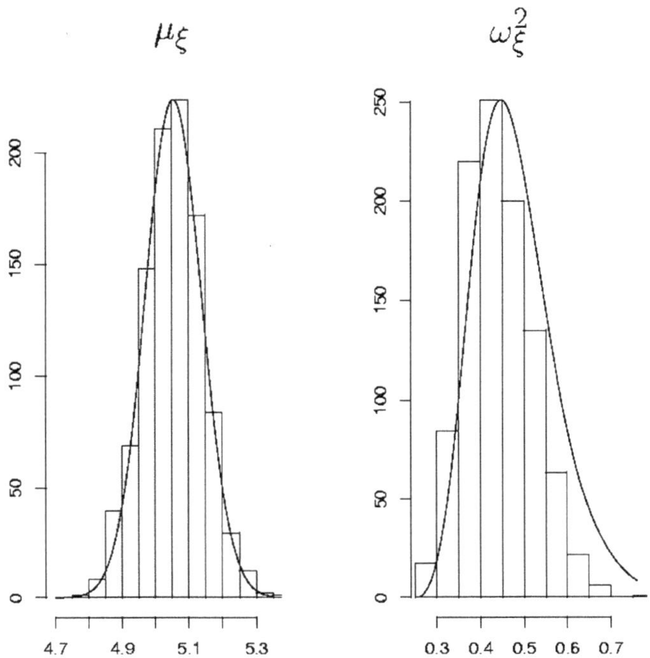

Fig. 4. Histograms of the posterior samples for μ_ξ (left) and ω^2_ξ (right) under the subjective priors scheme. The confidence intervals for the MLEs are represented with straight lines.

Table 2
Parameters of interest estimated under the frequentist (MLE) and Bayesian approach (summary of the posterior distributions)

	MLE		Centered MLE					Subjective				
			Prior beliefs		Posteriors			Prior beliefs		Posteriors		
	Mean	Sd	Mean	Sd	Median	Mean	Sd	Mean	Sd	Median	Mean	Sd
b_{21}	0.837	0.098	0.837	2.449	0.744	0.741	0.109	1	1.414	0.814	0.811	0.144
γ^{60}	1.483	0.399	1.483	2.449	1.455	1.454	0.313	1.5	1.414	1.083	1.077	0.209
γ^{65}	0.572	0.221	0.572	2.449	0.774	0.774	0.278	0.5	1.414	0.297	0.322	0.205
μ_ξ	5.054	0.084	5.054	2.5	5.054	5.053	0.087	5	1	5.040	5.035	0.098
ω^2_ξ	0.448	0.087	0.448	0.224	0.433	0.442	0.077	1	0.5	0.655	0.667	0.119

Table 3
R-square indicators for the frequentist model

R^2 estimates												
y_1	y_2	y_3	y_4	y_5	y_6	y_7	y_8	x_1	x_2	x_3	η^{60}	η^{65}
0.723	0.514	0.522	0.715	0.653	0.557	0.678	0.685	0.846	0.947	0.761	0.200	0.961

Table 4
Measures of the goodness of the predictive distribution

	$Country_{46}$	$Country_{61}$	$Country_{22}$
$RMSE_{Xi}$	0.94	0.24	0.323
MAE_{Xi}	0.13	0.33	0.384
$RMSE_{Yi}$	0.511	0.277	0.355
MAE_{Yi}	1	0.583	0.783
	IL_1	IL_2	IL_3
$RMSE_{Xj}$	0.029	0.032	0.075
MAE_{Xj}	0.21	0.214	0.54

	PDL_1	PDL_2	PDL_3	PDL_4	PDL_5	PDL_6	PDL_7	PDL_8
$RMSE_{Yj}$	0.096	0.193	0.172	0.116	0.131	0.134	0.131	0.114
MAE_{Yj}	0.689	1	1	0.818	0.892	0.912	0.931	0.765

measures computed as follows, $i = 1, \ldots, 75$ and $j = 1, \ldots, 8$,

$$RMSE_i = \frac{\sqrt{\sum_{j=1}^{8}(y_{ij} - E(\hat{y}_{ij}))^2}}{8}, \qquad RMSE_j = \frac{\sqrt{\sum_{i=1}^{75}(y_{ij} - E(\hat{y}_{ij}))^2}}{75},$$

$$MAE_i = \frac{\sum_{j=1}^{8}|y_{ij} - E(\hat{y}_{ij})|}{8}, \qquad MAE_j = \frac{\sum_{i=1}^{75}|y_{ij} - E(\hat{y}_{ij})|}{75},$$

where $E(\hat{y}_{ij}) = \frac{1}{MN}\sum_{l=1}^{M}\sum_{k=1}^{N}\hat{y}_{ij}^{lk}$ is the average of the posterior predictions for the jth PDL indicator and ith country. Those for the indicators of IL follow symmetrically. We report these estimates in Table 4, where countries $\{46, 61, 22\}$ are samples from each of the three industrialization clusters identified.

So far, we have presented summaries of the results obtained following both the frequentist's and Bayesian's approaches. However, there is more information available in the posterior than is present in marginal posterior summaries of the population parameters. In particular, the benefit of having posterior samples from the joint posterior distribution of the latent variables is large. They provide important information not contained in the measurement and structural models. We highlight these issues in the next section.

4.3. Democratization results

Recall that the main goal is to determine if the IL of a country has an impact on the change of its PDL. The average across countries of the posterior samples of IL in 1960 are summarized in Table 5.

In Figures 5 and 6 we show boxplots for the PDL in 1960 (gray boxes) and in 1965 (black boxes) for each country in the study, along with their IL (posterior mean) in

Table 5

Min.	1st qu.	Median	Mean	3rd qu.	Max	Sd
3.498	4.374	5.117	5.035	5.560	6.593	0.798

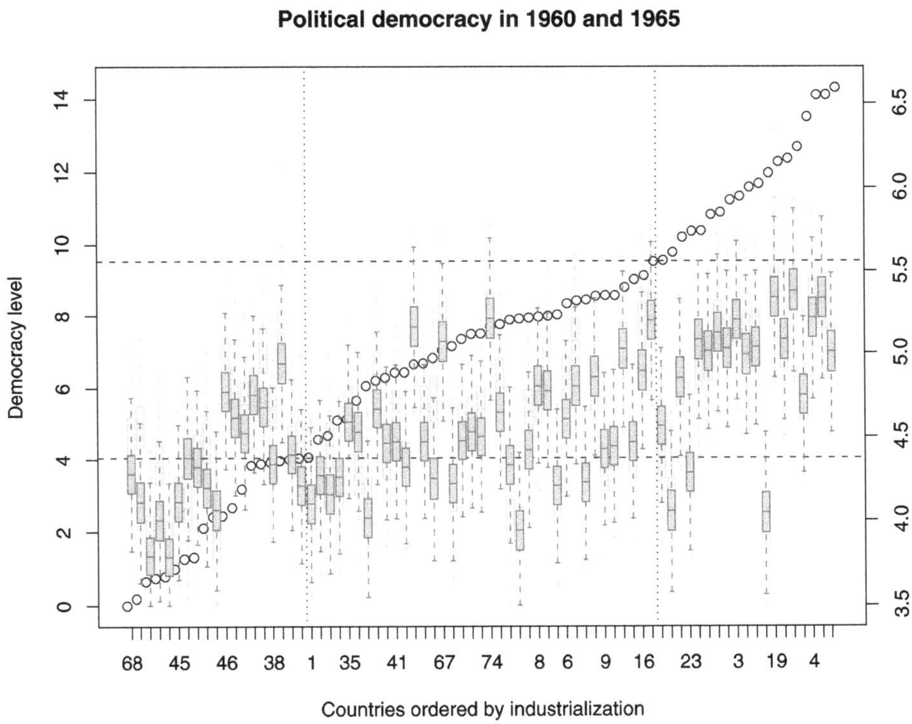

Fig. 5. Boxplots of the PDL in 1960, gray, and in 1965, black. The countries are sorted, black circles, by increasing posterior mean IL in 1960. The two vertical dotted lines separate the three clusters using IL as criteria. The two horizontal dashed lines show the first and fourth quartile for the posterior means of PDL in 1965.

1960 (black circles). To facilitate the interpretation, we have sorted the countries by increasing posterior mean of IL.

We notice a generalized (two thirds of the countries) reduction in PDL from 1960 to 1965 – in Figure 6 the black diamonds are mostly below the gray circles. In Figure 7 we show this behavior for the countries in the study, where the thin horizontal straight gray line represents the average across-countries of the PDL change; the PDL average reduced amount is 0.314. We plot also an horizontal straight black line at zero.

As a first approach, we have linearly regressed each posterior sample of the IL against the square of the PDL change for each country. We have estimated by least squares the slope of the regression line, finding that the posterior probability of having a negative slope is 0.94. This indicates that an increase in the IL will almost surely cause a positive

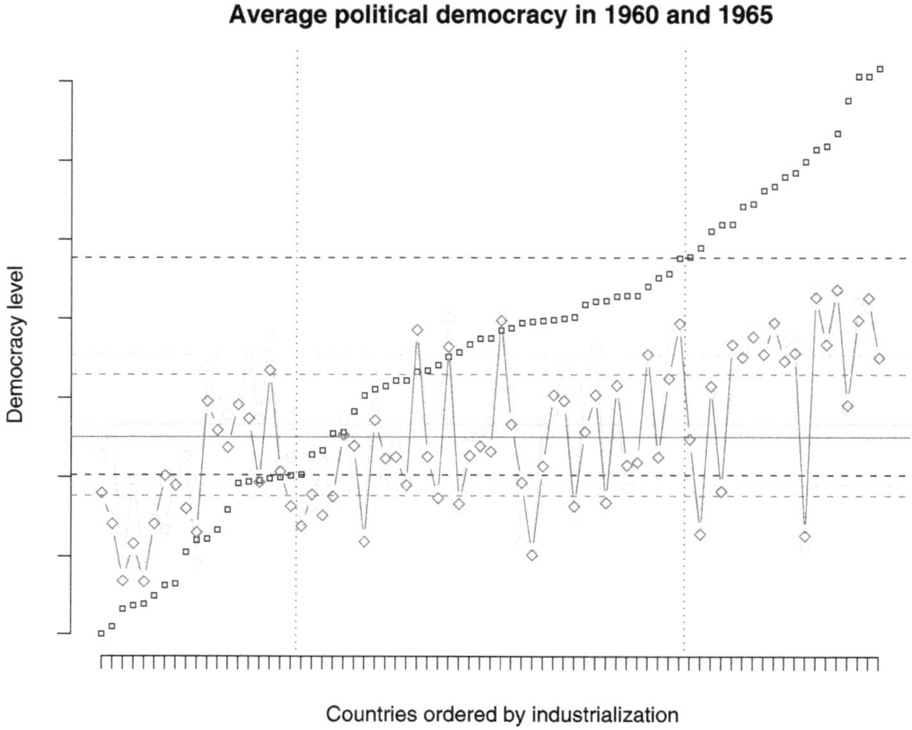

Fig. 6. Chart that shows the posterior mean of PDL in 1960, gray circles, and in 1965, black diamonds. The countries are sorted, black squares, by increasing IL (posterior mean) in 1960. Vertical dotted lines separate the three clusters using IL as criteria. The horizontal gray lines show the first and third quartile (dashed) and the median (straight) for the posterior means of PDL in 1960, and black dashed lines show these bands in 1965.

or negative change in the PDL. However, we notice that this behavior is not homogeneous among countries, and consequently further analysis is required. We define three clusters of countries, corresponding to: *poorly industrialized*, those countries in the first quartile; *mildly industrialized*, those in the second and third quartile; and *highly industrialized*, those in the fourth quartile. These clusters are represented with vertical dotted lines in Figures 5, 6 and 7, yielding 19, 37 and 19 countries, respectively, on each group.

The different behavior of the clusters is present in Figure 6, where for the poorly and highly industrialized groups, whenever their PDL in 1960 is within the (gray) bands, it will also remain within the (black) bands in 1965. No such pattern is detected in the case of the mildly industrialized countries, where the variability is considerably higher. In Figure 7 we also see the difference in the variability of PDL change among clusters: only 10.5% of the countries in the first cluster increased the PDL, whereas the 35.1 and 52.6% was recorded for the second and third clusters, respectively. Also, 42.1 and 27% are the percentages of countries that experienced an extreme (in the first quartile) reduction on PDL for the first and second clusters, respectively. This is not the case for the highly industrialized countries, where only 15.8% of the countries have a PDL

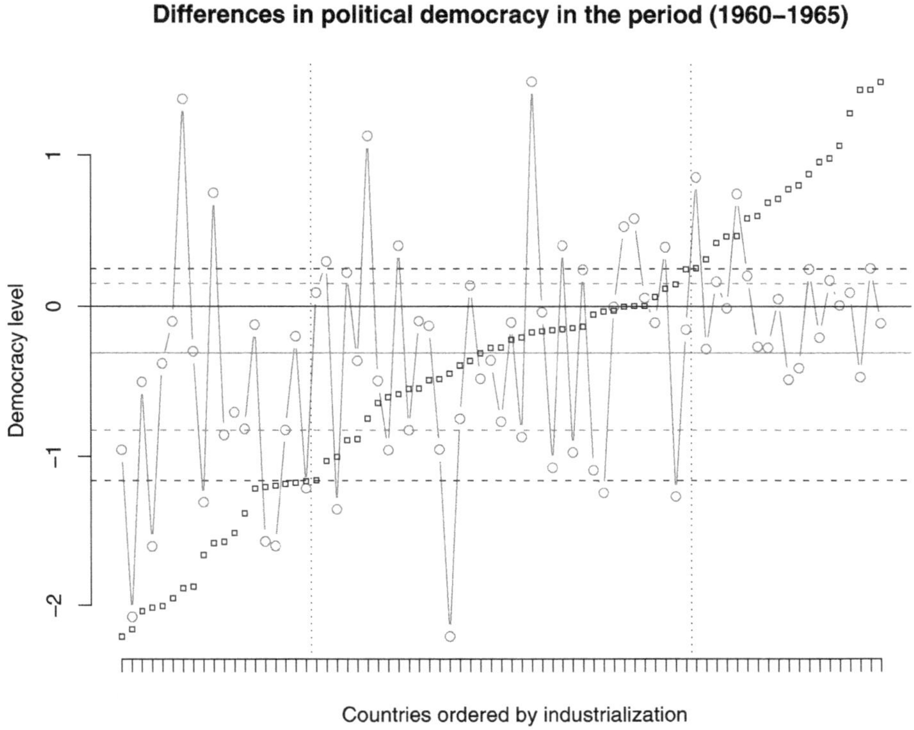

Fig. 7. Chart that shows the differences in PDL between 1960 and 1965. The countries are sorted, black squares, by increasing IL (posterior mean) in 1960. Vertical dotted lines separate the three clusters using IL as criteria, and horizontal gray lines represent the first and third quartile (dashed) and the median (straight) for the PDL reduction. The straight horizontal black line at zero separates the countries that increased the PDL in 1965.

change below the average, and none of them reaches the lower band. This indicates that low levels of industrialization cannot compensate the PDL reduction tendency over the period 1960–1965 for the countries of study.

5. Discussion and future research

This chapter has provided an overview of a Bayesian approach to structural equation modeling, highlighting some aspects of the Bayesian approach that have not been fully explored in previous articles on Bayesian SEMs. In particular, previous authors have not carefully considered the issue of parametrization, which really does have an enormous practical impact on Bayesian computation. The most important points to remember in conducting Bayesian SEM analysis are (1) use a centered parametrization allowing the latent variables to have free intercepts and variances; (2) do not use diffuse (high variance) or improper uniform priors for the latent variable variances; and (3) examine the posterior distributions of the latent variables, because they often provide additional information and insight not apparent from the estimated population parameters.

There is a need for additional research into computationally efficient algorithms for SEMs. There has been a lot of interest in computation for simpler variance component models, and a variety of approaches have been proposed, including not only centering but also clever parameter expansion techniques (refer to Gelman (2005), for a recent reference). The parameter expansion approach can potentially be applied directly to SEMs, but practical details remain to be worked out.

Another very important issue is the prior specification, particularly in cases in which limited information is available *a priori* or one wishes to choose a noninformative prior in conducting a reference analysis. However, special care must be taken with approaches that suggest, as a non-informative specification, e.g., Scheines et al. (1999), a uniform improper prior for the vector of parameters, including the variance components. These uniform improper priors on the latent variable variances, unfortunately, will result in an improper posterior, and, therefore, the results are not interpretable, since the posteriors are no longer a density function (it does not integrate to one). This problem is not solved by using highly diffuse inverse-gamma priors, because the posterior is then close to improper (i.e., it might as well be improper as far as MCMC behavior and interpretation). In addition, as illustrated by Gelman (2005) for simple variance component models, there can be enormous sensitivity to the prior variance chosen in the diffuse gamma prior. Better reference priors for variance component models were suggested by Gelman (2005) and similar specifications can be used in SEMs.

Additional areas in need of further research, include model selection/averaging and semiparametric methods, see Dunson (2006) for a semiparametric latent factor model. The Bayesian approach has the major advantage that it can allow for uncertainty in different aspects of the model specification in performing inferences about structural relations of interest. Raftery has developed Bayesian methods for model selection and averaging in SEMs in a series of papers, primarily based on the BIC and Laplace approximations to the marginal likelihood. However, due to the constraints involved with comparing models that have different variance component structures, more accurate approximations that also allow more flexibility in the prior specification need to be considered. Expanding the class of models considered to allow unknown latent variable and measurement error distributions can potentially be accomplished within a Bayesian approach using Dirichlet process priors, but again details remain to be worked out.

Appendix A. Prior specifications

We consider the following gamma and inverse-gamma formulations for the prior distribution of the precision and variance parameters respectively, where

$$\sigma^2 \sim InvGamma(\sigma^2; \alpha, \beta)$$

$$f(\sigma^2) = \frac{\beta^\alpha}{\Gamma(\alpha)}(\sigma^2)^{-(\alpha+1)} \exp\left(-\frac{\beta}{\sigma^2}\right),$$

$$\mu = \frac{\beta}{\alpha - 1}, \qquad \sigma^2 = \frac{\beta^2}{(\alpha - 1)^2(\alpha - 2)}$$

$$\sigma^{-2} \sim Gamma(\sigma^{-2}; \alpha, \beta)$$

$$f(\sigma^{-2}) = \frac{\beta^{\alpha}}{\Gamma(\alpha)}(\sigma^{-2})^{\alpha-1} \exp(-\beta\sigma^{-2})$$

$$\mu = \frac{\alpha}{\beta}, \qquad \sigma^2 = \frac{\alpha}{\beta^2}.$$

See Table A.1 for the prior parameters used.

Table A.1
Prior distributions for the model parameters and their corresponding ML estimates

	MLE centered parameters		Subjective parameters		MLE estimates	
$\sigma^2_{y1} \sim IGamma(\cdot,\cdot)$	6.0	9.455	10	36	1.891	0.444
$\sigma^2_{y2} \sim IGamma(\cdot,\cdot)$	6.0	36.865	10	36	7.373	1.374
$\sigma^2_{y3} \sim IGamma(\cdot,\cdot)$	6.0	25.335	10	36	5.067	0.952
$\sigma^2_{y4} \sim IGamma(\cdot,\cdot)$	6.0	15.74	10	36	3.148	0.739
$\sigma^2_{y5} \sim IGamma(\cdot,\cdot)$	6.0	11.755	10	36	2.351	0.480
$\sigma^2_{y6} \sim IGamma(\cdot,\cdot)$	6.0	24.77	10	36	4.954	0.914
$\sigma^2_{y7} \sim IGamma(\cdot,\cdot)$	6.0	17.155	10	36	3.431	0.713
$\sigma^2_{y8} \sim IGamma(\cdot,\cdot)$	6.0	16.27	10	36	3.254	0.695
$\sigma^2_{x1} \sim IGamma(\cdot,\cdot)$	6.0	0.41	6	1	0.082	0.019
$\sigma^2_{x2} \sim IGamma(\cdot,\cdot)$	6.0	0.6	6	1	0.120	0.070
$\sigma^2_{x3} \sim IGamma(\cdot,\cdot)$	6.0	2.335	6	1	0.467	0.090
$\mu_{\xi} \sim N(\cdot,\sigma^2)$	5.054	6.25	5	1	5.054	0.084
$\omega^2_{\xi} \sim IGamma(\cdot,\cdot)$	6.0	2.24	6	5	0.448	0.087
$\omega^{-1}_{\zeta 60} \sim Gamma(\cdot,\cdot)$	4.0	1.912	16	16	2.092	
$\omega^{-1}_{\zeta 65} \sim Gamma(\cdot,\cdot)$	4.0	43.977	4.0	93.023	0.091	
$\omega^2_{D15} \sim IGamma(\cdot,\cdot)$	6.0	3.12	6.0	5.0	0.624	0.358
$\omega^2_{D24} \sim IGamma(\cdot,\cdot)$	6.0	6.565	6.0	5.0	1.313	0.702
$\omega^2_{D26} \sim IGamma(\cdot,\cdot)$	6.0	10.765	6.0	5.0	2.153	0.734
$\omega^2_{D37} \sim IGamma(\cdot,\cdot)$	6.0	3.975	6.0	5.0	0.795	0.608
$\omega^2_{D48} \sim IGamma(\cdot,\cdot)$	6.0	1.74	6.0	5.0	0.348	0.442
$\omega^2_{D68} \sim IGamma(\cdot,\cdot)$	6.0	6.78	6.0	5.0	1.356	0.568
$\nu^y_2 \sim N(\cdot,\sigma^2)$	−2.611	6.0	0	1	−2.611	1.064
$\nu^y_3 \sim N(\cdot,\sigma^2)$	0.783	6.0	0	1	0.783	0.883
$\nu^y_4 \sim N(\cdot,\sigma^2)$	−2.459	6.0	0	1	−2.459	0.843

(continued on next page)

Table A.1
(*continued*)

	MLE centered parameters		Subjective parameters		MLE estimates	
$\nu_6^y \sim N(\cdot, \sigma^2)$	−3.112	6.0	0	1	−3.112	0.928
$\nu_7^y \sim N(\cdot, \sigma^2)$	−0.376	6.0	0	1	−0.376	0.878
$\nu_8^y \sim N(\cdot, \sigma^2)$	−2.459	6.0	0	1	−2.459	0.868
$\lambda_2^y \sim N(\cdot, \sigma^2)$	1.257	6.0	1	2	1.287	0.182
$\lambda_3^y \sim N(\cdot, \sigma^2)$	1.058	6.0	1	2	1.058	0.151
$\lambda_4^y \sim N(\cdot, \sigma^2)$	1.265	6.0	1	2	1.265	0.145
$\lambda_6^y \sim N(\cdot, \sigma^2)$	1.186	6.0	1	2	1.186	0.169
$\lambda_7^y \sim N(\cdot, \sigma^2)$	1.280	6.0	1	2	1.280	0.160
$\lambda_8^y \sim N(\cdot, \sigma^2)$	1.266	6.0	1	2	1.266	0.158
$\nu_2^x \sim N(\cdot, \sigma^2)$	−6.228	6.0	0	1	−6.228	0.705
$\nu_3^x \sim N(\cdot, \sigma^2)$	−5.634	6.0	0	1	−5.634	0.774
$\lambda_2^x \sim N(\cdot, \sigma^2)$	2.180	6.0	1	2	2.180	0.139
$\lambda_3^x \sim N(\cdot, \sigma^2)$	1.819	6.0	1	2	1.819	0.152
$\alpha^{60} \sim N(\cdot, \sigma^2)$	−2.031	6.0	1	2	−2.031	2.037
$\alpha^{65} \sim N(\cdot, \sigma^2)$	−2.332	6.0	1	2	−2.332	1.119
$b_{21} \sim N(\cdot, \sigma^2)$	0.837	6.0	1	2	0.837	0.098
$\gamma^{60} \sim N(\cdot, \sigma^2)$	1.483	6.0	1.5	2	1.483	0.399
$\gamma^{65} \sim N(\cdot, \sigma^2)$	0.572	6.0	0.5	2	0.572	0.221

Appendix B. Results: posterior parameters estimates (see Table B.1)

Table B.1

	MLE		Sub. priors			Centered MLE		
	Mean	Sd	Median	Mean	Sd	Median	Mean	Sd
α^{60}	−2.031	2.037	−0.021	−0.005	1.051	−1.865	−1.876	1.587
α^{65}	−2.332	1.119	−0.722	−0.763	0.944	−2.788	−2.806	1.287
b_{21}	0.837	0.098	0.814	0.811	0.144	0.744	0.741	0.109
γ^{60}	1.483	0.399	1.083	1.077	0.209	1.455	1.454	0.313
γ^{65}	0.572	0.221	0.297	0.322	0.205	0.774	0.774	0.278

(*continued on next page*)

Table B.1
(*continued*)

	MLE		Sub. priors			Centered MLE		
	Mean	Sd	Median	Mean	Sd	Median	Mean	Sd
ν_2^y	−2.611	1.064	−1.235	−1.205	0.708	−2.251	−2.289	0.945
ν_3^y	0.783	0.883	0.208	0.195	0.637	0.630	0.593	0.796
ν_4^y	−2.459	0.843	−1.415	−1.4	0.646	−2.249	−2.281	0.799
ν_6^y	−3.112	0.928	−1.292	−1.296	0.563	−2.555	−2.595	0.82
ν_7^y	−0.376	0.878	0.539	0.529	0.54	−0.014	−0.031	0.757
ν_8^y	−2.459	0.868	−0.841	−0.85	0.536	−2.024	−2.051	0.714
λ_2^y	1.257	0.182	1.039	1.04	0.131	1.186	1.193	0.165
λ_3^y	1.058	0.151	1.173	1.176	0.119	1.086	1.091	0.138
λ_4^y	1.265	0.145	1.111	1.106	0.118	1.218	1.228	0.14
λ_6^y	1.186	0.169	0.850	0.852	0.107	1.076	1.077	0.150
λ_7^y	1.280	0.160	1.095	1.094	0.1	1.207	1.208	0.135
λ_8^y	1.266	0.158	0.962	0.963	0.1	1.175	1.180	0.128
$\sigma_{y_1}^2$	1.891	0.444	1.486	1.509	0.215	0.777	0.799	0.178
$\sigma_{y_2}^2$	7.373	1.374	4.330	4.409	0.958	4.763	4.896	1.025
$\sigma_{y_3}^2$	5.067	0.952	3.534	3.590	0.7305	3.929	4.029	0.843
$\sigma_{y_4}^2$	3.148	0.739	2.581	2.620	0.55	2.061	2.118	0.537
$\sigma_{y_5}^2$	2.351	0.480	2.567	2.635	0.532	1.868	1.899	0.421
$\sigma_{y_6}^2$	4.954	0.914	2.801	2.869	0.619	2.662	2.739	0.647
$\sigma_{y_7}^2$	3.431	0.713	2.767	2.811	0.611	2.495	2.578	0.642
$\sigma_{y_8}^2$	3.254	0.695	2.422	2.467	0.479	1.937	2.012	0.497
ν_2^x	−6.228	0.705	−4.059	−4.059	0.48	−6.364	−6.405	0.624
ν_3^x	−5.634	0.774	−3.361	−3.380	0.536	−5.706	−5.734	0.732
λ_2^x	2.180	0.139	1.759	1.760	0.095	2.209	2.215	0.123
λ_3^x	1.819	0.152	1.382	1.381	0.105	1.836	1.838	0.144
$\sigma_{x_1}^2$	0.082	0.019	0.106	0.109	0.022	0.083	0.085	0.017
$\sigma_{x_2}^2$	0.120	0.070	0.172	0.179	0.056	0.113	0.118	0.04
$\sigma_{x_3}^2$	0.467	0.090	0.445	0.454	0.083	0.466	0.478	0.085
$\omega_{\zeta 60}^{-1}$	3.956	0.921	2.047	2.091	0.465	4.735	5.046	1.726
$\omega_{\zeta 65}^{-1}$	0.172	0.215	3.756	3.826	0.687	2.750	2.840	0.621
μ_ξ	5.054	0.084	5.040	5.035	0.098	5.054	5.053	0.087

(*continued on next page*)

Table B.1
(*continued*)

	MLE		Sub. priors			Centered MLE		
	Mean	Sd	Median	Mean	Sd	Median	Mean	Sd
ω_{ξ}^2	0.448	0.087	0.655	0.667	0.119	0.433	0.442	0.077
ω_{D15}^2	0.624	0.358	0.648	0.677	0.204	0.625	0.662	0.22
ω_{D24}^2	1.313	0.702	1.215	1.319	0.565	1.409	1.508	0.566
ω_{D26}^2	2.153	0.734	1.406	1.504	0.571	1.756	1.816	0.499
ω_{D37}^2	0.795	0.608	0.885	0.939	0.336	0.79	0.857	0.329
ω_{D48}^2	0.348	0.442	0.719	0.756	0.251	0.317	0.351	0.15
ω_{D68}^2	1.356	0.568	0.89	0.968	0.376	1.125	1.189	0.384

References

Ansari, A., Jedidi, K. (2000). Bayesian factor analysis for multilevel binary observations. *Psychometrika* **64**, 475–496.

Ansari, A., Jedidi, K., Jagpal, S. (2000). A hierarchical Bayesian methodology for treating heterogeneity in structural equation models. *Marketing Science* **19**, 328–347.

Arminger, G., Küsters, U. (1988). Latent trait models with indicators of mixed measurement level. In: Langeheine, R., Rost, J. (Eds.), *Latent Trait and Latent Class Models*. Plenum, New York, pp. 51–73.

Arminger, G., Muthén, B.O. (1998). A Bayesian approach to nonlinear latent variable models using the Gibbs sampler and the Metropolis–Hastings algorithm. *Psychometrika* **63**, 271–300.

Bauwens, L. (1984). *Bayesian Full Information Analysis of Simultaneous Equation Models using Integration by Monte Carlo*. Springer-Verlag, New York.

Bayarri, M.J., Berger, J.O. (2000). *P* values for composite null models. *Journal of the American Statistical Association* **95**, 1127–1142.

Bentler, P.M. (1992). *EQS Structural Equation Program Manual*. BMDP Statistical Software, Los Angeles.

Berger, J.O. (1985). *Statistical Decision Theory and Bayesian Analysis*, second ed. Springer-Verlag, New York.

Bernardo, J.M., Smith, A.F.M. (1994). *Bayesian Theory*. Wiley, Chichester.

Bollen, K.A. (1980). Issues in the comparative measurement of political democracy. *American Sociological Review* **80**, 370–390.

Bollen, K.A. (1989). *Structural Equations with Latent Variables*. Wiley, New York.

Bollen, K.A., Paxton, P. (1998). Two-stage least squares estimation of interaction effects. In: Schumacker, R.E., Marcoulides, G.A. (Eds.), *Interaction and Nonlinear Effects in Structural Equation Models*. Lawrence Erlbaum Associates, Mahwah, NJ, pp. 125–151.

Bollen, K.A., Stine, R. (1990). Direct and indirect effects: Classical and bootstrap estimates of variability. *Sociological Methodology* **20**, 115–140.

Bollen, K.A., Stine, R. (1993). Bootstrapping goodness-of-fit measures in structural equation models. In: Bollen, K.A., Long, J.S. (Eds.), *Testing Structural Equation Models*. Sage Publications, Newbury Park, CA.

Brooks, S.P., Gelman, A. (1998). General methods for monitoring convergence of iterative simulations. *Journal of Computational and Graphical Statistics* **7**, 434–455.

Brooks, S.P., Giudici, P. (2000). MCMC convergence assessment via two-way ANOVA. *Journal of Computational and Graphical Statistics* **9**, 266–285.

Browne, M.W. (1974). Generalized least squares estimations in the analysis of covariance structures. *South African Statistical Journal* **8**, 1–24.

Chen, M.H., Shao, Q.M., Ibrahim, J.G. (2000). *Monte Carlo Methods in Bayesian Computation*. Springer, New York.

Clinton, J., Jackman, S., Rivers, D. (2004). The statistical analysis of roll call data. *American Political Science Review* **98**, 355–370.

Croon, M., Bolck, A. (1997). On the use of factor scores in structural equations models. Tech. Report 97.10.102/7. Work and Organization Research Centre, Tilburg University.

Dunson, D.B. (2000). Bayesian latent variable models for clustered mixed outcomes. *Journal of the Royal Statistical Society B* **62**, 355–366.

Dunson, D.B. (2003). Dynamic latent trait models for multidimensional longitudinal data. *Journal of the American Statistical Association* **98**, 555–563.

Dunson, D.B. (2006). Bayesian dynamic modeling of latent trait distributions. *Biostatistics* **7**, 551–568.

Dunson, D.B., Chen, Z., Harry, J. (2003). A Bayesian approach for joint modeling of cluster size and subunit-specific outcomes. *Biometrics* **59**, 521–530.

Gelfand, A.E., Smith, A.F.M. (1990). Sampling-based approaches to calculate marginal densities. *Journal of the American Statistical Association* **85**, 398–409.

Gelman, A. (2005). Prior distributions for variance parameters in hierarchical models. *Bayesian Analysis* **1**, 1–19.

Geman, S., Geman, D. (1984). Stochastic relaxation, Gibbs distributions and the Bayesian restoration of images. *IEEE Trans. Pattern Anal. Machine Intell.* **6**, 721–741.

Gamerman, D. (1997). *Markov Chain Monte Carlo: Stochastic Simulation for Bayesian Inference*. Chapman and Hall, London.

Gilks, W.R., Richardson, S., Spiegelhalter, D.J. (1996). *Markov Chain Monte Carlo in Practice*. Chapman and Hall, London.

Herring, A.H., Dunson, D.B. (2004). Modeling the effects of a bi-directional latent predictor from multivariate questionnaire data. *Biometrics* **60**, 926–935.

Hills, S.E., Smith, A.F.M. (1992). Parametrization issues in Bayesian inference (with discussion). In: Bernardo, J.M., Berger, J.O., Dawid, A.P., Smith, A.F.M. (Eds.), *Bayesian Statistics*, vol. 4. Oxford University Press, pp. 227–246.

Jackman, S. (2004). What do we learn from graduate admissions committees? A multiple rater, latent variable model, with incomplete discrete and continuous indicators. *Political Analysis* **12**, 400–424.

Jedidi, K., Ansari, A. (2001). Bayesian structural equation models for multilevel data. In: Marcoulides, G.A., Schumacker, R.E. (Eds.), *New Developments and Techniques in Structural Equation Modeling*.

Jöreskog, K.G., Sörbom, D. (1985). *LISREL VI; Analysis of Linear Structural Relationships by Maximum Likelihood, Instrumental Variables, and Least Squares*. University of Uppsala, Uppsala.

Jöreskog, K.G., Sörbom, D. (1996). *LISREL 8; Structural Equation Modeling with the SIMPLIS Command Language*. Scientific Software International, Hove and London.

Jöreskog, K.G., Yang, F. (1996). Nonlinear structural equation models: the Kenny–Judd model with interaction effects. In: Marcoulides, G.A., Schumacker, R.E. (Eds.), *Advanced Structural Equation Modeling Techniques*. Lawrence Erlbaum Associates, Hillsdale, NJ, pp. 57–88.

Lee, S.Y. (1981). A Bayesian approach to confirmatory factor analysis. *Psychometrika* **46**, 153–160.

Lee, S.Y. (1992). Bayesian analysis of stochastic constraints in structural equation models. *British Journal of Mathematical and Statistical Psychology* **45**, 93–107.

Lee, S.Y., Shi, J.Q. (2000a). Bayesian analysis of structural equation model with fixed covariates. *Structural Equation Modeling: A Multidisciplinary Journal* **7**, 411–430.

Lee, S.Y., Shi, J.Q. (2000b). Joint Bayesian analysis of factor scores and structural parameters in the factor analysis model. *Annal of the Institute of Statistical Mathematics* **52**, 722–736.

Lee, S.Y., Song, X.Y. (2002). Bayesian selection on the number of factors in a factor analysis model. *Behaviormetrika* **27**, 23–39.

Lee, S.Y., Song, X.Y. (2003a). Bayesian model selection for mixtures of structural equation models with an unknown number of components. *British Journal of Mathematical and Statistical Psychology* **56**, 145–165.

Lee, S.Y., Song, X.Y. (2003b). Bayesian analysis of structural equation models with dichotomous variables. *Statistics in Medicine* **22**, 3073–3088.

Lee, S.Y., Song, X.Y. (2004). Bayesian model comparison of nonlinear structural equation models with missing continuous and ordinal data. *British Journal of Mathematical and Statistical Psychology* **57**, 131–150.

Liu, J.S., Wong, W.H., Kong, A. (1994). Covariance structure of the Gibbs sampler with applications to the comparisons of estimators and augmentation schemes. *Biometrika* **81**, 27–40.

Lopes, H., West, M. (2003). Bayesian model assessment in factor analysis. *Statistica Sinica* **14**, 41–67.

Martin, J.K., McDonald, R.P. (1975). Bayesian estimation in unrestricted factor analysis: A treatment for Heywood cases. *Psychometrika* **40**, 505–517.

Moustaki, I., Knott, M. (2000). Generalized latent trait models. *Psychometrika* **65**, 391–411.

Muthén, B.O. (1984). A general structural equation model with dichotomous, ordered categorical and continuous latent variable indicators. *Psychometrika* **49**, 115–132.

Muthén, L.K., Muthén, B.O. (1998). *Mplus User's Guide*. Muthén & Muthén, Los Angeles, CA.

Muthén, L.K., Muthén, B.O. (2003). *Mplus Version 2.13. Addendum to the Mplus User's Guide*. Muthén and Muthén, Los Angeles, CA. Downloadable from http://www.statmodel.com/support/download.

Raftery, A.E. (1993). Bayesian model selection in structural equation models. In: Bollen, K.A., Long, J.S. (Eds.), *Testing Structural Equation Models*. Sage, Newbury Park, CA.

Raftery, A.E., Lewis, S. (1992). How many iterations in the Gibbs sampler?. In: Bernardo, J.M., Berger, J.O., Dawid, A.P., Smith, A.F.M. (Eds.), *Bayesian Statistics*, vol. 4. Oxford University Press, pp. 763–773.

Robert, C.P., Casella, G. (2004). *Monte Carlo Statistical Methods*, second ed. *Springer Texts in Statistics*. Springer.

Roberts, G.O., Sahu, S.K. (1997). Updating schemes correlation structure blocking and parametrization for the Gibbs sampler. *Journal of the Royal Statistical Society B* **59**, 291–317.

Sammuel, M.D., Ryan, L.M., Legler, J.M. (1997). Latent variable models for mixed discrete and continuous outcomes. *Journal of the Royal Statistical Society B* **59**, 667–675.

Satorra, A., Bentler, P.M. (1988). Scaling corrections for chi-square statistics in covariance structure analysis. In: *Proceedings of the Business and Economic Statistics Section of the American Statistical Association*, pp. 308–313.

Scheines, R., Hoijtink, H., Boomsma, A. (1999). Bayesian estimation and testing of structural equation models. *Psychometrika* **64**, 37–52.

Song, X.Y., Lee, S.Y. (2004a). Bayesian analysis of two-level nonlinear structural equation models with continuous and polytomous data. *British Journal of Mathematical and Statistical Psychology* **57**, 29–52.

Song, X.Y., Lee, S.Y. (2004b). Local influence analysis of two-level latent variable models with continuous and polytomous data. *Statistica Sinica* **14**, 317–332.

Treier, S., Jackman, S. (2005). Democracy as a latent variable. Tech. Report. Available at http://jackman.stanford.edu/papers/download.php?i=2.

Wall, M.M., Amemiya, Y. (2000). Estimation for polynomial structural equation models. *Journal of the American Statistical Association* **95**, 929–940.

Spiegelhalter, D.J., Thomas, A., Best, N., Gilks, W. (2003). WinBUGS; Version 1.4 User Manual. MRC Biostatistics Unit. URL: http://weinberger.mrc-bsu.cam.ac.uk/bugs/Welcome.html.

Zhu, H.T., Lee, S.Y. (2001). Assessment of local influence for models with incomplete data. *Journal of the Royal Statistical Society, Series B* **63**, 111–126.

Handbook of Computing and Statistics with Applications, Vol. 1
ISSN: 1871-0301

DOI: 10.1016/S1871-0301(06)01009-2

9

The Analysis of Structural Equation Model with Ranking Data using Mx

Wai-Yin Poon

Abstract
The Thurstonian approach accompanying structural equation modeling is a useful approach in modeling and analyzing ranking data. The relationship between analyzing ranking data and ordinal categorical data is explored, with a view to making use of readily available structural equation modeling procedures for analyzing ordinal categorical data to analyze ranking data. The Mx program, which can be downloaded at no cost, is used to illustrate the implementation of the maximum likelihood estimation procedure. It will be demonstrated that the procedure is very flexible, enabling various kinds of Thurstonian model be analyzed in an easy, straightforward, and efficient way. Various identification constraints, across parameters constraints, mean structures as well as covariance and correlation structures can be incorporated into the analysis very easily. Moreover, models with partial ranking data can be analyzed.

Keywords: Ranking data; Thurstonian models; Partial ranking; Mx

1. Introduction

Ranking data are obtained when subjects are asked to rank p objects from 1 (most preferred) to p (least preferred). A number of approaches have been developed to model and analyze ranking data. Among others, researches on using Thurstonian models (Thurstone, 1927) remain active for more than 70 years (see, e.g., Critchlow and Fligner, 1991; Böckenholt, 1992, 1996, 2001; Chan and Bentler, 1998; Mayedu-Olivares, 1999), confirming their usefulness in a wide range of disciplines. Thurstonian models postulate that the ranking of the p objects are determined by a $p \times 1$ latent continuous random vector Y which is distributed as multivariate normal with mean μ and covariance matrix Σ. Different models are achieved by putting different restrictions on the elements in Σ. In particular, the use of structural equation models by imposing the structures on Σ substantially enriches the horizon of modeling ranking data and hence its practical applicability (see, e.g., Currim, 1982; Böckenholt, 1992; Elrod and Keane, 1995).

In effect, structural equation modeling remains an extremely active area of research. The interest derives from both the practical applicability of the approach in addressing and verifying substantive theories and the technical difficulties involving in modeling of various types of data together with the estimation of the resulted model. User-friendly structural equation modeling packages are widely available, such as PRELIS and LISREL (Jöreskog and Sörbom, 1996a, 1996b), EQS (Bentler and Wu, 1993), Mx (Neale et al., 1999) as well as Mplus (Muthén and Muthén, 1998). The capabilities of these packages continue to be extended and nowadays, they all provide options for analyzing ordinal categorical data in a convenient manner. Although ranking data and ordinal categorical data are different in nature and require different modeling and estimation techniques, the approaches adopted by popular packages for analyzing ordinal categorical data and the Thurstonian approach for analyzing ranking data both operate on the assumption that the observed variables are associated with some underlying continuous variables distributed as multivariate normal. With reference to this similarity, the current study establishes a relationship between the analysis of ordinal categorical data and ranking data with a view to making use of readily available structural equation modeling procedures for analyzing ordinal categorical data to analyze ranking data. Specifically, the implementation of the analysis of the Thurstonian models using options designated for analyzing ordinal categorical data in the widely available software program Mx (Neale et al., 1999) is addressed. Some initial effort along similar direction has been made by Mayedu-Olivares (1999). He has developed a procedure for analyzing ranking data using the package MECOSA (Arminger et al., 1996).

A description of the Thurstonian model of ranking data, the multivariate normal model for analyzing ordinal categorical data, and their similarity is given in Section 2. A summary of the Mx program together with its option for analyzing ordinal categorical data, and the procedure on using the option to implement the Thurstonian models are given in Section 3. It will be seen that the procedure is very flexible. Different identification and across parameters constraints together with various structures on the mean vector and on the covariance/correlation matrix can be incorporated in an efficient and easy manner. Some examples available in the literature are used for illustration. In effect, the procedure can also be applied to analyze models for partial ranking data (Böckenholt, 1992). Two applications are discussed in Section 4. The first application relates to an analysis of rankings of compound objects. The objective is to examine whether or not a mean score for a compound choice alternative consisting of two objects can be predicted by an additive combination of means scores obtained for each of the two objects separately. The second application relates to the analysis of a set of 8 soft drinks, and the rankings of the soft drinks are obtained via a balanced incomplete design. The paper is concluded with a discussion in Section 5.

2. Multivariate normal model for analyzing ranking and ordinal categorical data

2.1. The Thurstonian model of ranking data

Suppose that there are p alternatives which are ranked by N subjects without ties from 1 (most preferred) to p (least preferred), the class of Thurstonian models of ranking data operate on the assumption that there exists a $p \times 1$ random vector $Y = (Y_1, \ldots, Y_p)'$ which is distributed as multivariate normal with mean $\mu = (\mu_1, \ldots, \mu_p)'$ and covariance matrix $\Sigma = \{\sigma_{ij}\}$, and that a respondent ranks the object most preferred if his Y_i value is the largest, second preferred if his Y_i value is the second largest, and accordingly, least preferred if his Y_i value is the smallest.

Let A be a $(p-1) \times p$ matrix of contrasts given by

$$A = \begin{pmatrix} 1 & -1 & 0 & \ldots & 0 \\ 1 & 0 & -1 & \ldots & 0 \\ \vdots & \vdots & \vdots & \vdots & \vdots \\ 1 & \ldots & 0 & 0 & -1 \end{pmatrix}, \tag{1}$$

and let

$$Y^* = AY = \begin{pmatrix} Y_1 - Y_2 \\ Y_1 - Y_3 \\ \vdots \\ Y_1 - Y_p \end{pmatrix} = \begin{pmatrix} Y_1^* \\ Y_2^* \\ \vdots \\ Y_{p-1}^* \end{pmatrix}, \tag{2}$$

then the distribution of Y^* is multivariate normal with mean $\mu^* = A\mu = (\mu_1^*, \ldots, \mu_{p-1}^*)'$ and covariance matrix $\Sigma^* = A\Sigma A' = \{\sigma_{ij}^*\}$. Let $T = p!$ be the total number of possible ranking patterns, $(Y_{t(1)} > Y_{t(2)} > \cdots > Y_{t(p)})$ be the ordering of the elements of Y associated with the ranking pattern t, $t = 1, \ldots, T$, and π_t the probability of observing ranking pattern t, then it can be shown that (see, e.g., Chan and Bentler, 1998)

$$\begin{aligned} \pi_t &= \Pr(Y_{t(1)} > Y_{t(2)} > \cdots > Y_{t(p)}) \\ &= \Pr(Y_{t(1)} - Y_{t(2)} > 0, Y_{t(2)} - Y_{t(3)} > 0, \ldots, Y_{t(p-1)} - Y_{t(p)} > 0) \qquad (3) \\ &= \Phi_{p-1}\big(D_t S_t \mu^*; D_t S_t \Sigma^* S_t' D_t\big) \qquad (4) \\ &= \Phi_{p-1}\big(D_t S_t A\mu; D_t S_t A\Sigma A' S_t' D_t\big), \qquad (5) \end{aligned}$$

where $\Phi_{p-1}(z; R)$ denotes the distribution function of the $(p-1)$-dimensional standardized multivariate normal distribution with correlation matrix R, evaluating at z; S_t is a $(p-1) \times (p-1)$ selection matrix of 0's, 1's and -1's which transform Y^* to those contrasts involved in (3) specific to the pattern t, and $D_t = (\mathrm{diag}(S_t \Sigma^* S_t'))^{-1/2}$. Ignoring a constant term, the log-likelihood function is given by

$$L(\theta) = \sum_{t=1}^{T} f_t \log \pi_t, \tag{6}$$

where f_t is the observed frequency in pattern t and θ stores the unknown parameters in μ and Σ. Under the context of structural equation modeling where $\Sigma = \Sigma(\beta)$, θ consists of the unknown parameters in μ and the structural parameters in β. The model given in (6) is not identified. It is necessary to impose various constraints on μ and Σ or on the reduced form parameters μ^* and Σ^* to identify the model (see, e.g., Mayedu-Olivares, 1999). In the latter case, θ stores the unknown parameters in μ^* and Σ^* or its structural parameters.

2.2. The underlying multivariate normal model for ordinal categorical data

On the other hand, ordinal categorical variable is obtained when subjects are required to respond to a question on some five-point or seven-point scale. The analysis of ordinal categorical variables under the context of structural equation modeling has gained considerable attention over the past 15 years (see, e.g., Jöreskog, 1994; Lee et al., 1995) and many of these recently developed methods have been incorporated into widely available structural equation modeling software. These independently developed methods all operate on the same assumption that observed ordinal categorical variables are manifestations of some underlying continuous variables distributed as multivariate normal. Specifically, let $Z = (Z_1, \ldots, Z_q)'$ be a vector of the q observed ordinal categorical variables, and $X = (X_1, \ldots, X_q)'$ a random vector of the underlying continuous variables distributed as multivariate normal with mean μ_x and covariance matrix Σ_x, the relationship between Z_i and X_i is given by

$$Z_i = k_i - 1 \quad \text{if } \alpha_{i,k_i} < X_i < \alpha_{i,k_i+1} \tag{7}$$

for $i = 1, \ldots, q$ and $k_i = 1, \ldots, m_i$; where m_i is the number of categories for the ith variable with $\alpha_{i,1} = -\infty$ and $\alpha_{i,m_i+1} = \infty$.

The vector X is unobservable but a random sample of Z with size N_x is available, usually organized in the form of a q-way contingency table. Denote the observed frequency in the cell with multiple index $(k_1, \ldots, k_q)$ by $g_{k_1,\ldots,k_q}$ and the corresponding cell probability by $\xi_{k_1,\ldots,k_q}$, the probability that an observation falls into the cell is given by

$$\xi_{k_1,\ldots,k_q} = \Pr(\alpha_{1,k_1+1} < X_1 < \alpha_{1,k_1+2}, \ldots, \alpha_{q,k_q+1} < X_q < \alpha_{q,k_q+2}). \tag{8}$$

Ignoring a constant term, the log-likelihood function is given by

$$L_x(\gamma) = \sum_{k_1=1}^{m_1} \cdots \sum_{k_q=1}^{m_q} g_{k_1,\ldots,k_q} \log \xi_{k_1,\ldots,k_q}, \tag{9}$$

where γ consists of the unknown parameters in thresholds, in μ_x, and in Σ_x or its structural parameters. The model in (9) is not identified and it is again necessary to impose some constraints on γ to identify the model. Comparing (9) to (6) and (8) to (3), it is noted that the likelihood functions take the same form. Specifically, (3) is similar to (8) when $q = p - 1$, $m_i = 2$ and $\alpha_{i,2} = 0$ for all $i = 1, \ldots, q$. Given this similarity, it is possible to make use of procedures that are designed for finding the maximum likelihood estimate of γ to obtain the maximum likelihood estimate of θ.

3. Implementation by Mx

Although such similarities exist and a number of procedures have been developed to find the maximum likelihood estimate of γ, it is not straightforward to use the options that are available in structural equation modeling software for the analysis of ordinal categorical data to analyze ranking data. The reason lies in that those approaches adopted in the popular software for analyzing ordinal data are not full information maximum likelihood methods. Instead, they are some multistage methods built on the idea of partition maximum likelihood (Poon and Lee, 1987). The Mx program (Neale et al., 1999), on the other hand, uses the maximum likelihood approach in analyzing ordinal categorical data and is therefore able to perform maximum likelihood analysis of ranking data with reference to the aforementioned similarity between (6) and (9).

3.1. *The Mx program and its maximum likelihood analysis of raw ordinal data*

Mx (Neale et al., 1999) is a flexible software that is capable of doing advanced structural equation problems (Hamagami, 1997). When many of the fitting functions available in other structural equation modeling packages are available in Mx, it also allows users to define their own fit functions. Another important feature of Mx lies in its capability in conducting matrix operations. These features together with many others in Mx substantially facilitate its flexibility and applicability, enabling the analysis of many nonstandard models. The program can be mastered quite easily for those who are familiar with LISREL (Jöreskog and Sörbom, 1996b) or have some basic knowledge in structural equation modeling. Moreover, Mx can be downloaded at not cost, including program, documentation and examples (Neale et al., 1999; http://griffin.vcu.edu/mx/).

With regard to the analysis of ordinal categorical data, Mx has the option for maximum likelihood analysis of 2-way contingency tables (Neale et al., 1999, p. 85). For data of higher dimensions, it is necessary to use the raw ordinal data option (Neale et al., 1999, p. 83). As a result, for ranking data with 3 objects, the contingency table and raw ordinal data options can both be applied; but for ranking data with more than 3 objects, it is necessary to use the raw ordinal data option. We examine the use of the raw ordinal data option to analyze ranking data with reference to some ranking data sets available in the literature.

3.2. *Analysis of ranking data ($p = 4$) using Mx*

The first data set is the compact cars data set employed by Mayedu-Olivares (1999). 279 Spanish college students were asked to rank four compact cars $\{1 =$ Ford Fiesta, $2 =$ Opel Corsa, $3 =$ Peugeot 106, $4 =$ Volkswagen Polo$\}$ according to their purchase preferences. The ranking patterns' observed frequencies are provided in Table 1 where, for example, the pattern (2314) refers to Opel Corsa is the most preferred car, Peugeot is the second preferred, Ford Fiesta is the third and Volkswagen Polo is the least preferred one. We demonstrate how various analyses, including those presented in Mayedu-Olivares (1999), can be easily achieved using Mx. We first analyze the data using the basic Thurstonian model (model a), that is the model with mean vector μ and covariance matrix Σ, with the constraints $\mu_4 = 0$, $\sigma_{ii} = 1$ for all $i = 1, \ldots, 4$ and $\sigma_{43} = 0$; we then analyze the data with the additional structure $\Sigma = \Lambda\Lambda' + \Psi$ given by an one-factor factor analysis model (model b). Finally, we analyze the data set by estimating the reduced form parameters μ^* and Σ^* (model c). It will be seen that the Mx input scripts for analyzing these models can be prepared very easily by modifying the input script for the basic Thurstonian model. The sample input scripts for analyzing these three models are given in Appendices A, B and C.

Table 1
The observed frequencies of ranking pattern in the compact car's data

t	f_t	t	f_t	t	f_t	t	f_t
1234	16	2134	16	3124	22	4123	21
1243	16	2143	10	3142	14	4132	12
1324	14	2314	19	3214	14	4213	14
1342	2	2341	3	3241	3	4231	9
1423	4	2413	8	3412	8	4312	11
1432	9	2431	9	3421	11	4321	14

1 = Ford Fiesta, 2 = Opel Corsa, 3 = Peugeto 106, 4 = Volkswagen Polo.
From Mayedu-Olivares (1999, Table 1).

All inputs are in multiple group settings with 26 groups (or 25 groups when there are no functional constraints on parameters), which are divided into three parts. Part I consists of the first group only. This group is used to specify the data set and the model for analysis; it is therefore needed to be modified when different data set or different model is employed. Part II consists of 23 groups. This part remains the same for any data set and any analysis of ranking data with 4 objects. Once a template of this part is created, it can be used for any further analysis of ranking data with 4 objects. Part III consists of two groups, namely, group 25 and group 26. Group 25 is a calculation group that computes a component in the goodness-of-fit test statistic. The specification statements in group 25 are invariant for any analysis of ranking data with 4 objects. Group 26 is a constraint group which is used to specify functional constraints on parameters. Group 26 must be modified when applying to models with different functional constraints. Simple constraints equating a parameter to a constant can be specified in group 1. Group 26 is not required when there has no functional constraint on the parameters. Moreover, all statements after the mark "!" are for description.

Three external files have been used. The file "select.4" has $T \times (p-1) = 24 \times 3 = 72$ rows and $p - 1 = 3$ columns. It stores the 24 selection matrices, $S_t, t = 1, \ldots, 24$, each of size 3×3 (see (4)). This file can be used for all ranking data analysis with $p = 4$. The file "auto.obs" stores the frequencies of the observed ranking pattern, each frequency occupies a row. The file "auto.uni" is a file with all its entries equal to 1. It consists of $p - 1 = 3$ columns and $f_1 = 16$ rows, where f_1 is the observed frequency in the first pattern.

3.2.1. *Model a*

Some more explanations about the input file in Appendix A that is used to perform the first analysis for the basic Thurstonian model are given as follows:

Part I. This part consists of only group 1. It is used to compute the contribution of the observed frequency in the pattern $t = 1$ to the likelihood. We have specified in the "DATA" command line that "NI = 3" because $q = 3$ when $p = 4$. We use the "Ordinal" data option. Note that the data file "auto.uni" constructed using the aforementioned method consists of $f_1 = 16$ ordinal observations with 3 dichotomous items, and observations of all items fall into the second category ("0" represents the first category and

"1" the second), indicating they are greater than the threshold that is fixed at zero, the resulting fit function for group 1 is $-2f_1 \log \hat{\pi}_1$ (see (6) and (9)). The parameters in the basic Thurstonian model are respectively μ and Σ, they are stored in the matrices H and P, respectively, in the Mx program. It will be seen that for other models, only modification on input statements concerning these two matrices are required. We have indicated in the sample input script that many of the matrices are "invariant", such as the matrices S, J, U, L, T and I. These matrices can be specified in exactly the same way for all ranking data analysis with $p = 4$. Matrix A is a matrix of contrasts (see (1)). Whenever one works with the basic model with parameters μ and Σ or its structured form, this matrix is required and remains the same for all ranking data analysis with $p = 4$. However, when one is interested in the reduced form parameters μ^* and Σ^*, this matrix is no longer required. This point will be further addressed. Finally, the "Algebra" section is used to first transform μ and Σ to μ^* (stored in M) and Σ^* (stored in C), and then to $S_t\mu^*$ and $S_t\Sigma^* S_t'$ specific for the pattern $t = 1$ (see (5)).

Part II. This part consists of group 2 to group $T = 24$. Each group computes the contribution of the observation in the pattern t, $t = 2, \ldots, T$, to the likelihood. When the preparation of this part has been completed, it can be used for all analyses with $p = 4$ and no modification is required. For each group t, we compute $-2f_t \log \hat{\pi}_t$. Therefore, it is necessary to locate f_t and S_t respectively from the matrix O that stores the observed frequencies and the matrix S that stores S_t. Matrix K and E are used to point to the appropriate locations. For group t, the entries for matrix K are given by $\{(t-1) \times 3 + 1, 1, t \times 3, 3\}$, and those for matrix E are given by $\{1, t, 1, t\}$. The matrices O and E as well as other matrices that have been specified in group 1 will remain unchanged in group 2 to group T, unless otherwise specified. We then used the user-defined fit function to compute $-2f_t \log \hat{\pi}_t$, the contribution of group t to the overall fit function. Therefore, combining all contributions from group 2 to group T and added to the fit function in group 1 will result to an overall fit function given by

$$-2L(\hat{\theta}) = -2\sum_{t=1}^{T} f_i \log \hat{\pi}_t, \tag{10}$$

which is equivalent to the likelihood function given in (6).

Part III. Since there has no functional constraints on the parameters, Part III consists of only group 25. It is a calculation group that is used to produce a component of the likelihood ratio test statistic. Since the overall fit function is given by (10), an adjustment value equals to

$$2\sum_{t=1}^{T} f_i \log\left(\frac{f_i}{N}\right) \tag{11}$$

must be added to the fit function to produce the chi-squared distributed likelihood ratio test statistic G^2. The value of this component is computed by the specification of group 25 and is stored in the matrix F. This computation is invariant for all data sets and hence the input template for group 25 can be used for all ranking data analysis with $p = 4$.

The value of the fit function produced by this Mx input script is 1733.937 and the value produced by group 25 is −1707.38, resulting to a likelihood ratio test statistic $G^2 = 26.557$. G^2 is chi-squared distributed with 15 degrees of freedom. The maximum likelihood estimates of μ and Σ produced by the Mx are respectively given by

$$\hat{\mu} = \begin{pmatrix} 0.1628 \\ 0.1166 \\ 0.0889 \\ 0^* \end{pmatrix} \quad \text{and} \quad \widehat{\Sigma} = \begin{pmatrix} 1^* & 0.6633 & 0.5030 & 0.1688 \\ 0.6633 & 1^* & 0.4410 & 0.1310 \\ 0.5030 & 0.4410 & 1^* & 0^* \\ 0.1688 & 0.1310 & 0^* & 1^* \end{pmatrix}, \tag{12}$$

where parameters with a asterisk are fixed. The estimates are very close to those produced by Mayedu-Olivares (1999, Table 2).

3.2.2. *Model b*

Mayedu-Olivares (1999) has also applied an one-factor factor analysis model $\Sigma = \Lambda\Lambda' + \Psi$ to analyze the data set with the constraints $\mu = 0$, $\Lambda_4 = 0$, and Σ equals to a correlation matrix. Such an analysis can be conducted very easily by modifying the basic Mx input script already constructed in Appendix A. Specifications for group 2 through group 25 remain unchanged. However, modifications on group 1 are required to specify the factor structure and a constraint group (group 26) is included to specify the functional constraints. The details of the input for these two groups to achieve the analysis are presented in Appendix B. The parameter matrices are now H, F and G, storing, respectively, the mean vector μ, the factor loading matrix Λ and the matrix of error variance Ψ. In addition to the basic specifications on these parameter matrices, the statement "$P = F * F' + G$" in the "Algebra" section is included to specify the structure of the covariance matrix. It is worthy of note that in order to use Part II input as given in Appendix A, it is necessary to be consistent in notation. In other words, the matrices H and P should continue be used to store the mean and the covariance matrix. They will then be transformed using the contrast matrix A to the reduced form mean vector μ^* and covariance matrix Σ^* that is respectively denoted by M and C in Mx. Moreover, since Σ is constrained to be a correlation matrix, functional constraints on the parameters in F and G are required and are specified by group 26. The matrix H, on the other hand, is specified in group 1 as fixed at 0. The resulted likelihood ratio test statistic for this model is $G^2 = 32.36$ with 20 degrees of freedom. The estimates of Λ and the diagonal elements in Ψ are respectively given by

$$\hat{\Lambda}' = (0.8333, 0.7394, 0.5473, 0^*) \quad \text{and}$$

$$\text{Diag}(\widehat{\Psi}) = (0.3057, 0.4533, 0.7005, 1^*). \tag{13}$$

These estimates are again very close to those given by Mayedu-Olivares (1999, Table 3). It can be seen that once the input script such as that in Appendix A is available, the factor analysis model and any other structural models can be called for analysis in a very easy way.

3.2.3. *Model c*

In effect, any analysis of Thurstonian models of ranking data with 4 objects can be obtained by modifying the input of groups 1 and 26, including when one is interested in analyzing the reduced form parameters μ^* and Σ^* (see, e.g., Chan and Bentler, 1998). We present in Appendix C the input required in group 1 to produce a maximum likelihood analysis of μ^* and Σ^*. The basic parameter vectors have been changed to M and C and it is no longer necessary to use the contrast matrix A to transform the p-dimensional mean vector and covariance matrix into $(p-1)$-dimensional. As a result, the first two statements in the "Algebra" section of Appendix A and the specification for the matrix A have been deleted. It is worthy to recall that in order to make use of Part II input already constructed, the use of M and C to store the reduced form parameters μ^* and Σ^* should remain unchanged.

The Mx analysis gives $G^2 = 26.557$ with 15 degrees of freedom, the same as the unrestricted Thurstonian model, and the estimates of the parameters are given by

$$\hat{\mu}^* = \begin{pmatrix} 0.04^* \\ 0.0691 \\ 0.1520 \end{pmatrix} \quad \text{and} \quad \widehat{\Sigma}^* = \begin{pmatrix} 0.6060 & 0.2478 & 0.2691 \\ 0.2478 & 0.8909 & 0.2986 \\ 0.2691 & 0.2986 & 1.4778 \end{pmatrix}. \tag{14}$$

Three different analyses have been conducted for this example, all analyses can be implemented by Mx using the input script in Appendix A or its slight modifications. In effect, any analysis on ranking data with $p = 4$ can be produced by Mx using a simple modification of group 1 and group 26 of the standard input template given in Appendix A.

3.3. *Generalization*

When $p \neq 4$, it is necessary to create other input templates. Similar to the case of $p = 4$, the number of groups involved in the input is $p! + 2$. The first group is used to specify the data and the model, the next $p! - 1$ groups in Part II and the first group in Part III will remain the same for all ranking data analysis with p objects, and the final group is used to specify functional constraints on parameters. In effect, most of the specifications for other groups remain similar to those given in the preceding section for $p = 4$, only the specifications in the matrices K and E are required to be modified to reflect the change in dimensions. In order to facilitate future analysis, one can prepare some sample input templates for different p. As a result, any substantive researches on ranking data can be conducted by Mx with modifications on the first and the last groups of the standard templates. Moreover, it is also necessary to create the external file "select.p" that stores the $p!$ selection matrices $S_t, t = 1, \ldots, T$ (see (4)) each with dimension $(p-1) \times (p-1)$ for different p.

4. Applications

Once the standard templates have been created for analyzing ranking data with different p, they can also be used flexibly to analyze various models of ranking data. We present two applications on models with partial ranking data (Böckenholt, 1992).

Table 2
The observed frequencies of single ranking items in the gift's data

t	f_t	t	f_t	t	f_t	t	f_t
1234	13	2134	14	3124	4	4123	6
1243	3	2143	9	3142	4	4132	6
1324	10	2314	15	3214	7	4213	4
1342	7	2341	13	3241	9	4231	10
1423	5	2413	2	3412	8	4312	4
1432	4	2431	5	3421	9	4321	7

1 = Camera, 2 = Typewriter, 3 = Portable radio, 4 = Record player.
From (McKeon, 1961).

Table 3
The observed frequencies of composite ranking items in the gift's data set

Pattern	123	132	213	231	312	321
Frequency	48	17	31	25	19	38

1 = Camera & typewriter, 2 = Typewriter & portable radio, 3 = Portable radio & record player.
From (McKeon, 1961).

4.1. *Application: Ranking of compound objects*

The observed frequencies of the ranking patterns in Tables 2 and 3 are taken from Böckenholt (1992, Tables 1 and 2). The data set is originally from McKeon (1961). The objective is to examine whether or not a mean score for a compound choice alternative consisting of two objects, $O_{ij} = \{O_i, O_j\}$, can be predicted by an additive combination of mean scores obtained for each of the two objects separately. In other words, the relationship

$$\mu_{ij} = \mu_i + \mu_j, \tag{15}$$

is studied, where μ_{ij} represents the mean score for the compound package with individual items having means μ_i and μ_j, respectively. Table 2 presents the frequencies of ranking patterns of the single items {1: camera, 2: typewriter, 3: portable radio, 4: record player} and Table 3 presents the frequencies of ranking patterns of composite items {1: camera and typewriter, 2: typewriter and portable radio, 3: portable radio and radio player}. Böckenholt (1992, Table 3) applied four different models to analyze the data set, these models are summarized in Table 4, where μ_I and Σ_I denote the 4×1 mean vector and 4×4 covariance matrix for individual items, μ_{II} (3×1) and Σ_{II} (3×3) denote those for the compound packages, and I is the identify matrix of appropriate dimension. The matrix B is given by

$$B = \begin{pmatrix} 1 & 1 & 0 & 0 \\ 0 & 1 & 1 & 0 \\ 0 & 0 & 1 & 1 \end{pmatrix}, \tag{16}$$

Table 4
Mx analyses of the gift data set

	Model a	Model b	Model c	Model d
	$\Sigma_I = I$ $\Sigma_{II} = I$	$\Sigma_I = I$ $\Sigma_{II} = BB'$ $\mu_{II} = B\mu_I$	$\Sigma_I = I$ $\Sigma_{II} = BB'$	$\Sigma_I = \text{diag}(1, 1, \sigma_{33}^2, 1)$ $\Sigma_{II} = B\Sigma_I B'$ $\mu_{II} = B\mu_I$
	Parameter estimate			
$\hat{\mu}_{I_1}$	−0.0764	−0.0601	−0.0764	−0.0523
$\hat{\mu}_{I_2}$	0.2058	0.1755	0.2058	0.1681
$\hat{\mu}_{I_3}$	0.0989	0.0824	0.0989	0.0747
$\hat{\mu}_{I_4}$	−0.2283	−0.1978	−0.2284	−0.1905
$\hat{\mu}_{II_1}$	0.0190	0.1154	0.0341	0.1158
$\hat{\mu}_{II_2}$	0.1298	0.2579	0.1343	0.2428
$\hat{\mu}_{II_3}$	−0.1488	−0.1154	−0.1684	−0.1158
$\hat{\sigma}_{33}^2$	–	–	–	0.4417
Fit*1	1749.697	1732.726	1732.019	1724.474
Adj*2	−1704.994	−1704.994	−1704.994	−1704.994
G^2	44.703	27.732	27.025	19.480
df	23	25	23	24

*1 The overall fit function value.

*2 The adjustment value.

and the constraint that the sum of the mean scores equals to zero is also imposed in all models. These four models are analyzed using Mx. The input scripts for analysis can be easily constructed by combining the standard input templates for $p = 3$ and $p = 4$ with appropriate modifications on respectively their first and last groups. The resulting input script therefore consists of $(4! + 2) + (3! + 2) = 34$ groups. The likelihood ratio test statistic G^2 is obtained by adjusting the overall fit function value. The adjustment value is obtained as a total of the values produced respectively for $p = 3$ and $p = 4$ using (11), that is, the values produced by the first groups of Part III inputs in Mx. The results of the Mx analyses are presented in Table 4, they are nearly the same as those produced by Böckenholt (1992, Table 3).

4.2. Application: Balanced incomplete block design

Böckenholt (1992) discussed the analysis of partial ranking data. In particular, a balanced incomplete design has been employed to analyze a data set of 8 soft drinks. In order to reduce the number of possible response patterns that otherwise is equal to 8!, seven replications of two blocks each entails to the ranking of four objects have been used. As a result, the observed frequencies of 14 sets of ranking frequencies each with 24 patterns have been obtained. The details are available in Böckenholt (1992, Tables 5 and 6) and are not presented here. Since the objective function for analyzing such a data set is the sum of 14 functions each takes the form of the usual likelihood given in (6)

Table 5
Mx analyses of the soft drink data set

	Model a	Model b	Model c	Model d
	$\Sigma = I$	$\Sigma = \Lambda\Lambda' + \sigma^2 I$ (Λ given)	$\Sigma = \Lambda\Lambda' + \Psi$ Λ unrestricted $\Psi = I$	$\Sigma = \Lambda\Lambda' + \Psi$ Λ unrestricted $\Psi = I$ $\mu = \Lambda\nu$ $\nu_1^2 + \nu_2^2 = 1$
	Parameter estimate			
$\hat{\mu}_1$	0.6063	1.2545	1.1340	1.0995
$\hat{\mu}_2$	0.5544	1.1793	1.0437	0.9982
$\hat{\mu}_3$	0.3623	0.7161	0.6740	0.7496
$\hat{\mu}_4$	0.2833	0.5798	0.5251	0.5155
$\hat{\mu}_5$	−0.3699	−0.7412	−0.6591	−0.6191
$\hat{\mu}_6$	−0.3089	−0.6817	−0.5757	−0.5368
$\hat{\mu}_7$	−0.4827	−1.0043	−0.9554	−1.0557
$\hat{\mu}_8$	−0.6448	−1.3016	−1.1866	−1.1562
Fit*1	3270.857	3118.482	3090.168	3103.255
Adj*2	−2794.322	−2794.322	−2794.322	−2794.322
G^2	476.535	324.16	295.846	308.933
df	315	314	299	305

*1 The overall fit function value.
*2 The adjustment value.

with $p = 4$, it is possible to use the standard template with $p = 4$ as a basic component to construct the Mx input script. For simplicity, the adjustment on the likelihood ratio test statistic is computed using a separate program and only one constraint group is employed across the 14 sets to specify the constraints. As a result, the inputs in Part III are not required and the resulting input script consists of $14 \times 4! + 1 = 337$ groups. The first group in the input consists of all the major specifications. Specifically, since the objective is to estimate the mean vector and covariance matrix of a random vector of dimension 8, these two matrices of dimensions 8×1 and 8×8 are therefore specified in the first group as "Free" matrices. Moreover, in each of the first group in the 14 sets, it is necessary to construct a 4×8 selection matrix with reference to the 4 soft drinks available in the specific block and replication (see Böckenholt, 1992, Table 5) so as to obtain the corresponding 4-dimensional mean vector and covariance matrix. The other 23 groups are similar to those in Part II of the $p = 4$ standard template. Finally, a constraint group is used to specify the functional constraints. Four different models similar to those employed by Böckenholt (1992) have been analyzed. These models are summarized in Table 5 where in model b, the value of the factor loading matrix is fixed as

$$\Lambda' = \begin{pmatrix} 1 & 1 & 1 & 1 & -1 & -1 & -1 & -1 \\ 1 & 1 & -1 & -1 & 1 & 1 & -1 & -1 \end{pmatrix}. \tag{17}$$

Moreover, the constraint that the sum of the elements in the mean vector is zero has been imposed to models a, b and c. The results of Mx analyses are presented in Table 5. Some other parameter estimates not given in Table 5 are respectively given by $\hat{\sigma}^2 = 1.4387$ in model b;

$$\hat{\Lambda}' = \begin{pmatrix} 0.9782 & 0.6243 & 1.3836 & 0.8534 & -0.8028 & -0.7325 & -1.3376 & -0.9688 \\ 0.9453 & 1.2551 & -1.0653 & -1.0704 & 0.6028 & 0.9235 & -0.8359 & -0.7512 \end{pmatrix} \tag{18}$$

in model c; as well as $\hat{\nu}_1 = 0.9514$, $\hat{\nu}_2 = 0.3081$ and

$$\hat{\Lambda}' = \begin{pmatrix} 0.8427 & 0.6578 & 1.1438 & 0.8467 & -0.8268 & -0.8163 & -0.8503 & -1.0036 \\ 0.9665 & 1.2087 & -1.0991 & -0.9416 & 0.5438 & 0.7784 & -0.8010 & -0.6535 \end{pmatrix} \tag{19}$$

in model d.

It is worthy of note that since each respondent has been asked to provide rankings for two blocks of choice alternatives and most of the expected ranking frequencies are quite small, not too much attention should be paid to the likelihood ratio statistics. However, they can still be used to compare the goodness of fits of nested models (Böckenholt, 1992). From Table 5, it is found that the results lead to similar interpretations as those given in Böckenholt (1992). Specifically, similar preference structure for the soft drinks has been reviewed and the two underlying factors, namely cola-non-cola and diet-non-diet, are clearly recognized. Our likelihood ratio statistics are not the same as those produced by Böckenholt (1992), the differences may attribute to the fact that there are many empty cells for the ranking patterns, leading to inaccurate approximate of the normal probabilities.

5. Discussion

We have examined the implementation of the Thurstonian models of ranking data using the Mx program and have demonstrated that a wide range of models, including models based on partial ranking data, can be handled in a convenient manner. We have also proposed a unified method to prepare Mx input scripts. The script consists of many groups when p is large and appears to be cumbersome, but the price is the availability of a very flexible standard input template which can certainly be used for any other ranking data analyses. Only modification on the first group specifying the data set and model, and on the last group specifying the functional constraints is required. Moreover, although the standard input script consists of many groups, one can in fact prepare the standard template very easily due to the systematic and replicable nature of the command statements in different groups.

In effect, the long input script is a result of the fact that the procedure we employed in Mx is designated for analyzing raw ordinal data and users flexibility is hindered by various restrictions in the procedure, such as the format of data input and the limitations in using different available options. However, the software developers themselves possess a substantially larger flexibility than users. We have identified in this paper the similarities in analyzing ranking and ordinal categorical data, and have illustrated how the basic building blocks for analyzing ordinal categorical data can be used to analyze ranking data. Making use of these results, structural equation software developers can easily produce user-friendly options for handling ranking data, enhancing the practicability and applicability of the Thurstonian models.

It is well known that when p is large, the method of maximum likelihood for analyzing ranking data suffers from various practical and computational problems. Since the matrix function "mnor" in Mx can compute multiple integrals of the multivariate normal distribution for up to dimension 10, the method we proposed here can in theory analyze ranking data with the maximum likelihood method for up to 10 objects. However, for high dimensional data, Mx may require a long time to produce the solution that may not be stable. For example, we have studied the CPAI data set used by Chan and Bentler (1998) that consists of rankings of 6 objects. With good starting values, we were able to obtain the maximum likelihood estimates similar to those available in Chan and Bentler (1998); however, the same solution could not be achieved with some other starting values. Nevertheless, since the number of ranking patterns also increases drastically when p increases, leading to many practical problems; the use of models on partial ranking data (Böckenholt, 1992) seems a good alternative for analyzing ranking data. This alternative method can also be implemented using Mx in an easy and convenient manner.

Acknowledgements

The work described in this paper was partially supported by a grant from the Research Grants Council of the Hong Kong Special Administrative Region, China (Project No. CUHK4242/03H). The author is grateful to Professors Chih-Ping Chou and Sik-Yum Lee for their valuable suggestions for improving the paper, and Professor M. Neale and his associates for providing the Mx program at no cost.

Appendix A. Mx input script for $p = 4$, auto data set, basic Thurstonian model

```
! File auto4f.mx
! 2431 means object 2 most preferred, 4 the second ... 1 the least

#define ndata=279      ! define sample size

Group 1
DAta NG=25 NI=3
Ordinal file=auto.uni

Begin Matrices ;
H Full 4 1 Free     ! Mean vector, can be further modelled
P Stan 4 4 Free     ! Covariance mtx, can be further modelled
O Full 1 24         ! store the observed frequency, specification invariant
S Full 72 3         ! invariant, store the selection matrices $S_t$
A Full 3 4          ! stores the contrast matrix, not required for reduced form
J Full 1 4          ! invariant
U Full 1 3          ! invariant
L Full 1 3          ! invariant
```

```
T Full 1 3          ! invariant
I Full 1 1          ! invariant
End Matrices ;

Specify H 1 2 3 0                  ! specify free and fixed ("0") parameters
Specify P 4 5 6 7 8 0

Matrix H 0.0 0.0 0.0 0.0           ! starting values
Matrix P  0.0 0.0 0.0 0.0 0.0 0.0

Matrix O file=auto.obs             ! read in the data file

Matrix S file=select.4             ! invariant

Matrix A                           ! invariant, not required for reduced form
1 -1 0 0
1 0 -1 0
1 0 0 -1

Matrix J 1 1 3 3 ! for extracting $S_1$, the selection mtx of 1st pattern from $S$

Matrix U 100.0 100.0 100.0        ! invariant
Matrix L 0.0 0.0 0.0              ! invariant
Matrix T 1 1 1                    ! invariant
Matrix I -2.0                     ! invariant

Begin Algebra;             ! to compute $\mu_t$ and $\sigma_t$ for t=1
M=A*H ;                    ! N stores $\mu_t$ and
C=A*P*A' ;                 ! R stores $\sigma_t$
B=\part(S,J) ;
N=B*M ;
R=B*C*B' ;
End Algebra ;

Thresholds -N' ;            ! thresholds are fixed at zero
Covariance R ;
Options It=2000 Optimality=0.00000001 Function precision=0.00000001 End \

End

Group  2                    ! INVARIANT for any analyses of p=4
DA NI=0
Matrices=Group 1
K Full 1 4
E Full 1 4
End Matrices
Matrix K   4 1   6 3        ! for extract $S_2$ from $S$
Matrix E 1   2 1   2        ! for extract the frequency for pattern 2 from $O$
Begin Algebra ;             ! compute the user defined fit function for pattern 2
D=\part(S,K) ;
W=\part(O,E) ;
V=(D*M)' ;
Q=D*C*D' ;
X=\mnor((Q_V_U_L_T)) ;
Y=\ln(X) ;
Z=I*Y ;
End Algebra ;
Compute W*Z /
Option User defined NO Op=0.00000001 Fu=0.00000001 End /
End
.
.
.
Group 24                      ! INVARIANT for any analyses of p=4
DA NI=0
Matrices=Group 1
K Full 1 4
```

```
E Full 1 4
End Matrices
Matrix K  70 1  72 3          ! for extract $S_{24}$ from $S$
Matrix E 1  24 1  24          ! for extract the frequency for pattern 24 from $O$
Begin Algebra ;               ! compute the user defined fit function for pattern 24
D=\part(S,K) ;
W=\part(O,E) ;
V=(D*M)' ;
Q=D*C*D' ;
X=\mnor((Q_V_U_L_T)) ;
Y=\ln(X) ;
Z=I*Y ;
End Algebra ;
Compute W*Z /
Option User defined NO Op=0.00000001 Fu=0.00000001 End /
End


Group 25            ! for computation of the adjustment value
Ca

Begin Matrices ;
O Full 1 24=O1
N Full 1 1
I Full 1 1
R Full 1 2
End Matrices ;


Matrix N ndata
Matrix I 2


Begin Algebra;
G=\ln(N~*O) ;
Z=\sum(O.G) ;
F=I*Z ;
End Algebra;

Option Rs

End
```

Appendix B. Mx input script, auto data set, factor analysis model

```
! File auto4ffa.mx
! 2431 means object 2 most preferred, 4 the second ... 1 the least
! Factor Analysis Model

#define ndata=279

Group 1: pattern 123
DAta NG=26 NI=3
Ordinal file=auto.uni

Begin Matrices ;
H Full 4 1 Free        ! Mean vector, can be further modelled
F Full 4 1 Free        ! Factor loading matrix
G Diag 4 4 Free        ! Error variance matrix
O Full 1 24
S Full 72 3
A Full 3 4
J Full 1 4
U Full 1 3
```

```
L Full 1 3
T Full 1 3
I Full 1 1
End Matrices ;

Specify H 0 0 0 0
Specify F 4 5 6  0
Specify G 7 8 9 0

Matrix H 0.0 0.0 0.0 0.0
Matrix F  0.8 0.8 0.8  0.0
Matrix G 0.36 0.36 0.36 1.0

Matrix O file=auto.obs

Matrix S file=select.4

Matrix A
1 -1 0 0
1 0 -1 0
1 0 0 -1

Matrix J 1 1 3 3

Matrix U 100.0 100.0 100.0
Matrix L 0.0 0.0 0.0
Matrix T 1 1 1
Matrix I -2.0

Begin Algebra;
M=A*H ;
P=F*F'+G;           ! the structure of the covariance mtx
C=A*P*A' ;
B=\part(S,J) ;
N=B*M ;
R=B*C*B' ;
End Algebra ;

Thresholds -N' ;
Covariance R ;
Options It=2000 Optimality=0.00000001 Function precision=0.00000001 End \

End
:
:
Group 26 Constraint group          ! constrain diagonal elements of the
Constraint                         ! covariance matrix to 1

Matrices=Group 1
E Unit 1 4
End Matrices ;

Constraint E=\d2v(F*F'+G) ;

End
```

Appendix C. Mx input script, auto data set, model of reduced form parameters

```
! File auto4.mx
! 2431 means object 2 is most preferred, 4 the second ... 1 the most

#define ndata=279

Group 1: pattern 123
```

```
DAta NG=25 NI=3
Ordinal file=auto.uni

Begin Matrices ;
M Full 3 1 Free      ! Reduced form mean vector, can be further modelled
C Sym 3 3 Free     ! Reduced form covariance mtx, can be further modelled
O Full 1 24
S Full 72 3
J Full 1 4
U Full 1 3
L Full 1 3
T Full 1 3
I Full 1 1
End Matrices ;

Specify M 0 1 2
Specify C 3 4 5 6 7 8

Matrix M 0.04 0.0 0.0
Matrix C 1.0 0.0 1.0 0.0 0.0 1.0

Matrix O file=auto.obs

Matrix S file=select.4

Matrix J 1 1 3 3

Matrix U 100.0 100.0 100.0
Matrix L 0.0 0.0 0.0
Matrix T 1 1 1
Matrix I -2.0

Begin Algebra;
B=\part(S,J) ;
N=B*M ;
R=B*C*B' ;
End Algebra ;

Thresholds -N' ;
Covariance R ;
Options It=2000 Optimality=0.00000001 Function precision=0.00000001 End \

End
⋮
```

References

Arminger, G., Wittenberg, J., Schepers, A. (1996). *MECOSA 3 Users Guide*. Additive Gumbh, Friedrichsdorf.

Bentler, P.M., Wu, E.J.C. (1993). *EQS/Windows User's Guide, Version 4*. BMDP Statistical Software Inc., Los Angeles.

Böckenholt, U. (1992). Thurstonian representation for partial ranking data. *British Journal of Mathematical and Statistical Psychology* **45**, 31–49.

Böckenholt, U. (1996). Analyzing multiattribute ranking data: Joint and conditional approaches. *British Journal of Mathematical and Statistical Psychology* **49**, 57–78.

Böckenholt, U. (2001). Mixed-effects analyses of rank-ordered data. *Psychometrika* **66**, 45–62.

Chan, W., Bentler, P.M. (1998). Covariance structure analysis of ordinal ipsative data. *Psychometrika* **63**, 369–399.

Critchlow, D.E., Fligner, M.A. (1991). Paired comparison, triple comparison, and ranking experiments as generalized linear models, and their implementation on GLIM. *Psychometrika* **56**, 517–533.

Currim, I.S. (1982). Predictive testing of consumer choice models not subject to independence of irrelevant alternatives. *Journal of Marketing Research* **19**, 208–222.

Elrod, T., Keane, M.P. (1995). A factor analytic probit model for representing the market structure in panel data. *Journal of Marketing Research* **32**, 1–16.

Hamagami, F. (1997). A review of the Mx computer program for structural equation modeling. *Structural Equation Modeling* **4**, 157–175.

Jöreskog, K.G. (1994). On the estimation of polychoric correlations and their asymptotic covariance matrix. *Psychometrika* **59**, 381–389.

Jöreskog, K.G., Sörbom, D. (1996a). *LISREL 8: User's Reference Guide*. Scientific Software International, Chicago.

Jöreskog, K.G., Sörbom, D. (1996b). *PRELIS 2: User's Reference Guide*. Scientific Software International, Chicago.

Lee, S.Y., Poon, W.Y., Bentler, P.M. (1995). A two-stage estimation of structural equation models with continuous and polytomous variables. *British Journal of Mathematical and Statistical Psychology* **52**, 111–124.

Mayedu-Olivares, A. (1999). Thurstonian modeling of ranking data via mean and covariance structure analysis. *Psychometrika* **64**, 325–340.

McKeon, J.J. (1961). Measurement procedures based on comparative judgment. Unpublished doctoral dissertation. University of North Carolina, Chapel Hill.

Muthén, B., Muthén, L.K. (1998). *Mplus User's Guide*. Muthén and Muthén, Los Angeles.

Neale, M.C., Boker, S.M., Xie, G., Maes, H.H. (1999). *Mx: Statistical Modeling*, fifth ed. Department of Psychiatry, Box 126 MCV, Richmond, VA 23298.

Poon, W.Y., Lee, S.Y. (1987). Maximum likelihood estimation of multivariate polyserial and polychoric correlation coefficient. *Psychometrika* **52**, 409–430.

Thurstone, L.L. (1927). A law of comparative judgment. *Psychological Review* **34**, 273–286.

Handbook of Computing and Statistics with Applications, Vol. 1
ISSN: 1871-0301

DOI: 10.1016/S1871-0301(06)01010-9

10

Multilevel Structural Equation Modeling

Sophia Rabe-Hesketh, Anders Skrondal and Xiaohui Zheng

Abstract

In conventional structural equation models, all latent variables and indicators vary between units (typically subjects) and are assumed to be independent across units. The latter assumption is violated in multilevel settings where units are nested in clusters, leading to within-cluster dependence. Different approaches to extending structural equation models for such multilevel settings are examined. The most common approach is to formulate separate within-cluster and between-cluster models. An advantage of this set-up is that it allows software for conventional structural equation models to be 'tricked' into estimating the model. However, the standard implementation of this approach does not permit cross-level paths from latent or observed variables at a higher level to latent or observed variables at a lower level, and does not allow for indicators varying at higher levels. A multilevel regression (or path) model formulation is therefore suggested in which some of the response variables and some of the explanatory variables at the different levels are latent and measured by multiple indicators. The *Generalized Linear Latent and Mixed Modeling* (GLLAMM) framework allows such models to be specified by simply letting the usual structural part of the model include latent and observed variables varying at different levels. Models of this kind are applied to the U.S. sample of the Program for International Student Assessment (PISA) 2000 to investigate the relationship between the school-level latent variable 'teacher excellence' and the student-level latent variable 'reading ability', each measured by multiple ordinal indicators.

Keywords: Multilevel structural equation models; Generalized linear mixed models; Latent variables; Random effects; Hierarchical models; Item response theory; Factor models; Adaptive quadrature; Empirical Bayes; GLLAMM

1. Introduction

The popularity of multilevel modeling and structural equation modeling (SEM) is a striking feature of quantitative research in the medical, behavioral and social sciences.

Although developed separately and for different purposes, SEM and multilevel modeling have important communalities since both approaches include latent variables or random effects to induce, and therefore explain, correlations among responses.

Multilevel regression models are used when the data structure is hierarchical with elementary units at level 1 nested in clusters at level 2, which in turn may be nested in (super)clusters at level 3, and so on. The latent variables, or *random effects*, are interpreted as unobserved heterogeneity at the different levels which induce dependence among all lower-level units belonging to a higher-level unit. Random intercepts represent heterogeneity between clusters in the overall response and random coefficients represent heterogeneity in the relationship between the response and explanatory variables.

Structural equation models are used when the variables of interest cannot be measured perfectly. Instead, there are either sets of items reflecting a hypothetical construct (e.g., depression) or fallible measurements of a variable (e.g., calory intake) using different instruments. The latent variables, or *factors*, are interpreted as constructs, traits or 'true' variables, underlying the measured items and inducing dependence among them. The measurement model is sometimes of interest in its own right, but relations among the factors or between factors and observed variables (the structural part of the model) are often the focus of investigation.

Importantly, *multilevel structural equation modeling*, a synthesis of multilevel and structural equation modeling, is required for valid statistical inference when the units of observation form a hierarchy of nested clusters and some variables of interest are measured by a set of items or fallible instruments. Multilevel structural equation modeling also enables researchers to investigate exciting research questions which could not otherwise be validly addressed. For instance, in this chapter we will consider an important question in education: does student ability (a student-level latent trait) depend on teacher excellence (a school-level latent trait)?

Multilevel structural equation models could be specified using either multilevel regression models or structural equation models as the vantage point. An advantage of using the multilevel regression approach taken here is that the data need not be balanced and missing data are easily accommodated.

2. Response types

2.1. *Continuous responses*

Structural equation models were originally developed for continuous responses. In this case the 'response model' or 'measurement model' for subject j, relating the observed response vector $\mathbf{y}_j$ of manifest variables or indicators to the latent variables $\boldsymbol{\eta}_j$, the observed covariates $\mathbf{x}_j$, and the error terms $\boldsymbol{\varepsilon}_j$ (usually representing 'unique factors'), has the general form

$$\mathbf{y}_j = \boldsymbol{\nu}_j + \boldsymbol{\varepsilon}_j, \qquad \boldsymbol{\varepsilon}_j \sim \mathrm{N}(\mathbf{0}, \boldsymbol{\Theta}).$$

Here $\boldsymbol{\nu}_j$ are functions of $\boldsymbol{\eta}_j$ and $\mathbf{x}_j$ (see Section 3) and $\boldsymbol{\Theta}$ is the covariance matrix of $\boldsymbol{\varepsilon}_j$, usually specified as diagonal.

2.2. *Noncontinuous responses*

2.2.1. *Latent response formulation*

When the responses are dichotomous or ordinal, the same model as above can be specified for latent continuous responses $\mathbf{y}_j^*$ underlying the observed responses $\mathbf{y}_j$. A threshold model links the observed response for the ith indicator to the corresponding latent response,

$$y_{ij} = s \quad \text{if } \kappa_{is} < y_{ij}^* \leqslant \kappa_{i,s+1},$$
$$s = 0, \ldots, S-1, \ \kappa_{i0} = -\infty, \ \kappa_{iS} = \infty.$$

The threshold parameters κ_{is} (apart from κ_{i0} and κ_{iS}) can all be estimated if the mean and variance of $\mathbf{y}_j^*$ are fixed. Alternatively, two thresholds can be fixed (typically $\kappa_{i1} = 0$ and $\kappa_{i2} = 1$) for each response variable to identify the means and variances of $\mathbf{y}_j^*$.

Grouped or interval censored continuous responses can be modeled in the same way by constraining the threshold parameters to the limits of the censoring intervals. By allowing unit-specific right-censoring, this approach can be used for discrete time durations.

An advantage of the latent response formulation is that conventional models can be specified for the underlying continuous responses. By changing the distribution of $\boldsymbol{\varepsilon}_j$, the latent response formulation can also be used to specify logit models. Models for comparative responses such as rankings or pairwise comparisons can be formulated in terms of latent responses conceptualized as utilities or utility differences (e.g., Skrondal and Rabe-Hesketh, 2003).

2.2.2. *Generalized linear model formulation*

Unfortunately, the latent response formulation cannot be used to specify Poisson models for counts. Instead, a generalized linear model formulation is typically used where the conditional expectation of the response y_{ij} for indicator i given $\mathbf{x}_j$ and $\boldsymbol{\eta}_j$ is 'linked' to the linear predictor ν_{ij} via a link function $g(\cdot)$,

$$g\big(\mathrm{E}[y_{ij}|\mathbf{x}_j, \boldsymbol{\eta}_j]\big) = \nu_{ij}. \tag{1}$$

The linear model given above for continuous responses uses an identity link whereas the latent response model for dichotomous responses can be expressed as a generalized linear model with a probit or logit link. Other possible links are the log, reciprocal and complementary log–log.

The final component in the generalized linear model formulation is the conditional distribution of the response variable given the latent and explanatory variables. The conditional distribution is a member of the exponential family of distributions; a normal distribution is typically used for continuous responses, a Bernoulli distribution for dichotomous responses and a Poisson distribution for counts. In structural equation models with several latent variables, the measurement models for different latent variables may require different links and/or distributions.

For ordinal responses, the generalized linear model formulation is modified so that the link function is applied to cumulative probabilities instead of expectations,

$$g\big(\mathrm{P}[y_{ij} > s|\mathbf{x}_j, \boldsymbol{\eta}_j]\big) = \nu_{ij} - \kappa_{i,s+1}.$$

The threshold parameters $\kappa_{i,s+1}$ could alternatively be viewed as part of category-specific linear predictors ν_{ij}^{s} (treating multinomial responses as multivariate), but this will not be done here.

In structural equation modeling with categorical (dichotomous or ordinal) manifest variables, the latent response formulation is predominant. In contrast, item response models are invariably specified via the generalized linear model formulation (e.g., Mellenbergh, 1994). Although Takane and de Leeuw (1987) and Bartholomew (1987) pointed out the equivalence of the two formulations for many models, the literatures are still quite separate.

In the remainder of this chapter, we will use the generalized linear model formulation because it handles more response types. In most cases we are primarily interested in the form of the linear predictors ν_{ij} and view the choice of link functions and distributions as of secondary interest. For response types that can be modeled via a latent response formulation, the model for the latent responses can be written as $\nu_{ij} + \varepsilon_{ij}$.

3. Multilevel measurement models

3.1. Single-level factor models

Conventional single-level factor models can be specified as

$$\boldsymbol{\nu}_j = \boldsymbol{\beta} + \boldsymbol{\Lambda}\boldsymbol{\eta}_j, \quad \boldsymbol{\eta}_j \sim \mathrm{N}(\mathbf{0}, \boldsymbol{\Psi}).$$

For observed or latent continuous responses it follows that

$$\mathbf{y}_j^* = \boldsymbol{\beta} + \boldsymbol{\Lambda}\boldsymbol{\eta}_j + \boldsymbol{\varepsilon}_j, \quad \boldsymbol{\varepsilon}_j \sim \mathrm{N}(\mathbf{0}, \boldsymbol{\Theta}). \tag{2}$$

Here $\boldsymbol{\nu}_j$ and $\mathbf{y}_j^*$ are I-dimensional vectors with elements corresponding to the indicators, $\boldsymbol{\beta}$ is a vector of intercepts, $\boldsymbol{\Lambda}$ a matrix of factor loadings, $\boldsymbol{\eta}_j$ a m-dimensional vector of common factors and $\boldsymbol{\varepsilon}_j$ a vector of unique factors. The covariance structure of the latent responses becomes

$$\boldsymbol{\Sigma} \equiv \mathrm{Cov}(\mathbf{y}_j^*) = \boldsymbol{\Lambda}\boldsymbol{\Psi}\boldsymbol{\Lambda}' + \boldsymbol{\Theta}, \tag{3}$$

which is called a 'factor structure'. The factor model can be specified either directly as in (2) or via the above covariance structure.

An example of an 'independent clusters' two-factor model (where each indicator measures one and only one common factor) for $I = 6$ is

$$\underbrace{\begin{bmatrix} \nu_{1j} \\ \nu_{2j} \\ \nu_{3j} \\ \nu_{4j} \\ \nu_{5j} \\ \nu_{6j} \end{bmatrix}}_{\boldsymbol{\nu}_j} = \underbrace{\begin{bmatrix} \beta_1 \\ \beta_2 \\ \beta_3 \\ \beta_4 \\ \beta_5 \\ \beta_6 \end{bmatrix}}_{\boldsymbol{\beta}} + \underbrace{\begin{bmatrix} 1 & 0 \\ \lambda_{21} & 0 \\ \lambda_{31} & 0 \\ 0 & 1 \\ 0 & \lambda_{52} \\ 0 & \lambda_{62} \end{bmatrix}}_{\boldsymbol{\Lambda}} \underbrace{\begin{bmatrix} \eta_{1j} \\ \eta_{2j} \end{bmatrix}}_{\boldsymbol{\eta}_j},$$

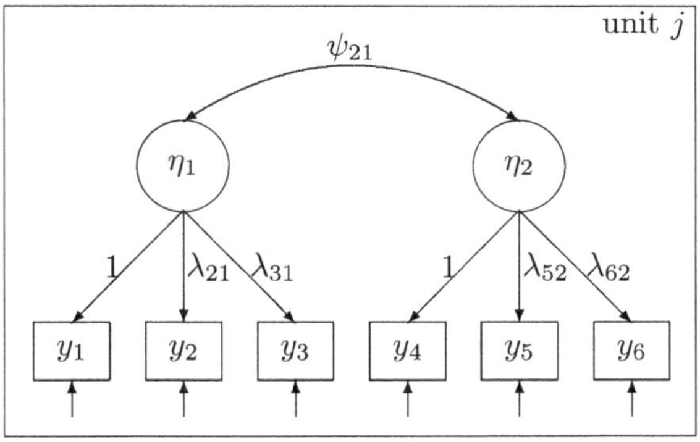

Fig. 1. Independent clusters two-factor model.

where the first factor is measured by the first three indicators and the second factor by the remaining indicators.

A path diagram for this model is given in Figure 1 where circles represent latent variables and rectangles observed variables. For continuous observed responses the long arrows represent linear relations between the responses and the common factors and the short arrows represent linear relations between the responses and the unique factors. For other response types the long arrows represent possibly nonlinear relations depending on the link function and the short arrows represent residual variability, following, for instance, a Bernoulli or Poisson distribution.

Factor models have a similar structure to random coefficient models as has been pointed out in the context of growth curve models (Skrondal, 1996), item response models (Rijmen et al., 2003; De Boeck and Wilson, 2004) and more generally in Skrondal and Rabe-Hesketh (2004). A two-level random coefficient model can be written as

$$\boldsymbol{\nu}_j = \mathbf{X}_j \boldsymbol{\beta} + \mathbf{Z}_j \boldsymbol{\eta}_j,$$

where the design matrix of known constants $\mathbf{Z}_j$ (varying over clusters j) takes the place of the parameter matrix of unknown factor loadings $\boldsymbol{\Lambda}$ (constant over clusters j).

Generalized linear latent and mixed models (GLLAMMs) (Rabe-Hesketh et al., 2004a, 2004b) unify factor models and random coefficient models by allowing each latent variable to multiply a term of the form $\mathbf{Z}_j \boldsymbol{\lambda}$, where $\mathbf{Z}_j$ is a design matrix and $\boldsymbol{\lambda}$ a parameter vector. The GLLAMM formulation of the independent clusters two-factor model discussed previously is as follows:

$$\underbrace{\begin{bmatrix} \nu_{1j} \\ \nu_{2j} \\ \nu_{3j} \\ \nu_{4j} \\ \nu_{5j} \\ \nu_{6j} \end{bmatrix}}_{\boldsymbol{\nu}_j} = \underbrace{\begin{bmatrix} \beta_1 \\ \beta_2 \\ \beta_3 \\ \beta_4 \\ \beta_5 \\ \beta_6 \end{bmatrix}}_{\boldsymbol{\beta}} + \eta_{1j} \underbrace{\begin{bmatrix} 1 & 0 & 0 \\ 0 & 1 & 0 \\ 0 & 0 & 1 \\ 0 & 0 & 0 \\ 0 & 0 & 0 \\ 0 & 0 & 0 \end{bmatrix}}_{\mathbf{Z}_{1j}} \underbrace{\begin{bmatrix} 1 \\ \lambda_{21} \\ \lambda_{31} \end{bmatrix}}_{\boldsymbol{\lambda}_1} + \eta_{2j} \underbrace{\begin{bmatrix} 0 & 0 & 0 \\ 0 & 0 & 0 \\ 0 & 0 & 0 \\ 1 & 0 & 0 \\ 0 & 1 & 0 \\ 0 & 0 & 1 \end{bmatrix}}_{\mathbf{Z}_{2j}} \underbrace{\begin{bmatrix} 1 \\ \lambda_{52} \\ \lambda_{62} \end{bmatrix}}_{\boldsymbol{\lambda}_2},$$

where $\mathbf{Z}_{1j}$ and $\mathbf{Z}_{2j}$ are design matrices containing only fixed constants whereas $\boldsymbol{\lambda}_1$ and $\boldsymbol{\lambda}_2$ are vectors of factor loadings. If the factor loadings are known constants, the products $\mathbf{Z}_{1j}\boldsymbol{\lambda}_1$ and $\mathbf{Z}_{2j}\boldsymbol{\lambda}_2$ become column vectors, giving a random coefficient model.

When viewing factor models as similar to random coefficient models, it is useful to describe the indicators as level-1 units and the subjects as level-2 units (or clusters). In the remainder of this chapter, we will therefore denote higher levels in which subjects are nested as level-3, level-4, etc.

3.2. Two-level factor models

Multilevel factor models are typically required if the subjects of interest are clustered in some way, for instance, students clustered in schools.

3.2.1. Within and between formulation

A two-level factor model for subjects j in clusters k is often formulated in terms of the within-cluster and between-cluster covariance matrices, $\boldsymbol{\Sigma}_W$ and $\boldsymbol{\Sigma}_B$, respectively (e.g., Longford and Muthén, 1992; Poon and Lee, 1992; Longford, 1993; Linda et al., 1993).

For continuous observed or latent responses, the following two-stage formulation can be used

$$\begin{aligned} \mathbf{y}^*_{jk} &\sim \mathrm{N}(\boldsymbol{\mu}_k, \boldsymbol{\Sigma}_W), \\ \boldsymbol{\mu}_k &\sim \mathrm{N}(\boldsymbol{\mu}, \boldsymbol{\Sigma}_B). \end{aligned} \tag{4}$$

Here, $\boldsymbol{\mu}$ is the overall intercept and $\boldsymbol{\mu}_k$ are cluster-specific intercepts. Factor structures of the form in (2) are then specified for the two covariance matrices

$$\boldsymbol{\Sigma}_W = \boldsymbol{\Lambda}^{(2)}\boldsymbol{\Psi}^{(2)}\boldsymbol{\Lambda}^{(2)\prime} + \boldsymbol{\Theta}^{(2)},$$

and

$$\boldsymbol{\Sigma}_B = \boldsymbol{\Lambda}^{(3)}\boldsymbol{\Psi}^{(3)}\boldsymbol{\Lambda}^{(3)\prime} + \boldsymbol{\Theta}^{(3)}.$$

Here we have used the superscript (2) to denote subject-level variables and parameters and (3) to denote the cluster-level counterparts. For consistency with the literature, we call the model a two-level factor model although we think of items as level-1 units, subjects as level-2 units and clusters as level-3 units.

The two-level factor model can alternatively be expressed more explicitly using a two-stage formulation with a within-model and a between-model:

$$\begin{aligned} \mathbf{y}^*_{jk} &= \boldsymbol{\mu}_k + \boldsymbol{\Lambda}^{(2)}\boldsymbol{\eta}^{(2)}_{jk} + \boldsymbol{\varepsilon}^{(2)}_{jk}, \\ \boldsymbol{\mu}_k &= \boldsymbol{\mu} + \boldsymbol{\Lambda}^{(3)}\boldsymbol{\eta}^{(3)}_{k} + \boldsymbol{\varepsilon}^{(3)}_{k}. \end{aligned} \tag{5}$$

The first equation for the latent responses $\mathbf{y}^*_{jk}$ represents a common factor model which includes random intercepts $\boldsymbol{\mu}_k$ that vary over clusters k. The second equation represents a common factor model for the random intercepts $\boldsymbol{\mu}_k$.

For the case of a single common factor at each level, a path diagram reflecting the above specification is given in Figure 2. Following the conventions used by Muthén and Muthén (2004), the models for the within and between covariance matrices are labeled

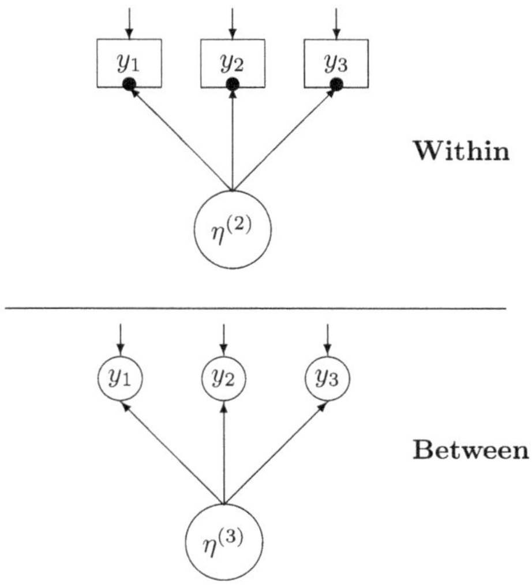

Fig. 2. Path diagram of two-level factor model in within and between formulation.

'within' and 'between'. The within-model shows the relationship between the observed responses and the common factor $\eta_1^{(2)}$ at the subject level. The solid circles attached to the responses indicate that the intercepts $\boldsymbol{\mu}_k$ of these responses vary randomly in the between-model. In the between-model, these random intercepts are shown as circles labeled with the names of the corresponding responses. These are modeled using a common and unique factors at the cluster level.

3.2.2. *Reduced-form formulation*

Substituting from the second line of (5) for $\boldsymbol{\mu}_k$ in the first line, we obtain the reduced form

$$\mathbf{y}^*_{jk} = \underbrace{\boldsymbol{\mu} + \boldsymbol{\Lambda}^{(3)}\boldsymbol{\eta}^{(3)}_k + \boldsymbol{\varepsilon}^{(3)}_k}_{\boldsymbol{\mu}_k} + \boldsymbol{\Lambda}^{(2)}\boldsymbol{\eta}^{(2)}_{jk} + \boldsymbol{\varepsilon}^{(2)}_{jk}.$$

A path diagram reflecting the reduced form is given in the left panel of Figure 3. Following the conventions in Rabe-Hesketh et al. (2004a, 2004b), nested frames represent the nested levels; variables located within the outer frame labeled 'cluster k' vary between clusters and have a k subscript and variables also inside the inner frame labeled 'unit j' vary between units within clusters and have both the j and k subscripts. Only common factors are enclosed in circles.

3.3. *Variance components factor model*

Instead of specifying separate factor models for the two levels, we could think of a single factor model defined for subjects in which the common factors have random intercepts

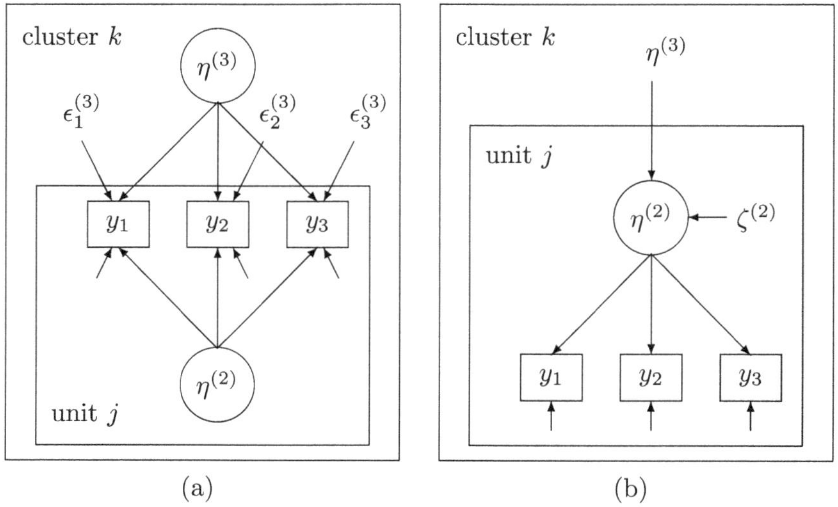

Fig. 3. (a) General two-level factor model and (b) variance components factor model (source: Skrondal and Rabe-Hesketh, 2004).

varying between the clusters. In the unidimensional case, with a single common factor $\eta_{jk}^{(2)}$ at level 2, the measurement model for this factor is combined with a structural model of the form

$$\eta_{jk}^{(2)} = \eta_k^{(3)} + \zeta_{jk}^{(2)}. \tag{6}$$

Such a *variance components factor model* is analogous to a MIMIC ('Multiple-Indicator Multiple-Cause') model (e.g., Jöreskog and Goldberger, 1975) except that the common factor is not regressed on observed covariates but on a random intercept representing the effects of unobserved covariates at a higher level. An obvious application is in item response models if, for example, children's mean latent abilities vary randomly between schools (see, e.g., Fox and Glas, 2001).

This model is a special case of the two-level factor model with the same number of common factors at both levels, no unique factors at level 3 and with factor loadings set equal across levels, $\mathbf{\Lambda}^{(2)} = \mathbf{\Lambda}^{(3)}$. Using the conventions of Muthén (e.g., Muthén and Muthén, 2004) the unidimensional variance components factor model would be depicted as in Figure 2 but without the short arrows in the 'between' model. Using our conventions, a natural representation is that given in Figure 3(b).

The cluster-level unique factors in the two-level factor model can be thought of as representing differential item functioning between clusters. In Longford and Muthén's (1992) application to test scores in eight areas of mathematics for students nested in classes, the unique factors were interpretable as representing the variability in emphases between classrooms, partly due to tracking.

Note that if the factor loadings are set to 1 the model simply becomes a multilevel regression model. Such a model has been used by Raudenbush and Sampson (1999).

4. Multilevel structural equation models

Just as for the single-level case, multilevel measurement models are sometimes of interest in their own right. However, it is often the nature of the relationships between latent variables at different levels that is the primary focus of the investigation.

4.1. Single-level models

The M latent variables $\boldsymbol{\eta}_j$ are defined via a measurement model as described in Section 3. The structural model for the latent variables then allows these latent variables to be regressed on each other and on observed covariates. This model often has the form (e.g., Muthén, 1984)

$$\boldsymbol{\eta}_j = \mathbf{B}\boldsymbol{\eta}_j + \boldsymbol{\Gamma}\mathbf{w}_j + \boldsymbol{\zeta}_j. \tag{7}$$

Here $\mathbf{B}$ is a regression parameter matrix for the relations among the latent variables $\boldsymbol{\eta}_j$, $\mathbf{w}_j$ is a vector of covariates, $\boldsymbol{\Gamma}$ is a parameter matrix for the regressions of the latent variables on the covariates, and $\boldsymbol{\zeta}_j$ is a vector of errors or disturbances. The relationships among the latent variables are recursive if the $\mathbf{B}$ matrix is strictly upper (or lower) triangular.

4.2. Multilevel structural equation models

Multilevel structural equation models can be specified in a number of different ways. The most common approach is the traditional two-stage approach described for factor models in Section 3.2.1. In this case separate structural equation models are specified for the within and between covariance matrices (e.g., Muthén, 1994; Lee and Shi, 2001). A recent application of this approach in education is described by Everson and Millsap (2004). In contrast, the approach advocated here is based on including latent variables in random coefficient models or generalized linear mixed models.

One possibility is to specify a conventional random coefficient model but let the *response variable* be a latent variable, for instance, ability. The intercept and possibly effects of covariates are then specified as varying randomly between clusters (e.g., Fox and Glas, 2001). This is an extension of the unidimensional variance components factor model to include covariates and possibly random coefficients of covariates. The model includes direct paths from cluster-level latent variables to subject-level latent variables as shown for the variance components factor model in Figure 3(b). While equivalent models can often be specified via separate models for the within and between covariance matrices, they require a large number of constraints, including nonlinear constraints (Rabe-Hesketh et al., 2004a). Furthermore, the simpler structure would not be apparent from separate diagrams for the within and between-models.

Remaining within the random coefficient framework, we can also let *covariates* be latent variables. If these covariates are cluster-specific, the model includes *responses varying at different levels*. This situation is accommodated within the framework suggested by Goldstein and McDonald (1988) and McDonald and Goldstein (1989) for

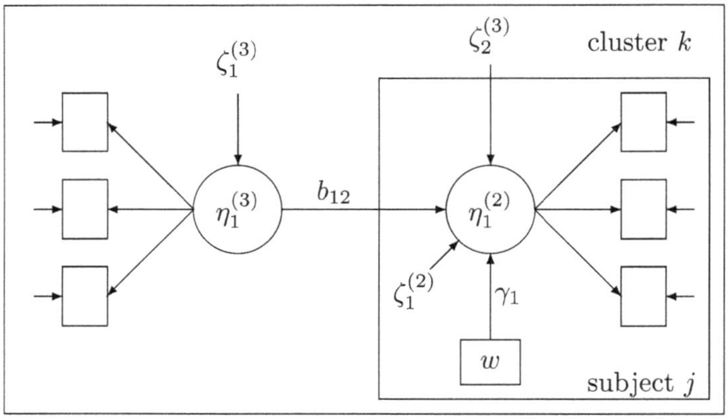

Fig. 4. Multilevel structural equation model with latent dependent variable and latent covariate at level 2.

continuous responses. Fox and Glas (2003) describe a model where both subject-level and cluster-level covariates are latent and where the measurement models are item response models. Unfortunately, the traditional two-stage formulations described in Section 3.2.1 cannot handle responses varying at different levels. This is a rather severe limitation for a multilevel structural equation model.

Rabe-Hesketh et al. (2004a, 2004b) develop the Generalized Linear Latent and Mixed Modeling (GLLAMM) framework consisting of a response model and a structural model. The response model has the form described in Section 3.1 but with L levels of nesting

$$\boldsymbol{\nu} = \mathbf{X}\boldsymbol{\beta} + \sum_{l=2}^{L}\sum_{m=1}^{M_l} \eta_m^{(l)} \mathbf{Z}_m^{(l)} \boldsymbol{\lambda}_m^{(l)}, \tag{8}$$

where we have omitted the indices for units at different levels for notational simplicity. This model allows specification of random coefficient models, measurement models or both, as well as hybrid models. The structural model has the same form as (7) for single-level models but is specified for the vector $\boldsymbol{\eta}_j$ of all latent variables for subject j. This allows lower-level latent variables to be regressed on same or higher-level latent and observed variables. This framework permits specification of random coefficient models with latent responses or covariates at different levels. In addition, models in the two-stage formulation can also be specified. One limitation is that it is not possible to have a random coefficient of a latent covariate as this would correspond to a (cross-level) interaction among latent variables.

In Section 6 we will apply a model of the kind shown in Figure 4. A subject-level latent variable is regressed on a cluster-level latent variable and has a cluster-level random intercept. Moreover, the subject-level latent variable is regressed on covariates.

5. Estimation

5.1. Continuous responses

5.1.1. Maximum likelihood

For the continuous case, Goldstein and McDonald (1988), McDonald and Goldstein (1989) and Lee (1990) derived theory and succinct expressions for the likelihood, allowing two-level structural equation models to be estimated. For unbalanced multilevel designs with missing items, Longford and Muthén (1992) proposed a Fisher scoring algorithm whereas Raudenbush (1995) and Poon and Lee (1998) suggested EM algorithms.

5.1.2. Ad-hoc methods

Because these approaches require specialized software, several two-stage alternatives have been proposed. Muthén (1989) suggests an approach which corresponds to maximum likelihood for balanced data where all clusters have the same size n. In this case, the empirical covariance matrix $\mathbf{S}_W$ of the cluster-mean centered responses is a consistent and unbiased estimator for $\boldsymbol{\Sigma}_W$,

$$\mathrm{E}(\mathbf{S}_W) = \boldsymbol{\Sigma}_W.$$

In contrast, the expectation of the empirical covariance matrix $\mathbf{S}_B$ of the cluster means is

$$\mathrm{E}(\mathbf{S}_B) = \boldsymbol{\Sigma}_B + \frac{1}{n}\boldsymbol{\Sigma}_W.$$

Within and between structural equation models are specified for $\boldsymbol{\Sigma}_W$ and $\boldsymbol{\Sigma}_B$. Since $\boldsymbol{\Sigma}_W$ contributes to both $\mathrm{E}(\mathbf{S}_B)$ and $\mathrm{E}(\mathbf{S}_W)$, both models must be fitted jointly to the empirical covariance matrices $\mathbf{S}_B$ and $\mathbf{S}_W$. This can be accomplished by treating the two matrices as if they corresponded to different groups of subjects and performing two-group analysis with the required constraints. If there are only a relatively small number of different cluster sizes, a multiple group approach (with more than two groups) can be used to obtain maximum likelihood estimates. These approaches as well as an ad-hoc solution for the completely unbalanced case are described in detail in Muthén (1994) and Hox (2002).

Goldstein (1987, 2003) suggests using multivariate multilevel modeling to estimate $\boldsymbol{\Sigma}_W$ and $\boldsymbol{\Sigma}_B$ consistently by either maximum likelihood or restricted maximum likelihood. Structural equation models can then be fitted separately to each estimated matrix. Advantages of this approach are that unbalanced data and missing values are automatically accommodated, and that it is straightforward to extend to more hierarchical levels and to models where levels are crossed instead of nested.

An alternative ad-hoc approach, similar to the work by Korn and Whittmore (1979), was proposed by Chou et al. (2000). Here, a factor or structural equation model is estimated separately for each cluster. The estimates are subsequently treated as responses in a between-model, typically a regression model with between-cluster covariates and an unstructured multivariate residual covariance matrix. This approach allows, and indeed requires, all parameters to vary between clusters, including factor loadings.

A common feature of these two-stage procedures is that standard errors provided from the second stage are incorrect since they treat the output from the first stage as data or as empirical covariance matrices.

5.2. Noncontinuous responses

For models with noncontinuous responses maximum likelihood estimation or Bayesian methods are typically used. Although computationally demanding, these methods automatically handle lack of balance and missing data and are straightforward to extend to include for instance mixed responses and nonlinear relations among latent variables. We note in passing that the ad-hoc approaches of Goldstein (2003, 1987) and Chou et al. (2000) discussed above can also be used for noncontinuous responses.

5.2.1. Maximum likelihood estimation

The major challenge in maximum likelihood estimation of multilevel latent variable models for noncontinuous responses is to integrate out the latent variables since closed form results typically do not exist. Thus, integration usually proceeds by either by Monte Carlo simulation or using numerical methods.

Lee and Shi (2001) and Lee and Song (2004) use Monte Carlo EM (MCEM) algorithms, employing Gibbs sampling to evaluate the integrals in the E-step. Rabe-Hesketh et al. (2004a, 2004b) suggest using Newton–Raphson where the latent variables are integrated out using adaptive quadrature, see also Rabe-Hesketh et al. (2005).

5.2.2. Mean posterior estimation

As in other areas of statistics, Markov Chain Monte Carlo (MCMC) methods have recently attracted considerable interest in multilevel structural equation modeling. Interestingly, very diffuse priors are almost invariably specified in practice. The mean of the posterior distribution is in this case often quite close to the mode of the likelihood. MCMC can thus be viewed as a convenient and powerful way of implementing maximum likelihood estimation for complex models.

MCMC methods have been used by Ansari and Jedidi (2000), Fox and Glas (2001) and Goldstein and Browne (2005) for binary responses and by Song and Lee (2004) for continuous and ordinal responses.

6. Application: Student ability and teacher excellence

To investigate whether student ability measured at the student-level depends on teacher excellence measured at the school-level we analyze data from the Program for International Student Assessment (PISA) 2000 Assessment of Reading (OECD, 2001) using multilevel structural equation modeling.

6.1. Data description

The data consist of student responses to a reading test and student background questionnaire and responses to a school questionnaire completed by principals.

At the student level, we focus on a unidimensional latent factor – the ability to interpret written information. We chose four items for this construct from the reading unit of the test. Three of the items have dichotomous responses and one item has ordinal responses. We included four observed covariates from the student questionnaire: Parents' education (one or both parents have higher education = 1, otherwise = 0), Male (male = 1, female = 0), Reading (some time spent on reading every day = 1, otherwise = 0), and English (English spoken at home = 1, otherwise = 0).

The school data include ordinal responses from principals to school survey questions regarding satisfaction with ten aspects of teacher excellence: teacher expectations, student–teacher relations, teacher turnover, teachers meeting individual students' needs, teacher absenteeism, teachers' strictness with students, teachers' morale, teachers' enthusiasm, teachers taking pride in the school, and teachers valuing academic achievement.

The sample comprises 2484 tenth grade students from 131 U.S. schools. School-level covariates were not included because this would have drastically reduced the number of schools due to missing data.

6.2. *Model specification*

In addition to developing measurement models for student interpretation ability and teacher excellence, we will estimate a structural equation model where student interpretation ability is regressed on student-level observed covariates and the school-level latent covariate teacher excellence. There is some doubt regarding the validity of the measurement of teacher excellence since this was based on a questionnaire completed by the principal. If teacher excellence is found to be predictive of student interpretation ability, this could be seen as supportive evidence for the validity of both instruments.

6.2.1. *Student-level model*

The single-level factor model discussed in Section 3.1 is estimated for the student data, where the common factor represents interpretation ability. The measurement model for interpretation ability $\eta_{1jk}^{(2)}$ can be written in terms of underlying continuous responses $\mathbf{y}_{jk}^{*}$. For item i for student j in school k we have

$$y_{ijk}^{*} = \beta_i + \lambda_i \eta_{1jk}^{(2)} + \varepsilon_{ijk}, \quad i = 1, 2, 3, 4. \tag{9}$$

Here β_i are item intercepts and λ_i factor loadings or discrimination parameters, and the ε_{ijk} have logistic distributions. Interpretation ability is measured by three dichotomous items and one ordinal item (item 2). For the dichotomous items ($i = 1, 3$ and 4),

$$y_{ijk} = \begin{cases} 1 & \text{if } y_{ijk}^{*} > 0, \\ 0 & \text{otherwise,} \end{cases}$$

and for the ordinal item ($i = 2$), the intercept β_2 is set to 0 and the threshold model is specified as

$$y_{2jk} = \begin{cases} 1 & \text{if } -\infty \leqslant y_{2jk}^{*} < \kappa_1, \\ 2 & \text{if } \kappa_1 \leqslant y_{2jk}^{*} < \kappa_2, \\ 3 & \text{if } \kappa_2 \leqslant y_{2jk}^{*} < \kappa_3, \\ 4 & \text{if } \kappa_3 \leqslant y_{2jk}^{*} < \infty. \end{cases}$$

This is a logistic graded response model (Samejima, 1969) where $-\beta_i$, $i = 1, 3, 4$, can be interpreted as the thresholds for the dichotomous items.

We regress student interpretation ability on the four student background covariates (Parents' education, Male, Reading and English):

$$\eta_{1jk}^{(2)} = \boldsymbol{\gamma}' \mathbf{w}_{jk} + \zeta_{1jk}^{(2)}.$$

Here $\mathbf{w}_{jk} = [w_{1jk}, w_{2jk}, w_{3jk}, w_{4jk}]'$ is a vector of the four covariates, $\boldsymbol{\gamma} = [\gamma_1, \gamma_2, \gamma_3, \gamma_4]'$ the corresponding regression parameter vector and $\zeta_{1jk}^{(2)}$ a vector of student-level disturbances.

6.2.2. School-level model

Teacher excellence is measured by ten ordinal items, with response categories 'dissatisfied', 'somewhat satisfied' and 'satisfied'. The following model was used for the underlying continuous responses for items i and schools k:

$$y_{ik}^{*} = \eta_{1k}^{(3)} + \varepsilon_{ik}, \tag{10}$$

where $\eta_{1k}^{(3)}$ represents teacher excellence and ε_{ik} has a logistic distribution. The ordinal responses are generated from the threshold model

$$y_{ik} = \begin{cases} 1 & \text{if } -\infty \leqslant y_{ik}^{*} < \alpha_1 + \tau_{i1}, \\ 2 & \text{if } \alpha_1 + \tau_{i1} \leqslant y_{ik}^{*} < \alpha_2 + \tau_{i2}, \\ 3 & \text{if } \alpha_2 + \tau_{i2} \leqslant y_{ik}^{*} < \infty, \end{cases}$$

where α_s $(s = 1, 2)$ is the sth threshold for item 1, whereas τ_{is} $(i = 2, \ldots, 10)$ is the difference in the sth threshold between item i and item 1. Thus, $\alpha_s + \tau_{is}$ corresponds to the threshold parameter κ_{is} for the ordinal responses as defined in Section 2.2.1. The model is a one-parameter version of the logistic graded response model which assumes that all items have the same discrimination.

The structural model is trivial if we do not wish to include school-level covariates:

$$\eta_{1k}^{(3)} = \zeta_{1k}^{(3)}.$$

6.2.3. Joint model

A joint model for the student data and school survey data combines the student-level and school-level models. In this example, students are the level-2 units and schools the level-3 units. Under the general response model in (8), the joint measurement model combines the item response model for the school survey data and a variance components factor model as discussed in Section 3.3 for the student data. A path diagram for this kind of model is shown in Figure 4.

We can write the model for the responses of a student j from school k and a principal from school k as:

$$\begin{bmatrix} y^*_{1jk} \\ y^*_{2jk} \\ y^*_{3jk} \\ y^*_{4jk} \\ \hline y^*_{1k} \\ \vdots \\ y^*_{10,k} \end{bmatrix} = \begin{bmatrix} 1 & 0 & 0 \\ 0 & 0 & 0 \\ 0 & 1 & 0 \\ 0 & 0 & 1 \\ \hline 0 & 0 & 0 \\ \vdots & \vdots & \vdots \\ 0 & 0 & 0 \end{bmatrix} \begin{bmatrix} \beta_1 \\ \beta_3 \\ \beta_4 \end{bmatrix} + \eta^{(2)}_{1jk} \begin{bmatrix} \mathbf{I}_{4\times 4} \\ \hline \mathbf{0}_{10\times 4} \end{bmatrix} \begin{bmatrix} 1 \\ \lambda^{(2)}_2 \\ \lambda^{(2)}_3 \\ \lambda^{(2)}_4 \end{bmatrix}$$

$$+ \eta^{(3)}_{1k} \begin{bmatrix} \mathbf{0}_{4\times 1} \\ \hline \mathbf{I}_{10\times 1} \end{bmatrix} 1 + \eta^{(3)}_{2k} \left[\mathbf{0}_{14\times 1}\right] 1 + \begin{bmatrix} \varepsilon_{1jk} \\ \varepsilon_{2jk} \\ \varepsilon_{3jk} \\ \varepsilon_{4jk} \\ \hline \varepsilon_{1k} \\ \vdots \\ \varepsilon_{10,k} \end{bmatrix}.$$

In the vectors and matrices of the above model, student-level elements are placed above the horizontal lines and school-level elements below the horizontal lines. $\eta^{(2)}_{1jk}$ represents student interpretation ability, $\eta^{(3)}_{1k}$ teacher excellence, and $\eta^{(3)}_{2k}$ the school-level random intercept of interpretation ability. The latter is multiplied by zero for each item since the random intercept does not affect the items directly.

In the structural model, teacher excellence becomes an explanatory variable for interpretation ability. Moreover, interpretation ability is regressed on student-level covariates and the school-level random intercept $\eta^{(3)}_{2k}$ which allows students' mean ability to vary randomly between schools after controlling for the covariates. The structural model can be written as

$$\boldsymbol{\eta} = \mathbf{B}\boldsymbol{\eta} + \boldsymbol{\Gamma}\mathbf{w} + \boldsymbol{\zeta}.$$

Specifically,

$$\begin{bmatrix} \eta^{(2)}_{1jk} \\ \eta^{(3)}_{1k} \\ \eta^{(3)}_{2k} \end{bmatrix} = \begin{bmatrix} 0 & b_{12} & 1 \\ 0 & 0 & 0 \\ 0 & 0 & 0 \end{bmatrix} \begin{bmatrix} \eta^{(2)}_{1jk} \\ \eta^{(3)}_{1k} \\ \eta^{(3)}_{2k} \end{bmatrix}$$

$$+ \begin{bmatrix} \gamma_1 & \gamma_2 & \gamma_3 & \gamma_4 \\ 0 & 0 & 0 & 0 \\ 0 & 0 & 0 & 0 \end{bmatrix} \begin{bmatrix} w_{1jk} \\ w_{2jk} \\ w_{3jk} \\ w_{4jk} \end{bmatrix} + \begin{bmatrix} \zeta^{(2)}_{1jk} \\ \zeta^{(3)}_{1k} \\ \zeta^{(3)}_{2k} \end{bmatrix}.$$

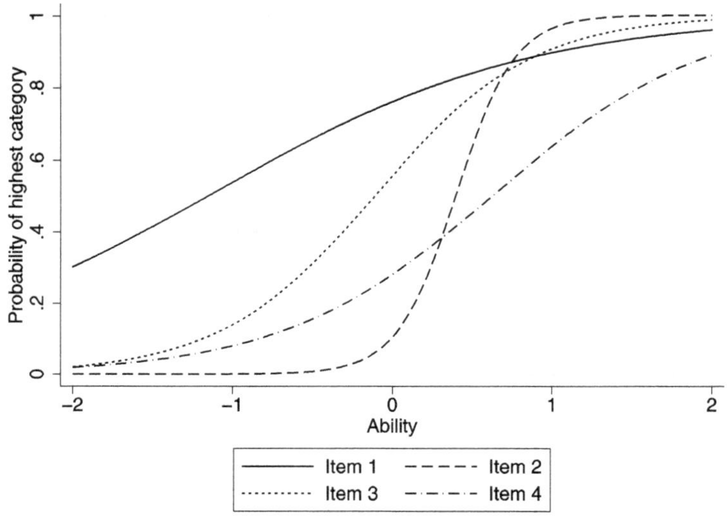

Fig. 5. Item characteristic curves for the four interpretation items.

In the structural model, the **B** matrix defines the relationship among the latent factors at different levels. In particular, the cross-level coefficient b_{12} represents the effect of school-level teacher excellence on student-level interpretation ability.

6.2.4. Results

Maximum likelihood estimates for the models considered above are given in Table 1. The estimates were obtained using *gllamm* (Rabe-Hesketh et al., 2004a, 2004b) which uses adaptive quadrature (Rabe-Hesketh et al., 2005) and runs in Stata (StataCorp, 2005).

In the student-level measurement model, the item difficulties for the dichotomous items are $-\beta_i/\lambda_i$. For the ordinal item, the κ_s represent the thresholds. Interpretation of these parameters is facilitated by inspecting the item characteristic curves shown in Figure 5.

Overall girls perform slightly better than boys as do students who read often or speak English at home. However, parents' education has a negligible estimated effect on student performance (not significant at the 5% level) which is somewhat surprising. This could be due to inaccurate reporting of students on their parents' education or insufficient reliability for interpretation ability due to the small number of items.

The school-level model includes threshold parameters for the ten ordinal items. In category 2 (somewhat satisfied), the principals find "teachers' strictness with students" (item 6) the easiest to endorse and "teachers meeting individual students' needs" (item 4) the most difficult. In category 3 (satisfied), "teachers valuing academic achievement" (item 10) is the easiest and "teachers meeting individual students' needs" is once again the most difficult to endorse.

Table 1
Maximum likelihood estimates for reading test data

Parameter		Student model Est	(SE)	School model Est	(SE)	Joint model Est	(SE)
Student-level:							
β_1	[Item 1, intercept]	1.16	(0.16)			1.15	(0.16)
κ_1	[Item 2, threshold 1]	−0.22	(0.57)			−0.25	(0.58)
κ_2	[Item 2, threshold 2]	−0.92	(0.61)			0.91	(0.62)
κ_3	[Item 2, threshold 3]	2.15	(0.79)			2.16	(0.77)
β_3	[Item 3, intercept]	0.23	(0.25)			0.25	(0.23)
β_4	[Item 4, intercept]	−0.94	(0.23)			−0.94	(0.22)
λ_1	[Item 1, loading]	1				1	
λ_2	[Item 2, loading]	5.44	(2.88)			5.07	(2.38)
λ_3	[Item 3, loading]	2.05	(0.71)			1.80	(0.59)
λ_4	[Item 4, loading]	1.51	(0.54)			1.40	(0.46)
γ_1	[Parents' education]	−0.02	(0.05)			−0.02	(0.05)
γ_2	[Male]	−0.11	(0.06)			−0.12	(0.06)
γ_3	[Reading]	0.16	(0.08)			0.16	(0.08)
γ_4	[English]	0.27	(0.13)			0.30	(0.13)
$\mathrm{var}(\zeta_{1jk}^{(2)})$	[Interpretation ability]	0.20	(0.12)			0.19	(0.10)
School-level:							
α_1	[Threshold parameters]			−1.86	(0.31)	−1.86	(0.31)
τ_{21}	[Threshold parameters]			−1.30	(0.47)	−1.30	(0.47)
τ_{31}	[Threshold parameters]			−0.84	(0.44)	−0.84	(0.43)
τ_{41}	[Threshold parameters]			0.39	(0.37)	0.39	(0.37)
τ_{51}	[Threshold parameters]			−0.41	(0.40)	−0.41	(0.40)
τ_{61}	[Threshold parameters]			−2.54	(0.67)	−2.54	(0.67)
τ_{71}	[Threshold parameters]			−0.33	(0.40)	−0.33	(0.40)
τ_{81}	[Threshold parameters]			−2.03	(0.56)	−2.03	(0.56)
τ_{91}	[Threshold parameters]			−2.29	(0.60)	−2.29	(0.60)
$\tau_{10,1}$	[Threshold parameters]			−2.30	(0.60)	−2.30	(0.60)
α_2	[Threshold parameters]			1.69	(0.31)	1.69	(0.31)
τ_{22}	[Threshold parameters]			0.06	(0.38)	0.06	(0.38)
τ_{32}	[Threshold parameters]			−0.56	(0.37)	−0.56	(0.37)
τ_{42}	[Threshold parameters]			0.86	(0.42)	0.86	(0.42)
τ_{52}	[Threshold parameters]			−0.36	(0.37)	−0.36	(0.37)
τ_{62}	[Threshold parameters]			−1.44	(0.36)	−1.43	(0.36)
τ_{72}	[Threshold parameters]			0.35	(0.40)	0.35	(0.40)
τ_{82}	[Threshold parameters]			0.29	(0.39)	0.29	(0.39)
τ_{92}	[Threshold parameters]			−0.67	(0.37)	−0.67	(0.37)
$\tau_{10,2}$	[Threshold parameters]			−1.79	(0.36)	−1.79	(0.36)
b_{12}	[Cross-level coefficient]			0.02	(0.03)		
$\mathrm{var}(\zeta_{1k}^{(3)})$	[Teacher excellence]			2.19	(0.42)	2.19	(0.42)
$\mathrm{var}(\zeta_{2k}^{(3)})$	[Intercept]					0.05	(0.04)

The school random intercept variance is estimated as 0.05 which is negligible for a logit model. The teacher excellence variance is estimated as 2.19. Somewhat surprisingly, the cross-level effect of teacher excellence on student interpretation ability appears to be negligible. One consequence of this is that the student-level and school-level parameters in the joint model do not differ much from those in the individual student and school models. The small estimated regression coefficient casts some doubt on the validity of the principal's assessment of teacher excellence based on the school questionnaire.

References

Ansari, A., Jedidi, K. (2000). Bayesian factor analysis for multilevel binary observations. *Psychometrika* **65**, 475–496.

Bartholomew, D.J. (1987). *Latent Variable Models and Factor Analysis*. Oxford University Press, Oxford.

Chou, C.-P., Bentler, P.M., Pentz, M.A. (2000). Two-stage approach to multilevel structural equation models: Application to longitudinal data. In: Little, T.D., Schnabel, K.U., Baumert, J. (Eds.), *Modeling Longitudinal and Multilevel Data: Practical Issues, Applied Approaches, and Specific Examples*. Erlbaum, Mahwah, NJ, pp. 33–49.

De Boeck, P., Wilson, M. (2004). *Explanatory Item Response Models: A Generalized Linear and Nonlinear Approach*. Springer, New York.

Everson, H.T., Millsap, R.E. (2004). Beyond individual differences: Exploring school effects on SAT scores. *Educational Psychologist* **39**, 157–172.

Fox, J.P., Glas, C.A.W. (2001). Bayesian estimation of a multilevel IRT model using Gibbs sampling. *Psychometrika* **66**, 271–288.

Fox, J.P., Glas, C.A.W. (2003). Bayesian modeling of measurement error in predictor variables using item response theory. *Psychometrika* **68**, 169–191.

Goldstein, H. (1987). Multilevel covariance component models. *Biometrika* **74**, 430–431.

Goldstein, H. (2003). *Multilevel Statistical Models*, third ed. Arnold, London.

Goldstein, H., Browne, W.J. (2005). Multilevel factor analysis models for continuous and discrete data. In: Maydeu-Olivares, A., McArdle, J.J. (Eds.), *Contemporary Psychometrics: A Festschrift for Roderick P. McDonald*. Erlbaum, Mahwah, NJ, pp. 453–475.

Goldstein, H., McDonald, R.P. (1988). A general model for the analysis of multilevel data. *Psychometrika* **53**, 455–467.

Hox, J. (2002). *Multilevel Analysis: Techniques and Applications*. Erlbaum, Mahwah, NJ.

Jöreskog, K.G., Goldberger, A.S. (1975). Estimation of a model with multiple indicators and multiple causes of a single latent variable. *Journal of the American Statistical Association* **70**, 631–639.

Korn, E.L., Whittmore, A.S. (1979). Methods for analyzing panel studies of acute health effects of air pollution. *Biometrics* **35**, 795–804.

Lee, S.-Y. (1990). Multilevel analysis of structural equation models. *Biometrika* **77**, 763–772.

Lee, S.-Y., Shi, J.-Q. (2001). Maximum likelihood estimation of two-level latent variable models with mixed continuous and polytomous data. *Biometrics* **57**, 787–794.

Lee, S.-Y., Song, X.-Y. (2004). Maximum likelihood analysis of a general latent variable model with hierarchically mixed data. *Biometrics* **60**, 624–636.

Linda, N.Y., Lee, S.-Y., Poon, W.-Y. (1993). Covariance structure analysis with three level data. *Computational Statistics Data Analysis* **15**, 159–178.

Longford, N.T. (1993). *Random Coefficient Models*. Oxford University Press, Oxford.

Longford, N.T., Muthén, B.O. (1992). Factor analysis for clustered observations. *Psychometrika* **57**, 581–597.

McDonald, R.P., Goldstein, H. (1989). Balanced and unbalanced designs for linear structural relations in two-level data. *British Journal of Mathematical and Statistical Psychology* **42**, 215–232.

Mellenbergh, G.J. (1994). Generalized linear item response theory. *Psychological Bulletin* **115**, 300–307.

Muthén, B.O. (1984). A general structural equation model with dichotomous, ordered categorical and continuous latent indicators. *Psychometrika* **49**, 115–132.

Muthén, B.O. (1989). Latent variable modeling in heterogeneous populations. *Psychometrika* **54**, 557–585.

Muthén, B.O. (1994). Multilevel covariance structure analysis. *Sociological Methods Research* **22**, 376–398.

Muthén, L.K., Muthén, B.O. (2004). *Mplus User's Guide*, third ed. Muthén & Muthén, Los Angeles, CA.

OECD (2001). *Knowledge and Skills for Life: First Results from PISA 2000*. OECD, Paris.

Poon, W.Y., Lee, S.Y. (1992). Maximum likelihood and generalized least squares analyses of two-level structural equation models. *Statistics and Probability Letters* **14**, 25–30.

Poon, W.Y., Lee, S.Y. (1998). Analysis of two-level structural equation models via EM-type algorithm. *Statistica Sinica* **8**, 749–766.

Rabe-Hesketh, S., Skrondal, A., Pickles, A. (2004a). Generalized multilevel structural equation modeling. *Psychometrika* **69**, 167–190.

Rabe-Hesketh, S., Skrondal, A., Pickles, A. (2004b). GLLAMM manual. Tech. rept. 160. U.C. Berkeley Division of Biostatistics Working Paper Series. Downloadable from http://www.bepress.com/ucbbiostat/paper160/.

Rabe-Hesketh, S., Skrondal, A., Pickles, A. (2005). Maximum likelihood estimation of limited and discrete dependent variable models with nested random effects. *Journal of Econometrics* **128**, 301–323.

Raudenbush, S.W. (1995). Maximum likelihood estimation for unbalanced multilevel covariance structure models via the EM algorithm. *British Journal of Mathematical and Statistical Psychology* **48**, 359–370.

Raudenbush, S.W., Sampson, R. (1999). Ecometrics: Toward a science of assessing ecological settings, with application to the systematic social observation of neighborhoods. In: Marsden, P.V. (Ed.), *Sociological Methodology*. Blackwell, Oxford, pp. 1–41.

Rijmen, F., Tuerlinckx, F., De Boeck, P., Kuppens, P. (2003). A nonlinear mixed model framework for item response theory. *Psychological Methods* **8**, 185–205.

Samejima, F. (1969). *Estimation of Latent Trait Ability Using A Response Pattern of Graded Scores. Psychometric Monograph*, vol. 17. Psychometric Society, Bowling Green, OH.

Skrondal, A. (1996). *Latent Trait, Multilevel and Repeated Measurement Modelling with Incomplete Data of Mixed Measurement Levels*. Section of Medical Statistics, University of Oslo, Oslo.

Skrondal, A., Rabe-Hesketh, S. (2003). Multilevel logistic regression for polytomous data and rankings. *Psychometrika* **68**, 267–287.

Skrondal, A., Rabe-Hesketh, S. (2004). *Generalized Latent Variable Modeling: Multilevel, Longitudinal, and Structural Equation Models*. Chapman Hall/CRC, Boca Raton, FL.

Song, X.Y., Lee, S.Y. (2004). Bayesian analysis of two-level structural equation models with continuous and polytomous data. *British Journal of Mathematical and Statistical Psychology* **57**, 29–52.

StataCorp (2005). *Stata Statistical Software: Release 9.0*. College Station, TX.

Takane, Y., de Leeuw, J. (1987). On the relationship between item response theory and factor analysis of discretized variables. *Psychometrika* **52**, 393–408.

Handbook of Computing and Statistics with Applications, Vol. 1
ISSN: 1871-0301

DOI: 10.1016/S1871-0301(06)01011-0

11

Statistical Inference of Moment Structures

Alexander Shapiro

Abstract

The statistical inference of moment structure models. Although the theory is presented in terms of general moment structures, the main emphasis is on the analysis of covariance structures. Identifiability and the minimum discrepancy function (MDF) approach to statistical analysis (estimation) of such models are discussed. Topics of the large samples theory, in particular, consistency, asymptotic normality of MDF estimators and asymptotic chi-squaredness of MDF test statistics are addressed. Results addressing asymptotic robustness of the normal theory based MDF statistical inference in the analysis of covariance structures are presented.

1. Introduction

Statistical inferences of moment structures where first and/or second population moments are hypothesized to have a parametric structure are discussed. Classical examples of such models are multinomial and covariance structure models. The presented theory is sufficiently general to handle various situations, however the main focus is on covariance structures. Theory and applications of covariance structures were motivated first by the factor analysis model and its various generalizations and later by the development of LISREL models (Jöreskog, 1977, 1981) (see also (Browne, 1982) for a thorough discussion of covariance structure modeling).

2. Moment structures models

In this section we discuss modeling issues in the analysis of moment structures, and, in particular, identifiability of such models. Let $\boldsymbol{\xi} = (\xi_1, \dots, \xi_m)'$ be a vector variable representing a parameter vector of some statistical population. For example, in the analysis of covariance structures, $\boldsymbol{\xi}$ will represent the elements of a $p \times p$ covariance matrix $\boldsymbol{\Sigma}$. That is, $\boldsymbol{\xi} := \text{vec}(\boldsymbol{\Sigma})$, where $\text{vec}(\boldsymbol{\Sigma})$ denotes the $p^2 \times 1$ vector formed by stacking elements of $\boldsymbol{\Sigma}$. (The notation ":=" means "equal by definition".) Of course, since matrix $\boldsymbol{\Sigma}$ is symmetric, vector $\operatorname{vec}(\boldsymbol{\Sigma})$ has duplicated elements. Therefore an alternative is to consider $\boldsymbol{\xi} := \operatorname{vecs}(\boldsymbol{\Sigma})$, where $\operatorname{vecs}(\boldsymbol{\Sigma})$ denotes the $p(p+1)/2 \times 1$ vector formed by stacking (nonduplicated) elements of $\boldsymbol{\Sigma}$ above and including the diagonal. Note that covariance matrices $\boldsymbol{\Sigma}$ are positive semidefinite, and hence the corresponding vectors $\boldsymbol{\xi}$ are restricted to a (convex) subset of $\mathbb{R}^m$. Therefore, we assume that $\boldsymbol{\xi}$ varies in a set $\Xi \subset \mathbb{R}^m$ representing a *saturated* model for the population vector $\boldsymbol{\xi}$. In the analysis of covariance structures we have a natural question whether to use vector $\boldsymbol{\xi} = \operatorname{vec}(\boldsymbol{\Sigma})$ or $\boldsymbol{\xi} = \operatorname{vecs}(\boldsymbol{\Sigma})$ for the saturated model. Since in both cases the dimension of the corresponding set Ξ is $p(p+1)/2$, it seems more advantageous to use $\boldsymbol{\xi} := \operatorname{vecs}(\boldsymbol{\Sigma})$. In that case the set Ξ has a nonempty interior. (The interior of Ξ is the set of points $\boldsymbol{\xi} \in \Xi$ such that Ξ contains a neighborhood of $\boldsymbol{\xi}$. For example, the interior of the set of positive semidefinite matrices is formed by its subset of positive definite matrices. A singular positive semidefinite matrix can be viewed as a boundary point of this set Ξ. A neighborhood of a point $\boldsymbol{\xi} \in \mathbb{R}^m$ is a subset of $\mathbb{R}^m$ containing a ball centered at $\boldsymbol{\xi}$ of a sufficiently small positive radius.) However, for actual calculations it is often more convenient to use $\boldsymbol{\xi} := \operatorname{vec}(\boldsymbol{\Sigma})$. When dealing with specific applications we will specify a choice of the corresponding vector $\boldsymbol{\xi}$.

A model for $\boldsymbol{\xi}$ is a subset Ξ_0 of Ξ. Of course, this definition is too abstract and one needs a constructive way of defining a model. There are two natural ways for constructing a model, namely either by imposing equations or by a parameterization. The parameterization approach suggests existence of an $m \times 1$ vector valued function $\boldsymbol{g}(\boldsymbol{\theta}) = (g_1(\boldsymbol{\theta}), \ldots, g_m(\boldsymbol{\theta}))$, and a parameter set $\Theta \subset \mathbb{R}^q$, which relates the parameter vector $\boldsymbol{\theta} = (\theta_1, \ldots, \theta_q)'$ to $\boldsymbol{\xi}$. That is,

$$\Xi_0 := \left\{\boldsymbol{\xi} \in \Xi \colon \boldsymbol{\xi} = \boldsymbol{g}(\boldsymbol{\theta}), \boldsymbol{\theta} \in \Theta\right\}. \tag{2.1}$$

We refer to $\boldsymbol{g}(\boldsymbol{\theta})$, $\boldsymbol{\theta} \in \Theta$, as a *structural* model for $\boldsymbol{\xi}$. We assume in the subsequent analysis that the mapping $\boldsymbol{g}(\boldsymbol{\theta})$ is sufficiently smooth. In particular, we always assume that $\boldsymbol{g}(\boldsymbol{\theta})$ is *twice continuously differentiable*. We associate with mapping $\boldsymbol{g}(\cdot)$ its $m \times q$ Jacobian matrix $\partial \boldsymbol{g}(\boldsymbol{\theta})/\partial \boldsymbol{\theta}' = [\partial g_i(\boldsymbol{\theta})/\partial \theta_j]_{i=1,\ldots,m,j=1,\ldots,q}$ of partial derivatives and use notation $\boldsymbol{\Delta}(\boldsymbol{\theta}) := \partial \boldsymbol{g}(\boldsymbol{\theta})/\partial \boldsymbol{\theta}'$.

REMARK 1. It should be noted that the same set Ξ_0 could be represented by different parameterizations in the form (2.1). For example, let Ξ be the set of all $p \times p$ symmetric positive semidefinite matrices (covariance matrices) and Ξ_0 be its subset of diagonal matrices with nonnegative diagonal elements. This model can be parameterized by the set $\Theta := \mathbb{R}^p_+$ and mapping $\boldsymbol{g}(\boldsymbol{\theta}) := \operatorname{diag}(\theta_1, \ldots, \theta_p)$. (The set $\mathbb{R}^p_+$ denotes the nonnegative orthant of the space $\mathbb{R}^p$, i.e., $\mathbb{R}^p_+ := \{\boldsymbol{\theta} \in \mathbb{R}^p \colon \theta_i \geqslant 0,\ i = 1, \ldots, p\}$.) Alternatively, it can be parameterized by $\Theta := \mathbb{R}^p$ and $\boldsymbol{g}(\boldsymbol{\theta}) := \operatorname{diag}(\theta_1^2, \ldots, \theta_p^2)$. Of course, in applications the considered parameters typically have an interpretation. For instance, in the above example of diagonal covariance matrices, in the first parameterization parameters θ_i represent the corresponding standard deviations while in the second parameterization these are the corresponding variances. Note that this set Ξ_0 can be also defined by equations by setting the off-diagonal elements of $\boldsymbol{\Sigma}$ to zero. In the subsequent analysis we mainly deal with the parameterization approach.

It is said that model $\boldsymbol{g}(\boldsymbol{\theta})$, $\boldsymbol{\theta} \in \Theta$, is (globally) *identified* at a point $\boldsymbol{\theta}_0 \in \Theta$ if $\boldsymbol{\theta}_0$ is a unique parameter vector corresponding to the value $\boldsymbol{\xi}_0 := \boldsymbol{g}(\boldsymbol{\theta}_0)$ of the population vector. It is said that the model is locally identified at $\boldsymbol{\theta}_0$ if such uniqueness holds in a neighborhood of $\boldsymbol{\theta}_0$. More formally, we have the following definition.

DEFINITION 2.1. It is said that structural model $\boldsymbol{g}(\boldsymbol{\theta})$, $\boldsymbol{\theta} \in \Theta$, is *identified* (*locally identified*) at a point $\boldsymbol{\theta}_0 \in \Theta$ if $\boldsymbol{g}(\boldsymbol{\theta}^*) = \boldsymbol{g}(\boldsymbol{\theta}_0)$ and $\boldsymbol{\theta}^* \in \Theta$ ($\boldsymbol{\theta}^*$ in a neighborhood of $\boldsymbol{\theta}_0$) implies that $\boldsymbol{\theta}^* = \boldsymbol{\theta}_0$.

Of course, (global) identifiability implies local identifiability. A well-known sufficient condition for local identifiability of $\boldsymbol{\theta}_0 \in \Theta$ is that the Jacobian matrix $\boldsymbol{\Delta}(\boldsymbol{\theta}_0)$, of $\boldsymbol{g}(\boldsymbol{\theta})$ at $\boldsymbol{\theta}_0$, has full column rank q (e.g., Fisher, 1966). In general, this condition is not necessary for local identifiability of $\boldsymbol{\theta}_0$ even if $\boldsymbol{\theta}_0$ is an interior point of Θ. Take, for example, $g(\theta) := \theta^3$, $\theta \in \mathbb{R}$. This model is locally (and globally) identified at $\theta = 0$, while $\partial g(0)/\partial\theta = 0$. This condition becomes necessary and sufficient under the following assumption of constant rank regularity which was used by several authors (e.g., Fisher, 1966; Rothenberg, 1971; Wald, 1950).

DEFINITION 2.2. We say that a point $\boldsymbol{\theta}_0 \in \Theta$ is *locally regular* if the Jacobian matrix $\boldsymbol{\Delta}(\boldsymbol{\theta})$ has the same rank as $\boldsymbol{\Delta}(\boldsymbol{\theta}_0)$ for every $\boldsymbol{\theta}$ in a neighborhood of $\boldsymbol{\theta}_0$.

If the mapping $\boldsymbol{g}(\boldsymbol{\theta})$ is independent of, say, last s parameters $\theta_{q-s+1}, \dots, \theta_q$, then, of course, these parameters are redundant and the model can be viewed as overparameterized. In that case the rank of the Jacobian matrix $\boldsymbol{\Delta}(\boldsymbol{\theta})$ is less than or equal to $q - s$ for any $\boldsymbol{\theta}$. In general, it is natural to view the structural model as being (locally) overparameterized, at a point $\boldsymbol{\theta}_0$ in the interior of Θ, if it can be reduced to the above case by a local transformation (reparameterization). More formally we have the following definition (Shapiro, 1986).

DEFINITION 2.3. We say that structural model $\boldsymbol{g}(\boldsymbol{\theta})$, $\boldsymbol{\theta} \in \Theta$, is *locally overparameterized*, at an interior point $\boldsymbol{\theta}_0$ of Θ, if the rank r of $\boldsymbol{\Delta}(\boldsymbol{\theta}_0)$ is less than q and there exists a local diffeomorphism $\boldsymbol{\theta} = \boldsymbol{h}(\boldsymbol{\gamma})$ such that the composite mapping $\boldsymbol{g}(\boldsymbol{h}(\boldsymbol{\gamma}))$ is independent of, say last, $q - r$ coordinates of $\boldsymbol{\gamma}$.

Mapping $\boldsymbol{h}(\boldsymbol{\gamma})$ from a neighborhood of $\boldsymbol{\gamma}_0 \in \mathbb{R}^q$ to a neighborhood of $\boldsymbol{\theta}_0$, with $\boldsymbol{h}(\boldsymbol{\gamma}_0) = \boldsymbol{\theta}_0$, is called local diffeomorphism if it is continuously differentiable, locally one-to-one and its inverse is also continuously differentiable. It can be shown that $\boldsymbol{h}(\boldsymbol{\gamma})$ is a local diffeomorphism if and only if it is continuously differentiable and the Jacobian matrix $\partial\boldsymbol{h}(\boldsymbol{\gamma}_0)/\partial\boldsymbol{\gamma}$ is nonsingular. We can assume, without loss of generality, that $\boldsymbol{\gamma}_0 = \boldsymbol{0}$.

The local diffeomorphism $\boldsymbol{h}(\boldsymbol{\gamma})$, in the above definition, can be viewed as a local reparameterization of the model. We do not need to construct such a reparameterization explicitly but rather to know about its existence since it gives us an information about a (local) structure of the model. Clearly, if the model is locally overparameterized at a point $\boldsymbol{\theta}_0$, then it is not locally identified at this point. Moreover, in the case of local overparameterization the set of points $\boldsymbol{\theta}$ such that $\boldsymbol{g}(\boldsymbol{\theta}) = \boldsymbol{g}(\boldsymbol{\theta}_0)$ forms a smooth manifold in a neighborhood of $\boldsymbol{\theta}_0$. Note that the rank of the Jacobian matrix of the composite mapping $\boldsymbol{g}(\boldsymbol{h}(\boldsymbol{\gamma}))$, at $\boldsymbol{\gamma}_0 = \boldsymbol{0}$, is the same as the rank of $\boldsymbol{\Delta}(\boldsymbol{\theta}_0)$. Therefore, in the reparameterized model the remaining r coordinates of $\boldsymbol{\gamma}$ are locally identified.

A relation between the concepts of local regularity and local overparameterization is clarified by the following result known as the Rank Theorem (Fisher, 1966).

PROPOSITION 2.1. *Let $\boldsymbol{\theta}_0$ be an interior point of the parameter set Θ. Then the following holds.*

(i) *Suppose that $\boldsymbol{\theta}_0$ is locally regular. Then the model is locally identified at $\boldsymbol{\theta}_0$ if and only if the rank r of $\boldsymbol{\Delta}(\boldsymbol{\theta}_0)$ is equal to q, otherwise if $r < q$, then the model is locally overparameterized at $\boldsymbol{\theta}_0$.*
(ii) *Conversely, if the model is locally overparameterized at $\boldsymbol{\theta}_0$, then $r < q$ and the point $\boldsymbol{\theta}_0$ is locally regular.*

The above results are not very useful for verification of (local) identifiability at an individual point $\boldsymbol{\theta}_0$. For one thing the population value of the parameter vector usually is unknown, and even if the value $\boldsymbol{\theta}_0$ is specified, it is not possible to calculate the rank of the corresponding Jacobian matrix numerically because of the round off errors. However, one can approach the identifiability problem from a generic point of view. Suppose that the mapping $\boldsymbol{g}(\cdot)$ is analytic, i.e., every coordinate function $g_i(\cdot)$ can be expanded into a power series in a neighborhood of every point $\boldsymbol{\theta} \in \Theta$. Suppose also that the set Θ is connected. Let ι be an index set of rows and columns of $\boldsymbol{\Delta}(\boldsymbol{\theta})$ defining its squared submatrix and $v_\iota(\boldsymbol{\theta})$ be the determinant of that submatrix. Clearly there is only a finite number of such submatrices and hence the corresponding determinant functions. Since $\boldsymbol{g}(\cdot)$ is analytic, every such determinant function $v_\iota(\boldsymbol{\theta})$ is also analytic. Consequently, either $v_\iota(\boldsymbol{\theta})$ is identically zero for all $\boldsymbol{\theta} \in \Theta$, or $v_\iota(\boldsymbol{\theta}) \neq 0$ for almost every $\boldsymbol{\theta} \in \Theta$. (The "almost every" statement here can be understood in the sense that it holds for all $\boldsymbol{\theta}$ in Θ except on a subset of Θ of Lebesgue measure zero.) These arguments lead to the following result.

PROPOSITION 2.2. *Suppose that the mapping $\boldsymbol{g}(\cdot)$ is analytic and the set Θ is connected. Then almost every point of Θ is locally regular with the same rank r of the Jacobian matrix $\boldsymbol{\Delta}(\boldsymbol{\theta})$, and, moreover, the rank of $\boldsymbol{\Delta}(\boldsymbol{\theta})$ is less than or equal to r for all $\boldsymbol{\theta} \in \Theta$.*

By the above proposition we have that with the model, defined by analytic mapping $\boldsymbol{g}(\boldsymbol{\theta})$, is associated an integer r equal to the rank of $\boldsymbol{\Delta}(\boldsymbol{\theta})$ almost everywhere. We refer to this number r as the *characteristic rank* of the model, and whenever talking about the characteristic rank we assume that the mapping $\boldsymbol{g}(\boldsymbol{\theta})$ is analytic and the parameter set Θ is connected (the concept of characteristic rank was introduced in Shapiro (1986)). It follows from the preceding discussion that either $r = q$ in which case the model is locally identified almost everywhere, or $r < q$ in which case the model is locally overparameterized almost everywhere.

We say that the model is identified (locally identified) in the *generic sense* if it is identified (locally identified) at almost every $\boldsymbol{\theta} \in \Theta$ (Shapiro, 1985a). By the above analysis we have that the model is locally identified in the generic sense if and only if its characteristic rank is equal q. Note that the characteristic rank is always less than or equal to the dimension m of the saturated model.

In situations where the model is (locally) overparameterized, the usual practice is to restrict the parameter space by imposing constraints. According to Definition 2.3, if the model is locally overparameterized, at a point $\boldsymbol{\theta}_0$, then it can be reparameterized such that the reparameterized model locally does not depend on the last $q - r$ coordinates $\gamma_{r+1}, \dots, \gamma_q$. Consequently by imposing the constraints $\gamma_i = 0$, $i = r + 1, \dots, q$, the reparameterized model becomes locally identified at $\boldsymbol{\gamma}_0 = \mathbf{0}$ while its image space Ξ_0 is not changed. For the original model this is equivalent to imposing (locally) the identifiability constraints $c_i(\boldsymbol{\theta}) = 0$, $i = r + 1, \dots, q$, where $\boldsymbol{\gamma} = \boldsymbol{c}(\boldsymbol{\theta})$ is the inverse of the mapping $\boldsymbol{\theta} = \boldsymbol{h}(\boldsymbol{\gamma})$.

EXAMPLE 2.1. Consider the factor analysis model:

$$\boldsymbol{\Sigma} = \boldsymbol{\Lambda}\boldsymbol{\Lambda}' + \boldsymbol{\Psi}, \tag{2.2}$$

which relates the $p \times p$ covariance $\boldsymbol{\Sigma}$ to the $p \times k$ matrix $\boldsymbol{\Lambda}$ of factor loadings and the $p \times p$ diagonal matrix $\boldsymbol{\Psi}$ of the residual variances. The corresponding parameter vector $\boldsymbol{\theta}$ is composed here from the elements of matrix $\boldsymbol{\Lambda}$ and diagonal elements of $\boldsymbol{\Psi}$, and hence has dimension $q = pk + p$. Note that the diagonal elements of the matrix $\boldsymbol{\Psi}$ should be nonnegative while there are no restrictions on the elements of $\boldsymbol{\Lambda}$.

By substituting $\boldsymbol{\Lambda}\boldsymbol{T}$ for $\boldsymbol{\Lambda}$, where $\boldsymbol{T}$ is an arbitrary $k \times k$ orthogonal matrix, we end up with the same matrix $\boldsymbol{\Sigma}$ (this is the so-called indeterminacy of the factor analysis model). Since the dimension of the (smooth) manifold of $k \times k$ orthogonal matrices is $k(k-1)/2$ and the dimension of the space of $p \times p$ symmetric matrices is $p(p+1)/2$, it is possible to show that the characteristic rank r of the factor analysis model (2.2) is

$$r = \min\{pk + p - k(k-1)/2,\, p(p+1)/2\}. \tag{2.3}$$

It follows that for $k > 1$ the model (2.2) is locally overparameterized. A way of dealing with this is to reduce the number of parameters given by matrix $\boldsymbol{\Lambda}$ by setting $k(k-1)/2$ appropriate elements of $\boldsymbol{\Lambda}$ to zero (Jennrich, 1987). Then the question of global (local) identifiability of the factor analysis model is reduced to the global (local) identifiability of the diagonal matrix $\boldsymbol{\Psi}$ (Anderson and Rubin, 1956). We have that a necessary condition for *generic* local identifiability of $\boldsymbol{\Psi}$ is that $pk + p - k(k-1)/2$ is less than or equal to $p(p+1)/2$, which is equivalent to $(p-k)(p-k+1)/2 \geqslant p$, and in turn is equivalent to $k \leqslant \phi(p)$, where

$$\phi(p) := \frac{2p + 1 - \sqrt{8p+1}}{2}. \tag{2.4}$$

The above function $\phi(p)$ corresponds to the so-called Ledermann bound (Ledermann, 1937). In the present case we have that $k \leqslant \phi(p)$ is a necessary and sufficient condition for local identifiability of the diagonal matrix $\boldsymbol{\Psi}$, of the factor analysis model, in the *generic* sense (Shapiro, 1985a).

It is more difficult to establish (global) identifiability of a considered model. We can also approach the (global) identifiability problem from the generic point of view. Of course, if a model is not locally identified it cannot be globally identified. Therefore, $k \leqslant \phi(p)$ is a necessary condition for (global) identifiability of the factor analysis model in the generic sense. It is known that matrix $\boldsymbol{\Psi}$, in the factor analysis model (2.2), is globally identified in the generic sense if and only if $k < \phi(p)$ (Bekker and ten Berge, 1997).

3. Minimum discrepancy function estimation approach

Let $\boldsymbol{\xi}_0 \in \Xi$ be a population value of the parameter vector of the saturated model. Recall that we refer to a subset Ξ_0 of Ξ as a model for $\boldsymbol{\xi}$. Unless stated otherwise it will be assumed that the model is structural, i.e., the set Ξ_0 is given in the parametric form (2.1). It is said that the model holds if $\boldsymbol{\xi}_0 \in \Xi_0$. Clearly this means that there exists $\boldsymbol{\theta}_0 \in \Theta$ such that $\boldsymbol{\xi}_0 = \boldsymbol{g}(\boldsymbol{\theta}_0)$. If the model is identified at $\boldsymbol{\theta}_0$, then this vector $\boldsymbol{\theta}_0$ is defined uniquely. In that case we refer to $\boldsymbol{\theta}_0$ as the *population* value of the parameter vector $\boldsymbol{\theta}$.

Suppose that we are given an estimator $\hat{\boldsymbol{\xi}}$ of $\boldsymbol{\xi}_0$, based on a sample of size n. We will be interested then in testing the hypothesis H_0: $\boldsymbol{\xi}_0 \in \Xi_0$, and consequently in estimation of the population value of the parameter $\boldsymbol{\theta}$. Consider the setting of the covariance structures with $\boldsymbol{\Sigma}$ being the covariance matrix of $p \times 1$ random vector $\boldsymbol{X}$, and let $\boldsymbol{\Sigma} = \boldsymbol{\Sigma}(\boldsymbol{\theta})$ be an associated structural model. Let $\boldsymbol{X}_1, \dots, \boldsymbol{X}_n$ be an iid (independent identically distributed) random sample drawn from a considered population. Then the standard estimator of the population value $\boldsymbol{\Sigma}_0$ of the covariance matrix is the sample covariance matrix

$$\boldsymbol{S} := \frac{1}{n-1} \sum_{i=1}^{n} (\boldsymbol{X}_i - \overline{\boldsymbol{X}})(\boldsymbol{X}_i - \overline{\boldsymbol{X}})', \tag{3.1}$$

where $\overline{\boldsymbol{X}} := n^{-1} \sum_{i=1}^{n} \boldsymbol{X}_i$. Suppose, further, that the population distribution is (multivariate) normal with mean vector $\boldsymbol{\mu}_0$ and covariance matrix $\boldsymbol{\Sigma}_0$. Then the corresponding log-likelihood function (up to a constant independent of the parameters) is

$$\ell(\boldsymbol{\mu}, \boldsymbol{\Sigma}) = -\frac{1}{2} n \ln |\boldsymbol{\Sigma}| - \frac{1}{2} \operatorname{tr}\left(\boldsymbol{\Sigma}^{-1} \sum_{i=1}^{n} (\boldsymbol{X}_i - \boldsymbol{\mu})(\boldsymbol{X}_i - \boldsymbol{\mu})' \right). \tag{3.2}$$

(By $|\boldsymbol{A}|$ and $\operatorname{tr}(\boldsymbol{A})$ we denote the determinant and the trace, respectively, of a (square) matrix $\boldsymbol{A}$.) The maximum likelihood (ML) estimator of $\boldsymbol{\mu}_0$ is $\overline{\boldsymbol{X}}$ and the ML estimator of $\boldsymbol{\Sigma}_0$, for the saturated model, is

$$\widehat{\boldsymbol{\Sigma}} = \frac{1}{n} \sum_{i=1}^{n} (\boldsymbol{X}_i - \overline{\boldsymbol{X}})(\boldsymbol{X}_i - \overline{\boldsymbol{X}})' = \frac{n-1}{n} \boldsymbol{S}. \tag{3.3}$$

Of course, for reasonably large values of n, the ML estimator $\widehat{\boldsymbol{\Sigma}}$ is "almost" equal to the unbiased estimator $\boldsymbol{S}$. Therefore, with some abuse of notation, we use $\boldsymbol{S} = \widehat{\boldsymbol{\Sigma}}$ as the estimator of the population covariance matrix.

It follows that two times log-likelihood ratio statistic for testing the null hypothesis $\boldsymbol{\Sigma}_0 = \boldsymbol{\Sigma}(\boldsymbol{\theta}_0)$ is given by $n\widehat{F}$, where

$$\widehat{F} := \min_{\boldsymbol{\theta}\in\Theta} F_{\mathrm{ML}}\big(\boldsymbol{S}, \boldsymbol{\Sigma}(\boldsymbol{\theta})\big), \tag{3.4}$$

with $F_{\mathrm{ML}}(\cdot, \cdot)$ being a function of two (matrix valued) variables defined by

$$F_{\mathrm{ML}}(\boldsymbol{S}, \boldsymbol{\Sigma}) := \ln|\boldsymbol{\Sigma}| - \ln|\boldsymbol{S}| + \mathrm{tr}(\boldsymbol{S}\boldsymbol{\Sigma}^{-1}) - p. \tag{3.5}$$

The corresponding ML estimator $\hat{\boldsymbol{\theta}}$ of $\boldsymbol{\theta}_0$ is a minimizer of $F_{\mathrm{ML}}(\boldsymbol{S}, \boldsymbol{\Sigma}(\boldsymbol{\theta}))$ over $\boldsymbol{\theta} \in \Theta$, i.e.,

$$\hat{\boldsymbol{\theta}} \in \arg\min_{\boldsymbol{\theta}\in\Theta} F_{\mathrm{ML}}\big(\boldsymbol{S}, \boldsymbol{\Sigma}(\boldsymbol{\theta})\big). \tag{3.6}$$

(By $\arg\min\{f(\boldsymbol{\theta})\colon \boldsymbol{\theta} \in \Theta\}$ we denote the set of all minimizers of $f(\boldsymbol{\theta})$ over $\boldsymbol{\theta} \in \Theta$; this set can be empty or contain more than one element.) Let us note at this point that the estimator $\hat{\boldsymbol{\theta}}$, defined in (3.6), can be calculated whether the model holds or not. We will discuss implications of this later.

Following Browne (1982), we say that a function $F(\boldsymbol{x}, \boldsymbol{\xi})$, of two vector variables $\boldsymbol{x}, \boldsymbol{\xi} \in \Xi$, is a *discrepancy* function if it satisfies the following conditions:

(i) $F(\boldsymbol{x}, \boldsymbol{\xi}) \geqslant 0$ for all $\boldsymbol{x}, \boldsymbol{\xi} \in \Xi$,
(ii) $F(\boldsymbol{x}, \boldsymbol{\xi}) = 0$ if and only if $\boldsymbol{x} = \boldsymbol{\xi}$,
(iii) $F(\boldsymbol{x}, \boldsymbol{\xi})$ is twice continuously differentiable jointly in $\boldsymbol{x}$ and $\boldsymbol{\xi}$.

Let $\boldsymbol{g}(\boldsymbol{\theta})$, $\boldsymbol{\theta} \in \Theta$, be a structural model being considered. Given an estimator $\hat{\boldsymbol{\xi}}$ of $\boldsymbol{\xi}_0$ and a discrepancy function $F(\boldsymbol{x}, \boldsymbol{\xi})$, we refer to the statistic $n\widehat{F}$, where

$$\widehat{F} := \min_{\boldsymbol{\theta}\in\Theta} F\big(\hat{\boldsymbol{\xi}}, \boldsymbol{g}(\boldsymbol{\theta})\big), \tag{3.7}$$

as the minimum discrepancy function (MDF) test statistic, and to

$$\hat{\boldsymbol{\theta}} \in \arg\min_{\boldsymbol{\theta}\in\Theta} F\big(\hat{\boldsymbol{\xi}}, \boldsymbol{g}(\boldsymbol{\theta})\big) \tag{3.8}$$

as the MDF estimator.

The function $F_{\mathrm{ML}}(\boldsymbol{S}, \boldsymbol{\Sigma})$ defined in (3.5), considered as a function of $\boldsymbol{s} = \mathrm{vec}(\boldsymbol{S})$ and $\boldsymbol{\sigma} = \mathrm{vec}(\boldsymbol{\Sigma})$, is an example of a discrepancy function. It is referred to as the maximum likelihood (ML) discrepancy function. Another popular choice of the discrepancy function in the analysis of covariance structures is

$$F_{\mathrm{GLS}}(\boldsymbol{S}, \boldsymbol{\Sigma}) := \frac{1}{2}\,\mathrm{tr}\big[(\boldsymbol{S} - \boldsymbol{\Sigma})\boldsymbol{S}^{-1}(\boldsymbol{S} - \boldsymbol{\Sigma})\boldsymbol{S}^{-1}\big]. \tag{3.9}$$

We refer to a function of the form

$$F(\boldsymbol{x}, \boldsymbol{\xi}) := (\boldsymbol{x} - \boldsymbol{\xi})'\big[\boldsymbol{V}(\boldsymbol{x})\big](\boldsymbol{x} - \boldsymbol{\xi}), \quad \boldsymbol{x}, \boldsymbol{\xi} \in \Xi, \tag{3.10}$$

as a generalized least squares (GLS) discrepancy function. Here $\boldsymbol{V}(\boldsymbol{x})$ is an $m \times m$ symmetric matrix valued function of $\boldsymbol{x} \in \Xi$. We assume that for any $\boldsymbol{x} \in \Xi$, the corresponding matrix $\boldsymbol{V}(\boldsymbol{x})$ is positive definite, and hence conditions (i) and (ii) hold, and that $\boldsymbol{V}(\boldsymbol{x})$ is twice continuously differentiable, and hence condition (iii) is satisfied.

The $F_{\mathrm{GLS}}(\boldsymbol{S}, \boldsymbol{\Sigma})$ function, defined in (3.9), is a particular example of GLS discrepancy functions with the weight matrix $\boldsymbol{V}(s) = \frac{1}{2}\boldsymbol{S}^{-1} \otimes \boldsymbol{S}^{-1}$ (by $\otimes$ we denote the Kronecker product of matrices).

We have the following basic result about the structure of discrepancy functions (Shapiro, 1985b).

PROPOSITION 3.1. *Let $F(\boldsymbol{x}, \boldsymbol{\xi})$ be a discrepancy function satisfying conditions* (i)–(iii). *Then there exists a continuous $m \times m$ symmetric matrix valued function $\boldsymbol{V}(\boldsymbol{x}, \boldsymbol{\xi})$ such that*

$$F(\boldsymbol{x}, \boldsymbol{\xi}) = (\boldsymbol{x} - \boldsymbol{\xi})'\big[\boldsymbol{V}(\boldsymbol{x}, \boldsymbol{\xi})\big](\boldsymbol{x} - \boldsymbol{\xi}) \tag{3.11}$$

for all $\boldsymbol{x}, \boldsymbol{\xi} \in \Xi$.

The above result shows that any discrepancy function can be represented in a form of an "almost" GLS function. A difference between the representation (3.11) and the general form (3.10) of GLS discrepancy functions is that the weight matrix in (3.11) can also depend on $\boldsymbol{\xi}$ as well as on $\boldsymbol{x}$.

Let $\boldsymbol{\xi}_0 \in \Xi$ be a given (say the population) value of vector $\boldsymbol{\xi}$. Consider matrix $\boldsymbol{V}_0 := \boldsymbol{V}(\boldsymbol{\xi}_0, \boldsymbol{\xi}_0)$ associated with the matrix valued function $\boldsymbol{V}(\cdot, \cdot)$ of representation (3.11). We can then write

$$F(\boldsymbol{x}, \boldsymbol{\xi}) = (\boldsymbol{x} - \boldsymbol{\xi})'\boldsymbol{V}_0(\boldsymbol{x} - \boldsymbol{\xi}) + \boldsymbol{r}(\boldsymbol{x}, \boldsymbol{\xi}), \tag{3.12}$$

where $\boldsymbol{r}(\boldsymbol{x}, \boldsymbol{\xi}) := (\boldsymbol{x} - \boldsymbol{\xi})'[\boldsymbol{V}(\boldsymbol{x}, \boldsymbol{\xi}) - \boldsymbol{V}_0](\boldsymbol{x} - \boldsymbol{\xi})$. We have that

$$\big|\boldsymbol{r}(\boldsymbol{x}, \boldsymbol{\xi})\big| \leqslant \|\boldsymbol{x} - \boldsymbol{\xi}\|^2 \big\|\boldsymbol{V}(\boldsymbol{x}, \boldsymbol{\xi}) - \boldsymbol{V}_0\big\|,$$

and $\boldsymbol{V}(\boldsymbol{x}, \boldsymbol{\xi})$ tends to $\boldsymbol{V}_0$ as $\boldsymbol{x} \to \boldsymbol{\xi}_0$ and $\boldsymbol{\xi} \to \boldsymbol{\xi}_0$. Consequently, for $(\boldsymbol{x}, \boldsymbol{\xi})$ near $(\boldsymbol{\xi}_0, \boldsymbol{\xi}_0)$ the remainder term $\boldsymbol{r}(\boldsymbol{x}, \boldsymbol{\xi})$ in (3.12) is of order

$$\boldsymbol{r}(\boldsymbol{x}, \boldsymbol{\xi}) = \mathrm{o}\big(\|\boldsymbol{x} - \boldsymbol{\xi}_0\|^2 + \|\boldsymbol{\xi} - \boldsymbol{\xi}_0\|^2\big).$$

The notation $\mathrm{o}(x)$ means that $\mathrm{o}(x)$ is a function of x such that $\mathrm{o}(x)/x$ tends to zero as $x \to 0$. For a sequence X_n of random variables the notation $X_n = \mathrm{o}_p(a_n)$ means that X_n/a_n converges in probability to zero. In particular, $X_n = \mathrm{o}_p(1)$ means that X_n converges in probability to zero.

This can be compared with a Taylor expansion of $F(\boldsymbol{x}, \boldsymbol{\xi})$ at $(\boldsymbol{\xi}_0, \boldsymbol{\xi}_0)$. We have that $F(\boldsymbol{\xi}_0, \boldsymbol{\xi}_0) = 0$, and since $F(\cdot, \boldsymbol{\xi}_0)$ attains its minimum (of zero) at $\boldsymbol{x} = \boldsymbol{\xi}_0$, we have that $\partial F(\boldsymbol{\xi}_0, \boldsymbol{\xi}_0)/\partial \boldsymbol{x} = 0$, and similarly $\partial F(\boldsymbol{\xi}_0, \boldsymbol{\xi}_0)/\partial \boldsymbol{\xi} = 0$. It follows that the second-order Taylor expansion of $F(\boldsymbol{x}, \boldsymbol{\xi})$ at $(\boldsymbol{\xi}_0, \boldsymbol{\xi}_0)$ can be written as follows

$$\begin{aligned} F(\boldsymbol{x}, \boldsymbol{\xi}) = {} & \frac{1}{2}(\boldsymbol{x} - \boldsymbol{\xi}_0)'\boldsymbol{H}_{xx}(\boldsymbol{x} - \boldsymbol{\xi}_0) + \frac{1}{2}(\boldsymbol{\xi} - \boldsymbol{\xi}_0)'\boldsymbol{H}_{\xi\xi}(\boldsymbol{\xi} - \boldsymbol{\xi}_0) \\ & + (\boldsymbol{x} - \boldsymbol{\xi}_0)'\boldsymbol{H}_{x\xi}(\boldsymbol{\xi} - \boldsymbol{\xi}_0) + \mathrm{o}\big(\|\boldsymbol{x} - \boldsymbol{\xi}_0\|^2 + \|\boldsymbol{\xi} - \boldsymbol{\xi}_0\|^2\big), \end{aligned} \tag{3.13}$$

where $\boldsymbol{H}_{xx} := \partial^2 F(\boldsymbol{\xi}_0, \boldsymbol{\xi}_0)/\partial \boldsymbol{x}\partial \boldsymbol{x}'$, $\boldsymbol{H}_{\xi\xi} := \partial^2 F(\boldsymbol{\xi}_0, \boldsymbol{\xi}_0)/\partial \boldsymbol{\xi}\partial \boldsymbol{\xi}'$, $\boldsymbol{H}_{x\xi} := \partial^2 F(\boldsymbol{\xi}_0, \boldsymbol{\xi}_0)/\partial \boldsymbol{x}\partial \boldsymbol{\xi}'$ are the corresponding Hessian matrices of second order partial derivatives. By comparing (3.12) and (3.13), we obtain that

$$\frac{\partial^2 F(\xi_0, \xi_0)}{\partial x \partial x'} = \frac{\partial^2 F(\xi_0, \xi_0)}{\partial \xi \partial \xi'} = -\frac{\partial^2 F(\xi_0, \xi_0)}{\partial x \partial \xi'} = 2V_0. \tag{3.14}$$

Both discrepancy functions $F_{\mathrm{ML}}(s, \sigma)$ and $F_{\mathrm{GLS}}(s, \sigma)$, defined in (3.5) and (3.9), respectively, have the same Hessian matrix

$$\frac{\partial^2 F_{\mathrm{ML}}(\sigma_0, \sigma_0)}{\partial s \partial s'} = \frac{\partial^2 F_{\mathrm{GLS}}(\sigma_0, \sigma_0)}{\partial s \partial s'} = \Sigma_0^{-1} \otimes \Sigma_0^{-1}, \tag{3.15}$$

and hence the same second-order Taylor approximation at (σ_0, σ_0).

REMARK 2. For technical reasons we also assume the following condition for discrepancy functions.

(iv) For any (fixed) $\bar{x} \in \Xi$, $F(x, \xi)$ tends to infinity as $x \to \bar{x}$ and $\|\xi\| \to \infty$.

It is not difficult to verify that the ML discrepancy function, defined in (3.5), and the GLS discrepancy functions satisfy this condition.

4. Consistency of MDF estimators

Let $\hat{\xi} = \hat{\xi}_n$ be a given estimator, based on a sample of size n, of the population value $\xi_0 \in \Xi$ of the parameter vector of the saturated model (we use the subscript n, in the notation $\hat{\xi}_n$, to emphasize that the estimator is a function of a sample of size n being considered). It is said that the estimator $\hat{\xi}_n$ is *consistent* if it converges with probability one (w.p.1) to ξ_0 as $n \to \infty$. For example, by the (strong) Law of Large Numbers we have that the sample covariance matrix S converges to Σ_0 w.p.1 as $n \to \infty$. For this to hold we only need to assume that the population distribution has finite second-order moments, and hence the covariance matrix Σ_0 does exist, and that the corresponding random sample is iid.

Let $F(x, \xi)$ be a chosen discrepancy function satisfying conditions (i)–(iv) specified in the previous section, and $\bar{\xi} = \bar{\xi}(x)$ be an optimal solution of the minimization problem:

$$\min_{\xi \in \Xi_0} F(x, \xi). \tag{4.1}$$

Note that since $F(x, \cdot)$ is continuous and because of condition (iv), such a minimizer always exists (provided that the set Ξ_0 is closed), although it may be not unique. Define

$$\hat{\xi}_n^* := \bar{\xi}(\hat{\xi}_n) \quad \text{and} \quad \xi^* := \bar{\xi}(\xi_0). \tag{4.2}$$

That is, $\hat{\xi}_n^*$ and ξ^* are minimizers of $F(\hat{\xi}_n, \cdot)$ and $F(\xi_0, \cdot)$, respectively, over Ξ_0. It could be noted that if the model holds, i.e., $\xi_0 \in \Xi_0$, then the minimizer ξ^* coincides with ξ_0 and is unique (because of the properties (i) and (ii) of the discrepancy function). It is possible to show that if the minimizer ξ^* is unique, then the function $\bar{\xi}(x)$ is continuous at $x = \xi_0$, i.e., $\bar{\xi}(x) \to \xi^*$ as $x \to \xi_0$. Together with consistency of $\hat{\xi}_n$ this implies the following result (Shapiro, 1984).

PROPOSITION 4.1. *Suppose that the discrepancy function satisfies conditions* (i)–(iv) *and* $\hat{\boldsymbol{\xi}}_n$ *is a consistent estimator of* $\boldsymbol{\xi}_0$. *Then* $\hat{\boldsymbol{\xi}}_n^*$ *converges to* $\boldsymbol{\xi}^*$ *w.p.1 as* $n \to \infty$, *provided that the minimizer* $\boldsymbol{\xi}^*$ *is unique. In particular, if* $\boldsymbol{\xi}_0 \in \Xi_0$, *then* $\hat{\boldsymbol{\xi}}_n^* \to \boldsymbol{\xi}_0$ *w.p.1 as* $n \to \infty$.

Similar analysis can be applied to studying consistency of the MDF estimators of the parameter vectors in Θ. For a given $\boldsymbol{x} \in \Xi$ consider the optimization (minimization) problem:

$$\min_{\boldsymbol{\theta} \in \Theta} F\big(\boldsymbol{x}, \boldsymbol{g}(\boldsymbol{\theta})\big). \tag{4.3}$$

Recall that the MDF estimator $\hat{\boldsymbol{\theta}}_n$ is an optimal solution of problem (4.3) for $\boldsymbol{x} = \hat{\boldsymbol{\xi}}_n$. Let $\boldsymbol{\theta}^*$ be an optimal solution of (4.3) for $\boldsymbol{x} = \boldsymbol{\xi}_0$, i.e.,

$$\boldsymbol{\theta}^* \in \arg\min_{\boldsymbol{\theta} \in \Theta} F\big(\boldsymbol{\xi}_0, \boldsymbol{g}(\boldsymbol{\theta})\big).$$

Of course, if $\boldsymbol{\xi}_0 = \boldsymbol{g}(\boldsymbol{\theta}_0)$ for some $\boldsymbol{\theta}_0 \in \Theta$ (i.e., the model holds), then $\boldsymbol{\theta}_0$ is an optimal solution of (4.3) for $\boldsymbol{x} = \boldsymbol{\xi}_0$, and we can take $\boldsymbol{\theta}^* = \boldsymbol{\theta}_0$. The optimal values of problems (4.1) and (4.3) are equal to each other and there is a one-to-one correspondence between the sets of optimal solutions of problems (4.1) and (4.3). That is, if $\bar{\boldsymbol{\theta}}$ is an optimal solution of (4.3), then $\bar{\boldsymbol{\xi}} = \boldsymbol{g}(\bar{\boldsymbol{\theta}})$ is an optimal solution of (4.1), and conversely if $\bar{\boldsymbol{\xi}}$ is an optimal solution of (4.1) and $\bar{\boldsymbol{\theta}} \in \Theta$ is a corresponding point of Θ, then $\bar{\boldsymbol{\theta}}$ is an optimal solution of (4.3). The relation between $\bar{\boldsymbol{\xi}}$ and $\bar{\boldsymbol{\theta}}$ is defined by the equation $\bar{\boldsymbol{\xi}} = \boldsymbol{g}(\bar{\boldsymbol{\theta}})$. If the model is identified at $\bar{\boldsymbol{\theta}}$, then the equation $\bar{\boldsymbol{\xi}} = \boldsymbol{g}(\bar{\boldsymbol{\theta}})$ defines the point $\bar{\boldsymbol{\theta}}$ uniquely.

It follows that, under the assumptions of Proposition 4.1, $\hat{\boldsymbol{\theta}}_n$ is a consistent estimator of $\boldsymbol{\theta}^*$, if the inverse of the mapping $\boldsymbol{g}(\cdot)$ is continuous at $\boldsymbol{\theta}^*$, i.e., if the following condition holds:

$$\begin{aligned} &\boldsymbol{g}(\boldsymbol{\theta}_n) \to \boldsymbol{g}(\boldsymbol{\theta}^*), \quad \text{for some sequence } \{\boldsymbol{\theta}_n\} \subset \Theta, \\ &\quad \text{implies that } \boldsymbol{\theta}_n \to \boldsymbol{\theta}^*. \end{aligned} \tag{4.4}$$

Note that the above condition (4.4) can only hold if the model is identified at $\boldsymbol{\theta}^*$. This leads to the following result (Kano, 1986; Shapiro, 1984).

PROPOSITION 4.2. *Suppose that the discrepancy function satisfies conditions* (i)–(iv) $\hat{\boldsymbol{\xi}}_n$ *is a consistent estimator of* $\boldsymbol{\xi}_0$, *and for* $\boldsymbol{x} = \boldsymbol{\xi}_0$ *problem* (4.3) *has unique optimal solution* $\boldsymbol{\theta}^*$ *and condition* (4.4) *holds. Then* $\hat{\boldsymbol{\theta}}_n$ *converges to* $\boldsymbol{\theta}^*$ *w.p.1 as* $n \to \infty$. *In particular, if* $\boldsymbol{\xi}_0 = \boldsymbol{g}(\boldsymbol{\theta}_0)$, *for some* $\boldsymbol{\theta}_0 \in \Theta$ *(i.e., the model holds), then* $\boldsymbol{\theta}_0 = \boldsymbol{\theta}^*$ *and the MDF estimator* $\hat{\boldsymbol{\theta}}_n$ *is a consistent estimator of* $\boldsymbol{\theta}_0$.

Note that uniqueness of the optimal solution $\boldsymbol{\theta}^*$ implies uniqueness of the corresponding optimal solution $\boldsymbol{\xi}^*$. Converse of that also holds if the model is identified at $\boldsymbol{\theta}^*$. As it was mentioned above, identifiability of $\boldsymbol{\theta}^*$ is a necessary condition for the property (4.4) to hold. It is also sufficient if the set Θ is compact (i.e., bounded and closed). For a noncompact set Θ, condition (4.4) prevents the MDF estimator from escaping to infinity.

The above proposition shows that if the model holds, then under mild regularity conditions (in particular, identifiability of the model at $\boldsymbol{\theta}_0$) the MDF estimator $\hat{\boldsymbol{\theta}}_n$ converges w.p.1 to the true (population) value $\boldsymbol{\theta}_0$ of the parameter vector. On the other hand, if the model does not hold, then $\hat{\boldsymbol{\theta}}_n$ converges to an optimal solution $\boldsymbol{\theta}^*$ of the problem (4.3). It could be noted that if the model does not hold, then such an optimal solution depends on a particular choice of the discrepancy function.

As we can see uniqueness of the (population) minimizer $\boldsymbol{\theta}^*$ is crucial for convergence of the MDF estimator $\hat{\boldsymbol{\theta}}_n$. If the model holds, then $\boldsymbol{\theta}^* = \boldsymbol{\theta}_0$ and uniqueness of $\boldsymbol{\theta}_0$ is equivalent to identifiability of the model at $\boldsymbol{\theta}_0$. Now if a point $\bar{\boldsymbol{\theta}} = \bar{\boldsymbol{\theta}}(\boldsymbol{x})$ is an optimal solution of problem (4.3) and is an interior point of the set Θ, then it satisfies the necessary optimality condition

$$\frac{\partial F(\boldsymbol{x}, \boldsymbol{g}(\boldsymbol{\theta}))}{\partial \boldsymbol{\theta}} = \mathbf{0}. \tag{4.5}$$

This condition can be viewed as a system of (nonlinear) equations. Consider a point $\boldsymbol{\theta}_0$ in the *interior* of the set Θ and let $\boldsymbol{\xi}_0 := \boldsymbol{g}(\boldsymbol{\theta}_0)$. By linearizing (4.5) at $\boldsymbol{x} = \boldsymbol{\xi}_0$ and $\boldsymbol{\theta} = \boldsymbol{\theta}_0$, we obtain the following (linear) system of equations:

$$\left[\frac{\partial^2 F(\boldsymbol{\xi}_0, \boldsymbol{g}(\boldsymbol{\theta}_0))}{\partial \boldsymbol{\theta} \partial \boldsymbol{x}'}\right](\boldsymbol{x} - \boldsymbol{\xi}_0) + \left[\frac{\partial^2 F(\boldsymbol{\xi}_0, \boldsymbol{g}(\boldsymbol{\theta}_0))}{\partial \boldsymbol{\theta} \partial \boldsymbol{\theta}'}\right](\boldsymbol{\theta} - \boldsymbol{\theta}_0) = \mathbf{0}. \tag{4.6}$$

Note that by (3.14) we have that

$$\frac{\partial^2 F(\boldsymbol{\xi}_0, \boldsymbol{g}(\boldsymbol{\theta}_0))}{\partial \boldsymbol{\theta} \partial \boldsymbol{x}'} = -2\boldsymbol{\Delta}_0' \boldsymbol{V}_0 \quad \text{and} \quad \frac{\partial^2 F(\boldsymbol{\xi}_0, \boldsymbol{g}(\boldsymbol{\theta}_0))}{\partial \boldsymbol{\theta} \partial \boldsymbol{\theta}'} = 2\boldsymbol{\Delta}_0' \boldsymbol{V}_0 \boldsymbol{\Delta}_0, \tag{4.7}$$

where $\boldsymbol{\Delta}_0 := \boldsymbol{\Delta}(\boldsymbol{\theta}_0)$. Since the matrix $\boldsymbol{V}_0$ is positive definite, we have that the matrix $\boldsymbol{\Delta}_0' \boldsymbol{V}_0 \boldsymbol{\Delta}_0$ is nonsingular iff the Jacobian matrix $\boldsymbol{\Delta}_0$ has full column rank q (recall that this is a sufficient condition for identifiability of $\boldsymbol{\theta}_0$). It follows then by the Implicit Function Theorem that:

> If $\boldsymbol{\Delta}_0$ has full column rank q, then for all $\boldsymbol{x}$ sufficiently close to $\boldsymbol{\xi}_0$ the system (4.5) has a unique solution $\bar{\boldsymbol{\theta}} = \bar{\boldsymbol{\theta}}(\boldsymbol{x})$ in a neighborhood of $\boldsymbol{\theta}_0$, and $\bar{\boldsymbol{\theta}}$ is the unique optimal solution of problem (4.3) in that neighborhood of $\boldsymbol{\theta}_0$.

The above is a local result. It implies that, under the specified conditions, if the estimator $\hat{\boldsymbol{\xi}}_n$ is sufficiently close to a point $\boldsymbol{\xi}_0 = \boldsymbol{g}(\boldsymbol{\theta}_0)$ satisfying the model, i.e., the fit is good enough, then the corresponding MDF estimator $\hat{\boldsymbol{\theta}}_n$ is unique, and can be obtained by solving Eq. (4.5) with $\boldsymbol{x} = \hat{\boldsymbol{\xi}}_n$, in a neighborhood of the point $\boldsymbol{\theta}_0 \in \Theta$. Of course, in practice it is impossible to say a priori when "sufficiently close" is close enough for the above to hold.

5. Asymptotic analysis of the MDF estimation procedure

In this section we discuss a basic theory of asymptotics of the MDF estimation procedure. We assume that the discrepancy function considered satisfies conditions (i)–(iv) specified in Section 3. We also assume that the estimator $\hat{\boldsymbol{\xi}}_n$ is *asymptotically normal*. That is, we assume that the sequence

$$\boldsymbol{Z}_n := n^{1/2}(\hat{\boldsymbol{\xi}}_n - \boldsymbol{\xi}_0),$$

of random vectors, with $\boldsymbol{\xi}_0$ being the population value of the parameter vector $\boldsymbol{\xi} \in \Xi$, converges in distribution to multivariate normal with mean vector zero and covariance matrix $\boldsymbol{\Gamma}$, i.e., $\boldsymbol{Z}_n \Rightarrow N(\boldsymbol{0}, \boldsymbol{\Gamma})$ (by "$\Rightarrow$" we denote convergence in distribution). For example, in the analysis of covariance structures we have that vector $\boldsymbol{s} := \text{vec}(\boldsymbol{S})$, associated with the sample covariance matrix $\boldsymbol{S}$, is asymptotically normal. This follows from the Central Limit Theorem provided that the population distribution has fourth-order moments and the sample is iid. Moreover, if the population distribution is normal, then $\boldsymbol{\Gamma} = \boldsymbol{\Gamma}_N$, where

$$\boldsymbol{\Gamma}_N := 2\boldsymbol{M}_p(\boldsymbol{\Sigma}_0 \otimes \boldsymbol{\Sigma}_0) \tag{5.1}$$

with $\boldsymbol{M}_p$ being an $p^2 \times p^2$ symmetric idempotent matrix of rank $p(p+1)/2$ with element in row ij and column kl given by $\boldsymbol{M}_p(ij, kl) = \frac{1}{2}(\delta_{ik}\delta_{jl} + \delta_{il}\delta_{jk})$ (Browne, 1974) (here $\delta_{ik} = 1$ if $i = k$, and $\delta_{ik} = 0$ if $i \neq k$). It follows that matrix $\boldsymbol{\Gamma}_N$ also has rank $p(p+1)/2$, provided that the covariance matrix $\boldsymbol{\Sigma}_0$ is nonsingular. We assume that the (asymptotic) covariance matrix $\boldsymbol{\Gamma}$, of $\boldsymbol{Z}_n$, has the *maximal rank*, which in the case of covariance structures is $p(p+1)/2$. It follows then that the linear space generated by columns of the Jacobian matrix $\boldsymbol{\Delta}(\boldsymbol{\theta})$ is contained in the linear space generated by columns of $\boldsymbol{\Gamma}$.

Denote by $\vartheta(\boldsymbol{x})$ the optimal value of problem (4.1), i.e., $\vartheta(\boldsymbol{x}) := \inf_{\boldsymbol{\xi} \in \Xi_0} F(\boldsymbol{x}, \boldsymbol{\xi})$. Recall that the optimal values of problems (4.1) and (4.3) are the same, and hence $\vartheta(\boldsymbol{x})$ is also the optimal value of problem (4.3), i.e., we can write $\vartheta(\boldsymbol{x}) = \inf_{\boldsymbol{\theta} \in \Theta} F(\boldsymbol{x}, \boldsymbol{g}(\boldsymbol{\theta}))$. Denote by $\bar{\boldsymbol{\theta}}(\boldsymbol{x})$ an optimal solution of problem (4.3), i.e. $\bar{\boldsymbol{\theta}}(\boldsymbol{x}) \in \arg\min_{\boldsymbol{\theta} \in \Theta} F(\boldsymbol{x}, \boldsymbol{g}(\boldsymbol{\theta}))$. By the definitions, we have that $\widehat{F} = \vartheta(\hat{\boldsymbol{\xi}}_n)$ and $\hat{\boldsymbol{\theta}}_n = \bar{\boldsymbol{\theta}}(\hat{\boldsymbol{\xi}}_n)$. Therefore it should not be surprising that asymptotic properties of the MDF test statistics and estimators are closely related to analytical properties of functions $\vartheta(\cdot)$ and $\bar{\boldsymbol{\theta}}(\cdot)$.

Suppose that the model holds, i.e., $\boldsymbol{\xi}_0 \in \Xi_0$ or equivalently $\boldsymbol{\xi}_0 = \boldsymbol{g}(\boldsymbol{\theta}_0)$ for some $\boldsymbol{\theta}_0 \in \Theta$. The second-order Taylor approximation of the discrepancy function, at the point $(\boldsymbol{x}, \boldsymbol{\xi}) = (\boldsymbol{\xi}_0, \boldsymbol{\xi}_0)$, can be written in the form (3.12). Suppose, further, that the set Ξ_0 can be approximated at the point $\boldsymbol{\xi}_0$ by a cone $\mathcal{T} \subset \mathbb{R}^m$ in the following sense:

$$\text{dist}(\boldsymbol{\xi}_0 + z, \Xi_0) = o(\|z\|), \quad z \in \mathcal{T}, \tag{5.2}$$

$$\text{dist}(\boldsymbol{\xi} - \boldsymbol{\xi}_0, \mathcal{T}) = o(\|\boldsymbol{\xi} - \boldsymbol{\xi}_0\|), \quad \boldsymbol{\xi} \in \Xi_0. \tag{5.3}$$

This definition of cone approximation goes back to Chernoff (1954). (By $\text{dist}(\boldsymbol{x}, A) := \inf_{z \in A} \|\boldsymbol{x} - \boldsymbol{z}\|$ we denote the distance from a point $\boldsymbol{x} \in \mathbb{R}^m$ to a set $A \subset \mathbb{R}^m$. A set $\mathcal{T} \subset \mathbb{R}^m$ is said to be a *cone* if for any $z \in \mathcal{T}$ and $t \geqslant 0$ it follows that $tz \in \mathcal{T}$.)

In particular, if Ξ_0 is a smooth manifold near $\boldsymbol{\xi}_0$, then it is approximated at $\boldsymbol{\xi}_0$ by a *linear space* referred to as its tangent space at $\boldsymbol{\xi}_0$. Suppose that $\boldsymbol{\theta}_0$ is an *interior* point of Θ and $\boldsymbol{\theta}_0$ is locally regular (see Definition 2.2). Denote $\boldsymbol{\Delta}_0 := \boldsymbol{\Delta}(\boldsymbol{\theta}_0)$. Then the image $\boldsymbol{g}(\mathcal{N}) := \{\boldsymbol{\xi}: \boldsymbol{\xi} = \boldsymbol{g}(\boldsymbol{\theta}), \boldsymbol{\theta} \in \mathcal{N}\}$ of the set Θ restricted to a neighborhood $\mathcal{N} \subset \Theta$ of

$\boldsymbol{\theta}_0$, is a smooth manifold with the tangent space at $\boldsymbol{\xi}_0$ given by

$$\mathcal{T} = \{\boldsymbol{\zeta} = \boldsymbol{\Delta}_0\boldsymbol{\beta}\colon\ \boldsymbol{\beta} \in \mathbb{R}^q\}. \tag{5.4}$$

Of course, $\boldsymbol{g}(\mathcal{N})$ is a subset of Ξ_0, restricted to a neighborhood of $\boldsymbol{\xi}_0$. The asymptotic analysis is local in nature. Therefore, there is no loss of generality here by restricting the set Θ to a neighborhood of $\boldsymbol{\theta}_0$.

DEFINITION 5.1. A point $\boldsymbol{\theta}_0 \in \Theta$ is said to be *regular* if $\boldsymbol{\theta}_0$ is locally regular and there exist a neighborhood $\mathcal{V}$ of $\boldsymbol{\xi}_0 = \boldsymbol{g}(\boldsymbol{\theta}_0)$ and a neighborhood $\mathcal{N} \subset \Theta$ of $\boldsymbol{\theta}_0$ such that $\Xi_0 \cap \mathcal{V} = \boldsymbol{g}(\mathcal{N})$.

In other words, regularity of $\boldsymbol{\theta}_0$ ensures that local structure of Ξ_0 near $\boldsymbol{\xi}_0$ is provided by the mapping $\boldsymbol{g}(\boldsymbol{\theta})$ defined in a neighborhood of $\boldsymbol{\theta}_0$. Regularity of $\boldsymbol{\theta}_0$ implies that Ξ_0 is a smooth manifold near $\boldsymbol{\xi}_0$ and is approximated at $\boldsymbol{\xi}_0$ by its tangent space $\mathcal{T}$ of the form (5.4).

In particular, if condition (4.4) holds and $\boldsymbol{\Delta}(\boldsymbol{\theta})$ has full column rank q for all $\boldsymbol{\theta}$ in a neighborhood of $\boldsymbol{\theta}_0$ (i.e., point $\boldsymbol{\theta}_0$ is locally regular of rank q), then Ξ_0 is a smooth manifold near $\boldsymbol{\xi}_0$ and its tangent space at $\boldsymbol{\xi}_0$ is given by (5.4). Note that $\boldsymbol{g}(\mathcal{N})$ is a smooth manifold even if the rank of the Jacobian matrix $\boldsymbol{\Delta}_0$ is less than q provided that the local regularity condition holds. Note also that if the tangent space $\mathcal{T}$ is given in the form (5.4), then its dimension, $\dim(\mathcal{T})$, is equal to the rank of the Jacobian matrix $\boldsymbol{\Delta}_0$, i.e., $\dim(\mathcal{T}) = \operatorname{rank}(\boldsymbol{\Delta}_0)$.

REMARK 3. Let us remark at this point that if the point $\boldsymbol{\theta}_0$ is a *boundary* point of Θ, then under certain regularity conditions the set Θ can be approximated at $\boldsymbol{\theta}_0$ by a cone $\mathcal{C} \subset \mathbb{R}^q$, rather than a linear space, and consequently Ξ_0 can be approximated by the cone

$$\mathcal{T} = \{\boldsymbol{\zeta} = \boldsymbol{\Delta}_0\boldsymbol{\beta}\colon\ \boldsymbol{\beta} \in \mathcal{C}\}. \tag{5.5}$$

Later we will discuss implications of this to the asymptotics of the MDF estimators (see Section 5.4).

The above discussion suggests the following approximation of the optimal value function $\vartheta(\boldsymbol{x})$ near $\boldsymbol{\xi}_0$ (Shapiro, 1985c, Lemma 3.1).

PROPOSITION 5.1. *Suppose that the set Ξ_0 can be approximated at $\boldsymbol{\xi}_0 \in \Xi_0$ by a cone $\mathcal{T} \subset \mathbb{R}^m$. Then*

$$\vartheta(\boldsymbol{\xi}_0 + \boldsymbol{z}) = \min_{\boldsymbol{\zeta}\in\mathcal{T}} (\boldsymbol{z} - \boldsymbol{\zeta})'\boldsymbol{V}_0(\boldsymbol{z} - \boldsymbol{\zeta}) + o(\|\boldsymbol{z}\|^2). \tag{5.6}$$

Suppose, further, that the cone $\mathcal{T}$ actually is a linear space, i.e., Ξ_0 can be approximated at $\boldsymbol{\xi}_0 \in \Xi_0$ by a linear space $\mathcal{T} \subset \mathbb{R}^m$. Then the main (first) term in the right-hand side of (5.6) is a quadratic function of $\boldsymbol{z}$ and can be written as $\boldsymbol{z}'\boldsymbol{Q}\boldsymbol{z}$ for some symmetric positive semidefinite matrix $\boldsymbol{Q}$. In particular, if the space $\mathcal{T}$ is given in the form (5.4), then this term can be written as

$$\min_{\boldsymbol{\beta}\in\mathbb{R}^q} (z - \boldsymbol{\Delta}_0\boldsymbol{\beta})' \boldsymbol{V}_0 (z - \boldsymbol{\Delta}_0\boldsymbol{\beta}) = z'\boldsymbol{Q}z, \tag{5.7}$$

where $\boldsymbol{Q} = \boldsymbol{V}_0 - \boldsymbol{V}_0\boldsymbol{\Delta}_0(\boldsymbol{\Delta}_0'\boldsymbol{V}_0\boldsymbol{\Delta}_0)^-\boldsymbol{\Delta}_0'\boldsymbol{V}_0$ (by $\boldsymbol{A}^-$ we denote a generalized inverse of matrix $\boldsymbol{A}$). It is also possible to write this matrix in the form

$$\boldsymbol{Q} = \boldsymbol{\Delta}_c\big(\boldsymbol{\Delta}_c'\boldsymbol{V}_0^{-1}\boldsymbol{\Delta}_c\big)^{-1}\boldsymbol{\Delta}_c', \tag{5.8}$$

where $\boldsymbol{\Delta}_c$ is an orthogonal complement of $\boldsymbol{\Delta}_0$, i.e., $\boldsymbol{\Delta}_c$ is an $m \times (m - \operatorname{rank}(\boldsymbol{\Delta}_0))$ matrix of full column rank such that $\boldsymbol{\Delta}_c'\boldsymbol{\Delta}_0 = 0$. This follows by the standard theory of linear models (see, e.g., (Seber, 1977, Sections 3.6 and 3.8)). Note that $\operatorname{rank}(\boldsymbol{Q}) = m - \operatorname{rank}(\boldsymbol{\Delta}_0)$.

We already discussed continuity properties of $\bar{\boldsymbol{\theta}}(\cdot)$ in Section 4. Suppose that the model is (globally) identified at $\boldsymbol{\theta}_0$, and hence $\bar{\boldsymbol{\theta}}(\boldsymbol{\xi}_0) = \boldsymbol{\theta}_0$ and is defined uniquely. Then under mild regularity conditions we have that $\bar{\boldsymbol{\theta}}(\cdot)$ is continuous at $\boldsymbol{\xi}_0$. Moreover, we have the following result (Shapiro, 1985c, Lemma 3.1).

PROPOSITION 5.2. *Suppose that the set Θ can be approximated at $\boldsymbol{\theta}_0 \in \Theta$ by a convex cone $\mathcal{C} \subset \mathbb{R}^q$, the Jacobian matrix $\boldsymbol{\Delta}_0$ has full column rank q and $\bar{\boldsymbol{\theta}}(\cdot)$ is continuous at $\boldsymbol{\xi}_0 = \boldsymbol{g}(\boldsymbol{\theta}_0)$. Then*

$$\bar{\boldsymbol{\theta}}(\boldsymbol{\xi}_0 + z) = \boldsymbol{\theta}_0 + \bar{\boldsymbol{\beta}}(z) + o(\|z\|), \tag{5.9}$$

where $\bar{\boldsymbol{\beta}}(z)$ is the optimal solution of the problem

$$\min_{\boldsymbol{\beta}\in\mathcal{C}} (z - \boldsymbol{\Delta}_0\boldsymbol{\beta})' \boldsymbol{V}_0 (z - \boldsymbol{\Delta}_0\boldsymbol{\beta}). \tag{5.10}$$

Note that since the approximating cone $\mathcal{C}$ is assumed to be *convex*, $\operatorname{rank}(\boldsymbol{\Delta}_0) = q$ and the matrix $\boldsymbol{V}_0$ is positive definite, the minimizer $\bar{\boldsymbol{\beta}}(z)$ is unique for any $z \in \mathbb{R}^m$. If, moreover, the point $\boldsymbol{\theta}_0$ is an interior point of Θ and hence $\mathcal{C} = \mathbb{R}^q$, then $\bar{\boldsymbol{\beta}}(\cdot)$ is a linear function and can be written explicitly as

$$\bar{\boldsymbol{\beta}}(z) = (\boldsymbol{\Delta}_0'\boldsymbol{V}_0\boldsymbol{\Delta}_0)^{-1}\boldsymbol{\Delta}_0'\boldsymbol{V}_0 z. \tag{5.11}$$

Now if $\boldsymbol{\xi}_0 \notin \Xi_0$, then the analysis becomes considerably more involved. It will be beyond the scope of this paper to give a detailed description of such theory. We refer the interested reader to Bonnans and Shapiro (2000) for a thorough development of that theory. We give below some, relatively simple, results which will be relevant for the statistical inference. In the optimization literature the following result, giving a first order approximation of the optimal value function, is often referred to as Danskin Theorem (Danskin, 1967).

PROPOSITION 5.3. *Let $\mathfrak{S}$ be the set of optimal solutions of problem* (4.1) *for $\boldsymbol{x} = \boldsymbol{\xi}_0$. Then*

$$\vartheta(\boldsymbol{\xi}_0 + z) = \vartheta(\boldsymbol{\xi}_0) + \min_{\xi\in\mathfrak{S}} \mathrm{g}_\xi' z + o(\|z\|), \tag{5.12}$$

where $\mathfrak{g}_{\boldsymbol{\xi}} := \partial F(\boldsymbol{\xi}_0, \boldsymbol{\xi})/\partial \boldsymbol{x}$. *In particular, if problem* (4.1) *has unique optimal solution* $\boldsymbol{\xi}^*$ *for* $\boldsymbol{x} = \boldsymbol{\xi}_0$, *then*

$$\frac{\partial \vartheta(\boldsymbol{\xi}_0)}{\partial \boldsymbol{x}} = \frac{\partial F(\boldsymbol{\xi}_0, \boldsymbol{\xi}^*)}{\partial \boldsymbol{x}}. \tag{5.13}$$

Of course, if $\boldsymbol{\xi}_0 \in \Xi_0$, then problem (4.1) has unique optimal solution $\boldsymbol{\xi}^* = \boldsymbol{\xi}_0$ and hence $\partial \vartheta(\boldsymbol{\xi}_0)/\partial \boldsymbol{x} = 0$. The following result is a consequence of the Implicit Function Theorem (Shapiro, 1983, Theorem 4.2).

PROPOSITION 5.4. *Suppose that*:

(i) *for* $\boldsymbol{x} = \boldsymbol{\xi}_0$ *problem* (4.3) *has unique optimal solution* $\boldsymbol{\theta}^*$,
(ii) *the point* $\boldsymbol{\theta}^*$ *is an interior point of* Θ,
(iii) $\bar{\boldsymbol{\theta}}(\cdot)$ *is continuous at* $\boldsymbol{\theta}^*$,
(iv) *the Hessian matrix* $\boldsymbol{H}_{\theta\theta} := \partial^2 F(\boldsymbol{\xi}_0, \boldsymbol{g}(\boldsymbol{\theta}^*))/\partial\boldsymbol{\theta}\partial\boldsymbol{\theta}'$ *is nonsingular.*

Then $\bar{\boldsymbol{\theta}}(\cdot)$ *is continuously differentiable and* $\vartheta(\cdot)$ *is twice continuously differentiable at* $\boldsymbol{\xi}_0$, *and*

$$\frac{\partial \bar{\boldsymbol{\theta}}(\boldsymbol{\xi}_0)}{\partial \boldsymbol{x}} = -\boldsymbol{H}_{\theta\theta}^{-1}\boldsymbol{H}_{\theta x}, \tag{5.14}$$

$$\frac{\partial^2 \vartheta(\boldsymbol{\xi}_0)}{\partial \boldsymbol{x}\partial \boldsymbol{x}'} = \boldsymbol{H}_{xx} - \boldsymbol{H}_{\theta x}'\boldsymbol{H}_{\theta\theta}^{-1}\boldsymbol{H}_{\theta x}, \tag{5.15}$$

where $\boldsymbol{H}_{\theta x} := \partial^2 F(\boldsymbol{\xi}_0, \boldsymbol{g}(\boldsymbol{\theta}^*))/\partial\boldsymbol{\theta}\partial\boldsymbol{x}'$ *and* $\boldsymbol{H}_{xx} := \partial^2 F(\boldsymbol{\xi}_0, \boldsymbol{g}(\boldsymbol{\theta}^*))/\partial\boldsymbol{x}\partial\boldsymbol{x}'$.

REMARK 4. If the model holds, and hence $\boldsymbol{\theta}^* = \boldsymbol{\theta}_0$, then $\boldsymbol{H}_{\theta\theta} = 2\boldsymbol{\Delta}_0'\boldsymbol{V}_0\boldsymbol{\Delta}_0$, $\boldsymbol{H}_{\theta x} = -2\boldsymbol{\Delta}_0'\boldsymbol{V}_0$ and $\boldsymbol{H}_{xx} = 2\boldsymbol{V}_0$ (compare with (4.7)). In that case formula (5.14) gives the same derivatives as (5.9) and (5.11), and (5.15) is equivalent to (5.7), and these formulas involve only first-order derivatives (i.e., the Jacobian matrix) of $\boldsymbol{g}(\cdot)$. On the other hand, if the model does not hold, and hence $\boldsymbol{\theta}^* \neq \boldsymbol{\theta}_0$, then these derivatives involve *second-order* derivatives of $\boldsymbol{g}(\cdot)$. Note also that the Hessian matrix $\boldsymbol{H}_{\theta\theta}$ can be nonsingular only if the Jacobian matrix $\boldsymbol{\Delta}(\boldsymbol{\theta}^*)$ has full column rank q and hence the model is locally identified at $\boldsymbol{\theta}^*$.

5.1. Asymptotics of MDF test statistics

Suppose that the model holds, i.e., $\boldsymbol{\xi}_0 \in \Xi_0$. Since $\widehat{F} = \vartheta(\hat{\boldsymbol{\xi}}_n)$, we obtain from the approximation (5.6) the following asymptotic expansion of the MDF test statistic (under the null hypothesis):

$$n\widehat{F} = \min_{\boldsymbol{\zeta}\in\mathcal{T}}(\boldsymbol{Z}_n - \boldsymbol{\zeta})'\boldsymbol{V}_0(\boldsymbol{Z}_n - \boldsymbol{\zeta}) + o_p(1). \tag{5.16}$$

Recall that $\boldsymbol{Z}_n \Rightarrow N(\boldsymbol{0}, \boldsymbol{\Gamma})$. It follows that

$$n\widehat{F} \Rightarrow \min_{\boldsymbol{\zeta}\in\mathcal{T}}(\boldsymbol{Z} - \boldsymbol{\zeta})'\boldsymbol{V}_0(\boldsymbol{Z} - \boldsymbol{\zeta}), \tag{5.17}$$

where $\boldsymbol{Z}$ is a random vector having normal distribution $N(\mathbf{0}, \boldsymbol{\Gamma})$. The optimal value of the right-hand side of (5.17) is a quadratic function of $\boldsymbol{Z}$. Under certain conditions this quadratic function $\boldsymbol{Z}'\boldsymbol{Q}\boldsymbol{Z}$ has a chi-square distribution (see, e.g., (Seber, 1977, Section 2.4)). In particular, this holds if $\boldsymbol{V}_0 = \boldsymbol{\Gamma}^{-1}$. As it was discussed earlier, nonsingularity of the covariance matrix $\boldsymbol{\Gamma}$ depends on a choice of the space where the saturated model is defined. In applications it is often convenient to take a larger space in which case $\boldsymbol{\Gamma}$ becomes singular. It is said that the discrepancy function is *correctly specified* if $\boldsymbol{V}_0$ is equal to a generalized inverse of $\boldsymbol{\Gamma}$, that is, $\boldsymbol{\Gamma}\boldsymbol{V}_0\boldsymbol{\Gamma} = \boldsymbol{\Gamma}$. Of course, if $\boldsymbol{\Gamma}$ is nonsingular, then this is the same as $\boldsymbol{V}_0 = \boldsymbol{\Gamma}^{-1}$. As it was mentioned earlier we assume that the asymptotic covariance matrix $\boldsymbol{\Gamma}$ has the maximal rank, e.g., in the case of covariance structures we assume that $\text{rank}(\boldsymbol{\Gamma}) = p(p+1)/2$. It follows then that each column vector of the Jacobian matrix $\boldsymbol{\Delta}(\boldsymbol{\theta})$ is contained in the linear space generated by columns of $\boldsymbol{\Gamma}$.

We have the following result (Browne, 1982; Shapiro, 1986) giving asymptotics of the null distribution of the MDF test statistic. Recall Definition 5.1 of a regular point.

THEOREM 5.1. *Suppose that the model holds, the discrepancy function is correctly specified and the point $\boldsymbol{\theta}_0$ is regular (and hence the set Ξ_0 is approximated at $\boldsymbol{\xi}_0 = \boldsymbol{g}(\boldsymbol{\theta}_0)$ by a linear space $\mathcal{T}$ of the form (5.4)). Then the MDF test statistic $n\widehat{F}$ converges in distribution to a (central) chi-square with*

$$\nu = \text{rank}(\boldsymbol{\Gamma}) - \dim(\mathcal{T}) = \text{rank}(\boldsymbol{\Gamma}) - \text{rank}(\boldsymbol{\Delta}_0) \tag{5.18}$$

degrees of freedom.

Suppose that the mapping $\boldsymbol{g}(\cdot)$ is analytic and let r be the *characteristic rank* of the model. The above results imply that if the discrepancy function is correctly specified, then under the null hypothesis, generically, the MDF test statistic $n\widehat{F}$ has asymptotically a chi-square distribution, with $\nu = \text{rank}(\boldsymbol{\Gamma}) - r$ degrees of freedom. Recall that "generically" means that this holds for almost every population value $\boldsymbol{\theta}_0 \in \Theta$ of the parameter vector. For example, consider the setting of covariance structures and suppose that the population distribution is normal. Then the covariance matrix $\boldsymbol{\Gamma}$ can be written in the form (5.1), has rank $p(p+1)/2$ and matrix $\boldsymbol{V}_0 := \frac{1}{2}\boldsymbol{\Sigma}_0^{-1} \otimes \boldsymbol{\Sigma}_0^{-1}$ is its generalized inverse. It follows that for a normally distributed population, both discrepancy functions F_{ML} and F_{GLS}, defined in (3.5) and (3.9), respectively, are correctly specified. Therefore, we have that:

> Under the null hypothesis, generically, the MDF test statistics, associated with F_{ML} and F_{GLS}, are asymptotically chi-square distributed, with $\nu = p(p+1)/2 - r$ degrees of freedom, provided that the population distribution is normal.

It is possible to extend this basic result in various directions. Suppose now that the model does not hold, i.e., $\boldsymbol{\xi}_0 \notin \Xi_0$. Let $\boldsymbol{\xi}^*$ be a minimizer of $F(\boldsymbol{\xi}_0, \cdot)$ over Ξ_0, i.e., $\boldsymbol{\xi}^*$ is an optimal solution of problem (4.1) for $\boldsymbol{x} = \boldsymbol{\xi}_0$. Since the model does not hold, we have here that $\boldsymbol{\xi}^* \neq \boldsymbol{\xi}_0$. Suppose, however, that the population value $\boldsymbol{\xi}_0$ is close to the model set Ξ_0, i.e., there is no big difference between $\boldsymbol{\xi}^*$ and $\boldsymbol{\xi}_0$. We can employ approximation (5.6) at the point $\boldsymbol{\xi}^*$, instead of $\boldsymbol{\xi}_0$, by taking $\boldsymbol{V}_0 := \boldsymbol{V}(\boldsymbol{\xi}^*, \boldsymbol{\xi}^*)$, i.e., making the second-order Taylor expansion of the discrepancy function at the point $(\boldsymbol{x}, \boldsymbol{\xi}) = (\boldsymbol{\xi}^*, \boldsymbol{\xi}^*)$, and using the tangent space $\mathcal{T}$ at $\boldsymbol{\xi}^*$. We obtain the following approximation of the MDF statistic:

$$n\widehat{F} = \min_{\zeta \in \mathcal{T}} (\boldsymbol{Z}_n^* - \boldsymbol{\zeta})' \boldsymbol{V}_0 (\boldsymbol{Z}_n^* - \boldsymbol{\zeta}) + o\big(\|\boldsymbol{Z}_n^*\|^2\big), \tag{5.19}$$

where

$$\boldsymbol{Z}_n^* := n^{1/2}(\hat{\boldsymbol{\xi}}_n - \boldsymbol{\xi}^*) = \underbrace{n^{1/2}(\hat{\boldsymbol{\xi}}_n - \boldsymbol{\xi}_0)}_{\boldsymbol{Z}_n} + \underbrace{n^{1/2}(\boldsymbol{\xi}_0 - \boldsymbol{\xi}^*)}_{\boldsymbol{\mu}_n}.$$

Recall that it is assumed that $\boldsymbol{Z}_n \Rightarrow N(\boldsymbol{0}, \boldsymbol{\Gamma})$. On the other hand, as n tends to infinity, the "deterministic" part $\boldsymbol{\mu}_n := n^{1/2}(\boldsymbol{\xi}_0 - \boldsymbol{\xi}^*)$ of $\boldsymbol{Z}_n^*$ grows indefinitely. However, the quadratic approximation, given by the right-hand side of (5.19), could be reasonable if the "stochastic" part $\boldsymbol{Z}_n$ is bigger than the "deterministic" part $\boldsymbol{\mu}_n$ (we will discuss this in more details later). In order to formulate this in a mathematically rigorous way, we make the so-called assumption of a *sequence of local alternatives*. That is, we assume that there is a sequence $\boldsymbol{\xi}_0 = \boldsymbol{\xi}_{0,n}$ of population values (local alternatives) converging to a point $\boldsymbol{\xi}^* \in \Xi_0$ such that $n^{1/2}(\boldsymbol{\xi}_{0,n} - \boldsymbol{\xi}^*)$ converges to a (deterministic) vector $\boldsymbol{\mu}$ (this assumption is often referred to as *Pitman drift*). It follows then that $\boldsymbol{Z}_n^* \Rightarrow N(\boldsymbol{\mu}, \boldsymbol{\Gamma})$ and the remainder term $o(\|\boldsymbol{Z}_n^*\|^2)$ in (5.19) converges in probability to zero. If, moreover, $\boldsymbol{V}_0 = \boldsymbol{\Gamma}^-$, then the quadratic term in (5.19) converges in distribution to noncentral chi-square with the same degrees of freedom ν and the noncentrality parameter

$$\delta = \min_{\zeta \in \mathcal{T}} (\boldsymbol{\mu} - \boldsymbol{\zeta})' \boldsymbol{V}_0 (\boldsymbol{\mu} - \boldsymbol{\zeta}). \tag{5.20}$$

Moreover, by (5.6) we have that

$$\delta = \lim_{n \to \infty} \Big[n \min_{\xi \in \Xi_0} F(\boldsymbol{\xi}_{0,n}, \boldsymbol{\xi}) \Big]. \tag{5.21}$$

This leads to the following result (Shapiro, 1983, Theorem 5.5; Steiger et al., 1985).

THEOREM 5.2. *Suppose that the assumption of a sequence of local alternatives* (*Pitman drift*) *holds, the discrepancy function is correctly specified and the set* Ξ_0 *is approximated at* $\boldsymbol{\xi}^* = \boldsymbol{g}(\boldsymbol{\theta}^*)$ *by a linear space* $\mathcal{T}$ *generated by the columns of the matrix* $\boldsymbol{\Delta}^* = \boldsymbol{\Delta}(\boldsymbol{\theta}^*)$. *Then the MDF test statistic* $n\widehat{F}$ *converges in distribution to a noncentral chi-square with* $\nu = \operatorname{rank}(\boldsymbol{\Gamma}) - \dim(\mathcal{T})$ *degrees of freedom and the noncentrality parameter* δ *given in* (5.20) *or, equivalently,* (5.21).

From the practical point of view it is important to understand when the noncentral chi-square distribution gives a reasonable approximation of the true distribution of the MDF test statistics. By the analysis of Section 4, we have that $\widehat{F}$ converges w.p.1 to the value

$$F^* := \min_{\xi \in \Xi_0} F(\boldsymbol{\xi}_0, \boldsymbol{\xi}) = F(\boldsymbol{\xi}_0, \boldsymbol{\xi}^*). \tag{5.22}$$

Recall that $\vartheta(\boldsymbol{x})$ denotes the optimal value of problem (4.1), and hence $\widehat{F} = \vartheta(\hat{\boldsymbol{\xi}}_n)$ and $F^* = \vartheta(\boldsymbol{\xi}_0)$. Suppose that $\boldsymbol{\xi}^*$ is the *unique* minimizer of $F(\boldsymbol{\xi}_0, \cdot)$ over Ξ_0. Then we have by Danskin theorem (see Proposition 5.3) that $\partial\vartheta(\boldsymbol{\xi}_0)/\partial\boldsymbol{x} = \partial F(\boldsymbol{\xi}_0, \boldsymbol{\xi}^*)/\partial\boldsymbol{x}$. It follows that

$$n^{1/2}(\widehat{F} - F^*) = \mathfrak{g}_0' \boldsymbol{Z}_n + o_p(1), \tag{5.23}$$

where $\mathfrak{g}_0 := \frac{\partial F(\boldsymbol{x}, \boldsymbol{\xi}^*)}{\partial \boldsymbol{x}}|_{\boldsymbol{x}=\boldsymbol{\xi}_0}$, which in turn implies the following asymptotic result (Shapiro, 1983, Theorem 5.3).

THEOREM 5.3. *Suppose that $\boldsymbol{\xi}^*$ is the unique minimizer of $F(\boldsymbol{\xi}_0, \cdot)$ over Ξ_0. Then*

$$n^{1/2}(\widehat{F} - F^*) \Rightarrow N(0, \mathfrak{g}_0' \boldsymbol{\Gamma} \mathfrak{g}_0). \tag{5.24}$$

If the model holds, then $F^* = 0$ and $\mathfrak{g}_0 = \boldsymbol{0}$. In that case the asymptotic result (5.24) degenerates into the trivial statement that $n^{1/2}\widehat{F}$ converges in probability to zero. And, indeed, as it was discussed above, under the null hypothesis one needs to scale $\widehat{F}$ by the factor of n, instead of $n^{1/2}$, in order to get meaningful asymptotics. However, as the distance between the population value $\boldsymbol{\xi}_0$ and the model set Ξ_0 becomes larger, the noncentral chi-square distribution approximation deteriorates and the normal distribution, with mean F^* and variance $n^{-1}\mathfrak{g}_0' \boldsymbol{\Gamma} \mathfrak{g}_0$, could become a better approximation of the distribution of $\widehat{F}$. The noncentral chi-square approximation is based on the distribution of the quadratic form

$$\begin{aligned}&\min_{\boldsymbol{\zeta} \in \mathcal{T}} (\boldsymbol{Z}_n^* - \boldsymbol{\zeta})' \boldsymbol{V}_0 (\boldsymbol{Z}_n^* - \boldsymbol{\zeta}) \\ &\quad = \boldsymbol{Z}_n^{*\prime} \boldsymbol{Q} \boldsymbol{Z}_n^* = \boldsymbol{Z}_n' \boldsymbol{Q} \boldsymbol{Z}_n + \boldsymbol{\mu}_n' \boldsymbol{Q} \boldsymbol{\mu}_n + 2\boldsymbol{\mu}_n' \boldsymbol{Q} \boldsymbol{Z}_n.\end{aligned} \tag{5.25}$$

Recall that $\boldsymbol{Z}_n^* = \boldsymbol{Z}_n + \boldsymbol{\mu}_n$ and $\boldsymbol{\mu}_n = n^{1/2}(\boldsymbol{\xi}_0 - \boldsymbol{\xi}^*)$. The first term, in the right-hand side of (5.25), has approximately a central chi-square distribution with $\nu = \operatorname{rank}(\boldsymbol{\Gamma}) - \dim(\mathcal{T})$ degrees of freedom. Suppose that $\boldsymbol{\xi}_0$ is close to Ξ_0. By (5.6), the second term in the right-hand side of (5.25) can be approximated as follows

$$\boldsymbol{\mu}_n' \boldsymbol{Q} \boldsymbol{\mu}_n = n \min_{\boldsymbol{\zeta} \in \mathcal{T}} (\boldsymbol{\xi}_0 - \boldsymbol{\xi}^* - \boldsymbol{\zeta})' \boldsymbol{V}_0 (\boldsymbol{\xi}_0 - \boldsymbol{\xi}^* - \boldsymbol{\zeta}) \approx nF^*. \tag{5.26}$$

Recall that, by (5.21), nF^* is approximately equal to the noncentrality parameter δ. We also have that

$$\begin{aligned}\mathfrak{g}_0 &= \left.\frac{\partial F(\boldsymbol{x}, \boldsymbol{\xi}^*)}{\partial \boldsymbol{x}}\right|_{\boldsymbol{x}=\boldsymbol{\xi}_0} \approx \left.\frac{\partial (\boldsymbol{x} - \boldsymbol{\xi}^*)' \boldsymbol{V}_0 (\boldsymbol{x} - \boldsymbol{\xi}^*)}{\partial \boldsymbol{x}}\right|_{\boldsymbol{x}=\boldsymbol{\xi}_0} \\ &= 2\boldsymbol{V}_0(\boldsymbol{\xi}_0 - \boldsymbol{\xi}^*).\end{aligned} \tag{5.27}$$

Moreover, by the first-order optimality conditions, the gradient vector $\mathfrak{g}_0$ is orthogonal to the space $\mathcal{T}$, and hence $\boldsymbol{V}_0(\boldsymbol{\xi}_0 - \boldsymbol{\xi}^*) \approx \boldsymbol{Q}(\boldsymbol{\xi}_0 - \boldsymbol{\xi}^*)$.

It follows that the sum of the second and third terms in the right-hand side of (5.25) has approximately a normal distribution with mean nF^* and variance $n\mathfrak{g}_0' \boldsymbol{\Gamma} \mathfrak{g}_0$. Therefore, for $\boldsymbol{\xi}_0$ close to Ξ_0 the difference between the noncentral chi-square and normal approximations, given in Theorems 5.2 and 5.3, respectively, is the first term in the right-hand side of (5.25). The expected value and the variance of a chi-square random variable with ν degrees of freedom is equal to ν and 2ν, respectively. Therefore, when the number of degrees of freedom $\nu = \operatorname{rank}(\boldsymbol{\Gamma}) - \dim(\mathcal{T})$ is bigger or comparable with the noncentrality parameter $\delta \approx nF^*$, the noncentral chi-square approximation, which is based on a second-order Taylor expansion at the point $\boldsymbol{\xi}^*$, should be better than the normal approximation, which is based on a first-order approximation at $\boldsymbol{\xi}_0$. On the other hand, if δ is significantly bigger than ν, then the first term in the right-hand side of (5.25) becomes negligible and the normal approximation could be reasonable. This is in agreement with the property that a noncentral chi-square distribution, with ν degrees of freedom and noncentrality parameter δ, becomes approximately normal if δ is much bigger than ν. In such a case the normal approximation can be used to construct a confidence interval, for F^*, of the form $\widehat{F} \pm \kappa\hat{\sigma}_F$. Here $\hat{\sigma}_F$ is an estimate of the standard deviation of $\widehat{F}$ and κ is a critical value. Recall that the expected value and variance of a noncentral chi-square random variable, with ν degrees of freedom and noncentrality parameter δ, is $\nu + \delta$ and $2\nu + 4\delta$, respectively. Therefore, the normal approximation could be reasonable if

$$\frac{n\widehat{F} - \nu}{\sqrt{4n\widehat{F} + 2\nu}} \geqslant \kappa, \tag{5.28}$$

where κ is a critical value, say $\kappa = 3$.

Let us finally remark that from a theoretical point of view one can obtain a better approximation of the distribution of the MDF test statistic by using a second-order Taylor expansion of the optimal value function at the population point $\boldsymbol{\xi}_0$. The corresponding first- and second-order derivatives are given in (5.13) and (5.15), respectively, provided that the optimal solution $\boldsymbol{\xi}^*$ is unique. Note, however, that in practical applications this will require an accurate estimation of the corresponding first- and second-order derivatives which could be a problem.

5.2. *Nested models*

Suppose now that we have two models $\Xi_1 \subset \Xi$ and $\Xi_2 \subset \Xi$ for the same parameter vector $\boldsymbol{\xi}$. It is said that the second model is *nested*, within the first model, if Ξ_2 is a subset of Ξ_1, i.e., $\Xi_2 \subset \Xi_1$. We refer to the models associated with the sets Ξ_1 and Ξ_2 as *full* and *restricted* models, respectively. If Ξ_1 is given in the parametric form

$$\Xi_1 := \{\boldsymbol{\xi} \in \Xi \colon \boldsymbol{\xi} = \boldsymbol{g}(\boldsymbol{\theta}), \boldsymbol{\theta} \in \Theta_1\}, \tag{5.29}$$

i.e., the full model is structural, then it is natural to define a nested model by restricting the parameter space Θ_1 to a subset Θ_2. Typically the subset $\Theta_2 \subset \Theta_1$ is defined by imposing constraints on the parameter vector $\boldsymbol{\theta}$. In this section we discuss asymptotics of the MDF test statistics $n\widehat{F}_i$, $i = 1, 2$, where

$$\widehat{F}_i := \min_{\boldsymbol{\xi} \in \Xi_i} F(\hat{\boldsymbol{\xi}}_n, \boldsymbol{\xi}). \tag{5.30}$$

Suppose that the population value $\boldsymbol{\xi}_0 \in \Xi_2$, i.e., the restricted model holds. Suppose, further, that the sets Ξ_1 and Ξ_2 are approximated at $\boldsymbol{\xi}_0$ by linear spaces $\mathcal{T}_1$ and $\mathcal{T}_2$, respectively. This holds if both Ξ_1 and Ξ_2 are smooth manifolds near $\boldsymbol{\xi}_0$ with respective tangent spaces $\mathcal{T}_1$ and $\mathcal{T}_2$. Note that $\mathcal{T}_2 \subset \mathcal{T}_1$ since $\Xi_2 \subset \Xi_1$. We have then (see (5.16)) that

$$n\widehat{F}_i = \min_{\boldsymbol{\zeta}\in\mathcal{T}_i} (\boldsymbol{Z}_n - \boldsymbol{\zeta})'\boldsymbol{V}_0(\boldsymbol{Z}_n - \boldsymbol{\zeta}) + o_p(1), \quad i = 1, 2. \tag{5.31}$$

Suppose, further, that the discrepancy function is *correctly specified*. Then by the analysis of the previous section we have that $n\widehat{F}_i$, $i = 1, 2$, converges in distribution to a (central) chi-square with $\nu_i = \operatorname{rank}(\boldsymbol{\Gamma}) - \dim(\mathcal{T}_i)$ degrees of freedom. Moreover, it follows from the representation (5.31) that $n\widehat{F}_1$ and $n\widehat{F}_2 - n\widehat{F}_1$ are asymptotically independent of each other. The corresponding arguments are analogous to derivations of the statistical inference of linear constraints in the theory of linear models (e.g., Seber, 1977, Section 4.5.1). This can be extended to the setting of a sequence of local alternatives, where there is a sequence $\boldsymbol{\xi}_{0,n}$ of population values converging to a point $\boldsymbol{\xi}^* \in \Xi_2$ such that the following limits exist

$$\delta_i = \lim_{n\to\infty}\Big[n \min_{\boldsymbol{\xi}\in\Xi_i} F(\boldsymbol{\xi}_{0,n}, \boldsymbol{\xi})\Big], \quad i = 1, 2. \tag{5.32}$$

Then the following asymptotic results hold (Steiger et al., 1985).

THEOREM 5.4. *Suppose that the assumption of a sequence of local alternatives holds, the discrepancy function is correctly specified and the sets Ξ_i, $i = 1, 2$, are approximated at $\boldsymbol{\xi}^* \in \Xi_2$ by respective linear spaces $\mathcal{T}_i$. Then the following holds*:

(i) *the MDF test statistics $n\widehat{F}_i$ converge in distribution to noncentral chi-square with respective degrees of freedom $\nu_i = \operatorname{rank}(\boldsymbol{\Gamma}) - \dim(\mathcal{T}_i)$ and noncentrality parameter δ_i given in* (5.32),

(ii) *the statistic $n\widehat{F}_2 - n\widehat{F}_1$ converges in distribution to a noncentral chi-square with $\nu_2 - \nu_1$ degrees of freedom and noncentrality parameter $\delta_2 - \delta_1$,*

(iii) *the statistics $n\widehat{F}_1$ and $n\widehat{F}_2 - n\widehat{F}_1$ are asymptotically independent of each other,*

(iv) *the ratio statistic $((\widehat{F}_2 - \widehat{F}_1)/(\nu_2 - \nu_1))/(\widehat{F}_1/\nu_1)$ converges in distribution to doubly noncentral F-distribution with noncentrality parameters $\delta_2 - \delta_1$ and δ_1 and with $\nu_2 - \nu_1$ and ν_1 degrees of freedom.*

It is straightforward to extend the above result to a sequence of nested models. Also we have that $n\widehat{F}_2 = n\widehat{F}_1 + (n\widehat{F}_2 - n\widehat{F}_1)$. Recall that the variance of a noncentral chi-square random variable with ν degrees of freedom and noncentrality parameter δ is $2\nu + 4\delta$. Therefore, under the assumptions of the above theorem, the asymptotic covariance between $n\widehat{F}_1$ and $n\widehat{F}_2$ is equal to the asymptotic variance of $n\widehat{F}_1$, which is equal to $2\nu_1 + 4\delta_1$. Consequently, the asymptotic correlation between the MDF statistics $n\widehat{F}_1$ and $n\widehat{F}_2$ is equal to $\sqrt{(\nu_1 + 2\delta_1)/(\nu_2 + 2\delta_2)}$ (Steiger et al., 1985).

5.3. *Asymptotics of MDF estimators*

In this section we discuss asymptotics of the MDF estimator $\hat{\boldsymbol{\theta}}_n$. Suppose that the model holds and $\hat{\boldsymbol{\theta}}_n$ is a consistent estimator of the population value $\boldsymbol{\theta}_0 \in \Theta$ (see Section 4 and Proposition 4.2 in particular). Since $\hat{\boldsymbol{\theta}}_n = \bar{\boldsymbol{\theta}}(\hat{\boldsymbol{\xi}}_n)$, we have by (5.9) that, under the assumptions of Proposition 5.2,

$$n^{1/2}(\hat{\boldsymbol{\theta}}_n - \boldsymbol{\theta}_0) = \bar{\boldsymbol{\beta}}(\boldsymbol{Z}_n) + o_p(1). \tag{5.33}$$

Recall that $\bar{\boldsymbol{\beta}}(z)$ is the optimal solution of (5.10) and note that $\bar{\boldsymbol{\beta}}(\cdot)$ is positively homogeneous, i.e., $\bar{\boldsymbol{\beta}}(tz) = t\bar{\boldsymbol{\beta}}(z)$ for any z and $t \geqslant 0$. This leads to the following asymptotics of the MDF estimator (Browne, 1974; Shapiro, 1983).

THEOREM 5.5. *Suppose that the model holds, $\hat{\boldsymbol{\theta}}_n$ is a consistent estimator of $\boldsymbol{\theta}_0$, the set Θ is approximated at $\boldsymbol{\theta}_0$ by a convex cone $\mathcal{C}$ and* $\operatorname{rank}(\boldsymbol{\Delta}_0) = q$. *Then $n^{1/2}(\hat{\boldsymbol{\theta}}_n - \boldsymbol{\theta}_0) \Rightarrow \bar{\boldsymbol{\beta}}(\boldsymbol{Z})$, where $\boldsymbol{Z} \sim N(0, \boldsymbol{\Gamma})$. If, furthermore, $\boldsymbol{\theta}_0$ is an interior point of Θ, and hence $\mathcal{C} = \mathbb{R}^q$, then $n^{1/2}(\hat{\boldsymbol{\theta}}_n - \boldsymbol{\theta}_0)$ converges in distribution to normal with mean vector zero and covariance matrix*

$$\boldsymbol{\Pi} = (\boldsymbol{\Delta}_0' \boldsymbol{V}_0 \boldsymbol{\Delta}_0)^{-1} \boldsymbol{\Delta}_0' \boldsymbol{V}_0 \boldsymbol{\Gamma} \boldsymbol{V}_0 \boldsymbol{\Delta}_0 (\boldsymbol{\Delta}_0' \boldsymbol{V}_0 \boldsymbol{\Delta}_0)^{-1}. \tag{5.34}$$

Moreover, if the discrepancy function is correctly specified, then $\boldsymbol{\Pi} = (\boldsymbol{\Delta}_0' \boldsymbol{V}_0 \boldsymbol{\Delta}_0)^{-1}$.

In particular, if in the setting of covariance structures the population distribution is normal and the employed discrepancy function is normal-theory correctly specified, then the asymptotic covariance matrix of $n^{1/2}(\hat{\boldsymbol{\theta}}_n - \boldsymbol{\theta}_0)$ can be written as

$$\boldsymbol{\Pi}_N = 2\big[\boldsymbol{\Delta}_0'(\boldsymbol{\Sigma}_0^{-1} \otimes \boldsymbol{\Sigma}_0^{-1})\boldsymbol{\Delta}_0\big]^{-1}. \tag{5.35}$$

Note that since it is assumed that the asymptotic covariance matrix $\boldsymbol{\Gamma}$ has the maximal rank, and hence the linear space generated by columns of $\boldsymbol{\Delta}_0$ is included in the linear space generated by columns of $\boldsymbol{\Gamma}$, we have here that matrix $\boldsymbol{\Delta}_0' \boldsymbol{\Gamma}^- \boldsymbol{\Delta}_0$ is independent of a particular choice of the generalized inverse $\boldsymbol{\Gamma}^-$ and is positive definite. In particular, if the discrepancy function is correctly specified, then $\boldsymbol{\Delta}_0' \boldsymbol{\Gamma}^- \boldsymbol{\Delta}_0 = \boldsymbol{\Delta}_0' \boldsymbol{V}_0 \boldsymbol{\Delta}_0$. It is possible to show (Browne, 1974, Proposition 3) that the inequality

$$\boldsymbol{\Pi} \succeq (\boldsymbol{\Delta}_0' \boldsymbol{\Gamma}^- \boldsymbol{\Delta}_0)^{-1} \tag{5.36}$$

always holds. (For $q \times q$ symmetric matrices $\boldsymbol{A}$ and $\boldsymbol{B}$ the inequality $\boldsymbol{A} \succeq \boldsymbol{B}$ is understood in the Loewner sense, i.e., that matrix $\boldsymbol{A} - \boldsymbol{B}$ is positive semidefinite.) Basically this is the Gauss–Markov theorem. That is, for a correctly specified discrepancy function, the asymptotic covariance matrix of the corresponding MDF estimator attains its lower bound given by the right-hand side of (5.36). Therefore, for a correctly specified discrepancy function the corresponding MDF estimator is *asymptotically efficient* within the class of MDF estimators.

The above asymptotics of MDF estimators were derived under the assumption of identifiability of the model. If the model is overparameterized, then it does not make sense to talk about distribution of the MDF estimators since these estimators are not uniquely defined. However, even in the case of overparameterization some of the parameters could be defined uniquely. Therefore it makes sense to consider the following concept of *estimable functions* borrowed from the theory of linear models (e.g., Seber, 1977, Section 3.8.2).

DEFINITION 5.2. Consider a continuously differentiable function $a(\boldsymbol{\theta})$. We say that $\alpha = a(\boldsymbol{\theta})$ is an *estimable function* (of the parameter vector $\boldsymbol{\theta}$) if $\boldsymbol{\theta}_1, \boldsymbol{\theta}_2 \in \Theta$ and $\boldsymbol{g}(\boldsymbol{\theta}_1) = \boldsymbol{g}(\boldsymbol{\theta}_2)$ imply that $a(\boldsymbol{\theta}_1) = a(\boldsymbol{\theta}_2)$. If this holds in a neighborhood of a point $\boldsymbol{\theta}_0 \in \Theta$, we say α is *locally estimable*, near $\boldsymbol{\theta}_0$.

In the analysis of covariance structures the above concept of estimable functions was discussed in Shapiro (1986, p. 146). By using local reparameterization (see Proposition 2.1) it is possible to show the following.

If $\boldsymbol{\theta}_0$ is a locally regular interior point of Θ, then a parameter $\alpha = a(\boldsymbol{\theta})$ is locally estimable, near $\boldsymbol{\theta}_0$, iff vector $\partial a(\boldsymbol{\theta})/\partial \boldsymbol{\theta}'$ belongs to the linear space generated by rows of $\boldsymbol{\Delta}(\boldsymbol{\theta})$ for all $\boldsymbol{\theta}$ in a neighborhood of $\boldsymbol{\theta}_0$.

Consider now vector $\boldsymbol{a}(\boldsymbol{\theta}) = (a_1(\boldsymbol{\theta}), \dots, a_s(\boldsymbol{\theta}))$ of locally estimable parameters, near a population point $\boldsymbol{\theta}_0$. Suppose that $\boldsymbol{\theta}_0$ is locally regular and let $\boldsymbol{\alpha}_0 := \boldsymbol{a}(\boldsymbol{\theta}_0)$ and $\hat{\boldsymbol{\alpha}}_n := \boldsymbol{a}(\hat{\boldsymbol{\theta}}_n)$. Note that, by local estimability of $\boldsymbol{\alpha}$, the estimator $\hat{\boldsymbol{\alpha}}_n$ is defined uniquely for $\hat{\boldsymbol{\theta}}_n$ sufficiently close to $\boldsymbol{\theta}_0$. We have then that $n^{1/2}(\hat{\boldsymbol{\alpha}}_n - \boldsymbol{\alpha}_0)$ converges in distribution to normal with mean vector zero and covariance matrix

$$\boldsymbol{A}_0(\boldsymbol{\Delta}_0'\boldsymbol{V}_0\boldsymbol{\Delta}_0)^{-}\boldsymbol{\Delta}_0'\boldsymbol{V}_0\boldsymbol{\Gamma}\boldsymbol{V}_0\boldsymbol{\Delta}_0(\boldsymbol{\Delta}_0'\boldsymbol{V}_0\boldsymbol{\Delta}_0)^{-}\boldsymbol{A}_0', \tag{5.37}$$

where $\boldsymbol{A}_0 := \partial \boldsymbol{a}(\boldsymbol{\theta}_0)/\partial \boldsymbol{\theta}'$ is $s \times q$ Jacobian matrix. Note that because of the local estimability of $\boldsymbol{\alpha}$, we have that row vectors of $\boldsymbol{A}_0$ belong to the linear space generated by rows of $\boldsymbol{\Delta}_0$, and hence the expression in (5.37) does not depend on a particular choice of the generalized inverse of $\boldsymbol{\Delta}_0'\boldsymbol{V}_0\boldsymbol{\Delta}_0$. In particular, for correctly specified discrepancy function this expression becomes $\boldsymbol{A}_0(\boldsymbol{\Delta}_0'\boldsymbol{V}_0\boldsymbol{\Delta}_0)^{-}\boldsymbol{A}_0'$.

We can also consider a situation when the model does not hold. Under the assumptions of Proposition 5.4, in particular, that $\boldsymbol{\theta}^*$ is the unique optimal solution of problem (4.3), we have that $\hat{\boldsymbol{\theta}}_n$ converges w.p.1 to $\boldsymbol{\theta}^*$ and, by (5.14), that (Shapiro, 1983, Theorem 5.4):

$$n^{1/2}(\hat{\boldsymbol{\theta}}_n - \boldsymbol{\theta}^*) \Rightarrow N\big(0, \boldsymbol{H}_{\theta\theta}^{-1}\boldsymbol{H}_{\theta x}\boldsymbol{\Gamma}\boldsymbol{H}_{\theta x}'\boldsymbol{H}_{\theta\theta}^{-1}\big). \tag{5.38}$$

5.4. *The situation where the population value of the parameter vector lies on the boundary of the parameter space*

In the previous sections we discussed asymptotics of MDF test statistics and estimators under the assumption that the set Ξ_0 can be approximated, at a considered point, by a *linear* space $\mathcal{T}$. In this section we consider a situation when $\boldsymbol{\theta}_0$ is a boundary point of the set Θ, and as a consequence the set Ξ_0 should be approximated at $\boldsymbol{\xi}_0 = \boldsymbol{g}(\boldsymbol{\theta}_0)$ by a (convex) cone rather than a linear space. This may happen if the set Θ is defined by inequality constraints and some of these inequality constraints are active at the population point. For instance, in Example 2.1 (Factor Analysis model) the diagonal entries of the matrix $\boldsymbol{\Psi}$ should be nonnegative. If the population value $\boldsymbol{\Psi}_0$ has zero diagonal entries (i.e., some residual variances are zeros), then the corresponding value of the parameter vector can be viewed as lying on the boundary of the feasible region. One can think about more sophisticated examples where, for instance, it is hypothesized that some of the residual variances (i.e., diagonal entries of $\boldsymbol{\Psi}$) are bigger than the others, or that elements of matrix $\boldsymbol{\Lambda}$ are nonnegative. Statistical theory of parameter estimation under inequality type constraints is often referred to as order restricted statistical inference. The interested reader can be referred to the recent comprehensive monograph (Silvapulle and Sen, 2005) for a thorough treatment of that theory. We give below a few basic results which are relevant for our discussion.

Suppose that the model holds, the Jacobian matrix $\boldsymbol{\Delta}_0$ has full column rank q and the set Θ is approximated at $\boldsymbol{\theta}_0$ by convex cone $\mathcal{C}$. We have then that

$$n\widehat{F} \Rightarrow \min_{\boldsymbol{\beta}\in\mathcal{C}}(\mathbf{Z} - \boldsymbol{\Delta}_0\boldsymbol{\beta})'\boldsymbol{V}_0(\mathbf{Z} - \boldsymbol{\Delta}_0\boldsymbol{\beta}), \tag{5.39}$$

where $\mathbf{Z} \sim N(\mathbf{0}, \boldsymbol{\Gamma})$. Let us look at the optimal value of the minimization problem in the right-hand side of (5.39). Suppose, for the sake of simplicity, that $\boldsymbol{\Delta}_0$ has full column rank q. Then we can decompose this optimal value into a sum of two terms as follows

$$\begin{aligned}&\min_{\boldsymbol{\beta}\in\mathcal{C}}(\mathbf{Z} - \boldsymbol{\Delta}_0\boldsymbol{\beta})'\boldsymbol{V}_0(\mathbf{Z} - \boldsymbol{\Delta}_0\boldsymbol{\beta})\\ &\quad = (\mathbf{Z} - \boldsymbol{\Delta}_0\tilde{\boldsymbol{\beta}})'\boldsymbol{V}_0(\mathbf{Z} - \boldsymbol{\Delta}_0\tilde{\boldsymbol{\beta}}) + \min_{\boldsymbol{\beta}\in\mathcal{C}}(\tilde{\boldsymbol{\beta}} - \boldsymbol{\beta})'\boldsymbol{\Delta}_0'\boldsymbol{V}_0\boldsymbol{\Delta}_0(\tilde{\boldsymbol{\beta}} - \boldsymbol{\beta}),\end{aligned} \tag{5.40}$$

where $\tilde{\boldsymbol{\beta}} = (\boldsymbol{\Delta}_0'\boldsymbol{V}_0\boldsymbol{\Delta}_0)^{-1}\boldsymbol{\Delta}_0'\boldsymbol{V}_0\mathbf{Z}$ is the corresponding unconstrained minimizer (compare with (5.11)). The term in the left-hand side of (5.40) can be viewed as a squared distance from $\mathbf{Z}$ to the cone $\{\boldsymbol{\Delta}_0\boldsymbol{\beta}\colon \boldsymbol{\beta} \in \mathcal{C}\}$, where the distance is defined with respect to the weight matrix $\boldsymbol{V}_0$. The first term in the right-hand side of (5.40) is the corresponding unconstrained minimum over $\boldsymbol{\beta} \in \mathbb{R}^q$, and can be viewed as the squared distance from $\mathbf{Z}$ to the linear space generated by $\boldsymbol{\Delta}_0$, and the second term can be considered as the squared distance from $\tilde{\boldsymbol{\beta}}$ to $\mathcal{C}$. In a sense the above decomposition (5.40) is just the Pythagoras Theorem. Suppose, further, that the discrepancy function is correctly specified. Recall that it is assumed that the asymptotic covariance matrix $\boldsymbol{\Gamma}$ has maximal rank. By reducing the saturated space, if necessary, we can assume here that $\boldsymbol{\Gamma}$ is nonsingular, and hence "correctly specified" means that $\boldsymbol{V}_0 = \boldsymbol{\Gamma}^{-1}$. It follows then that $\tilde{\boldsymbol{\beta}} \sim N(\mathbf{0}, (\boldsymbol{\Delta}_0'\boldsymbol{V}_0\boldsymbol{\Delta}_0)^{-1})$.

Assuming that the model holds and $\boldsymbol{V}_0 = \boldsymbol{\Gamma}^{-1}$, we have that the first term in the right-hand side of (5.40) has chi-square distribution, with $\nu = m-q$ degrees of freedom, and is distributed independently of the second term. The second term in the right-hand side of (5.40) follows a *mixture* of chi-square distributions (such distributions are called *chi-bar-squared* distributions). With various degrees of generality this result was derived in (Bartholomew, 1961; Kudô, 1963; Nüesch, 1966), it was shown in (Shapiro, 1985d) that this holds for any *convex* cone $\mathcal{C}$. Denote by $n\widetilde{F}$ the corresponding *unconstrained* MDF test statistic, i.e.,

$$\widetilde{F} := \min_{\boldsymbol{\theta}\in\mathbb{R}^q} F\big(\hat{\boldsymbol{\xi}}_n, \boldsymbol{g}(\boldsymbol{\theta})\big).$$

We have that $n\widetilde{F}$ is asymptotically equivalent to the first term in the right-hand side of (5.40). Under the above assumptions we obtain the following results:

(i) The unrestricted MDF test statistic $n\widetilde{F}$ converges in distribution to chi-square with $\nu = m - q$ degrees of freedom and is asymptotically independent of the difference statistic $n\widehat{F} - n\widetilde{F}$.

(ii) The difference statistic $n\widehat{F} - n\widetilde{F}$ is asymptotically equivalent to the second term in the right hand side of (5.40) and converges in distribution to a mixture of chi-square distributions, that is,

$$\lim_{n\to\infty} \operatorname{Prob}\{n\widehat{F} - n\widetilde{F} \geqslant c\} = \sum_{i=0}^{q} w_i \operatorname{Prob}\{\chi_i^2 \geqslant c\}, \tag{5.41}$$

where χ_i^2 is a chi-square random variable with i degrees of freedom, $\chi_0^2 \equiv 0$ and w_i are nonnegative weights such that $w_0 + \cdots + w_q = 1$.

Of course, the asymptotic distribution, given by the right-hand side of (5.41), depends on the weights w_i, which in turn depend on the covariance matrix of $\tilde{\boldsymbol{\beta}}$ and cone $\mathcal{C}$ (recall that, for correctly specified discrepancy function, the covariance matrix of $\tilde{\boldsymbol{\beta}}$ is $(\boldsymbol{\Delta}_0' \boldsymbol{V}_0 \boldsymbol{\Delta}_0)^{-1}$). A general property of these weights is that $\sum_{i=0}^{q}(-1)^i w_i = 0$ if at least two of these weights are nonzeros. If $\boldsymbol{\theta}_0$ is an interior point of Θ, and hence $\mathcal{C} = \mathbb{R}^q$, then $w_0 = 1$ and all other weights are zeros. In that case we have the same asymptotics of the MDF statistic $n\widehat{F}$ as given in Theorem 5.1. Often the set $\Theta \subset \mathbb{R}^q$ is defined by inequality constraints. Then, under mild regularity conditions (called *constraint qualifications*), the approximating cone $\mathcal{C}$ is obtained by linearizing the active at $\boldsymbol{\theta}_0$ inequality constraints. In particular, if only one inequality constraint is active at $\boldsymbol{\theta}_0$, then $\mathcal{C}$ is defined by one linear inequality constraint and hence is a half space of $\mathbb{R}^q$. In that case $w_0 = w_1 = 1/2$ and all other weights are zeros. If two inequality constraints are active at $\boldsymbol{\theta}_0$, then only weights w_0, w_1 and w_2 can be nonzeros, with $w_1 = 1/2$, etc. For a general discussion of how to calculate these weights we can refer, e.g., to (Shapiro, 1988; Silvapulle and Sen, 2005).

6. Asymptotic robustness of the MDF statistical inference

An important condition in the analysis of the previous section was the assumption of correct specification of the discrepancy function. In particular, the discrepancy functions F_{ML} and F_{GLS}, defined in (3.5) and (3.9), respectively, are motivated by the assumption that the underlying population has a normal distribution and are correctly specified in that case. Nevertheless, these discrepancy functions are often applied in situations where the normality assumption has no justification or even can be clearly wrong. It turns out, however, that the asymptotic chi-square distribution of MDF test statistics, discussed in Theorems 5.1 and 5.2, can hold under considerably more general conditions than correct specification of the discrepancy function. This was discovered in Amemiya and Anderson (1990) and Anderson and Amemiya (1988) for a class of factor analysis models, and in Browne (1987), Browne and Shapiro (1988), and Shapiro (1987) for general linear models, by using approaches based on different techniques. In this section we are going to discuss this theory following the Browne–Shapiro approach, which in a sense is more general although uses a slightly stronger assumption of existence of fourth-order moments. As in the previous section we assume that $n^{1/2}(\hat{\boldsymbol{\xi}}_n - \boldsymbol{\xi}_0) \Rightarrow N(\boldsymbol{0}, \boldsymbol{\Gamma})$.

Let us start with the following algebraic result (Shapiro, 1987, Theorem 3.1). It is based on a verification that the corresponding quadratic form has a chi-square distribution. (For the sake of notational convenience we drop here the superscript of the Jacobian matrix $\boldsymbol{\Delta}^*$.)

PROPOSITION 6.1. *Suppose that the assumption of a sequence of local alternatives holds and the set* Ξ_0 *is approximated at the point* $\boldsymbol{\xi}^* = \boldsymbol{g}(\boldsymbol{\theta}^*)$ *by a linear space* $\mathcal{T}$ *generated by the columns of the matrix* $\boldsymbol{\Delta} = \boldsymbol{\Delta}(\boldsymbol{\theta}^*)$ *(e.g., the point* $\boldsymbol{\theta}^*$ *is regular). Suppose, further, that the discrepancy function is correctly specified with respect to an* $m \times m$ *positive semidefinite matrix* $\boldsymbol{\Gamma}_0$ *of maximal rank. Then* $n\widehat{F}$ *is asymptotically chi-squared, with degrees of freedom* v *and the noncentrality parameter* δ *given in* (5.18) *and* (5.21)*, respectively, if and only if* $\boldsymbol{\Gamma}$ *is representable in the form*

$$\boldsymbol{\Gamma} = \boldsymbol{\Gamma}_0 + \boldsymbol{\Delta}\boldsymbol{C}' + \boldsymbol{C}\boldsymbol{\Delta}', \tag{6.1}$$

where $\boldsymbol{C}$ *is an arbitrary* $m \times q$ *matrix.*

In particular, we have that for the normal-theory discrepancy functions F_{ML} and F_{GLS} (in the analysis of covariance structures) the MDF test statistics are asymptotically chi-squared if and only if the corresponding $p^2 \times p^2$ asymptotic covariance matrix $\boldsymbol{\Gamma}$ can be represented in the form

$$\boldsymbol{\Gamma} = \boldsymbol{\Gamma}_N + \boldsymbol{\Delta}\boldsymbol{C}' + \boldsymbol{C}\boldsymbol{\Delta}', \tag{6.2}$$

where matrix $\boldsymbol{\Gamma}_N$ is defined in (5.1).

The representation (6.1) is slightly more general than the following representation

$$\boldsymbol{\Gamma} = \boldsymbol{\Gamma}_0 + \boldsymbol{\Delta}\boldsymbol{D}\boldsymbol{\Delta}', \tag{6.3}$$

where $\boldsymbol{D}$ is an arbitrary $q \times q$ symmetric matrix. Clearly, (6.3) is a particular form of the representation (6.1) with $\boldsymbol{C} := \frac{1}{2}\boldsymbol{\Delta}\boldsymbol{D}$. It turns out that under various structural assumptions, in the analysis of covariance structures, it is possible to show that the corresponding asymptotic covariance matrix $\boldsymbol{\Gamma}$ is of the form

$$\boldsymbol{\Gamma} = \boldsymbol{\Gamma}_N + \boldsymbol{\Delta}\boldsymbol{D}\boldsymbol{\Delta}', \tag{6.4}$$

and hence to verify that the normal-theory MDF test statistics have asymptotically chi-square distributions. We also have the following result about asymptotic robustness of MDF estimators (Shapiro, 1987, Corollary 5.4).

PROPOSITION 6.2. *Suppose that the model holds, the set* Ξ_0 *is approximated at the point* $\boldsymbol{\xi}_0$ *by a linear space* $\mathcal{T}$ *generated by the columns of the matrix* $\boldsymbol{\Delta} = \boldsymbol{\Delta}(\boldsymbol{\theta}_0)$*, the Jacobian matrix* $\boldsymbol{\Delta}$ *has full column rank* q*, the MDF estimator* $\hat{\boldsymbol{\theta}}_n$ *is a consistent estimator of* $\boldsymbol{\theta}_0$*, the discrepancy function is correctly specified with respect to an* $m \times m$ *positive semidefinite matrix* $\boldsymbol{\Gamma}_0$ *of maximal rank, and the representation* (6.3) *holds. Then* $n^{1/2}(\hat{\boldsymbol{\theta}}_n - \boldsymbol{\theta}_0)$ *converges in distribution to normal* $N(\boldsymbol{0}, \boldsymbol{\Pi})$*, the MDF estimator* $\hat{\boldsymbol{\theta}}_n$ *is asymptotically efficient within the class of MDF estimators, and*

$$\boldsymbol{\Pi} = \boldsymbol{\Pi}_0 + \boldsymbol{D}, \tag{6.5}$$

where $\boldsymbol{\Pi}_0 := (\boldsymbol{\Delta}'\boldsymbol{V}_0\boldsymbol{\Delta})^{-1}$.

We have that if the representation (6.3) holds, then the MDF test statistics designed for the asymptotic covariance matrix $\boldsymbol{\Gamma}_0$ still have asymptotically a chi-square distribution (under a sequence of local alternatives) and the asymptotic covariance matrix of the corresponding MDF estimators needs a simple correction given by formula (6.5). In the remainder of this section we discuss situations in the analysis of covariance structures which lead to the representation (6.4).

We assume below, in the remainder of this section, the setting of the analysis of covariance structures, with structural model $\boldsymbol{\Sigma} = \boldsymbol{\Sigma}(\boldsymbol{\theta})$ and with $\boldsymbol{\Gamma}$ being the $p^2 \times p^2$ asymptotic covariance matrix of $n^{1/2}(\boldsymbol{s} - \boldsymbol{\sigma}_0)$, where $\boldsymbol{s} := \text{vec}(\boldsymbol{S})$ and $\boldsymbol{\sigma}_0 := \text{vec}(\boldsymbol{\Sigma}_0)$. We assume that the underlying population has finite fourth-order moments, and hence the asymptotic covariance matrix $\boldsymbol{\Gamma}$ is well defined. As before, we denote by $\boldsymbol{\Gamma}_N$ and $\boldsymbol{\Pi}_N$ the normal-theory asymptotic covariance matrices given in (5.1) and (5.35), respectively. We also assume that the employed discrepancy function is *correctly specified with respect to a normal distribution* of the data, i.e., $\boldsymbol{V}_0$ is a generalized inverse of $\boldsymbol{\Gamma}_N$. Recall that the normal-theory discrepancy functions F_{ML} and F_{GLS} satisfy this property.

6.1. Elliptical distributions

In this section we assume that the underlying population has an *elliptical* distribution. We may refer to (Muirhead, 1982) for a thorough discussion of elliptical distributions. In the case of elliptical distributions the asymptotic covariance matrix $\boldsymbol{\Gamma}$ has the following structure:

$$\boldsymbol{\Gamma} = \alpha \boldsymbol{\Gamma}_N + \beta \boldsymbol{\sigma}_0 \boldsymbol{\sigma}_0', \tag{6.6}$$

where $\alpha = 1 + \kappa$, $\beta = \kappa$, and κ is the kurtosis parameter of a considered elliptical distribution. This basic asymptotic result was employed in the studies of Muirhead and Waternaux (1980), Tyler (1982, 1983) and Browne (1982, 1984).

It can be seen that the corrected covariance matrix $\alpha^{-1}\boldsymbol{\Gamma}$ has the structure specified in Eq. (6.4), provided that $\boldsymbol{\sigma}_0$ can be represented as a linear combination of columns of the Jacobian matrix $\boldsymbol{\Delta} = \boldsymbol{\Delta}(\boldsymbol{\theta}_0)$. If the point $\boldsymbol{\theta}_0$ is regular, and hence Ξ_0 is a smooth manifold near $\boldsymbol{\sigma}_0 = \boldsymbol{\sigma}(\boldsymbol{\theta}_0)$ and the tangent space $\mathcal{T}$, to Ξ_0 at $\boldsymbol{\sigma}_0$, is the linear space generated by the columns of $\boldsymbol{\Delta}$ (i.e., can be written in the form (5.4)), then this condition is equivalent to the condition that $\boldsymbol{\sigma}_0 \in \mathcal{T}$. This, in turn, holds if the set Ξ_0 is *positively homogeneous*, i.e., it satisfies the property that if $\boldsymbol{\sigma} \in \Xi_0$ and $t > 0$, then $t\boldsymbol{\sigma} \in \Xi_0$. For structural models, positive homogeneity of Ξ_0 can be formulated in the following form (this condition was introduced in Browne (1982) where models satisfying this condition were called *invariant under a constant scaling factor*):

(C) For every $t > 0$ and $\boldsymbol{\theta} \in \Theta$ there exists $\boldsymbol{\theta}^* \in \Theta$ such that $t\boldsymbol{\Sigma}(\boldsymbol{\theta}) = \boldsymbol{\Sigma}(\boldsymbol{\theta}^*)$.

The above condition (C) is easy to verify and it holds for many models used in applications. For example, it holds for the factor analysis model (2.2). By the above discussion we have the following results, which in somewhat different forms were obtained in Tyler (1983) and Browne (1982, 1984), and in the present form in Shapiro and Browne (1987).

THEOREM 6.1. *Suppose that the assumption of a sequence of local alternatives holds, the set Ξ_0 is approximated at the point $\boldsymbol{\sigma}^* = \boldsymbol{\sigma}(\boldsymbol{\theta}^*)$ by a linear space $\mathcal{T}$, the representation* (6.6) *holds with $\alpha > 0$ and that $\boldsymbol{\sigma}_0 \in \mathcal{T}$. Let $\hat{\alpha}$ be a consistent estimator of the parameter α. Then $\hat{\alpha}^{-1} n\widehat{F}$ has asymptotically a chi-squared distribution with ν degrees of freedom and the noncentrality parameter $\alpha^{-1}\delta$, where ν and δ are defined in* (5.18) *and* (5.21), *respectively.*

Recall that the condition "$\boldsymbol{\sigma}_0 \in \mathcal{T}$", used in the above theorem, holds if the set Ξ_0 is positively homogeneous, which in turn is implied by condition (C) (invariance under a constant scaling factor). We also have the following result about asymptotic robustness of the MDF estimators (Shapiro and Browne, 1987).

THEOREM 6.2. *Suppose that the model holds, the set Ξ_0 is approximated at the point $\boldsymbol{\sigma}_0 \in \Xi_0$ by a linear space $\mathcal{T}$ generated by the columns of the matrix $\boldsymbol{\Delta} = \boldsymbol{\Delta}(\boldsymbol{\theta}_0)$, the MDF estimator $\hat{\boldsymbol{\theta}}_n$ is a consistent estimator of $\boldsymbol{\theta}_0$, the representation* (6.6) *holds with $\alpha > 0$ and that $\boldsymbol{\sigma}_0 = \boldsymbol{\Delta}\boldsymbol{\zeta}$ for some $\boldsymbol{\zeta} \in \mathbb{R}^q$. Then $n^{1/2}(\hat{\boldsymbol{\theta}}_n - \boldsymbol{\theta}_0)$ converges in distribution to normal $N(\mathbf{0}, \boldsymbol{\Pi})$, the MDF estimator $\hat{\boldsymbol{\theta}}_n$ is asymptotically efficient within the class of MDF estimators, and*

$$\boldsymbol{\Pi} = \alpha\boldsymbol{\Pi}_N + \beta\boldsymbol{\zeta}\boldsymbol{\zeta}'. \tag{6.7}$$

The vector $\boldsymbol{\zeta}$ can be obtained by solving the system of linear equations $\boldsymbol{\Delta}\boldsymbol{\zeta} = \boldsymbol{\sigma}_0$, which is consistent if $\boldsymbol{\sigma}_0 \in \mathcal{T}$. If the model is invariant under a constant scaling factor, i.e., the above condition (C) holds, then we can view $\boldsymbol{\theta}^* = \boldsymbol{\theta}^*(\boldsymbol{\theta}, t)$ as a function of $\boldsymbol{\theta}$ and t. If, moreover, $\boldsymbol{\theta}^*(\boldsymbol{\theta}, t)$ is differentiable in t, then by differentiating both sides of the equation $\boldsymbol{\sigma}(\boldsymbol{\theta}^*(\boldsymbol{\theta}, t)) = t\boldsymbol{\sigma}(\boldsymbol{\theta})$ we obtain vector $\boldsymbol{\zeta}$ in the form

$$\boldsymbol{\zeta} = \left.\frac{\partial\boldsymbol{\theta}^*(\boldsymbol{\theta}_0, t)}{\partial t}\right|_{t=1}. \tag{6.8}$$

For example, if the model is linear in $\boldsymbol{\theta}$, then $\boldsymbol{\theta}^* = t\boldsymbol{\theta}$ and hence $\boldsymbol{\zeta} = \boldsymbol{\theta}_0$. Of course, in practice the (unknown) population value $\boldsymbol{\theta}_0$ should be replaced by its estimator $\hat{\boldsymbol{\theta}}$.

6.2. *Linear latent variate models*

Presentation of this section is based on Browne and Shapiro (1988). We assume here that the observed $p \times 1$ vector variate $\boldsymbol{X}$ can be written in the form

$$\boldsymbol{X} = \boldsymbol{\mu} + \sum_{i=1}^{s} \boldsymbol{A}_i \boldsymbol{z}_i, \tag{6.9}$$

where $\boldsymbol{\mu}$ is a $p \times 1$ mean vector, $\boldsymbol{z}_i$ is an (unobserved) $m_i \times 1$ vector variate and $\boldsymbol{A}_i$ is a (deterministic) $p \times m_i$ matrix of regression weights of $\boldsymbol{X}$ onto $\boldsymbol{z}_i$, $i = 1, \dots, s$. We assume that random vectors $\boldsymbol{z}_i$ and $\boldsymbol{z}_j$ are independently distributed for all $i \neq j$ and have finite fourth-order moments. The above model implies the following structure of the covariance matrix $\boldsymbol{\Sigma}$ of $\boldsymbol{X}$:

$$\boldsymbol{\Sigma} = \sum_{i=1}^{s} \boldsymbol{A}_i \boldsymbol{\Phi}_i \boldsymbol{A}_i', \tag{6.10}$$

where $\boldsymbol{\Phi}_i$ is the $m_i \times m_i$ covariance matrix of z_i, $i = 1, \ldots, s$.

For example, consider the factor analysis model:

$$\boldsymbol{X} = \boldsymbol{\mu} + \boldsymbol{\Lambda f} + \boldsymbol{u}, \tag{6.11}$$

where $\boldsymbol{\Lambda}$ is a $p \times k$ matrix of factor loadings, $\boldsymbol{f} = (f_1, \ldots, f_k)'$ is a $k \times 1$ common vector variate, and $\boldsymbol{u} = (u_1, \ldots, u_p)'$ is a $p \times 1$ unique factor vector variate. It is assumed that random variables $f_1, \ldots, f_k, u_1, \ldots, u_p$, are mutually independently distributed, and hence random vectors $\boldsymbol{f}$ and $\boldsymbol{u}$ are independent. This implies that the covariance matrices $\boldsymbol{\Phi}$ and $\boldsymbol{\Psi}$, of $\boldsymbol{f}$ and $\boldsymbol{u}$, respectively, are diagonal. Of course, we can write model (6.11) in the form

$$\boldsymbol{X} = \boldsymbol{\mu} + \sum_{i=1}^{k} \boldsymbol{\Lambda}_i f_i + \sum_{j=1}^{p} \boldsymbol{E}_j u_j, \tag{6.12}$$

where $\boldsymbol{\Lambda}_i$ is the ith column vector of $\boldsymbol{\Lambda}$ and $\boldsymbol{E}_j$ is the jth coordinate vector. This shows that this model is a particular case of model (6.9) with $s = k + p$ and $m_i = 1$, $i = 1, \ldots, s$. Now model (6.11) (or, equivalently, model (6.12)) generates the following covariance structures model:

$$\boldsymbol{\Sigma} = \boldsymbol{\Lambda \Phi \Lambda}' + \boldsymbol{\Psi}. \tag{6.13}$$

The only difference between the above model (6.13) and the model (2.2) is that in (2.2) the covariance matrix $\boldsymbol{\Phi}$ is assumed to be the identity matrix.

The linear model (6.9) implies the following structure of the asymptotic covariance matrix $\boldsymbol{\Gamma}$ (Browne and Shapiro, 1988, Theorem 2.1):

$$\boldsymbol{\Gamma} = \boldsymbol{\Gamma}_N + \sum_{i=1}^{s} (\boldsymbol{A}_i \otimes \boldsymbol{A}_i) \boldsymbol{C}_i (\boldsymbol{A}_i' \otimes \boldsymbol{A}_i'), \tag{6.14}$$

where $\boldsymbol{C}_i$ is the $m_i^2 \times m_i^2$ forth-order cumulant matrix of z_i, $i = 1, \ldots, s$.

Suppose now that the weight matrices have parametric structures $\boldsymbol{A}_i = \boldsymbol{A}_i(\boldsymbol{\upsilon})$, $i = 1, \ldots, s$, where $\boldsymbol{\upsilon} \in \Upsilon$ is a parameter vector varying in space $\Upsilon \subset \mathbb{R}^\ell$. Then (6.10) becomes the following covariance structures model

$$\boldsymbol{\Sigma}(\boldsymbol{\theta}) = \sum_{i=1}^{s} \boldsymbol{A}_i(\boldsymbol{\upsilon}) \boldsymbol{\Phi}_i \boldsymbol{A}_i(\boldsymbol{\upsilon})', \tag{6.15}$$

with the parameter vector $\boldsymbol{\theta} := (\boldsymbol{\upsilon}', \boldsymbol{\varphi}_1', \ldots, \boldsymbol{\varphi}_s')'$, where $\boldsymbol{\varphi}_i := \text{vec}(\boldsymbol{\Phi}_i)$, $i = 1, \ldots, s$. Note that the only restriction on the $m_i^2 \times 1$ parameter vectors $\boldsymbol{\varphi}_i$ imposed here is that the corresponding covariance matrix $\boldsymbol{\Phi}_i$ should be positive semidefinite. Note also that if $m_i > 1$ for at least one i, then such choice of the parameter vector results in over-parameterization of model (6.15) since $\boldsymbol{\varphi}_i$ will have duplicated elements. It is possible to include in the parameter vector $\boldsymbol{\theta}$ only nonduplicated elements of matrices $\boldsymbol{\Phi}_i$. This, however, is not essential at the moment.

By applying the 'vec' operator to both sides of Eq. (6.15), we can write this model in the form

$$\boldsymbol{\sigma}(\boldsymbol{\theta}) = \sum_{i=1}^{s} \big(\boldsymbol{A}_i(\boldsymbol{v}) \otimes \boldsymbol{A}_i(\boldsymbol{v})\big)\boldsymbol{\varphi}_i. \tag{6.16}$$

It can be seen that the model is linear in parameters $\boldsymbol{\varphi}_i$, $i = 1, \dots, s$, and

$$\frac{\partial \boldsymbol{\sigma}(\boldsymbol{\theta})}{\partial \boldsymbol{\varphi}_i'} = \boldsymbol{A}_i(\boldsymbol{v}) \otimes \boldsymbol{A}_i(\boldsymbol{v}). \tag{6.17}$$

That is, the corresponding Jacobian matrix can be written as

$$\boldsymbol{\Delta}(\boldsymbol{\theta}) = \big[\boldsymbol{\Delta}(\boldsymbol{v}), \boldsymbol{A}_1(\boldsymbol{v}) \otimes \boldsymbol{A}_1(\boldsymbol{v}), \dots, \boldsymbol{A}_s(\boldsymbol{v}) \otimes \boldsymbol{A}_s(\boldsymbol{v})\big]. \tag{6.18}$$

Together with (6.14) this implies that Eq. (6.4) holds with matrix

$$\boldsymbol{D} = \begin{bmatrix} \mathbf{0} & \mathbf{0} & \mathbf{0} & \cdots & \mathbf{0} \\ \mathbf{0} & \boldsymbol{C}_1 & \mathbf{0} & \cdots & \mathbf{0} \\ \mathbf{0} & \mathbf{0} & \boldsymbol{C}_2 & \cdots & \mathbf{0} \\ & \cdots & \cdots & \cdots & \\ \mathbf{0} & \mathbf{0} & \mathbf{0} & \cdots & \boldsymbol{C}_s \end{bmatrix}. \tag{6.19}$$

We obtain the following result (Browne and Shapiro, 1988, Proposition 3.3).

THEOREM 6.3. *Consider the linear latent variate model* (6.9) *and the corresponding covariance structures model* (6.15). *Suppose that random vectors* z_i, $i = 1, \dots, s$, *are mutually independently distributed, the assumption of a sequence of local alternatives (for the covariance structures model) holds, and* Ξ_0 *is a smooth manifold near the point* $\boldsymbol{\sigma}^*$. *Then the MDF test statistic* $n\widehat{F}$ *has asymptotically noncentral chi-squared distribution with* $\nu = p(p+1)/2 - \operatorname{rank}(\boldsymbol{\Delta}_0)$ *degrees of freedom and the noncentrality parameter* δ.

In particular, the above theorem can be applied to the factor analysis model (6.11). Note that the MDF test statistics for the model (6.13), with the covariance term $\boldsymbol{\Phi}$, and model (2.2), without this term, are the same since $\boldsymbol{\Phi}$ can be absorbed into $\boldsymbol{\Lambda}$ and hence the corresponding set Ξ_0 is the same. Note also that in order to derive the asymptotic chi-squaredness of the MDF test statistics we only used the corresponding independence condition, no other assumptions about distributions of $\boldsymbol{f}$ and $\boldsymbol{u}$ were made (except existence of fourth-order moments). For the factor analysis model this result was first obtained by Amemiya and Anderson (1990) by employing different techniques and without the assumption of finite forth-order moments.

It is also possible to give corrections for the asymptotic covariance matrix $\boldsymbol{\Pi}$ of $n^{1/2}(\hat{\boldsymbol{\theta}}_n - \boldsymbol{\theta}_0)$ (compare with formula (6.5) of Proposition 6.2). In order to do that we need to verify identifiability of the parameter vector $\boldsymbol{\theta}$. Let us replace now $\boldsymbol{\varphi}_i$ with parameter vector $\boldsymbol{\varphi}_i^* := \operatorname{vecs}(\boldsymbol{\Phi}_i)$, $i = 1, \dots, s$, i.e., $\boldsymbol{\varphi}_i^*$ is $m_i(m_i + 1)/2 \times 1$ vector formed from the nonduplicated elements of $\boldsymbol{\Phi}_i$. Eq. (6.4) still holds with the matrix $\boldsymbol{D}$

reduced to a smaller matrix $\boldsymbol{D}^*$ by replacing each $m_i^2 \times m_i^2$ matrix $\boldsymbol{C}_i$ by the corresponding $m_i(m_i+1)/2 \times m_i(m_i+1)/2$ matrix $\boldsymbol{C}_i^*$ formed by the nonduplicated rows and columns of $\boldsymbol{C}_i$. We have then that, under the above assumptions,

$$\boldsymbol{\Pi} = \boldsymbol{\Pi}_N + \boldsymbol{D}^*. \tag{6.20}$$

It follows that the asymptotic covariance matrix of the MDF estimator $\hat{\boldsymbol{\upsilon}}$ is independent of the particular distribution of the z_i, $i = 1, \dots, s$, while the asymptotic covariance matrix of the MDF estimator $\hat{\boldsymbol{\varphi}}_i^*$ needs the correction term $\boldsymbol{C}_i^*$ as compared with the normal case. Asymptotic covariances between $\hat{\boldsymbol{\upsilon}}$ and $\hat{\boldsymbol{\varphi}}_i^*$, and between $\hat{\boldsymbol{\varphi}}_i^*$ and $\hat{\boldsymbol{\varphi}}_j^*$, for $i \neq j$, are the same as in the normal case.

REMARK 5. Suppose, furthermore, that the population value $\boldsymbol{\upsilon}_0$, of the parameter vector $\boldsymbol{\upsilon}$, lies on the *boundary* of the parameter space Υ, and that Υ is approximated at $\boldsymbol{\upsilon}_0$ by convex cone $\mathcal{C}$ (recall that the parameter vectors $\hat{\boldsymbol{\varphi}}_i^*$, $i = 1, \dots, s$, are assumed to be unconstrained). Let $\tilde{\boldsymbol{\upsilon}}$ be the *unconstrained* MDF estimator of $\boldsymbol{\upsilon}_0$ (compare with the derivations of Section 5.4). Then, under the above assumptions, $n^{1/2}(\tilde{\boldsymbol{\upsilon}} - \boldsymbol{\upsilon}_0) \Rightarrow N(\boldsymbol{0}, \boldsymbol{U})$, where the asymptotic covariance matrix $\boldsymbol{U}$ is independent of the particular distribution of the z_i, $i = 1, \dots, s$. We also have then that the MDF test statistic $n\widehat{F}$ converges in distribution to the sum of two stochastically independent terms (compare with Eqs. (5.39) and (5.40)), one term having the usual chi-square distribution and the other term given by $\min_{\boldsymbol{\upsilon}\in\mathcal{C}}(\tilde{\boldsymbol{\upsilon}} - \boldsymbol{\upsilon})'\boldsymbol{U}^{-1}(\tilde{\boldsymbol{\upsilon}} - \boldsymbol{\upsilon})$. It follows that the asymptotic distribution of the MDF test statistic $n\widehat{F}$ is chi-bar-squared and is independent of the particular distribution of the z_i, $i = 1, \dots, s$. That is, under these assumptions, distribution of the MDF test statistic is again asymptotically robust.

Acknowledgements

Author is supported in part by the National Science Foundation award DMS-0510324. The author is indebted to Michael Browne for careful reading of the manuscript and many helpful comments and suggestions.

References

Amemiya, Y., Anderson, T.W. (1990). Asymptotic chi-square tests for a large class of factor analysis models. *Annals of Statistics* **18**, 1453–1463.

Anderson, T.W., Amemiya, Y. (1988). The Asymptotic normal distribution of estimators in Factor Analysis under general conditions. *Annals of Statistics* **16**, 759–771.

Anderson, T.W., Rubin, H. (1956). Statistical inference in factor analysis. In: Neyman, J. (Ed.), *Proceedings of the Third Berkley Symposium on Mathematical Statistics and Probability*, vol. 5. University of California Press, Berkeley, CA, pp. 111–150.

Bartholomew, D.J. (1961). Ordered tests in the analysis of variance. *Biometrika* **48**, 325–332.

Bekker, P.A., ten Berge, J.M.F. (1997). Generic global identification in factor analysis. *Linear Algebra and its Applications* **264**, 255–263.

Bonnans, J.F., Shapiro, A. (2000). *Perturbation Analysis of Optimization Problems*. Springer-Verlag, New York.

Browne, M.W. (1974). Generalized least squares estimators in the analysis of covariance structures. *South African Statistical Journal* **8**, 1–24.

Browne, M.W. (1982). Covariance structures. In: Hawkins, D.M. (Ed.), *Topics in Applied Multivariate Analysis*. Cambridge University Press.

Browne, M.W. (1984). Asymptotically distribution-free methods in the analysis of covariance structures. *British Journal of Mathematical and Statistical Psychology* **37**, 62–83.

Browne, M.W. (1987). Robustness in statistical inference in factor analysis and related models. *Biometrika* **74**, 375–384.

Browne, M.W., Shapiro, A. (1988). Robustness of normal theory methods in the analysis of linear latent variate models. *British Journal of Mathematical and Statistical Psychology* **41**, 193–208.

Chernoff, H. (1954). On the distribution of the likelihood ratio. *Ann. Math. Statist.* **25**, 573–578.

Danskin, J.M. (1967). *The Theory of Max-Min and its Applications to Weapon Allocation Problems*. *Econometrics and Operations Research*, vol. 5. Springer-Verlag, Berlin and New York.

Fisher, F.M. (1966). *The Identification Problem in Econometrics*. McGraw-Hill, New York.

Jennrich, R.I. (1987). Rotational equivalence of factor loading matrices with specified values. *Psychometrika* **43**, 421–426.

Jöreskog, K.G. (1977). Structural equation models in the social sciences: specification, estimation and testing. In: Krishnaiah, P.R. (Ed.), *Applications of Statistics*. North-Holland, Amsterdam, pp. 265–287.

Jöreskog, K.G. (1981). Analysis of covariance structures (with discussion). *Scandinavian Journal of Statistics* **8**, 65–92.

Kano, Y. (1986). Conditions on consistency of estimators in covariance structure model. *Journal of the Japan Statistical Society* **16**, 75–80.

Kudô, A. (1963). A multivariate analogue of the one-sided test. *Biometrika* **50**, 403–418.

Ledermann, W. (1937). On the rank of reduced correlation matrices in multiple factor analysis. *Psychometrika* **2**, 85–93.

Muirhead, R.J. (1982). *Aspects of Multivariate Statistical Theory*. Wiley, New York.

Muirhead, R.J., Waternaux, C.M. (1980). Asymptotic distribution in canonical correlation analysis and other multivariate procedures for nonnormal populations. *Biometrika* **67**, 31–43.

Nüesch, P.E. (1966). On the problem of testing location in multivariate populations for restricted alternatives. *Ann. Math. Stat.* **37**, 113–119.

Rothenberg, T.J. (1971). Identification in parametric models. *Econometrica* **39**, 577–591.

Seber, G.A.F. (1977). *Linear Regression Analysis*. Wiley, New York.

Shapiro, A. (1983). Asymptotic distribution theory in the analysis of covariance structures (a unified approach). *South African Statistical Journal* **17**, 33–81.

Shapiro, A. (1984). A note on consistency of estimators in the analysis of moment structures. *British Journal of Mathematical and Statistical Psychology* **37**, 84–88.

Shapiro, A. (1985a). Identifiability of factor analysis: some results and open problems. *Linear Algebra and its Applications* **70**, 1–7.

Shapiro, A. (1985b). Asymptotic equivalence of minimum discrepancy function estimators to GLS estimators. *South African Statistical Journal* **19**, 73–81.

Shapiro, A. (1985c). Second-order sensitivity analysis and asymptotic theory of parametrized nonlinear programs. *Mathematical Programming* **33**, 280–299.

Shapiro, A. (1985d). Asymptotic distribution of test statistics in the analysis of moment structures under inequality constraints. *Biometrika* **72**, 133–144.

Shapiro, A. (1986). Asymptotic theory of overparametrized structural models. *Journal of the American Statistical Association* **81**, 142–149.

Shapiro, A. (1987). Robustness properties of the MDF analysis of moment structures. *South African Statistical Journal* **21**, 39–62.

Shapiro, A. (1988). Towards a unified theory of inequality constrained testing in multivariate analysis. *International Statistical Review* **56**, 49–62.

Shapiro, A., Browne, M.W. (1987). Analysis of covariance structures under elliptical distributions. *Journal of the American Statistical Association* **82**, 1092–1097.

Silvapulle, M.J., Sen, P.K. (2005). *Constrained Statistical Inference: Order, Inequality, and Shape Constraints*. Wiley, New York.

Steiger, J.H., Shapiro, A., Browne, M.W. (1985). On the multivariate asymptotic distribution of sequential chi-square statistics. *Psychometrika* **50**, 253–264.
Tyler, E.D. (1982). Radial estimates and the test for sphericity. *Biometrika* **69**, 429–436.
Tyler, E.D. (1983). Robustness and efficiency properties of scatter matrices. *Biometrika* **70**, 411–420.
Wald, A. (1950). Note on the identification of economic relations. In: Koopmans, T.C. (Ed.), *Statistical Inference in Dynamic Economic Models*. John Wiley, New York.

Handbook of Computing and Statistics with Applications, Vol. 1
ISSN: 1871-0301

DOI: 10.1016/S1871-0301(06)01012-2

12

Meta-Analysis and Latent Variable Models for Binary Data

Jian Qing Shi

Abstract
Meta-analysis is widely used as a method of summarizing and combining results from individual research studies. Heterogeneity and publication bias are two major problems in meta-analysis for binary data. A latent meta-analysis model by using binomial distributions is proposed. This model allows for heterogeneity among different studies. Based on fitting a model to the funnel plot, a method for sensitivity analysis is discussed and is used to address the problem of publication bias. The maximum likelihood estimates based on the exact distributions are calculated by a Markov chain Monte Carlo EM algorithm. A meta-analysis of epidemiological studies on the effect of alcohol on the risk of breast cancer is used to illustrate the method.

Keywords: Heterogeneity; Latent variable model; Markov chain Monte Carlo EM algorithm; Meta-analysis; Publication bias; Sensitivity analysis; Trend estimation

1. Introduction

In epidemiology and other areas, binary data is usually recorded to study the relation between manifest and latent variables (the recent development on latent variable models can be found in other chapters in this book, and the most recent discussion on latent variable models with binary data is given by Lee and Tang (2006) and Song and Lee (2006)). Many individual studies may report small experimental or epidemiological investigations and fail to give any very firm conclusion individually. Meta-analysis is to summarise and combine results from those individual studies and collectively may suggest a clear overall result. *Heterogeneity* and *publication bias* are two major problems in meta-analysis. Different methods in experimental design, data collection and data analysis are used in different studies. All those result in the variation in the data used in meta-analysis. Furthermore, the sample collected is often selective and is usually biased in favour of large studies and in favour of small studies with positive outcomes. There are other small studies which have been carried out but which gave negative result and have not been published. In practice, publication bias is a very common problem. Sutton et al. (2000) assessed 48 meta-analyses and found that about half showed the signs of publication bias.

Using the terminology defined in Little and Rubin (2002), publication bias is a problem with nonignorable missing data. It is very difficult to correct publication bias without making very strong assumptions. For example, the 'trim and fill' method (Duval and Tweedie, 2000) is based on the strong symmetry assumption which is unverifiable. An alternative more cautious approach is *sensitivity analysis* which is proposed by Copas and Shi (2000, 2001). They draw conclusions from the meta-analysis under a variety of plausible possibilities for the extent of publication bias, and assess how different conclusions are drawn from one another and from the results of conventional approaches. This approach has been extended to 2×2 tables in Shi and Copas (2002) by using exact binomial distributions. They used a random effect model to address the problem of heterogeneity and used a selection model to address the problem of publication bias. A Markov chain Monte Carlo EM (MCMC-EM) algorithm was used to calculate maximum likelihood estimates. In this chapter, we will use a similar idea to analyse the relation between manifest and latent variables for binary data. A typical example is the trend estimation in epidemiological studies of the association between disease and exposure to some agent or hazard. It is often interested to know how much risk increases as exposure increases. For the example of alcohol use and breast cancer which we will discuss later in this chapter, we would like to estimate how much the risk increases as the amount of alcohol consumption increases.

Using the normal approximation for empirical log-odds ratio, Shi and Copas (2004) conducted a meta-analysis and sensitivity analysis for trend estimation. However, this method is inappropriate when the sample size is small or the related probability is close to 0 or 1. We will therefore assume exact binomial distributions for binary data in this chapter, and use a latent variable model to model the relation between the probability of being a case and the covariates. Those covariates measure the extent of the exposure and other factors in trend estimation, for example, the amount of alcohol use. We will focus on meta-analysis and trend estimation in this chapter, but there is no major difficulty to extend the method to a more general latent variable model to cover other types of categorical data.

Section 2 sets out the methodology for meta-analysis. Section 2.1 discusses a basic model for an individual study, in which an exact binomial distribution is used for binary data, with a logistic regression model for modelling the relation between the probability of having a disease and the extent of exposure. A meta-analysis model which allows for heterogeneity is introduced in Section 2.2. The implementation by using a MCMC-EM algorithm to calculate maximum likelihood estimates is discussed in Section 2.3. In Section 3, we will define a selection model with a meta-analysis model and discuss a method for sensitivity analysis, based on fitting a model to the funnel plot, to address the problem of publication bias. An illustrated example is discussed in Section 4. Some final comments and further development are given in Section 5.

2. Meta-analysis for binary data

2.1. A model for a single study

To study the association between an event and dose level, we usually observe a set of binomial outcomes from several exposure bands, including the baseline group with zero dose. To fix the notation, suppose that a typical study has $(n+1)$ exposure bands, and has z_j cases out of m_j subjects in the jth band for $j = 0, 1, \ldots, n$, where $j = 0$ stands for the baseline control group with zero dose. The binomial outcomes have the distribution:

$$\begin{aligned} z_0 &\sim Bin(m_0, \pi_0), \\ z_j &\sim Bin(m_j, \pi_j), \quad j = 1, \ldots, n. \end{aligned} \tag{1}$$

The related log-odds is

$$\eta_j = \log\left(\frac{\pi_j}{1-\pi_j}\right) \tag{2}$$

for $j = 0, 1, \ldots, n$. If the jth dose class can be assumed to have a single dose level x_j, we can define a logistic regression model by

$$\eta_j = \alpha + \beta x_j, \quad j = 0, 1, \ldots, n, \tag{3}$$

where β measures the association between the log-odds ratio and dose level. Parameter β is the parameter of interest in trend estimation, but α is a nuisance parameter. For the alcohol use and breast cancer example, z_j is the number of patients in group j in which the average alcohol consumption is x_j. The parameter β measures how the odds of being a case increases when the alcohol consumption increases. The likelihood for (α, β) given $z = (z_0, z_1, \ldots, z_n)^{\mathrm{T}}$ is

$$p(z|\alpha, \beta) = \prod_{j=0}^{n} \binom{m_j}{z_j} \pi_j^{z_j} (1-\pi_j)^{m_j - z_j} \tag{4}$$

$$\propto \frac{\exp(z_0\alpha)}{[1+\exp(\alpha)]^{m_0}} \prod_{j=1}^{n} \frac{\exp(z_j(\alpha+\beta x_j))}{[1+\exp(\alpha+\beta x_j)]^{m_j}}. \tag{5}$$

Maximum likelihood estimates of α and β can be calculated by maximising the above likelihood.

2.2. Meta-analysis

Suppose that there are K studies in the meta-analysis, in which the ith study has observation $z_i = (z_{i0}, z_{i1}, \ldots, z_{in_i})$ with sample size m_{i0} and m_{ij}, and probabilities π_{i0} and π_{ij} for $j = 1, \ldots, n_i$ as in distributions (1). Let η_{ij} and x_{ij} be the log-odds and the single dose level for ith study, as defined in Eqs. (2) and (3). If we allow for heterogeneity, we may define a random effect logistic regression model as follows

$$\eta_{ij} = \alpha_i + \beta_i x_{ij}, \tag{6}$$

$$\beta_i \sim N(\mu_\beta, \tau_\beta^2), \tag{7}$$

where the overall coefficient μ_β measures the association between log-odds ratio and dose level, and τ_β measures the magnitude of the variation between studies. The probability density function for the ith study now involves an integral

$$f(z_i|\boldsymbol{\theta}) = \int p(z_i|\alpha_i, \beta_i)\phi(\beta_i; \mu_\beta, \tau_\beta)\,\mathrm{d}\beta_i, \tag{8}$$

where $\boldsymbol{\theta} = (\boldsymbol{\alpha}, \mu_\beta, \tau_\beta)'$, $\boldsymbol{\alpha} = (\alpha_1, \dots, \alpha_K)$, $\phi(\cdot; \mu_\beta, \tau_\beta)$ is the density function of $N(\mu_\beta, \tau_\beta^2)$. The density function $p(z_i|\alpha_i, \beta_i)$ is given by (5). The log-likelihood for $\boldsymbol{\theta}$ is therefore

$$L(\boldsymbol{\theta}) = \sum_{i=1}^{K} \log\{f(z_i|\boldsymbol{\theta})\}. \tag{9}$$

The above log-likelihood involves integrals. Crouch and Spiegelman (1990) pointed out that it is not adequate to use Gaussian quadrature to approximate integrals (8), since that type of integrand is not well approximated by a polynominal function. An alternative way is to use Markov chain Monte Carlo EM (MCMC-EM) algorithm to calculate the maximum likelihood estimates directly, analogous to the method used in (Shi and Copas, 2002).

2.3. *Implementation*

The basic idea is to treat the latent variable $\boldsymbol{\beta} = (\beta_1, \dots, \beta_K)$ in (6) as missing and use an EM algorithm. Let $\boldsymbol{Z}$ be the collection of K vectors of z_i, the full log-likelihood for $\boldsymbol{\theta}$ given $(\boldsymbol{Z}, \boldsymbol{\beta})$ is

$$L(\boldsymbol{Z}, \boldsymbol{\beta}; \boldsymbol{\theta}) = \sum_{i=1}^{K} \big[\log\{p(z_i|\alpha_i, \beta_i)\} + \log\{\phi(\beta_i; \mu_\beta, \tau_\beta)\}\big], \tag{10}$$

where $p(z_i|\alpha_i, \beta_i)$ is given by (5) but (α, β) is replaced by (α_i, β_i). The EM algorithm involves the calculation of the expectation of the above full log-likelihood conditional on the current estimate of $\boldsymbol{\theta}$ and the observation $\boldsymbol{Z}$ in the E-step, and then updates the estimate of $\boldsymbol{\theta}$ by maximising this conditional expectation in the M-step. The details will be discussed in the rest of this section.

2.3.1. *MCMC-EM: E step*

In E-step of the $(r+1)$th iteration, we need to calculate the following conditional expectation

$$\begin{aligned} L(\boldsymbol{\theta}|\boldsymbol{\theta}^{(r)}) &= E\big[L(\boldsymbol{Z}, \boldsymbol{\beta}; \boldsymbol{\theta})|\boldsymbol{Z}, \boldsymbol{\theta}^{(r)}\big] \\ &= \sum_{i=1}^{K} \big\{E\big[\log(p(z_i|\alpha_i, \beta_i))|\boldsymbol{Z}, \boldsymbol{\theta}^{(r)}\big] \\ &\quad + E\big[\log(\phi(\beta_i; \mu_\beta, \tau_\beta))|\boldsymbol{Z}, \boldsymbol{\theta}^{(r)}\big]\big\}, \end{aligned} \tag{11}$$

where $\boldsymbol{\theta}^{(r)}$ is the current estimate after the rth iteration. The expectation is calculated in terms of $\boldsymbol{\beta}$. As there is no analytical form for Eq. (11), we use MCMC-EM algorithm; see, for example, Wei and Tanner (1990) and Booth and Hobert (1999).

To do this, we generate A number of random vectors $\{\boldsymbol{\beta}^a, a = 1, \dots, A\}$ from the conditional distribution $p(\boldsymbol{\beta}|\boldsymbol{Z}, \boldsymbol{\theta}^{(r)})$, and approximate (11) by

$$L_A\big(\boldsymbol{\theta}|\boldsymbol{\theta}^{(r)}\big) = \frac{1}{A}\sum_{a=1}^{A} L\big(\boldsymbol{Z}, \boldsymbol{\beta}^a; \boldsymbol{\theta}\big). \tag{12}$$

We will discuss how to update $\boldsymbol{\theta}$ by maximising the above log-likelihood in Section 2.3.2. Now, we discuss how to generate those random numbers.

Given $(\boldsymbol{Z}, \boldsymbol{\theta})$, the latent variable β_i's are conditional independent for $i = 1, \dots, K$. We can therefore generate random number for each component individually. For each component, its conditional density function is

$$p\big(\beta_i|z_i, \boldsymbol{\theta}^{(r)}\big) \propto p\big(z_i|\alpha_i^{(r)}, \beta_i\big)\phi\big(\beta_i; \mu_\beta^{(r)}, \tau_\beta^{(r)}\big), \tag{13}$$

where $p(z_i|\alpha_i^{(r)}, \beta_i)$ is given by (5). We can use Metropolis–Hastings algorithm (Metropolis et al., 1953; Hastings, 1970) to generate a random variate β_i from the above density function. Carlin and Louis (2000) discussed the details for Metropolis–Hastings algorithm. Shi and Copas (2004) gave the details how to define a transition density and calculate acceptance probability for a similar problem.

2.3.2. MCMC-EM: M-step

In the M-step, we need to update $\boldsymbol{\theta}$ by maximising the conditional likelihood (12). Since the unknown parameters (μ_β, τ_β) are involved in the second term only in the full log-likelihood function (10), the calculation of the maximum likelihood estimate is rather simple. This is to estimate (μ_β, τ_β) by maximising the following objective function

$$\sum_{i=1}^{K} \log\big\{\phi(\beta_i; \mu_\beta, \tau_\beta)\big\}.$$

It is actually equivalent to the problem that we have observed $\{\beta_1, \dots, \beta_K\}$ from the normal distribution $N(\mu_\beta, \tau_\beta)$ and want to calculate the maximum likelihood estimates of μ_β and τ_β. Their estimates are simply given by the sample mean and the sample standard error:

$$\bar{\beta} = \frac{1}{K}\sum \beta_i \quad \text{and} \quad V_\beta = \sqrt{\frac{1}{K}\sum(\beta_i - \bar{\beta})^2}.$$

Here, $\bar{\beta}$ and V_β are sufficient statistics for the parameters of (μ_β, τ_β). In the $(r+1)$th iteration, M-step is to update β and τ_β by their conditional expectations $E(\bar{\beta}|\boldsymbol{Z}, \boldsymbol{\theta}^{(r)})$ and $E(V_\beta|\boldsymbol{Z}, \boldsymbol{\theta}^{(r)})$. It is not possible to get an analytic form for those conditional expectations. We therefore use the random vectors $\{\boldsymbol{\beta}^1, \dots, \boldsymbol{\beta}^A\}$ generated from the conditional distribution $p(\boldsymbol{\beta}|\boldsymbol{Z}, \boldsymbol{\theta}^{(r)})$ in E-step, and approximate these conditional expectations by their sample means:

$$\big(\bar{\beta}^1 + \dots + \bar{\beta}^A\big)/A \quad \text{and} \quad \big(V_\beta^1 + \dots + V_\beta^A\big)/A,$$

where $\bar{\beta}^a$ and V_β^a are the sample mean and sample standard error of $\boldsymbol{\beta}^a = (\beta_1^a, \ldots, \beta_K^a)$, respectively.

The parameter α_i is updated by maximising $L_i = \log\{p(z_i|\alpha_i, \beta_i)\}$. There is no analytic solution. We use the following Newton method to approximate the estimate of α_i at the M-step. For simplifying the notation, we omit the index i here. We update α by the following subiteration

$$\alpha = \alpha_0 - \dot{L}(\alpha_0)/\ddot{L}(\alpha_0),$$

where α_0 is the current estimate of α, $\dot{L}(\alpha_0)$ and $\ddot{L}(\alpha_0)$ are the first two derivatives of $L_i = \log\{p(z_i|\alpha_i, \beta_i)\}$ in terms of α_i for the ith component. Bear in mind that we actually need to maximise the conditional expectation of $L = \log\{p(z|\alpha, \beta)\}$ given β. This can be approximated by the random numbers generated in MCMC-E step:

$$L = \frac{1}{A}\sum_{a=1}^{A} \log\{f(z|\alpha, \beta^a)\}.$$

The Newton method works very effectively as it is a univariate problem.

2.3.3. Average MCMC-EM algorithm and standard errors

As discussed in (Shi and Copas, 2002), the estimates by the MCMC-EM algorithm converges to the real maximum likelihood estimates when A is sufficiently large. The Monte Carlo error is mainly determined by the sample size A. To reduce the Monte Carlo error we need to take a sufficiently large A (see, e.g., Booth and Hobert, 1999). An alternative way is to use average MCMC-EM algorithm proposed in (Shi and Copas, 2002). Instead of increasing A, the average value of the estimates collected in the iterations after 'burn-in' is used. The average batch mean $\bar{\boldsymbol{\theta}}^{(r)}$ is defined as the sample mean of $\{\boldsymbol{\theta}^{(r-J+1)}, \boldsymbol{\theta}^{(r-J)}, \ldots, \boldsymbol{\theta}^{(r)}\}$, where $\boldsymbol{\theta}^{(r)}$ is the estimate obtained in the rth iteration. The estimates $\bar{\boldsymbol{\theta}}^{(r)}$ is roughly equivalent to the one calculated by using the MCMC-EM algorithm with Monte Carlo sample size JA, but the former is more efficient and much easier to implement than the latter.

The standard error of $\hat{\boldsymbol{\theta}}$ can be calculated quite easily for the MCMC-EM algorithm. The related observed information matrix can be calculated by Louis (1982)

$$I(\boldsymbol{\theta}) = -E\{\ddot{\boldsymbol{L}}(\boldsymbol{Z}, \boldsymbol{\beta}|\boldsymbol{Z})\} - E\{\dot{\boldsymbol{L}}(\boldsymbol{Z}, \boldsymbol{\beta}|\boldsymbol{Z}) \cdot \dot{\boldsymbol{L}}^{\mathrm{T}}(\boldsymbol{Z}, \boldsymbol{\beta}|\boldsymbol{Z})\},$$

where $\dot{\boldsymbol{L}}$ and $\ddot{\boldsymbol{L}}$ are the first two derivatives of the full log-likelihood (10) with respect to $\boldsymbol{\theta}$. The expectation is defined in terms of $\boldsymbol{\beta}$, which can be approximated by the random samples generated in MCMC-E step. In each iteration, we calculate

$$I(\boldsymbol{\theta}|\boldsymbol{\beta}) = -\ddot{\boldsymbol{L}}(\boldsymbol{Z}, \boldsymbol{\beta}) - \dot{\boldsymbol{L}}(\boldsymbol{Z}, \boldsymbol{\beta}) \cdot \dot{\boldsymbol{L}}^{\mathrm{T}}(\boldsymbol{Z}, \boldsymbol{\beta}),$$

evaluated at $\boldsymbol{\beta} = \boldsymbol{\beta}^a$ for each sample $\boldsymbol{\beta}^a$. The observed information matrix $I(\boldsymbol{\theta})$ is therefore approximated by the sample mean of $\{I(\boldsymbol{\theta}|\boldsymbol{\beta}^a),\ a = 1, \ldots, A\}$.

The final estimate of the information matrix can be calculated by the average 'batch mean' for average MCMC-EM algorithm. The variance of the estimates can be calculated from the inverse of the information matrix.

For the model we discussed in this section, we update the estimates of α_i's and (μ_β, τ_β) in the M-step separately. Thus, it is rather simple to calculate standard errors for those parameters since they can also be calculated separately. The expressions for the first two derivatives related to α_i are (the index i is omitted for simplifying the notation)

$$\dot{L}(\alpha|\beta) = \sum_{j=0}^{n}\left[z_j - m_j\frac{\exp(\alpha+\beta x_j)}{1+\exp(\alpha+\beta x_j)}\right],$$

$$\ddot{L}(\alpha|\beta) = -\sum_{j=0}^{n}\left[m_j\frac{\exp(\alpha+\beta x_j)}{(1+\exp(\alpha+\beta x_j))^2}\right],$$

with the related quantities evaluated for the ith component. The variance of α is therefore calculated by

$$\left[\frac{1}{A}\sum_{a=1}^{A}\{-\ddot{L}(\alpha|\beta^a) - \dot{L}^2(\alpha|\beta^a)\}\right]^{-1}.$$

The first two derivatives for (μ_β, τ_β) are given by

$$\dot{\boldsymbol{L}}(\mu_\beta, \tau_\beta|\boldsymbol{\beta}) = \begin{pmatrix} K(\bar{\boldsymbol{\beta}} - \mu_\beta)/\tau_\beta^2 \\ -K/\tau_\beta + \sum_{i=1}^{K}(\beta_i - \mu_\beta)^2/\tau_\beta^3 \end{pmatrix},$$

$$\ddot{\boldsymbol{L}}(\mu_\beta, \tau_\beta|\boldsymbol{\beta}) = \begin{pmatrix} -K/\tau_\beta^2 & -2K(\bar{\boldsymbol{\beta}} - \mu_\beta)/\tau_\beta^3 \\ -2K(\bar{\boldsymbol{\beta}} - \mu_\beta)/\tau_\beta^3 & K/\tau_\beta^2 - 3\sum_{i=1}^{m}(\beta_i - \mu_\beta)^2/\tau_\beta^4 \end{pmatrix},$$

where $\bar{\boldsymbol{\beta}}$ is the average of $\{\beta_1, \ldots, \beta_K\}$. The observed information matrix for (μ_β, τ_β) is calculated by

$$\sum_{a=1}^{A}\left[-\ddot{\boldsymbol{L}}(\mu_\beta, \tau_\beta|\boldsymbol{\beta}^a) - \dot{\boldsymbol{L}}(\mu_\beta, \tau_\beta|\boldsymbol{\beta}^a)\dot{\boldsymbol{L}}^{\mathrm{T}}(\mu_\beta, \tau_\beta|\boldsymbol{\beta}^a)\right]/A.$$

The final estimate of the information matrix can be approximated by the batch mean. The covariance matrix of (μ_β, τ_β) is the inverse of information matrix.

3. Publication bias and sensitivity analysis

3.1. Selection model

In meta-analysis, it is always that only a selection of studies is included. In most situations, there is selection bias or publication bias, meaning that the selected studies cannot represent the original population. Let $\hat{\beta}$ be the estimate of slope in a study and s is the standard error. We plotted $\hat{\beta}_i$ against s_i in Figure 1 for the breast cancer example. It shows that the small studies tend to have larger values of $\hat{\beta}_i$, recognising the publication bias that the studies with significant results and/or large sample sizes are more likely to be published. Let $\mathcal{S}$ be the event that a study is selected. To model the possibility

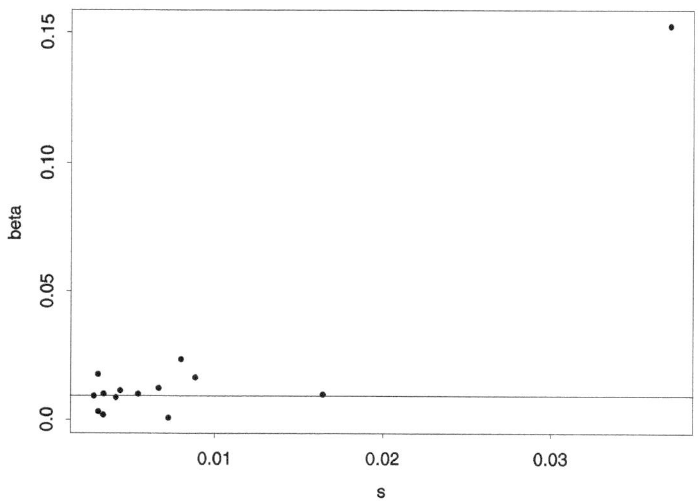

Fig. 1. Funnel plots of $\hat{\beta}_i$ against its standard error s_i, where $\hat{\beta}_i$ is calculated by using empirical odds ratio. The solid line gives the overall estimate $\hat{\mu}_\beta$ by using meta-analysis models (6) and (7) without assuming publication bias.

that the selection is biased in favour of larger studies (with smaller s_i), and in favour of studies having a more positive outcome (with larger $\hat{\beta}_i$ in this chapter), suppose that a study reporting estimate $\hat{\beta}_i$ and standard error s_i is selected with probability

$$q(z_i|\beta_i) = P(\mathcal{S}_i|\hat{\beta}_i, s_i, \beta_i) = \Phi\left(\frac{a + b/s_i + \rho(\hat{\beta}_i - \beta_i)/s_i}{(1-\rho^2)^{1/2}}\right), \tag{14}$$

where $b \geqslant 0$, $\rho \geqslant 0$, Φ is the cumulate distribution function of the standard normal distribution. Parameter ρ plays a very important role here, which models the association between the selection model and the study outcome $\hat{\beta}$. If $\rho = 0$, this is the model without publication bias. If $\rho > 0$, the selection probability is larger for the study with larger value of $\hat{\beta}$ (the study with more positive outcome) or smaller value of s (larger study). This models the phenomenon shown in Figure 1.

The selection probability defined in (14) is conditional on the true value β_i. Following the discussion in Shi and Copas (2004), the overall selection probability for a study with data $(\hat{\beta}_i, s_i)$ is given by

$$\begin{aligned} P(\mathcal{S}_i|\hat{\beta}_i, s_i) &= \int \Phi\left(\frac{a + b/s_i + \rho(\hat{\beta}_i - \beta_i)/s_i}{(1-\rho^2)^{1/2}}\right)\phi(\beta_i; \mu_\beta, \tau_\beta)\,\mathrm{d}\beta_i \\ &= \Phi\left(\frac{a + b/s_i + \tilde{\rho}(\hat{\beta}_i - \mu_\beta)/(\tau_\beta^2 + s_i^2)^{1/2}}{(1-\tilde{\rho}^2)^{1/2}}\right), \end{aligned} \tag{15}$$

where $\tilde{\rho} = \rho s_i/(\tau_\beta^2 + s_i^2)^{1/2}$. The marginal probability for a study with standard error s_i is

$$P(\mathcal{S}_i|s_i) = \Phi\left(a + \frac{b}{s_i}\right). \tag{16}$$

This gives the interpretation of the parameters a and b that a controls the overall proportion of studies selected and b (expected to be positive) controls how the chance of selection depends on study size. Some more explanation will be given in Section 4.

Publication bias is the problem with nonignorable missing data. The parameters a and b are used to measure the nonignorable missing-mechanism. Since we have no information about the studies in the population that have not been selected, a and b are not estimable without making strong assumptions. Adopted the same idea proposed in (Copas and Shi, 2000), we use a sensitivity analysis to deal with the problem of publication bias. We first give a range of different values of (a, b) and then monitor how sensitively the estimate $(\mu_\beta, \tau_\beta, \rho)$ and other quantities depend on the particular choice of these selection parameters.

3.2. Maximum likelihood estimates

Now we discuss how to obtain estimates for a given pair of (a, b). We still use $\boldsymbol{\theta}$ to denote all the unknown parameters $(\boldsymbol{\alpha}, \mu_\beta, \tau_\beta, \rho)$. For the meta-analysis of K studies, the log-likelihood is for those selected studies, and it is therefore given by

$$\begin{aligned} L(\boldsymbol{Z}; \boldsymbol{\theta}) &= \sum_{i=1}^{K} \log\{p(z_i | \mathcal{S}_i, \boldsymbol{\theta})\} \\ &= \sum_{i=1}^{K} \log\left\{\frac{p(z_i, \mathcal{S}_i | \boldsymbol{\theta})}{p(\mathcal{S}_i)}\right\} \\ &= \sum_{i=1}^{K} \left[\log\left\{\int p(z_i, \mathcal{S}_i | \boldsymbol{\theta}, \beta_i)\phi(\beta_i; \mu_\beta, \tau_\beta)\,\mathrm{d}\beta_i\right\} - \log\bigl[p(\mathcal{S}_i | s_i)\bigr]\right] \\ &= \sum_{i=1}^{K} \left[\log\left\{\int p(z_i | \alpha_i, \beta_i) q(z_i | \beta_i)\phi(\beta_i; \mu_\beta, \tau_\beta)\,\mathrm{d}\beta_i\right\}\right. \\ &\quad \left. - \log\bigl[p(\mathcal{S}_i | s_i)\bigr]\right], \end{aligned} \tag{17}$$

where $p(z_i | \alpha_i, \beta_i)$ is given by (5) and $q(z_i | \beta_i)$ is given by (14). We want to find the estimates of interesting parameters $(\mu_\beta, \mu_\tau, \rho)$ and nuisance parameters α_i's by maximising the above log-likelihood. The MCMC-EM algorithm discussed in Section 2 can be extended to cover this model.

As before, we still treat $\boldsymbol{\beta} = (\beta_1, \ldots, \beta_K)$ as missing data. The full log-likelihood of $(\boldsymbol{Z}, \boldsymbol{\beta})$ for $\boldsymbol{\theta} = (\boldsymbol{\alpha}, \mu_\beta, \tau_\beta, \rho)$ is

$$\begin{aligned} L(\boldsymbol{Z}, \boldsymbol{\beta}; \boldsymbol{\theta}) &= \sum_{i=1}^{K} \log\{p(z_i, \beta_i | \mathcal{S}_i, \boldsymbol{\theta})\} \\ &= \sum_{i=1}^{K} \Bigl[\log\{\phi(\beta_i; \mu_\beta, \tau_\beta)\} \end{aligned}$$

$$+\log\{q(z_i|\beta_i)\} + \log\{p(z_i|\alpha_i, \beta_i)\} - \log\{P(\mathcal{S}_i|s_i)\}\Big]$$
$$= L_1 + L_2 + L_3, \tag{18}$$

where

$$L_1 = \sum_{i=1}^{K} \log\{\phi(\beta_i; \mu_\beta, \tau_\beta)\},$$
$$L_2 = \sum_{i=1}^{K} \log\{p(z_i|\alpha_i, \beta_i)\}, \quad \text{and}$$
$$L_3 = \sum_{i=1}^{K} \big[\log\{q(z_i|\beta_i)\} - \log\{P(\mathcal{S}_i|s_i)\}\big].$$

Noting the fact that L_1 involves parameters (μ_β, τ_β) only, L_2 involves $\boldsymbol{\alpha}$ only and L_3 involves ρ only, we can update those three groups of parameters in the M-step by maximising L_1, L_2 and L_3 separately. L_1 and L_2 are the exact same as the related items in (10), we can use the same formulae as those discussed in Section 2.3 to update the estimates for (μ_β, τ_β) and $\boldsymbol{\alpha}$ in the M-step. The parameter ρ is updated by maximising L_3 alone. Since $P(\mathcal{S}_i|s_i)$ in L_3 is independent of any unknown parameters, the second term can be ignored. We use a subiteration of Newton method to update ρ analogous to the method updating α in M-step as in Section 2.3.2.

In MCMC-E step, we need to generate β_i from the following conditional distribution

$$p(\beta|z, \boldsymbol{\theta}) \propto q(z|\beta)p(z|\alpha, \beta)\phi(\beta; \mu_\beta, \tau_\beta), \tag{19}$$

where the related quantity is evaluated for ith component. We still use the Metropolis–Hasting algorithm as in the last section.

3.3. Sensitivity analysis

In trend estimation, we are mainly concerned with the estimate of the slope μ_β. We therefore develop a sensitivity analysis for μ_β allowing for a range of plausible values of (a, b). A rough procedure is described as follows.

(i) Select a range of possible values of (a, b). This can be helped by the marginal selection probabilities, especially the selection probabilities for the smallest study and the largest studies:

$$P_{\max} = \Phi(a + b/s_{\min}), \qquad P_{\min} = \Phi(a + b/s_{\max}),$$

where $s_{\min}$ and $s_{\max}$ are the standard error of β_i for the largest study and the smallest study, respectively. We may select $(P_{\max}, P_{\min})$ from a grid of $(0, 1) \times (0, 1)$, and then for each of the pair to calculate the values of a and b.

(ii) For each pair of (a, b), calculate the maximum likelihood estimates for all unknown parameters $\boldsymbol{\theta} = (\boldsymbol{\alpha}, \mu_\beta, \tau_\beta, \rho)$ by MCMC-EM algorithm discussed in the previous subsection.

(iii) Based on fitting to the funnel plots and other methods, develop sensitivity analysis about μ_β for the set of pairs of (a, b) selected in Step (i).

In sensitivity analysis, it is useful to calibrate (a, b) into a single or more statistical quantities. One method is to test the model for fit to the funnel plots. For example, the solid line shown in Figure 1 is the estimate of μ_β obtained from a meta-analysis model without assuming publication bias. It obvious not a good fit for the funnel plot. In sensitivity analysis, we test whether a meta-analysis model and a selection model with a particular pair of (a, b) give a good fit for the funnel plot. A formal approach can be given by constructing a hypothesis test. The meta-analysis model is given by Eqs. (6) and (7) and a selection model is given by (14) with the given pair of (a, b). We call this model as $M(a, b)$. Now, we replace (7) by

$$\beta_i \sim N\left(\mu_\beta + \gamma s_i, \tau_\beta^2\right), \tag{20}$$

and test the null hypothesis H_0: $\gamma = 0$ against H_1: $\gamma \neq 0$. If we reject the null hypothesis, it means that the model $M(a, b)$ does not give a good fit for the funnel plot. The related model and the pair of (a, b) are not acceptable.

In sensitivity analysis, we can therefore reject all the models $M(a, b)$ if the related P-value for the above hypothesis testing is less than say 0.05. We may also use other statistical quantities in sensitivity analysis, for example, the marginal selection probability, especially the selection probabilities for the smallest study and the largest study. We can also use the following quantity to give a rough guidance how many studies are missing

$$\sum_{i=1}^{K}\left(\frac{1 - P(S_i)}{P(S_i)}\right).$$

Some other statistics can also be used in sensitivity analysis (see, for example, Copas and Shi, 2000; Copas and Shi, 2001). In the next section, we will use a real example to detail the procedure for sensitivity analysis.

4. An illustrated example

4.1. Trend estimation for alcohol use and breast cancer

To study the association between breast cancer and alcohol consumption, a number of epidemiologic investigations have been conducted. Table 1 reports the results for such a study (Hiatt and Bawol, 1984). In this follow-up study, each row is correspond to an exposure band, including the baseline group with zero dose. The empirical odds ratio reported in the last column gives an estimate of the odds of being a case versus being a control. We apply the model defined in Section 2.1 for this data-set, and obtain the estimates $\hat{\alpha} = -4.6117$ and $\hat{\beta} = 0.0092$. The value of β measures the relation between the alcohol consumption and the risk of breast cancer. The estimate $\hat{\beta} = 0.0092$ implies that one extra drink daily (about 13 gram of alcohol) increases risk by about 12%.

Table 1
Follow-up data on alcohol use and breast cancer

Alcohol (g/day)	Assigned dose x	No. of cases	No. of controls	Empirical OR
0	0	252	24089	1.0
< 26	6.8	505	49432	1.024
39–65	46.34	68	3892	1.670
> 78	83.6	13	760	1.635

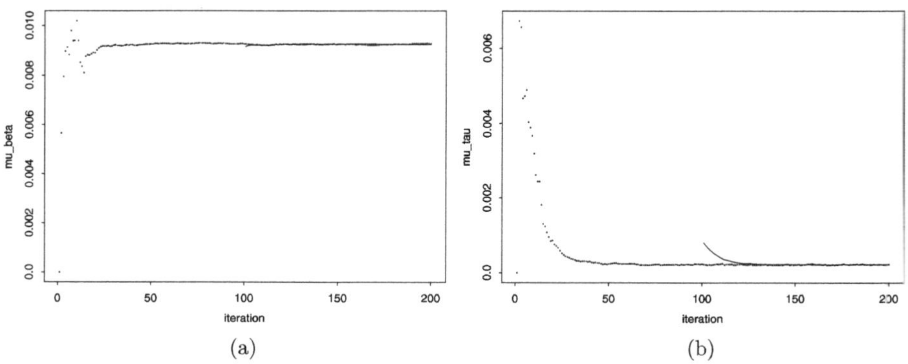

Fig. 2. The estimates of μ_β and τ_β in each iterations of MCMC-EM algorithm. The solid line gives the values of batch mean with $J = 100$.

Fourteen studies were collected by Greenland and Longnecker (1992). For each one of studies, we calculate the estimates of α_i and β_i. The values of $\hat{\beta}_i$ are presented in Figure 1. They ranges from 0.00084 (excess risk 1% for one extra drink daily) to 0.15232 (excess risk 198%). The epidemiologic findings regarding the relation between alcohol consumption and risk of breast cancer is inconsistent. It is therefore necessary to perform a meta-analysis to obtain an overall estimate for the slope β.

We use meta-analysis models (6) and (7) defined in Section 2.2 for those 14 studies. This model allows for heterogeneity between studies but without considering publication bias. The MCMC-EM algorithm is used to calculate maximum likelihood estimates. In MCMC-E step, we generate $A = 100$ random numbers and use them to calculate the conditional expectation of the log likelihood, and then update the estimates of the parameters in M-step. In each iteration, the estimates of unknown parameters $\boldsymbol{\theta}$ are calculated. The estimates of μ_β and τ_β in each iterations are plotted in Figure 2, which show that the MCMC-EM algorithm converges very fast. We use the average MCMC-EM algorithm, and calculate the batch mean with $J = 100$. The batch mean is plotted as solid line in Figure 2. They become stable after about 120 iterations. The automatic stopping rule proposed in Shi and Copas (2002) can be used here. The final results are

$$\hat{\mu}_\beta = 0.0093 \quad \text{and} \quad \hat{\tau}_\beta = 0.00022.$$

The overall estimate $\hat{\mu}_\beta = 0.0093$ indicates that the excess risk is about 12% for one extra drink daily, a quite high excess risk. In the next subsection, we will show that the risk is actually overestimated.

4.2. Publication bias and sensitivity analysis

Figure 1 gives a funnel plot of $\hat{\beta}_i$ against its standard error s_i. It shows that the small studies tend to give larger values of $\hat{\beta}_i$, showing the sign of publication bias. The solid line in this figure gives the estimate of μ_β, the estimate obtained from the model without considering publication bias, which is obvious not fitted to the funnel plot. This is also verified by using a formal hypothesis test. Using model (20) with meta-analysis model (6), we obtain estimates by MCMC-EM algorithm:

$$\mu_\beta = 0.006365, \quad \tau_\beta = 0.000153, \quad \rho = 0.979328, \quad \gamma = 0.666547.$$

A likelihood ratio test gives P-value = 0.0028 for testing H_0: $\gamma = 0$. We therefore reject the null hypothesis, indicating that the meta-analysis models (6) and (7) does not fit the funnel plot. The value of β_i may depend on the value of s_i, i.e., the size of study. There is publication bias for those 14 studies collected in this meta-analysis.

Now, we use the approach discussed in Section 3 to demonstrate how to do a sensitivity analysis. We first choose values for a pair of (a, b), for example, $a = -1.4$ and $b = 0.0073$. This corresponds to a maximum marginal selection probability of about 96% and a minimum marginal selection probability of about 30%. In this example, two studies are extremely small (see the studies with two largest values of s_i in Figure 1). By choosing the above values of a and b, the marginal selection probability corresponding to those two studies are quite small, but the selection probabilities corresponding to the other studies are actually not very small (larger than 53%). Thus the selection model with this pair of (a, b) assumes a moderate publication bias. It seems plausible for this example. By using MCMC-EM algorithm, we obtain the maximum likelihood estimates as follows

$$\hat{\mu}_\beta = 0.007, \quad \hat{\tau}_\beta = 0.002, \quad \hat{\rho} = 0.546.$$

This gives a 9% excess risk for one extra unit alcohol consumption daily, down a quarter comparing to the excess risk 12% ($\hat{\mu}_\beta = 0.009$) without considering publication bias.

By selecting $(P_{\max}, P_{\min})$ from a grid of $(0, 1) \times (0, 1)$, we get a set of pairs (a, b). For each pair, we calculate the maximum likelihood estimates for the unknown parameters, and also calculate other statistical quantities such as P-value for fit to the funnel plot discussed around Eq. (20). The results are presented in Figure 3. The numerical results for three typical pairs of (a, b) are reported in Table 2. Figure 3(i) presents contours of $\hat{\mu}_\beta$ against (a, b). The contours of the related maximum and minimum marginal selection probabilities are given in Figure 3 (ii) and (iii), respectively. The selection probabilities for both the smallest study and the largest study are close to one in upper right corner, thus it corresponds to the model with no selection bias or with very slight publication bias. The estimate of $\hat{\mu}_\beta$ is about 0.009. The first row in Table 2 gives a typical example in this area. The marginal selection probabilities for this example are 84

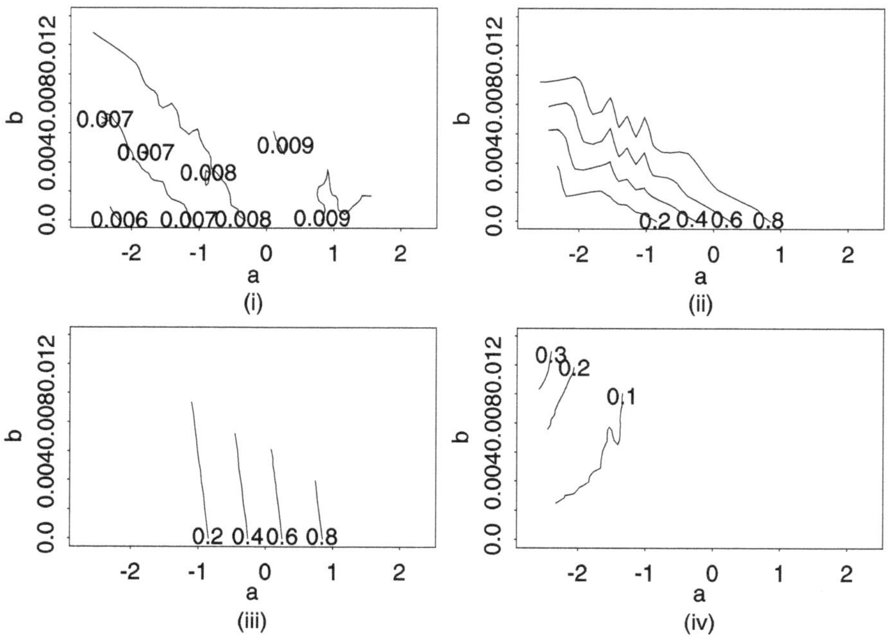

Fig. 3. Results of a sensitivity analysis: contours of different statistical quantities against (a, b), which are (i) the estimate of slope $\hat{\mu}_\beta$; (ii) selection probability for the largest study; (iii) selection probability for the smallest study; and (iv) P-value for the fit to funnel plot.

Table 2
The results of sensitivity analysis for three pairs of (a, b)

a	b	μ_β	τ_β	ρ	$P(s_{\max})^{*1}$	$P(s_{\min})^{*2}$	P-value*3
0.9292	0.0017	0.009027	0.000239	0.25739	0.8353	0.9384	0.0272
−1.4061	0.0073	0.007851	0.000225	0.54628	0.1131	0.8868	0.1075
−2.4769	0.0056	0.006544	0.000206	0.44224	0.0100	0.319	0.1858

*1 Marginal selection probability for the largest study.

*2 Marginal selection probability for the smallest study.

*3 P-value for fit to the funnel plot discussed around (20).

and 94% for the smallest and the largest studies, respectively. The P-value for fit to the funnel plot is less than 3%, meaning this pair (a, b) is not acceptable if we use 5% as a test level. The contours of the P-value for fit to the funnel plot are given in Figure 3(iv). It shows that the P-value is less than 0.05 in this corner.

In the middle of the contour plot, corresponding to moderate publication bias assumption, the estimate is about 0.007. Table 2 reports the numerical results for the pair discussed in the second paragraph in this subsection. The P-value for fit to the funnel plots is 11%. It indicates that the related model is acceptable.

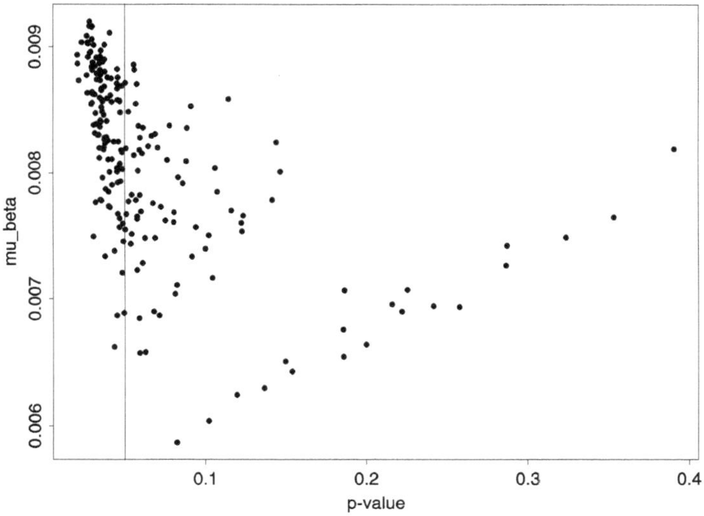

Fig. 4. The estimate $\hat{\mu}_\beta$ against P-values for fit to the funnel plot. The solid vertical line gives 0.05 test level.

The lower left corner corresponds to a severe publication bias. The estimates of $\hat{\mu}_\beta$ is down to about 0.005, i.e. the excess risk is only about 6%, half of the excess risk without considering publication bias. The third row in Table 2 gives a typical example in this corner. Although the P-value for fit to the funnel plot is 19%, the marginal selection probabilities are too extreme to be accepted.

In Figure 4, we plot the estimates of μ_β against the P-value for fit to the funnel plot for all pairs of (a, b). If we take 5% as the test level, then all the pairs of (a, b) with P-values less than 0.05 are not acceptable. The estimate of μ_β is therefore down to at most 0.0085 from 0.0093. Combining the above discussion for sensitivity analysis by using P-values and the marginal selection probabilities, it seems reasonable to assume a moderate publication bias for this example. The estimate of μ_β should be around 0.006 and 0.007. The conventional estimate $\hat{\mu}_\beta = 0.0093$ is overestimated.

5. Discussion and further development

In this chapter we have conducted a meta-analysis for binary data in trend estimation. A logistic regression model and a latent variable model are used to combine the data from a set of studies and give an overall estimation for the parameters of interest. This model allows for heterogeneity between studies. A MCMC-EM algorithm is used for implementation. We then use a sensitivity analysis approach to address the problem of publication bias by defining a selection model. Although we focus on the problem of trend estimation in this chapter, there is no major difficulty to extend the approach to a more general problem, for example, for multi-nominal data with a general latent model.

We use the unconditional likelihood approach in this chapter. In some circumstances, unconditional estimates may be biased (see, for example, (Cox and Snell, 1989; Shi and Copas, 2002)). If we use a conditional likelihood approach, fixed $Z_0+Z_1+\cdots+Z_n = t$, the conditional likelihood for a single study is

$$p(z_0, z_1, \dots, z_n | Z_0 + Z_1 + \dots + Z_n = t) = \frac{\binom{m_0}{z_0} \prod_{j=1}^{n} \binom{m_j}{z_j} \exp(\beta z_j x_j)}{\sum_{u_1,\dots,u_n} \binom{m_0}{t-u_1-\dots-u_n} \prod_{j=1}^{n} \binom{m_j}{u_j} \exp(\beta u_j x_j)},$$

for all $u_1, \dots, u_n$ satisfying

$$0 \leqslant u_j \leqslant m_j, \quad t - m_0 \leqslant u_1 + \dots + u_n \leqslant t.$$

Computation for such a likelihood is tedious. It needs an efficient algorithm or a good approximation.

In trend estimation, the exposure levels for each subject are often not recorded exactly but grouped into class intervals; see, for example, the first column in Table 1. We used an assigned value for each group in this chapter. But for the grouped dose level, we may suppose that exposure is an underlying continuous covariate belonging to observable intervals J_{ij}, so that each subject in the jth group in the ith study has $x \in J_{ij}$. Suppose the exposure levels of all individuals in a particular study are sampled from the same distribution with the probability density function $f(x)$. The probability of being a case, given dose x, is given by $\pi(x)$ in (1), (2) and (3), the probability that an individual in class interval J is a case is

$$\pi_J = \frac{\int_J \pi(x) f(x)\,\mathrm{d}x}{\int_J f(x)\,\mathrm{d}x} \tag{21}$$

where $\pi(x) = \exp(\alpha + \beta x)/\{1 + \exp(\alpha + \beta x)\}$. It is not trivial to extend the approach discussed in this chapter to the problem with grouped dose-level.

It is worth a further research for those problems.

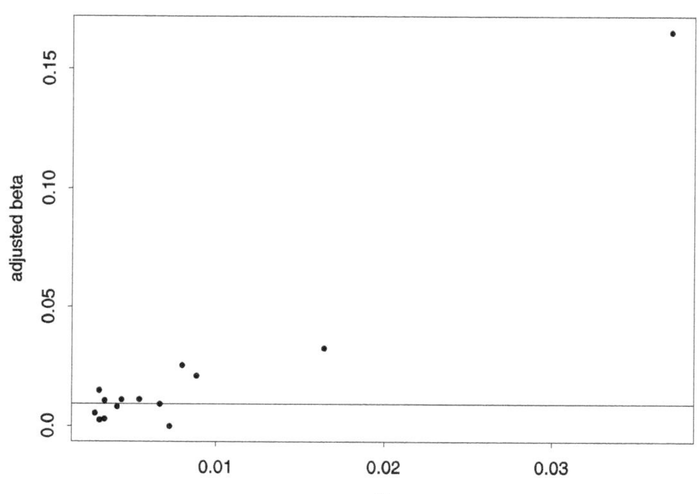

Fig. 5. Funnel plots of $\hat{\beta}_i$ against its standard error s_i, where $\hat{\beta}_i$ is calculated by using adjusted odds ratio.

In this chapter, we use the exact binomial distribution and the latent variable model to study the overall relation between alcohol use and breast cancer. If we believe there is moderate publication bias for the data collected (it seems plausible for this example), then the excess risk for one extra unit alcohol consumption is about 9%. By using normal approximation and the values of adjusted odds ratios, the funnel plot is given by Figure 5. It shows the sign of severer publication bias than Figure 1. Shi and Copas (2004) reported that the excess risk is only about 5% assuming a moderate publication bias. The detailed discussion for this example is given in Shi and Copas (2004). Unfortunately, we have not been able to extract the raw data which is used to calculate the adjusted odds ratio. Nevertheless, the results obtained in this chapter coincide with Shi and Copas' (2004) findings that the excess risk estimate 12% obtained by the conventional method is overestimated. The real risk may be much less than this percentage.

References

Booth, J.G., Hobert, J.P. (1999). Maximum generalised linear mixed model likelihoods with an automated Monte Carlo EM algorithm. *J. R. Statist. Soc. B* **61**, 265–285.

Carlin, B.P., Louis, T.A. (2000). *Bayes and Empirical Bayes Methods for Data Analysis*, second ed. Chapman & Hall.

Copas, J.B., Shi, J.Q. (2000). Meta-analysis, funnel plots and sensitivity analysis. *Biostatistics* **1**, 247–262.

Copas, J.B., Shi, J.Q. (2001). A sensitivity analysis for publication bias in systematic reviews. *Statistical Methods in Medical Research* **10**, 251–265.

Cox, D.R., Snell, E.J. (1989). *Analysis of Binary Data*, second ed. Chapman and Hall, London.

Crouch, E.A.C., Spiegelman, D. (1990). The evaluation of integrals of the form $\int_{-\infty}^{\infty} f(t)\exp(-t^2)\,dt$: application to logistic-normal models. *J. Am. Statist. Assoc.* **85**, 464–469.

Duval, S., Tweedie, R. (2000). A nonparametric 'Trim and Fill' methods of accounting for publication bias in meta-analysis. *J. Am. Stat. Assoc.* **95**, 89–98.

Greenland, S., Longnecker, M.P. (1992). Methods for trend estimation from summarized dose-response data, with applications to meta-analysis. *Am. J. Epidemiology* **135**, 1301–1309.

Hastings, W.K. (1970). Monte Carlo sampling methods using Markov chains and their applications. *Biometrika* **57**, 97–109.

Hiatt, R.A., Bawol, R.D. (1984). Alcoholic beverage consumption and breast cancer incidence. *Am. J. Epidemiology* **120**, 676–683.

Lee, S.Y., Tang, N.S. (2006). Bayesian analysis of structure equation models with mixed exponential family and ordered categorical data. Technical report. Department of Statistics, Chinese University of Hong Kong.

Little, R.J.A., Rubin, D.B. (2002). *Statistical Analysis with Missing Data*, second ed. John Wiley & Sons, New Jersey.

Louis, T.A. (1982). Finding the observed information matrix when using the EM algorithm. *J. R. Statist. Soc. B* **44**, 226–233.

Metropolis, N., Rosenbluth, A.W., Rosenbluth, M.N., Teller, A.H., Teller, E. (1953). Equations of state calculations by fast computing machines. *J. Chem. Phys.* **21**, 1087–1091.

Shi, J.Q., Copas, J.B. (2002). Publication bias and meta-analysis for 2 × 2 tables: an average Markov chain Monte Carlo EM algorithm. *J. R. Statist. Soc. B* **64**, 221–236.

Shi, J.Q., Copas, J.B. (2004). Meta-analysis for trend estimation. *Statistics in Medicine* **23**, 3–19.

Song, X.Y., Lee, S.Y. (2006). Bayesian analysis of latent variable models with nonignorable missing outcomes from exponential family. Technical report. Department of Statistics, Chinese University of Hong Kong.

Sutton, A.J., Duval, S.J., Tweedie, R.L., Abrams, K.R., Jones, D.R. (2000). Empirical assessment of effect of publication bias on meta-analyses. *Br. Med. J.* **320**, 1574–1577.

Wei, G.C.G., Tanner, M.A. (1990). A Monte Carlo implementation of the EM algorithm and the poor man's data augmentation algorithm. *J. Am. Statist. Assoc.* **85**, 699–704.

Handbook of Computing and Statistics with Applications, Vol. 1
ISSN: 1871-0301

DOI: 10.1016/S1871-0301(06)01013-4

13

Analysis of Multisample Structural Equation Models with Applications to Quality of Life Data

Xin-Yuan Song

Abstract

A maximum likelihood (ML) approach for analyzing a multisample structural equation model with missing ordered categorical outcomes is introduced. In the ML estimation, the observed data are augmented with the real missing data and the hypothetical missing data that involve the latent variables and the unobserved continuous responses underlying the ordered categorical data. The resulting problem is handled by the implementation of a Monte Carlo EM algorithm to produce ML estimates of the unknown parameters. A path sampling procedure is proposed to compute the complicated observed-data log-likelihood, and eventually the Bayesian information criterion for model comparison and hypothesis testing. The proposed methodologies are used to analyze a two-sample quality of life (QOL) data set to investigate the similarities and differences of QOL between people in Western countries and in China.

Keywords: Latent variables; Maximum likelihood; MCEM algorithm; Gibbs sampler; Path sampling; QOL

1. Introduction

Structural equation models (SEMs) are widely appreciated in behavioral, educational, sociological, and medical research in analyzing multivariate correlated data from latent variables. SEMs allow one to evaluate a series of simultaneous hypotheses about the relationships of some latent and manifest variables with other variables, on the basis of nonexperimental data. In the past quarter of a century, they have drawn a great deal of attention in psychometrics and sociometrics, both in terms of theoretical developments and practical applications (Bollen, 1989; Jöreskog and Sörbom, 1996; Bentler, 1992). Although not to the extent that they have been used in behavioral, educational, and social sciences, the LISREL (Jöreskog and Sörbom, 1996) or EQS (Bentler, 1992) programs have been widely used in other researches. In fact, research in various disciplines has outlined the usefulness of this kind of latent variable model with various types of data. For instance, in a review of methodology, Francis discussed the implications of structural equation models for necropsychological theory and practice (Francis, 1988). Bentler and Stein gave a very comprehensive review with many important medical applications (Bentler and Stein, 1992). In addition to those cited by Bentler and Stein (1992), SEMs have recently been applied in assessing the inter-relationships among components of metabolic syndrome (Chan et al., 1996); in analyzing ecological and evolutionary biology data (Pugesek et al., 2003); and in investigating the behavior of twins in genetics (Boomsma and Molenaar, 1987; Dolan et al., 1991).

More than a dozen standard programs use structural equation models to cope with the high demand in various fields. The statistical development of most software occurs under the assumption that the data are continuous measurements with a multivariate normal distribution. However, in practice studies, we frequently encounter dichotomous or ordered categorical variables. Typically, respondents are asked to select answers from 'yes' or 'no' about the existence of a symptom, 'strongly disagree', 'disagree', 'no opinion', 'agree', 'strongly agree' about a policy, 'feeling better', 'no change', 'worse' about the effect of a drug, or 'satisfactory', 'no opinion', 'unsatisfactory' about the performance of a staff, etc. Moreover, the empirical distribution of discrete observations is often skewed. Hence, the basic assumption that the data come from a continuous normal distribution is clearly violated. Maximum likelihood (ML) analysis of structural equation models with ordered categorical data is not straightforward. A difficult computational problem is encountered in evaluating the cell probabilities that are induced by the ordered categorical outcomes. Some two-stage methods have been proposed to reduce the computational burden in evaluating the high-dimensional integrals that are associated with the cell probabilities (Jöreskog and Sörbom, 1996; Lee et al., 1995). Shi and Lee (2000) pointed out that the two-stage estimators are not statistically optimal as the ML estimator, and need to invert at each iteration of its minimization procedure a huge matrix, the dimensions of which increase very rapidly with the number of manifest variables. They further developed a Monte Carlo EM type algorithm for ML analysis of a factor analysis model with mixed continuous and ordered categorical outcomes (see also, Wei and Tanner, 1990).

In this article, for handling more complex practical data, the ML method proposed in Shi and Lee (2000) in analyzing ordered categorical data is generalized in four directions. First, the factor analysis model is generalized to a more general SEM, for assessing the effect of the exogenous (independent) latent variables on the endogenous (dependent) variables. Second, the single sample model is extended to a multisample SEM, for analyzing the behavior of different treatment groups and/or cultural groups, etc. Third, the ML methodology is extended to accommodate data that are missing at random (MAR) (Little and Rubin, 1987). In most missing data sets in medical research, the number of missing patterns are large, and the sample size within some missing patterns can be small. Hence, the approach with existing software in structural equation models (Jöreskog and Sörbom, 1996; Bentler, 1992; Muthen and Muthen, 2001), which employs a multiple group analysis by treating observations in a missing pattern as an independent group, is not practical. Finally, hypothesis testing via the Bayesian information criterion (BIC) is proposed for comparisons across different groups.

The present work is motivated by the increasing recognition that measures of health related quality of life (QOL) have great value for clinical work and the planning and evaluation of health care. It has been generally accepted that QOL is a multidimensional concept that is best evaluated by a number of different latent domains (variables) such as health status, physical function, mental status, and social relationships (Staquet et al., 1998; Drota, 1998). Many questionnaires in QOL research are divided into several 'scales', each of which comprises several related items for measuring the corresponding latent variables. To improve consistency and efficiency, the related items are combined via the measurement equation of the SEM to form a latent variable (factor) for analyzing the corresponding latent domain. The relationship of these latent domains (variables) are assessed via the structural equation of the model. As the items in the questionnaire are often ordered categorical, most QOL data involve multidimensional discrete observations. Moreover, as missing data are frequently encountered and it is sometimes necessary to compare different groups (e.g., different treatment or cultural groups), it is necessary to develop ML methods for multisample SEMs, in the context of incomplete ordered categorical variables. To illustrate the newly developed ML methodologies, we will analyze the WHOQOL-BEEF (Power et al., 1999) data set, which contains twenty-six ordered categorical items, and is obtained from many Western countries and China. The underlying instrument has been shown to be useful to health professionals (Power et al., 1999). We note the presence of missing data and that several items are heavily skewed to the right. Treating these ordinal data as coming from a normal distribution is likely to produce biased results.

We will develop ML methods for analyzing a multisample SEM with missing ordered categorical outcomes, and apply these methods to study the similarities and differences of QOL in the different cultural groups. The statistical inferences include ML estimation and hypothesis testing via the BIC criterion. A MCEM algorithm is implemented to produce the ML estimate, and a path sampling (Geman and Geman, 1984) procedure is constructed to compute the complicated observed-data log-likelihood that involves intractable integrals.

The rest of the paper is organized as follows. Section 2 presents the multisample SEM, and describes the missing ordered categorical data. The ML approaches for estimation and model comparison (hypothesis testing) are discussed in Section 3. Section 4 describes the results that are obtained from the analysis of the WHOQOL data (Power et al., 1999). Section 5 provides a discussion, and technical details are given in Appendices A, B.

2. A multisample SEM with missing ordered categorical variables

2.1. The model and the data

We consider G independent groups of individuals that may represent populations of patients who receive different treatments, or are from different cultures, etc. For $g = 1$,

$\ldots, G$, let $\mathbf{v}_i^{(g)}$ be the random vector that contains $j = 1, \ldots, p$ manifest variables that correspond to the ith observation (subject) in the gth group, $i = 1, \ldots, n_g$. Similar to the LISREL model, $\mathbf{v}_i^{(g)}$ is related to a $q \times 1$ random vector of latent variables $\boldsymbol{\omega}_i$ by the following measurement equation:

$$\mathbf{v}_i^{(g)} = \boldsymbol{\mu}^{(g)} + \boldsymbol{\Lambda}^{(g)}\boldsymbol{\omega}_i^{(g)} + \boldsymbol{\varepsilon}_i^{(g)}, \quad (1)$$

where $\boldsymbol{\mu}^{(g)}$ is the vector of the intercepts, $\boldsymbol{\Lambda}^{(g)}$ is the parameter matrix of regression coefficients that reflects the relation of manifest variables in $\mathbf{v}_i^{(g)}$ with the latent variables in $\boldsymbol{\omega}_i^{(g)}$, and $\boldsymbol{\varepsilon}_i^{(g)}$ is a random vector of the measurement errors. It is assumed that $\boldsymbol{\omega}_i^{(g)}$ and $\boldsymbol{\varepsilon}_i^{(g)}$ are independent, and the distribution of $\boldsymbol{\varepsilon}_i^{(g)}$ is $N[\mathbf{0}, \boldsymbol{\Psi}^{(g)}]$, where $\boldsymbol{\Psi}^{(g)}$ is a diagonal covariance matrix. Let $\boldsymbol{\omega}_i^{(g)} = (\boldsymbol{\eta}_i^{(g)'}, \boldsymbol{\xi}_i^{(g)'})$, where $\boldsymbol{\eta}_i^{(g)}$ represents the vector of endogenous latent variables, and $\boldsymbol{\xi}_i^{(g)}$ represents the vector of the exogenous latent variables. An important component of the proposed model is the following structural equation that addresses relationships of $\boldsymbol{\eta}_i^{(g)}$ and $\boldsymbol{\xi}_i^{(g)}$:

$$\boldsymbol{\eta}_i^{(g)} = \mathbf{B}^{(g)}\boldsymbol{\eta}_i^{(g)} + \boldsymbol{\Gamma}^{(g)}\boldsymbol{\xi}_i^{(g)} + \boldsymbol{\delta}_i^{(g)}, \quad (2)$$

where $\mathbf{B}^{(g)}$ and $\boldsymbol{\Gamma}^{(g)}$ are parameter matrices of regression coefficients such that $\mathbf{I} - \mathbf{B}^{(g)}$ is nonsingular, and $\boldsymbol{\delta}_i^{(g)}$ is the random vector of error measurements that is independent of $\boldsymbol{\xi}_i^{(g)}$. It is further assumed that $\boldsymbol{\xi}_i^{(g)}$ is distributed as $N[\mathbf{0}, \boldsymbol{\Phi}^{(g)}]$, and $\boldsymbol{\delta}_i^{(g)}$ is distributed with $N[\mathbf{0}, \boldsymbol{\Psi}_\delta^{(g)}]$, where $\boldsymbol{\Psi}_\delta^{(g)}$ is a diagonal covariance matrix.

To handle the ordered categorical outcomes, suppose that $\mathbf{v}_i^{(g)} = (\mathbf{x}_i^{(g)'}, \mathbf{y}_i^{(g)'})'$, where $\mathbf{x}_i^{(g)}$ is an observable subvector of continuous responses, and $\mathbf{y}_i^{(g)}$ is a subvector of unobservable continuous responses, the information of which is reflected by an observable ordered categorical vector $\mathbf{z}_i$ with equal dimension. In a generic sense, an ordered categorical variable $z_h^{(g)}$ is defined with its underlying latent continuous random variable $y_h^{(g)}$ by:

$$z_h^{(g)} = k \quad \text{if } \alpha_{h,k}^{(g)} \leqslant y_h^{(g)} < \alpha_{h,k+1}^{(g)}, \ k = 1, \ldots, b_h^{(g)}, \quad (3)$$

where $\{-\infty = \alpha_{h,1}^{(g)} < \alpha_{h,2}^{(g)} < \cdots < \alpha_{h,b_h}^{(g)} < \alpha_{h,b_{h+1}}^{(g)} = \infty\}$ is the set of threshold parameters that define the categories, and $b_h^{(g)}$ is the number of categories for the ordered categorical variable $z_h^{(g)}$. Note that the categories can be unequally spaced under this formulation. To deal with the missing data, let $\mathbf{x}_i^{(g)} = \{\mathbf{x}_{io}^{(g)}, \mathbf{x}_{im}^{(g)}\}$ and $\mathbf{z}_i^{(g)} = \{\mathbf{z}_{io}^{(g)}, \mathbf{z}_{im}^{(g)}\}$, where $\mathbf{x}_{io}^{(g)}$ and $\mathbf{z}_{io}^{(g)}$ represent the observed data, and $\mathbf{x}_{im}^{(g)}$ and $\mathbf{z}_{im}^{(g)}$ represent the missing data. For a fully observed data point, $\mathbf{x}_{im}^{(g)}$ and $\mathbf{z}_{im}^{(g)}$ are empty. We assume that the missing data are missing at random (MAR); that is, the data mechanism that has caused the missing data depends only on the observed data and not on the missing data themselves (Little and Rubin, 1987). For any $i \neq j$, we allow that the dimensions of $\mathbf{x}_{io}^{(g)}$ and $\mathbf{z}_{io}^{(g)}$ can be different from the dimensions of $\mathbf{x}_{jo}^{(g)}$ and $\mathbf{z}_{jo}^{(g)}$. Therefore, the number of missing patterns can be large, and the sample sizes within some missing patterns can be small. Consequently, the multiple group approach that is commonly used by the common software in structural equation models by treating observations in a missing pattern as an independent group is not practical.

2.2. *Identification*

To tackle the identification problem, we have to pay attention to the following issues. There are two kind of indeterminacies for the multisample SEM with ordered categorical variables. First, the SEM that is defined by (1) and (2) is not identified. This indeterminacy can be solved by the common method of fixing appropriate elements in $\boldsymbol{\Lambda}^{(g)}$, $\mathbf{B}^{(g)}$, and/or $\boldsymbol{\Gamma}^{(g)}$ at preassigned values. The other indeterminacy is induced by the ordered categorical variable. Consider an ordered categorical variable $z_h^{(g)}$ that is defined by a set of thresholds $\alpha_{h,k}^{(g)}$ and an underlying latent continuous variable $y_h^{(g)}$ with a distribution $N[\mu_h^{(g)}, \sigma_h^{2(g)}]$. The indeterminacy is caused by the fact that $\alpha_{h,k}^{(g)}, \mu_h^{(g)}, \sigma_h^{2(g)}$ are not simultaneously estimable. For a given group g, a common method to solve this identification problem with respect to the hth ordered categorical variable is to fix $\alpha_{h,2}^{(g)}$ and $\alpha_{h,b_h}^{(g)}$ at preassigned values (Lee et al., 1995; Shi and Lee, 2000; Lee et al., 2005). For example, we may fix $\alpha_{h,2}^{(g)} = \Phi^{*-1}(f_{h,2}^{(g)})$, and $\alpha_{h,b_h}^{(g)} = \Phi^{*-1}(f_{h,b_h}^{(g)})$, where Φ^* is the cumulative function of $N[0, 1]$, $f_{h,2}^{(g)}$ and $f_{h,b_h}^{(g)}$ are the frequencies of the first category and the cumulative frequencies of categories with $z_h^{(g)} < b_h^{(g)}$. For analyzing multisample models with interest in group comparisons, it is important to impose conditions for identification of the ordered categorical variables such that the latent continuous variables have the same scale among the groups. To achieve this, we can choose the first group as the reference group, and identify its ordered categorical variables by fixing both end thresholds as above. Then, for any h and $g \neq 1$, we impose the following restrictions,

$$\alpha_{h,k}^{(g)} = \alpha_{h,k}^{(1)}, \quad k = 1, \ldots, b_h^{(g)}, \tag{4}$$

on the thresholds for every ordered categorical variable $z_h^{(g)}$. Under these identification conditions, the unknown parameters in the groups should be interpreted in a relative sense, compared over groups. Note that when different reference groups are used, relations over groups are unchanged. Hence, the statistical inferences are unaffected by the choice of the reference group. Clearly, the compatibility of the groups is reflected by the differences of the parameter estimates.

3. ML analysis

In this section, we describe a ML approach that produces consistent ML estimates with optimal statistical properties, standard error estimates, scores of the latent variables, the observed-data log-likelihood, and the BIC statistic for model comparison, on the basis of the fully and partially observed data of mixed continuous and ordered categorical outcomes. For $g = 1, \ldots, G$, let $\mathbf{X}_o^{(g)} = \{\mathbf{x}_{io}^{(g)}, i = 1, \ldots, n_g\}$,

$\mathbf{X}_m^{(g)} = \{\mathbf{x}_{im}^{(g)}, i = 1, \ldots, n_g\}$, $\mathbf{Z}_m^{(g)} = \{\mathbf{z}_{im}^{(g)}, i = 1, \ldots, n_g\}$, $\mathbf{Z}_o = \{\mathbf{Z}_o^{(1)}, \ldots, \mathbf{Z}_o^{(G)}\}$, and $\mathbf{Z}_m = \{\mathbf{Z}_m^{(1)}, \ldots, \mathbf{Z}_m^{(G)}\}$. Hence, $\mathbf{X}_o$ and $\mathbf{X}_m$ are the observed and the missing continuous data, and $\mathbf{Z}_o$ and $\mathbf{Z}_m$ are the observed and the missing ordered categorical data.

3.1. Estimation

Let $\boldsymbol{\theta}^{(g)}$ be the unknown parameter vector in the identified model that corresponds to the gth group. In multisample analysis, a certain type of parameter in $\boldsymbol{\theta}^{(g)}$ is often hypothesized to be equal to that type of parameter in $\boldsymbol{\theta}^{(h)}$. For example, we impose restrictions on the thresholds, and we often test $\boldsymbol{\Lambda}^{(1)} = \cdots = \boldsymbol{\Lambda}^{(G)}$, $\boldsymbol{\Phi}^{(1)} = \cdots = \boldsymbol{\Phi}^{(G)}$, and/or $\boldsymbol{\Gamma}^{(1)} = \cdots = \boldsymbol{\Gamma}^{(G)}$. Hence, we allow common parameters in $\boldsymbol{\theta}^{(1)}, \ldots, \boldsymbol{\theta}^{(G)}$. Let $\boldsymbol{\theta}$ be the vector that contains all unknown distinct parameters in $\boldsymbol{\theta}^{(1)}, \ldots, \boldsymbol{\theta}^{(G)}$, by definition, the ML estimate of $\boldsymbol{\theta}$ is a vector that maximizes the observed data log-likelihood $L_o(\mathbf{X}_o, \mathbf{Z}_o; \boldsymbol{\theta})$. However, due to the existence of missing data, and the discrete nature of the ordered categorical variables, $L_o(\mathbf{X}_o, \mathbf{Z}_o; \boldsymbol{\theta})$ involves very complicated multiple integrals. Maximizing this function for obtaining the ML estimate is difficult. To solve the problem, we augment the observed data with the real missing data and the hypothetical missing data that contain the latent variables and the latent continuous responses. Let $\mathbf{Y}_o$ and $\mathbf{Y}_m$ be the latent continuous data sets that correspond to $\mathbf{Z}_o$ and $\mathbf{Z}_m$, and let $\mathbf{V}_m = (\mathbf{X}_m, \mathbf{Y}_m)$. Moreover, let $\boldsymbol{\Omega} = \{\boldsymbol{\Omega}^{(1)}, \ldots, \boldsymbol{\Omega}^{(G)}\}$ be the collection of all latent variables, where $\boldsymbol{\Omega}^{(g)} = \{\boldsymbol{\omega}_i^{(g)}, i = 1, \ldots, n_g\}$. In the ML analysis, the observed data set $\mathbf{D}_o = (\mathbf{X}_o, \mathbf{Z}_o)$ is augmented with the missing data set $\mathbf{D}_m = (\mathbf{X}_m, \mathbf{Y}_o, \mathbf{Y}_m, \boldsymbol{\Omega}) = (\mathbf{Y}_o, \mathbf{V}_m, \boldsymbol{\Omega})$ to form a complete data set $\mathbf{D} = (\mathbf{D}_o, \mathbf{D}_m)$. Note that once $\mathbf{Y}_m$ is given, $\mathbf{Z}_m$ is not important. Let $L_c(\mathbf{D}; \boldsymbol{\theta})$ be the complete-data log-likelihood. The ML estimate of $\boldsymbol{\theta}$ is obtained by applying the Monte Carlo EM (MCEM) algorithm (Wei and Tanner, 1990). This algorithm is implemented as follows. At the rth iteration with a current value $\boldsymbol{\theta}_{(r)}$, it involves the following E-step and M-step.

E-step: Evaluate $Q(\boldsymbol{\theta}; \boldsymbol{\theta}_{(r)}) = E\{L_c(\mathbf{D}; \boldsymbol{\theta}) | \mathbf{D}_o, \boldsymbol{\theta}_{(r)}\}$, where the expectation is taken with respect to the joint conditional distribution of $\mathbf{D}_m$ given $\mathbf{D}_o$ at $\boldsymbol{\theta}_{(r)}$. It can be shown that the log-likelihood function based on the complete-data set $\mathbf{D}$ is given by

$$
\begin{aligned}
&L_c(\mathbf{D}; \boldsymbol{\theta}) \\
&\quad = \log p(\mathbf{D}|\boldsymbol{\theta}) \\
&\quad = -\frac{1}{2}\sum_{g=1}^{G}\sum_{i=1}^{n_g}\Big\{(p+q)\log(2\pi) + \log\big|\boldsymbol{\Psi}^{(g)}\big| + \log\big|\boldsymbol{\Psi}_\delta^{(g)}\big| + \log\big|\boldsymbol{\Phi}^{(g)}\big| \\
&\quad\quad + \boldsymbol{\xi}_i^{(g)\prime}\boldsymbol{\Phi}^{(g)^{-1}}\boldsymbol{\xi}_i^{(g)} + \big(\mathbf{v}_i^{(g)} - \boldsymbol{\mu}^{(g)} - \boldsymbol{\Lambda}^{(g)}\boldsymbol{\omega}_i^{(g)}\big)'\boldsymbol{\Psi}^{(g)^{-1}} \\
&\quad\quad \times \big(\mathbf{v}_i^{(g)} - \boldsymbol{\mu}^{(g)} - \boldsymbol{\Lambda}^{(g)}\boldsymbol{\omega}_i^{(g)}\big) \\
&\quad\quad \times \big(\boldsymbol{\eta}_i^{(g)} - \mathbf{B}^{(g)}\boldsymbol{\eta}_i^{(g)} - \boldsymbol{\Gamma}^{(g)}\boldsymbol{\omega}_i^{(g)}\big)'\boldsymbol{\Psi}_\delta^{(g)^{-1}}\big(\boldsymbol{\eta}_i^{(g)} - \mathbf{B}^{(g)}\boldsymbol{\eta}_i^{(g)} - \boldsymbol{\Gamma}^{(g)}\boldsymbol{\omega}_i^{(g)}\big)\Big\} \\
&\quad\quad + \frac{1}{2}\sum_{g=1}^{G}\sum_{i=1}^{n_g}\log I\big(\mathbf{y}_i^{(g)} \in \mathbf{A}_i^{(g)}\big),
\end{aligned}
\tag{5}
$$

where $I(y \in A)$ is a indicator function which takes the value 1 if $y \in A$ zero otherwise, and

$$\mathbf{A}_i^{(g)} = (\alpha_{1,z_{i1}^{(g)}}^{(g)}, \alpha_{1,z_{i1}^{(g)}+1}^{(g)}] \times \cdots \times (\alpha_{s,z_{is}^{(g)}}^{(g)}, \alpha_{s,z_{is}^{(g)}+1}^{(g)}],$$

where s is the number of ordered categorical variables, and $z_{ij}^{(g)}$ is the jth element of $\mathbf{z}_i^{(g)}$. It should be noted that the likelihood function based on the observed data is much more complicated than the function given in (5).

M-step: Update $\boldsymbol{\theta}_{(r)}$ to $\boldsymbol{\theta}_{(r+1)}$ by the maximum of $Q(\boldsymbol{\theta}; \boldsymbol{\theta}_{(r)})$.

The conditional expectations at the E-step are approximated by a large number of observations that are generated from the conditional distribution of $\mathbf{D}_m$ given $\mathbf{D}_o$ at $\boldsymbol{\theta}_{(r)}$ (Wei and Tanner, 1990). The idea is to approximate the expectation by the sample mean of a large sample of observations that are simulated from the target distribution. Hence, the main task is to simulate a sufficient amount of observations from the target conditional distribution. We use the well-known Markov chain Monte Carlo (MCMC) methods, namely the Gibbs sampler (Geman and Geman, 1984) and the Metropolis–Hastings (MH) algorithm (Metropolis et al., 1953; Hastings, 1970), to complete this task. Technical details are presented in Appendix A. Note that one component in the Gibbs sampler is sampling $\mathbf{V}_m$ from its conditional distribution given $\mathbf{D}_o$, $\mathbf{Y}_o$, and $\boldsymbol{\Omega}$ at $\boldsymbol{\theta}_{(r)}$. As observations in the independent groups are mutually independent, they can be simulated separately, one by one, from the corresponding conditional distribution (see Eq. (A.2) in Appendix A). Hence, the number of missing patterns, and the possible small sample sizes of the patterns do not create any problem in our MCMC procedure. Let $\{\mathbf{D}_{mj}; j = 1, \ldots, J\}$ be a sufficiently large sample that is simulated by the MCMC methods. Hence,

$$Q(\boldsymbol{\theta}; \boldsymbol{\theta}_{(r)}) \doteq \frac{1}{J} \sum_{j=1}^{J} L_c(\mathbf{D}_o, \mathbf{D}_{mj}; \boldsymbol{\theta}_{(r)}). \tag{6}$$

As observations that are obtained from the Monte Carlo simulation are used to complete the E-step, the algorithm is regarded as a Monte Carlo EM (MCEM) algorithm.

The M-step updates the unknown parameters by maximizing the conditional expectation of the complete-data log-likelihood that is obtained in the E-step. The unknown thresholds in $\boldsymbol{\alpha}$ are updated by the method that is given by Lee and Shi (2001). The other structural parameters are updated by solving the following system of equations:

$$\frac{\partial Q(\boldsymbol{\theta}; \boldsymbol{\theta}_{(r)})}{\partial \boldsymbol{\theta}} = 0. \tag{7}$$

The common Newton–Raphson or Fletcher–Powell algorithm can be applied. However, an attractive method is conditional maximization (Meng and Rubin, 1993).

The convergence of the MCEM algorithm is monitored by the method that was given by Shi and Copas (2002). This method has been found satisfactory in the ML approaches of a number of SEMs (Lee and Song, 2004a, 2004b). The estimates of the latent variable scores can be obtained from the sample of observations $\{\boldsymbol{\Omega}_j = (\boldsymbol{\Omega}_j^{(1)}, \ldots, \boldsymbol{\Omega}_j^{(G)}); j =$

$1, \ldots, J\}$ that are simulated in the E-step at the last iteration of the MCEM algorithm as follows:

$$\hat{\boldsymbol{\omega}}_i^{(g)} = \frac{1}{J}\sum_{j=1}^{J}\boldsymbol{\omega}_{ij}^{(g)}, \tag{8}$$

where $\boldsymbol{\omega}_{ij}^{(g)}$ is in $\boldsymbol{\Omega}_j^{(g)}$.

The selection of the sample size, say J, in the Monte Carlo E-step is an issue in the MCEM algorithm. To decrease the Monte Carlo error at the E-step, J should be large. As it is inefficient to start with a large J when $\hat{\boldsymbol{\theta}}^{(j)}$ is far from the ML estimate, it has been suggested that J should be increased from one iteration to the next (Wei and Tanner, 1990; Booth and Hobert, 1999). One method is to take $J = J_1 + J_2 j$, for some positive integers J_1 and J_2. As J_1 is not related to j, its size does not have a significant impact. The choice of J_2 depends on the complexity of the underlying problem, and should be approached on a problem-by-problem basis. Convergence can either be monitored by ratios of the observed-data likelihood values at consecutive iterations (Meng and Schilling, 1996), computed via bridge sampling (Meng and Wong, 1996), or computed by absolute or relative error of the parameter estimates. Alternatively, Shi and Copas (2002) proposed the following scheme without iteratively increasing J. After the mth iteration, they computed

$$\bar{\boldsymbol{\theta}}^{(j)} = \frac{1}{m}(\boldsymbol{\theta}^{(j-m+1)} + \cdots + \boldsymbol{\theta}^{(j)}), \tag{9}$$

and monitored convergence via the stopping rule: for given small values δ_1 and δ_2 (e.g., 0.001), the procedure is stopped if

$$\frac{\|\bar{\boldsymbol{\theta}}^{(j)} - \bar{\boldsymbol{\theta}}^{(j-\gamma_0)}\|}{\|\bar{\boldsymbol{\theta}}^{(j-\gamma_0)}\| + \delta_1} \tag{10}$$

is smaller than some predetermined small value δ_2. To avoid the danger of premature stopping, a value of $\gamma_0 = 5$ is suggested. Convergence is claimed after the stopping rule is satisfied for several consecutive iterations; $\bar{\boldsymbol{\theta}}^{(j)}$ is then taken to be the ML estimate. Shi and Copas (2002) argued that for a sufficiently large m, the average of the Monte Carlo errors in (9) is negligible. To reduce the bias, one should take an appropriate m which can control the Monte Carlo errors within a bearable limit. In the example given in Section 4, we take $m = 50$. One can use the method in Shi and Copas (2002), or take $J = J_1 + J_2 j$, or use a combination of both as in our example.

Finally, standard error estimates are produced by approximating the Louis (1982) formulae at the ML estimates via a sufficiently large number of observations that are obtained at the E-step of the final iteration, together with newly generated observations, if necessary.

Standard errors estimates of the ML estimates can be obtained by inverting the Hessian matrix of the observed-data log-likelihood function $L_o(\mathbf{D}_o|\boldsymbol{\theta})$. However, this matrix is generally not in closed form. Hence, we use an identity of Louis (1982) and random samples generated from $p(\mathbf{D}_m|\mathbf{D}_o, \hat{\boldsymbol{\theta}})$ to obtain standard error estimates. It follows from Louis (1982) that

$$-\frac{\partial^2 L_o(\mathbf{D}_o|\boldsymbol{\theta})}{\partial\boldsymbol{\theta}\partial\boldsymbol{\theta}'} = E_{\mathbf{D}_m}\left\{-\frac{\partial^2 L_c(\mathbf{D}_o, \mathbf{D}_m|\boldsymbol{\theta})}{\partial\boldsymbol{\theta}\partial\boldsymbol{\theta}'}\right\} - Var_{\mathbf{D}_m}\left\{\frac{\partial L_c(\mathbf{D}_o, \mathbf{D}_m|\boldsymbol{\theta})}{\partial\boldsymbol{\theta}}\right\}, \tag{11}$$

where expectations involved in (11) are taken with respect to the conditional distribution of $\mathbf{D}_m$ given $\mathbf{D}_o$ and $\boldsymbol{\theta}$, and the whole expression is evaluated at $\hat{\boldsymbol{\theta}}$. These expectations are difficult to evaluate analytically, but they can be approximated respectively by the sample mean and the sample covariance matrix of the random samples $\{\mathbf{D}_{mt}, t = 1, \ldots, T\}$ generated from $p(\mathbf{D}_m|\mathbf{D}_o, \hat{\boldsymbol{\theta}})$ using the Gibbs sampler algorithm. Thus, the right-hand side of (11) can be estimated by

$$T^{-2}\left(\sum_{t=1}^{T}\frac{\partial L_c(\mathbf{D}_o, \mathbf{D}_{mt}|\boldsymbol{\theta})}{\partial\boldsymbol{\theta}}\right)\left(\sum_{t=1}^{T}\frac{\partial L_c(\mathbf{D}_o, \mathbf{D}_{mt}|\boldsymbol{\theta})}{\partial\boldsymbol{\theta}}\right)'\Bigg|_{\boldsymbol{\theta}=\hat{\boldsymbol{\theta}}} + T^{-1}\sum_{t=1}^{T}\left[-\frac{\partial^2 L_c(\mathbf{D}_o, \mathbf{D}_{mt}|\boldsymbol{\theta})}{\partial\boldsymbol{\theta}\partial\boldsymbol{\theta}'} - \left(\frac{\partial L_c(\mathbf{D}_o, \mathbf{D}_{mt}|\boldsymbol{\theta})}{\partial\boldsymbol{\theta}}\right) \times \left(\frac{\partial L_c(\mathbf{D}_o, \mathbf{D}_{mt}|\boldsymbol{\theta})}{\partial\boldsymbol{\theta}}\right)'\right]_{\boldsymbol{\theta}=\hat{\boldsymbol{\theta}}}. \tag{12}$$

Explicit formulae for the partial derivatives can be obtained via standard matrix differentiation.

3.2. Model comparison and hypothesis testing

In analyzing the multisample SEM one important statistical inference beyond estimation is on testing whether some types of parameters are invariant over the groups. For instance, the hypotheses of interest may be $\boldsymbol{\Lambda}^{(1)} = \cdots = \boldsymbol{\Lambda}^{(G)}$, $\boldsymbol{\Phi}^{(1)} = \cdots = \boldsymbol{\Phi}^{(G)}$, etc., which specify certain kinds of constraints on some parameters among groups. One common approach to hypothesis testing is to use the significance tests on the basis of p-values that are determined by the asymptotic chi-square statistic. For the current situation, the asymptotic test is not available. Hence, the BIC is used. This statistic has been widely applied to model comparison in the ML analysis of models in statistics, and structural equation models (Lee and Song, 2004b; Raftery, 1993; Song and Lee, 2005, 2006).

Let M_1 and M_2 be two nested or non-nested multisample SEMs under comparison. For $k = 1, 2$, let $\hat{\boldsymbol{\theta}}_k$ be the ML estimate of the parameter vector $\boldsymbol{\theta}_k$ that contains the distinct unknown parameters under M_k, let d_k be the dimension of $\boldsymbol{\theta}_k$, and let n be the sample size. The BIC for comparing M_1 and M_2 is defined as

$$BIC_{12} = -2\{\log p(\mathbf{D}_o; \hat{\boldsymbol{\theta}}_1, M_1) - \log p(\mathbf{D}_o; \hat{\boldsymbol{\theta}}_2, M_2)\} + (d_1 - d_2)\log n, \tag{13}$$

where $p(\mathbf{D}_o; \hat{\boldsymbol{\theta}}_k, M_k)$ is the observed-data likelihood evaluated at $\hat{\boldsymbol{\theta}}_k$ under M_k. The following criterion can be used for interpretation of BIC_{12} (see Kass and Raftery, 1995):

BIC_{12}	< 0	0 to 2	2 to 6	> 6
	Support M_1	No conclusion	Support M_2	Strongly support M_2

This statistic can be applied to test various hypotheses about the invariance of some types of parameters over the groups. For example, let M_1 be the model without any constraint on the parameters, and M_2 be the model that associates with the constraints $\boldsymbol{\Lambda}^{(1)} = \cdots = \boldsymbol{\Lambda}^{(G)}$. We can use the BIC_{12} to assess the hypothesis H_1: $\boldsymbol{\Lambda}^{(1)} \neq \cdots \neq \boldsymbol{\Lambda}^{(G)}$ against H_2: $\boldsymbol{\Lambda}^{(1)} = \cdots = \boldsymbol{\Lambda}^{(G)}$. The statistic BIC_{12} gives precise evidence about which model is better. Moreover, it can be applied to compare non-nested models. For the current multisample SEM with missing continuous and ordered categorical data, the computation of $\log p(\mathbf{D}_o; \hat{\boldsymbol{\theta}}_k, M_k)$ in the BIC is rather involved. Some MCMC methods, such as importance sampling (Newton and Raftery, 1994) or bridge sampling (Meng and Wong, 1996), can be applied to compute $\log p(\mathbf{D}_o; \hat{\boldsymbol{\theta}}_k, M_k)$ (see, for example, Song et al., 2001). Recently, Gelman and Meng (1998) developed an efficient method, known as path sampling, and showed that it is a generalization of importance sampling and bridge sampling. Hence, this method is expected to produce more accurate results. In this article, the method is applied to compute the observed data log-likelihoods in (13), and then the BIC. A description of how to use the method is given in Appendix B.

4. Illustrative example: analysis of multisample synthetic QOL data

As the latent constructs of QOL can be naturally regarded as latent variables that are reflected by the related items (observed variables) in the questionnaire, factor analysis and structural equation models have been used in analyzing QOL data. For instance, Power et al. (1999) applied a multisample model to investigate whether the WHOQOL instrument is structurally comparable in different cultures, and Menleners et al. (2003) applied a LISREL model to analyze the QOL for adolescents. However, the above cited work, as well as most applications of the factor analysis model to QOL, are based on fully observed continuous data with a normality assumption.

The WHOQOL-BREF (Power et al., 1999) instrument was taken from the WHOQOL-100 instrument by selecting twenty-six ordered categorical items out of 100 original items. The observations were taken from 15 international field centers, one of which is China, and most of the rest are Western countries, such as the United Kingdom, Italy, and Germany. The first two items are the overall QOL and general health, the next seven items address physical health, the next six items address psychological health, the three items that follow are for social relationships, and the last eight items address the

Table 1
Frequencies of the ordered categorical scores of the items

WHOQOL items		Group 1					No. of missing obs.	Group 2					No. of missing obs.
		1	2	3	4	5		1	2	3	4	5	
Q1	Overall QOL	3	40	95	186	75	1	6	27	191	165	8	1
Q2	Overall health	29	109	89	132	41	0	13	59	116	196	7	7
Q3	Pain and discomfort	22	57	93	125	103	0	21	53	135	137	44	8
Q4	Medical treatment dependence	22	60	79	90	147	2	18	63	139	110	64	4
Q5	Energy and fatigue	17	59	128	101	90	5	10	44	198	121	23	2
Q6	Mobility	17	38	72	114	154	5	11	31	123	189	43	1
Q7	Sleep and rest	29	71	85	139	76	0	8	45	104	216	24	1
Q8	Daily activities	10	69	76	184	59	2	9	26	108	235	19	1
Q9	Work capacity	23	85	95	135	57	5	9	31	107	233	18	0
Q10	Positive feeling	10	25	114	191	56	4	7	25	206	129	28	3
Q11	Spirituality/personal beliefs	9	35	115	165	74	2	6	20	204	141	26	1
Q12	Memory and concentration	5	30	172	154	36	3	5	23	211	137	22	0
Q13	Bodily image/appearance	5	34	126	125	107	3	2	17	235	102	41	1
Q14	Self-esteem	9	47	121	173	50	0	4	20	110	249	14	1
Q15	Negative feeling	4	40	105	196	52	3	8	38	135	126	80	11
Q16	Personal relationship	6	19	70	195	110	0	4	8	104	259	17	6
Q17	Sexual activity	28	51	120	111	58	32	7	18	119	143	8	103
Q18	Social support	9	6	87	191	105	2	6	19	130	223	11	9
Q19	Physical safety and security	4	23	160	152	52	2	10	25	186	162	12	3
Q20	Physical environment	8	23	160	152	52	5	29	36	197	123	7	6
Q21	Financial resources	17	41	162	103	75	2	32	67	227	61	8	3
Q22	Daily life information	7	26	120	175	68	4	22	72	223	67	5	9
Q23	Leisure activity participation	16	92	120	124	48	0	13	94	171	104	13	3
Q24	Living condition	6	16	42	203	133	0	19	73	115	179	7	5
Q25	Health accessibility	4	23	71	241	60	1	12	70	122	180	2	12
Q26	Transportation	7	19	56	213	103	2	16	76	117	178	7	4

environment. All of the items are measured with a 5-point scale (1 = 'not at all/very dissatisfied'; 2 = 'a little/dissatisfied'; 3 = 'moderate/neither'; 4 = 'very much/satisfied'; and 5 = 'extremely/very satisfied').

To illustrate the developed methodology, we use a synthetic two-sample data set that mimic the QOL study with the same items as mentioned above for each sample. The two data for analysis are presented in Table 1. The sample sizes for the first and the second groups are 388 and 400, respectively. We note from these datasets that several items, especially those in relation to group 1 tend to take maximum values for most patients, and hence seriously skew to the right. Treating these discrete data as coming from a normal distribution may lead to a misleading conclusion.

We apply the multisample SEM as defined in (1) and (2) with $G = 2$ to analyze the data. In the ML analysis, we identify the ordered categorical variables by the method described in Section 2.2, using the Western countries as the reference group ($g = 1$). Based on the meaning of the questions, we use the following non-overlapping $\boldsymbol{\Lambda}^{(g)}$ for clear interpretation of latent variables: For $g = 1, 2$

$$\boldsymbol{\Lambda}^{(g)'} = \begin{bmatrix} 1 & \lambda^{(g)}_{2,1} & 0 & 0 & \cdots & 0 & 0 & 0 & \cdots & 0 & 0 & 0 & 0 & 0 & 0 & \cdots & 0 \\ 0 & 0 & 1 & \lambda^{(g)}_{4,2} & \cdots & \lambda^{(g)}_{9,2} & 0 & 0 & \cdots & 0 & 0 & 0 & 0 & 0 & 0 & \cdots & 0 \\ 0 & 0 & 0 & 0 & \cdots & 0 & 1 & \lambda^{(g)}_{11,3} & \cdots & \lambda^{(g)}_{15,3} & 0 & 0 & 0 & 0 & 0 & \cdots & 0 \\ 0 & 0 & 0 & 0 & \cdots & 0 & 0 & 0 & \cdots & 0 & 1 & \lambda^{(g)}_{17,4} & \lambda^{(g)}_{18,4} & 0 & 0 & \cdots & 0 \\ 0 & 0 & 0 & 0 & \cdots & 0 & 0 & 0 & \cdots & 0 & 0 & 0 & 0 & 1 & \lambda^{(g)}_{20,5} & \cdots & \lambda^{(g)}_{26,5} \end{bmatrix}, \tag{14}$$

where 1's and 0's are fixed parameters. Hence, the latent variables in $\boldsymbol{\omega}_i^{(g)'} = (\boldsymbol{\eta}_i^{(g)}, \boldsymbol{\xi}_{1i}^{(g)}, \boldsymbol{\xi}_{2i}^{(g)}, \boldsymbol{\xi}_{3i}^{(g)}, \boldsymbol{\xi}_{4i}^{(g)})$ are interpreted as 'health related QOL, η', 'physical health, ξ_1', 'psychological health, ξ_2', 'social relationship, ξ_3', and 'environment, ξ_4'. The measurement equation in the model is given by

$$\mathbf{v}_i^{(g)} = \boldsymbol{\mu}^{(g)} + \boldsymbol{\Lambda}^{(g)}\boldsymbol{\omega}_i^{(g)} + \boldsymbol{\varepsilon}_i^{(g)}, \quad g = 1, 2, \tag{15}$$

with $\boldsymbol{\Lambda}^{(g)}$ defined as above. The following structural equation is used to assess the effects of the latent constructs in $\boldsymbol{\xi}_i^{(g)}$ to the health related QOL, $\eta^{(g)}$:

$$\eta_i^{(g)} = \gamma_1^{(g)}\xi_{1i}^{(g)} + \gamma_2^{(g)}\xi_{2i}^{(g)} + \gamma_3^{(g)}\xi_{3i}^{(g)} + \gamma_4^{(g)}\xi_{4i}^{(g)} + \delta_i^{(g)}. \tag{16}$$

In the MCEM algorithm for computing the ML estimates under the competing models, we generate $J = 50+20j$ observations to approximate the conditional expectations at the E-step of the jth iteration. Convergence was be monitored via the method of Shi and Copas (2002) by using (10) with $\gamma_0 = 5$, $\delta_1 = \delta_2 = 0.001$, and $\bar{\boldsymbol{\theta}}^{(j)}$ is computed via (9) with $m = 50$. Convergence is claimed if the stopping rule is satisfied for five consecutive iterations. To reveal convergence in analyzing M_0, plots of some parameters in the second group against iterations are displayed in Figure 1. Based on the above stopping rule, the algorithm stops at the 89-th MCEM iteration; and $\bar{\boldsymbol{\theta}}^{(89)}$ is taken as the ML estimate. In the path sampling procedure, we take $S = 20$ and $J = 1000$.

In the multisample analysis, we employ the BIC to study the relations of the parameters in the model, and try to identify a comparatively good model. We consider the model M_0 as the general model that is defined by (15) and (16) without any constraints among the parameters in group 1 and group 2. This model is compared with M_1, which is the model defined by (15) and (16) together with the constraint $\boldsymbol{\mu}^{(1)} = \boldsymbol{\mu}^{(2)}$, and with M_2, which is also defined by (15) and (16) but with the constraint $\boldsymbol{\Lambda}^{(1)} = \boldsymbol{\Lambda}^{(2)}$. Using the path sampling procedure for computing $p(\mathbf{D}_o; \hat{\boldsymbol{\theta}}_k, M_k)$, we obtain the observed-data log-likelihoods that correspond to M_0, M_1 and M_2 as -24308.22, -25199.66, and -24741.68. Hence, it follows from (13) that $BIC_{01} = -1609.10$, and $BIC_{02} = -726.59$. Hence, the hypothesis $\boldsymbol{\mu}^{(1)} = \boldsymbol{\mu}^{(2)}$ is rejected, and the hypothesis $\boldsymbol{\Lambda}^{(1)} = \boldsymbol{\Lambda}^{(2)}$ is also rejected.

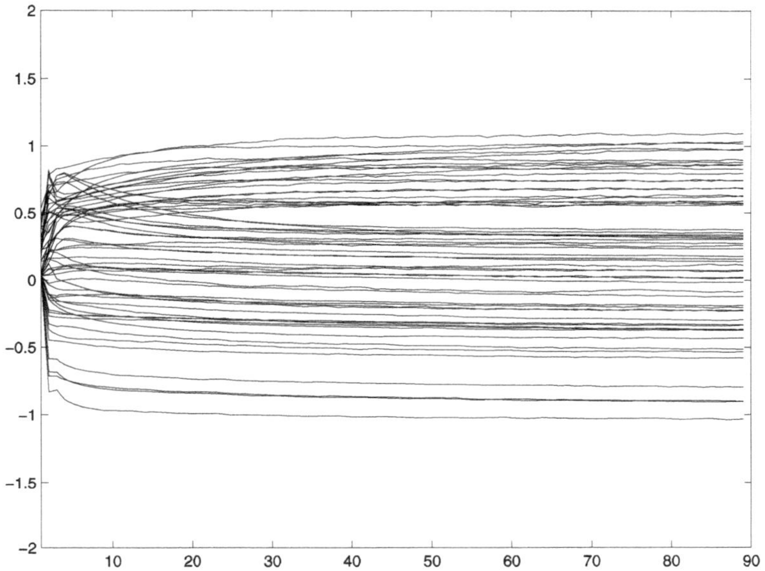

Fig. 1. Average estimates of $\boldsymbol{\mu}^{(2)}$, $\boldsymbol{\Lambda}^{(2)}$, $\boldsymbol{\Gamma}^{(2)}$, and $\boldsymbol{\Phi}^{(2)}$ in M_0.

As the hypothesis $\boldsymbol{\mu}^{(1)} = \boldsymbol{\mu}^{(2)}$ is rejected, we conclude that the latent means of the QOL items corresponding to these two groups are different. From the ordered categorical outcomes of these two groups, which are presented in Table 1, we observe that people from group 1 have high scores in most of the QOL items (the empirical distribution skewed more to the right). As the hypothesis H_0: $\boldsymbol{\Lambda}^{(1)} = \boldsymbol{\Lambda}^{(2)}$ is rejected, the factor loading matrices of these two groups are different. What is the interpretation of this result? From the structure of $\boldsymbol{\Lambda}^{(g)}$ (see Eq. (14)) and the corresponding QOL items (indicators), we have a rather clear interpretation of the latent variables $\eta^{(g)}, \xi_1^{(g)}, \xi_2^{(g)}, \xi_3^{(g)}$, and $\xi_4^{(g)}$ as the health related QOL, physical health, etc. Based on (15), we see that the associations between the latent variables and their respective indicators (QOL items) are clearly indicated by the corresponding elements in $\boldsymbol{\Lambda}^{(g)}$. The rejection of the null hypothesis H_0 reveals that the associations in group 1 are different from those in group 2. A common practice in multisample analysis of structural equation models is to continue the analysis by testing more strict hypotheses in a hierarchical order if H_0 is not rejected, and stop if H_0 is rejected (Bollen, 1989). Hence, we can just report the ML estimates and stop. The reason for ending further analysis may be due to the belief that rejection of H_0 implies that the scale of measurements of the latent variables in group 1 are different from the scale of measurements of the latent variables in group 2. However, based on the significance test with p-value, the conclusion that H_0 is not rejected does not imply that H_0 is true, and it does not imply that the scales of these two groups are the same. Hence, the justification that is based on the result of hypothesis testing for continuing the analysis is not very strong. Moreover, for ordered categorical variables, as the underlying continuous responses are not observed, it is difficult to draw a conclusion on the scales of the latent variables on the basis of the hypothesis testing

Table 2
ML estimates of unknown parameters in the 2-group SEM

	Western countries	China		Western countries	China		Western countries	China
μ_1	0.001	−0.404	λ_y	1.226	0.909	γ_1	0.560	0.560
μ_2	0.006	0.185	λ_{12}	0.936	0.552	γ_2	0.267	0.267
μ_3	0.003	−0.180	λ_{13}	1.128	0.905	γ_3	−0.054	−0.054
μ_4	0.006	−0.303	λ_{14}	1.173	0.855	γ_4	0.083	0.083
μ_5	−0.002	−0.152	λ_{15}	0.770	0.758	ψ_1	0.407	0.150
μ_6	0.000	−0.329	λ_{16}	1.340	1.098	ψ_2	0.350	0.164
μ_7	0.004	0.054	λ_{17}	1.194	0.951	ψ_3	0.593	0.533
μ_8	0.007	0.030	λ_{29}	0.768	0.835	ψ_4	0.601	0.368
μ_9	0.014	0.220	$\lambda_{2,10}$	0.690	0.899	ψ_5	0.408	0.149
μ_{10}	0.010	−0.310	$\lambda_{2,11}$	0.733	0.559	ψ_6	0.352	0.094
μ_{11}	−0.000	−0.246	$\lambda_{2,12}$	1.001	0.741	ψ_7	0.730	0.184
μ_{12}	0.008	−0.068	$\lambda_{2,13}$	0.780	0.577	ψ_8	0.178	0.093
μ_{13}	0.004	−0.308	$\lambda_{3,15}$	0.326	0.607	ψ_9	0.346	0.084
μ_{14}	0.008	0.113	$\lambda_{3,16}$	1.130	0.988	ψ_{10}	0.440	0.180
μ_{15}	0.012	0.050	$\lambda_{4,18}$	0.876	1.043	ψ_{11}	0.635	0.167
μ_{16}	−0.004	−0.340	$\lambda_{4,19}$	0.851	0.711	ψ_{12}	0.695	0.365
μ_{17}	0.006	0.091	$\lambda_{4,20}$	0.874	0.837	ψ_{13}	0.663	0.282
μ_{18}	0.004	−0.501	$\lambda_{4,21}$	0.834	0.647	ψ_{14}	0.369	0.190
μ_{19}	0.006	−0.190	$\lambda_{4,22}$	1.028	0.651	ψ_{15}	0.609	1.290
μ_{20}	0.004	−0.467	$\lambda_{4,23}$	0.761	0.814	ψ_{16}	0.459	0.107
μ_{21}	0.007	−0.546	$\lambda_{4,24}$	0.883	0.756	ψ_{17}	0.956	0.244
μ_{22}	0.005	−0.856	ϕ_{11}	0.405	0.405	ψ_{18}	0.474	0.173
μ_{23}	0.006	−0.170	ϕ_{12}	0.367	0.367	ψ_{19}	0.532	0.546
μ_{24}	0.019	−1.007	ϕ_{13}	0.186	0.186	ψ_{20}	0.653	0.472
μ_{25}	−0.001	−0.756	ϕ_{14}	0.280	0.280	ψ_{21}	0.680	0.391
μ_{26}	0.011	−0.864	ϕ_{22}	0.587	0.587	ψ_{22}	0.660	0.293
			ϕ_{23}	0.310	0.310	ψ_{23}	0.687	0.428
			ϕ_{24}	0.354	0.354	ψ_{24}	0.565	0.347
			ϕ_{33}	0.377	0.377	ψ_{25}	0.755	0.428
			ϕ_{34}	0.294	0.294	ψ_{26}	0.671	0.348
			ϕ_{44}	0.419	0.419			
			ψ_δ	0.079	0.117			

result. From the non-overlapping structure of $\boldsymbol{\Lambda}^{(g)}$, these latent variables can be clearly interpreted as the latent constructs in relation to health related QOL, physical health, etc., although the associations of these latent constructs and their indicators are not the same in groups 1 and 2. Hence, we think that it is desirable to conduct a complementary analysis to obtain a deeper understanding of these latent constructs about QOL. This optional complementary analysis involves comparisons with following models.

M_3: The model that is given in (15) and (16), and $\boldsymbol{\Gamma}^{(1)} = \boldsymbol{\Gamma}^{(2)}$,

M_4: The model that is given in (15) and (16), and $\boldsymbol{\Phi}^{(1)} = \boldsymbol{\Phi}^{(2)}$,

M_5: The model that is given in (15) and (16), and $\boldsymbol{\Psi}_\varepsilon^{(1)} = \boldsymbol{\Psi}_\varepsilon^{(2)}$,

M_6: The model that is given in (15) and (16), and $\boldsymbol{\Psi}_\delta^{(1)} = \boldsymbol{\Psi}_\delta^{(2)}$.

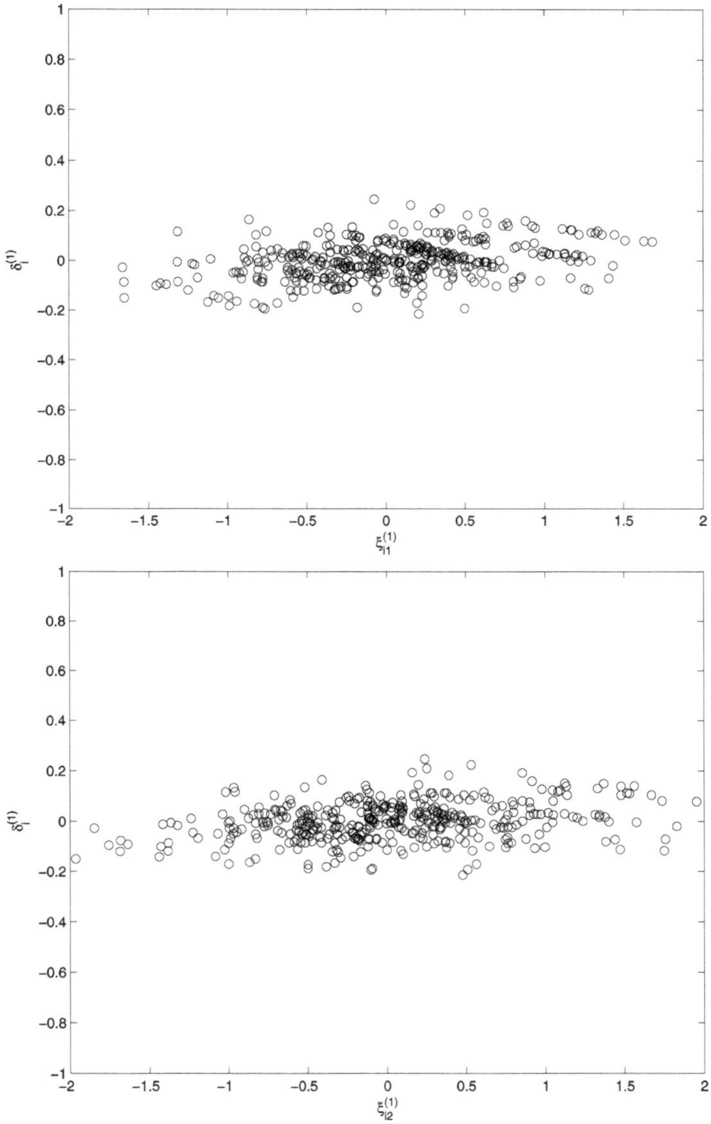

Fig. 2. Plots of residuals $\hat{\delta}_i^{(1)}$ against (a) $\hat{\xi}_{i1}^{(1)}$, (b) $\hat{\xi}_{i2}^{(1)}$, (c) $\hat{\xi}_{i3}^{(1)}$, and (d) $\hat{\xi}_{i4}^{(1)}$.

The observed-data log-likelihoods computed via the path sampling procedure for M_3, M_4, M_5, and M_6 are equal to -24314.56, -24322.02, -24859.40, and -24311.80. It follows from (13) that BIC_{03}, BIC_{04}, BIC_{05}, and BIC_{06} are equal to 14.05, 39.23, -928.61, and -0.47. From BIC_{03} and BIC_{04}, we conclude that M_3 and M_4 are better than M_0. Hence, the sample data give evidence of support to the null hypothesis H_0: $\boldsymbol{\Gamma}^{(1)} = \boldsymbol{\Gamma}^{(2)}$, and H_0: $\boldsymbol{\Phi}^{(1)} = \boldsymbol{\Phi}^{(2)}$. To cross validate this conclusion, we consider

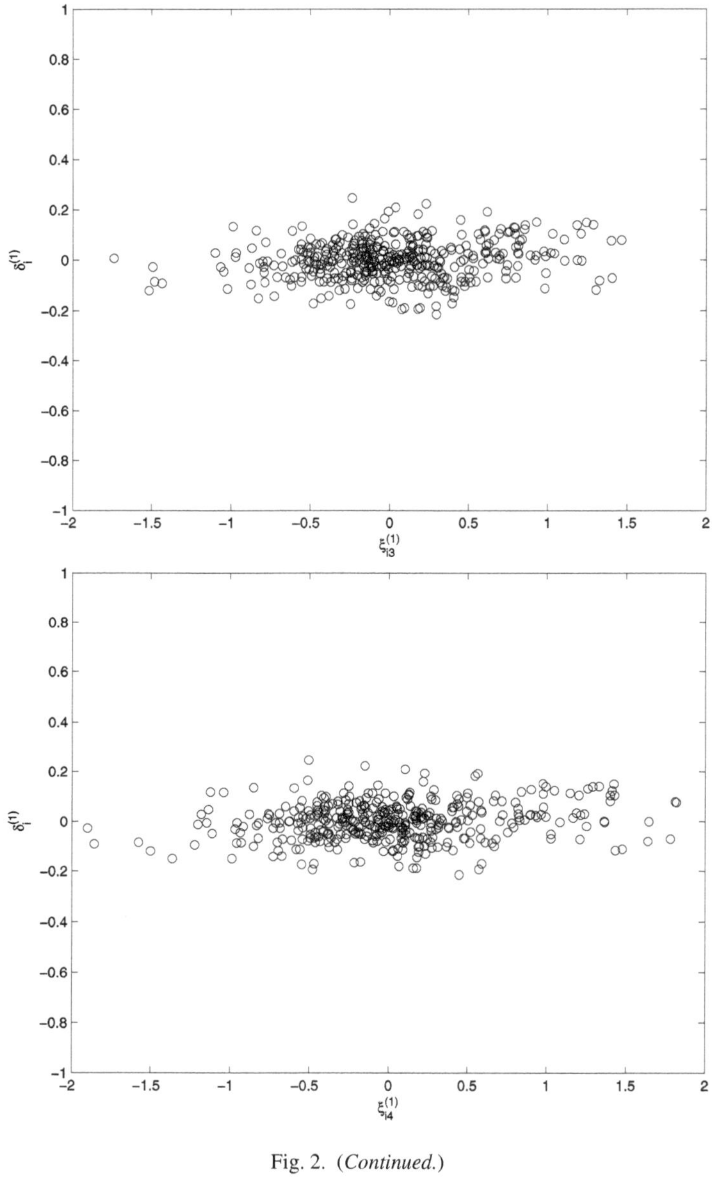

Fig. 2. (*Continued.*)

the following M_7.

$$M_7\text{: The model given by (15) and (16), and } \boldsymbol{\Gamma}^{(1)} = \boldsymbol{\Gamma}^{(2)}, \text{ and } \boldsymbol{\Phi}^{(1)} = \boldsymbol{\Phi}^{(2)}.$$

The observed-data log-likelihood under M_7 is -24334.51. Again, it follows from (13) that $BIC_{37} = 26.92$, and $BIC_{47} = 1.74$. Hence, the sample data give evidence of support to the fact that M_7 is better than M_3 or M_4. The ML estimates of the parameters in the selected M_7 are presented in Table 2. Inspired by model checking techniques in

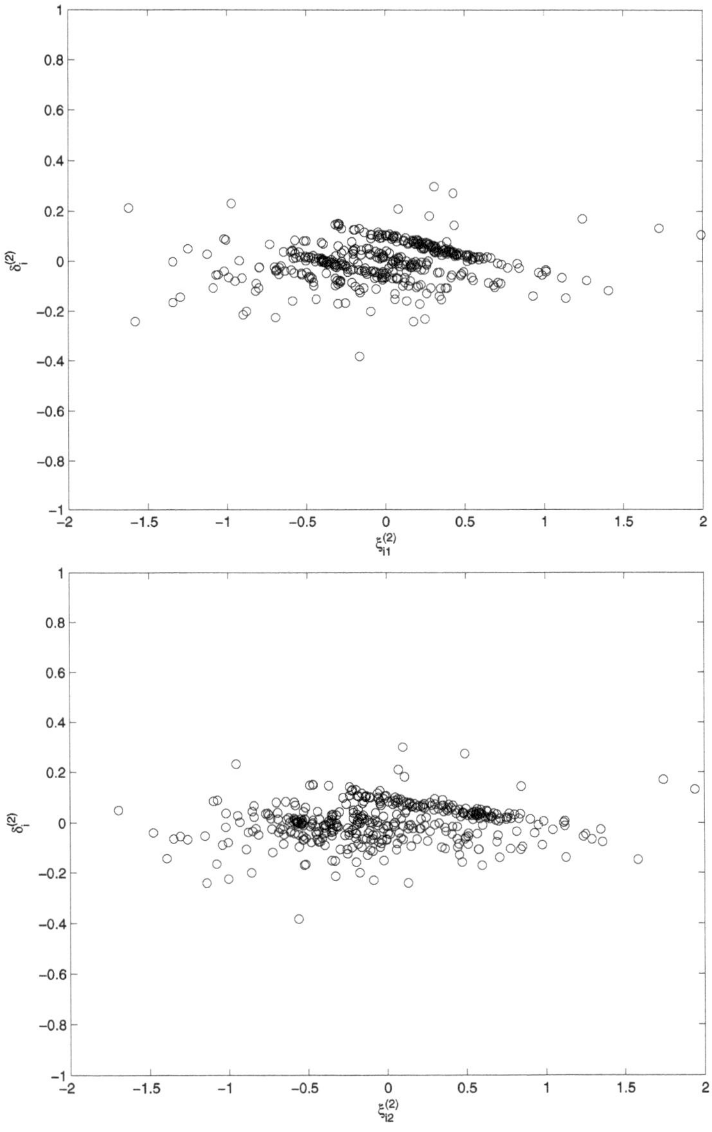

Fig. 3. Plots of residuals $\hat{\delta}_i^{(2)}$ against (a) $\hat{\xi}_{i1}^{(2)}$, (b) $\hat{\xi}_{i2}^{(2)}$, (c) $\hat{\xi}_{i3}^{(2)}$, and (d) $\hat{\xi}_{i4}^{(2)}$.

regression, we use the following estimated residuals to reveal the goodness-of-fit of M_7 to the empirical data with the latent structure: $\hat{\delta}_i^{(g)} = \hat{\eta}_i^{(g)} - \hat{\gamma}_1^{(g)}\hat{\xi}_{1i}^{(g)} - \hat{\gamma}_2^{(g)}\hat{\xi}_{2i}^{(g)} - \hat{\gamma}_3^{(g)}\hat{\xi}_{3i}^{(g)} - \hat{\gamma}_4^{(g)}\hat{\xi}_{4i}$, $\hat{\boldsymbol{\varepsilon}}_i^{(g)} = \hat{\mathbf{v}}_i^{(g)} - \hat{\boldsymbol{\mu}}^{(g)} - \hat{\boldsymbol{\Lambda}}^{(g)}\hat{\boldsymbol{\omega}}_i^{(g)}$, where estimates of the latent vectors $\boldsymbol{\omega}_i^{(g)}$ and latent quantities $\mathbf{v}_i^{(g)}$ are obtained from the observations that are simulated by the Gibbs sampler in the E-step of the final MCEM iterations. The estimated residuals

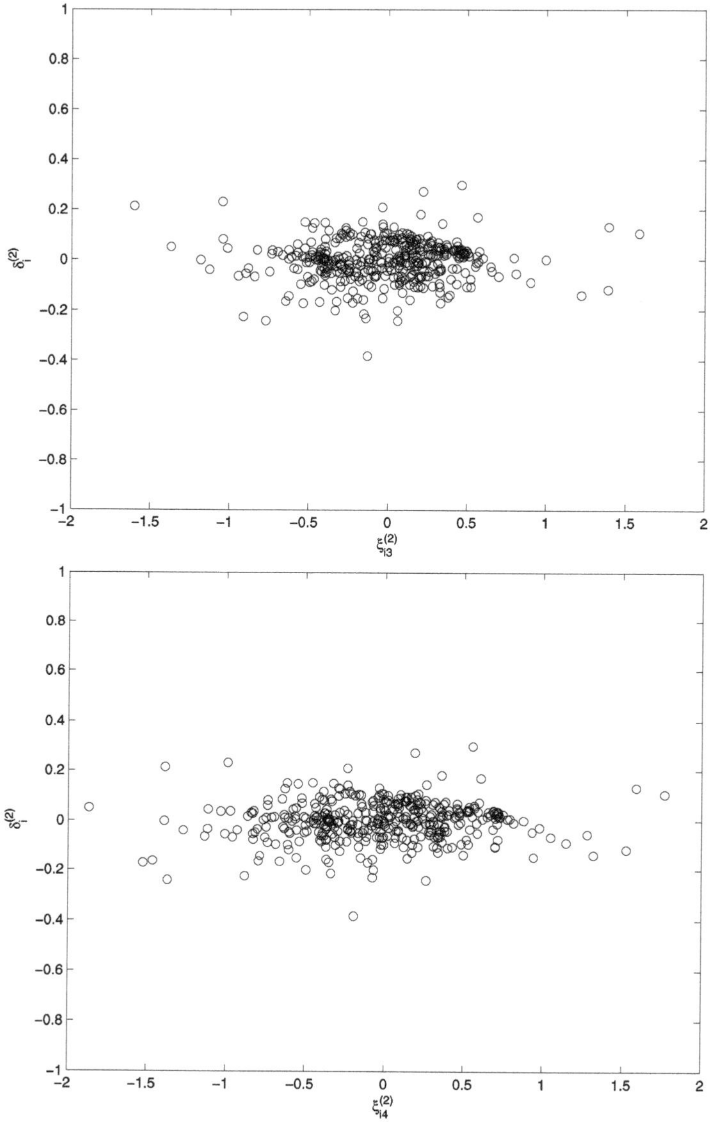

Fig. 3. (*Continued.*)

$\hat{\boldsymbol{\epsilon}}_i^{(g)}$ and $\hat{\delta}_i^{(g)}$ are computed. To save space, only plots of $\hat{\delta}_i^{(g)}$ against $\hat{\xi}_{i1}^{(g)}$, $\hat{\xi}_{i2}^{(g)}$, $\hat{\xi}_{i3}^{(g)}$, and $\hat{\xi}_{i4}^{(g)}$ are displayed in Figure 2 ($g = 1$) and Figure 3 ($g = 2$). The residual plots that correspond to each component in $\hat{\boldsymbol{\epsilon}}_i^{(g)}$ against the latent variables are similar. These plots lie within two parallel horizontal lines that are centered at zero, and no linear or quadratic trends are detected. This roughly indicates that the proposed measurement and structural equations at both groups are adequate.

We have the following interpretations from the results in Table 2.

(1) Comparing the relative mean estimates of the items that correspond to group 1 with those correspond to group 2, it seems that the averages of most item scores of the observation in group 1 are lower than those in group 2, including the overall QOL (see $\hat{\mu}_1^{(1)}$ and $\hat{\mu}_1^{(2)}$). However, by comparing $\hat{\mu}_2^{(1)} = 0.006$ with $\hat{\mu}_2^{(2)} = 0.185$, it seems that the subjects in group 1 consider themselves as having a better health than the subjects in group 2.

(2) All of the loadings are high, which indicates strong associations between each of the latent constructs and their respective items, in both group. Note that most of the $\hat{\lambda}_{i,j}^{(1)}$ are close to $\hat{\lambda}_{i,j}^{(2)}$, only $\hat{\lambda}_{4,2}^{(1)}$, $\hat{\lambda}_{17,4}^{(1)}$, and $\hat{\lambda}_{24,5}^{(1)}$ are substantially different from the corresponding estimates in group 2. This finding gives more support to conducting the complementary analysis.

(3) As M_7 is selected, the relationships of the latent constructs that are identified by the measurement equations (keeping in mind that three $\hat{\lambda}_{i,j}^{(1)}$ and $\hat{\lambda}_{i,j}^{(2)}$ are different) to the latent health-related QOL in these two groups are the same. Hence, the effects of physical health, psychological health, social relationships, and the environment on the health-related QOL of patients in these two different groups are the same. From the magnitudes of $\hat{\gamma}_1^{(g)}, \ldots, \hat{\gamma}_4^{(g)}$, it can be concluded that physical health and psychological health have stronger effects than social relationships and the environment, and that the effect of the last two latent constructs are not substantial. These results seem theoretically reasonable and logical.

(4) The sample data give evidence of support to the hypothesis that the correlations of the latent constructs in these two groups are not different. As expected, these latent constructs are positively correlated.

(5) The estimated residual variables, $\hat{\psi}_\delta^{(1)} = 0.075$ and $\hat{\psi}_\delta^{(2)} = 0.117$, are quite small. This indicates that the fitting of the health-related QOL by its latent constructs via the structural equation is quite good.

5. Discussion

Latent variables are frequently encountered in substantive research. A clear understanding of the inter-relationships among these latent variables and their relationships with manifest variables is important in making correct decisions. Structural equation modeling is an important multivariate method of achieving this purpose, and its standard model with the normality assumption has been widely applied to practical problems. However, to analyze multisample data from different treatment groups, cultural groups, etc., it is necessary to extend the single sample model to a multisample model. Moreover, as missing data, and/or heavily skewed ordered categorical data are frequently encountered in medical research, the developed statistical methods should be able to handle this kind of missing discrete data. The main objective of this article is to establish an ML approach for analyzing multisample SEMs with missing ordered categorical variables, by means of the standard tools in statistical computing, such as the MCEM algorithm and path sampling. Based on the statistics produced from the proposed ML approach, we can perform other statistical analyses. For instance, the observed-data log-likelihood can be applied to discriminant analysis.

Most probably due to the complexity of the different data and the problem, there are a number of research directions of analyzing various kind of QOL data. The current development provides an alternative for analyzing multisample ordered categorical QOL data with missing entries.

An advantage of the current SEM approach is that it can be naturally generalized to more complex models and data structures to cope with the real complicated situations in QOL or other medical research. In future research, it will be desirable to establish multilevel SEMs for analyzing hierarchically structural data, and mixture SEMs for analyzing heterogenous data. Finally, the methodology that is provided in this paper provides a solid foundation for developing longitudinal SEMs, which will be important for investigating time effects on QOL analysis, and other medical research.

Acknowledgements

This research was partially supported by a Hong Kong UGC Earmarked grant CUHK 4243/03H and a Direct Grant 2060279 from the Chinese University of Hong Kong.

Appendix A

The Gibbs sampler is implemented as follows. At the jth iteration with a current value $\mathbf{D}_{mj} = (\mathbf{V}_{mj}, \mathbf{Y}_{oj}, \boldsymbol{\Omega}_j)$, we simulate in turn:

$$\begin{aligned} &\mathbf{V}_{m,j+1} \quad \text{from } p[\mathbf{V}_m | \mathbf{Y}_{oj}, \boldsymbol{\Omega}_j, \mathbf{D}_o, \boldsymbol{\theta}_{(r)}], \\ &\mathbf{Y}_{o,j+1} \quad \text{from } p[\mathbf{Y}_o | \mathbf{V}_{m,j+1}, \boldsymbol{\Omega}_j, \mathbf{D}_o, \boldsymbol{\theta}_{(r)}], \\ &\boldsymbol{\Omega}_{j+1} \quad \text{from } p[\boldsymbol{\Omega} | \mathbf{V}_{m,j+1}, \mathbf{Y}_{o,j+1}, \mathbf{D}_o, \boldsymbol{\theta}_{(r)}], \end{aligned} \tag{A.1}$$

where $p[\cdot|\cdots]$ denotes the corresponding conditional distribution. The above cycle is iterated many times. As j tends to infinity, it can be shown by standard theory in statistical computing that the joint distribution of $\mathbf{D}_{mj}$ converges in distribution to the joint conditional distribution of $\mathbf{D}_m$ given $\mathbf{D}_o$ at $\boldsymbol{\theta}_{(r)}$. Hence, there is a sufficiently large integer J^*, such that for $j > J^*$, $\mathbf{D}_{mj}$ can be regarded as an observation from the conditional distribution $(\mathbf{D}_m|\mathbf{D}_o)$ at $\boldsymbol{\theta}_{(r)}$. The conditional distributions that are involved in (A.1) are given follows.

(i) Conditional distribution of $p[\mathbf{V}_m|\cdot]$. Let $\mathbf{V}_m = \{\mathbf{v}_{im}^{(g)} = (\mathbf{x}_{im}^{(g)'}, \mathbf{y}_{im}^{(g)'})', g = 1, \ldots, G; i = 1, \ldots, n_g\}$. For $i = 1, \ldots, n_g$, as $\mathbf{v}_i^{(g)}$ are mutually independent, $\mathbf{v}_{im}^{(g)}$ are also mutually independent. As $\boldsymbol{\Psi}^{(g)}$ is diagonal, $\mathbf{v}_{im}^{(g)}$ is conditionally independent with $\mathbf{y}_{io}^{(g)}$ given $\boldsymbol{\omega}_i^{(g)}$. Hence, it follows from (1) that

$$p(\mathbf{V}_m|\mathbf{Y}_o, \boldsymbol{\Omega}, \mathbf{D}_o, \boldsymbol{\theta}) = \prod_{g=1}^{G}\prod_{i=1}^{n_g} p\big(\mathbf{v}_{im}^{(g)}|\boldsymbol{\omega}_i^{(g)}, \boldsymbol{\theta}^{(g)}\big), \quad \text{and}$$

$$\big[\mathbf{v}_{im}^{(g)}|\boldsymbol{\omega}_i^{(g)}, \boldsymbol{\theta}^{(g)}\big] \stackrel{D}{=} N\big[\boldsymbol{\mu}_{i,m}^{(g)} + \boldsymbol{\Lambda}_{i,m}^{(g)}\boldsymbol{\omega}_i^{(g)}, \boldsymbol{\Psi}_{i,m}^{(g)}\big], \tag{A.2}$$

where $\boldsymbol{\mu}_{i,m}^{(g)}$ is a $p_i^{(g)} \times 1$ subvector of $\boldsymbol{\mu}^{(g)}$, $\boldsymbol{\Lambda}_{i,m}^{(g)}$ is a $p_i^{(g)} \times q$ submatrix of $\boldsymbol{\Lambda}^{(g)}$ with rows that correspond to observed components deleted, and $\boldsymbol{\Psi}_{i,m}^{(g)}$ is a $p_i^{(g)} \times p_i^{(g)}$ submatrix of $\boldsymbol{\Psi}^{(g)}$ with the appropriate rows and columns deleted. In general, the structure of $\mathbf{V}_m$ may be very complicated with a large number of missing patterns having different positions of missing entries. However, the corresponding conditional distribution only involves a product of very simple normal distributions. The computational burden for simulating $\mathbf{V}_m$ is light.

(ii) Conditional distribution of $p[\mathbf{Y}_o|\cdot]$. Suppose the dimension of $\mathbf{y}_i^{(g)}$ is t. Based on the definition and properties of the current model, it can be shown that:

$$
\begin{aligned}
&p(\mathbf{Y}_o|\mathbf{V}_m, \boldsymbol{\Omega}, \mathbf{D}_o, \boldsymbol{\theta}) \\
&\quad = p(\mathbf{Y}_o|\boldsymbol{\Omega}, \mathbf{Z}_o, \boldsymbol{\theta}) = \prod_{g=1}^{G}\prod_{i=1}^{n_g}\prod_{h=1}^{t} p\big(y_{ih}^{(g)}|\boldsymbol{\omega}_i^{(g)}, z_{ih}^{(g)}, \boldsymbol{\theta}^{(g)}\big), \\
&p\big(y_{ih}^{(g)}|\boldsymbol{\omega}_i^{(g)}, z_{ih}^{(g)}, \boldsymbol{\theta}^{(g)}\big) \\
&\quad \stackrel{D}{=} N\big[\mu_h^{(g)} + \boldsymbol{\Lambda}_h^{(g)}\boldsymbol{\omega}_i^{(g)}, \psi_{hh}^{(g)}\big] I\big(y_{ih}^{(g)} \in \big(\alpha_{h,z_{ih}^{(g)}}^{(g)}, \alpha_{h,z_{ih}^{(g)}+1}^{(g)}\big]\big),
\end{aligned}
\tag{A.3}
$$

where $\boldsymbol{\Lambda}_h^{(g)}$ is the hth row of $\boldsymbol{\Lambda}^{(g)}$; $\psi_{hh}^{(g)}$ is the hth diagonal element of $\boldsymbol{\Psi}^{(g)}$; $z_{ih}^{(g)}$ is hth element of $\mathbf{z}_i^{(g)}$ associated with $y_{ih}^{(g)}$, and I is an indicator function.

(iii) Conditional distribution of $p[\boldsymbol{\Omega}|\cdot]$.

$$
\begin{aligned}
&p(\boldsymbol{\Omega}|\mathbf{V}_m, \mathbf{Y}_o, \mathbf{D}_o, \boldsymbol{\theta}) = \prod_{g=1}^{G}\prod_{i=1}^{n_g} p\big(\boldsymbol{\omega}_i^{(g)}|\mathbf{v}_i^{(g)}, \boldsymbol{\theta}^{(g)}\big), \quad \text{where} \\
&p\big(\boldsymbol{\omega}_i^{(g)}|\mathbf{v}_i^{(g)}, \boldsymbol{\theta}^{(g)}\big) \propto p\big(\boldsymbol{\omega}_i^{(g)}|\boldsymbol{\theta}^{(g)}\big) p\big(\mathbf{v}_i^{(g)}|\boldsymbol{\omega}_i^{(g)}, \boldsymbol{\theta}_i^{(g)}\big) \\
&\qquad \times \exp\bigg[-\frac{1}{2}\big\{\big(\mathbf{v}_i^{(g)} - \boldsymbol{\mu}^{(g)} - \boldsymbol{\Lambda}^{(g)}\boldsymbol{\omega}_i^{(g)}\big)'\boldsymbol{\Psi}^{(g)^{-1}} \\
&\qquad \times \big(\mathbf{v}_i^{(g)} - \boldsymbol{\mu}^{(g)} - \boldsymbol{\Lambda}^{(g)}\boldsymbol{\omega}_i^{(g)}\big) + \boldsymbol{\omega}_i^{(g)'}\boldsymbol{\Sigma}_\omega^{(g)^{-1}}\boldsymbol{\omega}_i^{(g)}\big\}\bigg].
\end{aligned}
\tag{A.4}
$$

It can be shown that

$$
p\big(\boldsymbol{\omega}_i^{(g)}|\cdot\big) \stackrel{D}{=} N\big(\boldsymbol{\Sigma}_g^{-1}\boldsymbol{\Lambda}^{(g)'}\boldsymbol{\Psi}^{(g)^{-1}}(\mathbf{v}_i^{(g)} - \boldsymbol{\mu}^{(g)}), \boldsymbol{\Sigma}_g\big),
$$

where

$$
\begin{aligned}
\boldsymbol{\Sigma}_g^{-1} &= \boldsymbol{\Sigma}_\omega^{(g)^{-1}} + \boldsymbol{\Lambda}^{(g)'}\boldsymbol{\Psi}^{(g)^{-1}}\boldsymbol{\Lambda}^{(g)}, \quad \text{and} \\
\boldsymbol{\Sigma}_\omega^{(g)} &= \begin{bmatrix} \mathbf{A}^{(g)^{-1}}(\boldsymbol{\Gamma}^{(g)}\boldsymbol{\Phi}^{(g)}\boldsymbol{\Gamma}^{(g)'} + \boldsymbol{\Psi}_\delta^{(g)})\mathbf{A}^{(g)} & \mathbf{A}^{(g)}\boldsymbol{\Gamma}^{(g)}\boldsymbol{\Phi}^{(g)} \\ \boldsymbol{\Phi}^{(g)}\boldsymbol{\Gamma}^{(g)'}\mathbf{A}^{(g)'} & \boldsymbol{\Phi}^{(g)} \end{bmatrix}, \\
\mathbf{A}^{(g)} &= (\mathbf{I} - \mathbf{B}^{(g)})^{-1}.
\end{aligned}
$$

Simulating observations from the conditional distributions that are given in (A.3) involves the univariate truncated normal distribution, and this is done by the inverse distribution method proposed by Devroye (1985). Conditional distribution in (A.4) is normal, simulating observations from it is straightforward.

Appendix B

We describe the application of path sampling (Gelman and Meng, 1998) to compute $\log p(\mathbf{D}_o; \hat{\boldsymbol{\theta}}_1, M_1)$. To facilitate the computation, we choose a model M_0 with a parameter vector $\boldsymbol{\theta}_0$ that is nested in M_1, such that $\boldsymbol{\theta}_1 = (\boldsymbol{\theta}_0, \boldsymbol{\theta}_1^*)$ and $\boldsymbol{\theta}_1$ reduces to $\boldsymbol{\theta}_0$ under M_0. Then we define a path to link M_1 and M_0 by a linked model M_t with a parameter vector $t\boldsymbol{\theta}_1 = (\boldsymbol{\theta}_0, t\boldsymbol{\theta}_1^*)$ via a continuous path t in [0, 1], such that $M_t = M_1$ if $t = 1$, and $M_t = M_0$ if $t = 0$. Let $p(\mathbf{D}; \boldsymbol{\theta})$ be the complete-data likelihood function, $U(\mathbf{D}; t\hat{\boldsymbol{\theta}}_1) = d \log p(\mathbf{D}; t\hat{\boldsymbol{\theta}}_1)/dt$, and

$$\lambda_{10} = \log\left[p(\mathbf{D}_o; \hat{\boldsymbol{\theta}}_1, M_1)/p(\mathbf{D}_o; \hat{\boldsymbol{\theta}}_0, M_0)\right].$$

It can be shown that (Gelman and Meng, 1998)

$$\lambda_{10} = \frac{1}{2}\sum_{s=1}^{S}(t_{(s+1)} - t_{(s)})(\overline{U}_{(s+1)} + \overline{U}_{(s)}),$$

where $\{t_{(s)}\colon s = 0, \ldots, S\}$ are grids in [0, 1] such that $0 = t_{(0)} < t_{(1)} < \cdots < t_{(s+1)} = 1$, $\overline{U}_{(s)} = J^{-1}\sum_{j=1}^{J} U(\mathbf{D}_o, \mathbf{D}_{mj}; t_{(s)}\hat{\boldsymbol{\theta}}_1)$, with $\{\mathbf{D}_{mj}\colon j = 1, \ldots, J\}$ are observations that are simulated from the conditional distribution of $\mathbf{D}_m$ given $\mathbf{D}_o$ at $t_{(s)}\hat{\boldsymbol{\theta}}_1$. The MCMC method at the E-step of the MCEM algorithm can be directly applied to simulate the above sample of observations for computing λ_{10}. If we can find a simple M_0 such that $p(\mathbf{D}_o; \hat{\boldsymbol{\theta}}_0, M_0)$ can be evaluated, then

$$\log p(\mathbf{D}_o; \hat{\boldsymbol{\theta}}_1, M_1) = \lambda_{10} + \log p(\mathbf{D}_o; \hat{\boldsymbol{\theta}}_0, M_0).$$

In practice, M_0 can be chosen as $\mathbf{v}_i^{(g)} = \boldsymbol{\mu}^{(g)} + \boldsymbol{\varepsilon}_i^{(g)}$, so that its observed-data log-likelihood can be obtained via single integration (note that $\boldsymbol{\Psi}^{(g)}$ is a diagonal), see (Lee and Song, 2004b).

References

Bentler, P.M. (1992). *EQS: Structural Equation Program Manual*. BMDP Statistical Software, Los Angeles.

Bentler, P.M., Stein, J.A. (1992). Structural equation models in medical research. *Statistical Methods in Medical Research* **1**, 159–181.

Bollen, K.A. (1989). *Structural Equations with Latent Variables*. John Wiley and Sons, New York.

Boomsma, D.I., Molenaar, P.C.M. (1987). Constrained maximum likelihood analysis of familiar resemblance of twins and their parents. *Acta Genet. Med. Gemellol.* **36**, 29–39.

Booth, J.G., Hobert, J.P. (1999). Maximum generalized linear model likelihoods with an automated Monte Carlo EM algorithm. *Journal of the Royal Statistical Society, Series B* **61**, 265–285.

Chan, C.N., Chan, Y.W., Cheung, C.K., Swaminathan, R., Lan, M.C., Cockrama, S., Wooa, J. (1996). The metabolic syndrome in Hong Kong Chinese: The interrelationship among its components analyzed by structural equation modeling. *Diabetes Care* **19**, 953–959.

Devroye, L. (1985). *Non-Uniform Random Variate Generation*. Springer-Verlag, New York.
Dolan, C.V., Molenaar, P.C.M., Boomsma, D.I. (1991). Simultaneous genetic analysis of longitudinal means and covariance structure in the simplex model using twin data. *Behavior Genetics* **21**, 49–65.
Drota, D. (Ed.) (1998). *Measuring Health-Related Quality of Life in Children and Adolescents: Implications for Research and Practice*. Lawrence Erlbaum Associates, Mahwah, NJ.
Francis, D.J. (1988). An introduction to structural equation models. *Journal of Clinical and Experimental Neuropsychology* **10**, 623–639.
Gelman, A., Meng, X.L. (1998). Simulating normalizing constant: From importance sampling to bridge sampling to path sampling. *Statistical Science* **13**, 163–185.
Geman, S., Geman, D. (1984). Stochastic relaxation, Gibbs distributions, and the Bayesian restoration of images. *IEEE Transactions on Pattern Analysis and Machine Intelligence* **6**, 721–741.
Hastings, W.K. (1970). Monte Carlo sampling methods using Markov chains and their application. *Biometrika* **57**, 97–109.
Jöreskog, K.G., Sörbom, D. (1996). *LISREL 8: Structural Equation Modeling with the SIMPLIS Command Language*. Scientific Software International, Hove and London.
Kass, R.E., Raftery, A.E. (1995). Bayes factors. *Journal of the American Statistical Association* **90**, 773–795.
Lee, S.Y., Shi, J.Q. (2001). Maximum likelihood estimation of two-level latent variable models with mixed continuous and polytomous data. *Biometrics* **57**, 787–794.
Lee, S.Y., Song, X.Y. (2004a). Maximum likelihood estimation and model comparison for mixtures of structural equation models with ignorable missing data. *Journal of Classification* **20**, 221–255.
Lee, S.Y., Song, X.Y. (2004b). Maximum likelihood analysis of a general latent variable model with hierarchically mixed data. *Biometrics* **60**, 624–636.
Lee, S.Y., Poon, W.Y., Bentler, P.M. (1995). A 2-stage estimation of structural equation models with continuous and polytomous variables. *British Journal of Mathematical and Statistical Psychology* **48**, 339–358.
Lee, S.Y., Song, X.Y., Skevington, S., Hao, Y.T. (2005). Application of structural equation models to quality of life. *Structural Equation Modeling: A Multidisciplinary Journal* **12**, 435–453.
Little, R.J.A., Rubin, D.B. (1987). *Statistical Analysis with Missing Data*. Wiley, New York.
Louis, T.A. (1982). Finding the observed information matrix when using EM algorithm. *Journal of the Royal Statistical Society, Series B* **44**, 226–233.
Meng, X.L., Rubin, D.B. (1993). Maximum likelihood estimation via the ECM algorithm: A general framework. *Biometrika* **80**, 267–278.
Meng, X.L., Schilling, S. (1996). Fitting full-information item factor models and an empirical investigation of bridge sampling. *Journal of American Statistical Association* **91**, 1254–1267.
Meng, X.L., Wong, H.W. (1996). Simulating ratios of normalizing constants via a simple identity: a theoretical exploration. *Statistica Sinica* **6**, 831–860.
Menleners, L.B., Lee, A.H., Binns, C.W., Lower, A. (2003). Quality of life for adolescents: Assessing measurement properties using structural equation modeling. *Quality of Life Research* **12**, 283–290.
Metropolis, N., Rosenbluth, A.W., Rosenbluth, M.N., Teller, A.H., Teller, E. (1953). Equations of state calculations by fast computing machine. *Journal of Chemical Physics* **21**, 1087–1091.
Muthen, L., Muthen, B. (2001). *Mplus User's Guide*. Muthen and Muthen, Los Angeles, CA.
Newton, M.A., Raftery, A.E. (1994). Approximate Bayesian inference by the weighted likelihood bootstrap (with discussion). *Journal of the Royal Statistical Society, Series B* **56**, 3–48.
Power, M., Bullingen, M., Hazper, A., WHOQOL Group (1999). The World Health Organization WHOQOL-100: Tests of the universality of quality of life in 15 different cultural groups worldwide. *Health Psychology* **18** (5), 495–505.
Pugesek, B.H., Tomer, T.A., von Eye, A. (2003). *Structural Equation Modeling Applications in Ecological and Evolutionary Biology*. Cambridge Univ. Press, New York.
Raftery, A.E. (1993). Bayesian model selection in structural equation models. In: Bollen, K.A., Long, J.S. (Eds.), *Testing Structural Equation Models*. New Delhi, London.
Shi, J.Q., Copas, J. (2002). Publication bias and meta-analysis for 2×2 tables: an average Markov chain Monte Carlo EM algorithm. *Journal of the Royal Statistical Society, Series B* **64**, 221–236.
Shi, J.Q., Lee, S.Y. (2000). Latent variable models with mixed continuous and polytomous data. *Journal of the Royal Statistical Society, Series B* **62**, 77–87.

Song, X.Y., Lee, S.Y. (2005). A multivariate probit latent variable model for analyzing dichotomous responses. *Statistica Sinica* **15**, 645–664.

Song, X.Y., Lee, S.Y. (2006). Model comparison of generalized linear mixed models. *Statistics in Medicine* **25**, 1685–1698.

Song, X.Y., Lee, S.Y., Zhu, H.T. (2001). Model selection in structural equation models with continuous and polytomous data. *Structural Equation Modeling – A Multidisciplinary Journal* **8**, 378–396.

Staquet, M.J., Hayes, R.D., Fayes, P.M. (Eds.) (1998). *Quality of Life Assessment in Clinical Trials*. Oxford University Press, New York.

Wei, G.C.G., Tanner, M.A. (1990). A Monte Carlo implementation of the EM algorithm and the poor man's data augmentation algorithm. *Journal of the American Statistical Association* **85**, 699–704.

Handbook of Computing and Statistics with Applications, Vol. 1
ISSN: 1871-0301

DOI: 10.1016/S1871-0301(06)01014-6

14

The Set of Feasible Solutions for Reliability and Factor Analysis

Jos M.F. Ten Berge and Gregor Sočan

Abstract
The mathematical model underlying factor analysis has the same structure as classical test theory. Both in factor analysis and in test theory, a convex set of "feasible tautologies" has been defined. In factor analysis, the set contains all nonnegative diagonal matrices $\boldsymbol{\Psi}$ of unique variances which entail a reduced covariance matrix $\boldsymbol{\Sigma} - \boldsymbol{\Psi}$ with no negative eigenvalues. In reliability theory, the set contains all nonnegative diagonal matrices of error variances which entail a true score covariance matrix with no negative eigenvalues. Points in the feasible set are considered which have direct psychometric interpretations. The most reliable point in test theory has Principal Component Analysis as a factor analysis counterpart. Similarly, the least reliable point, associated with the greatest lower bound to reliability, corresponds to Minimum Trace Factor Analysis. Other factor solutions in the set are Minimum Rank Factor Analysis (minimizing the unexplained common variance for different numbers of retained common factors), and a novel quadratic variant of it. The latter method minimizes the same function as Least Squares Factor Analysis, but does constrain the solution to be in the feasible set. Minimum Rank Factor Analysis solutions have no direct counterparts in test theory, except when only one common factor is extracted. The single factor solution corresponds to the most unidimensional (most congeneric) point in test theory. To evaluate the reliability of congeneric test, the greatest lower bound is argued to be preferable to reliability measures based on a single factor hypothesis. The greatest lower bound is also argued to be superior to reliability based on multiple factor solutions.

For many decades, factor analysis has upheld intimate relationships with classical test theory. It is not just that practitioners use factor analysis to construct reliable scales. The very mathematical model underlying factor analysis has the same structure as classical test theory. Both in factor analysis and in reliability theory, a "set of feasible tautologies" has been defined. In factor analysis, the set contains all nonnegative diagonal matrices $\boldsymbol{\Psi}$ of unique variances such that the common parts covariance matrix $\boldsymbol{\Sigma} - \boldsymbol{\Psi}$, holding communalities in the diagonal, has no negative eigenvalues. Each point in the set refers to one particular factor analysis solution in a tautological manner, e.g., Browne (1969). In test theory, the set contains all nonnegative diagonal matrices of error variances which entail a true score covariance matrix that has no negative eigenvalues. Each point in the set now defines a possible set of error variances, and, by implication, a possible value of the reliability. The focus of this paper is on points in the feasible set which have direct psychometric interpretations.

After an introduction of the feasible set in the factor analysis context, a historical overview is given of Ledermann's bound to the number of factors needed for perfect fit. Then the feasible set is discussed from a reliability point of view, by reviewing the "greatest lower bound" (glb) to reliability (Jackson and Agunwamba, 1977). Next, the set is again considered from a factor analysis point of view. Minimum Trace Factor Analysis (Bentler and Woodward, 1980; Shapiro, 1982) and Minimum Rank Factor Analysis (Ten Berge and Kiers, 1991) are reviewed, and a new type of least squares factor analysis which is also in the feasible set is introduced. The concepts of test unidimensionality and test reliability are discussed and compared in terms of an example. Finally, reliability based on multiple factor solutions is discussed. Both for the single and the multiple factor situation, the glb is argued to be the superior estimate of reliability.

1. Introduction

Factor analysis of a set of variables $x_1, \ldots, x_m$, decomposes each variable x_j into a common part c_j, giving rise to correlation between x_j and x_k, $j \neq k$, and a unique part u_j, defined to be uncorrelated with $c_1, \ldots, c_m$ and with u_k, $k \neq j$. Upon writing

$$x_j = c_j + u_j, \tag{1}$$

$j = 1, \ldots, m$, and using the property that $u_1, \ldots, u_m$ are uncorrelated with $c_1, \ldots, c_m$ and among themselves, we have

$$\boldsymbol{\Sigma} = \mathbf{FF}' + \boldsymbol{\Psi}, \tag{2}$$

where $\boldsymbol{\Sigma}$ is the covariance or correlation matrix of the variables, $\boldsymbol{\Psi}$ the diagonal matrix of unique variances, and $\mathbf{FF}'$ the covariance matrix of the common parts c_j, $j = 1, \ldots, m$, of the variables. The variances of these common parts are in the diagonal of $\mathbf{FF}'$. They are the communalities of the variables.

Factor analysis aims at a decomposition (2) with $\boldsymbol{\Sigma} - \boldsymbol{\Psi}$ of low rank r, which can be factored as $\boldsymbol{\Sigma} - \boldsymbol{\Psi} = \mathbf{FF}'$, with $\mathbf{F}$ an $m \times r$ matrix. However, in practice, the smallest value of r for which (2) can be solved exactly tends to be prohibitively high, an issue to be reviewed in the next section. For practical purposes, we shall have to allow that some common factors will be discarded, and decompose $\boldsymbol{\Sigma}$, for some small value of r, as

$$\boldsymbol{\Sigma} = \mathbf{F}_1\mathbf{F}_1' + \mathbf{F}_2\mathbf{F}_2' + \boldsymbol{\Psi}, \tag{3}$$

where $\mathbf{F}_1$ is the $m \times r$ matrix of loadings on r retained common factors, and $\mathbf{F}_2$, with columns orthogonal to those of $\mathbf{F}_1$, holds the loadings on common factors to be discarded. This means that the variances of the variables (diagonal elements of $\boldsymbol{\Sigma}$) are decomposed into explained common variances (diagonal elements of $\mathbf{F}_1\mathbf{F}_1'$), unexplained common variances (diagonal elements of $\mathbf{F}_2\mathbf{F}_2'$), and unique variances (diagonal elements of $\boldsymbol{\Psi}$). It is essential to note that (3) still requires that both $\boldsymbol{\Sigma} - \boldsymbol{\Psi}$ and $\boldsymbol{\Psi}$ are at least positive semidefinite. Negative elements in $\boldsymbol{\Psi}$, known as Heywood cases, have drawn a lot of attention. However, when $\boldsymbol{\Sigma} - \boldsymbol{\Psi}$, the covariance matrix for the common parts of the variables, would appear to be indefinite, that would be equally incompatible with (3).

A small example may be instructive. Suppose we have observed the covariance matrix

$$\boldsymbol{\Sigma} = \begin{bmatrix} 1 & 0.73 & 0.36 & 0.36 \\ 0.73 & 1 & 0.36 & 0.36 \\ 0.36 & 0.36 & 1 & 0.52 \\ 0.36 & 0.36 & 0.52 & 1 \end{bmatrix}. \tag{4}$$

When $\boldsymbol{\Psi}$ has the diagonal elements 0.27, 0.27, 0.48, and 0.48, respectively, we obtain a matrix $\boldsymbol{\Sigma} - \boldsymbol{\Psi}$ of rank 2, with communalities 0.73, 0.73, 0.52 and 0.52 in the diagonal cells. Its eigenvalues are 2.0, 0.50, 0, and 0, so $\boldsymbol{\Psi}$ is indeed in the feasible set. We can factor $\boldsymbol{\Sigma} - \boldsymbol{\Psi}$ as $\mathbf{FF}'$, where

$$\mathbf{F} = \begin{bmatrix} 0.8 & -0.3 \\ 0.8 & -0.3 \\ 0.6 & 0.4 \\ 0.6 & 0.4 \end{bmatrix}. \tag{5}$$

Because the common factors account for all covariances between the variables, we have found a solution for (2). The total common variance is 2.5, the sum of communalities. It is fully explained by the two common factors, which means that 100% of the common variance is explained by the two factors. Because of the perfect fit, any particular method of common factor analysis would produce the same solution.

The situation changes when it is desired to determine only a single common factor. One possible solution would be to preserve the elements of $\boldsymbol{\Psi}$, and determine $\mathbf{F}_1$ and $\mathbf{F}_2$ as columns 1 and 2 of $\mathbf{F}$, respectively. The common variance to be explained remains 2.5, but now only 2.0 is explained by the single factor, which amounts to 80% of explained common variance.

The above rank-one solution is not the standard solution from least squares factor analysis. The latter method yields $\mathbf{F}_1 = [0.7905\ 0.7905\ 0.5509\ 0.5509]'$, and it defines the diagonal elements of $\boldsymbol{\Psi}$ as 0.375, 0.375, 0.697, and 0.697. Now $\boldsymbol{\Sigma} - \boldsymbol{\Psi}$ has eigenvalues 1.8566, 0.3217, −0.1052, and −0.2165. Dividing the first eigenvalue (explained common variance) by the sum of all eigenvalues (common variance to be explained) would reveal that 100% of the common variance is explained by a single factor. Because there is no perfect fit, this value is of no use at all. This is not an idiosyncrasy of the miniature data. In fact, when for any r-factor solution, the percentage of explained common variance is evaluated as the sum of the r largest eigenvalues of $\boldsymbol{\Sigma} - \boldsymbol{\Psi}$, divided by the sum of all its eigenvalues, least squares factor analysis (regardless of the particular algorithm used) invariably yields 100% of explained common variance, because the last $m - r$ eigenvalues (0.3217, −0.1052, and −0.2165 in the example above) sum to zero (Harman, 1976, p. 182). Incidentally, as Harman has explained, the same goes for maximum-likelihood factor analysis. When perfect fit is abandoned, and approximate least squares or maximum likelihood solutions are adopted instead, the concept of explained common variance is sacrificed. This is because these methods imply matrices $\boldsymbol{\Psi}$ which do not belong to the feasible set. If it is desired to preserve the concept of explained common variance, we shall have to resort to methods which do stay in the feasible set. Such methods will be discussed in due course.

2. The Ledermann bound

As we have seen above, the unexplained common variance entailed by (3) will be zero when $\mathbf{F}_2$ vanishes, that is, when r, the number of common factors to be retained, equals the rank of $\boldsymbol{\Sigma} - \boldsymbol{\Psi}$, which we shall call the reduced rank of $\boldsymbol{\Sigma}$ in the sequel. Accordingly, the question is to what extent the rank of $\boldsymbol{\Sigma} - \boldsymbol{\Psi}$ can be reduced by an appropriate choice of unique variances in $\boldsymbol{\Psi}$, or, equivalently, of communalities to put in the diagonal of $\boldsymbol{\Sigma}$. The answer has an intriguing history, revolving around Ledermann's bound. Ledermann (1937) proposed that the rank of $\boldsymbol{\Sigma} - \boldsymbol{\Psi}$ could always be reduced to the smallest integer equal to or above the function

$$\varphi(m) = \left[2m + 1 - \sqrt{8m + 1}\,\right]/2. \tag{6}$$

For instance, when $m = 6$, we would need at most three factors to solve (2), and when $m = 10$, six factors would be enough. For $m = 4$, the Ledermann bound is 1.63, indicating that 2 factors would be enough.

Ledermann (1937) inferred that the function $\varphi(m)$ given in (6) should be seen as an upper bound to r, the number of factors needed to solve (2), by counting the equations and parameters involved in solving (2). This optimistic view was shattered by counterexamples presented by Wilson and Worcester (1939) and Guttman (1958). Guttman showed that the universal upper bound to r is as high as $m - 1$, also see Bekker and De Leeuw (1987). Ledermann's bound seemed history, albeit that the bound did keep a role in the number of degrees of freedom for the chi-square test of the factor analysis model, e.g., Jöreskog (1967).

A surprising remake of Ledermann's bound took place in the eighties. Contrary to what Ledermann believed, Shapiro (1982) showed that the bound is almost surely *a lower bound* to the number of factors needed in factor analysis. Specifically, Shapiro has shown that the set of covariance matrices $\boldsymbol{\Sigma}$ the rank of which can be reduced to a value below the Ledermann bound by changing the diagonal elements has Lebesgue measure zero. This means that, when $m = 4$, we almost surely need *at least* two factors to solve (2), when $m = 6$ we almost surely need *at least* three, and when $m = 10$, we need *at least* six. For the covariance matrix given in (4), where $m = 4$, we have seen that two factors are enough to solve (2), but cases where we need three factors also arise with positive probability. For instance, if the covariance between variables 1 and 4 in (4) is changed to 0.55, we need three factors to solve (2).

Shapiro (1982) also proved that the minimum rank is unstable below the Ledermann bound. This means that, when the rank of a given $\boldsymbol{\Sigma}$ can be reduced to a value below the Ledermann bound (for instance, when such a matrix is artificially constructed), any change, no matter how small, of the off-diagonal elements of $\boldsymbol{\Sigma}$ will increase the minimum reduced rank of $\boldsymbol{\Sigma}$. Furthermore, Shapiro (1985) has shown that any $\boldsymbol{\Psi}$ that solves (2), with $\boldsymbol{\Sigma} - \boldsymbol{\Psi}$ positive semidefinite of rank r, is almost surely non-unique (meaning that other choices of $\boldsymbol{\Psi}$, entailing the same reduced rank r, exist) when r is above the Ledermann bound, and almost surely locally unique at and below the Ledermann bound. Local uniqueness means that in a neighborhood of any $\boldsymbol{\Psi}$ that satisfies (2) no other solutions exist. Shapiro (1985) also conjectured that $\boldsymbol{\Psi}$ is almost surely globally unique when r is strictly below the Ledermann bound. This conjecture has been proven correct (Bekker and Ten Berge, 1997).

The results on Ledermann's bound are important in that they provide answers to long-standing questions. However, from a practical point of view, the results are not what one might have hoped for. In cases of perfect fit, uniqueness of $\boldsymbol{\Psi}$ holds only below (sometimes at) Ledermann's bound, but such cases do not arise in practice. Accordingly, we shall have to resort to solutions for (3) rather than (2), to obtain an $\mathbf{F}_2$ which is small in some sense, and can be discarded without losing much information. Methods for obtaining such solutions will be discussed below.

Incidentally, Ledermann's bound also plays a role in the so-called inverse principal component problem of a correlation matrix. That is, when the number of retained principal components (possibly rotated) is at or above the Ledermann bound, the entire correlation matrix can be retrieved from the retained loadings (Ten Berge and Kiers, 1999). For instance, the loadings on three principal components of a correlation matrix of six variables carry enough information to allow retrieving that correlation matrix.

The fact that solving (2) is not possible for a small number of factors r has led the mainstream of factor analysis to turn to approximations where some function of $\boldsymbol{\Sigma} - \mathbf{F}\mathbf{F}'$, with $\mathbf{F}$ a tall matrix, is minimized subject to either no constraints, or subject to the constraint that the unique variances, evaluated as the diagonal elements of $\boldsymbol{\Sigma} - \mathbf{F}\mathbf{F}'$, are nonnegative. We have seen an example of Least Squares Factor Analysis above. It yields a $\boldsymbol{\Psi}$ outside the feasible set. It is well known that Maximum Likelihood Factor Analysis does satisfy the constraint that $\boldsymbol{\Sigma} - \mathbf{F}_1\mathbf{F}_1'$ is a Gramian matrix (Browne, 1969), but it yields a $\boldsymbol{\Sigma} - \boldsymbol{\Psi}$ that cannot be written as $\mathbf{F}_1\mathbf{F}_1' + \mathbf{F}_2\mathbf{F}_2'$ because it has one or more negative eigenvalues. So it is likewise outside the feasible set. In fact, there are very few methods which decompose $\boldsymbol{\Sigma}$ by (2) with both $\boldsymbol{\Sigma} - \boldsymbol{\Psi}$ and $\boldsymbol{\Psi}$ Gramian. Until the nineties, the only exception was Constrained Minimum Trace Factor Analysis (CMTFA, Bentler and Woodward, 1980; Shapiro, 1982), which has its roots in the framework of classical test theory. That framework will be discussed next.

3. Reliability theory and a convex set of possible solutions for (2)

Classical test theory starts from the axiom that, for a test X composed of test parts $x_1, \ldots, x_m$, each test part x_j consists of a true score t_j and error e_j:

$$x_j = t_j + e_j, \tag{7}$$

$j = 1, \ldots, m$, the error e_j being uncorrelated with $t_1, \ldots, t_m$ and with e_k, $k \neq j$. The reliability of test X is defined as

$$\mathrm{Var}(T)/\,\mathrm{Var}(X) = 1 - \mathrm{Var}(E)/\,\mathrm{Var}(X), \tag{8}$$

where $X = x_1 + \cdots + x_m$, $T = t_1 + \cdots + t_m$, and $E = e_1 + \cdots + e_m$. The reliability can be evaluated as the correlation between the test and a parallel test. However, parallel tests are more often than not absent, whence one may have to settle for lower bounds to the reliability. By far the best known amid these is Guttman's lower bound λ_3 (Guttman, 1945), usually referred to as Cronbach's alpha. The relation between alpha and the reliability can be expressed by the identity

$$alpha + \frac{1}{(m-1)\,\mathrm{Var}(X)} \sum_{j<k} \mathrm{Var}(t_j - t_k) = reliability, \tag{9}$$

see Ten Berge and Sočan (2004). This formula reveals at once that alpha is a lower bound to the reliability, as we know from Guttman (1945), and that alpha equals the reliability if and only if all true score differences $t_j - t_k$ have variance zero, as we know from Novick and Lewis (1967).

Jackson and Agunwamba (1977) noted that all lower bounds to reliability use *part of the information* implicit in the classical definitions. For instance, alpha uses nonnegativity of all variances of true score differences $t_j - t_k$, as is immediate from (9). Jackson and Agunwamba, elaborating on pioneering work by Bentler (1972), asked the fundamental question of how to use *all information* implied by the classical assumptions. The zero covariance of error terms with true scores and with errors of other test parts implies that the observed covariance matrix $\boldsymbol{\Sigma}$ can be decomposed as

$$\boldsymbol{\Sigma} = \boldsymbol{\Sigma}_T + \boldsymbol{\Sigma}_E, \tag{10}$$

where $\boldsymbol{\Sigma}_E$, the covariance matrix of $e_1, \ldots, e_m$, is diagonal, and both $\boldsymbol{\Sigma}_E$ and $\boldsymbol{\Sigma}_T$ (the covariance matrix of $t_1, \ldots, t_m$), are positive semidefinite. Jackson and Agunwamba thus arrived at the set of all feasible states of nature: It is the set of all nonnegative diagonal matrices $\boldsymbol{\Sigma}_E$ for which $\boldsymbol{\Sigma} - \boldsymbol{\Sigma}_E$ has no negative eigenvalue. The set is convex, and the $m = 2$ case can be pictured as in Figure 1.

The area at and below the curve represents the set of all cases where $\boldsymbol{\Sigma} - \boldsymbol{\Sigma}_E$ is positive semidefinite and positive definite, respectively. The positive quadrant contains all nonnegative diagonal matrices $\boldsymbol{\Sigma}_E$, and the shaded area is the intersection of the two sets. This intersection defines all possible solutions for $\boldsymbol{\Sigma}_E$. It contains all diagonal matrices $\boldsymbol{\Sigma}_E$ that are compatible with the classical axioms. The points in the set that are on the curve are the boundary points of the set. They correspond to $\boldsymbol{\Sigma} - \boldsymbol{\Sigma}_E$ singular, and all interior points correspond to $\boldsymbol{\Sigma} - \boldsymbol{\Sigma}_E$ nonsingular, assuming that $\boldsymbol{\Sigma}$ is nonsingular to begin with. It may be noted that we can move from each interior point to a boundary point by adding $\lambda_m \mathbf{I}_m$ to $\boldsymbol{\Sigma}_E$, where λ_m is the smallest eigenvalue of $\boldsymbol{\Sigma} - \boldsymbol{\Sigma}_E$. The new point $\boldsymbol{\Sigma}_E + \lambda_m \mathbf{I}_m$ corresponds to a matrix $\boldsymbol{\Sigma} - \boldsymbol{\Sigma}_E - \lambda_m \mathbf{I}_m$ which is positive semidefinite and singular.

At this point, it is instructive to consider points in the set that can be identified at once. First, because $\boldsymbol{\Sigma}_X - \mathbf{O} = \boldsymbol{\Sigma}_X$ is positive semidefinite, there is the origin. It

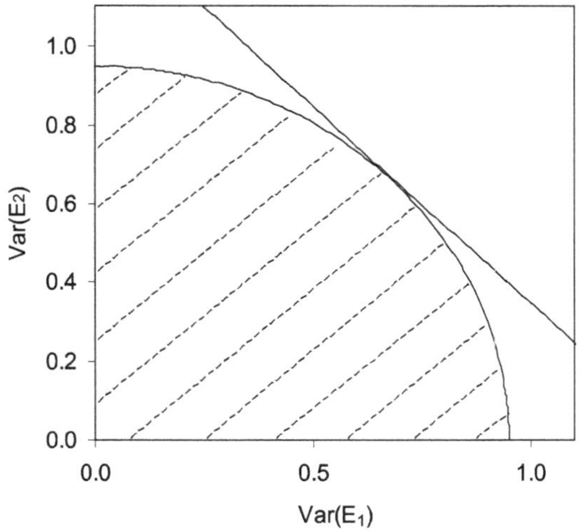

Fig. 1. An example of the feasible set ($m = 2$).

represents the point where all error variances are zero. Next, there are m boundary points which have one positive coordinate, and all other coordinates zero. They correspond to the covariance matrices that remain when variable x_j is replaced by its multiple linear regression on all other $m - 1$ variables, $j = 1, \ldots, m$, viz. the points (0.955, 0) and (0, 0.951) in Figure 1. The nonzero coordinate (that which is subtracted from the variance of variable j) is the residual variance associated with the regression on the other $m - 1$ variables. Clearly, the resulting $\boldsymbol{\Sigma}_E$ is a boundary point with $\boldsymbol{\Sigma} - \boldsymbol{\Sigma}_E$ singular.

Furthermore, we have a boundary point with all coordinates equal. It is the point where $\boldsymbol{\Sigma}_E$ is $\lambda_m \mathbf{I}_m$, where λ_m is the smallest eigenvalue of $\boldsymbol{\Sigma}$. When that eigenvalue is zero, either the entire feasible set collapses into the origin, or one or more coordinates vanish. For instance, when variable x_m is the sum of variables $x_1, \ldots, x_{m-1}$, the origin is the only point in the feasible set; when, however, there is a linear dependency in just a subset of variables, the coordinates of points in the feasible set vanish for those variables. In both cases, removing one or more variables to restore linear independence is indicated.

Finally, the set has a unique boundary point (Ten Berge et al., 1981; Della Riccia and Shapiro, 1982) that has the largest sum of coordinates. This is the point where the tangent hyperplane orthogonal to the vector of ones, that is, the hyperplane defined by a constant sum of error variances, touches the feasible set. When $m = 2$, the hyperplane becomes the tangent line with angle $-45°$, portrayed in Figure 1.

We now consider implications of the feasible set for reliability theory. By mapping the set of possible $\boldsymbol{\Sigma}_E$ into the interval [0, 1] by the reliability function $r_{XX} = 1 - \mathrm{tr}(\boldsymbol{\Sigma}_E)/\mathrm{Var}(X)$, that interval is split in two areas of possible versus impossible values for the reliability. Clearly, because $\boldsymbol{\Sigma}_E = \mathbf{O}$ (the origin) belongs to the set, a reliability as high as 1 can never be ruled out on the basis of a single test administration. On the other hand, the worst possible situation for reliability is when the sum of error variances is a maximum over the set of possible $\boldsymbol{\Sigma}_E$. It defines the worst case scenario for reliability. It has the largest possible sum of error variances under full consideration of all information. Accordingly, the *greatest lower bound* (glb) to reliability is defined as the reliability value associated with that point (Jackson and Agunwamba, 1977). Hence, in the [0, 1] interval, the possible reliability values are in the range [glb, 1], and the impossible reliability values are in the range [0, glb). All lower bounds (except when they happen to coincide with the glb) are in the interval [0, glb) of impossible reliability values, or they are impossible because they are negative.

An efficient computational method to evaluate the glb as the worst possible case in the feasible set has been proposed by Bentler and Woodward (1980), also see Ten Berge et al. (1981). It does not just yield the maximum value of $\mathrm{tr}(\boldsymbol{\Sigma}_E)$, but also identifies the maximizing $\boldsymbol{\Sigma}_E$. Hence, the method offers a possible solution for (10). But then it also offers a possible solution for (2): Upon replacing $\boldsymbol{\Sigma}_E$ by $\boldsymbol{\Psi}$, and $\boldsymbol{\Sigma}_T$ by $\mathbf{FF}'$, it provides the solution to (2) with the maximum sum of unique variances. It has been christened Constrained Minimum Trace Factor Analysis (CMTFA) by Shapiro (1982). It yields the solution to (2) having the smallest possible trace for the reduced covariance matrix $\boldsymbol{\Sigma} - \boldsymbol{\Psi}$, subject to the constraints that both $\boldsymbol{\Sigma} - \boldsymbol{\Psi}$ and $\boldsymbol{\Psi}$ (diagonal) be at least positive semidefinite. However, although CMTFA does solve (2) subject to its constraints, it has no appeal whatsoever as a method of factor analysis, because it is not aimed at reducing $\boldsymbol{\Sigma}$ to a matrix $\boldsymbol{\Sigma} - \boldsymbol{\Psi}$ of (approximately) low rank, as will be shown next.

4. Minimizing the sum and the sum of squares of unexplained common variances

The purpose of CMTFA is to find the solution for (2) which minimizes $\mathrm{tr}(\boldsymbol{\Sigma} - \boldsymbol{\Psi})$ or, equivalently, $\sum_{j=1}^{m} \lambda_j(\boldsymbol{\Sigma} - \boldsymbol{\Psi})$ (the sum of *all* eigenvalues of $\boldsymbol{\Sigma} - \boldsymbol{\Psi}$) subject to its constraints. Because there is no separation here between explained and unexplained common variance as detailed in (3), CMTFA has no link to the purpose of factor analysis. It is not aimed at (approximate) rank reduction. In fact, CMTFA often yields all reduced eigenvalues nonzero, except the last. In other words, CMTFA is not aimed at finding a decomposition (3) with $\mathbf{F}_1$ "big" in some sense, and $\mathbf{F}_2$ "small". To obtain a method more germane to the purpose of factor analysis, Ten Berge and Kiers (1991) proposed a generalization of CMTFA, called Minimum Rank Factor Analysis (MRFA), which does aim at finding a "small" $\mathbf{F}_2$ in (3). For any given number r of common factors to be retained, MRFA minimizes the sum of the smallest $m-r$ reduced eigenvalues. Equivalently, it decomposes $\boldsymbol{\Sigma}$ in such a way that, for a fixed r, the sum of squares of $\mathbf{F}_2$ is minimized. In terms of reduced eigenvalues, MRFA minimizes

$$f_1(\boldsymbol{\Psi}) = \sum_{j=r+1}^{m} \lambda_j(\boldsymbol{\Sigma} - \boldsymbol{\Psi}), \tag{11}$$

subject to the constraint $\lambda_1 \geqslant \lambda_2 \geqslant \cdots \geqslant \lambda_m \geqslant 0$, with $\boldsymbol{\Psi}$ a nonnegative diagonal matrix. It may be noted that CMTFA is the special case of MRFA when $r = 0$.

In earlier work (e.g., Shapiro, 1982), MRFA used to be associated with factor solutions that satisfy (2) exactly with r as small as possible. However, Ten Berge and Kiers (1991) have relaxed this definition of MRFA to include approximate solutions.

Table 1
Three methods targeted at small reduced eigenvalues

	Eigenvalue criterion minimized	ssq minimized	Heywood case protection	Last $m - r$ eigenvalues	Explained common variance
MRFA	Sum of last $m - r$ eigenvalues	$\text{ssq}(\mathbf{F}_2)$	Automatic	Nonnegative	Percentage available
MRFA-Q	Sum of squares of last $m - r$ eigenvalues	$\text{ssq}(\mathbf{F}_2\mathbf{F}_2')$	Automatic	Nonnegative	Percentage available
ULS	Sum of squares of last $m - r$ eigenvalues	$\text{ssq}(\mathbf{F}_2\mathbf{F}_2')$	Available	Summing to 0	Percentage not available

The practical interpretation of MRFA is that it offers the very solution to (3) that minimizes the common variance left unexplained when as few as r factors are used. This is fully compatible with the purpose of factor analysis. However, MRFA does not just minimize the unexplained common variance: It also reveals how small it is. This is possible because MRFA preserves the distinction between communalities and explained common variances: The former are the diagonal elements of $\boldsymbol{\Sigma} - \boldsymbol{\Psi}$, and the latter are the row sums of squares of $\mathbf{F}_1$. In terms of eigenvalues of $\boldsymbol{\Sigma} - \boldsymbol{\Psi}$, the total common variance is $\lambda_{1+\dots+}\lambda_m$ and the unexplained part of this is $f_1(\boldsymbol{\Psi})$, whence the percentage of explained common variance is $100 \times (\lambda_1 + \cdots + \lambda_r)/(\lambda_1 + \cdots + \lambda_m)$.

Although f_1 seems a meaningful criterion for factor analysis, alternatives are easily conceived of. In particular, we might minimize

$$f_2(\boldsymbol{\Psi}) = \sum_{j=r+1}^{m} \lambda_j^2(\boldsymbol{\Sigma} - \boldsymbol{\Psi}), \tag{12}$$

the *sum of squares* of the smallest $m - r$ eigenvalues of $\boldsymbol{\Sigma} - \boldsymbol{\Psi}$, subject to the same constraints as (11). This minimizes the sum of squares (rather than the plain sum) of unexplained common variances. We shall refer to it as quadratic MRFA (MRFA-Q). Research into the practical performance of MRFA-Q, using a numerical method by Prof. Arkadi Nemirovski, is in progress, e.g., Sočan and Ten Berge (2005). It may be noted that MRFA-Q is a protected version of Least Squares Factor Analysis, for which we shall use the acronym ULS (Unweighted Least Squares) henceforth. The latter method also minimizes f_2, but with $\boldsymbol{\Sigma} - \boldsymbol{\Psi}$ indefinite except in cases of perfect fit. Incidentally, ULS refers to one particular way (Jöreskog, 1967) of minimizing (12). Other approaches (such as Minres and Principal Axis Factoring) can be found in Harman and Jones (1966). The properties of MRFA, MRFA-Q, and ULS are summarized in Table 1.

Maximum Likelihood Factor Analysis (MLFA) also has an interpretation in terms of reduced eigenvalues, very similar to ULS (Jöreskog, 1967; Ten Berge, 1998). In simulation studies by Briggs and MacCallum (2003) and Sočan (2003), ULS appeared slightly superior to MLFA in retrieving underlying factors from samples.

It is tempting to believe that minimizing the unexplained common variance, as is done by MRFA, is tantamount to maximizing the explained common variance, but that is not the case. The eigenvalues of a symmetric matrix are monotonic functions of the diagonal elements. Hence, the maximum of the sum of the r largest eigenvalues of $\boldsymbol{\Sigma} - \boldsymbol{\Psi}$ is obtained when $\boldsymbol{\Psi}$ vanishes. This would take us right back to PCA. It is the point in the feasible set with the minimum sum of unique variances. It might be called *maximum trace* factor analysis, maximizing $\mathrm{tr}(\boldsymbol{\Sigma} - \boldsymbol{\Psi})$.

To see how useful a percentage of explained common variance can be, consider the ULS analysis of nine intelligence tests by Carroll (1993, pp. 96–98). Carroll reports that a four factor solution is sufficient. When MRFA is applied with $r = 3$ and $r = 4$, we find solutions for $\boldsymbol{\Psi}$ with percentages of explained common variance of 93.16 and 99.10, respectively. This is very clear evidence that four factors are indeed enough. In terms of the resulting factor solutions, the MRFA loadings for $r = 4$ are very similar to those reported by Carroll. The main difference is that MRFA (or MRFA-Q, for that matter) yields, as a bonus, the percentage of explained common variance associated with any number of factors.

A study on the sample sizes, needed to infer population factors from MRFA factors in samples, was done by Sočan (2003). The results indicate that, except for cases of low average communalities (which may be expected when factoring, for example, test items), samples of $n = 200$ are more than enough to obtain reliable estimates of loadings and explained common variances. Asymptotic theory for MRFA has been developed by Shapiro and Ten Berge (2002).

Sočan (2003) also found that standardizing variables before the analysis has hardly any impact at all on retrieving underlying factors from samples. This is reassuring because MRFA is not scale-free.

5. The feasible set from two perspectives

The feasible set of test theory has two points of particular interest: The glb point is the maximum error variance point (minimum reliability point), and the origin is the perfect reliability point. In terms of factor analysis, these points correspond to CMTFA (the maximum unique variance point) and PCA, the minimum unique variance point, respectively. In terms of reduced eigenvalues (the eigenvalues of $\boldsymbol{\Sigma} - \boldsymbol{\Psi}$), CMTFA and PCA are the points where the sum of reduced eigenvalues is a minimum, and a maximum, respectively. Additional points can now be identified in the context of factor analysis.

Before turning to such points, it is important to remember that the covariance matrix $\boldsymbol{\Sigma}$ does not admit a $\boldsymbol{\Psi}$ such that the rank of $\boldsymbol{\Sigma} - \boldsymbol{\Psi}$ is below the Ledermann bound, almost surely (Shapiro, 1982). Low ranks occur, for that matter, neither inside nor outside the feasible set. Since approximations of low rank are mandatory in applied factor analysis, we may either seek points outside the feasible set, implying that one or more reduced eigenvalues are negative (ULS, MLFA), or we may seek solutions inside the set (MRFA, MRFA-Q). Using MRFA for a given value of r produces the point inside the set where the sum of the last $m - r$ reduced eigenvalues is a minimum, and MRFA-Q produces the point where the sum of squares of those eigenvalues is a minimum. In both cases, the explained common variance can meaningfully be evaluated as sum of the first r reduced eigenvalues divided by the sum of all m reduced eigenvalues. It follows that the feasible set has a best rank-r MRFA (and MRFA-Q) point, $r = 1, 2, 3, \ldots, m - 2$. All these points are boundary points, and they often are very close. The points will typically be unique when at least one of the smallest $m - r$ reduced eigenvalues is positive, and non-unique otherwise.

The $r = 1$ MRFA solution can be interpreted as the most unidimensional point in the set, the $r = 2$ MRFA solution is the most two-dimensional point, and so on. Counterparts in test theory are not obvious except when $r = 1$. This point corresponds to the most "congeneric" point in test theory.

Clearly, all mathematical results on the feasible set apply equally to test theory and factor analysis. For instance, Roff (1936) noted that the residual variances of a test in the regression analysis on all $m - 1$ other tests are upper bounds to the unique variances. So they are also upper bounds to the error variances in test theory. When Ihara and Kano (1986) proposed a new estimator of the uniqueness in factor analysis, they contributed a new estimator for error variance in test theory. When Yanai and Ichikawa (1990) developed new bounds for communalities in factor analysis, they contributed new bounds for true score variances.

The above discussion of factor analysis and reliability from the perspective of the feasible set is not meant to imply that points inside the set are necessarily "good" and those outside are "bad". After all, all lower bounds to reliability (Guttman, 1945; Jackson and Agunwamba, 1977) except the glb are outside the set (unless they happen to coincide with the glb) and one of these, coefficient alpha, has been extremely useful for applied research. Likewise, ULS has an excellent record of retrieving underlying factors from samples in simulation studies (Briggs and MacCallum, 2003; Sočan and Ten Berge, 2005). The fact that ULS solutions are outside the feasible set does not detract from that at all.

It should be clear at this point that, although factor analysis and test theory can be described in terms of the same mathematical framework of a "feasible set", they refer to entirely different points in that set. In fact, because we treat specific variance as part of the reliable variance, evaluating reliability on the basis of factor solutions is impossible. This will be further explained for single and multiple factor solutions, respectively.

6. Reliability measures derived from a single factor solution

When a test is congeneric, i.e., made up of test parts having perfectly correlated true scores, there is a decomposition $\boldsymbol{\Sigma} = \boldsymbol{\Sigma}_T + \boldsymbol{\Sigma}_E$, see (10), with $\boldsymbol{\Sigma}_T$ of rank one. So $\boldsymbol{\Sigma}_T$ can be factored as $\mathbf{f}_1\mathbf{f}'_1$, for some vector $\mathbf{f}_1$. The reliability is the sum of elements of $\boldsymbol{\Sigma}_T$, divided by $\mathrm{Var}(X)$. This gives

$$(\mathbf{1}'\boldsymbol{\Sigma}_T\mathbf{1})/\mathrm{Var}(X) = (\mathbf{1}'\mathbf{f}_1\mathbf{f}'_1\mathbf{1})/\mathrm{Var}(X) = (\mathbf{1}'\mathbf{f}_1)^2/\mathrm{Var}(X), \tag{13}$$

e.g., McDonald, 1970, also see Zinbarg et al., 2005. In practice, we know that rank one is not possible when $m > 3$. But we may still run a factor analysis program with $r = 1$, and evaluate (13). For example, Ten Berge and Sočan (2004) have done this for six political survey items described by De Leeuw (1983). These items are meant

Table 2
Six variables of De Leeuw (1983)

	alpha	glb	(13)	ECV
$n = 100$	0.836	0.893	0.892	77.48
$n = 250$	0.839	0.888	0.888	79.40
$n = 500$	0.840	0.887	0.886	79.76
$n = 1000$	0.839	0.886	0.885	79.92
Population	0.840	0.885	0.885	80.10

to measure the same trait, and therefore should be close to congeneric. The correlation matrix is

$$\boldsymbol{\Sigma} = \begin{bmatrix} 1.000 & & & & & \\ 0.446 & 1.000 & & & & \\ 0.462 & 0.380 & 1.000 & & & \\ 0.398 & 0.241 & 0.589 & 1.000 & & \\ 0.583 & 0.536 & 0.569 & 0.459 & 1.000 & \\ 0.516 & 0.483 & 0.417 & 0.403 & 0.514 & 1.000 \end{bmatrix}. \tag{14}$$

Following De Leeuw (1983), Ten Berge and Sočan treated this correlation matrix (based on $n = 119$ members of parliament) as if it was based on the population, and constructed sets of 500 samples of sizes 100, 250, 500, and 1000, respectively, under the assumption of multivariate normality. For each sample, they evaluated *alpha*, glb, and the single-factor based reliability coefficient (13), evaluated from the $r = 1$ MRFA solution. Also, they included the "unidimensionality" measure ECV (percentage of Explained Common Variance with one factor) associated with the latter solution. Table 2 gives average results (over 500 replications) for 4 different sample sizes, and for the population.

In general, the sampling bias of the glb was quite small for the correlation matrix (14), yet it was slightly larger (in absolute size) than for alpha. As expected, the glb was much higher than alpha both in population and in samples. Remarkably, (13) behaved very similarly to the glb in every respect. In fact, it tends to display the same sampling bias for which the glb is renowned. Ten Berge and Sočan (2004) gave the following explanation:

For correlation matrices like (14), with all elements positive, the true score variances implied by the glb tend to be very close to the communalities of the $r = 1$ MRFA solution. This means that $\boldsymbol{\Sigma}_T$ of the glb will be nearly identical to $\boldsymbol{\Sigma} - \boldsymbol{\Psi}$ of MRFA with $r = 1$. Upon factoring $\boldsymbol{\Sigma}_T$ as $\boldsymbol{\Sigma}_T = \mathbf{f}_1\mathbf{f}_1' + \mathbf{F}_2\mathbf{F}_2'$, where $\mathbf{f}_1$ is the vector of loadings on the first common factor, the glb $\mathbf{1}'\boldsymbol{\Sigma}_T\mathbf{1}/\mathbf{1}'\boldsymbol{\Sigma}\mathbf{1}$ will be very close to $\mathbf{1}'(\mathbf{f}_1\mathbf{f}_1' + \mathbf{F}_2\mathbf{F}_2')\mathbf{1}/\mathbf{1}'\boldsymbol{\Sigma}\mathbf{1}$, whereas (13) yields $\mathbf{1}'\mathbf{f}_1\mathbf{f}_1'\mathbf{1}/\mathbf{1}'\boldsymbol{\Sigma}\mathbf{1}$. Because all elements of $\boldsymbol{\Sigma}_T$ are positive, all loadings in $\mathbf{f}_1$ have the same sign. Because the columns of $\mathbf{F}_2$ are orthogonal to $\mathbf{f}_1$, the column sums of $\mathbf{F}_2$ are close to zero, hence $\mathbf{1}'\mathbf{F}_2\mathbf{F}_2'\mathbf{1}$ is near zero. This means that, although the factors associated with $\mathbf{F}_2$ do explain some variance, they do not contribute to reliability, because positive and negative contributions to true scores cancel. This explains why the glb must be close to (13) for data like (14). However, whereas (13) is based on an assumption of unidimensionality, the glb is not. Therefore, it seems safer to use the glb.

7. Reliability derived from multiple factor analysis

To obtain a reliability estimate involving multiple common factors, (13) needs to be generalized to

$$(\mathbf{1}'\mathbf{FF}'\mathbf{1})/\,\mathrm{Var}(X), \tag{15}$$

e.g., McDonald (1970), where **F** contains the loadings on r common factors. Bentler (2004) has, just like Ten Berge and Sočan (2004) challenged the idea of reliability based on a single factor because the single factor hypothesis is untenable. Whereas the latter authors adopted the glb instead, Bentler insisted that (15) should be used. Interestingly, Bentler showed that the solution **F** entering (15) can be rotated such that one factor will have a loading vector $\mathbf{f}^*$ with the maximum possible sum of loadings, and loadings on the other factors sum to zero. Accordingly, either **F** or $\mathbf{f}^*$ can be entered in (15), yielding the multifactorial reliability estimate

$$(\mathbf{1}'\mathbf{FF}'\mathbf{1})/\,\mathrm{Var}(X) = (\mathbf{1}'\mathbf{f}^*\mathbf{f}^{*\prime}\mathbf{1})/\,\mathrm{Var}(X) = (\mathbf{1}'\mathbf{f}^*)^2/\,\mathrm{Var}(X). \tag{16}$$

Because it defines the reliable variance on the basis of more than one factor, it will usually exceed (13). Before discussing the merits of (16), it may be instructive to consider an example.

Bentler (2004, p. 20), computed (16) for Harman's (1976) nine psychological tests (Table 3), using maximum likelihood factor analysis with $r = 1$ and $r = 3$, respectively. Because **F** or $\mathbf{f}^*$ can be based on any method of factor analysis, it is instructive to complement Bentler's analysis with the $r = 1$ and $r = 3$ solutions of MRFA and ULS, respectively.

In fact, we first consider the MRFA factor solutions for $r = 0, 1, 2,$ and 3, respectively. The $r = 0$ solution is obtained from the MRFA program by running it with zero common factors. Because MRFA minimizes the sum of the last $m - r$ reduced eigenvalues, it will minimize the sum of all reduced eigenvalues when $r = 0$, which means

Table 3
Correlation matrix of Harman's nine psychological tests

1.00	0.75	0.78	0.44	0.45	0.51	0.21	0.30	0.31
0.75	1.00	0.72	0.52	0.53	0.58	0.23	0.32	0.30
0.78	0.72	1.00	0.47	0.48	0.54	0.28	0.37	0.37
0.44	0.52	0.47	1.00	0.82	0.82	0.33	0.33	0.31
0.45	0.53	0.48	0.82	1.00	0.74	0.37	0.36	0.36
0.51	0.58	0.54	0.82	0.74	1.00	0.35	0.38	0.38
0.21	0.23	0.28	0.33	0.37	0.35	1.00	0.45	0.52
0.30	0.32	0.37	0.33	0.36	0.38	0.45	1.00	0.67
0.31	0.30	0.37	0.31	0.36	0.38	0.52	0.67	1.00

Table 4
MRFA communalities (h^2), eigenvalues (λ), and percentage of ECV

	$r=0$		$r=1$		$r=2$		$r=3$	
	h^2	λ	h^2	λ	h^2	λ	h^2	λ
	0.85	4.543	0.85	4.543	0.84	4.543	0.85	4.543
	0.75	1.083	0.75	1.083	0.75	1.087	0.75	1.086
	0.77	0.882	0.77	0.882	0.77	0.882	0.77	0.885
	0.96	0.058	0.96	0.057	0.96	0.058	0.97	0.058
	0.77	0.048	0.77	0.048	0.77	0.048	0.77	0.048
	0.77	0.029	0.77	0.030	0.76	0.029	0.76	0.030
	0.41	0.016	0.41	0.016	0.41	0.013	0.41	0.012
	0.60	0	0.60	0	0.59	0	0.59	0
	0.80	0	0.80	0	0.81	0	0.81	0
Sum	6.658	6.658	6.658	6.658	6.660	6.660	6.662	6.662
ECV			68.23%		84.52%		97.79%	

that it yields the error variances defining the glb, or, equivalently, the unique variances defining CMTFA. Communalities and reduced eigenvalues are reported in Table 4.

The communalities in Table 4 reveal that, for the data of Table 3, the best $r = 1$ point, the best $r = 2$ point, and the best $r = 3$ point of MRFA are very close to the CMTFA ($r = 0$) point. The percentage of ECV is 97.79 for three factors, indicating that a three-factor solution gives an almost perfect fit. The glb is $1 - \text{tr}(\boldsymbol{\Sigma}_E)/\mathbf{1}'\boldsymbol{\Sigma}_X\mathbf{1} = 1 - (9 - 6.658)/42.3 = 0.9446$, much higher than alpha (0.886). Incidentally, Table 4 may suggest that the exact minimum rank is 7, but it is 6. This fact can not be seen from the results presented here, but rather from inspection of MRFA solutions for $r = 5$ and $r = 6$.

Next, we turn to factor based reliability, using MLFA, MRFA and ULS. Table 5 gives the loadings $\mathbf{f}$ for $r = 1$ and $\mathbf{f}^*$ for $r = 3$, for each of these three methods, along with the implied reduced eigenvalues λ_j for ULS. The bottom lines of the table report the sums of loadings, and the associated values of (15). In the $r = 1$ cases, this boils down to (13).

It is clear from Table 5 that the $r = 1$ loading vectors $\mathbf{f}$ differ widely between the three methods, and so do the associated reliability estimates, ranging from 0.8347 to 0.9398. In fact, the MLFA value of 0.8347 is even smaller than alpha (0.886), just as Bentler has noted, reporting 0.880 for (15) instead. This shows that, when the single factor model is blatantly inadequate (it is always inadequate when $m > 3$, but sometimes may get close), it matters quite a bit which particular method of factor analysis is used.

When turning to the more realistic $r = 3$ case, the loading vectors $\mathbf{f}^*$ become very similar, and so do the associated values of (15), now ranging from 0.9389 to 0.9446. The explanation for this is that, when $r = 3$, the negative reduced eigenvalues of ULS (see Table 5) and MLFA are very close to zero for the data under consideration. That is, the solutions are still outside the feasible set (they always will be unless there is perfect fit), yet they are very close to the boundary of that set. Since the MRFA rank-3 solution

Table 5
Loadings of MLFA, MRFA and ULS, and ULS eigenvalues (λ)

	MLFA		MRFA		ULS			
	$r=1$	$r=3$	$r=1$	$r=3$	$r=1$		$r=3$	
	f	**f***	**f**	**f***	**f**	λ	**f***	λ
	0.636	0.727	0.742	0.728	0.706	4.297	0.726	4.519
	0.697	0.738	0.760	0.743	0.750	0.746	0.740	1.062
	0.677	0.754	0.764	0.755	0.750	0.614	0.755	0.858
	0.867	0.789	0.814	0.793	0.774	−0.094	0.788	0.027
	0.844	0.767	0.785	0.771	0.780	−0.163	0.769	0.018
	0.879	0.803	0.816	0.800	0.824	−0.201	0.802	0.009
	0.424	0.492	0.472	0.497	0.461	−0.241	0.494	−0.002
	0.466	0.597	0.562	0.597	0.535	−0.266	0.596	−0.021
	0.462	0.635	0.589	0.638	0.537	−0.395	0.635	−0.031
Sum	5.942	6.302	6.305	6.321	6.117	4.297	6.304	6.439
(15)	0.8347	0.9389	0.9398	0.9446	0.8846		0.9395	

is on that very boundary, the closeness of all three solutions is no surprise. Also, the glb solution happens to be close to the rank-3 MRFA solution, as we know from Table 4.

Unfortunately, the tendency of the three methods to yield increasingly similar loadings as the fit gets better does not extend to the situation of perfect fit. The reason is that perfect fit will only arise in situations where the number of factors is equal to or above the Ledermann bound. Specifically, for the data of Table 3, the minimum rank is 6. Because 6 is strictly above the Ledermann bound, there is an infinite number of rank-6 solutions, each having their own value of (15). It is not just a problem that different methods may produce different solutions: The methods themselves admit an infinite number of solutions. The feasible set for the data of Table 3 has an infinite number of rank-6 solutions on its boundary.

The situation is less dramatic but still problematic when the minimum rank is exactly on the Ledermann bound. Then there is only a finite number of solutions. For instance, consider the data of Wilson and Worcester (1939, p. 74), who constructed the correlation matrix of Table 6. This correlation matrix admits two different rank-3 solutions. That is, the feasible set contains two distinct boundary points, each entailing a 3-factor solution with perfect fit. The associated values of (15) are 0.8666 and 0.8548, whereas the glb is 0.8542. Because the former two values are in the feasible set, they are "possible values of the reliability", and therefore can no longer be smaller than the glb.

Undoubtedly, Bentler's suggestion to derive reliability estimates from (15) or (16), taking multiple factors into account, is on a much better footing than adhering to the single factor hypothesis. Also, it is likely to bring the reliability estimate closer to the glb. Still, it is not clear what "reliability on the basis of multiple factor analysis" has to offer in addition to glb.

First, consider the case of close but imperfect fit, where r is less than the minimum reduced rank of $\boldsymbol{\Sigma}$. Regardless of the method we use, when evaluating (15), $\boldsymbol{\Sigma} - \boldsymbol{\Psi}$ will merely be *approximated* by a rank-r matrix $\mathbf{FF}'$, and that rank-r matrix still does not

Table 6
Correlation matrix of Wilson and Worcester

1.00	0.56	0.16	0.48	0.24	0.64
0.56	1.00	0.20	0.66	0.51	0.86
0.16	0.20	1.00	0.18	0.07	0.23
0.48	0.66	0.18	1.00	0.30	0.72
0.24	0.51	0.07	0.30	1.00	0.41
0.64	0.86	0.23	0.72	0.41	1.00

correspond to a point in the feasible set because the feasible set does not contain low rank points (almost surely). Nevertheless, the associated value of (15) can be expected to be close to the glb. Because points outside the set may still imply reliability values in the range [glb, 1], the value of (15) may or may not be a possible value of the reliability. Incidentally, the numerical closeness of (15) and the glb implies that the former are likely to have the same upward sampling bias as the latter, see (Ten Berge and Sočan, 2004). The sampling bias is indeed a problem for the glb (Shapiro and Ten Berge, 2000; Li and Bentler, 2004), but it affects (15) just as much.

Next, suppose we do base (15) on a perfectly fitting solution. Now every factor solution corresponds to a $\boldsymbol{\Psi}$ inside the feasible set, hence every value of (15) derived from such a solution is *a possible value of the reliability*. As was demonstrated for the data of Table 6, there typically is no unique factor solution, even when a fixed method of factor analysis is adopted. For the sake of the argument, suppose we were able to determine, among all perfectly fitting the solutions, the one ($\rho_{\max}$) which maximizes and the one ($\rho_{\min}$) which minimizes (15). Then the range [glb, 1] of possible reliability values could be narrowed down to the range $[\rho_{\min}, \rho_{\max}]$. However, this would still be based on a the fundamental assumption that unique variance and error variance of the variables coincide. In view of the fact that the glb does not rely on such an assumption, and typically will be very close to (15) in cases of imperfect but close fit, it seems that the glb is to be preferred. As said before, the glb does have a fierce sampling bias problem, but there is every reason to believe that (15) will have exactly the same problem.

8. Discussion

The set of feasible tautologies allows treating reliability and factor analysis in the same framework. The set contains all possible states of nature, but there is no way of telling which is true. As for reliability, we only know that the true reliability of a test is somewhere between the glb (the least reliable point) and 1, the most reliable point. As for unidimensionality, we only know that the true unidimensionality is somewhere between the percentage of variance explained by the first principal component (the least unidimensional point) and the percentage of common variance explained by a single-factor MRFA (the most unidimensional point).

McDonald (1970) proposed reliability estimates based on either single or multiple factor solutions. Bentler (2004) has insisted that only the latter should be used. Although this is indeed much more realistic, the method still has no clear advantages over using the glb. It remains based on the assumption that unique variance and error variance coincide. This assumption seems unlikely, because it implies that a variable has no reliable specific variance and, besides, it resists verification, whence it seems that the glb is to be preferred.

The discussion of the feasible set of solutions for error variances in test theory or unique variances in factor analysis may have given the impression that solutions inside the set are necessarily good and those outside the set are always evil. However, that impression would not be warranted, as we have explained. The main reason for adhering to the feasible set in factor analysis rests in the percentage of explained common variance. This concept will be lost when some of the reduced eigenvalues are allowed to be negative. The main reason for adhering to it in the reliability context is the glb. Numerically, the glb behaves almost exactly like reliability estimates based on multiple factor analysis, sampling bias included. Conceptually, however, the glb is more elegant in that it neither relies on the hypothesis that specific variances and error variances of the variables coincide nor does its numerical value depend on the specific choice of a factor analysis method.

A Pascal program for MRFA can be downloaded from http://www.ppsw.rug.nl/~kiers/. A Matlab code for MRFA-Q is available upon request.

References

Bekker, P.A., De Leeuw, J. (1987). The rank of reduced dispersion matrices. *Psychometrika* **52**, 125–135.

Bekker, P.A., Ten Berge, J.M.F. (1997). Generic global identification in factor analysis. *Linear Algebra and its Applications* **264**, 255–264.

Bentler, P.M. (1972). A lower-bound method for the dimension-free measurement of reliability. *Social Science Research* **1**, 343–357.

Bentler, P.M. (2004). Maximal reliability for unit-weighted components. Statistical preprint 405. UCLA. http://preprints.stat.ucla.edu/405.

Bentler, P.M., Woodward, J.A. (1980). Inequalities among lower bounds to reliability: With applications to test construction and factor analysis. *Psychometrika* **45**, 249–267.

Briggs, N.E., MacCallum, R.C. (2003). Recovery of weak common factors by maximum likelihood and ordinary least squares estimation. *Multivariate Behavioral Research* **38**, 25–56.

Browne, M.W. (1969). Fitting the factor analysis model. *Psychometrika* **34**, 375–394.

Carroll, J.B. (1993). *Human Cognitive Abilities: A Survey of Factor Analytic Research.* Cambridge University Press, New York.

De Leeuw, J. (1983). Models and methods for the analysis of correlation coefficients. *Journal of Econometrics* **22**, 113–137.

Della Riccia, G., Shapiro, A. (1982). Minimum rank and minimum trace of covariance matrices. *Psychometrika* **47**, 443–448.

Guttman, L. (1945). A basis for analyzing test-retest reliability. *Psychometrika* **10**, 255–282.

Guttman, L. (1958). To what extent can communalities reduce rank? *Psychometrika* **23**, 297–308.

Harman, H.H. (1976). *Modern Factor Analysis*, third ed. University of Chicago Press, Chicago.

Harman, H.H., Jones, W.H. (1966). Factor analysis by minimizing residuals (Minres). *Psychometrika* **31**, 351–368.

Ihara, M., Kano, Y. (1986). A new estimator of the uniqueness in factor analysis. *Psychometrika* **51**, 563–566.

Jackson, P.H., Agunwamba, C.C. (1977). Lower bounds for the reliability of the total score on a test composed of non-homogeneous items: I. Algebraic lower bounds. *Psychometrika* **42**, 567–578.
Jöreskog, K.G. (1967). Some contributions to maximum likelihood factor analysis. *Psychometrika* **31**, 443–482.
Ledermann, W. (1937). On the rank of reduced correlation matrices in multiple factor analysis. *Psychometrika* **2**, 85–93.
Li, L., Bentler, P.M. (2004). The greatest lower bound to reliability: Corrected and resampling estimates. Paper presented at the 82nd Symposium of the Behaviormetric Society of Japan, Tokyo, August 25.
McDonald, R.P. (1970). The theoretical foundations of principal factor analysis, canonical factor analysis, and alpha factor analysis. *British Journal of Mathematical and Statistical Psychology* **23**, 1–21.
Novick, M.R., Lewis, C. (1967). Coefficient alpha and the reliability of composite measurements. *Psychometrika* **32**, 1–13.
Roff, M. (1936). Some properties of the communality in multiple factor theory. *Psychometrika* **1**, 1–6.
Shapiro, A. (1982). Rank reducibility of a symmetric matrix and sampling theory of minimum trace factor analysis. *Psychometrika* **47**, 187–199.
Shapiro, A. (1985). Identifiability of factor analysis: Some results and open problems. *Linear Algebra and its Applications* **70**, 1–7.
Shapiro, A., Ten Berge, J.M.F. (2000). The asymptotic bias of minimum trace factor analysis, with applications to the greatest lower bound to reliability. *Psychometrika* **65**, 413–425.
Shapiro, A., Ten Berge, J.M.F. (2002). Statistical inference of minimum rank factor analysis. *Psychometrika* **67**, 79–94.
Sočan, G. (2003). The incremental value of MRFA. PhD thesis. University of Groningen, The Netherlands.
Sočan, G., Ten Berge, J.M.F. (2005). The link between MINRES and Minimum Rank Factor Analysis. Paper presented at the 14th International Meeting of the Psychometric Society, Tilburg.
Ten Berge, J.M.F. (1998). Some recent developments in factor analysis and the search for proper communalities. In: Rizzi, A., Vichi, M., Bock, H.-H. (Eds.), *Advances in Data Science and Classification*. Springer, Berlin, pp. 325–334.
Ten Berge, J.M.F., Kiers, H.A.L. (1991). A numerical approach to the exact and the approximate minimum rank of a covariance matrix. *Psychometrika* **56**, 309–315.
Ten Berge, J.M.F., Kiers, H.A.L. (1999). Retrieving the correlation matrix from a truncated PCA solution: The inverse principal component problem. *Psychometrika* **64**, 317–324.
Ten Berge, J.M.F., Sočan, G. (2004). The greatest lower bound to reliability of a test and the hypothesis of unidimensionality. *Psychometrika* **69**, 613–625.
Ten Berge, J.M.F., Snijders, T.A.B., Zegers, F.E. (1981). Computational aspects of the greatest lower bound to reliability and constrained minimum trace factor analysis. *Psychometrika* **46**, 357–366.
Wilson, E.B., Worcester, J. (1939). The resolution of six tests into three general factors. *Proc. National Academy of Sciences* **25**, 73–77.
Yanai, H., Ichikawa, M. (1990). New lower and upper bounds for communality in factor analysis. *Psychometrika* **55**, 405–410.
Zinbarg, R.E., Revelle, W., Yovel, I., Li, W. (2005). Cronbach's α, Revelle's β, and McDonald's ω_H: Their relations with each other and two alternative conceptualizations of reliability. *Psychometrika* **70**, 123–133.

Handbook of Computing and Statistics with Applications, Vol. 1
ISSN: 1871-0301

DOI: 10.1016/S1871-0301(06)01015-8

15

Nonlinear Structural Equation Modeling as a Statistical Method

Melanie M. Wall and Yasuo Amemiya

Abstract
Structural equation analysis allows exploring and modeling relationships among latent variables. The most traditional analysis deals only with linear relationships. However, in applied problems, relevant research questions can be associated with some form of nonlinear relations. Also, from a statistical modeling or data analysis point of view, capabilities to address unrestricted nonlinear relations would be welcome. A review is given for the existing approaches that have been developed and designed for specific nonlinear models. Then, a statistical formulation of general nonlinear structural equation analysis is introduced, and a general model fitting procedure applicable under weak assumptions on latent variable distributions is developed. An example with a nonpolynomial nonlinear structural model is discussed using the new method.

1. Introduction

Structural equation modeling originated (Jöreskog (1973); Bentler (1980); Bollen (1989)) as a method for modeling linear relations among observed and hypothesized latent variables. Despite limitations inherent in the linearity assumption of traditional structural equation modeling, it has indeed provided a revolutionary and popular framework for addressing research questions in the social, psychological and behavioral sciences where latent variables are quite common. In order to expand the flexibility and thus applicability of this already useful statistical modeling method, a natural extension is to include the possibility of modeling nonlinear relations among the latent variables in addition to linear relations.

There has been a growing literature (some of which described later in this paper) developing different kinds of nonlinear structural equation models and estimation methods for them. Generally, the estimation methods in this literature can be described as either making and relying on distributional assumptions for the underlying latent variables or instead leaving the distribution unspecified. Furthermore, the methods can be described as being either tailor-made to a specific sort of nonlinear structural model, i.e., polynomial or specifically low-dimensional polynomial, or else being applicable to a more general nonlinear structural model.

In this paper we present a general nonlinear structural equation model and estimation methods for it. Section 2 presents the general nonlinear structural equation model as an extension of the linear structural equation model. Section 2 also describes special cases of the nonlinear structural equation model including those that have been considered in the literature. Section 3 presents an estimation method for the general model which does not make strong distributional assumptions about the latent variables and can be implemented using a pseudo-likelihood approach combined with the Monte Carlo Expectation Maximization algorithm (MCEM). Section 4 presents an example motivated by an investigation of cystic fibrosis patients where treatment adherence is examined in relation to social, familial and personal factors, and a nonlinear structural equation model is specified and the estimation method described herein is used. Section 5 provides some discussion.

2. General nonlinear structural equation model

2.1. Linear structural equation models

The traditional linear structural equation model is typically made up of two parts: the measurement model describing the relationships between the observed and latent variables and the structural model describing the relationships between the latent variables. Given a vector of p observed variables $\mathbf{Z}_i$ for the ith individual in a sample of size n and a vector of q latent variables $\boldsymbol{f}_i$, the linear structural equation model system can be written:

$$\mathbf{Z}_i = \boldsymbol{\mu} + \boldsymbol{\Lambda} \boldsymbol{f}_i + \boldsymbol{\varepsilon}_i, \tag{1}$$

$$\mathbf{b}_0 + \mathbf{B}_0 \boldsymbol{f}_i = \boldsymbol{\delta}_{0i}, \tag{2}$$

where in the measurement model, the matrices $\boldsymbol{\mu}$ $(p \times 1)$ and $\boldsymbol{\Lambda}$ $(p \times q)$ contain fixed or unknown scalars describing the linear relation between the observations $\mathbf{Z}_i$ and the common latent factors $\boldsymbol{f}_i$, and $\boldsymbol{\varepsilon}_i$ represents the $(p \times 1)$ vector of random measurement error independent of $\boldsymbol{f}_i$ such that $E(\boldsymbol{\varepsilon}_i) = \mathbf{0}$ and $\mathrm{Var}(\boldsymbol{\varepsilon}_i) = \boldsymbol{\Psi}$ with fixed and unknown scalars in $\boldsymbol{\Psi}$; and in the structural model, the matrices $\mathbf{b}_0$ $(d \times 1)$ and $\mathbf{B}_0$ $(d \times q)$ contain fixed or unknown scalars defining d different additive linear simultaneous structural equations relating the factors to one another plus the $(d \times 1)$ vector of random equation error $\boldsymbol{\delta}_{0i}$, where $E(\boldsymbol{\delta}_{0i}) = \mathbf{0}$ and $\mathrm{Var}(\boldsymbol{\delta}_{0i}) = \boldsymbol{\Delta}_0$ with fixed and unknown scalars in $\boldsymbol{\Delta}_0$.

The simultaneous linear structural model as written in (2) is very general. For many practical research questions which can be addressed by simultaneous structural models, it is useful to model specific variables in terms of the rest of the variables, i.e., it is useful to consider some of the latent variables as endogenous and others as exogenous, where endogenous variables are those that are functions of other endogenous and exogenous variables. Let $\boldsymbol{f}_i = (\boldsymbol{\eta}_i', \boldsymbol{\xi}_i')'$ where $\boldsymbol{\eta}_i$ are the d endogenous latent variables and $\boldsymbol{\xi}_i$ are the $q - d$ exogenous latent variables. Then a commonly used form for the structural model (2) becomes:

$$\boldsymbol{\eta}_i = \mathbf{b} + \mathbf{B}\boldsymbol{\eta}_i + \boldsymbol{\Gamma}\boldsymbol{\xi}_i + \boldsymbol{\delta}_i, \tag{3}$$

where it is assumed the equation errors $\boldsymbol{\delta}_i$ have $E(\boldsymbol{\delta}_i) = \mathbf{0}$, $\text{Var}(\boldsymbol{\delta}_i) = \boldsymbol{\Delta}$ and are independent of the $\boldsymbol{\xi}_i$ as well as independent of $\boldsymbol{\varepsilon}_i$ in (1), and the matrices $\mathbf{b}$ $(d \times 1)$, $\mathbf{B}$ $(d \times d)$, $\boldsymbol{\Gamma}$ $(d \times (q - d))$, and $\boldsymbol{\Delta}$ $(d \times d)$ are fixed or unknown scalars. The structural model (3) is said to be in **implicit form**, implicit because it has endogenous variables on both sides of the equations, i.e., it is not "solved" for the endogenous variables. It is assumed that the diagonal of $\mathbf{B}$ is zero so that no element of $\boldsymbol{\eta}_i$ is a function of itself. A sufficient condition for solving (3) is that $(\mathbf{I} - \mathbf{B})$ is invertible, then (3) can be solved for the endogenous variables and written as

$$\boldsymbol{\eta}_i = \mathbf{b}^* + \boldsymbol{\Gamma}^*\boldsymbol{\xi}_i + \boldsymbol{\delta}_i^*, \tag{4}$$

where $\mathbf{b}^* = (\mathbf{I} - \mathbf{B})^{-1}\mathbf{b}$, $\boldsymbol{\Gamma}^* = (\mathbf{I} - \mathbf{B})^{-1}\boldsymbol{\Gamma}$, and $\text{Var}(\boldsymbol{\delta}_i^*) = (\mathbf{I} - \mathbf{B})^{-1}\boldsymbol{\Delta}(\mathbf{I} - \mathbf{B})^{-1'}$. The structural model (4) is said to be in **reduced form** as the $\boldsymbol{\eta}_i$ now appears only on the left-hand side of the equation. It is important to note the assumption that the equation errors $\boldsymbol{\delta}_i$ were additive and independent of the $\boldsymbol{\xi}_i$ in the implicit form (3) results in the equation errors $\boldsymbol{\delta}_i^*$ in the reduced form (4) also being additive and independent of the $\boldsymbol{\xi}_i$.

Given p, q and d, additional restrictions must be placed on $\boldsymbol{\mu}$, $\boldsymbol{\Lambda}$, $\boldsymbol{\Psi}$, $\mathbf{b}_0$, $\mathbf{B}_0$, and $\boldsymbol{\Delta}_0$ in (1)–(2) in order to make all the unknown parameters identifiable. The assumption that (2) can be written in reduced form (4) is the typical restriction placed on the structural model. Additionally, a common restriction placed on the measurement model (1) is the errors-in-variables parameterization where q of the observed variables are each fixed to be equal to one of the q different latent variables plus measurement error. For a thorough discussion of identifiability in linear structural equation models see, e.g., Bollen (1989). Finally, it should be noted that there is no inherent distributional assumptions needed for $\boldsymbol{\varepsilon}_i$, $\boldsymbol{\delta}_{0i}$, nor $\boldsymbol{f}_i$ at this point of model specification although distributional assumptions may be added eventually to perform estimation.

2.2. *Extension to nonlinear structural models*

A natural way to examine many scientific theories empirically is by measuring some variables on a sample of a population then examining several possible relationships between the variables. The two parts of the structural equation model (1)–(2) match this idea where (1) is measuring the latent variables and (2) is relating the latent variables to one another. A straightforward extension then is to assume that the way the variables are measured in (1) is reasonable, but that there may be more complicated relationships between the latent variables of interest than just linear ones. Thus the general nonlinear structural equation model we introduce retains the linear measurement model but considers nonlinear structural relations:

$$\mathbf{Z}_i = \boldsymbol{\mu} + \boldsymbol{\Lambda}\boldsymbol{f}_i + \boldsymbol{\varepsilon}_i, \tag{5}$$

$$\mathbf{H}_0(\boldsymbol{f}_i; \boldsymbol{\beta}_0) = \boldsymbol{\delta}_{0i}, \tag{6}$$

where the general simultaneous nonlinear structural model system is described by the $(d \times 1)$ vector function $\mathbf{H}_0$ which is a known function of $\boldsymbol{f}_i$ with unknown parameters $\boldsymbol{\beta}_0$. The parameters, $\boldsymbol{\mu}$, $\boldsymbol{\Lambda}$, and the specification of the errors $\boldsymbol{\varepsilon}_i$ and $\boldsymbol{\delta}_{0i}$ to have zero expectation and variances $\boldsymbol{\Psi}$ and $\boldsymbol{\Delta}_0$, respectively, are the same as for the linear structural equation model (1)–(2) above.

Motivated by the desire (as in the linear structural model) to model systems of structural models where certain sets of variables are written as functions of other variables plus error, we consider again endogenous and exogenous $\boldsymbol{\eta}_i$ and $\boldsymbol{\xi}_i$ and introduce the following class of nonlinear structural equation model

$$\boldsymbol{\eta}_i = \mathbf{H}(\boldsymbol{\eta}_i, \boldsymbol{\xi}_i; \boldsymbol{\beta}) + \boldsymbol{\delta}_i, \tag{7}$$

where $\mathbf{H}$ is a $(d \times 1)$ vector function with unknown parameters $\boldsymbol{\beta}$, and $\boldsymbol{\delta}_i$ is random equation error independent of $\boldsymbol{\xi}_i$ and $\boldsymbol{\varepsilon}_i$ with $E(\boldsymbol{\delta}_i) = 0$ and $\mathrm{Var}(\boldsymbol{\delta}_i) = \boldsymbol{\Delta}$ such that $\boldsymbol{\Delta}$ is a $(d \times d)$ matrix of fixed or unknown scalars. Note that since $\mathbf{H}$ is a function of both $\boldsymbol{\eta}_i$ and $\boldsymbol{\xi}_i$ we refer to this simultaneous nonlinear model (7) as being in **implicit form**. It is assumed that $\mathbf{H}$ is such that there are no elements of $\boldsymbol{\eta}_i$ which are functions of themselves.

In order that the parameters in (7) are identifiable, it is important that the model is written in an unambiguous way. One way of doing this is to focus on models that can be written in an explicit **reduced form**. In the linear structural model, choosing models that had reduced form meant restricting to the subset of models in (3) that had $(\mathbf{I}-\mathbf{B})$ invertible. Here in the nonlinear case, the rules for knowing when (7) can be solved explicitly for $\boldsymbol{\eta}_i$ and thus written in reduced form are not so simple. A substantial literature from econometrics investigates the solvability of systems of nonlinear simultaneous equations (where the $\boldsymbol{\eta}_i$ and $\boldsymbol{\xi}$ would be considered observed), see, e.g., Benkard and Berry (2005) and no general set of rules is available. Nevertheless we continue the general development of the nonlinear structural model by assuming that the model of interest can be written in reduced form and then (in the next subsection) we describe useful subclasses of implicit form models which can be written in reduced form.

The general **reduced form** simultaneous nonlinear structural model of (7) (when a reduced form exists) can then be written

$$\boldsymbol{\eta}_i = \mathbf{h}(\boldsymbol{\xi}_i, \boldsymbol{\delta}_i; \boldsymbol{\beta}^*), \tag{8}$$

where $\mathbf{h}$ is a $(d \times 1)$ vector function with unknown parameters $\boldsymbol{\beta}^*$ and $\boldsymbol{\eta}_i$, $\boldsymbol{\xi}_i$, and $\boldsymbol{\delta}_i$ are as in (7). Note that in general, solving a nonlinear implicit form (7) results in the equation error term $\boldsymbol{\delta}_i$ entering the reduced form function $\mathbf{h}$ nonlinearly. Thus, additive equation error independent of $\boldsymbol{\xi}_i$ in the implicit form of the nonlinear structural model does not necessarily result in additive error in the reduced form.

2.3. *Subclasses of nonlinear structural models*

In the following we present several classes of simultaneous nonlinear structural models each of which can be written in reduced form and hence are subclasses of the general nonlinear structural model (8).

Linear endogenous, nonlinear exogenous. This name describes a class of nonlinear structural models that is linear in the endogenous variables but possibly nonlinear in the exogenous variables, that is,

$$\boldsymbol{\eta}_i = \mathbf{B}\boldsymbol{\eta}_i + \mathbf{g}(\boldsymbol{\xi}_i, \boldsymbol{\gamma}) + \boldsymbol{\delta}_i, \tag{9}$$

where $\mathbf{g}$ is a vector function of $\boldsymbol{\xi}_i$ with unknown parameters $\boldsymbol{\gamma}$ and the matrix $(\mathbf{I} - \mathbf{B})$ is invertible. The equation errors $\boldsymbol{\delta}_i$ are as before in (7). Note that because of the linearity in the endogenous variables, nonrecursive models are also included here. It is straightforward to see how this model can be written in reduced form by multiplying both sides by the inverse of $(\mathbf{I}-\mathbf{B})$. We note that the reduced form has separable (additive) equation error.

Nonlinear recursive. This name describes a class of nonlinear structural models that is possibly nonlinear in both the endogenous and exogenous variables but where the system of equations is recursive (i.e., one equation can be substituted into the next), that is,

$$\eta_{1i} = g_1(\boldsymbol{\xi}_i, \boldsymbol{\beta}_1) + \delta_{1i}, \tag{10}$$

$$\eta_{2i} = g_2(\eta_{1i}, \boldsymbol{\xi}_i, \boldsymbol{\beta}_2) + \delta_{2i}, \tag{11}$$

$$\vdots$$

$$\eta_{di} = g_d(\eta_{1i}, \dots, \eta_{(d-1)i}, \boldsymbol{\xi}_i, \boldsymbol{\beta}_d) + \delta_{di}, \tag{12}$$

where $g_1 \dots g_d$ are nonlinear functions of the corresponding latent variables and unknown parameters $\boldsymbol{\beta}_1, \dots \boldsymbol{\beta}_d$ respectively and the vector of equation errors formed by taking $(\delta_{1i}, \delta_{2i} \dots \delta_{di})'$ is treated as $\boldsymbol{\delta}_i$ in (7). As a result of the triangular recursive form, it is straightforward to see how the model can be written in reduced form by substitution, but we note that the equation error will not necessarily be separable in the reduced form.

Linear endogenous, additive nonlinear exogenous. This class restricts the model (9) so that $\mathbf{g}(\boldsymbol{\xi}_i, \boldsymbol{\gamma})$ is an additive function of possibly nonlinear terms involving only $\boldsymbol{\xi}_i$. This model could also be described as linear in parameters, but nonlinear in the exogenous latent variables, that is,

$$\boldsymbol{\eta}_i = \mathbf{B}\boldsymbol{\eta}_i + \boldsymbol{\Gamma}\mathbf{g}(\boldsymbol{\xi}_i) + \boldsymbol{\delta}_i, \tag{13}$$

where $\boldsymbol{\Gamma}$ is a $(d \times r)$ matrix of fixed or unknown scalars and $\mathbf{g}(\boldsymbol{\xi}_i) = (g_1(\boldsymbol{\xi}_i), g_2(\boldsymbol{\xi}_i), \dots g_r(\boldsymbol{\xi}_i))'$ is a $(r \times 1)$ vector function of known functions of the exogenous variables.

This class of nonlinear structural models and particularly its subsets below for the polynomial and specifically the second order model is the one almost exclusively examined in the literature up to this point. Assuming normality for $\boldsymbol{\xi}_i$, Arminger and Muthén (1998) and Zhu and Lee (1999) described the Bayesian method for (13) with a linear measurement model (5) while Lee and Zhu (2002) describe the full maximum likelihood method for it. The nonlinear structural model (13) has also been examined by Lee and Zhu (2000), Lee and Song (2003a, 2003b), Song and Lee (2002), Lee and Lu (2003), and Lee et al. (2003). An estimation method for (13) not relying on distribution assumptions for $\boldsymbol{\xi}_i$ was developed by Bollen (1995, 1996) using a two-stage least squares. The method uses the instrumental variable technique where instruments are formed by taking functions of the observed indicators. One difficulty of the method comes from finding an appropriate instrument. Bollen (1995) and Bollen and Paxton (1998) show that the method works for the quadratic and interaction model but for general $\mathbf{g}(\boldsymbol{\xi}_i)$ it may be impossible to find appropriate instruments.

General polynomial. Further restricting (13) so that $\mathbf{g}(\boldsymbol{\xi}_i)$ is taken to be all the pure powers and all the multi-way interactions of those powers of the elements in $\boldsymbol{\xi}_i$ results in the polynomial structural equation model. An estimation method for the general order polynomial structural equation model was described by Wall and Amemiya (2000, 2003). The two stage method of moments estimator produces consistent estimators for the structural model parameters for virtually any distribution of the observed indicator variables where the linear measurement model holds. The procedure uses factor score estimates and estimates of their measurement error in a form of nonlinear errors-in-variables regression and produces closed-form method of moments type estimators as well as asymptotically correct standard errors.

Quadratic and interactions. Finally, if the general polynomial model is restricted to simply the second order model, we have the quadratic and/or interaction structural equation model. In particular,

$$\eta_i = \gamma_0 + \gamma_1\xi_i + \gamma_2\xi_i^2 + \delta_i \quad \text{or} \tag{14}$$

$$\eta_i = \gamma_0 + \gamma_1\xi_{1i} + \gamma_2\xi_{2i} + \gamma_3\xi_{1i}\xi_{2i} + \delta_i, \tag{15}$$

each taken as the structural model underlying its own linear measurement model are the models presented and estimated by the pioneering paper of Kenny and Judd (1984). In fact, these came to be known by some literature as the "Kenny and Judd model" and attracted much methodological discussions and alterations by a number of papers, including Hayduck (1987), Ping (1996), Jaccard and Wan (1995), Jöreskog and Yang (1996, 1997), Schumacker and Marcoulides (1998), Li et al. (1998) and within growth curve modeling Li et al. (2000) and Wen et al. (2002). The method of estimation proposed by Kenny and Judd (1984) involved taking products of the observed indicators $\mathbf{Z}_i$ and treating these products as themselves indicators of the quadratic or interaction terms. This results in many (tedious) constraints on the model covariance matrix but nevertheless is possible to implement in existing *linear* structural equation modeling software programs (e.g., LISREL). The Kenny and Judd (1984) method relied on the normality assumption for $\boldsymbol{\xi}_i$ and was shown to produce inconsistent estimators when the observed indicators are not normally distributed (Wall and Amemiya, 2001). Building on the products of indicators method, Wall and Amemiya (2001) developed an estimation method practical for the quadratic and interaction model that produces consistent estimators without assuming any distributional form for the underlying factors or errors. Comparisons via simulation study between several different approaches for the interaction model were examined in Marsh et al. (2004) and Lee et al. (2004).

3. Pseudo-likelihood estimation for the general nonlinear structural equation model

3.1. Motivation and setup

Estimation for the parameters in the general nonlinear structural model comprised of the linear measurement model (5) and the reduced form nonlinear structural model (8) will be the aim of this section. The error terms $\boldsymbol{\varepsilon}_i$, and $\boldsymbol{\delta}_i$ will be assumed to be normally distributed which can often be considered reasonable in most applications. The distribution of the latent variables $\boldsymbol{\eta}_i$ and $\boldsymbol{\xi}_i$ on the other hand will not be specified as normal. Because the structural model can be written in reduced form where $\boldsymbol{\eta}_i$ is a direct function of $\boldsymbol{\xi}_i$ and $\boldsymbol{\delta}_i$, only the distribution of $\boldsymbol{\xi}_i$ remains unspecified. The aim is to develop an estimator of $\boldsymbol{\beta}^*$ and $\boldsymbol{\Delta}$ that work well for weakly specified distributions of $\boldsymbol{\xi}_i$. The method presented here follows closely the work of Amemiya and Zhao (2001, 2002).

For individual i, the joint distribution of the observed data and the latent variables can be written under the nonlinear structural equation model (5)–(8)

$$\begin{aligned} P(\mathbf{Z}_i, \boldsymbol{f}_i; \boldsymbol{\theta}) &= P(\mathbf{Z}_i | \boldsymbol{f}_i; \boldsymbol{\theta}_m) P(\boldsymbol{f}_i; \boldsymbol{\theta}_s) \\ &= P(\mathbf{Z}_i | \boldsymbol{\eta}_i, \boldsymbol{\xi}_i; \boldsymbol{\theta}_m) P(\boldsymbol{\eta}_i, \boldsymbol{\xi}_i; \boldsymbol{\theta}_s) \\ &= P(\mathbf{Z}_i | \boldsymbol{\eta}_i, \boldsymbol{\xi}_i; \boldsymbol{\theta}_m) P(\boldsymbol{\eta}_i | \boldsymbol{\xi}_i; \boldsymbol{\theta}_1) P(\boldsymbol{\xi}_i; \boldsymbol{\theta}_\xi), \end{aligned} \tag{16}$$

where $\boldsymbol{\theta}_m$ represents the measurement model parameters, $\boldsymbol{\theta}_m = \{\boldsymbol{\mu}, \boldsymbol{\Lambda}, \boldsymbol{\Psi}\}$, and $\boldsymbol{\theta}_s$ represents the structural model parameters which are made up of the parameters in the nonlinear structural function (8), i.e., $\boldsymbol{\theta}_1 = \{\boldsymbol{\beta}^*, \boldsymbol{\Delta}\}$ and the parameters $\boldsymbol{\theta}_\xi$ describing the distribution of $\boldsymbol{\xi}$. Note that the parameters in the three parts are all distinct.

Given a known distribution for $P(\boldsymbol{\xi}_i; \boldsymbol{\theta}_\xi)$ with nice form, estimation for all the parameters given data could proceed via maximum likelihood, or with the addition of prior information proceed within a fully Bayesian setting. Treating the latent variables as missing data, the expectation maximization algorithm can be used for full maximum likelihood estimation although difficulty arises in the integration of the E-step since no closed form is available. Taking the distribution of $P(\boldsymbol{\xi}_i; \boldsymbol{\theta}_\xi)$ to be normally distributed, Amemiya and Zhao (2001) performed the full maximum likelihood for the general nonlinear model using the Monte Carlo EM algorithm.

Very commonly it is assumed that the distribution of the exogenous variables, $P(\boldsymbol{\xi}_i; \boldsymbol{\theta}_\xi)$ are normally distributed only for computational convenience. This restrictive assumption will be weakened in the current method by taking the hypothetically assumed distribution for $\boldsymbol{\xi}_i$ to be a multivariate normal mixture, i.e.,

$$\boldsymbol{\xi}_i \sim \sum_{j=1}^{J} \pi_j N(\boldsymbol{\mu}_j, \boldsymbol{\Sigma}_\xi), \tag{17}$$

where $\boldsymbol{\mu}_j$ is the $((q-d)\times 1)$ mean vectors for the jth component of the mixture and the covariance matrix $\boldsymbol{\Sigma}_\xi$ is assumed to be the same for all components. The normal mixture is considered as it can approximate a large class of distributions reasonably well and it is practical. That is, essential aspects of the estimation method, latent variable distribution deconvolution and Monte Carlo simulation from an estimated density, can be carried out readily using the normal mixture form.

While it would be possible to consider a full likelihood or fully Bayesian approach incorporating the finite mixture distribution for $P(\boldsymbol{\xi}_i; \boldsymbol{\theta}_\xi)$, the current paper presents an estimate for $\boldsymbol{\theta}_1$ utilizing the pseudo maximum likelihood estimation procedure proposed by Gong and Samaniego (1981) and Parke (1986). In this approach, instead of maximizing the likelihood with respect to $\boldsymbol{\theta}_1$, $\boldsymbol{\theta}_m$, and $\boldsymbol{\theta}_\xi$, some consistent estimators of the nuisance parameters $\boldsymbol{\theta}_m$, and $\boldsymbol{\theta}_\xi$ are substituted into the likelihood, and the resulting function is maximized only with respect to $\boldsymbol{\theta}_1$. The pseudo-likelihood approach is computationally simpler than full likelihood while not loosing the ability to consider flexible distributions for $\boldsymbol{\xi}_i$.

3.2. Estimating the nuisance parameters

Estimation of the structural model parameter $\boldsymbol{\theta}_1$ is of primary interest, thus $\boldsymbol{\theta}_m$ and $\boldsymbol{\theta}_\xi$ are considered nuisance parameters and will be estimated separately. The goal is to use estimators for these nuisance parameters which are consistent under weak distributional assumptions for the latent variables. The approach presented here is similar to that presented by Amemiya and Zhao (2002).

We start with describing our estimator $\hat{\boldsymbol{\theta}}_m$ of $\boldsymbol{\theta}_m$. It has been shown that the maximum normal likelihood estimators of the factor loadings and error variances in the linear factor analysis are consistent and have nice properties for nearly any unspecified distribution of the factor vector. See, e.g., Amemiya et al. (1987), Anderson and Amemiya (1988), and Brown and Shapiro (1988). Hence, we apply the maximum likelihood estimation to the linear measurement models (5) treating $\boldsymbol{f}_i$ normal with an unrestricted covariance matrix, and obtain $\hat{\boldsymbol{\theta}}_m$.

Now, given estimates for $\boldsymbol{\theta}_m$ we focus on estimating the parameters $\boldsymbol{\theta}_\xi$ describing the distribution of $\boldsymbol{\xi}_i$. To obtain an estimate of $\boldsymbol{\theta}_\xi$ in the latent variable normal mixture distribution, we use a method referred to as a measurement error deconvolution. This method starts with obtaining the so-called factor score estimator $\hat{\boldsymbol{\xi}}_i$ of each $\boldsymbol{\xi}_i$ based on the measurement model and its estimated parameters. The factor score estimator for $\boldsymbol{f}_i$ is

$$\hat{\boldsymbol{f}}_i = [\hat{\boldsymbol{\Lambda}}'\widehat{\boldsymbol{\Psi}}^{-1}\hat{\boldsymbol{\Lambda}}]^{-1}\hat{\boldsymbol{\Lambda}}'\widehat{\boldsymbol{\Psi}}^{-1}[\boldsymbol{Z}_i - \hat{\boldsymbol{\mu}}] \tag{18}$$

and $\hat{\boldsymbol{\xi}}_i$ is taken as the corresponding subset of elements from $\hat{\boldsymbol{f}}_i$. To use these factor score estimates for making inference about the distribution of $\boldsymbol{\xi}_i$, it is necessary to have an estimate of the measurement error that exists in $\hat{\boldsymbol{f}}_i$ as a measure of $\boldsymbol{f}_i$. Ignoring the errors of $O_p(n^{-1/2})$ in estimation of $\boldsymbol{\theta}_m$ and recalling the $\boldsymbol{\varepsilon}_i$ are normally distributed and independent of $\boldsymbol{f}_i$, we have

$$\hat{\boldsymbol{f}}_i = \boldsymbol{f}_i + \mathbf{r}_i, \tag{19}$$

where $\mathbf{r}_i \sim N(0, \boldsymbol{\Sigma}_r)$, and $\boldsymbol{\Sigma}_{r\xi}$ denotes the elements of $\boldsymbol{\Sigma}_r$ corresponding to $\mathrm{Var}(\hat{\boldsymbol{\xi}}_i - \boldsymbol{\xi}_i)$, and a consistent estimator $\widehat{\boldsymbol{\Sigma}}_r$ of $\boldsymbol{\Sigma}_r$ is

$$\widehat{\boldsymbol{\Sigma}}_r = [\hat{\boldsymbol{\Lambda}}\widehat{\boldsymbol{\Psi}}^{-1}\hat{\boldsymbol{\Lambda}}']^{-1}. \tag{20}$$

An alternate form of (18) and (20) are given in Wall and Amemiya (2000, 2003) for the case when $\widehat{\boldsymbol{\Psi}}$ is singular. Denote the elements of $\widehat{\boldsymbol{\Sigma}}_r$ corresponding to $\boldsymbol{\Sigma}_{r\xi}$ as $\widehat{\boldsymbol{\Sigma}}_{r\xi}$. It follows from (19) and (17) that

$$\hat{\boldsymbol{\xi}}_i \sim \sum_{j=1}^{J} \pi_j N(\boldsymbol{\mu}_j, \boldsymbol{\Sigma}_{\hat{\xi}}), \tag{21}$$

where

$$\boldsymbol{\Sigma}_{\hat{\xi}} = \boldsymbol{\Sigma}_{\xi} + \boldsymbol{\Sigma}_{r\xi} \tag{22}$$

and so by fitting a normal mixture to $\hat{\boldsymbol{\xi}}$ via maximum likelihood following, for example, a standard EM algorithm for normal mixtures from McLachlan and Peel (2000) to obtain $\{\hat{\pi}_j, j = 1 \ldots J\}$, $\{\hat{\boldsymbol{\mu}}_j, j = 1 \ldots J\}$, and $\widehat{\boldsymbol{\Sigma}}_{\hat{\xi}}$, we can then obtain an estimate of $\boldsymbol{\Sigma}_{\xi}$ by subtraction (deconvolution), i.e. $\widehat{\boldsymbol{\Sigma}}_{\xi} = \widehat{\boldsymbol{\Sigma}}_{\hat{\xi}} - \widehat{\boldsymbol{\Sigma}}_{r\xi}$. However, such a difference estimator may not be a proper covariance matrix, i.e., a nonnegative definite matrix. Also, to assure meaningful estimation of $\boldsymbol{\theta}_1$ at this model fitting stage, we need to have a strictly positive definite estimate of $\boldsymbol{\Sigma}_{\xi}$. One practical way to address this difficulty is to use an adjustment by eigenvalues in the error-matrix metric, described in, e.g., Amemiya (1985). Consider the eigenvalue-eigenvector decomposition

$$\widehat{\boldsymbol{\Sigma}}_{r\xi}^{-1/2} \widehat{\boldsymbol{\Sigma}}_{\hat{\xi}} \widehat{\boldsymbol{\Sigma}}_{r\xi}^{-1/2} = \mathbf{QDQ}.$$

Then, an estimator of $\boldsymbol{\Sigma}_{\xi}$ based on the difference $\widehat{\boldsymbol{\Sigma}}_{\hat{\xi}} - \widehat{\boldsymbol{\Sigma}}_{r\xi}$ guaranteed to be positive definite is

$$\widehat{\boldsymbol{\Sigma}}_{r\xi}^{1/2} \mathbf{M} \widehat{\boldsymbol{\Sigma}}_{r\xi}^{1/2},$$

where the ith element of a diagonal $\mathbf{M}$ is

$$m_i = \max\left\{d_i - 1, \frac{c}{n}\right\},$$

where d_i is the ith diagonal element of $\mathbf{D}$ and c is a positive constant.

3.3. MCEM for the pseudo-likelihood

Given consistent estimators $\hat{\boldsymbol{\theta}}_m$ and $\hat{\boldsymbol{\theta}}_{\xi}$ for the nuisance parameters, the pseudo maximum likelihood estimator (PMLE) for $\boldsymbol{\theta}_1$ is obtained by maximizing the likelihood evaluated at $\hat{\boldsymbol{\theta}}_m$ and $\hat{\boldsymbol{\theta}}_{\xi}$ with respect to $\boldsymbol{\theta}_1$. Since the likelihood function does not have an explicit expression, we consider performing the maximization using a Monte Carlo EM (MCEM) algorithm. The complete data pseudo-likelihood is

$$\begin{aligned} L_c &= \prod_{i=1}^{n} P\big(\mathbf{Z}_i, \boldsymbol{f}_i; \boldsymbol{\theta}_1, \hat{\boldsymbol{\theta}}_m, \hat{\boldsymbol{\theta}}_{\xi}\big) \\ &= \prod_{i=1}^{n} P\big(\mathbf{Z}_i | \boldsymbol{\eta}_i, \boldsymbol{\xi}_i; \hat{\boldsymbol{\theta}}_m\big) P(\boldsymbol{\eta}_i | \boldsymbol{\xi}_i; \boldsymbol{\theta}_1) P\big(\boldsymbol{\xi}_i; \hat{\boldsymbol{\theta}}_{\xi}\big). \end{aligned}$$

Then the E-step obtains the expectation of the complete data pseudo-likelihood given the observations and the current parameter estimates $\boldsymbol{\theta}_1^{(t)}$

$$E\big(\log L_c|\mathbf{Z}_1\ldots\mathbf{Z}_n;\boldsymbol{\theta}_1^{(t)},\hat{\boldsymbol{\theta}}_m,\hat{\boldsymbol{\theta}}_\xi\big) \tag{23}$$

$$=\sum_{i=1}^{n}\int\log P\big(\mathbf{Z}_i,\boldsymbol{f}_i;\boldsymbol{\theta}_1,\hat{\boldsymbol{\theta}}_m,\hat{\boldsymbol{\theta}}_\xi\big)P\big(\boldsymbol{f}_i|\mathbf{Z}_i,\boldsymbol{\theta}_1^{(t)},\hat{\boldsymbol{\theta}}_m,\hat{\boldsymbol{\theta}}_\xi\big)\,\mathrm{d}\boldsymbol{f}_i \tag{24}$$

$$=\sum_{i=1}^{n}E\Bigg(\log P(\mathbf{Z}_i,\boldsymbol{f}_i;\boldsymbol{\theta}_1,\hat{\boldsymbol{\theta}}_m,\hat{\boldsymbol{\theta}}_\xi)\times\frac{P(\mathbf{Z}_i|\boldsymbol{f}_i;\boldsymbol{\theta}_1^{(t)},\hat{\boldsymbol{\theta}}_m,\hat{\boldsymbol{\theta}}_\xi)}{\int P(\mathbf{Z}_i|\boldsymbol{f}_i;\boldsymbol{\theta}_1^{(t)},\hat{\boldsymbol{\theta}}_m,\hat{\boldsymbol{\theta}}_\xi)P(\boldsymbol{f}_i;\boldsymbol{\theta}_1^{(t)},\hat{\boldsymbol{\theta}}_m,\hat{\boldsymbol{\theta}}_\xi)\,\mathrm{d}\boldsymbol{f}_i}\Bigg) \tag{25}$$

$$\equiv g_{\boldsymbol{\theta}_1^{(t)}}\big(\boldsymbol{\theta}_1;\mathbf{Z},\hat{\boldsymbol{\theta}}_m,\hat{\boldsymbol{\theta}}_\xi\big), \tag{26}$$

where the expectation is taken with respect to the random latent variables $\boldsymbol{f}_i$. The Monte Carlo method can then be used to approximate this expectation. Given the current $\boldsymbol{\theta}_1^{(t)}$ along with $\hat{\boldsymbol{\theta}}_\xi$, a Monte Carlo sample $(\boldsymbol{f}_i^1,\ldots\boldsymbol{f}_i^M)$ is generated. The mth sample for individual i is generating as follows:

$$\boldsymbol{\xi}_i^m\sim P\big(\boldsymbol{\xi}_i;\hat{\boldsymbol{\theta}}_\xi\big), \tag{27}$$

$$\boldsymbol{\delta}_i^m\sim N(\mathbf{0},\boldsymbol{\Delta}^{(t)}), \tag{28}$$

then take

$$\boldsymbol{\eta}_i^m=\mathbf{h}\big(\boldsymbol{\xi}_i^m,\boldsymbol{\delta}_i^m;\boldsymbol{\beta}^{*(t)}\big) \tag{29}$$

so $\boldsymbol{f}_i^m=(\boldsymbol{\eta}_i^m,\boldsymbol{\xi}_i^m)'$. The expectation can then be approximated as

$$g_{\boldsymbol{\theta}_1^{(t)}}\big(\boldsymbol{\theta}_1;\mathbf{Z},\hat{\boldsymbol{\theta}}_m,\hat{\boldsymbol{\theta}}_\xi\big)\approx\sum_{i=1}^{n}\frac{1}{M}\sum_{m=1}^{M}\big[\log P(\mathbf{Z}_i,\boldsymbol{f}_i^m;\boldsymbol{\theta}_1,\hat{\boldsymbol{\theta}}_m,\hat{\boldsymbol{\theta}}_\xi)\big]W_i^m$$

$$\equiv g^{\mathrm{MC}}_{\boldsymbol{\theta}_1^{(t)}}\big(\boldsymbol{\theta}_1;\mathbf{Z},\hat{\boldsymbol{\theta}}_m,\hat{\boldsymbol{\theta}}_\xi\big) \tag{30}$$

where

$$W_i^m=\frac{P(\mathbf{Z}_i|\boldsymbol{f}_i^m;\boldsymbol{\theta}_1^{(t)},\hat{\boldsymbol{\theta}}_m,\hat{\boldsymbol{\theta}}_\xi)}{\frac{1}{M}\sum_{m=1}^{M}P(\mathbf{Z}_i|\boldsymbol{f}_i^m;\boldsymbol{\theta}_1^{(t)},\hat{\boldsymbol{\theta}}_m,\hat{\boldsymbol{\theta}}_\xi)}$$

is a weight which can be calculated straightforwardly. The value for $P(\mathbf{Z}_i|\boldsymbol{f}_i^m;\boldsymbol{\theta}_1^{(t)},\hat{\boldsymbol{\theta}}_m,\hat{\boldsymbol{\theta}}_\xi)$ can be calculated directly from the multivariate normal distribution. Note that if the normally distributed equation error $\boldsymbol{\delta}_i$ is not additive in the reduced form for $\mathbf{h}$ in (8), then the probability $P(\boldsymbol{\eta}_i^m|\boldsymbol{\xi}_i^m;\boldsymbol{\beta}^{*(t)},\boldsymbol{\Delta}^{(t)})$ will not follow a multivariate normal distribution. In cases where the structural model has the nonlinear recursive structure described in Section 2.3 with independent equation errors, then it will be possible to calculate $P(\boldsymbol{\eta}_i^m|\boldsymbol{\xi}_i^m;\boldsymbol{\beta}^{*(t)},\boldsymbol{\Delta}^{(t)})$ by appropriate recursive conditioning on the sequential endogenous variables using a product of conditional normal distributions. Generally, the probability can be calculated by taking the multivariate normal probability for $\boldsymbol{\delta}_i^m$ appropriately scaled by the Jacobian of $\boldsymbol{\delta} = \mathbf{h}^{-1}(\boldsymbol{\eta})$ taken with respect to $\boldsymbol{\eta}$. That is

$$P\big(\boldsymbol{\eta}_i^m|\boldsymbol{\xi}_i^m; \boldsymbol{\beta}^{*(t)}, \boldsymbol{\Delta}^{(t)}\big) = P\big(\boldsymbol{\delta}_i^m; \boldsymbol{\Delta}^{(t)}\big)\left|\frac{\partial \boldsymbol{\delta}}{\partial \boldsymbol{\eta}}\bigg|_{\boldsymbol{\eta}_i^m}\right|.$$

The M-step is to maximize $g^{\mathrm{MC}}_{\boldsymbol{\theta}_1^{(t)}}(\boldsymbol{\theta}_1; \mathbf{Z}, \hat{\boldsymbol{\theta}}_m, \hat{\boldsymbol{\theta}}_\xi)$ with respect to $\boldsymbol{\theta}_1$. From (16), we note that $\boldsymbol{\theta}_1$ only appears in the term in $P(\boldsymbol{\eta}_i|\boldsymbol{\xi}_i; \boldsymbol{\theta}_1)$, hence to maximize $g^{\mathrm{MC}}_{\boldsymbol{\theta}_1^{(t)}}(\boldsymbol{\theta}_1; \mathbf{Z}, \hat{\boldsymbol{\theta}}_m, \hat{\boldsymbol{\theta}}_\xi)$ we only need to maximize

$$G_{\boldsymbol{\theta}_1^{(t)}}(\boldsymbol{\theta}_1; \mathbf{Z}, \hat{\boldsymbol{\theta}}_m, \hat{\boldsymbol{\theta}}_\xi) = \sum_{i=1}^{n} \frac{1}{M} \sum_{m=1}^{M} \big[\log P(\boldsymbol{\eta}_i^m|\boldsymbol{\xi}_i^m; \boldsymbol{\theta}_1)\big] W_i^m.$$

Given the general nonlinear structural model $\boldsymbol{\eta}_i = \mathbf{h}(\boldsymbol{\xi}_i, \boldsymbol{\delta}_i; \boldsymbol{\beta}^*)$ and given the equation errors $\boldsymbol{\delta}_i$ are multivariate normal with $\operatorname{Var} \boldsymbol{\delta}_i = \boldsymbol{\Delta}$, the maximization to obtain $\boldsymbol{\theta}_1^{(t+1)} = (\boldsymbol{\beta}^{*(t+1)}, \boldsymbol{\Delta}^{(t+1)})$ can be accomplished by multivariate nonlinear weighted least squares. Note that for simpler forms of the nonlinear structural model, less computationally involved methods of maximization may be possible to implement. For example, multivariate linear regression can be used when the nonlinear structural model has the linear in endogenous, additive nonlinear in exogenous form as in (13).

The MCEM algorithm will iterate between the E-step and the M-step until the parameters converge according to some criteria. It has been pointed out that it is inefficient to choose a large Monte Carlo sample size M when theta is far from the ML estimate (Wei and Tanner, 1990; Booth and Hobert, 1999) and that it is preferable to start with a small M and increase it for each iteration by some fixed number. The convergence of the EM algorithm can be monitored by plotting theta versus the iteration number.

3.4. Standard errors estimation

The computation of the estimated covariance matrix for the PMLE was discussed in Parke (1986). Let $\boldsymbol{\theta}_2 = (\boldsymbol{\theta}_m, \boldsymbol{\theta}_\xi)$ represent all the nuisance parameters. Let $(\boldsymbol{\theta}_1^0, \boldsymbol{\theta}_2^0)$ be the true values for $(\boldsymbol{\theta}_1, \boldsymbol{\theta}_2)$, and let the information matrix for $(\boldsymbol{\theta}_1, \boldsymbol{\theta}_2)$ at $(\boldsymbol{\theta}_1^0, \boldsymbol{\theta}_2^0)$ for the full likelihood be denoted by

$$\boldsymbol{\Sigma} = \begin{pmatrix} \boldsymbol{\Sigma}_{11} & \boldsymbol{\Sigma}_{12} \\ \boldsymbol{\Sigma}_{21} & \boldsymbol{\Sigma}_{22} \end{pmatrix}, \tag{31}$$

partitioned corresponding to $(\boldsymbol{\theta}_1, \boldsymbol{\theta}_2)$. Parke (1986) showed that if

$$\sqrt{n}\big(\hat{\boldsymbol{\theta}}_2 - \boldsymbol{\theta}_2^0\big) \xrightarrow{\mathcal{L}} N(\mathbf{0}, \boldsymbol{\Upsilon}), \quad \text{as } n \to \infty,$$

then the PMLE $\hat{\boldsymbol{\theta}}_1$ satisfies

$$\sqrt{n}\big(\hat{\boldsymbol{\theta}}_1 - \boldsymbol{\theta}_1^0\big) \xrightarrow{\mathcal{L}} N(\mathbf{0}, \boldsymbol{\Xi}),$$

where n is the sample size, and

$$\boldsymbol{\Xi} = \boldsymbol{\Sigma}_{11}^{-1} + \boldsymbol{\Sigma}_{11}^{-1}\boldsymbol{\Sigma}_{12}\boldsymbol{\Upsilon}\boldsymbol{\Sigma}_{21}\boldsymbol{\Sigma}_{11}^{-1}.$$

While an estimator for the part of $\boldsymbol{\Upsilon}$ corresponding to $\hat{\boldsymbol{\theta}}_m$ may be readily available from standard software packages that fit confirmatory factor analysis models, the part of $\boldsymbol{\Upsilon}$ corresponding to $\hat{\boldsymbol{\theta}}_\xi$ and the covariance between the two are not readily available from canned software. Thus an estimator $\widehat{\boldsymbol{\Upsilon}}$ of $\boldsymbol{\Upsilon}$ can be obtained using a nonparametric bootstrap covariance matrix to estimate $\boldsymbol{\Upsilon}$. To avoid identifiability problems bootstrapping the estimates for the mixture model parameters $\boldsymbol{\theta}_\xi$, it is recommended to use the same starting values for estimation on each bootstrap sample (McLachlan and Peel (2000), p. 70). To estimate $\boldsymbol{\Sigma}_{11}$ and $\boldsymbol{\Sigma}_{12}$ in (31), we use an approximation to the expected information matrix, as described in McLachlan and Krishnan (1997, pp. 120–122). The observed data log-likelihood is

$$\sum_{i=1}^{n} \log P(\mathbf{Z}_i; \boldsymbol{\theta}) = \sum_{i=1}^{n} \log \int P(\mathbf{Z}_i | \boldsymbol{f}_i; \boldsymbol{\theta}) P(\boldsymbol{f}_i; \boldsymbol{\theta})\, \mathrm{d}\boldsymbol{f}_i,$$

and the corresponding individual score vector is

$$\mathbf{s}(\mathbf{Z}_i, \boldsymbol{\theta}) = \partial \log P(\mathbf{Z}_i; \boldsymbol{\theta}) / \partial \boldsymbol{\theta}.$$

We propose to use an estimator of the form

$$\sum_{i=1}^{n} \mathbf{s}(\mathbf{Z}_i; \hat{\boldsymbol{\theta}}) \mathbf{s}'(\mathbf{Z}_i; \hat{\boldsymbol{\theta}}),$$

using our estimator $\hat{\boldsymbol{\theta}}$, and to extract $\widehat{\boldsymbol{\Sigma}}_{11}$ and $\widehat{\boldsymbol{\Sigma}}_{12}$ parts. It can be shown that $\mathbf{s}(\mathbf{Z}_i; \boldsymbol{\theta}) = E\{\partial l_{ci}(\boldsymbol{\theta}) / \partial \boldsymbol{\theta} | \mathbf{Z}; \boldsymbol{\theta}\}$, where $l_{ci}(\boldsymbol{\theta}) = \log P(\mathbf{Z}_i, \boldsymbol{f}_i; \boldsymbol{\theta})$. Thus $\hat{\mathbf{s}}(\mathbf{Z}_i; \hat{\boldsymbol{\theta}})$ can be computed using Monte Carlo method, with $\{(\eta_i^m, \xi_i^m)\colon m = 1, 2, \ldots, M\}$ and $\{W_i^m\colon m = 1, 2, \ldots, M\}$ obtained in the last step of the MCEM algorithm. Then, $\widehat{\boldsymbol{\Sigma}}_{11}$ and $\widehat{\boldsymbol{\Sigma}}_{12}$ can be obtained in

$$\sum_{i=1}^{n} \hat{\mathbf{s}}(\mathbf{Z}_i; \hat{\boldsymbol{\theta}}) \hat{\mathbf{s}}'(\mathbf{Z}_i; \hat{\boldsymbol{\theta}}).$$

Combining $\widehat{\boldsymbol{\Sigma}}_{11}$, $\widehat{\boldsymbol{\Sigma}}_{12}$, and $\widehat{\boldsymbol{\Upsilon}}$, we obtain our estimate of the asymptotic covariance matrix of the PMLE $\hat{\boldsymbol{\theta}}_1$ as

$$n^{-1}\big[\widehat{\boldsymbol{\Sigma}}_{11}^{-1} + \widehat{\boldsymbol{\Sigma}}_{11}^{-1} \widehat{\boldsymbol{\Sigma}}_{12} \widehat{\boldsymbol{\Upsilon}} \widehat{\boldsymbol{\Sigma}}_{21} \widehat{\boldsymbol{\Sigma}}_{11}^{-1}\big]. \tag{32}$$

4. Example

4.1. The data

The data used in this section is not real life data (due to data privacy issues) but is instead computer generated data mimicking and motivated by a real life study. The data is motivated by a study of children with cystic fibrosis (CF) which was interested in examining the influences that stressors in the child's life, self esteem, and feelings of dejection have on the child's adherence to the treatment regimes. The data have been generated directly from the nonlinear structural equation model described below and are used only to provide an example of the kind of nonlinear structural equations that might be considered and how to apply the pseudo-likelihood approach for inference. The results are not intended to represent or even reflect the results in the motivating study.

Suppose we have data collected from a self-report questionnaire asking adolescents who have cystic fibrosis about the strains and stresses they encounter and feel, their self esteem, their feelings of dejection, and their frequency of skipping (nonadhering) to their treatments. The model considered of interest is shown in Figure 1 where

- *parental/youth strain* is measured by 3 items (Z1–Z3), e.g., You get into hassles/fights with your parents. Denote this latent factor as ξ_1.
- *peer/youth strain* is measured by 3 items (Z4–Z6), e.g., None of your friends seem to understand what having CF is like. Denote this latent factor as ξ_2.
- *personal worries and strains* is measured by 5 items (Z7–Z11), e.g., You worry about the future or You stay at home when you really do not want to. Denote this latent factor as ξ_3.
- *self-esteem* is measured by 6 items (Z12–Z17), e.g., I feel that I have a number of good qualities. Denote this latent factor as η_1.
- *feelings of dejection* is measured by 2 items (Z18–Z19), e.g., You get so sick of all you have to do to take care of yourself that you just want to give up. Denote this latent factor as η_2.
- *nonadherence* is a score (Z20) created as a frequency of not adhering to a number of items including, e.g., You skip doing chest physical therapy treatments. This observed variable will be treated as an observed latent variable, denoted η_3.

From Figure 1 we see that there are three exogenous and three endogenous variables of interest. The explicit relationships specified among these variables are described below. Denote $\mathbf{Z} = (Z1, \ldots Z20)$ as in Figure 1. Then the nonlinear structural equations model considered is

$$\mathbf{Z} = \boldsymbol{\mu} + \boldsymbol{\Lambda}(\xi_1, \xi_2, \xi_3, \eta_1, \eta_2, \eta_3)' + \boldsymbol{\varepsilon}, \tag{33}$$

$$\eta_1 = \beta_{10} + \beta_{11}\xi_1 + \beta_{12}\xi_2 + \beta_{13}\xi_3 + \delta_1, \tag{34}$$

$$\eta_2 = \beta_{20} + \beta_{21}\exp(\beta_{22}\eta_1 + \beta_{23}\xi_1 + \beta_{24}\xi_2 + \beta_{25}\xi_3) + \delta_2, \tag{35}$$

$$\eta_3 = \beta_{30} + \beta_{31}\eta_1 + \beta_{32}\eta_2 + \beta_{33}\eta_1\eta_2 + \delta_3, \tag{36}$$

$$\boldsymbol{\mu}' = \left(0\ \mu_1\ \mu_2\ 0\ \mu_3\ \mu_4\ 0\ \mu_5\ \mu_6\ \mu_7\ \mu_8\ 0\ \mu_9\ \mu_{10}\ \mu_{11}\ \mu_{12}\ \mu_{13}\ 0\ \mu_{14}\ 0\right),$$

$$\boldsymbol{\Lambda}' = \begin{pmatrix} 1 & \lambda_{11} & \lambda_{12} & 0 & 0 & 0 & 0 & 0 & 0 & 0 & 0 & 0 & 0 & 0 & 0 & 0 & 0 & 0 & 0 & 0 \\ 0 & 0 & 0 & 1 & \lambda_{21} & \lambda_{22} & 0 & 0 & 0 & 0 & 0 & 0 & 0 & 0 & 0 & 0 & 0 & 0 & 0 & 0 \\ 0 & 0 & 0 & 0 & 0 & 0 & 1 & \lambda_{31} & \lambda_{32} & \lambda_{33} & \lambda_{34} & 0 & 0 & 0 & 0 & 0 & 0 & 0 & 0 & 0 \\ 0 & 0 & 0 & 0 & 0 & 0 & 0 & 0 & 0 & 0 & 0 & 1 & \lambda_{41} & \lambda_{42} & \lambda_{43} & \lambda_{44} & \lambda_{45} & 0 & 0 & 0 \\ 0 & 0 & 0 & 0 & 0 & 0 & 0 & 0 & 0 & 0 & 0 & 0 & 0 & 0 & 0 & 0 & 0 & 1 & \lambda_{51} & 0 \\ 0 & 0 & 0 & 0 & 0 & 0 & 0 & 0 & 0 & 0 & 0 & 0 & 0 & 0 & 0 & 0 & 0 & 0 & 0 & 1 \end{pmatrix}$$

and the last element of $\boldsymbol{\varepsilon}$ is set equal to zero since the endogenous variable η_3 = nonadherence is treated as directly observed by $Z20$. The data shown in Figure 2 and

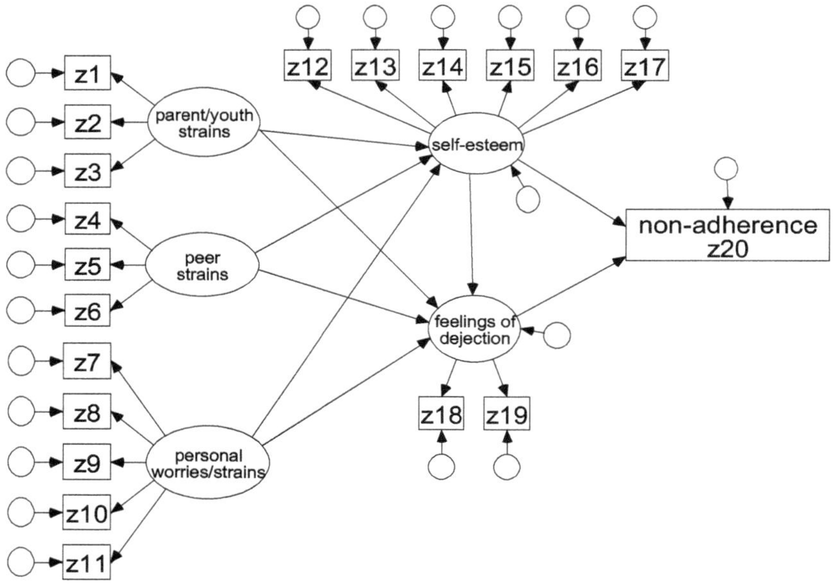

Fig. 1. Structural equation model – Unobserved variables represented by ovals or circles (circles for the errors), rectangles represent observed variables. Nonlinear relationships are not explicitly represented in the figure.

that used for the rest of this example were generated from the model above as follows: $(\xi_1, \xi_2, \xi_3)' = sqrt(\exp(x_1, x_2, x_3))$, where $(x_1, x_2, x_3)'$ is multivariate normal with all means zero, variances equal to 1 and $\mathrm{Cov}(x_1, x_2) = 0.5$, $\mathrm{Cov}(x_1, x_3) = 0.6$, $\mathrm{Cov}(x_2, x_3) = 0.4$; $\varepsilon_1 \ldots \varepsilon_{19}$, δ_1, δ_2, δ_3 are independent and normally distributed each with mean zero (recall $\varepsilon_{20} = 0$); variances of ε's are equal to 0.25, variance of δ_1 is 1, and variances of δ_2 and δ_3 are 0.25; all unknown values in $\boldsymbol{\mu}$ are set at zero and all unknown elements of $\boldsymbol{\Lambda}$ are set at one for data generation; and $\beta_{10} = 6$, $\beta_{11} = -0.5$, $\beta_{12} = -0.5$, $\beta_{13} = -0.5$; $\beta_{20} = 0$, $\beta_{21} = 1$, $\beta_{22} = -0.05$, $\beta_{23} = 0.25$, $\beta_{24} = 0.25$, $\beta_{25} = 0.25$; $\beta_{30} = 6$, $\beta_{31} = -0.7$, $\beta_{32} = 1$, $\beta_{33} = -0.125$.

Note that the exogenous variables (ξ_1, ξ_2, ξ_3) are not normally distributed due to the transformation taken. One dataset with 1000 independently sampled vectors $\mathbf{Z}$ was generated under these specifications.

4.2. *Description of the nonlinearities*

The particular nonlinearity considered between the strains, self-esteem, feelings of dejection and the unhealthy behavior of nonadhering to treatment extends the types of models usually considered for relating stress, self-esteem, and unhealthy behaviors. Many studies have considered linear relationships between stress, self-esteem and different unhealthy behaviors, e.g., related to suicide, Wilburn and Smith (2005); related to smoking, Byrne and Mazanov (2001); while some others have considered interaction effects between stress and self-esteem, e.g., Roberts and Kassel (1997), and Abel

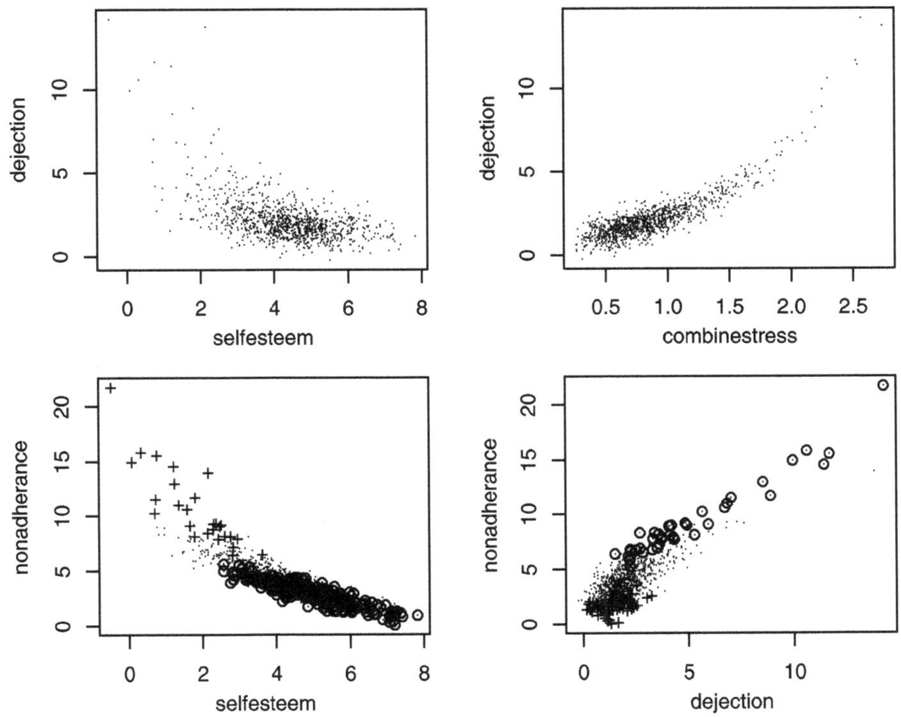

Fig. 2. Relationships between the true generated latent variables. In upper right figure, "combinestress" is the true sum of the three strain variables. To aid in viewing the interaction effect, plus signs and circles in bottom two figures represent those observations with high or low (respectively) values of the other nonplotted variable predicting nonadherence, i.e. dejection on the left and self-esteem on the right.

(1996). In contrast to the studies given as examples here where the measurement of the latent variables was considered to be done exactly using an observed scale treated with no measurement error, the full nonlinear structural equation model not only considers more general nonlinear relationships but also takes into account the measurement error inherent in the latent variables through the measurement model.

A description of the theoretical reasons for considering the nonlinear structural models (35) and (36), for η_2, i.e., feelings of dejection and η_3 nonadherence, is best given by examining the behavior of the functions seen for the generated data found in Figure 2. Dejection (Parrott, 2001) is a state of sadness in particular describing a feeling of being defeated. As self-esteem decreases and as overall stress increases, feelings of dejection increase. But the nonlinearities suggest that dramatically increased feelings of dejection come at a sort of tipping point or breaking point. That is, after a certain level of stress or a certain lack of self-esteem, the feelings of dejection are much higher. This can be modelled by the exponential model. The interaction term in the model for nonadherence is motivated by the fact that higher self-esteem is expected to weaken the effect that feelings of dejection have on nonadherence and likewise high levels of feelings of dejection would be expected to weaken the protective effect that self-esteem has on nonadherence. This is seen, for example, in the bottom left figure in Figure 2 where the for low levels of dejection (represented with circles) the increase of nonadherence as self-esteem decreases is much milder than when dejection is high (represented with plus signs).

4.3. Estimation

The pseudo maximum likelihood method described in Section 3 will be used to fit model (33)–(36) to the generated data.

First the measurement model is fit using SAS Proc Calis and the $\hat{\boldsymbol{\mu}}$, $\hat{\boldsymbol{\Lambda}}$, and $\widehat{\boldsymbol{\Psi}}$ are obtained. Then using Eqs. (18) and (20), the $\hat{\boldsymbol{\xi}}$ are obtained and also $\widehat{\boldsymbol{\Sigma}}_{r\xi}$. Then using the *EMclust* function from the *mclust* library (Fraley and Raftery (2002)) in the R statistical package, several finite mixture models were considered for fitting the model (17) to $\hat{\boldsymbol{\xi}}$. Based on the BIC criterion, a mixture model with 5 components and spherical covariance matrix $\boldsymbol{\Sigma}_{\hat{\xi}}$ fit best. Thus the estimates for $\{\hat{\pi}_j, j = 1\ldots5\}$, $\{\hat{\boldsymbol{\mu}}_j, j = 1\ldots5\}$, and $\widehat{\boldsymbol{\Sigma}}_{\hat{\xi}}$, are obtained and $\widehat{\boldsymbol{\Sigma}}_{\xi}$ can be obtained by subtracting $\widehat{\boldsymbol{\Sigma}}_{r\xi}$. Appendix A presents all the estimated nuisance parameters. Figures 3 and 4 compare the marginal and bivariate distributions of the true underlying non normally distributed exogenous latent variables with the estimated distributions fitted with the mixture model. Marginally (Figure 3) we see that the mixture model with 5 components appears to adequately capture the positive skew. Bivariately, the major fanning feature between the true latent variables is captured by the mixture model via the two distant components in opposite directions from the rest. It is possible that a mixture model allowing different volumes (i.e., different $\boldsymbol{\Sigma}_{\hat{\xi}}$) for each of the 5 components would capture the distribution more closely, although the much simpler model fit here with common $\boldsymbol{\Sigma}_{\hat{\xi}}$ appears reasonable. Furthermore we note that after the adjustment for the differential measurement error in the $\hat{\boldsymbol{\xi}}_i$ by subtracting $\widehat{\boldsymbol{\Sigma}}_{r\xi}$, the estimate $\widehat{\boldsymbol{\Sigma}}_{\xi}$ has different variances along the diagonal.

Given the estimates for the nuisance parameters (shown in Appendix A), we can proceed with the MCEM for doing maximum pseudo-likelihood. Taking advantage of the recursive nature of the nonlinear structural model considered and the equation errors being independent we have

$$\begin{aligned} P\big(\boldsymbol{\eta}_i^m|\boldsymbol{\xi}_i^m;\boldsymbol{\theta}_1\big) &\equiv P\big(\eta_{1i}^m,\eta_{2i}^m,\eta_{3i}^m|\boldsymbol{\xi}_i^m;\boldsymbol{\theta}_1\big) \\ &= P\big(\eta_{3i}^m|\eta_{2i}^m,\eta_{1i}^m;\boldsymbol{\beta}_3\big)P\big(\eta_{2i}^m|\eta_{1i}^m,\boldsymbol{\xi}_i^m;\boldsymbol{\beta}_2\big)P\big(\eta_{1i}^m|\boldsymbol{\xi}_i^m;\boldsymbol{\beta}_1\big),\end{aligned}$$

where each of the three components has a univariate normal distribution and is a function of separate parameters. Hence calculation of the weights in the E-step is straightforward as a product of normal densities and maximization in the M-step can be performed separately for each of the three parts using least squares. For the $P(\eta_{1i}^m|\boldsymbol{\xi}_i^m;\boldsymbol{\beta}_1)$ and $P(\eta_{3i}^m|\eta_{2i}^m,\eta_{1i}^m;\boldsymbol{\beta}_3)$, ordinary linear least squares is used since the equations are linear in the parameters $\boldsymbol{\beta}_1$ and $\boldsymbol{\beta}_3$. For $P(\eta_{2i}^m|\eta_{1i}^m,\boldsymbol{\xi}_i^m;\boldsymbol{\beta}_2)$ a nonlinear least squares is necessary but is easily implemented, for example, using the *nlm* maximization function in R. A program for the MCEM pseudo-likelihood approach for this model has been implementing in R and is available from the authors.

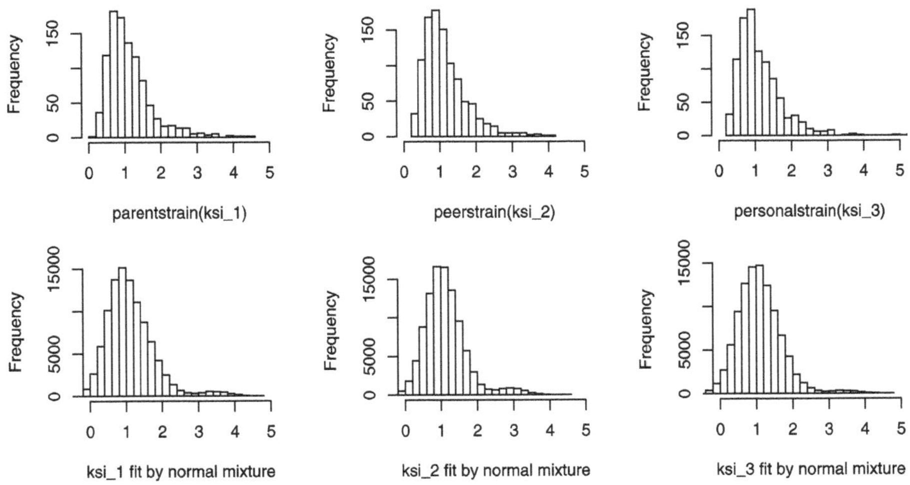

Fig. 3. Marginal distribution of true underlying exogenous latent variables and respective estimated distributions for them using a normal mixture with 5 components.

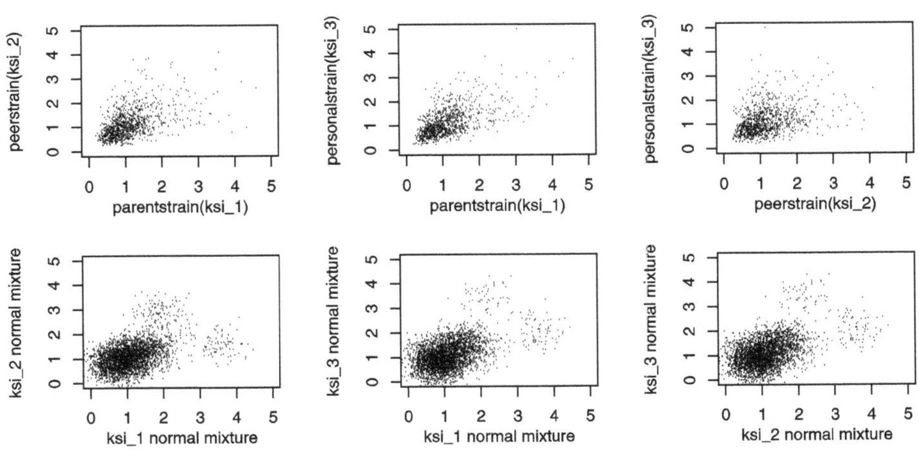

Fig. 4. Bivariate distribution of true underlying exogenous latent variables and respective estimated distributions for them using a normal mixture with 5 components.

The standard errors as described in Section 3.4 are also computed. The bootstrap method was used to obtain $\widehat{\boldsymbol{\Upsilon}}$. This involved obtaining 5000 bootstrap samples for which $\hat{\boldsymbol{\theta}}_m^{(B)}$ $B = 1 \ldots 5000$ were computed for the measurement model from SAS PROC CALIS, and then using the same bootstrap samples, $\hat{\boldsymbol{\theta}}_\xi^{(B)}$ was computed using the *Mclust* function in R. Denoting $\hat{\boldsymbol{\theta}}_2^{(B)} = (\hat{\boldsymbol{\theta}}_m^{(B)\prime}, \hat{\boldsymbol{\theta}}_\xi^{(B)\prime})'$ and taking $mean(\hat{\boldsymbol{\theta}}_2^{(B)}) = \frac{1}{5000}\sum_{B=1}^{5000}\hat{\boldsymbol{\theta}}_2^{(B)}$, then the estimator $\widehat{\boldsymbol{\Upsilon}}$ is computed as $\sum_{B=1}^{5000}(\hat{\boldsymbol{\theta}}_2^{(B)} - mean(\hat{\boldsymbol{\theta}}_2^{(B)}))(\hat{\boldsymbol{\theta}}_2^{(B)} -$

Table 1
Estimates based on pseudo maximum likelihood for $\boldsymbol{\theta}_1$

Parameter	Truth	Estimate	Standard error
β_{10}	6	5.979	0.113
β_{11}	−0.5	−0.540	0.122
β_{12}	−0.5	−0.483	0.089
β_{13}	−0.5	−0.484	0.114
δ_1	1	0.994	0.072
β_{20}	0	0.301	0.265
β_{21}	1	0.674	0.214
β_{22}	−0.05	−0.032	0.013
β_{23}	0.25	0.314	0.039
β_{24}	0.25	0.270	0.048
β_{25}	0.25	0.316	0.036
δ_2	0.25	0.218	0.029
β_{30}	6	5.803	0.200
β_{31}	−0.7	−0.655	0.044
β_{32}	1	1.095	0.061
β_{33}	−0.125	−0.149	0.016
δ_3	0.25	0.295	0.036

$mean(\hat{\boldsymbol{\theta}}_2^{(B)}))'$. For obtaining standard errors of the structural model parameters $\hat{\boldsymbol{\theta}}_1$, the estimator $\widehat{\boldsymbol{\Upsilon}}$ is combined with results from the MCEM. In particular, the estimators $\hat{\mathbf{s}}(\mathbf{Z}_i; \hat{\boldsymbol{\theta}})$ are obtained from the last step of the MCEM and the estimated asymptotic covariance is calculated for $\hat{\boldsymbol{\theta}}_1$ using (32). Note that the distribution for $\boldsymbol{\xi}_i$ used in forming the complete data likelihood for calculating $\hat{\mathbf{s}}(\mathbf{Z}_i; \hat{\boldsymbol{\theta}})$ was taken to be a 5 component normal mixture with common diagonal covariance matrix in each component. The standard errors are taken as the square root of the diagonal and 95% confidence intervals can be formed based on an assumption of asymptotic normality. Results are shown in Table 1. Not surprisingly since this is simulated data with a sample of size 1000 the resulting estimates are close to the true values.

5. Discussion

In this paper, the nonlinear structural equation model was described generally and an estimation procedure was presented using a flexible mixture model for the underlying exogenous factors and a pseudo-likelihood approach for estimating the parameters of interest. Particular attention was placed on distinguishing between techniques that make strong distributional assumptions for the underlying factors (i.e., normally distributed) verses those techniques not relying on (or more robust) to these assumptions. Sound statistical methods should generally aim to be practicably feasible and applicable to many problems while minimizing the number of uncheckable assumptions. That is the aim of the model and method presented in this paper.

A related modeling area to the one presented in this paper is nonlinear measurement error models, e.g., Fuller (1987), and Carroll et al. (1995). When the unknown measurements are considered random, it is then sometimes called nonlinear structural models, e.g., Patefield (2002). Generally a nonlinear measurement error model considers a nonlinear relationship between latent variables (variables that cannot be observed directly) similar to those in the current paper, but a confirmatory factor analysis type model is not typically used as the measurement model. Instead, only one variable is typically used to measure each latent variable and there is some assumptions made or external information used about the magnitude of the measurement error. Estimation methods including nonparametrics are well developed in this related field and may provide insight into methods useful for nonlinear structural equation modeling.

The nonlinear structural equation model presented in this paper could be made more general in two natural ways. One would be to allow for a measurement model with categorical observed variables. That is, a deviation from the linear measurement model. One of the possible difficulties with this is related to the desire to keep the underlying exogenous factor distribution flexible via the finite mixture model. It is not clear that the good results expected for estimating a flexible distribution for $\boldsymbol{\xi}$ will work well when the measurement model is not linear. The other extension is to include observed covariates into the structural model. In fact, this can already be considered within the model presented, although it may not be immediately obvious. An observed covariate can be included directly into the structural model by taking it to be a perfect measure of its own exogenous latent variable with error equal to zero (similar to what was done for Z20 in the example). Then the observed variable can be considered as a special element in $\boldsymbol{\xi}$ that is always equal to itself and never generated within the MCEM.

While general nonlinear relationships between latent variables may be natural to consider, there appears to be a lack of current theories motivating their use in the behavioral sciences (based on the authors' own experience and reading of literature). This may be do in part to the small signal to noise ratio often expected in data collected in the social sciences which does not lend itself to thinking of any more than just linear relationships, but it may also be due to a lack (until more recently) of models and methods for fitting nonlinear relationships between latent variables. The hope is that if there are nonlinear theories out there just waiting for a method, that the already existing methods (including that in this paper) will be discovered and become implemented.

Appendix A

```
> Lambda
       [,1]   [,2]   [,3]   [,4]   [,5] [,6]
 [1,] 1.0000 0.0000 0.0000 0.0000 0.0000    0
 [2,] 0.9431 0.0000 0.0000 0.0000 0.0000    0
 [3,] 0.9901 0.0000 0.0000 0.0000 0.0000    0
 [4,] 0.0000 1.0000 0.0000 0.0000 0.0000    0
```

```
 [5,] 0.0000 0.9654 0.0000 0.0000 0.0000    0
 [6,] 0.0000 0.9691 0.0000 0.0000 0.0000    0
 [7,] 0.0000 0.0000 1.0000 0.0000 0.0000    0
 [8,] 0.0000 0.0000 1.0058 0.0000 0.0000    0
 [9,] 0.0000 0.0000 0.9731 0.0000 0.0000    0
[10,] 0.0000 0.0000 1.0439 0.0000 0.0000    0
[11,] 0.0000 0.0000 0.9804 0.0000 0.0000    0
[12,] 0.0000 0.0000 0.0000 1.0000 0.0000    0
[13,] 0.0000 0.0000 0.0000 0.9815 0.0000    0
[14,] 0.0000 0.0000 0.0000 1.0145 0.0000    0
[15,] 0.0000 0.0000 0.0000 1.0165 0.0000    0
[16,] 0.0000 0.0000 0.0000 1.0040 0.0000    0
[17,] 0.0000 0.0000 0.0000 0.9898 0.0000    0
[18,] 0.0000 0.0000 0.0000 0.0000 1.0000    0
[19,] 0.0000 0.0000 0.0000 0.0000 0.9631    0
[20,] 0.0000 0.0000 0.0000 0.0000 0.0000    1

> mu
 [1]  0.00000  0.08254  0.03744  0.00000  0.07588  0.07115
 [7]  0.00000  0.04293  0.06978 -0.01494  0.02725  0.00000
      0.07329 -0.07574
[15] -0.07303 -0.00933  0.04260  0.00000  0.05085  0.00000

> diag(Psi)
 [1] 0.22879 0.25550 0.26387 0.22457 0.24857 0.25313 0.24940
 [8] 0.24991 0.26225 0.23446 0.26237 0.24384 0.23845 0.25896
     0.25449
[16] 0.24356 0.25510 0.23300 0.28825 0.00000

> output$mu, mu_1...mu_5 for mixture model using 5 components
            1        2        3         4        5
[1,] 1.450040 2.158594 3.445283 0.7947789 1.891052
[2,] 1.321064 1.770537 1.559521 0.8865848 2.919759
[3,] 1.431786 3.415500 2.113866 0.8422715 1.368160

> output$pro, pi_1...pi_5 for mixture model
[1] 0.27121709 0.01595621 0.02719309 0.64279488 0.04283873

> output$sigma, Sigma_ksihat for mixture model for ksihat
          [,1]      [,2]      [,3]
[1,] 0.2304556 0.0000000 0.0000000
[2,] 0.0000000 0.2304556 0.0000000
[3,] 0.0000000 0.0000000 0.2304556

> Sigrksihat, from equation (21)
           [,1]       [,2]       [,3]
[1,] 0.08645236 0.00000000 0.00000000
[2,] 0.00000000 0.08394503 0.00000000
[3,] 0.00000000 0.00000000 0.05005088
```

```
> Sigma_ksi = Sigma_ksihat - Sigrksihat
          [,1]      [,2]      [,3]
[1,] 0.1440033 0.0000000 0.0000000
[2,] 0.0000000 0.1465106 0.0000000
[3,] 0.0000000 0.0000000 0.1804047
```

References

Abel, M.H. (1996). Self-esteem: Moderator or mediator between perceived stress and expectancy of success?. *Psychological Reports* **79**, 635–641.

Amemiya, Y. (1985). What shold be done when an estimated between-group covaraince matrix is not non-negative definite?. *The American Statistician* **30**, 112–117.

Amemiya, Y., Zhao, Y. (2001). Estimation for nonlinear structural equation system with an unspecified distribution. In: *Proceedings of Business and Economic Statistics Section, The Annual Meeting of the American Statistical Association* (CD-ROM).

Amemiya, Y., Zhao, Y. (2002). Pseudo likelihood approach for nonliear and non-normal structural equation analysis. In: *Proceedings of Business and Economic Statistics Section, The Annual Meeting of the American Statistical Association* (CD-ROM).

Amemiya, Y., Fuller, W.A., Pantula, S.G. (1987). The asymptotic distributions of some estimators for a factor analysis model. *Journal of Multivariate Analysis* **22**, 51–64.

Anderson, T.W., Amemiya, Y. (1988). The asymptotic normal distribution of estimators in factor analysis under general conditions. *Annals of Statistics* **16**, 759–771.

Arminger, G., Muthén, B. (1998). A Bayesian approach to nonlinear latent variable models using the Gibbs sampler and the Metropolis–Hastings algorithm. *Psychometrika* **63** (3), 271–300.

Benkard, C.L., Berry, S. (2005). On the nonparametric identification of nonlinear simultaneous equations models: Comment on B. Browne (1983) and Roehrig (1988). Technical Report. http://www.stanford.edu/~lanierb/research/emanote041805.pdf.

Bentler, P.M. (1980). Multivariate analysis with latent variables: Causal modeling. *Annual Review of Psychology* **31**, 419–456.

Bollen, K.A. (1989). *Structural Equations with Latent Variables*. Wiley, New York.

Bollen, K.A. (1995). Structural equation models that are nonlinear in latent variables: A least squares estimator. *Sociological Methodology* **25**, 223–251.

Bollen, K.A. (1996). An alternative two stage least squares (2SLS) estimator for latent variable equation. *Psychometrika* **61**, 109–121.

Bollen, K.A., Paxton, P. (1998). Interactions of latent variables in structural equation models. *Structural Equation Modeling* **5**, 267–293.

Booth, J.G., Hobert, J.H. (1999). Maximizing generalized linear mixed model likelihoods with an automated Monte Carlo EM algorithm. *Journal of the Royal Statistical Society Series B* **62**, 265–285.

Brown, M.W., Shapiro, A. (1988). Robustness of normal theory methods in the analysis of linear latent variate models. *British Journal of Mathematical and Statistical Psychology* **41**, 193–208.

Byrne, D.G., Mazanov, J. (2001). Self-esteem, stress and cigarette smoking in adolescents. *Stress and Health* **17** (2), 105–110.

Carroll, R.J., Ruppert, D., Stefanski, L.A. (1995). *Measurement Error in Nonlinear Models*. Chapman and Hall, London.

Fraley, C., Raftery, A.E. (2002). MCLUST: Software for model-based clustering, density estimation and discriminant analysis. Technical Report, Department of Statistics, University of Washington. See URL: http://www.stat.washington.edu/mclust.

Fuller, W.A. (1987). *Measurement Error Models*. John Wiley, New York.

Gong, G., Samaniego, F.J. (1981). Pseudo maximum likelihood estimation: theory and applications. *The Annals of Statistics* **9** (4), 861–869.

Hayduck, L.A. (1987). *Structural Equation Modeling with LISREL: Essentials and Advances*. The Johns Hopkins University Press, Baltimore.

Jaccard, J., Wan, C.K. (1995). Measurement error in the analysis of interaction effects between continuous predictors using multiple regression: Multiple indicator and structural equation approaches. *Psychological Bulletin* **117** (2), 348–357.

Jöreskog, K.G. (1973). A general method for estimating a linear structural equation system. In: Goldberger, A.S., Duncan, O.D. (Eds.), *Structural Equation Models in the Social Sciences*. Academic Press, New York, pp. 85–112.

Jöreskog, K.G., Yang, F. (1996). Non-linear structural equation models: The Kenny–Judd model with interaction effects. In: Marcoulides, G.A., Schumacker, R.E. (Eds.), *Advanced Structural Equation Modeling: Issues and Techniques*, pp. 57–88.

Jöreskog, K.G., Yang, F. (1997). Estimation of interaction models using the augmented moment matrix: Comparison of asymptotic standard errors. In: Bandilla, W., Faulbaum, F. (Eds.), *SoftStat '97. Advances in Statistical Software* 6, pp. 467–478.

Kenny, D.A., Judd, C.M. (1984). Estimating the nonlinear and interactive effects of latent variables. *Psychological Bulletin* **96** (1), 201–210.

Lee, S.Y., Lu, B. (2003). Case-deletion diagnostics for nonlinear structural equation models. *Multivariate Behavioral Research* **38** (3), 375–400.

Lee, S.Y., Song, X.Y. (2003a). Maximum likelihood estimation and model comparison of nonlinear structural equation models with continuous and polytomous variables. *Computational Statistics and Data Analysis* **44**, 125–142.

Lee, S.Y., Song, X.Y. (2003b). Model comparison of nonlinear structural equation models with fixed covariates. *Psychometrika* **68** (1), 27–47.

Lee, S.Y., Zhu, H.T. (2000). Statistical analysis of nonlinear structural equation models with continuous and polytomous data. *British Journal of Mathematical and Statistical Psychology* **53**, 209–232.

Lee, S.Y., Zhu, H.T. (2002). Maximum likelihood estimation of nonlinear structural equation models. *Psychometrika* **67** (2), 189–210.

Lee, S.Y., Song, X.Y., Lee, J.C.K. (2003). Maximum likelihood estimation of nonlinear structural equation models with ignorable missing data. *Journal of Educational and Behavioral Statistics* **28** (2), 111–134.

Lee, S.Y., Song, X.Y., Poon, W.Y. (2004). Comparison of approaches in estimating interaction and quadratic effects of latent variables. *Multivariate Behavioral Research* **39**, 37–67.

Li, F., Harmer, P., Duncan, T., Duncan, S., Acock, A., Boles, S. (1998). Approaches to testing interaction effects using structural equation modeling methodology. *Multivariate Behavioral Research* **33** (1), 1–39.

Li, F., Duncan, T., Acock, A. (2000). Modeling interaction effects in latent growth curve models. *Structural Equation Modeling* **7**, 497–533.

Marsh, H.W., Wen, Z., Hau, K.T. (2004). Structural equation models of latent interactions: Evaluation of alternative estimation strategies and indicator construction. *Psychological Methods* **9** (3), 275–300.

McLachlan, G., Krishnan, T. (1997). *The EM Algorithm and Extensions*. John Wiley and Sons, New York.

McLachlan, G., Peel, D. (2000). *Finite Mixture Models*. John Wiley and Sons, New York.

Parke, W.R. (1986). Pseudo maximum likelihood estimation: the asymptotic distribution. *The Annals of Statistics* **14** (1), 355–357.

Parrott, W. (2001). *Emotions in Social Psychology*. Psychology Press, Philadelphia.

Patefield, M. (2002). Fitting non-linear structural relationships using SAS procedure NLMIXED. *The Statistician* **51** (3), 355–366.

Ping, R.A. (1996). Estimating latent variable interactions and quadratics: The state of this art. *Journal of Management* **22**, 163–183.

Roberts, J.E., Kassel, J.D. (1997). Labile self-esteem, life stress, and depressive symptoms: prospective data testing a model of vulnerability. *Cognitive Therapy and Research* **21** (5), 569–589.

Schumacker, R., Marcoulides, G. (Eds.) (1998). *Interaction and Nonlinear Effects in Structural Equation Modeling*. Lawrence Erlbaum Associates, Mahwah, NJ.

Song, X.Y., Lee, S.Y. (2002). A Bayesian approach for multigroup nonlinear factor analysis. *Structural Equation Modeling* **9** (4), 523–553.

Wall, M.M., Amemiya, Y. (2000). Estimation for polynomial structural equation models. *Journal of the American Statistical Association* **95**, 929–940.

Wall, M.M., Amemiya, Y. (2001). Generalized appended product indicator procedure for nonlinear structural equation analysis. *Journal of Educational and Behavioral Statistics* **26** (1), 1–29.

Wall, M.M., Amemiya, Y. (2003). A method of moments technique for fitting interaction effects in structural equation models. *British Journal of Mathematical and Statistical Psychology* **56**, 47–63.
Wei, G.C.G., Tanner, M.A. (1990). A Monte Carlo implementation of the EM algorithm and the poor man's data augmentation algorithm. *Journal of the American Statistical Association* **85**, 699–704.
Wen, Z., Marsh, H.W., Hau, K.T. (2002). Interaction effects in growth modeling: A full model. *Structural Equation Modeling* **9** (1), 20–39.
Wilburn, V., Smith, D. (2005). Stress, self-esteem, and suicidal ideation in late adolescents. *Adolescence* **40** (157), 33–45.
Zhu, H.T., Lee, S.Y. (1999). Statistical analysis of nonlinear factor analysis models. *British Journal of Mathematical and Statistical Psychology* **52**, 225–242.

Handbook of Computing and Statistics with Applications, Vol. 1
ISSN: 1871-0301

DOI: 10.1016/S1871-0301(06)01016-X

16

Matrix Methods and their Applications to Factor Analysis

Haruo Yanai and Yoshio Takane

Abstract
Since the introduction of Spearman's two factor model in 1904, a number of books and articles on theories of factor analysis have been published. During the same period, a number of matrix methods have also been developed, particularly in the theory of g-inverses and projection matrices. These two lines of developments, matrix methods and some important topics of factor analysis are integrated, and some of the earlier theories of factor analysis extended. A wide range of topics of factor analysis are covered including identifiability conditions, communality problems, analysis of image and anti-image variables, estimation of factor scores, and equivalence conditions on canonical factor analysis. In particular, the conditions under which the SMC of a variable is equal to the communality of the variable are developed, and some equivalent conditions under which the eigenvalues resulting from canonical factor analysis are either 1 or 0 are discussed. Methods for estimating factor score matrices when the unique variance matrix is singular are also introduced.

1. Introduction

Over the past hundred years since the introduction of Spearman's two-factor model of intelligence (in 1904), a number of books and articles have been published on theories of factor analysis (Harman, 1967; see also Yanai and Ichikawa (2007)). During the same period of time, there have been a number of interesting developments in matrix theory, particularly in the theory of g-inverses and projectors. In this chapter we attempt to integrate these two lines of developments, matrix methods and some important topics of factor analysis such as identifiability conditions, communality problems with special reference to squared multiple correlation (SMC), image and anti-image analysis, estimation of factor scores, equivalence conditions on canonical factor analysis, etc. Through this exercise, we also attempt to generalize some of the earlier theories. Throughout this chapter, we emphasize the use of g-inverse and projection matrices, which have been proven useful (Takeuchi et al., 1982; see also Takane (2004)) in explicating some intricate concepts underlying factor analysis models as well as other multivariate data analysis techniques. All matrices considered in this paper are real matrices.

2. Fundamentals of matrix methods

2.1. General definitions of g-inverse matrices and orthogonal projectors

Let A be a matrix of order $n \times m$, and let X be a matrix of order $m \times n$. Consider the following four equations:

$$\text{(i) } AXA = A, \quad \text{(ii) } XAX = A, \quad \text{(iii) } (AX)' = AX, \quad \text{(iv) } (XA)' = XA. \tag{1}$$

Matrix X satisfying (i) is called g-inverse of A and is generally denoted as A^-, while X satisfying both (i) and (iii) is called least squares g-inverse of A, and X satisfying both (i) and (iv) is called minimum norm g-inverse. These three types of g-inverses are not uniquely determined. Matrix X satisfying all of the above four conditions is called Moore–Penrose (g)-inverse matrix and is generally denoted as A^+. The Moore–Penrose inverse is uniquely determined. (See (c) below.)

We give some basic properties of g-inverses and orthogonal projectors:

(a) Let $X = A_\ell^-$ be a least squares g-inverse of A. Then,

$$AX = AA_\ell^- = A(A'A)^-A' = P_A, \tag{2}$$

where P_A is the orthogonal projector onto Sp(A), space spanned by the column vectors of A.

(b) Let $X = A_m^-$ be a minimum norm g-inverse of A. Then,

$$XA = A_m^-A = A'(AA')^-A = P_{A'}, \tag{3}$$

where $P_{A'}$ is the orthogonal projector onto Sp(A'), space spanned by the row vectors of A. Observe that P_A and $P_{A'}$ are symmetric and invariant over any choice of g-inverses of $A'A$ and AA', respectively, and for any choice of vectors spanning Sp(A) and Sp(A'), respectively.

(c) Let X_1 and X_2 be two Moore–Penrose inverse matrices of A. Then,

$$X_1 = (X_1A)X_1 = (X_2A)X_1 = X_2(AX_1) = X_2(AX_2) = X_2 \tag{4}$$

due to the relationships given in (2) and (3). This shows the uniqueness of the Moore–Penrose inverse matrix.

2.2. Decompositions of the orthogonal projector

Let A and B be $n \times p$ and $n \times q$ matrices, respectively, and let Sp(A) and Sp(B) represent subspaces spanned by the column vectors of A and B. Let I_n be the identity matrix of order n. Then, $Q_A = I_n - P_A$ and $Q_B = I_n - P_B$ are the orthogonal projectors onto Sp$(A)^\perp$ and Sp$(B)^\perp$, respectively, where Sp$(A)^\perp$ and Sp$(B)^\perp$ are the ortho-complement subspaces of Sp(A) and Sp(B). Obviously, $P_AQ_A = Q_AP_A = P_BQ_B = Q_BP_B = O$.

We introduce three important properties of the orthogonal projectors:

PROPERTY 1. (Rao and Yanai, 1979) *Let* $\text{Sp}(A, B)$ *represent the space spanned by column vectors of matrix* $[A, B]$. *Let* $P_{[A,B]}$ *be the orthogonal projector onto* $\text{Sp}(A, B)$. *Then,*

$$P_{[A,B]} = P_A + P_{Q_A B} = P_B + P_{Q_B A}. \tag{5}$$

PROPERTY 2. (Yanai and Puntanen, 1993) *Let* $Q_{[A,B]}$ *be the orthogonal projector onto the ortho-complement subspace of* $\text{Sp}(A, B)$, *that is,* $\text{Sp}(A, B)^{\perp}$. *Then,*

$$Q_{[A,B]} = Q_{Q_A B} Q_A = Q_{Q_B A} Q_B. \tag{6}$$

PROPERTY 3. (Baksalary, 1987) *Let* A *and* B *be* $n \times p$ *and* $n \times q$ *matrices, respectively. Further, let* P_A *and* P_B *be orthogonal projectors onto* $\text{Sp}(A)$ *and* $\text{Sp}(B)$. *Then the following eight statements are equivalent*:

(1) $P_A P_B = P_B P_A$.
(2) $A'B = A' P_B P_A B$.
(3) $(P_A P_B)^2 = P_A P_B$.
(4) $Q_B P_A B = O$.
(5) $Q_A P_B A = O$.
(6) $P_{[A,B]} = P_A + P_B - P_A P_B$.
(7) $\text{rank}(Q_B A) = \text{rank}(A) - \text{rank}(A'B)$.
(8) $A' Q_B Q_A B = O$.

Baksalary (1987, Theorem 1) provides thirty eight other equivalent conditions.

PROPERTY 4. (Baksalary and Styan, 1990) *Let* A, B, P_A, P_B, Q_A *and* Q_B *be matrices as defined in Properties* 1, 2 *and* 3. *Then,*

$$\text{rank}(A'B) = \text{rank}(A' Q_B Q_A B) + \text{rank}(A) + \text{rank}(B) - \text{rank}(A, B). \tag{7}$$

A straightforward proof of the equivalence between (7) and (8) in Property 3 can be given by combining Property 4 and the following rank formula:

$$\text{rank}(A, B) = \text{rank}(A) + \text{rank}(Q_A B) = \text{rank}(B) + \text{rank}(Q_B A). \tag{8}$$

2.3. *Image and anti-image vectors*

Let $X = [x_1, \ldots, x_p]$ be a column-wise centered data matrix of order $n \times p$. Further, let $X_{(j)} = [x_1, \ldots, x_{j-1}, x_{j+1}, \ldots, x_p]$ be the n by $p - 1$ matrix excluding the jth column vector, x_j, from X. Then, using an orthogonal projector we can write the squared multiple correlation, $\text{SMC}(j)$, obtained by regressing x_j onto $X_{(j)}$ as

$$\text{SMC}(j) \equiv R^2_{j/(j)} = \|P_{X_{(j)}} x_j\|^2 / \|x_j\|^2. \tag{9}$$

We also have

$$x_j = P_{X_{(j)}} x_j + Q_{X_{(j)}} x_j \tag{10}$$

for $i = 1, \ldots, p$, where $P_{X_{(j)}} x_j$ and $Q_{X_{(j)}} x_j$ are called image vector of x_j on $\text{Sp}(X_{(j)})$ and anti-image vector of x_j on $\text{Sp}(X_{(j)})^{\perp}$, respectively. Note that the image and anti-image vectors of x_j are orthogonal.

Observe that (9) ensures that SMC(j) can be computed even if $X'_{(j)}X_{(j)}$ is singular. Let

$$X_I = [P_{X_{(1)}}x_1, \ldots, P_{X_{(p)}}x_p], \tag{11}$$

and

$$X_A = [Q_{X_{(1)}}x_1, \ldots, Q_{X_{(p)}}x_p]. \tag{12}$$

Then, it follows from (10) that $X = X_I + X_A$. Assume that X is columnwise standardized. Then, $R = (1/n)X'X$, where R is the correlation matrix.

PROPERTY 5. (Yanai and Mukherjee, 1987, Theorem 1)

$$\frac{1}{n}X'_A X = \begin{bmatrix} d_1 & 0 & \ldots & 0 \\ 0 & d_2 & \ldots & 0 \\ \vdots & \vdots & \ddots & \vdots \\ 0 & 0 & \ldots & d_p \end{bmatrix} \equiv D, \tag{13}$$

where $1 - d_j = R^2_{j/(j)}$ *(the latter having been defined in* (9)*), and*

$$X_A = P_X X_A = XR^- D, \tag{14}$$

$$\frac{1}{n}X'_A X_A = DR^- D, \tag{15}$$

$$\frac{1}{n}X'_I X_I = (R - D)R^-(R - D), \tag{16}$$

and

$$\frac{1}{n}X'_A X_I = D - DR^- D. \tag{17}$$

PROOF. (13) follows immediately by noting that $x'_j Q_{X_{(j)}} x_i = 0$ because $x_i \in \mathrm{Sp}(X_{(j)})$ for $i \neq j$. (14) follows, since $\mathrm{Sp}(X_A) \subset \mathrm{Sp}(X)$ and $X_A = P_X X_A = X((1/n)X'X)^-(1/n)X'X_A = XR^- D$. (15), (16) and (17) are direct consequences of (13) and (14). To prove (15), note that $(1/n)X'_A X_A = (1/n)X'_A XR^- D = DR^- D$, observing that $X = X_A + X_I$. □

The above results are extensions of Kaiser (1976) in that R^+ (the Moore–Penrose inverse of R) is replaced by a weaker g-inverse R^-.

Now, from X one can construct an $n \times j$ matrix of the form $X_{[j]} = [x_1, \ldots, x_j]$. Further, let $P_{Y|X} = P_{Q_X Y}$. Then, Properties 1 and 2 can be extended to the following lemmas.

LEMMA 1. (Rao and Yanai, 1979) *Let* $V_i = \mathrm{Sp}(X_i)$, *and* $X = (X_1, X_2, \ldots, X_q)$ *and let* $V_{i|i-1} = \{x \mid x = Q_{X_{[i-1]}}y,\ y \in V_i\}$. *Then,*

$$P_X = P_{X_1} + P_{X_2|X_1} + \cdots + P_{X_j|X_{[j-1]}} + \cdots + P_{X_p|X_{[p-1]}}, \tag{18}$$

where $P_{X_i|X_{[i-1]}}$ *is the orthogonal projector onto* $V_{i|i-1}$.

LEMMA 2.

$$Q_X = Q_{X_1} Q_{X_2|X_1} \cdots Q_{X_j|X_{[j-1]}} \cdots Q_{X_p|X_{[p-1]}}. \tag{19}$$

2.4. Matrix inequalities

PROPERTY 6. (Beckenbach and Bellman, 1961) *If A and B are positive-semidefinite (PSD) matrices of order p, such that $A - B$ is also PSD, then*

$$\rho_j(A) \geqslant \rho_j(B) \tag{20}$$

for $1 \leqslant j \leqslant p$, where $\rho_j(A)$ is the jth largest eigenvalue of A.

PROPERTY 7 *(Poincaré Separation Theorem). Let A be a symmetric matrix of order p, and let B be a $p \times m$ matrix such that $B'B = I_m$. Then,*

$$\rho_{p-m+i}(A) \leqslant \rho_i(B'AB) \leqslant \rho_i(A) \tag{21}$$

for $i = 1, \ldots, p$.

PROPERTY 8. (Anderson and Gupta, 1963) *If A and B are symmetric matrices of order p,*

$$\rho_i(A + B) \leqslant \rho_j(A) + \rho_k(B) \tag{22}$$

for $j + k \leqslant i + 1$.

2.5. Miscellaneous properties of matrix and its rank

PROPERTY 9. (Yanai, 1990) *Let A and B be matrices of order $p \times m$ and $q \times m$, respectively. Then,* $\operatorname{rank}(AB') = \operatorname{rank}(A)$ *is necessary and sufficient for*

$$B'(AB')^- AB' = B'. \tag{23}$$

Further, let $\operatorname{rank}(AB') = \operatorname{rank}(A) = \operatorname{rank}(B)$. *Then, $B'(AB')^- A$ is the projector onto* $\operatorname{Sp}(B')$ *along* $\operatorname{Ker}(A)$.

PROPERTY 10. (Kristof, 1970) *Let T_j $(j = 1, \ldots, m)$ denote orthogonal matrices of order p, and let D_j $(j = 1, \ldots, m)$ denote diagonal matrices of order p with nonnegative diagonal elements. Then,* $\operatorname{tr}(\prod_{j=1}^m T_j D_j) \leqslant \operatorname{tr}(\prod_{j=1}^m D_j)$.

When $m = 1$, Property 10 reduces to the following.

PROPERTY 11. (ten Berge, 1993) *Let T and X be $n \times p$ matrices. Let $T'T = I_p$, and let the singular value decomposition of X be given by $X = V \Delta U'$, where $V'V = U'U = UU' = I_p$. Then,* $\operatorname{tr}(T'X) \leqslant \operatorname{tr}(\Delta)$, *and the equality is attained when*

$$T = VU' = X(X'X)^{-1/2}. \tag{24}$$

PROOF. Let $H = U'T'V$ denote a square matrix of order p. Then, $H'H = V'TUU'T'V = V'P_TV \leqslant V'V = I_p$, where $V'P_TV \leqslant V'V$ indicates $V'V - V'P_TV$ is PSD. Since $\sum_{k=1}^{p} h_{jk}^2 \leqslant 1$ implies $h_{jj}^2 \leqslant 1$, where $H = (h_{jk})$, we obtain $0 \leqslant h_{jj} \leqslant 1$, establishing

$$\operatorname{tr}(T'X) = \operatorname{tr}(T'V\Delta U') = \operatorname{tr}\big((U'T'V)\Delta\big) = \operatorname{tr}(H\Delta)$$
$$= \sum_{j=1}^{p} h_{jj}\delta_j \leqslant \sum_{j=1}^{p} \delta_j = \operatorname{tr}(\Delta),$$

where δ_j is the jth diagonal element of Δ. The equality holds when $H = I_p$ yielding $U'T'V = I_p$, which implies $T = VU'$. □

3. Applications of matrix methods to factor analysis

3.1. Lower bounds for communalities

Let X denote an $n \times p$ columnwise standardized data matrix, and consider the following traditional factor analysis model:

$$X = F\Lambda' + E\Psi^{1/2}, \tag{25}$$

where $F = [f_1, \ldots, f_m]$ is the $n \times m$ common factor score matrix, Λ is the $p \times m$ factor loading matrix, $E = [e_1, \ldots, e_p]$ is the $n \times p$ unique factor score matrix, and Ψ is the positive-definite diagonal matrix of order p of unique variances. We typically assume $(1/n)E'E = I_p$, and $F'E = O$. In an orthogonal solution, we additionally assume $(1/n)F'F = I_m$, so that

$$R = \Lambda\Lambda' + \Psi, \tag{26}$$

where $R = (1/n)X'X$ is a correlation matrix, and Ψ is a diagonal matrix whose jth diagonal element, ψ_j, is the unique variance of the jth variable, x_j. Let h_j^2 denote the communality of this variable satisfying $h_j^2 + \psi_j = 1$ for $j = 1, \ldots, p$.

We first give a property that allows to represent the communality in terms of orthogonal projector onto $\operatorname{Sp}(F)$, the space spanned by column vectors of F.

PROPERTY 12. *Let h_j^2 denote the communality of the jth observed variable, x_j, and let P_F denote the orthogonal projector onto* $\operatorname{Sp}(F)$. *Then,*

$$h_j^2 = \|P_F x_j\|^2 / \|x_j\|^2. \tag{27}$$

The relationships among x_j, $\operatorname{Sp}(F)$, $P_F x_j$, and $h_j = \|P_F x_j\|$ are depicted in Figure 1.

Observe that $0 \leqslant h_j^2 \leqslant 1$, where the first equality holds if $x_j \in \operatorname{Sp}(F)^\perp$ and the second equality holds if $x_j \in \operatorname{Sp}(F)$. The unique variance, ψ_j, is obtained by

$$\psi_j = \|Q_F x_j\|^2 / \|x_j\|^2. \tag{28}$$

We next give a well-known property on the relationship between communality and SMC.

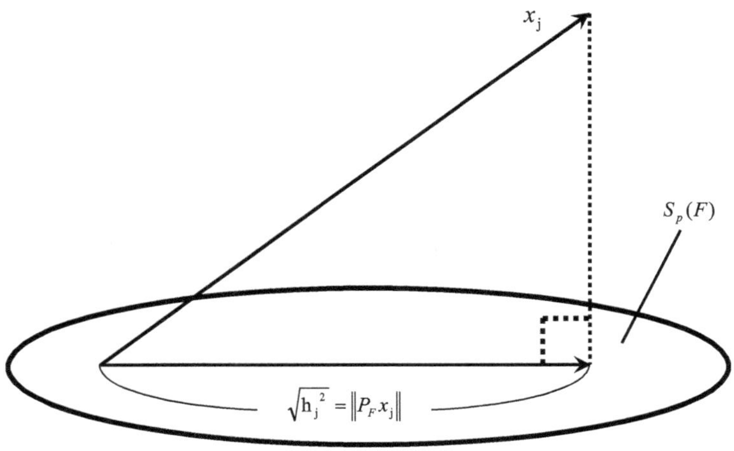

Fig. 1. Representation of communality h_j^2 of a vector x_j in terms of orthogonal projection assuming $\|x_j\| = 1$.

PROPERTY 13. (Roff, 1936) *For* $1 \leqslant j \leqslant p$, SMC($j$) *is a lower bound to communality* h_j^2, *i.e.*,

$$\text{SMC}(j) \leqslant h_j^2. \tag{29}$$

Using Property 1, we are in a position to give a straightforward proof of Property 13 and look into the conditions under which the equality in (29) holds.

THEOREM 1. *Let* $X_{(j)} = [x_1, \ldots, x_{j-1}, x_{j+1}, \ldots, x_p]$ *be a columnwise standardized* n *by* $p-1$ *matrix obtained by eliminating* x_j *from* X. *Then,* (29) *holds, and the equality in* (29) *holds in the following two cases*:

$$\text{SMC}(j) = 1 \quad (\textit{Case } 1), \tag{30}$$

and

$$\text{SMC}(j) \neq 1, \quad \textit{and} \quad r^{ji}\psi_i = 0 \quad \textit{for any } i \neq j \quad (\textit{Case } 2), \tag{31}$$

where r^{ji} *is the* (j, i)*th element of* R^-.

PROOF. By Property 1, we have

$$P_F + P_{Q_F X_{(j)}} = P_{X_{(j)}} + P_{Q_{X_{(j)}} F}. \tag{32}$$

By pre- and postmultiplying the above equation by x_j' and x_j, respectively, we obtain from (26) and $(1/n)F'F = I_m$ that

$$\begin{aligned} \frac{1}{n} x_j' Q_F X_{(j)} &= \frac{1}{n} x_j' X_{(j)} - \left(\frac{1}{n} x_j' F\right)\left(\frac{1}{n} F' X_{(j)}\right) \\ &= r_{j/(j)}' - \lambda_j' (\Lambda_{(j)})' = 0, \end{aligned} \tag{33}$$

where $\Lambda_{(j)}$ is the factor loading matrix with $p-1$ variables excluding x_j, λ_j is the vector of factor loadings of the jth variable, and $r_{j/(j)}$ is the vector of correlation coefficients between x_j and the remaining $p-1$ variables. It follows from (27) and (33) that

$$h_j^2 = \mathrm{SMC}(j) + x_j' P_{Q_{X_{(j)}}F} x_j, \tag{34}$$

which implies (29), since $x_j' P_{Q_{X_{(j)}}F} x_j$ is nonnegative.

We now look for conditions under which the equality holds in (29). If (33) is true, then $x_j' P_{Q_{X_{(j)}}F} x_j = 0$, leading to $x_j' Q_{X_{(j)}} F = 0'$, which implies that anti-image vector $Q_{X_{(j)}} x_j$ ($j = 1, \dots, p$) is orthogonal to $\mathrm{Sp}(F)$. Noting that $\mathrm{Sp}(Q_{X_{(j)}}) \subset \mathrm{Sp}(X)$, we obtain

$$(P_X Q_{X_{(j)}} x_j)' F = x_j' Q_{X_{(j)}} X R^- \Lambda = O. \tag{35}$$

By postmultiplying (35) by Λ', and using (26), we obtain from (13) that

$$\begin{aligned}
&(Q_{X_{(j)}} x_j)' X\big((1/n)X'X\big)^- \{(1/n)X'X - \Psi\} \\
&\quad = (Q_{X_{(j)}} x_j)' X\big(I_p - ((1/n)X'X)^- \Psi\big) \\
&\quad = \big(0, \dots, 0, 1 - \mathrm{SMC}(j), 0, \dots, 0\big)(I_p - R^- \Psi) = 0',
\end{aligned}$$

which implies

$$\begin{aligned}
&\big(1 - \mathrm{SMC}(j)\big) r^{ji} \psi_i = 0 \quad (i \neq j), \quad \text{and} \\
&\big(1 - \mathrm{SMC}(j)\big)\big(1 - r^{jj} \psi_j\big) = 0.
\end{aligned} \tag{36}$$

This completes the proof of Theorem 1. □

We provide an example of Theorem 1, assuming that $r^{ji} \neq 0$, which implies $\psi_i = 0$ $(i \neq j)$, and $h_i^2 = 1$ $(i \neq j)$. We will discuss the case in which $r^{ji} = 0$ $(i \neq j)$ later.

EXAMPLE 1. Suppose that the correlation matrix among four variables, x_1, x_2, x_3, and x_4, is given by

$$R = \begin{bmatrix} 1 & 0 & a & a \\ 0 & 1 & a & -a \\ a & a & 1 & 0 \\ a & -a & 0 & 1 \end{bmatrix},$$

where $2a^2 \leqslant 1$. The SMC of x_1 can be computed as

$$1 - \det(R)/\det \begin{bmatrix} 1 & a & -a \\ a & 1 & 0 \\ -a & 0 & 1 \end{bmatrix} = 1 - \frac{(1-2a^2)^2}{1-2a^2} = 2a^2,$$

provided that $2a^2 \neq 1$. Similarly, it can be shown that the SMC's of all the four variables are equal to $2a^2$.

The following factor loading matrix Λ and the unique variance matrix Ψ,

$$\Lambda = \begin{bmatrix} 1 & 0 \\ 0 & 1 \\ a & a \\ a & -a \end{bmatrix}, \quad \text{and} \quad \Psi = \begin{bmatrix} 0 & 0 & 0 & 0 \\ 0 & 0 & 0 & 0 \\ 0 & 0 & 1-2a^2 & 0 \\ 0 & 0 & 0 & 1-2a^2 \end{bmatrix},$$

on the other hand, satisfy the factor analysis model, (26). The communalities of the four variables can be computed as $(1, 1, 2a^2, 2a^2)$. Thus, the SMC's are equal to the communalities for variables 3 and 4, while the SMC's are smaller than (or equal to) the communalities for variables 1 and 2. Since the communalities of variables 1 and 2 are unity, factors f_1 and f_2 can be rotated to coincide with them. Since the SMC's are equal to the squared length of the projection of x_3 and x_4 onto the factor space which are now spanned by x_1 and x_2, it can be easily seen that the SMC's of variables x_3 and x_4 coincide with their communalities. In terms of the factor analysis model, we can write

$$\psi_1 = \psi_2 = 0, \quad \text{and} \quad \psi_3 = \psi_4 = 1 - 2a^2. \tag{37}$$

This result covers Case 2 in (31). It also covers Theorem 3 of Roff (1936, p. 5), which states that SMC(j) is equal to the communality of variable x_j, if variables contain m $(m < p)$ statistically independent variables each with unit communality (where p is the number of variables and m is the number of common factors).

In (31) it is important to consider the case in which $\psi_i = 0$ $(i \neq j)$ does not hold. In such a case, $r^{ji} = 0$ $(i \neq j)$ should be true. Since r^{ji} is the (j, i)th element of R^-, it follows from (8) of Property 3, that

$$(Q_{X_{(j)}} x_j)'(Q_{X_{(i)}} x_i) = 0 \quad (i \neq j), \tag{38}$$

provided that $R_{j/(j)} \neq 1$, which implies that the diagonal matrix D as defined by (13) is nonsingular. (38) implies that the anti-image of variable x_j is uncorrelated with that of variable x_i. It is interesting to note that (38) is closely related to Theorem 4 of Guttman (1953), which states that if a common-factor space of dimensionality m is determinate for an infinitely large universe of content, then there is no other determinate common factor space. In this case, the communalities are uniquely determined and are equal to the corresponding total norms, and in addition the common-factor scores are the total image scores, and the unique factor scores are the total ant-images. If (38) holds for any combination of i and j, then anti-image variable $Q_{X_{(j)}} x_j$ behaves like the unique factor, e_j, corresponding to the jth variable, x_j.

NOTE 1. If Sp(F) is a subspace of Sp$(X_{(j)})$, then $P_{X_{(j)}} F = F$, leading to $X'_{(j)} Q_{X_{(j)}} F = O$. In case of orthogonal factor analysis, F is columnwise orthogonal. Then, if m vectors in $X_{(j)}$ are orthogonal, Sp(F) can be embedded in a subspace of Sp$(X_{(j)})$. Thus, the equality in (29) holds.

NOTE 2. Let $X_1 = [x_1, \ldots, x_k]$ and $X_2 = [x_{k+1}, \ldots, x_p]$, which satisfies the following factor analysis model:

$$[X_1, X_2] = F[\Lambda'_1, \Lambda'_2] + E \begin{bmatrix} \Psi_1^{1/2} & O \\ O & \Psi_2^{1/2} \end{bmatrix},$$

where F is the $n \times m$ matrix of common factor scores, Λ_1 and Λ_2 are $k \times m$ and $(p-k)\times m$ factor loading matrices corresponding to X_1 and X_2, respectively, and Ψ_1 and Ψ_2 are diagonal matrices of uniqueness variances of order k and $p-k$, corresponding to X_1 and X_2, respectively. Then,

$$\Lambda_j \Lambda'_j \geqslant X'_j P_{X_i} X_j \quad (j, i = 1, 2,\ j \neq i), \tag{39}$$

where $\Lambda_j \Lambda'_j \geqslant X'_j P_{X_i} X_j$ means $\Lambda_j \Lambda'_j - X'_j P_{X_i} X_j$ is PSD.

From Property 1, we have

$$P_{[F, X_i]} = P_F + P_{Q_F X_i} = P_{X_i} + P_{Q_{X_i} F}. \tag{40}$$

Premultiplying (40) by P_{X_j} and noting that $(1/n)X'_j Q_F X_i = (1/n)X'_j X_i - \Lambda_j \Lambda'_i = O$, we obtain $X'_j P_F X_j = X'_j P_{X_i} X_j + X'_j P_{Q_{X_i} F} X_j$, establishing (39). The term on the left side of (39) represents generalized forms of communalities for variables X_j, and the term on the right may be called generalized SMC's.

3.2. *Stronger upper and lower bounds for communalities*

In this section, we consider the random model of factor analysis as opposed to the traditional model of factor analysis introduced earlier in (25). The random model of factor analysis is written as

$$x = \Lambda f + e \tag{41}$$

with $\mathrm{E}(f) = 0$, $\mathrm{E}(e) = 0$, $\mathrm{Cov}(f, e) = \mathrm{E}(fe') = O$, $\mathrm{V}(f) = \Phi$, and $\mathrm{V}(e) = \Psi$, where E, V, and Cov are expectation, variance, and covariance operators, respectively. The corresponding representation of the factor analysis model in terms of a correlation matrix can be expressed as

$$\Sigma = \Lambda \Phi \Lambda' + \Psi, \tag{42}$$

where Σ is the population correlation matrix. We have

PROPERTY 14. (Yanai and Ichikawa, 1990)

(a) *Let $h^2_{(j)}$ denote the jth largest communality among the p variables. Then, for $1 \leqslant j \leqslant p$,*

$$h^2_{(j)} \geqslant 1 - \rho_{p+1-j}(\Sigma), \tag{43}$$

where $\rho_{p+1-j}(\Sigma)$ is the $(p+1-j)$th largest eigenvalue of Σ.

(b) *For any positive definite correlation matrix Σ with distinct eigenvalues, we have*

$$1 - \rho_p(\Sigma) \geqslant \mathrm{SMC}(j) \quad \textit{for } 1 \geqslant j \geqslant p. \tag{44}$$

(c) *For any $1 \geqslant j \geqslant p$,*

$$h^2_{(j)} \leqslant 1 - \rho_{(p+m+1-j)}(\Sigma). \tag{45}$$

These results can be proved by the matrix inequalities given by (20), (21), *and* (22).

Table 1
Communalities, SMC's, and upper and lower bounds for communalities in the numerical example (Yanai and Ichikawa, 1990)

Variable	Communality	NLB	UB	SMC
1	0.770	0.735	–	0.654
2	0.680	0.645	–	0.573
3	0.650	0.568	–	0.518
4	0.650	0.510	0.735	0.498
5	0.560	–	0.645	0.453
6	0.450	–	0.568	0.384

EXAMPLE 2. Suppose we have the following factor loading matrix, Λ, and the unique variance matrix, Ψ:

$$\Lambda = \begin{bmatrix} 0.6 & 0.5 & 0.4 \\ 0.6 & 0.4 & 0.4 \\ 0.2 & 0.6 & 0.5 \\ 0.2 & 0.5 & 0.6 \\ 0.4 & 0.6 & 0.2 \\ 0.5 & 0.4 & 0.2 \end{bmatrix} \quad \text{and} \quad \Psi = \operatorname{diag}\begin{pmatrix} 0.23 \\ 0.32 \\ 0.35 \\ 0.35 \\ 0.44 \\ 0.55 \end{pmatrix},$$

which yield

$$\Sigma = \Lambda\Lambda' + \Psi = \begin{bmatrix} 1 & 0.72 & 0.62 & 0.61 & 0.62 & 0.58 \\ 0.72 & 1 & 0.56 & 0.56 & 0.56 & 0.54 \\ 0.62 & 0.56 & 1 & 0.64 & 0.54 & 0.44 \\ 0.61 & 0.56 & 0.64 & 1 & 0.50 & 0.42 \\ 0.62 & 0.56 & 0.54 & 0.50 & 1 & 0.48 \\ 0.58 & 0.54 & 0.44 & 0.42 & 0.48 & 1 \end{bmatrix}.$$

With some calculations, the eigenvalues of Σ are found to be $\rho_1 = 3.810$, $\rho_2 = 0.648$, $\rho_3 = 0.490$, $\rho_4 = 0.432$, $\rho_5 = 0.355$, and $\rho_6 = 0.265$, from which we obtain the new lower bounds (NLB), and the upper bounds (UB) summarized in Table 1. For comparison we also give SMC's in the last column of the table. The NLB's for variables 1, 2, 3, and 4 improve upon SMC's used as lower bounds of communalities. It seems that there are generally more than one variable in which the NLB is better than the SMC.

EXAMPLE 3. Let Σ_k ($k = p - 1$ and p) be a correlation matrix of order k with all the correlation coefficients being equal to a ($0 < a < 1$). The SMC's of all p variables are computed by

$$\begin{aligned} \mathrm{SMC}(j) &= 1 - \det(\Sigma_p)/\det(\Sigma_{p-1}) \\ &= 1 - \{(1 + (p-1)a)(1-a)^p\}/\{(1 + (p-2)a)\}(1-a)^{p-1} \\ &= a^2/(a + (1-a)/(p-1)), \end{aligned} \tag{46}$$

and the eigenvalues of Σ_p are:

$$\rho_1(\Sigma_p) = 1 + (p-1)a, \quad \text{and} \quad \rho_j(\Sigma_p) = 1 - a \quad \text{for } 2 \leqslant j \leqslant p.$$

Note that the last $p-1$ eigenvalues are equal to a. Furthermore, from (a) of Property 14, a gives a stronger lower bound of communality for each of the $p-1$ variables, since

$$a - a^2/\big(a + (1-a)/(p-1)\big) = a(1-a)/\big((1-a) + a(p-1)\big) > 0.$$

It is interesting to note that as p approaches infinity, SMC(j) computed by (46) approaches a which coincides with the communalities of the p variables. This is consistent with the suggestion first made by Roff (1936) and later proved by Guttman (1940).

3.3. Variable selection in factor analysis

It is recommended that some rotation methods be applied to the factor loading matrix derived by some initial factor extraction method to construct some psychological scales such as personality, vocational interest, and so on. In some cases, a number of items load highly on some factors, while smaller numbers of items load highly on other factors. In such cases, it is important to check whether a particular variable is a suitable indicator of a factor extracted. We present a stepwise variable selection method in factor analysis, following Yanai (1980).

Let $X = [x_1, \ldots, x_p]$ denote a standardized data matrix, and assume that the factor score matrix, $F = [f_1, \ldots, f_m]$, consists of m orthogonal factors. Let R^2_{X/f_j} denote the squared multiple correlation obtained by regressing f_j onto X. Then, from Lemma 1 we obtain

$$\begin{aligned} s &= R^2_{X/f_1} + \cdots + R^2_{X/f_m} \\ &= \sum_{j=1}^{m} (f_j' P_X f_j)/(f_j' f_j) = \mathrm{tr}\left(P_X \sum_{j=1}^{m} P_{f_j}\right) = \mathrm{tr}(P_X P_F), \end{aligned} \tag{47}$$

using the relationship, $P_F = P_{f_1} + \cdots + P_{f_m}$, which follows from Lemma 1 and the orthogonality of F. Note that s defined in (47) is the sum of the squared canonical correlation coefficients between F and X representing the relationship between the extracted factors and observed data. We propose a forward inclusion method for stepwise selection of variables in factor analysis by employing the following decomposition of $\mathrm{tr}(P_X P_F)$:

$$\begin{aligned} s &= \mathrm{tr}(P_X P_F) = \mathrm{tr}(P_{x_1} P_F) + \mathrm{tr}(P_{x_2|x_1} P_F) + \cdots + \mathrm{tr}(P_{x_p|X_{[p-1]}} P_F) \\ &= s_1 + s_2 + \cdots + s_p. \end{aligned} \tag{48}$$

Then, the proposed procedure of stepwise selection can be described as:

Step 1: Select a variable x_j with the largest communality h_j^2, since $\mathrm{tr}(P_{x_j} P_F) = \|P_F x_j\|^2/\|x_j\|^2 = h_j^2$ follows from (27).

Step 2: Suppose that variable x_j is selected. Then, select variable x_k ($k \neq j$) with the largest value of

Table 2
Results of the stepwise selection method in principal factor analysis

Step	Scale	Factor 1	Factor 2	Factor 3	Factor 4	Communality	s_j	$s_{(j)}$
1	D	**0.684**	−0.298	0.254	−0.326	0.728	0.728	0.728
2	A	0.183	**0.821**	−0.040	0.098	0.715	0.698	1.426
3	Co	**0.751**	0.186	−0.189	0.199	0.674	0.451	1.877
4	C	0.466	−0.102	**0.534**	−0.018	0.512	0.265	2.412
5	R	0.032	**0.457**	0.209	0.438	0.445	0.190	2.333
6	G	−0.082	**0.677**	−0.030	−0.117	0.479	0.170	2.503
7	S	−0.135	**0.795**	0.012	0.030	0.651	0.099	2.602
8	N	**0.808**	−0.157	0.045	0.059	0.684	0.088	2.690
9	Ag	−0.056	0.404	**0.407**	0.026	0.332	0.057	2.747
10	O	**0.837**	0.073	−0.020	0.011	0.707	0.053	2.800
11	T	0.157	0.092	−0.171	**0.471**	0.284	0.034	2.834
12	I	0.383	**−0.535**	0.185	0.000	0.461	0.020	2.854

$$
\begin{aligned}
s_k &= \operatorname{tr}(P_{x_k|x_j} P_F) = \operatorname{tr}(P_{Q_{x_j} x_k} P_F) \\
&= \left(h_j^2 r_{jk}^2 + h_k^2 - 2r_{jk}\sqrt{h_j^2 h_k^2}\right) / \left(1 - r_{jk}^2\right),
\end{aligned} \tag{49}
$$

where r_{jk} is the correlation coefficient between x_j and x_k.

Step 3: In earlier $j-1$ steps, suppose, for simplicity, that $j-1$ variables $X_{[j-1]} = [x_1, \ldots, x_{j-1}]$ are selected. (This is just for notational convenience.) Then, select a variable x_k ($k \geqslant j$) with the largest value of

$$
s_k = \operatorname{tr}(P_{x_k|X_{[j-1]}} P_F) = \|b_k\|^2 / \left(1 - R^2_{X_{[j-1]}/x_k}\right), \tag{50}
$$

where $b_k = \lambda_k - \Lambda'_{[j-1]} R^-_{[j-1],[j-1]} r_{(j-1)/k}$.

EXAMPLE 4. We first performed principal factor analysis and extracted four common factors from the data with twelve scales in Yatabe–Guilford Personality Inventory. (This is the most popular personality inventory in Japan.) We show the result of stepwise selection of the variables in Table 2, in which scales are arranged in descending order of s_j for $j = 1, \ldots, p$. (The list of the twelve scales is given in Table 3.) We then computed the communalities for the twelve scales. It turned out that the Depression scale (Scale D, for short) had the largest communality of 0.728 among the twelve scales. In the second step, Scale A with the s_k value (defined in (49)) of 0.698 was selected. Interestingly, Scale D had the highest factor loading on the first factor, while Scale A had the highest factor loading on the second factor. Continuing this way, we came to the final step where Scale I was selected with the s_k value (defined in (50)) of only 0.020. In reference to the values of $s_{(j)} = s_1 + \cdots + s_j$ given in the last column of Table 2, we may say that only five or six scales are sufficient for explaining the information contained in the four common factors. As an alternative method of stepwise selection in factor analysis, Kano and Harada (2000) developed SEFA (Stepwise variable selection in Exploratory Factor Analysis) by employing several goodness-of-fit measures used in structural equation modeling.

Table 3
List of twelve scales

No.	Symbol	Scale
1)	D	Depression
2)	A	Ascendance
3)	Co	Lack of cooperativeness
4)	C	Cyclic tendency
5)	R	Rhathymia
6)	G	General activity
7)	S	Social extraversion
8)	N	Nervousness
9)	Ag	Lack of agreeableness
10)	O	Lack of objectivity
11)	T	Thinking extraversion
12)	I	Inferiority feelings

3.4. Representation of SMC when the correlation matrix may be singular

Let R denote a correlation matrix with three variables, x_1, x_2, and x_3, with correlation coefficients $r_{x_1x_2} = r_{x_1x_2} = a$ $(a \neq \pm 1)$ and $r_{x_2x_3} = 1$. We write R, and a g-inverse of R, denoted by $P = R^-$, as

$$R = \begin{bmatrix} 1 & a & a \\ a & 1 & 1 \\ a & 1 & 1 \end{bmatrix}, \quad \text{and}$$

$$P = \big(1/(1-a^2)\big)\begin{bmatrix} 1 & -wa & -(1-w)a \\ -xa & t_1 & t_2 \\ -(1-x)a & t_3 & 1-t_1-t_2-t_3 \end{bmatrix},$$

where $-1 \leqslant a \leqslant 1$, and t_1, t_2, t_3, w, and x are arbitrary. Let $I_3 - RP = [g_1, g_2, g_3]$. Then with some computations, we obtain $g_1 = 0$, and $g_j \neq 0$ $(j = 2, 3)$. According to Theorem 1 of Kahtri (1976), $\text{SMC}(1) = a^2$, and $\text{SMC}(j) = 1$ for $j = 2, 3$. Thus, SMC's can be computed for all variables, even if R is singular.

Furthermore, let $W = [X, Y]$ be a matrix of order $n \times (p+q)$, where $X = [x_1, \dots, x_p]$ and $Y = [y_1, \dots, y_q]$ are matrices of orders $n \times p$ and $n \times q$, respectively. Let $R = R_{WW}$ be the correlation matrix, and let $P = R^-$ denote a g-inverse of R. Let R and P be partitioned analogously:

$$R = \begin{bmatrix} R_{XX} & R_{XY} \\ R_{YX} & R_{YY} \end{bmatrix}, \quad \text{and} \quad P = \begin{bmatrix} P^{XX} & P^{XY} \\ P^{YX} & P^{YY} \end{bmatrix}. \tag{51}$$

Then, with some computations, we obtain

$$R_{XX.Y} P^{XX} R_{XX.Y} = R_{XX.Y}, \quad \text{where}$$
$$R_{XX.Y} = R_{XX} - R_{XY} R_{YY}^- R_{YX}. \tag{52}$$

Let

$$\mathrm{SMC}(X|Y) = \begin{bmatrix} R^2_{x_1|Y} & R_{x_1x_2|Y} & \dots & R_{x_1x_p|Y} \\ R_{x_2x_1|Y} & R^2_{x_2|Y} & \dots & R_{x_2x_p|Y} \\ \vdots & \vdots & \ddots & \vdots \\ R_{x_px_1|Y} & R_{x_px_2|Y} & \dots & R^2_{x_p|Y} \end{bmatrix}$$

a square matrix of order p, where

$$R^2_{x_j|Y} = \|P_Y x_j\|^2, \quad \text{and} \quad R_{x_ix_j|Y} = x_i' P_Y x_j.$$

Then, we have the following lemma.

LEMMA 3. *Let R and P be matrices defined in* (51). *Further, let*

$$H = RP = \begin{bmatrix} H_{XX} & H_{XY} \\ H_{YX} & H_{YY} \end{bmatrix}, \quad \textit{and} \quad B = \begin{bmatrix} H_{XX} - I_p \\ H_{XY} \end{bmatrix}$$

be matrices of orders $p+q$ and $(p+q) \times p$, respectively. If $B = O$, then

$$\mathrm{SMC}(X|Y) = R_{XY} R_{YY}^- R_{YX} = R_{XX} - \left(P^{XX}\right)^{-1}, \tag{53}$$

and if $\mathrm{rank}(B) = p$, *then* $\mathrm{SMC}(X|Y) = R_{XX}$.

PROOF. It follows from Lemma 4 of Kahtri (1976), that $\mathrm{rank}(B) = p - \mathrm{rank}(X' Q_Y X)$. If $B = O$, then $\mathrm{rank}(X' Q_Y X) = p$, indicating $R_{XY} R_{YY}^- R_{YX}$ is nonsingular. Then, $R_{XX.Y} P^{XX} = I_p$ follows from (52), establishing (53). If $\mathrm{rank}(B) = p$, then $X' Q_Y X = O$, which implies $\mathrm{SMC}(X|Y) = X' P_Y X = X' X$. □

The term $\mathrm{SMC}(X|Y)$ defined in Lemma 3 coincides with the right-hand side of (39), which we call generalized SMC, since it reduces to the communality of a variable when X consists of a single vector. The above result represents an extension of Kahtri (1976).

3.5. Interpretation of communalities from a regularization perspective

A major difference between principal factor analysis (PFA) and principal component analysis (PCA) is that the former obtains the eigen-decomposition of $R - \Psi$ (assuming that Ψ is tentatively known), whereas the latter obtains that of R. The analysis of $R - \Psi$ may be justified from a regularization perspective. In the ridge type of regularization (Hoerl and Kennard, 1970) estimates of regression coefficients in linear regression models are obtained by

$$\tilde{b} = (X'X + \kappa I_p)^{-1} X' y, \tag{54}$$

where X is an $n \times p$ matrix of predictor variables, y is an n-component vector of criterion variable, and κ is a ridge parameter, which typically takes a small positive value. The prediction vector is obtained by

$$X\tilde{b} = P(\kappa) y, \tag{55}$$

where $P(\kappa) = X(X'X + \kappa I_p)^{-1}X'$ is called ridge operator. The ridge estimates of regression coefficients are usually biased, but are associated with smaller MSE (mean square errors; Hoerl and Kennard, 1970).

Takane and Yanai (2005) recently introduced the following metric,

$$M(\kappa) = I_n + \kappa(XX')^+, \tag{56}$$

in an effort to generalize the ridge type of regularization to other techniques of multivariate analysis. Using $M(\kappa)$, they could rewrite $P(\kappa)$ as $P(\kappa) = X(X'M(\kappa)X)^- X'$, where $X'M(\kappa)X = X'X + \kappa P_{X'}$, and $P_{X'}$ is the orthogonal projector onto $\mathrm{Sp}(X')$, which reduces to I_p when X is columnwise nonsingular. Note that $P(\kappa)$ is invariant over the choice of g-inverse of $X'M(\kappa)X$, and that $(X'X + \kappa I_p)^{-1} \in \{(X'M(\kappa)X)^-\}$. Takane and Yanai (2005) further extended the metric matrix to:

$$M^{(L)}(\kappa) = I_n + \kappa(XL^- X')^+, \tag{57}$$

where L is PSD with $\mathrm{Sp}(L) = \mathrm{Sp}(X')$. With this generalized metric matrix, we obtain $X'M^{(L)}(\kappa)X = X'X + \kappa L$.

The ridge estimation generally has the effect of shrinking the estimates toward zero by adding $\kappa P_{X'}$ or κL to $X'X$ on the predictor side. Presumably, a similar shrinkage effect can be obtained by subtracting the same from the criterion side. Let

$$M^{(\Psi)}(-1) = I_n - \left(X\Psi^{-1}X'\right)^+. \tag{58}$$

Then,

$$(1/n)X'M^{(\Psi)}(-1)X = R - \Psi, \tag{59}$$

which, as noted earlier, is the matrix whose eigen-decomposition is taken in PFA. (59) may be seen from $(X\Psi^{-1}X')^+ = X\Psi^{-1/2}((\Psi^{-1/2}X'X\Psi^{-1/2})^+)^2\Psi^{-1/2}X' = X(X'X\Psi^{-1}X'X)^+X'$. This indicates that in PFA we are sort of obtaining shrinkage estimates (of factor loadings) relative to PCA loadings. This leads to the idea that the estimate of Ψ is chosen in such a way that it reproduces an R that cross-validates best.

3.6. *Methods of estimating factor scores*

It is useful to estimate factor scores of individual subjects. A number of methods of estimating factor scores have been proposed so far.

The first estimator starts from the parametric model of factor analysis, $x = \Lambda f + e$, where $\mathrm{E}(e) = 0$, and $\mathrm{V}(e) = \Psi$ is a nonsingular diagonal matrix of unique variances. It is assumed that the factor loading matrix, Λ, and the unique variance matrix, Ψ, are known, and only e is a vector of random variables analogous to the disturbance terms in linear regression models. The generalized least squares estimate of f, which we denote by f_1, minimizing

$$(x - \Lambda f)'\Psi^{-1}(x - \Lambda f) \tag{60}$$

is given by

$$f_1 = \left(\Lambda'\Psi^{-1}\Lambda\right)^{-1}\Lambda'\Psi^{-1}x. \tag{61}$$

It can be easily verified that this estimator is unbiased and its covariance matrix is given by $\mathrm{V}(f_1) = (\Lambda'\Psi^{-1}\Lambda)^{-1}$. This is called Bartlett estimator (Bartlett, 1937).

NOTE 3. If we neglect Ψ, the minimization of $(x - \Lambda f)'(x - \Lambda f)$ with respect to f yields

$$f_2 = (\Lambda'\Lambda)^{-1}\Lambda' x, \tag{62}$$

which was first derived by Horst (1965). Note that f_2 defined above is also unbiased.

We now consider an estimation of f when Ψ is possibly singular.

LEMMA 4. (Rao and Yanai, 1979) *Under the Gauss–Markov model,* $(y, X\beta, \alpha^2 G)$ *where* G *may be singular, the BLUE (the best linear unbiased estimator) of* $X\beta$ *can be expressed as*

$$Xb = Py, \tag{63}$$

where P *satisfies both* (i) $PX = X$, *and* (ii) $PGZ = O$, *where* $Z = Q_X$ *is the orthogonal projector onto* $\mathrm{Sp}(X)^{\perp}$. *If* $\mathrm{Sp}(X)$ *and* $\mathrm{Sp}(GZ)$ *cover the entire space of* E^n, P *is the projector onto* $\mathrm{Sp}(X)$ *along* $\mathrm{Sp}(GZ)$, *and it can be expressed in the following three forms*:

(1) $X(X'Q_{GZ}X)^{-}X'Q_{GZ}$.
(2) $I_n - GZ(ZGZ)^{-}Z$.
(3) $X(X'T^{-1}X)^{-}X'T^{-1}$, *where* $T = XUX' + G$ *and* $\mathrm{rank}(T) = \mathrm{rank}(X, G)$.

We attempt to minimize (60) when Ψ is singular. To deal with this problem, Bentler and Yuan (1997) minimized $(x - \Lambda f)'\Psi^{+}(x - \Lambda f)$. Our solution, on the other hand, is based on Lemma 4.

LEMMA 5. *If* $E^n = \mathrm{Sp}(\Lambda) + \mathrm{Sp}(\Psi)$, *and* $W = Q_\Lambda$ *is the orthogonal projector onto* $\mathrm{Sp}(\Lambda)^{\perp}$, *the BLUE of* f *can be expressed in the following three equivalent forms*:

(1) $\Lambda(\Lambda' Q_{\Psi W}\Lambda)^{-}\Lambda' Q_{\Psi W} x$.
(2) $(I_n - \Psi W(W\Psi W)^{-}W)x$.
(3) $\Lambda(\Lambda' T^{-1}\Lambda)^{-}\Lambda' T^{-1} x$, *where* $T = \Lambda U \Lambda' + \Psi$ *and* $\mathrm{rank}(T) = \mathrm{rank}(\Lambda, \Psi)$.

Note that in the parametric model of factor analysis, a factor score vector and a raw data vector can be defined for each of n individual subjects. Let $f_{(j)}$ and $x_{(j)}$ denote these vectors for the jth subject. These vector may be represented collectively by matrices $F' = [f_{(1)}, \ldots, f_{(n)}]$ and $X' = [x_{(1)}, \ldots, x_{(n)}]$. Anderson and Rubin (1956) obtained an estimate of F which minimizes

$$(1/n)\sum_{j=1}^{n}(x_{(j)} - \Lambda f_{(j)})'\Psi^{-1}(x_{(j)} - \Lambda f_{(j)}) \tag{64}$$

subject to $(1/n)F'F = (1/n)\sum_{j=1}^n f_{(j)}f_{(j)}' = \Phi$, where Φ is the matrix of correlations among m factors and thus is positive-definite (PD). Observe that $(1/n)\sum_{j=1}^n \mathrm{tr}(f_{(j)}'\Lambda' \times \Psi^{-1}\Lambda f_{(j)}) = \mathrm{tr}(\Lambda'\Psi^{-1}\Lambda((1/n)\sum_{j=1}^n f_{(j)}f_{(j)}')) = \mathrm{tr}(\Lambda'\Psi^{-1}\Lambda\Phi)$. Thus, the minimization of (64) is equivalent to maximizing $\sum_{j=1}^n f_{(j)}'\Lambda'\Psi^{-1}x_{(j)} = \mathrm{tr}(F\Lambda'\Psi^{-1}X') = \mathrm{tr}(F\Phi^{-1/2}(X\Psi^{-1}\Lambda\Phi^{1/2})')$ subject to $(1/n)F'F = \Phi$. Note that $(1/n)F'F = \Phi$ is equivalent to $\Phi^{-1/2}(1/n)F'F\Phi^{-1/2} = I_m$. We obtain from Property 11 that

$$F = X\Psi^{-1}\Lambda\Phi^{1/2}\big(\Phi^{1/2}\Lambda'\Psi^{-1}X'X\Psi^{-1}\Lambda\Phi^{1/2}\big)^{-1/2}\Phi^{1/2}, \tag{65}$$

which yields

$$f_{(j)} = \Phi^{1/2}(\Phi^{1/2}\Lambda'\Psi^{-1}X'X\Psi^{-1}\Lambda\Phi^{1/2})^{-1/2}\Phi^{1/2}\Lambda'\Psi^{-1}x_{(j)} \quad (j = 1, \ldots, n). \tag{66}$$

Note that

$$\begin{aligned}\Phi^{1/2}\Lambda'\Psi^{-1}X'X\Psi^{-1}\Lambda\Phi^{1/2} &= \Phi^{1/2}\Lambda'\Psi^{-1}(\Lambda\Phi\Lambda' + \Psi)\Psi^{-1}\Lambda\Phi^{1/2} \\ &= (\Phi^{1/2}\Lambda'\Psi^{-1}\Lambda\Phi^{1/2})^2 + \Phi^{1/2}\Lambda'\Psi^{-1}\Lambda\Phi^{1/2}.\end{aligned}$$

By denoting $L = \Phi^{1/2}\Lambda'\Psi^{-1}\Lambda\Phi^{1/2}$, we may rewrite (66) as

$$f_{(j)} = \Phi^{1/2}(L^2 + L)^{-1/2}\Psi^{1/2}\Lambda'\Psi^{-1}x_{(j)}. \tag{67}$$

We denote (67) by f_3 for any j. Obviously, $(1/n)F'F = \Phi$ holds. This estimator was further discussed by Rao (1979) and ten Berge (1999).

Next, let us consider a random effect model of the form, $x = \Lambda f + e$, where Λ is a factor loading matrix of order $p \times m$, x and e are p-dimensional random vectors, the latter satisfying $\mathrm{E}(fe') = O$. Let P denote a square matrix of order p and define Px as an estimate of Λf where f is assumed to be random. Differentiating

$$\begin{aligned} g(P) &= \mathrm{tr}\big(\mathrm{E}(Px - \Lambda f)(Px - \Lambda f)'\big) \\ &= \mathrm{tr}\big(\mathrm{E}(Pxx'P' - Pxf'\Lambda' - \Lambda fx'P' + \Lambda ff'\Lambda')\big) \\ &= \mathrm{tr}(P\Sigma P' - P\Lambda\Phi\Lambda' - \Lambda\Phi\Lambda'P' + \Lambda\Phi\Lambda') \end{aligned} \tag{68}$$

with respect to P, we obtain

$$P\Sigma = \Lambda\Phi\Lambda'. \tag{69}$$

If Σ is nonsingular, we have

$$\Lambda f_4 = Px = \big(\Lambda\Phi\Lambda'\Sigma^{-1}\big)x = \Lambda\Phi\Lambda'(\Lambda\Phi\Lambda' + \Psi)^{-1}x,$$

yielding

$$f_4 = \Phi\Lambda'\Sigma^{-1}x = \Phi\Lambda'(\Lambda\Phi\Lambda' + \Psi)^{-1}x, \tag{70}$$

which coincides with the regression estimator of f on x first introduced by Thurstone (1935) and further discussed by Thomson (1946).

If Σ is singular, we obtain from (69) that

$$P = \Lambda\Phi\Lambda'\Sigma^{-} + Z(I - \Sigma\Sigma^{-}), \tag{71}$$

where Z is arbitrary. Let $S = \Sigma\Sigma^{-}x - x$. Then, $\mathrm{E}(SS') = O$, which implies $\Sigma\Sigma^{-}x = x$. Postmultiplying (71) by x, we obtain $\Lambda f_4 = Px = \Lambda\Phi\Lambda'\Sigma^{-}x$, yielding

$$f_4 = \Phi\Lambda'\Sigma^{-}x. \tag{72}$$

The relationships among the four methods of estimating factor scores were discussed in McDonald and Burr (1967).

Now we consider the relationship between f_1 and f_4. With some derivations, it follows that

$$\begin{aligned} f_4 &= \Phi\Lambda'\Sigma^{-1}x = \Phi\Lambda'(\Psi + \Lambda'\Phi\Lambda)^{-1}x \\ &= \Phi\Lambda'\big(\Psi^{-1} - \Psi^{-1}\Lambda(\Lambda'\Psi^{-1}\Lambda + \Phi^{-1})^{-1}\Lambda'\Psi^{-1}\big)x \\ &= \Phi\big(I_m - \Lambda'\Psi^{-1}\Lambda(\Lambda'\Psi^{-1}\Lambda + \Phi^{-1})^{-1}\big)\Lambda'\Psi^{-1}x \\ &= (\Lambda'\Psi^{-1}\Lambda + \Phi^{-1})^{-1}\Lambda'\Psi^{-1}x \\ &= \big(I + \Phi^{-1}(\Lambda'\Psi^{-1}\Lambda)^{-1}\big)^{-1}(\Lambda'\Psi^{-1}\Lambda)^{-1}\Lambda'\Psi^{-1}x \\ &= \big(I + \Phi^{-1}(\Lambda'\Psi^{-1}\Lambda)^{-1}\big)^{-1}f_1. \end{aligned} \tag{73}$$

Assume that $\Phi = I_m$. Then, it follows from Anderson (2003, Section 14.7) that the mean square errors of f_4 given by

$$\mathrm{E}\big[(f_4 - f)(f_4 - f)'\big] = \big(I_m + \Lambda'\Psi^{-1}\Lambda\big)^{-1}$$

are smaller than the variances of the unbiased estimator, f_1, given by $\mathrm{V}(f_1) = (\Lambda'\Psi^{-1}\Lambda)^{-1}$. The above result indicates that f_4 is a linear combination of f_1.

3.7. Application of Property 3 to canonical factor analysis

Let F denote a matrix of common factor scores, and let X denote a standardized data matrix of p variables. Further, let $f = Xw$ denote a linear composite score vector. Then, maximizing

$$\|P_F f\|^2/\|f\|^2 \tag{74}$$

with respect to w yields

$$(X'P_F X)w = \lambda(X'X)w, \tag{75}$$

leading to

$$Rw = \bar{\lambda}\Psi w, \quad \text{where } \bar{\lambda} = 1/(1+\lambda)$$

in view of $\Lambda\Lambda' = R - \Psi$. (75) is the eigen-equation resulting from canonical factor analysis introduced by Rao (1955). Note that the sum of eigenvalues obtained from (75) coincides with $\mathrm{tr}(P_X P_F)$, as defined by (47).

Using Property 3, we can establish:

THEOREM 2. *Let Λ denote a factor loading matrix of order $p \times m$. Then, the following seven statements are all equivalent*:

(1) $\lambda_j = 1$ *or* 0 *for* $j = 1, \ldots, m$.

(2) $P_X P_F = P_F P_X$.

(3) $\Lambda\Lambda'(\frac{1}{n}X'X)^-\Lambda = \Lambda$.

(4) $\Psi(X'X)^-\Lambda = O$.

(5) $((\frac{1}{n}X'X)^-\Lambda\Lambda')^2 = (\frac{1}{n}X'X)^-\Lambda\Lambda'$.

(6) $\operatorname{rank}(X) = \operatorname{rank}(\Lambda) + \operatorname{rank}(\Psi)$.

(7) $(X - F\Lambda')'Q_X F = O$.

PROOF. Equivalence between (1) and (2) is well known. To show (2) implies (3), we note $\Lambda = (1/n)F'X$. To show (3) implies (4), we note $\Lambda\Lambda'((1/n)X'X)^-\Lambda = \Lambda$, which implies $(R - \Psi)R^-\Lambda = RR^-\Lambda - \Psi R^-\Lambda = \Lambda$. This establishes the desired result, since $RR^-\Lambda = (1/n)(X'X)(X'X)^-X'F = (1/n)X'F = \Lambda$. To show (4) implies (3), we have $((1/n)X'X - \Lambda\Lambda')((1/n)X'X)^-(1/n)X'F = \Lambda - \Lambda\Lambda'((1/n)X'X)^-\Lambda = O$. To show (7) implies (3), $\Lambda F'Q_X F = \Lambda F'(I_n - X(X'X)^-X')F = n(\Lambda - \Lambda\Lambda' \times ((1/n)X'X)^-\Lambda) = O$. Equivalence between (3) and (5) is easy to establish, since (5) immediately follows from (3), while (3) follows from (5) by pre- and postmultiplying both sides of (5) by $(1/n)X'X$ and $(\Lambda\Lambda')^-\Lambda$, respectively. To show (4) implies (6), observe that $\operatorname{Sp}(\Lambda\Lambda' + \Psi) = \operatorname{Sp}(\Lambda, \Psi^{1/2})$. Further, suppose that $\Lambda\Lambda'\alpha + \Psi\beta = 0$. By premultiplying both sides by $\Lambda'(X'X)^-$, we obtain $\Lambda\Lambda'\alpha = 0$. Thus, $\operatorname{Sp}(\Lambda)$ and $\operatorname{Sp}(\Psi)$ are disjoint, establishing $\operatorname{rank}(X'X) = \operatorname{rank}(X) = \operatorname{rank}(\Lambda) + \operatorname{rank}(\Psi)$. Equivalence between (6) and (7) is established immediately, following a similar line of proof to that of the equivalence between (7) and (8) in Property 3. We set $A = X$ and $B = F$ and note that $\operatorname{rank}(Q_F X) = \operatorname{rank}((1/n)X'Q_F X) = \operatorname{rank}(R - \Lambda\Lambda') = \operatorname{rank}(\Psi)$. □

3.8. Some extension of the identifiability condition

It is well known that a sufficient condition for the matrix of unique variances to be uniquely determined is that there exists at least two disjoint square matrices both non-singular and of rank m in the factor loading matrix Λ when any one row vector is deleted from Λ (Anderson and Rubin, 1956). Ihara and Kano (1986), and Kano (1989) gave some extensions to Anderson and Rubin's result. We give an alternative extension using Property 9.

LEMMA 6. *Suppose that the factor loading matrix, Λ, of order $(p_1 + p_2 + r) \times m$, where $p_1 \geqslant m$, $p_2 \geqslant m$, and $r \leqslant \min(p_1, p_2)$, is partitioned as $\Lambda' = [\Lambda_1', \Lambda_2', \Lambda_3']$, where Λ_1, Λ_2, and Λ_3 are of orders $p_1 \times m$, $p_2 \times m$, and $r \times m$, respectively. The population correlation (covariance) matrix, Σ, is then expressed in a partitioned form as*

$$\Sigma = \begin{bmatrix} \Sigma_{11} & \Sigma_{12} & \Sigma_{13} \\ \Sigma_{21} & \Sigma_{22} & \Sigma_{23} \\ \Sigma_{31} & \Sigma_{32} & \Sigma_{33} \end{bmatrix} = \begin{bmatrix} \Lambda_1\Lambda_1' + \Psi_1 & \Lambda_1\Lambda_2' & \Lambda_1\Lambda_3' \\ \Lambda_2\Lambda_1' & \Lambda_2\Lambda_2' + \Psi_2 & \Lambda_2\Lambda_3' \\ \Lambda_3\Lambda_1' & \Lambda_3\Lambda_2' & \Lambda_3\Lambda_3' + \Psi_3 \end{bmatrix}$$
$$= \Lambda\Lambda' + \Psi.$$

Assume further that $\operatorname{rank}(\Lambda_1\Lambda_2') = \operatorname{rank}(\Lambda_1) = \operatorname{rank}(\Lambda_2)$, *and* $\operatorname{Sp}(\Lambda_3') \subset \operatorname{Sp}(\Lambda_2')$. *Then, Ψ is determined uniquely.*

PROOF. $\mathrm{Sp}(\Lambda_3') \subset \mathrm{Sp}(\Lambda_2')$ implies $\Lambda_3' = \Lambda_2' W$ for some W. $\mathrm{rank}(\Lambda_1\Lambda_2') = \mathrm{rank}(\Lambda_1)$ $= \mathrm{rank}(\Lambda_2)$ implies $\Lambda_2'(\Lambda_1\Lambda_2')^-\Lambda_1$ is the projector onto $\mathrm{Sp}(\Lambda_2')$ along $\mathrm{Ker}(\Lambda_1)$ (Property 9), and it is thus invariant over any choice of g-inverse of $\Lambda_1\Lambda_2'$. We then have

$$\begin{aligned}\Sigma_{32}\Sigma_{12}^-\Sigma_{13} &= \Lambda_3\Lambda_2'(\Lambda_1\Lambda_2')^-\Lambda_1\Lambda_3' = \Lambda_3\Lambda_2'(\Lambda_1\Lambda_2')^-\Lambda_1\Lambda_2'W \\ &= \Lambda_3\Lambda_3' = \Sigma_{33} - \Psi_3, \end{aligned} \tag{76}$$

establishing $\Psi_3 = \Sigma_{33} - \Sigma_{32}\Sigma_{12}^-\Sigma_{13}$, which is invariant over any choice of g-inverse of Σ_{12}. □

This establishes the desired result. Observe that Lemma 6 covers the result of Anderson and Rubin (1956), where it was assumed that $\mathrm{rank}(\Lambda_1\Lambda_2') = \mathrm{rank}(\Lambda_1) = \mathrm{rank}(\Lambda_2) = m$, which automatically implies $\mathrm{Sp}(\Lambda_3') \subset \mathrm{Sp}(\Lambda_2')$. Lemma 6 also covers Kano (1989), since

$$m = \mathrm{rank}(P_{\Lambda_1'}P_{\Lambda_2'}) \leqslant \mathrm{rank}(\Lambda_1\Lambda_2') \leqslant \mathrm{rank}(\Lambda_j) = m$$

for $j = 1, 2$. Note that $\mathrm{rank}(P_{\Lambda_1'}) = \mathrm{rank}(P_{\Lambda_2'}) = m$, and that $\mathrm{Sp}(\Lambda_3') \subset \mathrm{Sp}(\Lambda_2')$ also holds.

NOTE 4. As a reviewer of this manuscript has pointed out (see also Takeuchi et al., 1982, Section 7.2.2), there are in general an infinite number of possible Ψ's that satisfy the factor analysis model if Λ is allowed to change in such a way that $\mathrm{Sp}(\Lambda_3') \subset \mathrm{Sp}(\Lambda_2')$ no longer holds.

Acknowledgements

We are deeply indebted to an anonymous reviewer of this article who read the manuscript carefully and gave us some useful comments which lead to Note 4.

References

Anderson, T.W. (2003). *A Introduction to Multivariate Statistical Analysis*, third ed. Wiley, New York.

Anderson, T.W., Gupta, D. (1963). Some inequalities on characteristic roots of matrices. *Biometrika* **59**, 522–524.

Anderson, T.W., Rubin, H. (1956). Statistical inferences in factor analysis. In: Neyman, J. (Ed.), *Proceedings of the Third Berkeley Symposium on Mathematical Statistics and Probability*, vol. 5. University of California Press, Berkeley, pp. 111–150.

Baksalary, J.K. (1987). Algebraic characterizations and statistical implications of the commutativity of orthogonal projectors. In: Pukkila, T., Puntanen, S. (Eds.), *Proceedings of the Second International Tampere Conference in Statistics*. University of Tampere, Finland, pp. 113–142.

Baksalary, J.K., Styan, G.P.H. (1990). Around a formula for the rank of a matrix product with some statistical applications. In: Rees, R.S. (Ed.), *Graphs, Matrices, and Designs*. Marcel Dekker, New York, pp. 1–18.

Bartlett, M.S. (1937). The statistical conception of mental factors. *British Journal of Psychology* **28**, 97–104.

Beckenbach, E.F., Bellman, R. (1961). *Inequalities*. Springer, New York.

Bentler, P.M., Yuan, K.H. (1997). Optimal conditionally equivalent factor score estimation. In: Berkane, M. (Ed.), *Latent Variable Modeling and Application to Causality*. Springer, New York, pp. 259–281.

Guttman, L. (1940). Multiple rectilinear prediction and the resolution into components. *Psychometrika* **5**, 75–99.
Guttman, L. (1953). Image theory for the structure of quantitative variates. *Psychometrika* **18**, 277–296.
Harman, H.H. (1967). *Modern Factor Analysis*. University of Chicago Press, Chicago (revised 1976).
Hoerl, A.F., Kennard, R.W. (1970). Ridge regression: Biased estimation for nonorthogonal problems. *Technometrics* **12**, 55–67.
Horst, P. (1965). *Factor Analysis of Data Matrices*. Holt, Rinehart and Winston, New York.
Ihara, M., Kano, Y. (1986). A new estimator of the uniqueness in factor analysis. *Psychometrika* **51**, 563–566.
Kahtri, C.G. (1976). A note on multiple and canonical correlation matrix of a singular covariance matrix. *Psychometrika* **41**, 465–470.
Kaiser, H.F. (1976). Image and anti-image covariance matrices from a correlation matrix that may be singular. *Psychometrika* **41**, 295–300.
Kano, Y. (1989). A new estimation procedure using g-inverse matrix in factor analysis. *Mathematica Japonica* **34**, 43–52.
Kano, Y., Harada, A. (2000). Stepwise variable selection in factor analysis. *Psychometrika* **65**, 7–22.
Kristof, W. (1970). A theorem on the trace of certain matrix product and some application. *Journal of Mathematical Psychology* **7**, 515–530.
McDonald, R.P., Burr, E.J. (1967). A comparison of four methods constructing factor scores. *Psychometrika* **32**, 381–401.
Rao, C.R. (1955). Estimation and tests of significance in factor analysis. *Psychometrika* **20**, 93–111.
Rao, C.R. (1979). Separation theorem for singular values of matrices and their applications in multivariate analysis. *Journal of Multivariate Analysis* **9**, 362–377.
Rao, C.R., Yanai, H. (1979). General definition and decomposition of projectors and some applications to statistical problems. *Journal of Statistical Planning and Inferences* **3**, 1–17.
Roff, M. (1936). Some properties of the communality in multiple factor theory. *Psychometrika* **1**, 1–6.
Takane, Y. (2004). Matrices with special reference to applications in psychometrics. *Linear Algebra and its Applications C* **388**, 341–361.
Takane, Y., Yanai, H. (2005). On ridge operators. Submitted for publication.
Takeuchi, K., Yanai, H., Mukherjee, B.N. (1982). *The Foundation of Multivariate Analysis*. Wiley Eastern/Halsted Press, New Delhi/New York.
ten Berge, J.M.F. (1993). *Least Squares Optimization in Multivariate Analysis*. DSWO Press, Leiden.
ten Berge, J.M.F. (1999). Some new results on correlation-preserving factor score prediction methods. *Linear Algebra and its Applications* **289**, 311–318.
Thomson, G.H. (1946). *The Factorial Analysis of Human Ability*, second ed. Houghton Mifflin, New York.
Thurstone, L.L. (1935). *Vector of Mind*. The University of Chicago, Chicago.
Yanai, H. (1980). A proposition of generalized method for forward selection of variables. *Behaviormetrika* **7**, 95–107.
Yanai, H. (1990). Some generalized forms of least squares g-inverses, minimum norm g-inverses, and Moore–Penrose inverse matrices. *Computational Statistics and Data Analysis* **10**, 251–260.
Yanai, H., Ichikawa, M. (1990). New lower and upper bounds for communality in factor analysis. *Psychometrika* **55**, 405–409.
Yanai, H., Ichikawa, M. (2007). Factor analysis. In: Rao, R.S., Sinharay, S. (Eds.), *Psychometrics*. In: *Handbook of Statistics*, vol. 26. North-Holland, pp. 257–296 (Chapter 9).
Yanai, H., Mukherjee, B.N. (1987). A generalized method of image analysis from an intercorrelation matrix which may be singular. *Pychometrika* **52**, 555–564.
Yanai, H., Puntanen, S. (1993). Partial canonical correlation associated with symmetric reflexive generalized inverses of the dispersion matrix. In: Matsushita, K., Puri, M.L., Hayakawa, T. (Eds.), *Statistical Sciences and Data Analysis*. VSP, Utrecht, The Netherlands, pp. 253–264.

Handbook of Computing and Statistics with Applications, Vol. 1
ISSN: 1871-0301

DOI: 10.1016/S1871-0301(06)01017-1

17

Robust Procedures in Structural Equation Modeling*

Ke-Hai Yuan and Peter M. Bentler

Abstract

The classical procedure for structural equation modeling was developed under the assumption of normally distributed data. In practice, data are seldom normally distributed, and often possess heavy tails. When the normality assumption is slightly violated, the normal distribution based maximum likelihood (ML) procedure still generates consistent parameter estimates. When data come from a distribution with severe heavy tails, parameter estimates by ML may no longer be consistent. Standard errors and test statistics based on modeling the sample means and covariances may not be valid either. This chapter systematically introduces three types of robust procedures. Statistical properties of each procedure are reviewed, and their strengths and weaknesses as well as scope of applicability are discussed. Examples are provided to contrast the properties of these procedures. While each of the robust procedures improves the ML procedure to certain extent, only those that downweight the effect of outlying cases are really robust. The ML procedure is not recommended for use with non-normally distributed data in practice although it may possess asymptotic robustness.

1. Introduction

Structural equation modeling (SEM) has been widely used in social and behavioral sciences. The most commonly used method for estimation and testing in SEM is the normal theory based maximum likelihood (ML). In this method, parameter estimates are obtained by maximizing the likelihood function derived from the multivariate normal distribution. Standard errors (SE's) of the maximum likelihood estimators (MLE) are based on the covariance matrix that is obtained by inverting the associated information matrix. Overall model evaluation is accomplished by referring the likelihood ratio (LR) statistic to a chi-square distribution. In practice, however, data, including survey data that are often analyzed with SEM, may never be normally distributed (see Micceri, 1989;

*This research was supported by NSF grant DMS04-37167, the James McKeen Cattell Fund, and grants DA01070 and DA00017 from the National Institute on Drug Abuse.

Geary, 1947). It is possible that the ML procedure can still provide reliable inference when data do not follow a normal distribution, but this cannot be taken for granted. In cases where the ML methodology does not provide reliable model inference, improved procedures are necessary. This chapter discusses various robust procedures that generally lead to more reliable inference, i.e., procedures that can give reasonable inference when the underlying distribution of the population is unknown or when data are contaminated. We will discuss the theoretical/asymptotic properties as well as the empirical performance of each approach, so that readers will have a good overall picture of the field.

In the SEM literature, the term "robust" is used very broadly. We will discuss three classes of robust procedures. The first class involves extending the ML methodology in which parameter estimates remain the MLE but the SE's are asymptotically correct for arbitrary distributions with finite 4th-order moments. An associated rescaled LR statistic is used for overall model evaluation. In certain situations, the rescaled statistic asymptotically follows a chi-square distribution and can lead to more reliable model inference than the LR statistic. The second class is based on generalized least squares (GLS), where weight matrices involving 4th-order moments are explicitly involved in the estimation process. This class of procedures enjoys many nice analytical properties, and some also enjoy good finite sample properties. The third class is more consistent with the standard statistical literature on robustness (e.g., Huber, 1981; Hampel et al., 1986), where a proper weight is assigned to each case, and those lying far from the center of the data cloud or from the model get smaller weights. Because problematic cases receive smaller weights, their effect is minimized not only on SE's and model evaluation but also on parameter estimates. Most properties of these procedures are studied under the assumption of a correctly specified model; limited results relating to misspecified models also are available. The three classes of procedures in Sections 2 to 4 are developed under the assumption that the model is correctly specified. Section 5 will introduce the parallel results covering situations where the model is misspecified. Examples contrasting these procedures will be presented in Section 6.

Although the ML methodology in SEM is very sensitive to bad data, the LR statistic may still asymptotically follow a chi-square distribution when the population distribution of the sample and the model satisfy certain structural specifications. Similarly, asymptotically correct SE's for some parameter estimates may be obtained as well. These strengths of ML have been designated as *asymptotic robustness* in the literature of covariance structure analysis. We will discuss asymptotic robustness in the context of the ML procedure and its extensions. Furthermore, robust procedures have been developed mainly to deal with covariance structure analysis (CSA). We will present robust methods in the more general framework of mean and covariance structure analysis. In the remainder of this section, we introduce notation that will facilitate the technical development in later sections.

Let $\mathbf{x}$ represent the p observed variables whose population mean vector and covariance matrix are $E(\mathbf{x}) = \boldsymbol{\mu}_0$ and $\mathrm{Cov}(\mathbf{x}) = \boldsymbol{\Sigma}_0$. In SEM we want to model $\boldsymbol{\mu}_0$ and $\boldsymbol{\Sigma}_0$ by interesting structures $\boldsymbol{\mu}(\boldsymbol{\theta})$ and $\boldsymbol{\Sigma}(\boldsymbol{\theta})$, where $\boldsymbol{\theta}$ is a vector of free parameters. As described in several other chapters in this volume, free parameters may include factor loadings, latent variable regression coefficients, latent means, and so on, depending on the particular SEM structure being studied. Ideally, one would like to have

$$H_0\colon \boldsymbol{\mu}_0 = \boldsymbol{\mu}(\boldsymbol{\theta}_0) \quad \text{and} \quad \boldsymbol{\Sigma}_0 = \boldsymbol{\Sigma}(\boldsymbol{\theta}_0), \tag{1}$$

where $\boldsymbol{\theta}_0$ is the population value of $\boldsymbol{\theta}$ corresponding to the correctly specified models. Because $\boldsymbol{\mu}_0$ and $\boldsymbol{\Sigma}_0$ are generally unknown, we have to use their sample counterparts to evaluate whether (1) is achievable, and to obtain an estimate of $\boldsymbol{\theta}$ that is close to $\boldsymbol{\theta}_0$ if (1) holds. All the procedures will assume a random sample $\mathbf{x}_1, \ldots, \mathbf{x}_n$ with the sample mean vector $\bar{\mathbf{x}}$ and sample covariance matrix $\mathbf{S}$. For a $p \times p$ symmetric matrix $\mathbf{A}$, let vec($\mathbf{A}$) be the operator that stacks the columns of $\mathbf{A}$, and let vech($\mathbf{A}$) stack only the elements on and below the diagonal. Then vec($\mathbf{A}$) contains duplicated elements of vech($\mathbf{A}$), and there exists a matrix $\mathbf{D}_p$ such that $\text{vec}(\mathbf{A}) = \mathbf{D}_p\, \text{vech}(\mathbf{A})$. We denote $\mathbf{s} = \text{vech}(\mathbf{S})$, $\boldsymbol{\sigma} = \text{vech}(\boldsymbol{\Sigma})$, $\mathbf{t}_i = (\mathbf{x}_i', \text{vech}'\{(\mathbf{x}_i - \bar{\mathbf{x}})(\mathbf{x}_i - \bar{\mathbf{x}})'\})'$, $\bar{\mathbf{t}} = (\bar{\mathbf{x}}', \mathbf{s}')'$, $\boldsymbol{\beta} = (\boldsymbol{\mu}', \boldsymbol{\sigma}')'$,

$$\mathbf{W}_c = 2^{-1}\mathbf{D}_p'(\boldsymbol{\Sigma}^{-1} \otimes \boldsymbol{\Sigma}^{-1})\mathbf{D}_p, \quad \text{and}$$

$$\mathbf{W} = \text{diag}(\boldsymbol{\Sigma}^{-1}, \mathbf{W}_c) = \begin{pmatrix} \boldsymbol{\Sigma}^{-1} & \mathbf{0} \\ \mathbf{0} & \mathbf{W}_c \end{pmatrix}.$$

We will use a dot on top of a vector function to imply derivatives or a Jacobian matrix as in $\dot{\boldsymbol{\beta}}(\boldsymbol{\theta}) = \partial\boldsymbol{\beta}(\boldsymbol{\theta})/\partial\boldsymbol{\theta}'$ and double dots on top of a function to imply the matrix of second derivatives as in $\ddot{\mu}_i(\boldsymbol{\theta}) = \partial^2\mu_i(\boldsymbol{\theta})/\partial\boldsymbol{\theta}\partial\boldsymbol{\theta}'$. We may omit the argument of the function when it is evaluated at the population value corresponding to a correctly specified model as in $\boldsymbol{\beta} = \boldsymbol{\beta}(\boldsymbol{\theta}_0)$ and $\mathbf{W} = \mathbf{W}(\boldsymbol{\theta}_0)$. We will use $\mathbf{I}$ for the identity matrix and $\mathbf{0}$ for a vector or matrix of zeros, and subscripts will be used when their dimensions are not obvious. Convergence in probability and distribution will be denoted by $\xrightarrow{P}$ and $\xrightarrow{\mathcal{L}}$, respectively.

The regularity conditions for SEM are

(C1) $\boldsymbol{\theta}_0$ is an interior point of $\boldsymbol{\Theta}$ which is a compact subset of R^q;
(C2) $\boldsymbol{\beta}(\boldsymbol{\theta}) = \boldsymbol{\beta}(\boldsymbol{\theta}_0)$ only when $\boldsymbol{\theta} = \boldsymbol{\theta}_0$;
(C3) $\boldsymbol{\beta}(\boldsymbol{\theta})$ is twice continuously differentiable;
(C4) $\dot{\boldsymbol{\beta}}$ is full rank; and
(C5) the vector $(\mathbf{x}_i', \text{vech}'\{(\mathbf{x}_i - \boldsymbol{\mu}_0)(\mathbf{x}_i - \boldsymbol{\mu}_0)'\})'$ has a covariance matrix that is full rank.

Conditions (C1) and (C3) are the standard regularity conditions and are generally satisfied in practice. Condition (C2) implies that the model structure is identified. If the structural model is properly parameterized, condition (C4) will be satisfied. Conditions (C1) and (C2) are needed for consistency of parameter estimates. Conditions (C3) and (C4) are needed to establish the asymptotic normality of parameter estimates. (C5) is needed in order for parameter estimates or statistics for the overall model evaluation to have proper asymptotic distributions. (C5) will be satisfied when $\mathbf{x}_i \sim N(\boldsymbol{\mu}_0, \boldsymbol{\Sigma}_0)$ and $\boldsymbol{\Sigma}_0$ is full rank. When directly modeling $\mathbf{S}$, $\text{Cov}[\text{vech}\{(\mathbf{x}_i - \boldsymbol{\mu}_0)(\mathbf{x}_i - \boldsymbol{\mu}_0)'\}]$ needs to exist. When the outlying cases are downweighted as in Section 4, only $\boldsymbol{\Sigma}_0$ needs to exist. Further discussions of regularity conditions can be found in Browne (1984), Shapiro (1984), Kano (1986), and Yuan and Bentler (1997a). We will implicitly assume these conditions throughout the chapter.

2. Normal theory ML and related procedures

When the distribution of $\mathbf{x}$ is known, the ML methodology based on the true distribution of $\mathbf{x}$ has many nice asymptotic properties. Because we do not know the true distribution of $\mathbf{x}$, we might use $\mathbf{x} \sim N(\boldsymbol{\mu}, \boldsymbol{\Sigma})$ as a working assumption. Then the log likelihood function of $\boldsymbol{\theta}$ is given by

$$l(\boldsymbol{\theta}) = -\frac{n}{2}\ln\left|\boldsymbol{\Sigma}(\boldsymbol{\theta})\right| - \frac{n}{2}\operatorname{tr}\left\{\boldsymbol{\Sigma}^{-1}(\boldsymbol{\theta})\left(\mathbf{S} + \left[\bar{\mathbf{x}} - \boldsymbol{\mu}(\boldsymbol{\theta})\right]\left[\bar{\mathbf{x}} - \boldsymbol{\mu}(\boldsymbol{\theta})\right]'\right)\right\}. \tag{2}$$

Let the MLE that maximizes $l(\boldsymbol{\theta})$ be $\hat{\boldsymbol{\theta}}$. When all the elements in $\boldsymbol{\mu}$ and $\boldsymbol{\Sigma}$ are free parameters, their MLE's are given respectively by $\hat{\boldsymbol{\mu}} = \bar{\mathbf{x}}$ and $\widehat{\boldsymbol{\Sigma}} = \mathbf{S}$ with $\hat{\boldsymbol{\beta}} = (\bar{\mathbf{x}}', \mathbf{s}')'$. When evaluated at $\hat{\boldsymbol{\mu}} = \bar{\mathbf{x}}$ and $\widehat{\boldsymbol{\Sigma}} = \mathbf{S}$,

$$l(\bar{\mathbf{x}}, \mathbf{S}) = -\frac{n}{2}\ln|\mathbf{S}| - \frac{np}{2}. \tag{3}$$

It follows from (2) and (3) that

$$2\left[l(\bar{\mathbf{x}}, \mathbf{S}) - l(\boldsymbol{\theta})\right] = nD_{\mathrm{ML}}\left[\bar{\mathbf{x}}, \mathbf{S}, \boldsymbol{\mu}(\boldsymbol{\theta}), \boldsymbol{\Sigma}(\boldsymbol{\theta})\right],$$

where

$$\begin{aligned} D_{\mathrm{ML}}\left[\bar{\mathbf{x}}, \mathbf{S}, \boldsymbol{\mu}(\boldsymbol{\theta}), \boldsymbol{\Sigma}(\boldsymbol{\theta})\right] = {} & \left[\bar{\mathbf{x}} - \boldsymbol{\mu}(\boldsymbol{\theta})\right]'\boldsymbol{\Sigma}^{-1}(\boldsymbol{\theta})\left[\bar{\mathbf{x}} - \boldsymbol{\mu}(\boldsymbol{\theta})\right] + \operatorname{tr}\left[\mathbf{S}\boldsymbol{\Sigma}^{-1}(\boldsymbol{\theta})\right] \\ & - \log\left|\mathbf{S}\boldsymbol{\Sigma}^{-1}(\boldsymbol{\theta})\right| - p \end{aligned} \tag{4}$$

is commonly called the normal theory based ML discrepancy function. Because $l(\bar{\mathbf{x}}, \mathbf{S})$ is a constant, the MLE $\hat{\boldsymbol{\theta}}$ also minimizes $D_{\mathrm{ML}}[\bar{\mathbf{x}}, \mathbf{S}, \boldsymbol{\mu}(\boldsymbol{\theta}), \boldsymbol{\Sigma}(\boldsymbol{\theta})]$. It is obvious that

$$T_{\mathrm{ML}} = nD_{\mathrm{ML}}\left[\bar{\mathbf{x}}, \mathbf{S}, \boldsymbol{\mu}(\hat{\boldsymbol{\theta}}), \boldsymbol{\Sigma}(\hat{\boldsymbol{\theta}})\right]$$

is the LR statistic. When $\mathbf{x} \sim N(\boldsymbol{\mu}_0, \boldsymbol{\Sigma}_0)$ and (1) holds, standard large sample theory (Ferguson, 1996; Rao, 1973) tells us that

$$T_{\mathrm{ML}} \xrightarrow{\mathcal{L}} \chi^2_{df},$$

where $df = p(p+3)/2 - q$ and q is the number of free parameters in $\boldsymbol{\theta}$. Thus, a significantly large T_{ML} when referred to the given chi-square distribution implies that the model structure is most likely misspecified. Under $\mathbf{x} \sim N(\boldsymbol{\mu}_0, \boldsymbol{\Sigma}_0)$ and (1), standard asymptotic theory also implies that $\hat{\boldsymbol{\theta}}$ is consistent and

$$\sqrt{n}(\hat{\boldsymbol{\theta}} - \boldsymbol{\theta}_0) \xrightarrow{\mathcal{L}} N(\mathbf{0}, \boldsymbol{\Omega}_{\mathrm{ML}}), \tag{5}$$

where

$$\boldsymbol{\Omega}_{\mathrm{ML}} = (\dot{\boldsymbol{\beta}}'\mathbf{W}\dot{\boldsymbol{\beta}})^{-1}.$$

When the mean structure $\boldsymbol{\mu}(\boldsymbol{\theta})$ is saturated, (4) becomes

$$D_{\mathrm{ML}c}\left[\mathbf{S}, \boldsymbol{\Sigma}(\boldsymbol{\theta})\right] = \operatorname{tr}\left[\mathbf{S}\boldsymbol{\Sigma}^{-1}(\boldsymbol{\theta})\right] - \log\left|\mathbf{S}\boldsymbol{\Sigma}^{-1}(\boldsymbol{\theta})\right| - p, \tag{6}$$

where the subscript c is used to denote CSA only. We will still use $\boldsymbol{\theta}$ to denote the vector of free parameters whose dimension is now q_c, and the corresponding test statistic is

$T_{\mathrm{ML}c}$. For normally distributed data and a correctly specified $\boldsymbol{\Sigma}(\boldsymbol{\theta})$, there exist

$$T_{\mathrm{ML}c} \xrightarrow{\mathcal{L}} \chi^2_{df_c}, \tag{7}$$

with $df_c = p(p+1)/2 - q_c$ and

$$\sqrt{n}(\hat{\boldsymbol{\theta}} - \boldsymbol{\theta}_0) \xrightarrow{\mathcal{L}} N(\mathbf{0}, \boldsymbol{\Omega}_{\mathrm{ML}c}), \tag{8}$$

where $\boldsymbol{\Omega}_{\mathrm{ML}c} = (\dot{\boldsymbol{\sigma}}'\mathbf{W}_c\dot{\boldsymbol{\sigma}})^{-1}$. In the context of CSA with non-normally distributed data, $T_{\mathrm{ML}c}$ enjoys some nice properties that are not shared by T_{ML}. We will discuss these properties first and then turn to limited results for mean and covariance structure analysis.

When data are not normally distributed, the MLE $\hat{\boldsymbol{\theta}}$ is still consistent as long as $\boldsymbol{\beta}(\boldsymbol{\theta})$ is identified and correctly specified. It follows from the central limit theorem that

$$\sqrt{n}(\bar{\mathbf{t}} - \boldsymbol{\beta}_0) \xrightarrow{\mathcal{L}} N(\mathbf{0}, \boldsymbol{\Pi}), \tag{9}$$

where

$$\boldsymbol{\Pi} = \begin{pmatrix} \boldsymbol{\Sigma}_0 & \boldsymbol{\Delta} \\ \boldsymbol{\Delta}' & \boldsymbol{\Gamma} \end{pmatrix}$$

with $\boldsymbol{\Delta} = \mathrm{Cov}(\mathbf{x}, \mathrm{vech}[(\mathbf{x} - \boldsymbol{\mu}_0)(\mathbf{x} - \boldsymbol{\mu}_0)'])$ and $\boldsymbol{\Gamma} = \mathrm{Cov}[\mathrm{vech}\{(\mathbf{x} - \boldsymbol{\mu}_0)(\mathbf{x} - \boldsymbol{\mu}_0)'\}]$. Let

$$\mathbf{U}_c = \mathbf{W}_c - \mathbf{W}_c\dot{\boldsymbol{\sigma}}(\dot{\boldsymbol{\sigma}}'\mathbf{W}_c\dot{\boldsymbol{\sigma}})^{-1}\dot{\boldsymbol{\sigma}}'\mathbf{W}_c$$

and κ_j, $j = 1, \ldots, df_c$, be the nonzero eigenvalues of $\mathbf{U}_c\boldsymbol{\Gamma}$. Under the condition of a correctly specified model, by using a Taylor expansion we will obtain (see the appendix of Yuan et al. (2002a) for detail)

$$T_{\mathrm{ML}c} = \sum_{j=1}^{df_c} \kappa_j z_j^2 + o_p(1), \tag{10}$$

where z_j^2's are independent and each follows a chi-square distribution with 1 degree of freedom, and $o_p(1)$ denotes a random term that approaches zero in probability as $n \to \infty$. Eq. (10) implies that the asymptotic distribution of $T_{\mathrm{ML}c}$ is determined by the fourth-order moments of $\mathbf{x}$ as well as the model structure. Notice that, when $\mathbf{x} \sim N(\boldsymbol{\mu}, \boldsymbol{\Sigma})$,

$$\boldsymbol{\Gamma} = 2\mathbf{D}_p^+(\boldsymbol{\Sigma} \otimes \boldsymbol{\Sigma})\mathbf{D}_p^{+\prime} = \mathbf{W}_c^{-1},$$

where $\mathbf{D}_p^+ = (\mathbf{D}_p'\mathbf{D}_p)^{-1}\mathbf{D}_p'$ is the generalized inverse of $\mathbf{D}_p$. Thus, $\kappa_1 = \kappa_2 = \cdots = \kappa_{df_c} = 1$ and (7) holds. We next explore the structure of $\boldsymbol{\Gamma}$ and asymptotic robustness of several ML statistics within a class of non-normal distributions introduced in Yuan and Bentler (1999a).

Let $\xi_1, \ldots, \xi_m$ be independent random variables with $E(\xi_j) = 0$, $E(\xi_j^2) = 1$, $E(\xi_j^3) = \zeta_j$, $E(\xi_j^4) = \varphi_j$, and $\boldsymbol{\xi} = (\xi_1, \ldots, \xi_m)'$. Let r be a random variable that is independent of $\boldsymbol{\xi}$, $E(r^2) = 1$, $E(r^3) = \gamma$, and $E(r^4) = \tau$. Also, let $m \geqslant p$ and

$\mathbf{L} = (l_{ij}) = (\mathbf{l}_1, \ldots, \mathbf{l}_m)$ be a $p \times m$ matrix of rank p such that $\mathbf{L}\mathbf{L}' = \boldsymbol{\Sigma}$, where $\mathbf{l}_j = (l_{1j}, \ldots, l_{pj})'$. Then the random vector

$$\mathbf{x} = \boldsymbol{\mu} + r\mathbf{L}\boldsymbol{\xi} \tag{11}$$

follows a non-normal distribution with $E(\mathbf{x}) = \boldsymbol{\mu}$ and $\mathrm{Cov}(\mathbf{x}) = \boldsymbol{\Sigma}$. Different distributions are obtained by choosing a different set of ξ_j's, $\mathbf{L}$ and r. Yuan and Bentler (1999a) obtained the fourth-order moment matrix $\boldsymbol{\Gamma} = \mathrm{Cov}\{\mathrm{vech}[(\mathbf{x} - \boldsymbol{\mu})(\mathbf{x} - \boldsymbol{\mu})']\}$ as

$$\begin{aligned}\boldsymbol{\Gamma} &= 2\tau\mathbf{D}_p^+(\boldsymbol{\Sigma} \otimes \boldsymbol{\Sigma})\mathbf{D}_p^{+\prime} + (\tau - 1)\boldsymbol{\sigma}\boldsymbol{\sigma}' \\ &\quad + \tau\sum_{j=1}^{m}(\varphi_j - 3)\,\mathrm{vech}(\mathbf{l}_j\mathbf{l}_j')\,\mathrm{vech}'(\mathbf{l}_j\mathbf{l}_j').\end{aligned} \tag{12}$$

It was noted by Yuan and Bentler (1999a) that for a given matrix $\mathbf{L}$, different marginal skewnesses will not affect the $\boldsymbol{\Gamma}$ matrix in (12). When all the φ_j's equal 3, then the $\boldsymbol{\Gamma}$ in (12) reduces to that of an elliptical distribution for $\mathbf{x}$ (see Browne, 1984). Yuan and Bentler (1999a) called the corresponding distribution of $\mathbf{x}$ in (11) a *pseudo-elliptical distribution* because it may still have arbitrary skewness. When $\tau = 1$ in addition to $\varphi_j = 3$, then $\boldsymbol{\Gamma}$ reduces to that of $\mathbf{x} \sim N(\boldsymbol{\mu}, \boldsymbol{\Sigma})$. Yuan and Bentler (1999a) called the corresponding distribution of $\mathbf{x}$ in (11) a *pseudo-normal distribution*. Due to allowing skewness, the class of pseudo-elliptical distributions is much larger than the class of elliptical distributions (see Fang et al., 1990). Similarly, $N(\boldsymbol{\mu}, \boldsymbol{\Sigma})$ is only a member of the class of pseudo-normal distributions.

It is obvious that in CSA $T_{\mathrm{ML}c}$ and $\hat{\boldsymbol{\theta}}$ are functions of $\mathbf{S}$ only. Thus, (7) and (8) hold within the class of pseudo-normal distributions. Of course, statistics in other multivariate procedures such as correlations, principal components and exploratory factor models also enjoy the same asymptotic distributions within the class of pseudo-normal distributions as well (Yuan and Bentler, 2000a, 2002).

It follows from (10) that any asymptotic robustness conditions for $T_{\mathrm{ML}c}$ must lead to $\kappa_1 = \kappa_2 = \cdots = \kappa_{df_c} = 1$. When the κ_j's are not equal, $T_{\mathrm{ML}c}$ will not asymptotically follow a chi-square distribution. When $\kappa_1 = \kappa_2 = \cdots = \kappa_{df_c} = \kappa \neq 1$, $T_{\mathrm{ML}c}$ will not approach a chi-square variate either. However, using a consistent estimate $\hat{\kappa}$ for κ, we can rescale the LR statistic to $T_{\mathrm{RML}c} = \hat{\kappa}^{-1}T_{\mathrm{ML}c}$ and

$$T_{\mathrm{RML}c} \xrightarrow{\mathcal{L}} \chi^2_{df_c}. \tag{13}$$

Thus, conditions leading to $\kappa_1 = \kappa_2 = \cdots = \kappa_{df_c} = \kappa$ are also of great interest. Several conditions are necessary to characterize properties of $T_{\mathrm{RML}c}$. For a covariance structure $\boldsymbol{\Sigma}(\boldsymbol{\theta})$, if for any parameter vector $\boldsymbol{\theta}$ and positive constant α there exists a parameter vector $\boldsymbol{\theta}^*$ such that $\boldsymbol{\Sigma}(\boldsymbol{\theta}^*) = \alpha\boldsymbol{\Sigma}(\boldsymbol{\theta})$, then $\boldsymbol{\Sigma}(\boldsymbol{\theta})$ is invariant with respect to a constant scaling factor (ICSF). As noted by Browne (1984), nearly all structural models in current use are ICSF. Let $\mathcal{R}(\dot{\boldsymbol{\sigma}})$ represent the range space spanned by the column vectors of $\dot{\boldsymbol{\sigma}}$. If $\boldsymbol{\sigma} \in \mathcal{R}(\dot{\boldsymbol{\sigma}})$, then the model is said to be quasi-linear (QL). Because ICSF implies QL (Satorra and Bentler, 1986), condition QL is also satisfied by almost all of the models in current use.

Under the condition of ICSF, Browne (1984) and Shapiro and Browne (1987) showed that the κ_j's are equal within the class of elliptical symmetric distributions. Earlier work in this direction by Muirhead and Waternaux (1980) and Tyler (1983) showed that the LR statistics in canonical correlation and several simple covariance structures can be rescaled to asymptotically follow chi-square distributions. The correction factor $\hat{\kappa}$ they used is based on Mardia's (1970) measure of multivariate kurtosis. Satorra and Bentler (1988) observed that $\mathrm{tr}(\mathbf{U}_c\boldsymbol{\Gamma}) = \sum_{j=1}^{df_c} \kappa_j$. They suggested

$$\hat{\kappa} = \mathrm{tr}(\widehat{\mathbf{U}}_c\widehat{\boldsymbol{\Gamma}})/df_c \tag{14}$$

in constructing $T_{\mathrm{RML}c}$. With (14), Yuan and Bentler (1999a) noted that (13) holds when $\mathbf{x}$ follows (11) and $\mathrm{vech}(\mathbf{l}_j\mathbf{l}_j') \in \mathcal{R}(\dot{\boldsymbol{\sigma}})$. When κ is estimated using Mardia's multivariate kurtosis and the condition of QL is satisfied, (13) holds within the class of pseudo-elliptical distributions. So (13) is valid within a much larger class of non-normal distributions when (14) is used in $T_{\mathrm{RML}c}$.

Similarly, there exist conditions for using (8) to provide consistent SE's for a subset of $\hat{\boldsymbol{\theta}}_c$ (see Anderson and Amemiya, 1988; Yuan and Bentler, 1999b). As noted above, the property that $T_{\mathrm{ML}c}$ or a subset of $\hat{\boldsymbol{\theta}}$ enjoys the same asymptotic distribution for non-normally distributed data as for normally distributed data is commonly called asymptotic robustness. Early characterizations of this type of robustness were made by Browne (1987), Anderson and Amemiya (1988) and Amemiya and Anderson (1990) for confirmatory factor models (CFM). The results were generalized in various direction by Browne and Shapiro (1988), Kano (1992), Mooijaart and Bentler (1991), Satorra (1992), Satorra and Bentler (1990), Yuan and Bentler (1999a, 1999b).

Simulation studies indicate that $T_{\mathrm{RML}c}$ performs quite robustly under a variety of conditions (Curran et al., 1996; Hu et al., 1992; Yuan and Bentler, 1998a). While exceptions have been studied, most data generation in these studies satisfies the asymptotic robustness condition for $T_{\mathrm{RML}c}$ (see Yuan and Bentler, 1999a). In general, however, $T_{\mathrm{RML}c}$ does not approach a variate possessing a chi-square distribution. Instead, it only approaches a variate T with $E(T) = df$. It is likely that the distributional shape of T is far from chi-square. In such cases, $T_{\mathrm{RML}c}$ will not behave like a chi-square variate. Limited simulation results in Yuan and Bentler (1998a) and Bentler and Yuan (1999) indicate that, when the κ_j's are not equal, $T_{\mathrm{RML}c}$ may behave worse as n increases. When sample size is small, $T_{\mathrm{RML}c}$ as well as $T_{\mathrm{ML}c}$ may not behave like a chi-square variate even for normally distributed data. It should be noted that most conditions for asymptotic robustness are not verifiable in practice. It is misleading to blindly trust that the statistic $T_{\mathrm{ML}c}$ or $T_{\mathrm{RML}c}$ will asymptotically follow a chi-square distribution before the necessary conditions can be verified.

In contrast to CSA, little or no results exist on the asymptotic robustness of T_{ML} when simultaneously modeling mean and covariance structures. Yuan and Bentler (2006) noted that the asymptotic robustness condition in CSA will not make T_{ML} asymptotically follow a chi-square distribution. But, for correctly specified models, we can still write

$$T_{\mathrm{ML}} = \sum_{j=1}^{df} \kappa_j z_j^2 + o_p(1),$$

where κ_j's are the nonzero eigenvalues of $\mathbf{U}\boldsymbol{\Pi}$ with

$$\mathbf{U} = \mathbf{W} - \mathbf{W}\dot{\boldsymbol{\beta}}(\dot{\boldsymbol{\beta}}'\mathbf{W}\dot{\boldsymbol{\beta}})^{-1}\dot{\boldsymbol{\beta}}'\mathbf{W}, \tag{15}$$

and where z_j^2, $j = 1, \ldots, df$, are independent χ_1^2 variates. The following rescaled statistic with $\hat{\kappa} = \operatorname{tr}(\widehat{\mathbf{U}}\widehat{\boldsymbol{\Pi}})/df$ has been proposed in different contexts (Satorra, 1992; Satorra and Bentler, 1994; Yuan and Bentler, 2000b, 2001a)

$$T_{\mathrm{RML}} = \hat{\kappa}^{-1} T_{\mathrm{ML}},$$

but few or no empirical studies have been conducted to evaluate its performance.

When data are not normally distributed, the asymptotic distribution of $\hat{\boldsymbol{\theta}}$ is characterized by

$$\sqrt{n}(\hat{\boldsymbol{\theta}} - \boldsymbol{\theta}_0) \xrightarrow{\mathcal{L}} N(\mathbf{0}, \boldsymbol{\Omega}_{\mathrm{SW}}), \tag{16}$$

where

$$\boldsymbol{\Omega}_{\mathrm{SW}} = (\dot{\boldsymbol{\beta}}'\mathbf{W}\dot{\boldsymbol{\beta}})^{-1}(\dot{\boldsymbol{\beta}}'\mathbf{W}\boldsymbol{\Pi}\mathbf{W}\dot{\boldsymbol{\beta}})(\dot{\boldsymbol{\beta}}'\mathbf{W}\dot{\boldsymbol{\beta}})^{-1}$$

is commonly called the sandwich-type covariance matrix. Like T_{RML}, $\boldsymbol{\Omega}_{\mathrm{SW}}$ makes an adjustment for distributional violations of the normal theory based ML procedure. When data are normally distributed, $\mathbf{W} = \boldsymbol{\Pi}^{-1}$, so (16) reduces to (5). The sandwich-type covariance matrix was first proposed by Huber (1967). It is now widely used in SEM (Bentler, 1983; Bentler and Dijkstra, 1985; Browne, 1984; Browne and Arminger, 1995; Satorra and Bentler, 1994; Shapiro, 1983; Yuan and Bentler, 1997b, 1998b, 1998c). Comparing $\boldsymbol{\Omega}_{\mathrm{SW}}$ with $\boldsymbol{\Omega}_{\mathrm{ML}}$, one has to estimate the extra matrix $\boldsymbol{\Pi}$. Simulation results in CSA (Yuan and Bentler, 1997b) indicate that standard errors based on (16) match those of empirical ones very well for normally as well as non-normally distributed data. Yuan and Bentler (1997b) recommended using (16) as the default formula for calculating SE's of $\hat{\boldsymbol{\theta}}$.

3. Generalized Least Squares (GLS) procedures

With typical non-normal data in the social and behavioral sciences (Micceri, 1989), the ideal is to have a statistic that approximately follows a chi-square distribution regardless of the underlying distribution of the data. One of the original proposals in this direction was a GLS procedure made by Browne (1984) in the context of CSA. Because CSA alone does not enjoy extra properties for GLS procedures, we will discuss mean and covariance structure models, without paying special attention to CSA. Let $\mathbf{S}_t$ be the sample covariance matrix of $\mathbf{t}_i$. Then $\widehat{\boldsymbol{\Pi}} = \mathbf{S}_t$ is consistent for $\boldsymbol{\Pi}$. When $\widehat{\boldsymbol{\Pi}}$ is nonsingular, which needs the sample size n to be greater than $p(p+3)/2$ at least, the GLS estimator $\tilde{\boldsymbol{\theta}}$ will be obtained by minimizing

$$D_{\mathrm{GLS}}\big[\bar{\mathbf{x}}, \mathbf{S}, \boldsymbol{\mu}(\boldsymbol{\theta}), \boldsymbol{\Sigma}(\boldsymbol{\theta})\big] = \big[\bar{\mathbf{t}} - \boldsymbol{\beta}(\boldsymbol{\theta})\big]'\widehat{\boldsymbol{\Pi}}^{-1}\big[\bar{\mathbf{t}} - \boldsymbol{\beta}(\boldsymbol{\theta})\big]. \tag{17}$$

The corresponding statistic for overall model evaluation is

$$T_{\mathrm{GLS}} = n D_{\mathrm{GLS}}\big[\bar{\mathbf{x}}, \mathbf{S}, \boldsymbol{\mu}(\tilde{\boldsymbol{\theta}}), \boldsymbol{\Sigma}(\tilde{\boldsymbol{\theta}})\big] \tag{18}$$

and is referred to χ^2_{df} for significance. For the asymptotic distribution of $\tilde{\boldsymbol{\theta}}$, we have

$$\sqrt{n}(\tilde{\boldsymbol{\theta}} - \boldsymbol{\theta}_0) \xrightarrow{\mathcal{L}} N(\mathbf{0}, \boldsymbol{\Omega}_{\mathrm{GLS}}), \tag{19}$$

where

$$\boldsymbol{\Omega}_{\mathrm{GLS}} = (\dot{\boldsymbol{\beta}}' \boldsymbol{\Pi}^{-1} \dot{\boldsymbol{\beta}})^{-1}.$$

It is obvious that, for normally distributed data, $\boldsymbol{\Pi} = \mathbf{W}^{-1}$ and $\boldsymbol{\Omega}_{\mathrm{GLS}} = \boldsymbol{\Omega}_{\mathrm{ML}} = \boldsymbol{\Omega}_{\mathrm{SW}}$. For non-normally distributed data, $\boldsymbol{\Omega}_{\mathrm{GLS}} \leqslant \boldsymbol{\Omega}_{\mathrm{SW}}$, thus $\tilde{\boldsymbol{\theta}}$ is asymptotically at least as efficient as $\hat{\boldsymbol{\theta}}$. However, empirically $\tilde{\boldsymbol{\theta}}$ can be much less efficient than $\hat{\boldsymbol{\theta}}$ even when data are non-normally distributed. Also, SE's of $\tilde{\boldsymbol{\theta}}$ based on (19), with $\widehat{\boldsymbol{\Pi}} = \mathbf{S}_t$, are too optimistic compared to those obtained by Monte-Carlo. Yuan and Bentler (1997b) proposed a corrected estimator of $\boldsymbol{\Omega}_{\mathrm{GLS}}$ by

$$\widehat{\boldsymbol{\Omega}}_{\mathrm{CGLS}} = \frac{n-1}{n - p(p+3)/2 - 2} \big(\dot{\boldsymbol{\beta}}'(\tilde{\boldsymbol{\theta}}) \mathbf{S}_t^{-1} \dot{\boldsymbol{\beta}}(\tilde{\boldsymbol{\theta}})\big)^{-1}, \tag{20}$$

which is also consistent for $\boldsymbol{\Omega}_{\mathrm{GLS}}$. Although SE's based on (20) are still smaller than the Monte-Carlo SE's, the biases are much smaller compared to those directly based on (19), especially when p is large.

Because T_{GLS} asymptotically follows χ^2_{df} as long as $\boldsymbol{\Pi}$ exists – without requiring any specific distribution assumption – the GLS procedure is commonly referred to as the asymptotically distribution free (ADF) method in the context of covariance structure analysis (Browne, 1984). The ADF property is desirable. However, the distribution of T_{GLS} may be far from χ^2_{df} for typical sample sizes encountered in practice. Most correctly specified models are rejected when T_{GLS} is referred to χ^2_{df} (Yuan and Bentler, 1997a). With T_{GLSc} being the corresponding statistic for CSA, Hu et al. (1992) showed that the mean and variance of T_{GLSc} are also much greater than those of $\chi^2_{df_c}$.

Note that, when $\mathbf{x} \sim N(\boldsymbol{\mu}, \boldsymbol{\Sigma})$, both $2\mathbf{D}_p^+(\mathbf{S} \otimes \mathbf{S})\mathbf{D}_p^{+\prime}$ and the sample covariance matrix, $\mathbf{S}_z$, of $\mathbf{z}_i = \mathrm{vech}\{(\mathbf{x}_i - \bar{\mathbf{x}})(\mathbf{x}_i - \bar{\mathbf{x}})'\}$, $i = 1, \ldots, n$, are consistent for $\boldsymbol{\Gamma}$. In the literature of CSA, the GLS procedure with weight given by $2^{-1}[\mathbf{D}_p^+(\mathbf{S} \otimes \mathbf{S})\mathbf{D}_p^{+\prime}]^{-1}$ is commonly called the normal theory GLS procedure, while the one with weight given by $\mathbf{S}_z^{-1}$ is commonly called the ADF procedure.

In an effort to find statistics that perform better in rejection rates with smaller n's, Yuan and Bentler (1997a) compared mean and covariance structure analysis with multivariate regression and suggested using

$$\widetilde{\boldsymbol{\Pi}} = \frac{1}{n} \sum_{i=1}^{n} \big(\mathbf{t}_i - \boldsymbol{\beta}(\tilde{\boldsymbol{\theta}})\big)\big(\mathbf{t}_i - \boldsymbol{\beta}(\tilde{\boldsymbol{\theta}})\big)'$$

in constructing a GLS discrepancy function. Because $\widetilde{\boldsymbol{\Pi}} \geqslant \widehat{\boldsymbol{\Pi}} = \mathbf{S}_t$, the greater type I error in T_{GLS} will be corrected when using $\widetilde{\boldsymbol{\Pi}}^{-1}$ as the weight matrix in formulating a

GLS procedure. Let the corrected GLS discrepancy function be

$$D_{\mathrm{CGLS}}\big[\bar{\mathbf{x}}, \mathbf{S}, \boldsymbol{\mu}(\boldsymbol{\theta}), \boldsymbol{\Sigma}(\boldsymbol{\theta})\big] = \big[\bar{\mathbf{t}} - \boldsymbol{\beta}(\boldsymbol{\theta})\big]'\widetilde{\boldsymbol{\Pi}}^{-1}\big[\bar{\mathbf{t}} - \boldsymbol{\beta}(\boldsymbol{\theta})\big]. \tag{21}$$

Because (21) involves $\widetilde{\boldsymbol{\Pi}}$, it seems that one has to obtain $\tilde{\boldsymbol{\theta}}$ before minimizing (21). Yuan and Bentler (1997a) showed that $\tilde{\boldsymbol{\theta}}$ also minimizes (21) and the relation

$$T_{\mathrm{CGLS}} = nD_{\mathrm{CGLS}}\big[\bar{\mathbf{x}}, \mathbf{S}, \boldsymbol{\mu}(\tilde{\boldsymbol{\theta}}), \boldsymbol{\Sigma}(\tilde{\boldsymbol{\theta}})\big] = T_{\mathrm{GLS}}/(1 + T_{\mathrm{GLS}}/n) \tag{22}$$

holds. Because T_{GLS}/n approaches zero, T_{CGLS} asymptotically follows χ^2_{df} as long as $\boldsymbol{\Pi}$ exists. Simulation results in Yuan and Bentler (1997a) imply that the mean of T_{CGLS} approximately equals df for all sample sizes across many distribution conditions. However, at smaller sample sizes T_{CGLS} tends to over-correct the behavior of T_{GLS} by having a type I error smaller than the nominal level.

Notice that T_{GLS} is in the form of Hotelling's T^2 statistic. It is well-known that T^2 approaches a chi-square distribution as $n \to \infty$, but using an F-distribution better describes the distribution of T^2. Motivated by this, Yuan and Bentler (1999c) further proposed the following F-statistic

$$T_{\mathrm{F}} = \frac{(n - df)T_{\mathrm{GLS}}}{(n - 1)df} \tag{23}$$

and suggested referring T_{F} to $F_{df,n-df}$, the F-distribution with df and $n-df$ degrees of freedom. It is obvious that T_{CGLS}, T_{GLS} and T_{F} are asymptotically equivalent. Actually, they also perform similarly when n is very large (Yuan and Bentler, 1999c). When n is small, the performance of T_{F} lies between that of T_{CGLS} and T_{GLS}, with a slight over-rejection on average, but with quite satisfactory performance overall.

Both T_{F} and T_{CGLS} are asymptotically distribution free and perform quite reliably when n is not too small. However, the GLS procedure has a drawback of nonconvergences at smaller sample sizes (see Hu et al., 1992; Yuan and Bentler, 1997a). When a nonconvergence occurs, $\tilde{\boldsymbol{\theta}}$ is not available nor is T_{F} or T_{CGLS}. In the context of covariance structure analysis, Browne (1984) also proposed a residual-based statistic. Extending this statistic to mean and covariance structure analysis yields

$$\begin{aligned} T_{\mathrm{RGLS}} = n\big[\bar{\mathbf{t}} - \boldsymbol{\beta}(\hat{\boldsymbol{\theta}})\big]'\big(\widehat{\boldsymbol{\Pi}}^{-1} - \widehat{\boldsymbol{\Pi}}^{-1}\dot{\boldsymbol{\beta}}(\hat{\boldsymbol{\theta}})\big[\dot{\boldsymbol{\beta}}'(\hat{\boldsymbol{\theta}})\widehat{\boldsymbol{\Pi}}^{-1}\dot{\boldsymbol{\beta}}(\hat{\boldsymbol{\theta}})\big]^{-1}\dot{\boldsymbol{\beta}}'(\hat{\boldsymbol{\theta}})\widehat{\boldsymbol{\Pi}}^{-1}\big) \\ \times \big[\bar{\mathbf{t}} - \boldsymbol{\beta}(\hat{\boldsymbol{\theta}})\big], \end{aligned} \tag{24}$$

where $\hat{\boldsymbol{\theta}}$ can be the MLE, a GLS or any other consistent estimator that is easier to obtain. Like T_{GLS}, T_{RGLS} approaches χ^2_{df} as n tends to infinity. However, T_{RGLS} also behaves like T_{GLS} for finite n. It over-rejects correct models too often unless n is very large. Parallel to T_{CGLS}, Yuan and Bentler (1998a) proposed T_{CRGLS} whose performance under H_0 is almost the same as T_{CGLS}, with an empirical mean approximately equal to df and some under-rejection of the correct model with small sample sizes (Bentler and Yuan, 1999; Yuan and Bentler, 1998a). Yuan and Bentler (1998a) also proposed a residual-based F-statistic T_{RF} based on T_{RGLS}; it also performs similarly to T_{F}.

The GLS procedures enjoy better asymptotic properties than the ML procedure. The statistics T_{RF} and T_{CRGLS} also enjoy nice finite sample properties. Because T_{RML} does not approach a chi-square distribution in general, conditions exist for T_{RML} to yield poor inferences. These have not been systematically explored.

4. Real robust procedures

Although the ML and GLS procedures enjoy some nice properties, they are based on modeling $\bar{\mathbf{x}}$ and $\mathbf{S}$, which are efficient only when $\mathbf{x} \sim N(\boldsymbol{\mu}, \boldsymbol{\Sigma})$. When a data set possesses moderately heavy tails, $\bar{\mathbf{x}}$ and $\mathbf{S}$ will be inefficient estimates of their population counterparts (Kano et al., 1993; Tyler, 1983). When the heavy tails are severe, the population 4th-order moment matrix $\boldsymbol{\Gamma}$ may not exist, and the previously obtained $\hat{\boldsymbol{\theta}}$ and $\tilde{\boldsymbol{\theta}}$ have unbounded covariance matrices. Heavy tails can be caused by outliers. Since one outlier in a sample can move an element of $\bar{\mathbf{x}}$ or $\mathbf{S}$ to an arbitrary value, results obtained by ML or GLS procedures may have little meaning. When the sample contains outliers or $\mathbf{x}$ has heavy tails, a robust procedure can provide better parameter estimates and more reliable model evaluation.

Following Huber (1964), many robust statistical methods have been developed (e.g., Huber, 1981; Hampel et al., 1986; Hubert et al., 2004; Rousseeuw and Leroy, 1987; Wilcox, 2004). Maronna (1976) obtained the properties of M-estimators for the population mean vector and covariance matrix. Lopuhaä (1989, 1991) and Rocke (1996) studied other estimators that can allow the sample to contain nearly 50% contaminated observations. We will mainly use M-estimators for robust SEM procedures, aiming for samples containing a proportion of contaminated but not extreme observations.[1] There exist two main approaches to robust SEM. One is to first get robust estimates $\hat{\boldsymbol{\mu}}$ and $\widehat{\boldsymbol{\Sigma}}$, then treat these estimates as $\bar{\mathbf{x}}$ and $\mathbf{S}$ and use ML or GLS procedures for further analysis. The other fits the structural model to raw data directly by a robust method without explicitly estimating the saturated model. We will call the first one the two-stage procedure and the second the direct procedure. We will also discuss related procedures for robustness and robust procedures for related models.

4.1. *Two-stage procedures*

In a robust estimation process for $\hat{\boldsymbol{\mu}}$ and $\widehat{\boldsymbol{\Sigma}}$, cases are differentiated by their distances to the center of the majority of the data, as measured by

$$d(\mathbf{x}, \boldsymbol{\mu}, \boldsymbol{\Sigma}) = \left[(\mathbf{x} - \boldsymbol{\mu})'\boldsymbol{\Sigma}^{-1}(\mathbf{x} - \boldsymbol{\mu})\right]^{1/2}.$$

The farther $\mathbf{x}_i$ is from the center, the less weight it will get. Let $u_1(\cdot)$ and $u_2(\cdot)$ be decreasing weight functions. Robust M-estimates $\hat{\boldsymbol{\mu}}$ and $\widehat{\boldsymbol{\Sigma}}$ can be obtained by iteratively solving

$$\boldsymbol{\mu} = \sum_{i=1}^{n} u_1(d_i)\mathbf{x}_i \Big/ \sum_{i=1}^{n} u_1(d_i), \tag{25a}$$

$$\boldsymbol{\Sigma} = \sum_{i=1}^{n} u_2\left(d_i^2\right)(\mathbf{x}_i - \boldsymbol{\mu})(\mathbf{x}_i - \boldsymbol{\mu})'/n, \tag{25b}$$

[1] When a large proportion of the sample is contaminated, the sample may come from two or several distributions. Then a mixture model might be more appropriate (Arminger et al., 1999; Hoshino, 2001; Lee and Song, 2003; Muthén, 2001; Yung, 1997). This topic is beyond the scope of this chapter.

where $d_i = d(\mathbf{x}_i, \boldsymbol{\mu}, \boldsymbol{\Sigma})$. Two commonly used classes of weight functions for M-estimators are Huber-type weights and weights based on a multivariate t-distribution. Let ρ be the percentage of influential cases one wants to control, and r be a constant determined by ρ through $P(\chi_p^2 > r^2) = \rho$. The Huber-type weights are given by

$$u_1(d) = \begin{cases} 1, & \text{if } d \leqslant r, \\ r/d, & \text{if } d > r \end{cases} \tag{26}$$

and $u_2(d^2) = \{u_1(d)\}^2/\varrho$, where ϱ is a constant determined by ρ through $E\{\chi_p^2 u_2(\chi_p^2)\} = p$. The purpose of ϱ is to make $\widehat{\boldsymbol{\Sigma}}$ unbiased for $\boldsymbol{\Sigma}$ when $\mathbf{x} \sim N(\boldsymbol{\mu}, \boldsymbol{\Sigma})$. The weights corresponding to a p-variate t-distribution with m degrees of freedom are given by

$$u_1(d_i) = u_2(d_i^2) = (p+m)/(m+d_i^2). \tag{27}$$

Notice that the only tuning parameter in using (26) is ρ and that in using (27) is m.

Motivated by the formula for calculating the sample covariance matrix, Campbell (1980) defined another form of M-estimator by solving

$$\boldsymbol{\mu} = \sum_{i=1}^n u(d_i)\mathbf{x}_i / \sum_{i=1}^n u_i \tag{28a}$$

and

$$\boldsymbol{\Sigma} = \sum_{i=1}^n u^2(d_i)(\mathbf{x}_i - \boldsymbol{\mu})(\mathbf{x}_i - \boldsymbol{\mu})' \Big/ \left(\sum_{i=1}^n u^2(d_i) - 1\right), \tag{28b}$$

where $u(d) = w(d)/d$ with

$$w(d) = \begin{cases} d, & \text{if } d \leqslant d_0, \\ d_0 \exp\{-\frac{1}{2}(d-d_0)^2/b_2^2\}, & \text{if } d > d_0, \end{cases} \tag{29}$$

$d_0 = \sqrt{p} + b_1/\sqrt{2}$, b_1 and b_2 are constants. So there are two tuning parameters in (29). Based on empirical experience, Campbell (1980) suggested $b_1 = 2$ and $b_2 = \infty$ or $b_1 = 2$ and $b_2 = 1.25$. When $b_1 = 2$ and $b_2 = \infty$, (28) defines a Huber-type M-estimator; when $b_1 = 2$, $b_2 = 1.25$, (28) defines Hampel-type redescending M-estimator (Hampel, 1974); (28) leads to $\hat{\boldsymbol{\mu}} = \bar{\mathbf{x}}$ and $\widehat{\boldsymbol{\Sigma}} = \mathbf{S}$ when choosing $b_1 = \infty$.

Robust M-estimation for $\boldsymbol{\mu}_0$ and $\boldsymbol{\Sigma}_0$ was motivated by ML within the class of elliptical distributions (see Maronna, 1976). However, $\widehat{\boldsymbol{\Sigma}}$ generally does not approach $\boldsymbol{\Sigma}_0 = \text{Cov}(\mathbf{x})$ within the class of elliptical distributions. Instead, it approaches $\alpha\boldsymbol{\Sigma}_0$ for a scalar $\alpha > 0$. When the population elliptical distribution is known, one can theoretically calculate α and rescale $\widehat{\boldsymbol{\Sigma}}$ to make it consistent for $\boldsymbol{\Sigma}_0$. Because essentially all commonly used covariance structure models are ICSF, such an inconsistency will not cause any problems for statistical inference when treating $\alpha\boldsymbol{\Sigma}$ as the covariance matrix. See Yuan and Bentler (1998c) and Yuan et al. (2004a) for further discussion on this aspect. Thus we may still use $\boldsymbol{\Sigma}_0$ instead of $\alpha\boldsymbol{\Sigma}_0$ in the following development.

When replacing $\bar{\mathbf{x}}$ and $\mathbf{S}$ in (2), (4), (6), (17) or (21) by $\hat{\boldsymbol{\mu}}$ and $\widehat{\boldsymbol{\Sigma}}$, we will obtain the corresponding robust estimates for $\boldsymbol{\theta}$. Let these robust estimates be denoted by $\hat{\boldsymbol{\theta}}$ or $\tilde{\boldsymbol{\theta}}$, as before. Then all the robust properties of $\hat{\boldsymbol{\mu}}$ and $\widehat{\boldsymbol{\Sigma}}$ will be inherited by $\hat{\boldsymbol{\theta}}$ or $\tilde{\boldsymbol{\theta}}$ (see Yuan and Bentler, 1998b, 1998c). In order to have proper SE's for parameter estimates and test statistics for overall model evaluation, we need a consistent estimate for the covariance matrix of $\hat{\boldsymbol{\beta}} = (\hat{\boldsymbol{\mu}}', \hat{\boldsymbol{\sigma}}')'$, where $\hat{\boldsymbol{\sigma}} = \text{vech}(\widehat{\boldsymbol{\Sigma}})$. Yuan and Bentler (1998c) proposed to use the estimating equation approach to characterize the asymptotic distribution of $\hat{\boldsymbol{\beta}}$ and to estimate its covariance matrix. It is obvious that $\hat{\boldsymbol{\mu}}$ and $\hat{\boldsymbol{\sigma}}$ satisfy

$$\frac{1}{n}\sum_{i=1}^{n} \mathbf{g}_1(\mathbf{x}_i, \hat{\boldsymbol{\mu}}, \widehat{\boldsymbol{\Sigma}}) = \mathbf{0} \quad \text{and} \quad \frac{1}{n}\sum_{i=1}^{n} \mathbf{g}_2(\mathbf{x}_i, \hat{\boldsymbol{\mu}}, \widehat{\boldsymbol{\Sigma}}) = \mathbf{0},$$

where

$$\mathbf{g}_1(\mathbf{x}, \boldsymbol{\mu}, \boldsymbol{\Sigma}) = u_1\big\{d(\mathbf{x}, \boldsymbol{\mu}, \boldsymbol{\Sigma})\big\}(\mathbf{x} - \boldsymbol{\mu})$$

and

$$\mathbf{g}_2(\mathbf{x}, \boldsymbol{\mu}, \boldsymbol{\Sigma}) = u_2\big\{d^2(\mathbf{x}, \boldsymbol{\mu}, \boldsymbol{\Sigma})\big\}\,\text{vech}\big[(\mathbf{x} - \boldsymbol{\mu})(\mathbf{x} - \boldsymbol{\mu})'\big] - \boldsymbol{\sigma}$$

corresponding to (25); and

$$\mathbf{g}_1(\mathbf{x}, \boldsymbol{\mu}, \boldsymbol{\Sigma}) = u\big\{d(\mathbf{x}, \boldsymbol{\mu}, \boldsymbol{\Sigma})\big\}(\mathbf{x} - \boldsymbol{\mu})$$

and

$$\mathbf{g}_2(\mathbf{x}, \boldsymbol{\mu}, \boldsymbol{\Sigma}) = u^2\big\{d(\mathbf{x}, \boldsymbol{\mu}, \boldsymbol{\Sigma})\big\}\,\text{vech}\big\{(\mathbf{x} - \boldsymbol{\mu})(\mathbf{x} - \boldsymbol{\mu})' - \boldsymbol{\Sigma}\big\} + \frac{1}{n}\boldsymbol{\sigma}$$

corresponding to (28). Let $\mathbf{g} = (\mathbf{g}_1', \mathbf{g}_2')'$ and $\dot{\mathbf{g}} = \partial\mathbf{g}/\partial\boldsymbol{\beta}'$, then (see Yuan and Bentler, 1998c)

$$\sqrt{n}\begin{pmatrix} \hat{\boldsymbol{\mu}} - \boldsymbol{\mu}_0 \\ \hat{\boldsymbol{\sigma}} - \boldsymbol{\sigma}_0 \end{pmatrix} \xrightarrow{\mathcal{L}} N(\mathbf{0}, \mathbf{V}), \tag{30a}$$

where $\mathbf{V} = \mathbf{H}^{-1}\mathbf{B}\mathbf{H}'^{-1}$ with

$$\mathbf{H} = E\big\{\dot{\mathbf{g}}(\mathbf{x}, \boldsymbol{\mu}_0, \boldsymbol{\sigma}_0)\big\} \quad \text{and} \quad \mathbf{B} = E\big\{\mathbf{g}(\mathbf{x}, \boldsymbol{\mu}_0, \boldsymbol{\sigma}_0)\mathbf{g}'(\mathbf{x}, \boldsymbol{\mu}_0, \boldsymbol{\sigma}_0)\big\}. \tag{30b}$$

In contrast to (9), we only require $\boldsymbol{\Sigma}_0$ to exist in order for (30) to hold. A consistent estimate of $\mathbf{V}$ can be obtained by using consistent estimates for $\mathbf{H}$ and $\mathbf{B}$; these are given by

$$\widehat{\mathbf{H}} = \frac{1}{n}\sum_{i=1}^{n} \dot{\mathbf{g}}(\mathbf{x}_i, \hat{\boldsymbol{\mu}}, \hat{\boldsymbol{\sigma}}) \quad \text{and} \quad \widehat{\mathbf{B}} = \frac{1}{n}\sum_{i=1}^{n} \mathbf{g}(\mathbf{x}_i, \hat{\boldsymbol{\mu}}, \hat{\boldsymbol{\sigma}})\mathbf{g}'(\mathbf{x}_i, \hat{\boldsymbol{\mu}}, \hat{\boldsymbol{\sigma}}).$$

Now, with $\bar{\mathbf{x}}$, $\mathbf{S}$ and $\widehat{\boldsymbol{\Pi}}$ being replaced by $\hat{\boldsymbol{\mu}}$, $\widehat{\boldsymbol{\Sigma}}$ and $\widehat{\mathbf{V}}$, respectively, all the procedures in the previous two subsections can be applied in a robust manner.

Let $\mathbf{V}_{11}$ and $\mathbf{V}_{22}$ be the submatrices of $\mathbf{V}$ corresponding to asymptotic covariance matrices of $\hat{\boldsymbol{\mu}}$ and $\hat{\boldsymbol{\sigma}}$, respectively. When data follow an elliptical distribution, $\hat{\boldsymbol{\mu}}$ and $\hat{\boldsymbol{\sigma}}$ are asymptotically independent and (see Maronna, 1976; Tyler, 1983; Yuan and Bentler, 1998b)

$$\mathbf{V}_{11} = \tau_1\boldsymbol{\Sigma}_0, \quad \text{and} \quad \mathbf{V}_{22} = 2\tau_2\mathbf{D}_p^{+}(\boldsymbol{\Sigma}_0 \otimes \boldsymbol{\Sigma}_0)\mathbf{D}_p^{+'} + \tau_3\boldsymbol{\sigma}_0\boldsymbol{\sigma}_0', \tag{31}$$

where the scalars τ_1 to τ_3 are related to the underlying distribution of the data and the weights used in the estimation. A consistent estimator for $\mathbf{V}$ can also be obtained based on (31) when data are truly elliptically distributed. The one based on (30) is more robust against violation of distribution assumptions (Yuan and Jennrich, 1998). The value of (31) is that, when replacing $\mathbf{S}$ by $\widehat{\boldsymbol{\Sigma}}$ and $\widehat{\boldsymbol{\Gamma}}$ by $\widehat{\mathbf{V}}_{22}$, the rescaled statistic $T_{\mathrm{RML}c}$ for CSA asymptotically follows $\chi^2_{df_c}$. Parallel to modeling $\bar{\mathbf{x}}$ and $\mathbf{S}$ in mean and covariance structure analysis, the rescaled statistic T_{RML} when modeling $\hat{\boldsymbol{\mu}}$ and $\widehat{\boldsymbol{\Sigma}}$ will only approach a distribution whose mean equals df. Of course, we can also obtain the MLE when the specific elliptical distribution form of $\mathbf{x}$ is known (see Fang et al., 1990; Kano et al., 1993). However, in real data analysis, it is unlikely that we would know the exact distributional form for a given data set.

The Huber-type weight, the redescending weight, and the weight based on a multivariate t-distribution, can all effectively control the influence of heavy tails or outliers. However, differences exist among them. Based on empirical experience, Yuan and Bentler (1998b, 1998c) found that, in Huber-type estimators, the effect of abnormal cases is down-weighted but not eliminated. If data are nearly normal, the estimators based on Huber-type weights are still highly efficient. So Huber-type weights are better used for data sets whose distributions are not too far away from normal. By using redescending weights, the effect of outlying cases can be minimized; this is almost equivalent to outlier removal. But the estimators will lose efficiency as compared to those based on Huber-type weights when data are approximately normal. Yuan and Bentler (1998c) also found that the tuning parameters for the redescending M-estimator recommended by Campbell ($b_1 = 2$ and $b_2 = 1.5$) leads to many cases with essentially zero weights, resulting in near singular $\widehat{\boldsymbol{\Sigma}}$ and $\widehat{\mathbf{V}}$. Thus, b_2 has to be adjusted upwards to have a proper redescending effect. The weights based on a multivariate t-distribution are best used for a data set whose spread can be approximately described by the t-distribution, or for a data set with heavy tails but with no obvious outliers. In addition to these rough guidelines, statistical procedures have also been proposed for choosing proper weights. These have been studied mainly in the context of CSA.

For choosing the tuning parameter ρ in using the Huber-type weights of (26), let $u_{2i} = u_2\{d^2(\mathbf{x}_i, \hat{\boldsymbol{\mu}}, \widehat{\boldsymbol{\Sigma}})\}$ and

$$\mathbf{x}_i^{(\rho)} = \left\{\sqrt{u_{2i}}(\mathbf{x}_i - \hat{\boldsymbol{\mu}})\right\} \tag{32}$$

at the convergence of (25). Then we can rewrite (25b) as

$$\widehat{\boldsymbol{\Sigma}} = \frac{1}{n}\sum_{i=1}^{n} \mathbf{x}_i^{(\rho)} \mathbf{x}_i^{(\rho)\prime},$$

which is just the sample covariance matrix of the $\mathbf{x}_i^{(\rho)}$. Yuan et al. (2000) proposed using (32) as a transformation formula. Working with several practical data sets, they found that the transformed samples $\mathbf{x}_i^{(\rho)}$ are much better approximated by a multivariate normal distribution than the original sample $\mathbf{x}_i$, as measured by Mardia's (1970) multivariate skewness

$$M_1 = \frac{1}{n^2}\sum_{i,j=1}^{n} \left\{(\mathbf{x}_i - \bar{\mathbf{x}})'\mathbf{S}^{-1}(\mathbf{x}_j - \bar{\mathbf{x}})\right\}^3$$

and kurtosis

$$M_2 = \frac{1}{n}\sum_{i=1}^{n}\{(\mathbf{x}_i - \bar{\mathbf{x}})'\mathbf{S}^{-1}(\mathbf{x}_i - \bar{\mathbf{x}})\}^2.$$

Considering that the sample covariance matrix is most efficient when data are normally distributed, they further proposed using

$$z_{M_2} = [M_2 - p(p+2)]/[8p(p+2)/n]^{1/2} \sim N(0, 1)$$

as a criterion in choosing the tuning parameter ρ. This can be done by starting at 0 and incrementing by a step of size 0.05 until z_{M_2} corresponding to $\mathbf{x}_i^{(\rho)}$ is no longer significant. By transforming (32) further into

$$\mathbf{x}_{i0}^{(\rho)} = \boldsymbol{\Sigma}^{1/2}(\hat{\boldsymbol{\theta}})\widehat{\boldsymbol{\Sigma}}^{-1/2}\mathbf{x}_i^{(\rho)}$$

and resampling from the sample represented by $\mathbf{x}_{i0}^{(\rho)}$, Yuan and Hayashi (2003) proposed to use a bootstrap procedure to choose ρ. They discussed a rationale for choosing ρ that makes the statistic $T_{\text{ML}c}$ approximately following $\chi^2_{df_c}$. Using real data sets, they showed that $T_{\text{ML}c}$ applied to a properly transformed sample can closely follow $\chi^2_{df_c}$, while the behavior of $T_{\text{GLS}c}$ and $T_{\text{RML}c}$ remains significantly different from $\chi^2_{df_c}$ when applied to the sample $\mathbf{x}_{i0}^{(\rho)}$. It is obvious that these approaches to selecting ρ are equally applicable to selecting m when using the weights based on a multivariate t-distribution. They can also be applied to selecting b_1 and b_2 in applying (28) and (29) by defining

$$u_{2i} = nu_i^2 \Big/ \left(\sum_{i=1}^{n} u_i^2 - 1\right)$$

in (32).

Because efficiency of parameter estimates is one of the most important concerns in promoting a statistical methodology, Yuan et al. (2004a) proposed to use the empirical efficiency of $\hat{\boldsymbol{\theta}}$ in selecting a particular transformation or weight among different downweighting procedures. For such a purpose, we need to focus on a set of invariant parameters. Within the context of LISREL models (Jöreskog and Sörbom, 1996, pp. 1–3), when factor loadings are fixed at 1.0 to identify the scales of latent variables, all the free coefficients in the measurement and structural models are invariant parameters. Working with several real data sets, Yuan et al. (2004a) found that Huber-type weights in (26) often lead to the most efficient parameter estimates.

4.2. *Direct procedures*

In the two-stage approaches, cases are judged according to their distances from $\boldsymbol{\mu}$. In mean and covariance structure analysis, the model structure reflects a substantive theory. A case lying far from the saturated $\boldsymbol{\mu}$ may not necessarily be far from the model structure. This parallels regression where not all leverage points are necessarily outliers. Thus, it is more reasonable to weight cases according to their distances from the structural model. Let $\mathbf{y}_i = (\mathbf{x}_i', \text{vech}'(\mathbf{x}_i\mathbf{x}_i'))'$ and

$$\boldsymbol{\nu}(\boldsymbol{\theta}) = \begin{pmatrix} \boldsymbol{\mu}(\boldsymbol{\theta}) \\ \text{vech}[\boldsymbol{\Sigma}(\boldsymbol{\theta}) + \boldsymbol{\mu}(\boldsymbol{\theta})\boldsymbol{\mu}'(\boldsymbol{\theta})] \end{pmatrix}.$$

Then mean and covariance structure analysis can be expressed as the regression model

$$\mathbf{y}_i = \boldsymbol{\nu}(\boldsymbol{\theta}) + \mathbf{e}_i, \quad i = 1, 2, \ldots, n,$$

where $\mathbf{e}_i$ is the error term. Parallel to robust regression (see e.g., Holland and Welsch, 1977), we may define a robust estimator $\hat{\boldsymbol{\theta}}$ by solving

$$\sum_{i=1}^{n} \dot{\boldsymbol{\nu}}'(\boldsymbol{\theta})\mathbf{W}_i(\boldsymbol{\theta})\big(\mathbf{y}_i - \boldsymbol{\nu}(\boldsymbol{\theta})\big) = \mathbf{0}, \tag{33}$$

where $\mathbf{W}_i(\boldsymbol{\theta})$ is a weight matrix with dimension $p(p+3)/2$. In order for the estimator $\hat{\boldsymbol{\theta}}$ to be robust, the $\mathbf{W}_i$ has to explicitly account for the distance of the ith case from the model. Yuan and Bentler (2000c) proposed

$$\mathbf{W}_i = u(d_i)\mathbf{W},$$

where $\mathbf{W}$ does not depend on i, $u(\cdot)$ is a decreasing function, and

$$d_i^2 = \big(\mathbf{y}_i - \boldsymbol{\nu}(\boldsymbol{\theta})\big)'\mathbf{W}\big(\mathbf{y}_i - \boldsymbol{\nu}(\boldsymbol{\theta})\big). \tag{34}$$

As with the two-stage procedures, there are several ways to choose the function $u(\cdot)$ in controlling the influence of $\mathbf{y}_i$ on $\hat{\boldsymbol{\theta}}$. We may choose $u(\cdot)$ to be the Huber-type weight as in (26) or the weight based on a multivariate t-distribution as in (27). Yuan and Bentler (2000c) used $u(d) = w(d)/d$ with $w(\cdot)$ being given in (29), d being given in (34) and

$$\mathbf{W}(\boldsymbol{\theta}) = \left[\sum_{i=1}^{n} u_i^2\big(\mathbf{y}_i - \boldsymbol{\nu}(\boldsymbol{\theta})\big)\big(\mathbf{y}_i - \boldsymbol{\nu}(\boldsymbol{\theta})\big)' \Big/ \left(\sum_{i=1}^{n} u_i^2 - 1\right)\right]^{-1},$$

where $u_i = u(d_i)$. Once the $u(\cdot)$ and the $\mathbf{W}$ are chosen, $\hat{\boldsymbol{\theta}}$ can be obtained through the following iterative procedure:

Step 1. Choosing a positive definite matrix $\mathbf{W}^{(1)}$ and initial estimator $\boldsymbol{\theta}^{(1)}$.

Step 2. With the jth-step estimates $\boldsymbol{\theta}^{(j)}$ and $\mathbf{W}^{(j)}$, the $(j+1)$th-step estimates are given by

$$\boldsymbol{\theta}^{(j+1)} = \boldsymbol{\theta}^{(j)} + \Delta\boldsymbol{\theta}^{(j)}$$

with

$$\Delta\boldsymbol{\theta}^{(j)} = \left(\sum_{i=1}^{n} u_{ij}\dot{\boldsymbol{\nu}}^{(j)\prime}\mathbf{W}^{(j)}\dot{\boldsymbol{\nu}}^{(j)}\right)^{-1} \sum_{i=1}^{n} u_{ij}\dot{\boldsymbol{\nu}}^{(j)\prime}\mathbf{W}^{(j)}\big(\mathbf{y}_i - \boldsymbol{\nu}^{(j)}\big), \tag{35}$$

and

$$\mathbf{W}^{(j+1)} = \left[\frac{1}{(\sum_{i=1}^{n} u_{ij}^2 - 1)} \sum_{i=1}^{n} u_{ij}^2\big(\mathbf{y}_i - \boldsymbol{\nu}^{(j)}\big)\big(\mathbf{y}_i - \boldsymbol{\nu}^{(j)}\big)'\right]^{-1}, \tag{36}$$

where $\boldsymbol{\nu}^{(j)} = \boldsymbol{\nu}(\boldsymbol{\theta}^{(j)})$, $\dot{\boldsymbol{\nu}}^{(j)} = \dot{\boldsymbol{\nu}}(\boldsymbol{\theta}^{(j)})$, $u_{ij} = u(d_{ij})$ with d_{ij} being (34) evaluated at $\boldsymbol{\theta}^{(j)}$ and $\mathbf{W}^{(j)}$.

Step 3. For a prespecified small number ε, repeat step 2 until $\|\Delta\boldsymbol{\theta}^{(j)}\| < \varepsilon$.

Since $\mathbf{W}^{(j)}$ does not change from case to case in each iteration, we can also write (35) as

$$\Delta\boldsymbol{\theta}^{(j)} = \left(\dot{\boldsymbol{\nu}}^{(j)\prime}\mathbf{W}^{(j)}\dot{\boldsymbol{\nu}}^{(j)}\right)^{-1}\dot{\boldsymbol{\nu}}^{(j)\prime}\mathbf{W}^{(j)}\left(\tilde{\mathbf{y}} - \boldsymbol{\nu}^{(j)}\right), \tag{37}$$

where $\tilde{\mathbf{y}} = \sum_{i=1}^{n} u_{ij}\mathbf{y}_i/(\sum_{i=1}^{n} u_{ij})$. Eqs. (35) and (37) are parallel to (28a), and (36) is parallel to (28b). The main difference between the two-stage and the direct approaches is that in (35) and (36) each case gets its weight based on its distance to the structural model $\boldsymbol{\nu}(\boldsymbol{\theta})$ rather than to the saturated mean as in (28). Actually, (35) and (36) do not generate $\hat{\boldsymbol{\mu}}$ and $\widehat{\boldsymbol{\Sigma}}$ corresponding to the saturated model. Steps 1 to 3 constitute the well-known iteratively reweighted least squares (IRLS) algorithm. The convergence properties of this algorithm were studied by Holland and Welsch (1977), Rubin (1983) and Green (1984).

The IRLS procedure leads to a robust estimate $\hat{\boldsymbol{\theta}}$, but it does not provide a way to obtain the SE's of $\hat{\boldsymbol{\theta}}$ or to evaluate the overall model structure. Similar problems also exist in robust regression. Although the regression model is a saturated model where a statistic for overall model evaluation is not relevant, SE's for robust regression coefficients are necessary and have been well-discussed in the literature (see Huber, 1973; Gross, 1977; Carroll, 1979; Birch and Myers, 1982). Motivated by SE's in robust regression, Yuan and Bentler (2000c) proposed to obtain SE's of $\hat{\boldsymbol{\theta}}$ similar to that for the GLS procedures in the previous section. Denote the estimated weights at convergence of (35) and (36) by $\hat{u}_i$ and $\widehat{\mathbf{W}}$ with $\widehat{\mathbf{W}}_i = \hat{u}_i\widehat{\mathbf{W}}$. Then, $\hat{\boldsymbol{\theta}}$ satisfying (33) corresponds to minimizing the GLS function

$$\sum_{i=1}^{n}\left(\mathbf{y}_i - \boldsymbol{\nu}(\boldsymbol{\theta})\right)'\widehat{\mathbf{W}}_i\left(\mathbf{y}_i - \boldsymbol{\nu}(\boldsymbol{\theta})\right).$$

However, $n_1 = \sum_{i=1}^{n}\hat{u}_i$ and $n_2 = \sum_{i=1}^{n}\hat{u}_i^2$ play a similar role as sample size, and hence they need to be properly accounted for in obtaining SE's and formulating a test statistic for overall model evaluation. Comparing (33) and the IRLS procedure to those in robust regression (see Rousseeuw and Leroy, 1987, pp. 44–45), Yuan and Bentler (2000c) suggested using

$$\sqrt{n_1}(\hat{\boldsymbol{\theta}} - \boldsymbol{\theta}_0) \xrightarrow{\mathcal{L}} N(\mathbf{0}, \boldsymbol{\Omega}),$$

where

$$\boldsymbol{\Omega} = \frac{(n_2 - 1)}{n_1}(\dot{\boldsymbol{\nu}}'\mathbf{W}\dot{\boldsymbol{\nu}})^{-1},$$

for the SE's of $\hat{\boldsymbol{\theta}}$. They also suggested referring

$$T_{\mathrm{IRLS}} = \frac{n_1^2}{(n_2 - 1)}(\tilde{\mathbf{y}} - \hat{\boldsymbol{\nu}})'\mathbf{W}(\tilde{\mathbf{y}} - \hat{\boldsymbol{\nu}})$$

to χ^2_{df} for the overall model evaluation.

The above IRLS procedure is totally decided by $u(\cdot)$, d_i and $\mathbf{W}$. When $u_i = 1$ and $\mathbf{W} = \mathbf{S}_y^{-1}$ with $\mathbf{S}_y$ being the sample covariance matrix of $\mathbf{y}_i$, then the above inference procedure yields the GLS procedures discussed in Section 3. The statistic T_{IRLS} will be equivalent to T_{CGLS}. In the context of covariance structure analysis, Yuan et al. (2004b) proposed to use $\mathbf{W}(\boldsymbol{\theta}) = [\mathrm{Cov}(\mathbf{y}_i)]^{-1}$ when assuming $\mathbf{x}_i \sim N(\boldsymbol{\mu}, \boldsymbol{\Sigma}(\boldsymbol{\theta}))$, $u(\cdot)$ as in (26), and a d_i based on residuals from predicting the factor score. Examples show that the residual based d_i can effectively identify cases that deviate from a theoretical model structure. When these cases are downweighted according to a Huber-type weight, the model fits the data better, and reliability coefficients of a measurement scale can increase.

The principle for selecting turning parameters as discussed for two-stage approaches also applies in the direct approach. The bootstrap can be used to compare the empirical efficiency of a set of invariant parameter estimates. With given tuning parameters, one may also use the bootstrap to obtain the SE's of $\hat{\boldsymbol{\theta}}$ and to test model hypothesis based on T_{IRLS}.

4.3. *Other approaches to robustness and related procedures*

Techniques for identifying outliers or influential cases are also useful when facing data containing heavy tails or outliers. Berkane and Bentler (1988), Bollen and Arminger (1991), Lee and Wang (1996), Poon and Poon (2002), and Poon and Wong (2004) provided various procedures for identifying outliers. With multiple outliers, masking effect may cause difficulty in identifying the true ones (Fung, 1993; Poon et al., 2000; Rousseeuw and van Zomeren, 1990). When true outliers are not obvious, the idea of using empirical efficiency to compare different procedures can also be used to decide whether to keep or remove certain cases or outliers. Evidence in Yuan and Hayashi (2003) and Yuan et al. (2004a) implies that outlier removal is not as efficient as a proper robust procedure.

Model comparison is of fundamental interest in statistics and SEM as well, and cross-validation is a universally accepted criterion for model selection (Stone, 1974). When a sample contains heavy tails or outliers, the model that fits the majority of the data may not cross-validate well. Cross-validation combined with a robust procedure will give a model the proper merit it deserves. Yuan et al. (2002b) discussed cross-validation with robust covariance matrices. The same idea can be applied to mean and covariance structure analysis using two-stage as well as direct robust procedures.

The success of robust estimation requires having a weighting scheme so that cases lying far from the center of the data cloud or from the model are properly downweighted. When the underlying distribution of $\mathbf{x}$ is known, then the ML procedure is generally preferred. Similar to robust estimation, an ML procedure based on a distribution having heavy tails can also properly control the effect of influential cases. Recent development in Markov chain Monte Carlo can allow $\mathbf{x}$ to have rather complicated distribution where ML methodology is still feasible (see Lee and Xia, In press). When the chosen distribution is misspecified, MLE other than those based on $\mathbf{x} \sim N(\boldsymbol{\mu}, \boldsymbol{\Sigma})$ may not be consistent (Gourieroux et al., 1984), but they may still control the effect of influential cases.

Due to the advance of computing power, robust procedures have been developed and applied to almost every aspect of statistics and data analysis[2] (see e.g., Maddala and Rao, 1997). This chapter has focused mainly on robust procedures for SEM. Robust procedures for two closely related procedures, principal components and exploratory factor analysis, are also well developed. Some useful sources on robust principal components are Devlin et al. (1981), Ruymgaart (1981), Cui et al. (2003), Ibazizen and Dauxois (2003), and Hubert et al. (2005); and on factor analysis are Yuan et al. (2002a) and Pison et al. (2003).

Statistical theory for robust estimation has been developed primarily within the class of elliptical distributions (Huber, 1981; Hampel et al., 1986), mainly because analyzing $\widehat{\boldsymbol{\Sigma}}$ and $\mathbf{S}$ leads to the same substantive conclusion. In practice, data might contain outliers which will make a true symmetric distribution skewed at the sample level. In such a situation, a robust procedure is definitely preferred (Yuan and Bentler, 2001b; Yuan et al., 2000). If one believes that the true distribution of $\mathbf{x}$ is skewed, then the results corresponding to analyzing $\hat{\boldsymbol{\mu}}$ and $\widehat{\boldsymbol{\Sigma}}$ may not be substantively equivalent to those of analyzing $\bar{\mathbf{x}}$ and $\mathbf{S}$. Hampel et al.'s (1986, p. 401) discussion implies that robust procedures might still be preferred even when $\mathbf{x}$ has a skewed distribution. We recommend that robust and classical procedures be compared when having a data set with a highly significant z_{M_2}. We illustrate this comparison in Section 6.

5. Misspecified models

We have discussed the properties of parameter estimates and test statistics with correctly specified models. When models are misspecified, the results presented in Sections 2 to 4 may no longer hold. This section will present parallel results for misspecified models. Because most technical developments for misspecified models are based on ML, we will mainly discuss the ML procedure. Most results also hold for the GLS and the robust procedures. We will discuss consistency and asymptotic normality of the parameter estimates, consistency of standard errors, the distribution of the LR and rescaled statistics.

Whether or not the model is correctly specified, there exists

$$D_{\mathrm{ML}}\big[\bar{\mathbf{x}}, \mathbf{S}, \boldsymbol{\mu}(\boldsymbol{\theta}), \boldsymbol{\Sigma}(\boldsymbol{\theta})\big] \xrightarrow{P} D_{\mathrm{ML}}\big[\boldsymbol{\mu}_0, \boldsymbol{\Sigma}_0, \boldsymbol{\mu}(\boldsymbol{\theta}), \boldsymbol{\Sigma}(\boldsymbol{\theta})\big]. \tag{38}$$

Let $\boldsymbol{\theta}_*$ be the vector that minimizes $D_{\mathrm{ML}}[\boldsymbol{\mu}_0, \boldsymbol{\Sigma}_0, \boldsymbol{\mu}(\boldsymbol{\theta}), \boldsymbol{\Sigma}(\boldsymbol{\theta})]$. Then, under standard regularity conditions $\hat{\boldsymbol{\theta}}$ will converge to $\boldsymbol{\theta}_*$ according to (38) (see Kano, 1986; Shapiro, 1984). In general, $\boldsymbol{\theta}_*$ does not equal $\boldsymbol{\theta}_0$. Thus, $\hat{\boldsymbol{\theta}}$ is no longer consistent for $\boldsymbol{\theta}_0$. The asymptotic bias in $\hat{\boldsymbol{\theta}}$ is given by $\boldsymbol{\theta}_* - \boldsymbol{\theta}_0$. However, this does not mean that all the parameter estimates are inconsistent. Yuan et al. (2003) and Yuan and Bentler (2006) showed that many parameter estimates in CSA and SEM are still consistent even when a model is misspecified. An intuitive approach to identifying parameters that are not affected by misspecification was recently provided by Yuan et al. (2005). We might

[2] One may find an overwhelming list of books, monographs and conference proceedings on robust statistics by searching the web (e.g., http://www.amazon.com).

need to emphasize that when the model is misspecified, different procedures result in different $\boldsymbol{\theta}_*$'s, and statistics T_{ML}, T_{RML} and T_{GLS} are no longer equivalent (see Yuan and Chan, 2005).

When the model is misspecified, the MLE $\hat{\boldsymbol{\theta}}$ is still asymptotically normally distributed. We need to introduce additional notation to characterize its distribution. Let

$$l_i(\boldsymbol{\theta}) = -\frac{1}{2}\ln\left|\boldsymbol{\Sigma}(\boldsymbol{\theta})\right| - \frac{1}{2}\left[\mathbf{x}_i - \boldsymbol{\mu}(\boldsymbol{\theta})\right]'\boldsymbol{\Sigma}^{-1}(\boldsymbol{\theta})\left[\mathbf{x}_i - \boldsymbol{\mu}(\boldsymbol{\theta})\right].$$

Then the $l(\boldsymbol{\theta})$ in (2) can be rewritten as

$$l(\boldsymbol{\theta}) = \sum_{i=1}^{n} l_i(\boldsymbol{\theta}).$$

Let $\mathbf{A} = E[\ddot{l}_i(\boldsymbol{\theta}_*)]$ and $\mathbf{B} = E[\dot{l}_i(\boldsymbol{\theta}_*)\dot{l}_i'(\boldsymbol{\theta}_*)]$, where the expectations are with respect to the true distribution of the data. Then (see Arminger and Schoenberg, 1989; Browne and Arminger, 1995; Gourieroux et al., 1984; Shapiro, 1983; Vuong, 1989; White, 1982)

$$\sqrt{n}(\hat{\boldsymbol{\theta}} - \boldsymbol{\theta}_*) \xrightarrow{\mathcal{L}} N(\mathbf{0}, \mathbf{A}^{-1}\mathbf{B}\mathbf{A}^{-1}). \tag{39}$$

Consistent estimates of $\mathbf{A}$ and $\mathbf{B}$ are obtained by

$$\hat{\mathbf{A}} = \frac{1}{n}\sum_{i=1}^{n}\ddot{l}_i(\hat{\boldsymbol{\theta}}) \quad \text{and} \quad \widehat{\mathbf{B}} = \frac{1}{n}\sum_{i=1}^{n}\dot{l}_i(\hat{\boldsymbol{\theta}})\dot{l}_i'(\hat{\boldsymbol{\theta}}).$$

Note that when the model is misspecified, (39) is different from (5) even when $\mathbf{x} \sim N(\boldsymbol{\mu}, \boldsymbol{\Sigma})$; (39) is also different from (16) unless the model is correctly specified. SE's in standard software are commonly based on either (5) or (16), which are not consistent in general (see Yuan and Hayashi, 2006). When the model is saturated, (39) is a special case of (30).

Misspecified models not only affect parameter estimates but also statistics for overall model evaluation. An early technical development in CSA was given by Satorra and Saris (1985) and Steiger et al. (1985), who proposed to use the noncentral chi-square to describe the distribution of $T_{\mathrm{ML}c}$. Actually, T_{ML} as well as other LR statistics also asymptotically follow noncentral chi-square distributions (Wald, 1943) under certain conditions. The key condition behind this development is the concept of a sequence of local alternative hypotheses. For the given model structures $\boldsymbol{\mu}(\boldsymbol{\theta})$ and $\boldsymbol{\Sigma}(\boldsymbol{\theta})$, this assumption specifies a sequence of population mean vectors $\boldsymbol{\mu}_0^n$ and covariance matrices $\boldsymbol{\Sigma}_0^n$ that satisfy

$$\boldsymbol{\mu}_0^n = \boldsymbol{\mu}\left(\boldsymbol{\theta}_*^n\right) + \mathrm{O}(1/\sqrt{n}) \quad \text{and} \quad \boldsymbol{\Sigma}_0^n = \boldsymbol{\Sigma}\left(\boldsymbol{\theta}_*^n\right) + \mathrm{O}(1/\sqrt{n}). \tag{40}$$

Condition (40) makes $nD_{\mathrm{ML}}[\boldsymbol{\mu}_0^n, \boldsymbol{\Sigma}_0^n, \boldsymbol{\mu}(\boldsymbol{\theta}_*^n), \boldsymbol{\Sigma}(\boldsymbol{\theta}_*^n)]$ have a limit that does not depend on n. This limit value is the commonly used noncentrality parameter δ so that

$$T_{\mathrm{ML}} \xrightarrow{\mathcal{L}} \chi_{df}^2(\delta). \tag{41}$$

It can be shown that, if $T \xrightarrow{\mathcal{L}} \chi_{df}^2$ under (1), then T will approach a noncentral chi-square distribution under (40). For example, Shapiro and Browne (1987) showed that $T_{\mathrm{RML}c}$ approaches a noncentral chi-square distribution within the class of elliptical distributions. When a robust covariance matrix is modeled instead of **S**, Yuan et al. (2004a) showed that the noncentrality parameter in the noncentral chi-square distribution is greater, as is the power of $T_{\mathrm{RML}c}$.

In practice, one has a sample of size n whose population means and covariance matrix are unknown but fixed. So the limiting noncentral chi-square distribution in (41) is not valid with fixed alternatives (see Stroud, 1972). Actually, in many contexts, LR statistics have been shown to asymptotically follow normal distributions (Sugiura, 1969; Vuong, 1989; Yanagihara et al., 2005). Under fixed alternatives, Shapiro (1983) showed that test statistics in CSA generally follow normal distributions. By extending the results of Shapiro (1983) and Vuong (1989) to mean and covariance structure models, Yuan et al. (In press) provided a normal distribution description for T_{ML}. Let $\boldsymbol{\mu}_* = \boldsymbol{\mu}(\boldsymbol{\theta}_*)$, $\boldsymbol{\Sigma}_* = \boldsymbol{\Sigma}(\boldsymbol{\theta}_*)$, $\dot{\boldsymbol{\beta}}_* = \dot{\boldsymbol{\beta}}(\boldsymbol{\theta}_*)$, $\ddot{\sigma}_{ij*} = \ddot{\sigma}_{ij}(\boldsymbol{\theta}_*)$, $\ddot{\mu}_{i*} = \ddot{\mu}_i(\boldsymbol{\theta}_*)$,

$$
\begin{aligned}
\mathbf{W}_{c*} &= 2^{-1}\mathbf{D}_p'\left(\boldsymbol{\Sigma}_*^{-1} \otimes \boldsymbol{\Sigma}_*^{-1}\right)\mathbf{D}_p, \\
\boldsymbol{\eta} &= \boldsymbol{\Sigma}_*^{-1}(\boldsymbol{\mu}_0 - \boldsymbol{\mu}_*) = (\eta_1, \ldots, \eta_p)', \\
\boldsymbol{\Upsilon} &= \boldsymbol{\Sigma}_*^{-1}\left[\boldsymbol{\Sigma}_0 + (\boldsymbol{\mu}_0 - \boldsymbol{\mu}_*)(\boldsymbol{\mu}_0 - \boldsymbol{\mu}_*)'\right]\boldsymbol{\Sigma}_*^{-1}, \\
\mathbf{H} &= (h_{ij}) = \boldsymbol{\Sigma}_*^{-1} - \boldsymbol{\Upsilon}, \\
\mathbf{M}_* &= \begin{pmatrix} \boldsymbol{\Sigma}_*^{-1} & 2^{-1}(\boldsymbol{\eta}' \otimes \boldsymbol{\Sigma}_*^{-1})\mathbf{D}_p \\ 2^{-1}\mathbf{D}_p'(\boldsymbol{\eta} \otimes \boldsymbol{\Sigma}_*^{-1}) & \mathbf{W}_{c*} \end{pmatrix} \dot{\boldsymbol{\beta}}_*, \\
\mathbf{Q}_* &= \dot{\boldsymbol{\beta}}_*' \begin{pmatrix} \boldsymbol{\Sigma}_*^{-1} & (\boldsymbol{\eta}' \otimes \boldsymbol{\Sigma}_*^{-1})\mathbf{D}_p \\ \mathbf{D}_p'(\boldsymbol{\eta} \otimes \boldsymbol{\Sigma}_*^{-1}) & \mathbf{D}_p'[(\boldsymbol{\Upsilon} - 2^{-1}\boldsymbol{\Sigma}_*^{-1}) \otimes \boldsymbol{\Sigma}_*^{-1}]\mathbf{D}_p \end{pmatrix} \dot{\boldsymbol{\beta}}_* \\
&\quad + \frac{1}{2}\sum_{i=1}^{p}\sum_{j=1}^{p} h_{ij}\ddot{\sigma}_{ij*} - \sum_{i=1}^{p}\eta_i\ddot{\mu}_{i*}, \\
\mathbf{W}_{*0} &= \operatorname{diag}\left(\boldsymbol{\Sigma}_*^{-1}, \mathbf{W}_{c0}\right), \qquad \mathbf{U}_* = \mathbf{W}_{*0} - \mathbf{M}_*\mathbf{Q}_*^{-1}\mathbf{M}_*'
\end{aligned}
$$

and

$$
\boldsymbol{\Gamma}_{\mathcal{N}} = 2\mathbf{D}_p^+(\boldsymbol{\Sigma}_0 \otimes \boldsymbol{\Sigma}_0)\mathbf{D}_p^{+\prime}.
$$

The result of Yuan et al. (In press) is

$$
\sqrt{n}(T_{\mathrm{ML}}/n - \mu_*) \xrightarrow{\mathcal{L}} N\left(0, \omega_*^2\right), \tag{42}
$$

where

$$
\mu_* = D_{\mathrm{ML}}[\boldsymbol{\mu}_0, \boldsymbol{\Sigma}_0, \boldsymbol{\mu}_*, \boldsymbol{\Sigma}_*] + \frac{\operatorname{tr}(\mathbf{U}_*\boldsymbol{\Pi}]}{n}
$$

and

$$
\begin{aligned}
\omega_*^2 &= 4(\boldsymbol{\mu}_0 - \boldsymbol{\mu}_*)'\boldsymbol{\Sigma}_*^{-1}\boldsymbol{\Sigma}_0\boldsymbol{\Sigma}_*^{-1}(\boldsymbol{\mu}_0 - \boldsymbol{\mu}_*) + 2\operatorname{tr}\left[\left(\boldsymbol{\Sigma}_*^{-1}\boldsymbol{\Sigma}_0 - \mathbf{I}_p\right)^2\right] \\
&\quad + \operatorname{tr}\left\{\left[\mathbf{D}_p'\left(\boldsymbol{\Sigma}_*^{-1} - \boldsymbol{\Sigma}_0^{-1}\right) \otimes \left(\boldsymbol{\Sigma}_*^{-1} - \boldsymbol{\Sigma}_0^{-1}\right)\mathbf{D}_p\right](\boldsymbol{\Gamma} - \boldsymbol{\Gamma}_{\mathcal{N}})\right\} \\
&\quad + 4\operatorname{tr}\left(\left\{\left[(\boldsymbol{\mu}_0 - \boldsymbol{\mu}_*)'\boldsymbol{\Sigma}_*^{-1}\right] \otimes (\boldsymbol{\Sigma}_*^{-1} - \boldsymbol{\Sigma}_0^{-1})\right\}\mathbf{D}_p\boldsymbol{\Delta}\right).
\end{aligned}
$$

Notice that (42) is valid for both normal and non-normally distributed data. When the model is approximately correct, $\text{tr}(\mathbf{U}_*\boldsymbol{\Pi}) \approx \text{tr}(\mathbf{U}\boldsymbol{\Pi})$ with $\mathbf{U}$ being given by (15). For normally distributed data, $\text{tr}(\mathbf{U}\boldsymbol{\Pi}) = df$. So the term $\text{tr}(\mathbf{U}_*\boldsymbol{\Pi})/n$ in $\boldsymbol{\mu}_*$ accounts for model complexity. Empirical results in Yuan et al. (In press) indicated that, for normally distributed data, (42) better describes the distribution of T_{ML} than (41) unless (1) is trivially violated; for non-normally distributed data (42) is better than (41) even for a small effect size in a mean comparison of latent variables.

6. Illustration

Sections 2 to 4 discussed ML, GLS, and their extensions, as well as robust procedures. Each is insensitive to certain distributional violations. This section compares these procedures based on a real data set and with some artificial contamination. Our interest is to compare the main strengths of these procedures, not to illustrate every feature.

Holzinger and Swineford (1939) contains test scores on the following subtests or variables: Visual Perception, Cubes, Lozenges, Paragraph Comprehension, Sentence Completion, Word Meaning, Addition, Counting Dots, Straight-Curved Capitals. The first three variables were designed to measure spatial ability, the next three variables were designed to measure verbal ability, and the last three variables were administered with a limited time and were designed to measure a speed factor in performing the tasks. Let $\mathbf{x} = (x_1, x_2, \ldots, x_9)'$ represent the observed variables, f_1, f_2, and f_3 represent respectively the spatial, verbal, and speed latent scores. Then, with $\mathbf{f} = (f_1, f_2, f_3)'$, Holzinger and Swineford's design can be represented by the following confirmatory factor model

$$\mathbf{x} = \boldsymbol{\mu} + \boldsymbol{\Lambda}\mathbf{f} + \mathbf{e}, \tag{43a}$$

where $\boldsymbol{\mu} = E(\mathbf{x})$, $E(\mathbf{f}) = \mathbf{0}$, $E(\mathbf{e}) = \mathbf{0}$,

$$\boldsymbol{\Lambda} = \begin{pmatrix} \lambda_{11} & \lambda_{21} & \lambda_{31} & 0 & 0 & 0 & 0 & 0 & 0 \\ 0 & 0 & 0 & \lambda_{42} & \lambda_{52} & \lambda_{62} & 0 & 0 & 0 \\ 0 & 0 & 0 & 0 & 0 & 0 & \lambda_{73} & \lambda_{83} & \lambda_{93} \end{pmatrix}'.$$

Let $\boldsymbol{\Phi} = (\phi_{ij}) = \text{Corr}(\mathbf{f})$ be a correlation matrix, $\boldsymbol{\Psi} = \text{Cov}(\mathbf{e})$ be a diagonal matrix and assume $\mathbf{f}$ and $\mathbf{e}$ are uncorrelated. Then $\boldsymbol{\Sigma}_0 = \text{Cov}(\mathbf{x})$ can be modeled by

$$\boldsymbol{\Sigma}(\boldsymbol{\theta}) = \boldsymbol{\Lambda}\boldsymbol{\Phi}\boldsymbol{\Lambda}' + \boldsymbol{\Psi}, \tag{43b}$$

where $\boldsymbol{\theta}$ contains the unknown elements in $\boldsymbol{\Lambda}$, $\boldsymbol{\Phi}$ and $\boldsymbol{\Psi}$. Because standard deviations of the nine variables differ a lot, we divide them respectively by 6, 4, 8, 3, 4, 7, 23, 20, 36 to keep each marginal standard deviation between 1 and 2. This change of scale has no substantive effect on the model evaluation but makes the iterative convergence faster.

The normalized Mardia's kurtosis for this data set is $z_{M_2} = 3.037$. Compared to many data sets in the social sciences,[3] the sample only has slightly heavier tails than those of a normal distribution. Based on empirical efficiency of parameter estimates,

[3] We have seen many z_{M_2}'s that are greater than 100 in students' dissertations and research projects.

Table 1
Test statistics for the overall model evaluation

	ML					GLS			H(0.25)
	$T_{\mathrm{ML}c}$	$T_{\mathrm{RML}c}$	$T_{\mathrm{RGLS}c}$	$T_{\mathrm{CRGLS}c}$	$T_{\mathrm{RF}c}$	$T_{\mathrm{GLS}c}$	$T_{\mathrm{CGLS}c}$	$T_{\mathrm{F}c}$	$T_{\mathrm{RML}c}$
(I)									
T	51.187	49.381	64.261	44.527	2.250	57.916	41.386	2.028	41.879
P	0.001	0.002	0.000	0.007	0.002	0.000	0.015	0.007	0.013
(II)									
T	83.148	51.630	74.735	49.317	2.617	64.355	44.572	2.253	42.597
P	0.000	0.001	0.000	0.002	0.000	0.000	0.007	0.002	0.011
(III)									
T	28.099	27.831	31.501	25.879	1.160	30.124	24.942	1.110	23.302
P	0.212	0.222	0.111	0.307	0.294	0.146	0.353	0.345	0.443
(IV)									
T	52.432	34.387	46.047	34.948	1.696	35.246	28.354	1.298	24.494
P	0.000	0.060	0.003	0.053	0.035	0.049	0.203	0.183	0.377

(I) raw data, model 1; (II) contaminated data, model 1; (III) raw data, model 2; (IV) contaminated data, model 2.
Data from Holzinger and Swineford (1939).

Yuan and Hayashi (2003) recommended Huber-type weights as in (26) with $\rho = 0.25$, denoting it by H(0.25). They also showed that, when fitting the model in (43) to $\mathbf{x}_{i0}^{(0.25)}$, $T_{\mathrm{ML}c}$ approximately follows χ^2_{24}. Yuan et al. (2002b) found that H(0.25) applied to this sample is supported by cross-validation. So, in addition to the ML and GLS procedures, we will apply the two-stage robust procedure by minimizing $D_{\mathrm{ML}c}(\widehat{\boldsymbol{\Sigma}}, \boldsymbol{\Sigma}(\boldsymbol{\theta}))$ for parameter estimates and using the rescaled statistic for model evaluation.

Fitting the sample covariance matrix $\mathbf{S}$ to the model in (43) (model 1), nine statistics for overall model evaluation are presented in the first line of Table 1: These are respectively the LR statistic and the rescaled statistic discussed in Section 2; the residual based GLS statistic for CSA parallel to (24) with $\hat{\boldsymbol{\theta}}$ being the MLE; the corrected residual-based statistic parallel to (22); the residual based F-statistic parallel to (23); the GLS statistic as in (18) for CSA; the corrected GLS statistic as in (22) for CSA; the F-statistic as in (23) for CSA; and the rescaled statistic in the robust procedure H(0.25). The second line of numbers are the corresponding p-values when these statistics are referred to either χ^2_{24} or $F_{24,121}$. The largest p-value, corresponding to the corrected GLS statistic $T_{\mathrm{CGLS}c}$, is still highly significant. So none of the statistics endorses the model in (43).

Being available to the public, the covariance structure of the nine variables has been examined by many authors using the ML procedure (e.g., Jöreskog, 1969; Sörbom, 1989; Yuan et al., 2003) and robust procedures (Yuan and Bentler, 1998c). One of the recommended models is to let variable x_9 load on factor f_1 in (43), that is, free λ_{91} in the factor loading matrix. Adding this parameter to (43) results in a different model (model 2) whose corresponding nine statistics are given in part (III) of Table 1. The p-values are obtained by referring these statistic to either χ^2_{23} or $F_{23,122}$.

The smallest p-value (0.111) corresponds to the residual based GLS statistic $T_{\text{RGLS}c}$, which still indicates that the model fits the data reasonably well. The largest p-value (0.443) corresponds to the rescaled statistic based on H(0.25).

As has been discussed, all the procedures may give reasonable inference when a sample comes from a distribution that is approximately normally distributed. To see the reactions of these methods to bad data, we change the last five cases in the Holzinger and Swineford (1939) data file by

$$\mathbf{x}_i = r_i \mathbf{x}_i, \quad i = 141 \text{ to } 145,$$

where $r_i = \exp(z_i)$ and z_i is generated by the function *normal(seed)* in SAS IML with the initial seed given by 1111111111. With $n = 145$, about 3% observations are contaminated. The contaminated sample has a much larger multivariate kurtosis, $z_{M_2} = 55.377$, and thus is more comparable to many non-normal data sets in practice.

Fitting model 1 and model 2 to the contaminated sample, the resulting statistics are in parts (II) and (IV) of Table 1, respectively. Comparing the statistics under (I) with those under (II), all the statistics increased for the contaminated sample, so model 1 is still a poor model. Among these statistics, $T_{\text{ML}c}$ is most sensitive to the data contamination; $T_{\text{RML}c}$ under H(0.25) is least sensitive, since the influence of the contaminated observations are automatically downweighted. Because Huber-type weights keep a balance between efficiency and robustness, the effect of contaminated data are not totally removed by H(0.25). Actually, in H(0.25), cases with numbers 143, 144, 145, 24, 7 get the smallest weights in the contaminated sample, while cases numbered 24, 106, 7, 77, 40 get the smallest weights before the contamination. Compared to $T_{\text{ML}c}$, the other statistics in Table 1 also enjoy some robustness. Turning to part (IV) for the contaminated data with model 2, $T_{\text{ML}c}$ implies that the model is far from being adequate. Other statistics under ML also indicate that the model is marginal or not adequate. This implies that the ML and related statistics are not robust enough. Under the GLS heading, $T_{\text{GLS}c}$ implies that model 2 is marginal. On the other hand, $T_{\text{CGLS}c}$ and $T_{\text{F}c}$ indicate the model fits the sample pretty well, which indicates that the improved statistics for the GLS procedure not only are asymptotically distribution free but also enjoy some finite sample robustness properties. As with model 1, the least sensitive statistic is still $T_{\text{RML}c}$ under H(0.25), reflecting its real robustness to data contaminations.

Next we examine the parameter estimates and their SE's for some of these procedures in the original data as well as in the contaminated sample. Table 2 contains parameter estimates and their associated SE's by ML, GLS and H(0.25) for model 1, with the left side containing information on the original data and the right side containing the parallel information for the contaminated data. The SE's for the MLE $\hat{\boldsymbol{\theta}}$ are based on the sandwich-type covariance matrix as in (16), the SE's for the GLS estimates $\tilde{\boldsymbol{\theta}}$ are based on (20), the SE's for $\hat{\boldsymbol{\theta}}$ under H(0.25) are based on (16) with $\widehat{\boldsymbol{\Pi}} = \widehat{\mathbf{V}}$ being estimated according to (30). Comparing results on the right to those on the left, the MLE changes the most. For example, $\hat{\lambda}_{11}$ changes from 0.779 to 2.001; $\hat{\lambda}_{93}$ changes from 0.720 to 3.222, reflecting the lack of robustness of the ML procedure with data contamination. The SE's associated with the MLE also change substantially, reflecting the heavier tails of the contaminated sample. The GLS estimates, as well as estimates by H(0.25), are much more stable. Although the GLS estimates are based on modeling $\mathbf{S}$, because the

Table 2
Parameter estimates and the associated SEs for model 1, based on data from Holzinger and Swineford (1939) and with some contaminations

	Raw data			Contaminated		
	ML	GLS	H(0.25)	ML	GLS	H(0.25)
θ	$\hat{\theta}$ (SE[1])	$\tilde{\theta}$ (SE[2])	$\hat{\theta}$ (SE[3])	$\hat{\theta}$ (SE)	$\tilde{\theta}$ (SE)	$\hat{\theta}$ (SE)
λ_{11}	0.779 (0.116)	0.734 (0.113)	0.783 (0.107)	2.001 (0.766)	0.792 (0.119)	0.852 (0.111)
λ_{21}	0.574 (0.094)	0.322 (0.076)	0.536 (0.098)	2.304 (0.901)	0.450 (0.159)	0.666 (0.120)
λ_{31}	0.721 (0.091)	0.560 (0.092)	0.690 (0.095)	0.871 (0.285)	0.577 (0.105)	0.664 (0.091)
λ_{42}	0.974 (0.084)	0.940 (0.086)	0.928 (0.089)	1.769 (0.565)	0.972 (0.086)	0.971 (0.086)
λ_{52}	0.964 (0.083)	1.005 (0.086)	0.965 (0.087)	2.606 (0.991)	1.061 (0.133)	1.042 (0.101)
λ_{62}	0.938 (0.082)	0.961 (0.085)	0.891 (0.086)	2.012 (0.786)	0.961 (0.087)	0.906 (0.085)
λ_{73}	0.682 (0.080)	0.710 (0.079)	0.681 (0.089)	1.804 (0.647)	0.794 (0.090)	0.741 (0.088)
λ_{83}	0.837 (0.089)	0.794 (0.079)	0.789 (0.085)	2.645 (1.020)	0.917 (0.138)	0.908 (0.110)
λ_{93}	0.720 (0.086)	0.825 (0.069)	0.667 (0.083)	3.222 (1.465)	0.900 (0.135)	0.831 (0.104)
ϕ_{21}	0.541 (0.094)	0.899 (0.103)	0.578 (0.090)	0.926 (0.059)	0.886 (0.103)	0.627 (0.087)
ϕ_{31}	0.522 (0.100)	0.680 (0.109)	0.479 (0.107)	0.965 (0.031)	0.755 (0.116)	0.619 (0.099)
ϕ_{32}	0.335 (0.115)	0.608 (0.093)	0.342 (0.112)	0.918 (0.066)	0.628 (0.109)	0.453 (0.106)
ψ_{11}	0.721 (0.168)	0.513 (0.146)	0.617 (0.149)	0.736 (0.141)	0.495 (0.158)	0.581 (0.148)
ψ_{22}	0.905 (0.140)	0.892 (0.138)	0.858 (0.129)	1.043 (0.199)	0.961 (0.145)	0.891 (0.136)
ψ_{33}	0.560 (0.108)	0.658 (0.099)	0.580 (0.117)	0.913 (0.122)	0.690 (0.111)	0.627 (0.114)
ψ_{44}	0.317 (0.066)	0.273 (0.067)	0.307 (0.066)	0.487 (0.071)	0.258 (0.069)	0.308 (0.066)
ψ_{55}	0.422 (0.072)	0.374 (0.067)	0.376 (0.070)	0.294 (0.080)	0.358 (0.070)	0.386 (0.073)
ψ_{66}	0.409 (0.076)	0.318 (0.078)	0.388 (0.077)	0.535 (0.098)	0.309 (0.077)	0.384 (0.074)
ψ_{77}	0.604 (0.080)	0.566 (0.086)	0.594 (0.091)	0.776 (0.108)	0.541 (0.087)	0.611 (0.089)
ψ_{88}	0.402 (0.112)	0.189 (0.080)	0.321 (0.091)	0.602 (0.125)	0.198 (0.082)	0.365 (0.088)
ψ_{99}	0.540 (0.085)	0.360 (0.081)	0.510 (0.082)	0.691 (0.285)	0.284 (0.079)	0.441 (0.076)

[1] Based on the sandwich-type covariance matrix as in (16);
[2] based on the correct estimator as in (20);
[3] based on the sandwich-type covariance matrix as in (16) with $\boldsymbol{\Pi} = \mathbf{V}$ being estimated according to (30).

weight matrix $\mathbf{S}_t^{-1}$ automatically adjusts for the heavy tails of the contaminated sample, $\tilde{\boldsymbol{\theta}}$ changes little between the two samples.

Parameter estimates and standard errors for the three procedures applied to model 2 are given in Table 3. As in Table 2, the MLE as well as their associated SE's change most when data are contaminated, especially, $\hat{\lambda}_{93}$, $\hat{\psi}_{88}$, $\hat{\lambda}_{91}$ in the contaminated sample are no longer significant at the 0.05 level when referring $z = \hat{\theta}/\mathrm{SE}$ to $N(0, 1)$. The GLS estimates and the estimates by H(0.25) are very stable. The GLS estimate $\tilde{\psi}_{88}$ is negative, an improper solution or a Heywood case in both raw and contaminated samples. This could happen due to a large sampling error, ψ_{88} being small in the population, or due to a misspecified model (Anderson and Gerbing, 1984; Boomsma, 1985; Chen et al., 2001; Kano, 1998; Rindskopf, 1984; van Driel, 1978). Because $\tilde{\psi}_{88}$ is not statistically significant, the improper solution here is most likely caused by a small ψ_{88} together with a large sampling error. Actually, the GLS estimates can be quite inefficient at $n = 145$. The SE's in the table, based on (20), might be too optimistic (see Yuan and Bentler, 1997b). In addition to $\tilde{\psi}_{88}$ not being significant, $\tilde{\phi}_{32}$ is also not significant in

Table 3
Parameter estimates and the associated SEs for model 2, based on data from Holzinger and Swineford (1939) and with some contaminations

	Raw data			Contaminated		
	ML	GLS	H(0.25)	ML	GLS	H(0.25)
θ	$\hat{\theta}$ (SE)	$\tilde{\theta}$ (SE)	$\hat{\theta}$ (SE)	$\hat{\theta}$ (SE)	$\tilde{\theta}$ (SE)	$\hat{\theta}$ (SE)
λ_{11}	0.819 (0.109)	0.845 (0.105)	0.831 (0.101)	1.994 (0.767)	0.879 (0.118)	0.882 (0.106)
λ_{21}	0.543 (0.093)	0.557 (0.097)	0.501 (0.098)	2.282 (0.907)	0.714 (0.136)	0.641 (0.119)
λ_{31}	0.688 (0.088)	0.678 (0.088)	0.656 (0.090)	0.866 (0.284)	0.688 (0.098)	0.647 (0.087)
λ_{42}	0.975 (0.084)	0.892 (0.091)	0.930 (0.089)	1.771 (0.564)	0.933 (0.095)	0.972 (0.086)
λ_{52}	0.964 (0.083)	0.909 (0.090)	0.965 (0.087)	2.602 (0.993)	0.943 (0.131)	1.041 (0.100)
λ_{62}	0.937 (0.083)	0.858 (0.092)	0.890 (0.087)	2.016 (0.784)	0.862 (0.099)	0.905 (0.085)
λ_{73}	0.708 (0.084)	0.521 (0.084)	0.703 (0.095)	1.832 (0.640)	0.586 (0.100)	0.770 (0.090)
λ_{83}	0.900 (0.098)	1.158 (0.134)	0.833 (0.097)	2.733 (0.993)	1.147 (0.140)	0.976 (0.113)
λ_{91}	0.460 (0.103)	0.603 (0.099)	0.416 (0.106)	2.355* (1.220)	0.667 (0.126)	0.451 (0.121)
λ_{93}	0.453 (0.093)	0.251 (0.080)	0.452 (0.101)	0.902* (0.579)	0.295 (0.104)	0.524 (0.121)
ϕ_{21}	0.554 (0.091)	0.536 (0.101)	0.582 (0.088)	0.938 (0.050)	0.564 (0.117)	0.627 (0.085)
ϕ_{31}	0.392 (0.112)	0.393 (0.104)	0.355 (0.115)	0.933 (0.052)	0.484 (0.134)	0.489 (0.110)
ϕ_{32}	0.240 (0.117)	0.068* (0.109)	0.259 (0.116)	0.870 (0.097)	0.148* (0.172)	0.362 (0.115)
ψ_{11}	0.657 (0.160)	0.455 (0.134)	0.540 (0.140)	0.764 (0.121)	0.491 (0.142)	0.529 (0.140)
ψ_{22}	0.940 (0.142)	0.920 (0.137)	0.894 (0.133)	1.146 (0.235)	0.952 (0.143)	0.924 (0.139)
ψ_{33}	0.607 (0.096)	0.599 (0.100)	0.626 (0.105)	0.922 (0.113)	0.610 (0.111)	0.649 (0.106)
ψ_{44}	0.315 (0.066)	0.335 (0.072)	0.304 (0.065)	0.481 (0.072)	0.326 (0.078)	0.304 (0.055)
ψ_{55}	0.422 (0.072)	0.322 (0.068)	0.377 (0.070)	0.316 (0.082)	0.323 (0.069)	0.387 (0.073)
ψ_{66}	0.411 (0.077)	0.402 (0.081)	0.390 (0.077)	0.519 (0.095)	0.388 (0.080)	0.386 (0.074)
ψ_{77}	0.568 (0.083)	0.584 (0.088)	0.562 (0.092)	0.676 (0.097)	0.550 (0.089)	0.568 (0.089)
ψ_{88}	0.293 (0.130)	−0.241* (0.272)	0.249 (0.124)	0.131* (0.144)	−0.127* (0.198)	0.238 (0.122)
ψ_{99}	0.479 (0.076)	0.459 (0.083)	0.445 (0.079)	0.747 (0.199)	0.404 (0.080)	0.423 (0.075)

* Not significant at 0.05 level.

both samples. This again shows the stability of the GLS procedure. As for estimates by H(0.25), the z score for $\hat{\psi}_{88} = 0.238$ within the contaminated sample is $z = 1.957$, corresponding to a p-value of 0.025 or 0.050 using one-sided or two-sided test, so it is marginally significant. With the raw data, $\hat{\psi}_{88} = 0.249$, corresponding to a z score of 2.001, is also marginally significant. Improper solutions do not happen in either the MLE or estimates by H(0.25).

In summary, although there exist conditions for asymptotic robustness, the ML procedure is quite sensitive to data contamination. The GLS procedure together with the improved statistics is not sensitive to data contaminations. By giving a proper weight to each case, the robust procedure is designed to handle heavy tails, or contamination and outliers, and thus, as expected, it performs very stably.

The robustness properties illustrated in the example should be distinguished from the performance of these methods when samples are drawn from the same distribution with finite 4th-order moments. In such situations, the ML procedure with rescaled or residual-based statistics and SE's based on the sandwich-type covariance matrix performs reasonably well. Compared to ML, the GLS procedure may encounter more nonconvergences, more improper solutions, as well as less accurate SE's. In practice, of course, a sample may be from a heavy-tailed population, contain data contamination, or both. From this point of view, the robust procedure in Section 4 should be preferred in general.

The statistics T_{RML}, T_{CRGLS}, T_{RF}, T_{CGLS}, T_{F} as well as T_{RML} based on the robust procedure with weights given by (29) are available in EQS 6.0 (Bentler, 2007). SE's based on the sandwich-type covariance matrix in (16), based on robust estimates, and based on the GLS estimators (20), are also available in EQS 6.0.

References

Amemiya, Y., Anderson, T.W. (1990). Asymptotic chi-square tests for a large class of factor analysis models. *Annals of Statistics* **18**, 1453–1463.

Anderson, T.W., Amemiya, Y. (1988). The asymptotic normal distribution of estimators in factor analysis under general conditions. *Annals of Statistics* **16**, 759–771.

Anderson, J.C., Gerbing, D.W. (1984). The effects of sampling error on convergence, improper solutions and goodness-of-fit indices for maximum likelihood confirmatory factor analysis. *Psychometrika* **49**, 155–173.

Arminger, G., Schoenberg, R. (1989). Pseudo maximum likelihood estimation and a test for misspecification in mean and covariance structure models. *Psychometrika* **54**, 409–426.

Arminger, G., Stein, P., Wittenberg, J. (1999). Mixtures of conditional mean- and covariance-structure models. *Psychometrika* **64**, 475–494.

Bentler, P.M. (1983). Some contributions to efficient statistics in structural models: Specification and estimation of moment structures. *Psychometrika* **48**, 493–517.

Bentler, P.M. (2007). *EQS 6 Structural Equations Program Manual*. Multivariate Software. Encino, CA.

Bentler, P.M., Dijkstra, T. (1985). Efficient estimation via linearization in structural models. In: Krishnaiah, P.R. (Ed.), *Multivariate Analysis VI*. North-Holland, Amsterdam, pp. 9–42.

Bentler, P.M., Yuan, K.-H. (1999). Structural equation modeling with small samples: Test statistics. *Multivariate Behavioral Research* **34**, 181–197.

Berkane, M., Bentler, P.M. (1988). Estimation of contamination parameters and identification of outliers in multivariate data. *Sociological Methods & Research* **17**, 55–64.

Birch, J.B., Myers, R.H. (1982). Robust analysis of covariance. *Biometrics* **38**, 699–713.

Bollen, K.A., Arminger, G. (1991). Observational residuals in factor analysis and structural equation models. *Sociological Methodology* **21**, 235–262.

Boomsma, A. (1985). Nonconvergence, improper solutions, and starting values in LISREL maximum likelihood estimation. *Psychometrika* **50**, 229–242.

Browne, M.W. (1984). Asymptotic distribution-free methods for the analysis of covariance structures. *British Journal of Mathematical and Statistical Psychology* **37**, 62–83.

Browne, M.W. (1987). Robustness of statistical inference in factor analysis and related models. *Biometrika* **74**, 375–384.

Browne, M.W., Arminger, G. (1995). Specification and estimation of mean and covariance structure models. In: Arminger, G., Clogg, C.C., Sobel, M.E. (Eds.), *Handbook of Statistical Modeling for the Social and Behavioral Sciences*. Plenum, New York, pp. 185–249.

Browne, M.W., Shapiro, A. (1988). Robustness of normal theory methods in the analysis of linear latent variate models. *British Journal of Mathematical and Statistical Psychology* **41**, 193–208.

Campbell, N.A. (1980). Robust procedures in multivariate analysis I: Robust covariance estimation. *Applied Statistics* **29**, 231–237.

Carroll, R.J. (1979). On estimating variances of robust estimators when the errors are asymmetric. *Journal of the American Statistical Association* **74**, 674–679.

Chen, F., Bollen, K.A., Paxton, P., Curran, P., Kirby, J. (2001). Improper solutions in structural equation models: Causes, consequences, and strategies. *Sociological Methods & Research* **29**, 468–508.

Cui, H., He, X., Ng, K.W. (2003). Asymptotic distributions of principal components based on robust dispersions. *Biometrika* **90**, 953–966.
Curran, P.J., West, S.G., Finch, J.F. (1996). The robustness of test statistics to nonnormality and specification error in confirmatory factor analysis. *Psychological Methods* **1**, 16–29.
Devlin, S.J., Gnanadesikan, R., Kettenring, J.R. (1981). Robust estimation of dispersion matrices and principal components. *Journal of the American Statistical Association* **76**, 354–362.
Fang, K.-T., Kotz, S., Ng, K.W. (1990). *Symmetric Multivariate and Related Distributions*. Chapman Hall, London.
Ferguson, T. (1996). *A Course in Large Sample Theory*. Chapman Hall, London.
Fung, W.K. (1993). Unmasking outliers and leverage points: A confirmation. *Journal of the American Statistical Association* **88**, 515–519.
Geary, R.C. (1947). Testing for normality. *Biometrika* **34**, 209–242.
Gourieroux, C., Monfort, A., Trognon, A. (1984). Pseudo maximum likelihood methods: Theory. *Econometrica* **52**, 681–700.
Green, P.J. (1984). Iteratively reweighted least squares for maximum likelihood estimation, and some robust and resistant alternatives (with discussion). *Journal of the Royal Statistical Society B* **46**, 149–192.
Gross, A.M. (1977). Confidence intervals for bisquare regression estimates. *Journal of the American Statistical Association* **72**, 341–354.
Hampel, F.R. (1974). The influence curve and its role in robust estimation. *Journal of the American Statistical Association* **69**, 383–393.
Hampel, F.R., Ronchetti, E.M., Rousseeuw, P.J., Stahel, W.A. (1986). *Robust Statistics: The Approach Based on Influence Functions*. Wiley, New York.
Holland, P.W., Welsch, R.E. (1977). Robust regression using iteratively reweighted least-squares. *Communications in Statistics – Theory and Methods A* **6**, 813–827.
Holzinger, K.J., Swineford, F. (1939). *A Study in Factor Analysis: The Stability of a Bi-Factor Solution. Supplementary Educational Monographs*, vol. 48. University of Chicago.
Hoshino, T. (2001). Bayesian inference for finite mixtures in confirmatory factor analysis. *Behaviormetrika* **28**, 37–63.
Hu, L.T., Bentler, P.M., Kano, Y. (1992). Can test statistics in covariance structure analysis be trusted?. *Psychological Bulletin* **112**, 351–362.
Huber, P.J. (1964). Robust estimation of a location parameter. *Annals of Mathematical Statistics* **35**, 73–101.
Huber, P.J. (1967). The behavior of maximum likelihood estimates under nonstandard conditions. In: *Proceedings of the Fifth Berkeley Symposium on Mathematical Statistics and Probability*. University of California Press, Berkeley, CA, pp. 221–233.
Huber, P.J. (1973). Robust regression: Asymptotics, conjectures and Monte Carlo. *Annals of Statistics* **1**, 799–821.
Huber, P.J. (1981). *Robust Statistics*. Wiley, New York.
Hubert, M., Pison, G., Struyf, A., Van Aelst, S. (Eds.) (2004). *Theory and Applications of Recent Robust Methods*. Springer-Verlag, Berlin.
Hubert, M., Rousseeuw, P.J., Vanden Branden, K. (2005). ROBPCA: A new approach to robust principal component analysis. *Technometrics* **47**, 64–79.
Ibazizen, M., Dauxois, J. (2003). A robust principal component analysis. *Statistics* **37**, 73–83.
Jöreskog, K.G. (1969). A general approach to confirmatory maximum likelihood factor analysis. *Psychometrika* **34**, 183–202.
Jöreskog, K.G., Sörbom, D. (1996). *LISREL 8 Users's Reference Guide*. Scientific Software International, Chicago.
Kano, Y. (1986). Conditions on consistency of estimators in covariance structure model. *Journal of the Japan Statistical Society* **16**, 75–80.
Kano, Y. (1992). Robust statistics for test-of-independence and related structural models. *Statistics & Probability Letters* **15**, 21–26.
Kano, Y. (1998). Improper solutions in exploratory factor analysis: Causes and treatments. In: Rizzi, A., Vichi, M., Bock, H. (Eds.), *Advances in Data Sciences and Classification*. Springer-Verlag, Berlin, pp. 375–382.
Kano, Y., Berkane, M., Bentler, P.M. (1993). Statistical inference based on pseudo-maximum likelihood estimators in elliptical populations. *Journal of the American Statistical Association* **88**, 135–143.

Lee, S.-Y., Song, X.Y. (2003). Bayesian model selection for mixtures of structural equation models with an unknown number of components. *British Journal of Mathematical and Statistical Psychology* **56**, 145–165.
Lee, S.-Y., Wang, S.J. (1996). Sensitivity analysis of structural equation models. *Psychometrika* **61**, 93–108.
Lee, S.-Y., Xia, Y.-M. (In press). Maximum likelihood methods in treating outliers and symmetrically heavy-tailed distributions for nonlinear structural equation models with missing data. *Psychometrika*.
Lopuhaä, H.P. (1989). On the relation between S-estimators and M-estimators of multivariate location and covariances. *Annals of Statistics* **17**, 1662–1683.
Lopuhaä, H.P. (1991). τ-estimators for location and scatter. *Canadian Journal of Statistics* **19**, 307–321.
Maddala, G.S., Rao, C.R. (Eds.) (1997). *Handbook of Statistics 15: Robust Inference*. Elsevier, Amsterdam.
Mardia, K.V. (1970). Measures of multivariate skewness and kurtosis with applications. *Biometrika* **57**, 519–530.
Maronna, R.A. (1976). Robust M-estimators of multivariate location and scatter. *Annals of Statistics* **4**, 51–67.
Micceri, T. (1989). The unicorn, the normal curve, and other improbable creatures. *Psychological Bulletin* **105**, 156–166.
Mooijaart, A., Bentler, P.M. (1991). Robustness of normal theory statistics in structural equation models. *Statistica Neerlandica* **45**, 159–171.
Muirhead, R.J., Waternaux, C.M. (1980). Asymptotic distributions in canonical correlation analysis and other multivariate procedures for nonnormal populations. *Biometrika* **67**, 31–43.
Muthén, B. (2001). Latent variable mixture modeling. In: Marcoulides, G.A., Schumacker, R.E. (Eds.), *New Developments and Techniques in Structural Equation Modeling*. Erlbaum, Mahwah, NJ, pp. 1–34.
Pison, G., Rousseeuw, P.J., Filzmoser, P., Croux, C. (2003). Robust factor analysis. *Journal of Multivariate Analysis* **84**, 145–172.
Poon, W.Y., Poon, Y.S. (2002). Influential observations in the estimation of mean vector and covariance matrix. *British Journal of Mathematical and Statistical Psychology* **55**, 177–192.
Poon, W.Y., Wong, Y.K. (2004). A forward search procedure to identifying influential observations in the estimation of covariance matrix. *Structural Equation Modeling* **11**, 357–374.
Poon, W.Y., Lew, S.F., Poon, Y.S. (2000). A local influence approach to identify multiple multivariate outliers. *British Journal of Mathematical and Statistical Psychology* **53**, 255–273.
Rao, C.R. (1973). *Linear Statistical Inference and its Applications*, second ed. Wiley, New York.
Rindskopf, D. (1984). Structural equation models: Empirical identification, Heywood cases, and related problems. *Sociological Methods & Research* **13**, 109–119.
Rocke, D.M. (1996). Robustness properties of S-estimators of multivariate location and shape in high dimension. *Annals of Statistics* **24**, 1327–1345.
Rousseeuw, P.J., Leroy, A.M. (1987). *Robust Regression and Outlier Detection*. Wiley, New York.
Rousseeuw, P.J., van Zomeren, B.C. (1990). Unmasking multivariate outliers and leverage points (with discussion). *Journal of the American Statistical Association* **85**, 633–651.
Rubin, D.B. (1983). Iteratively reweighted least squares. In: Johnson, N.L., Kotz, S. (Eds.), *Encyclopedia of Statistical Sciences*, vol. 4. Wiley, New York, pp. 272–275.
Ruymgaart, F.H. (1981). A robust principal component analysis. *Journal of Multivariate Analysis* **11**, 485–497.
Satorra, A. (1992). Asymptotic robust inferences in the analysis of mean and covariance structures. *Sociological Methodology* **22**, 249–278.
Satorra, A., Bentler, P.M. (1986). Some robustness properties of goodness of fit statistics in covariance structure analysis. In: *American Statistical Association 1986 Proceedings of Business and Economics Sections*. American Statistical Association, pp. 549–554.
Satorra, A., Bentler, P.M. (1988). Scaling corrections for chi-square statistics in covariance structure analysis. In: *American Statistical Association 1988 Proceedings of Business and Economics Sections*. American Statistical Association, pp. 308–313.
Satorra, A., Bentler, P.M. (1990). Model conditions for asymptotic robustness in the analysis of linear relations. *Computational Statistics & Data Analysis* **10**, 235–249.
Satorra, A., Bentler, P.M. (1994). Corrections to test statistics and standard errors in covariance structure analysis. In: von Eye, A., Clogg, C.C. (Eds.), *Latent Variables Analysis: Applications for Developmental Research*. Sage, Thousand Oaks, CA, pp. 399–419.

Satorra, A., Saris, W. (1985). Power of the likelihood ratio test in covariance structure analysis. *Psychometrika* **50**, 83–90.
Shapiro, A. (1983). Asymptotic distribution theory in the analysis of covariance structures (a unified approach). *South African Statistical Journal* **17**, 33–81.
Shapiro, A. (1984). A note on the consistency of estimators in the analysis of moment structures. *British Journal of Mathematical and Statistical Psychology* **37**, 84–88.
Shapiro, A., Browne, M. (1987). Analysis of covariance structures under elliptical distributions. *Journal of the American Statistical Association* **82**, 1092–1097.
Sörbom, D. (1989). Model modification. *Psychometrika* **54**, 371–384.
Steiger, J.H., Shapiro, A., Browne, M.W. (1985). On the multivariate asymptotic distribution of sequential chi-square statistics. *Psychometrika* **50**, 253–264.
Stone, M. (1974). Cross-validation choice and assessment of statistical predictions. *Journal of the Royal Statistical Society B* **36**, 44–47.
Stroud, T.W.F. (1972). Fixed alternatives and Wald's formulation of the noncentral asymptotic behavior of the likelihood ratio statistic. *Annals of Mathematical Statistics* **43**, 447–454.
Sugiura, N. (1969). Asymptotic expansions of the distributions of the likelihood ratio criteria for covariance matrix. *Annals of Mathematical Statistics* **40**, 2051–2063.
Tyler, D.E. (1983). Robustness and efficiency properties of scatter matrices. *Biometrika* **70**, 411–420.
van Driel, O.P. (1978). On various causes of improper solutions in maximum likelihood factor analysis. *Psychometrika* **43**, 225–243.
Vuong, Q.H. (1989). Likelihood ratio tests for model selection and nonnested hypotheses. *Econometrica* **57**, 307–333.
Wald, A. (1943). Tests of statistical hypotheses concerning several parameters when the number of observations is large. *Transactions of the American Mathematical Society* **54**, 426–482.
White, H. (1982). Maximum likelihood estimation of misspecified models. *Econometrica* **50**, 1–25.
Wilcox, R.R. (2004). *Introduction to Robust Estimation and Hypothesis Testing*, second ed. Academic Press, San Diego.
Yanagihara, H., Tonda, T., Matsumoto, C. (2005). The effects of nonnormality on asymptotic distributions of some likelihood ratio criteria for testing covariance structures under normal assumption. *Journal of Multivariate Analysis* **96**, 237–264.
Yuan, K.-H., Bentler, P.M. (1997a). Mean and covariance structure analysis: Theoretical and practical improvements. *Journal of the American Statistical Association* **92**, 767–774.
Yuan, K.-H., Bentler, P.M. (1997b). Improving parameter tests in covariance structure analysis. *Computational Statistics and Data Analysis* **26**, 177–198.
Yuan, K.-H., Bentler, P.M. (1998a). Normal theory based test statistics in structural equation modeling. *British Journal of Mathematical and Statistical Psychology* **51**, 289–309.
Yuan, K.-H., Bentler, P.M. (1998b). Robust mean and covariance structure analysis. *British Journal of Mathematical and Statistical Psychology* **51**, 63–88.
Yuan, K.-H., Bentler, P.M. (1998c). Structural equation modeling with robust covariances. *Sociological Methodology* **28**, 363–396.
Yuan, K.-H., Bentler, P.M. (1999a). On normal theory and associated test statistics in covariance structure analysis under two classes of nonnormal distributions. *Statistica Sinica* **9**, 831–853.
Yuan, K.-H., Bentler, P.M. (1999b). On asymptotic distributions of normal theory MLE in covariance structure analysis under some nonnormal distributions. *Statistics & Probability Letters* **42**, 107–113.
Yuan, K.-H., Bentler, P.M. (1999c). F-tests for mean and covariance structure analysis. *Journal of Educational and Behavioral Statistics* **24**, 225–243.
Yuan, K.-H., Bentler, P.M. (2000a). Inferences on correlation coefficients in some classes of nonnormal distributions. *Journal of Multivariate Analysis* **72**, 230–248.
Yuan, K.-H., Bentler, P.M. (2000b). Three likelihood-based methods for mean and covariance structure analysis with nonnormal missing data. *Sociological Methodology* **30**, 167–202.
Yuan, K.-H., Bentler, P.M. (2000c). Robust mean and covariance structure analysis through iteratively reweighted least squares. *Psychometrika* **65**, 43–58.
Yuan, K.-H., Bentler, P.M. (2001a). A unified approach to multigroup structural equation modeling with nonstandard samples. In: Marcoulides, G.A., Schumacker, R.E. (Eds.), *Advanced Structural Equation Modeling: New Developments and Techniques*. Lawrence Erlbaum Associates, Mahwah, NJ, pp. 35–56.

Yuan, K.-H., Bentler, P.M. (2001b). Effect of outliers on estimators and tests in covariance structure analysis. *British Journal of Mathematical and Statistical Psychology* **54**, 161–175.
Yuan, K.-H., Bentler, P.M. (2002). On robustness of the normal-theory based asymptotic distributions of three reliability coefficient estimates. *Psychometrika* **67**, 251–259.
Yuan, K.-H., Bentler, P.M. (2006). Mean comparison: Manifest variable versus latent variable. *Psychometrika* **71**, 139–159.
Yuan, K.-H., Chan, W. (2005). On nonequivalence of several procedures of structural equation modeling. *Psychometrika* **70**, 791–798.
Yuan, K.-H., Hayashi, K. (2003). Bootstrap approach to inference and power analysis based on three statistics for covariance structure models. *British Journal of Mathematical and Statistical Psychology* **56**, 93–110.
Yuan, K.-H., Hayashi, K. (2006). Standard errors in covariance structure models: Asymptotics versus bootstrap. *British Journal of Mathematical and Statistical Psychology* **59**, 397–417.
Yuan, K.-H., Jennrich, R.I. (1998). Asymptotics of estimating equations under natural conditions. *Journal of Multivariate Analysis* **65**, 245–260.
Yuan, K.-H., Chan, W., Bentler, P.M. (2000). Robust transformation with applications to structural equation modeling. *British Journal of Mathematical and Statistical Psychology* **53**, 31–50.
Yuan, K.-H., Marshall, L.L., Bentler, P.M. (2002a). A unified approach to exploratory factor analysis with missing data, nonnormal data, and in the presence of outliers. *Psychometrika* **67**, 95–122.
Yuan, K.-H., Marshall, L.L., Weston, R. (2002b). Cross-validation through downweighting influential cases in structural equation modeling. *British Journal of Mathematical and Statistical Psychology* **55**, 125–143.
Yuan, K.-H., Marshall, L.L., Bentler, P.M. (2003). Assessing the effect of model misspecifications on parameter estimates in structural equation models. *Sociological Methodology* **33**, 241–265.
Yuan, K.-H., Bentler, P.M., Chan, W. (2004a). Structural equation modeling with heavy tailed distributions. *Psychometrika* **69**, 421–436.
Yuan, K.-H., Fung, W.K., Reise, S. (2004b). Three Mahalanobis-distances and their role in assessing unidimensionality. *British Journal of Mathematical and Statistical Psychology* **57**, 151–165.
Yuan, K.-H., Kouros, C.D., Kelley, K. (2005). Diagnosis for covariance structure models by analyzing the path. Under review.
Yuan, K.-H., Hayashi, K., Bentler, P.M. (In press). Normal theory likelihood ratio statistic for mean and covariance structure analysis under alternative hypotheses. *Journal of Multivariate Analysis*.
Yung, Y.-F. (1997). Finite mixtures in confirmatory factor-analysis models. *Psychometrika* **62**, 297–330.

Handbook of Computing and Statistics with Applications, Vol. 1
ISSN: 1871-0301

DOI: 10.1016/S1871-0301(06)01018-3

18

Stochastic Approximation Algorithms for Estimation of Spatial Mixed Models

Hongtu Zhu, Faming Liang, Minggao Gu and Bradley S. Peterson

Abstract

A class of spatial mixed models is introduced first. Spatial mixed models include latent Markov random fields, which make their likelihood functions complex. This complexity in turn makes statistical inferences (e.g., parameter estimates and prediction of latent fields) prohibitively difficult. Therefore, two algorithms are also introduce by integrating recent developments in stochastic approximation algorithms and Monte Carlo methods. The first of these algorithms, a stochastic approximation expectation-maximization (SAEM) algorithm, is developed to estimate the strength of spatial regularization in latent Markov random fields and other parameters. The second algorithm, an annealing stochastic approximation Monte Carlo (ASAMC) algorithm, is proposed to compute optimal estimates of latent fields, which are the global maxima of the likelihood functions of complete data. These algorithms are applied to data sets of the distribution of vegetation species and simulated images to demonstrate their effectiveness.

Keywords: Expectation-maximization; Multicanonical algorithm; Spatial mixed models; Stochastic approximation; Vegetation

1. Introduction

Spatial mixed models (SMM) are natural extensions of generalized linear models and allow for additional components of variability that account for unobservable latent processes. SMMs have wide applications in image analysis, ecology, psychology, physics, and biophysics. For instance, a number of fundamental processes in image analysis, including image restoration, segmentation, and edge-preserving filtering, have been modeled by using SMMs since the seminal work by Besag (1974) and Geman and Geman (1984). See, for example, Zhang (1993), Jalobeanu et al. (2002), Saquib et al. (1998), and Lakshmanan and Derin (1989), among many others. SMMs also include generalized linear mixed models (GLMM) (Breslow and Clayton, 1993; Zhu and Lee, 2002) and latent variable models (LVM) (Bentler and Dudgeon, 1996), both of which can be used to accommodate overdispersion and correlation among outcomes (Zeger et al., 1988) and to predict and interpolate (or smooth) spatial data (Diggle et al., 1998; Zhang, 2002). Thus, these models have applications in biomedical and educational research, the social sciences, and other fields that investigate complex multivariate, longitudinal, and family data.

However, SMMs are highly complex because of the latent Markov random fields they contain. This complexity makes calculating the maximum likelihood estimates and estimating latent fields prohibitively difficult and therefore poses a major challenge in the applications of SMMs. This challenge can be divided into three distinct issues.

The first issue is that likelihood functions of observed data are often represented by high-dimensional integrals in order to account for latent variables, and these integrals may become irreducibly complicated. Most of the existing procedures for locating maxima of likelihood functions include the expectation-maximization (EM) algorithm (Dempster et al., 1977), the Monte Carlo EM algorithm (Wei and Tanner, 1990; Booth and Hobert, 1999), Monte Carlo Newton–Raphson algorithm (McCulloch, 1997), and stochastic approximation algorithm (Gu and Kong, 1998; Delyon et al., 1999). However, these optimization algorithms only work for some SMMs (such as GLMMs and LVMs) but not for others, because they strongly depend on a simple likelihood function of complete data (including both latent variables and observed data).

Second, latent Markov random fields (MRF) also add to the complexity of the likelihood functions of SMMs. MRFs have been recently used in a range of fields, such as ecology and image processing, to model spatial and geographical correlation among observations (Winkler, 1995; Li, 2001). For example, ecologists use MRFs to describe the spatially correlated distribution of single or multiple species within a given geographical area (Huffer and Wu, 1998; He et al., 2003). Unknown parameters of MRFs in SMMs usually control the granularity of latent fields, and the corresponding normalizing factor (or partition function) of MRFs, as a function of these unknown control parameters, is not known analytically. In practice, the values of these unknown control parameters are either arbitrarily set or heuristically tuned to particular datasets; maximum likelihood estimates (MLE) of control parameters for MRFs in SMMs are rarely calculated because of the considerable computational burden that is involved. Moreover, most existing optimization algorithms for computing MLEs (Geyer and Thompson, 1992; Gu and Zhu, 2001) are based on observed MRFs and therefore are inappropriate for latent MRFs in SMMs. Several approximation methods, such as mean-field approximation, have been used to find approximate estimates of control parameters for latent MRFs (Jalobeanu et al., 2002; Qian and Titterington, 1991; Vasconcelos and Lippman, 2001). Recently, a stochastic approximation EM algorithm has been proposed, and its convergence has been established under some conditions (Zhu et al., 2005a, 2005b). However, performance of this SAEM algorithm when applied to many important applications, such as distributions of vegetation data in ecology and imaging analysis, has not yet been investigated.

The third issue in the use of SMMs is how to find optimal estimates of latent fields. This is the central issue of many research questions in ecology, psychology, and image analysis that involve the prediction of latent variables within a data set. For instance, latent field in image segmentation is a set of labels that represents the identities of individual voxels/pixels. In some cases, estimating latent fields is equivalent to minimizing/maximizing a complicated energy function with a large number of variables and is therefore nearly infeasible computationally. Several optimization methods, such as the iterated conditional modes (ICM), can only give a local optimal solution. Stochastic algorithms, such as the simulating annealing and genetic algorithms, have been proposed to search for the globally optimal estimates of latent fields (Kirkpatrick et al., 1983; Holland, 1975); however, these stochastic algorithms converge very slowly and have a high probability of missing the global minimum (Liang, 2005c). Recently, advanced Monte Carlo algorithms, including annealing stochastic approximation and contour Monte Carlo, have been proposed that are efficient for complex simulation and optimization (Liang, 2004, 2005a, 2005b, 2005c). We will apply the annealing stochastic approximation Monte Carlo (ASAMC) algorithm to find optimal latent fields by maximizing the complete-data likelihood functions given MLEs.

In this paper, we formally introduce SMMs and discuss some examples in the fields of ecology. We then propose two advanced stochastic approximation algorithms (SAEM and ASAMC) for calculating MLE and optimal estimates of latent fields in SMMs. Finally, we evaluate the performance of these algorithms using real-world examples, including distributions of vegetation species and image restoration. Throughout the discussion, we will address the three computational issues of SMMs discussed above.

2. Spatial mixed models

We consider stochastic processes $\boldsymbol{f} = \{f(s)\colon s \in S\}$, $\boldsymbol{X} = \{X(s)\colon s \in S\}$, and $\boldsymbol{Y} = \{Y(s)\colon s \in S\}$, where $S = \{s_i\colon i = 1, \ldots, n\}$ is a known discrete index set in R^d. We define SMMs as follows:

(i) conditional on $(\boldsymbol{f}, \boldsymbol{x})$, the components of $\boldsymbol{Y}$ are mutually independent, and the conditional density of $Y(s)$ given $(\boldsymbol{f}, \boldsymbol{x})$ is $p(y(s)|\boldsymbol{f}, \boldsymbol{x}; \alpha)$, where α is an unknown parameter vector;

(ii) latent field $\boldsymbol{f} = \{f(s_i)\colon i = 1, \ldots, n\}$ is said to be an MRF with respect to a neighborhood system $\mathcal{N} = \{\mathcal{N}_i\colon i = 1, \ldots, n\}$, which is characterized by a Gibbs distribution:

$$p(\boldsymbol{f}|\tau) = \exp\bigl\{-U(\boldsymbol{f}, \tau) - \log C(\tau)\bigr\}, \tag{1}$$

where $U(\boldsymbol{f}, \tau)$ is a potential (or energy) function, which exhibits the interaction between components of $\boldsymbol{f}$ (Besag, 1974). In addition, the normalizing constant $C(\tau)$ is a partition function having the form

$$C(\tau) = \int_{S_f} \exp\bigl\{-U(\boldsymbol{f}, \tau)\bigr\} m(\mathrm{d}\boldsymbol{f}), \tag{2}$$

where S_f is the minimal sample space of $\boldsymbol{f}$ and $m(\mathrm{d}\boldsymbol{f})$ is either the Dirac's delta measure or $\mathrm{d}\boldsymbol{f}$ according to whether $\boldsymbol{f}$ takes discrete or continuous values, respectively.

The above SMMs include many statistical models as special cases. For instance, GLMM is a special class of the SMMs (Breslow and Clayton, 1993) in which $\boldsymbol{f}$ are random effects. For linear LV models, we have $\mu(s) = \mathrm{E}[y(s)|\boldsymbol{f}, \mathbf{x}] = \mu + \Lambda f(s)$ with $\boldsymbol{f}$ following a multivariate normal distribution (Bentler and Dudgeon, 1996), where Λ is a factor loading matrix. SMMs also include more general LV models (Lee and Zhu, 2000, 2002).

Let us study two examples from image analysis: image restoration and segmentation.

EXAMPLE 1 *(Image restoration).* Let s be a pixel-site (or line-site) in a pixelated image, $\boldsymbol{f}$ the true scene and $\boldsymbol{y}$ the observed image, which is a noisy version of $\boldsymbol{f}$. SMMs have been used to characterize image construction and restoration. A particular example for image restoration is defined by

$$\boldsymbol{y} = \boldsymbol{H}\boldsymbol{f} + \varepsilon, \tag{3}$$

where $\boldsymbol{H}$ is the convolution matrix and $\varepsilon \sim N(\mathbf{0}, \phi^{-1}\boldsymbol{I}_n)$, in which $\boldsymbol{I}_n$ is an identity matrix. In this case, we have $\mu(s) = \mathrm{E}[y(s)|\boldsymbol{f}] = \boldsymbol{H}(s)\boldsymbol{f}$. Furthermore, we will assume that the true image $\boldsymbol{f}$ follows a Gaussian random field (GRF) given by

$$p(\boldsymbol{f}|\boldsymbol{B}) = \text{const} \times |\boldsymbol{B}|^{1/2} \exp\{-0.5\sigma^{-2}(\boldsymbol{f} - \mu)^{\mathrm{T}}\boldsymbol{B}(\boldsymbol{f} - \mu)\},$$

where $\boldsymbol{B} = (b(s_i, s_j))$, the inverse matrix of the covariance matrix of $\boldsymbol{f}$, controls the spatial dependence structure of $\boldsymbol{f}$ (Besag, 1974). Therefore, we have

$$U(\boldsymbol{f}, \tau) = 0.5\sigma^{-2}\boldsymbol{f}^{\mathrm{T}}\boldsymbol{B}\boldsymbol{f} - \sigma^{-2}\mu^{\mathrm{T}}\boldsymbol{B}\boldsymbol{f} + 0.5\sigma^{-2}\mu^{\mathrm{T}}\boldsymbol{B}\mu,$$

where τ represents all unknown parameters in $(\boldsymbol{B}, \mu, \sigma)$. In particular, evaluating $|\boldsymbol{B}|^{1/2}$ is computationally prohibit, because $\boldsymbol{B}$ is an $n \times n$-dimensional matrix (e.g., a 2048×2048 matrix corresponding to a 64×64 lattice) (Rue, 2001; Gu and Zhu, 2001). For edge-preserving image recovery, we further consider a generalized GRF (Bouman and Sauer, 1993) defined as follow:

$$p(\boldsymbol{f}|\boldsymbol{B}) = \frac{1}{\sigma^N C(\boldsymbol{B}, p_0)} \exp\left\{-\frac{1}{p_0\sigma^{p_0}} \sum_{s_i \sim s_j} b(s_i, s_j)\left|f(s_i) - f(s_j)\right|^{p_0}\right\},$$

where the summation is taken over all nearest-neighbor pairs $(s_i \sim s_j)$, $p_0 \in (1, 2]$, and the normalized constant $C(\boldsymbol{B}, p_0)$ depends on both $b(s_i, s_j)$ and p_0.

EXAMPLE 2 *(Image segmentation).* Image segmentation is used to classify an image into a set of nonoverlapping regions $\{R_1, \ldots, R_K\}$. We consider a special case of SMMs as follows. The observation at a particular pixel s can be written as

$$y(s) = \sum_{k=1}^{K} \Phi(s, \beta_k) f_k(s) + \varepsilon(s), \tag{4}$$

where $\varepsilon(s) \sim N(0, \phi^{-1})$, $\Phi(\cdot, \cdot)$ is a parametric model, and β_k is the parameter vector for R_k. In addition, $f(s) = (f_1(s), \ldots, f_K(s))$, $f_k(s) \in \{0, 1\}$, $\sum_{k=1}^{K} f_k(s) = 1$, and $f_k(s) = 1$ if and only if $s \in R_k$. Thus, $\mu(s) = \mathrm{E}[y(s)|\boldsymbol{f}] = \sum_{k=1}^{K} \Phi(s, \beta_k) f_k(s)$. We further assume that the joint distribution of the label field $\boldsymbol{f} = \{f(s)\colon s = 1, \ldots, n\}$ is given by

$$p(\boldsymbol{f}|\tau) = \exp\left\{\tau \sum_{s_i \sim s_j} \delta\big(f(s_i), f(s_j)\big) - \log C(\tau)\right\},$$

where the summation is taken over all nearest-neighbor pairs ($s_i \sim s_j$), $\delta(x, z)$ is the Kronecker function equaling to 1 when $x = z$ and 0 otherwise, and τ is the parameter controlling the granularity of the field. In addition, $C(\tau)$ is obtained by summing over all possible configurations $\boldsymbol{f}$ (e.g., n^M terms).

3. Estimation procedure

Much effort has been devoted to developing procedures for estimating the parameters and latent fields of SMMs. See, for example, Marroquin et al. (2003), Lakshmanan and Derin (1989), Jalobeanu et al. (2002), Saquib et al. (1998), Qian and Titterington (1991), Zhu et al. (2005a), and Younes (1989). An approach proposed by Lakshmanan and Derin (1989) is based on jointly maximizing the unknown parameters and the latent fields, but the estimates of parameters under this approach may not be consistent statistically (Neyman and Scott, 1948). For instance, for GLMM, specific conditions are required for validity of this approach (Jiang et al., 2001). To avoid such a pitfall, we take an alternative approach by calculating MLE of $\xi = (\tau, \alpha)$ first and then computing a maximum a posteriori (MAP) estimate of latent field $\boldsymbol{f}$. In particular, MLE of ξ is a consistent estimate under certain conditions (Guyon, 1995). Thus, our estimation procedure consists of two key steps as follows:

Stage 1: compute MLE of ξ, denoted by $\hat{\xi}$, by using the SAEM algorithm;
Stage 2: given $\hat{\xi}$ obtained in Stage 1, we calculate the MAP estimate of $\boldsymbol{f}$ by using the ASAMC algorithm.

3.1. Stochastic approximation expectation-maximization algorithm

The MLE $\hat{\xi} = (\hat{\tau}, \hat{\alpha})$ is defined by

$$L(\hat{\xi}; \boldsymbol{y}_o) = \max_{\xi} L(\xi; \boldsymbol{y}_o), \tag{5}$$

where $\boldsymbol{y}_o$ denotes the observed data, and the likelihood function of observed-data $L(\xi; \boldsymbol{y}_o)$ is given by

$$L(\xi; \boldsymbol{y}_o) = \int \left[\prod_{i=1}^{n} p\big(y(s_i)|\boldsymbol{f}, \mathbf{x}, \alpha\big)\right] \exp\big\{-U(\boldsymbol{f}, \tau) - \log C(\tau)\big\} m(\mathrm{d}\boldsymbol{f}). \tag{6}$$

The integration above is usually of very high dimension, making direct numerical evaluation difficult even for today's computers. In addition, $C(\tau)$ may involve a large matrix as in Example 1, a huge summation as in Example 2, and so on. In order to obtain $\hat{\xi}$, we assume throughout the paper that $L(\xi; \mathbf{y}_o)$ is sufficiently smooth and $\hat{\xi}$ always exists and is unique throughout the paper.

To calculate MLE, we approximate the first-order and second-order derivatives of the log-likelihood function of observed data. From the missing information principle, it follows that the first-order derivative of log-likelihood function can be written as

$$s_\xi(\xi; \mathbf{y}_o) = \partial_\xi \log L(\xi; \mathbf{y}_o) = \mathrm{E}\big[S_\xi(\xi; \boldsymbol{f})|\mathbf{y}_o, \xi\big], \tag{7}$$

where $\partial_\xi = \partial/\partial\xi$, $\mathrm{E}[\cdot|\mathbf{y}_o, \xi]$ denotes the expectation taken with respect to the conditional distribution $\boldsymbol{f}$ given the observed data, and $S_\xi(\xi; \boldsymbol{f})$ is the first derivative of complete-data log-likelihood function. Here, the complete-data log-likelihood function $l_c(\xi; \boldsymbol{f}, \mathbf{y}_o)$ is given by

$$\sum_{s\in S} \log p\big(y(s)|\boldsymbol{f}, \mathbf{x}, \alpha\big) - U(\boldsymbol{f}, \tau) - \log C(\tau). \tag{8}$$

To calculate the second-order derivative of the log-likelihood function, we apply Louis's (1982) formula and obtain

$$-\partial_\xi^2 \log L(\xi; \mathbf{y}_o) = \mathrm{E}\big[I_{\xi\xi}(\xi; \boldsymbol{f}) - S_\xi(\xi; \boldsymbol{f})^{\otimes 2}|\mathbf{y}_o, \xi\big] + s_\xi(\xi; \mathbf{y}_o)^{\otimes 2}, \tag{9}$$

where for vector $\mathbf{a}$, $\mathbf{a}^{\otimes 2} = \mathbf{a}\mathbf{a}^{\mathrm{T}}$, and $I_{\xi\xi}(\xi; \boldsymbol{f}) = -\partial_\xi^2 l_c(\xi; \boldsymbol{f}, \mathbf{y}_o)$ denotes the complete data information matrix.

For SMMs, we need to approximate the first-order and second-order derivatives of the complex $C(\tau)$. Following Gu and Zhu (2001), we can show that

$$\begin{aligned} \partial_\tau \log C(\tau) &= -\mathrm{E}_\tau\big[\partial_\tau U(\boldsymbol{f}, \tau)\big], \\ \partial_\tau^2 \log C(\tau) &= -\mathrm{E}_\tau\big[J(\tau; \boldsymbol{f})\big] - \big\{\partial_\tau \log C(\tau)\big\}^{\otimes 2}, \end{aligned} \tag{10}$$

where $J(\tau; \boldsymbol{f}) = \partial_\tau^2 U(\boldsymbol{f}, \tau) - [\partial_\tau U(\boldsymbol{f}, \tau)]^{\otimes 2}$ and E_τ is taken with respect to MRF (1). Based on (10), we approximate $\partial_\tau \log C(\tau)$ and $\partial_\tau^2 \log C(\tau)$ by using certain Markov chain Monte Carlo (MCMC) methods, such as the hybrid Markov chain, the birth-and-death process, and the Metropolis–Hastings (MH) algorithm. See, for example, Metropolis et al. (1953), Hastings (1970), Liu (2001), Möller (1999), and Robert and Casella (1999), among others. An alternative approach is to use numerical integration, but it usually gives unstable estimates except in some special cases.

We can approximate the first-order and second-order derivatives of the likelihood functions of observed data by using Eqs. (7), (8), and (10). The $\partial_\xi \log L(\xi; \mathbf{y}_o)$ can be approximated by $([S_{\tau,1} - S_{\tau,2}]^{\mathrm{T}}, S_\alpha(\xi; \boldsymbol{f})^{\mathrm{T}})^{\mathrm{T}}$, where $S_{\tau,2} = \partial_\tau \log C(\tau)$ and $S_{\tau,1} = -\mathrm{E}_\xi[\partial_\tau U(\boldsymbol{f}, \tau)|\mathbf{y}_o]$. We define

$$I_1(\xi; \boldsymbol{f}) = \begin{pmatrix} \partial_\tau^2 U(\boldsymbol{f}, \tau) & 0 \\ 0 & I_{\alpha\alpha}(\xi; \boldsymbol{f}) \end{pmatrix}$$

and

$$I_2(\xi; \boldsymbol{f}) = -\begin{pmatrix} -\partial_\tau U(\boldsymbol{f}, \tau) \\ S_\alpha(\xi; \boldsymbol{f}) \end{pmatrix}^{\otimes 2}.$$

The information matrix $-\partial_\xi^2 \log L(\xi; \mathbf{y}_o)$ can be approximated by

$$\mathrm{E}_\xi\big[I_1(\xi; \boldsymbol{f})|\mathbf{y}_o\big] + \begin{pmatrix} -\mathrm{E}_\tau[J(\tau; \boldsymbol{f})] - (S_{\tau,2})^{\otimes 2} & 0 \\ 0 & 0 \end{pmatrix} + \mathrm{E}_\xi\big[I_2(\xi; \boldsymbol{f})|\mathbf{y}_o\big]$$
$$+ \begin{pmatrix} -(S_{\tau,2})^{\otimes 2} + S_{\tau,1}S_{\tau,2}^{\mathrm{T}} + S_{\tau,2}S_{\tau,1}^{\mathrm{T}} & \mathbf{0} \\ \mathbf{0} & 0 \end{pmatrix} + s_\xi(\xi, \mathbf{y}_o)^{\otimes 2}. \qquad (11)$$

3.1.1. Basic steps of the SAEM algorithm

We introduce the SAEM algorithm for SMMs as follows. We adaptively update seven estimates: ξ^k, the current estimate of $\hat{\xi}$; $S_{\tau,1}^k$, the current estimate of $\mathrm{E}_{\hat{\xi}}[-\partial_\tau U(\boldsymbol{f}, \tau)|\mathbf{y}_o]$; $S_{\tau,2}^k$, the current estimate of $-\partial_\tau \log C(\hat{\tau})$; $\mathbf{h}^k$, the current estimate of $s_\xi(\hat{\xi}, \mathbf{y}_o)$; $\boldsymbol{\Gamma}_1^k$, the current estimate of $\mathrm{E}_{\hat{\xi}}[I_1(\hat{\xi}; \boldsymbol{f})|\mathbf{y}_o]$; $\boldsymbol{\Gamma}_2^k$, the current estimate of $\mathrm{E}_{\hat{\xi}}[I_2(\hat{\xi}; \boldsymbol{f})|\mathbf{y}_o]$; and $\boldsymbol{\Gamma}_3^k$, the current estimate of $\mathrm{E}_{\hat{\tau}}[J(\hat{\tau}; \boldsymbol{f})]$. Let $\Pi_\tau(\cdot, \cdot)$ denote the Markov transition probability of the MH algorithm for simulating $\boldsymbol{f}$ from MRF (1), and let $\Pi_{\mathbf{y}_o,\xi}(\cdot,\cdot)$ denote the transition probability of the MH algorithm for simulating $\boldsymbol{f}$ conditional on $\mathbf{y}_o$.

Step 1. At the kth iteration, set $\boldsymbol{f}_{k,0} = \boldsymbol{f}_{k-1,N_{k-1}}$ and $\boldsymbol{f}_{y,k,0} = \boldsymbol{f}_{y,k-1,N_{k-1}}$. For $i = 1, \ldots, N_k$, generate $\boldsymbol{f}_{k,i}$ and $\boldsymbol{f}_{y,k,i}$ from the transition probability $\Pi_{\tau^{k-1}}(\boldsymbol{f}_{k,i-1}, \cdot)$ and $\Pi_{\mathbf{y}_o,\xi^{k-1}}(\boldsymbol{f}_{y,k,i-1}, \cdot)$, respectively.

Step 2. Update the seven estimates as follows:

$$\begin{cases} \xi^k = \xi^{k-1} + \gamma_k[\boldsymbol{\Gamma}^k]^{-1}\overline{H}(\xi^{k-1}; \boldsymbol{f}_k, \boldsymbol{f}_{y,k}), \\ \mathbf{h}^k = \mathbf{h}^{k-1} + \gamma_k\big(\overline{H}(\xi^{k-1}; \boldsymbol{f}_k, \boldsymbol{f}_{y,k}) - \mathbf{h}^{k-1}\big), \\ \boldsymbol{\Gamma}_1^k = \boldsymbol{\Gamma}_1^{k-1} + \gamma_k\big(\bar{I}_1(\xi^{k-1}; \boldsymbol{f}_{y,k}) - \boldsymbol{\Gamma}_1^{k-1}\big), \\ \boldsymbol{\Gamma}_2^k = \boldsymbol{\Gamma}_2^{k-1} + \gamma_k\big(\bar{I}_2(\xi^{k-1}; \boldsymbol{f}_{y,k}) - \boldsymbol{\Gamma}_2^{k-1}\big), \\ \boldsymbol{\Gamma}_3^k = \boldsymbol{\Gamma}_3^{k-1} + \gamma_k\big(\bar{J}(\tau^{k-1}; \boldsymbol{f}_k) - \boldsymbol{\Gamma}_3^{k-1}\big), \\ S_{\tau,1}^k = S_{\tau,1}^{k-1} + \gamma_k\big(-\partial_\tau\overline{U}(\boldsymbol{f}_{y,k}, \tau^{k-1}) - S_{\tau,1}^{k-1}\big), \\ S_{\tau,2}^k = S_{\tau,2}^{k-1} + \gamma_k\big(\partial_\tau\widehat{U}(\boldsymbol{f}_k, \tau^{k-1}) - S_{\tau,2}^{k-1}\big), \end{cases} \qquad (12)$$

where $\mathbf{h}^{k\mathrm{T}} = (\mathbf{h}_\tau^{k\mathrm{T}}, \mathbf{h}_\alpha^{k\mathrm{T}})$, $\boldsymbol{f}_k = (\boldsymbol{f}_{k,1}, \ldots, \boldsymbol{f}_{k,N_k})$ and $\boldsymbol{f}_{y,k} = (\boldsymbol{f}_{y,k,1}, \ldots, \boldsymbol{f}_{y,k,N_k})$,

$$\bar{I}_1(\xi; \boldsymbol{f}_{y,k}) = \sum_{i=1}^{N_k} I_1(\xi; \boldsymbol{f}_{y,k,i})/N_k,$$

$$\bar{I}_2(\xi; \boldsymbol{f}_{y,k}) = \sum_{i=1}^{N_k} I_2(\xi; \boldsymbol{f}_{y,k,i})/N_k,$$

$$\bar{J}(\tau; \boldsymbol{f}_k) = \sum_{i=1}^{N_k} J(\tau; \boldsymbol{f}_{k,i})/N_k,$$

$$\boldsymbol{\Gamma}^k = \boldsymbol{\Gamma}_1^k + [\mathbf{h}^k]^{\otimes 2} + \begin{pmatrix} -\boldsymbol{\Gamma}_3^k - (S_{\tau,2}^k)^{\otimes 2} & 0 \\ 0 & 0 \end{pmatrix},$$

$$\partial_\tau \overline{U}(\boldsymbol{f}_{y,k}, \tau) = \frac{1}{N_k}\sum_{i=1}^{N_k} \partial_\tau U(\boldsymbol{f}_{y,k,i}, \tau),$$

$$\partial_\tau \widehat{U}(\boldsymbol{f}_k, \tau) = \frac{1}{N_k}\sum_{i=1}^{N_k} \partial_\tau U(\boldsymbol{f}_{k,i}, \tau),$$

$$\overline{H}(\xi; \boldsymbol{f}_k, \boldsymbol{f}_{y,k}) = \left(\left[-\partial_\tau \overline{U}(\boldsymbol{f}_{y,k}, \tau) + \partial_\tau \widehat{U}(\boldsymbol{f}_k, \tau)\right]^{\mathrm{T}}, \frac{1}{N_k}\sum_{i=1}^{N_k} S_\alpha(\xi; \boldsymbol{f}_{y,k,i})^{\mathrm{T}}\right)^{\mathrm{T}}.$$

3.1.2. *Gain constants*

Gain constants play an essential role in ensuring the convergence of stochastic approximation algorithms. For fixed N_k, the gain constants sequence $\{\gamma_k\}$ must satisfy the following conditions:

$$0 \leqslant \gamma_k \leqslant 1 \text{ for all } k, \quad \sum_{k=1}^{\infty} \gamma_k = \infty, \quad \text{and} \quad \sum_{k=1}^{\infty} \gamma_k^2 < \infty. \tag{13}$$

In practice, gain constants are usually defined by $\gamma_k = b_1/(k^{a_1}+b_1-1)$, $k = 1, \dots, K_1$, where integer b_1 and real number $a_1 \in (1/2, 1]$ are preassigned and K_1 is determined by some random criteria (Gu and Zhu, 2001; Zhu et al., 2005a). For a given sequence γ_k, the SAEM algorithm iterates Steps 1 and 2 as described above. At the beginning of the SAEM algorithm, we suggest to choosing a small a_1 so that the SAEM algorithm will move quickly towards to the feasible region. When the algorithm starts to stabilize near the neighborhood of MLE, we set a_1 to be close to 1, and a small integer is chosen for b_1, say, $a_1 = 0.8$ and $b_1 = 2$. At the same time, an averaging procedure is used, with $\tilde{\xi}^0 = \xi^0$, $\tilde{\mathbf{h}}^0 = \mathbf{h}^0$, $\widetilde{S}^0_{\tau,m} = S^0_{\tau,m}$, and $\widetilde{\boldsymbol{\Gamma}}^0_{m'} = \boldsymbol{\Gamma}^0_{m'}$,

$$\tilde{\xi}^k = \tilde{\xi}^{k-1} + (\xi^k - \tilde{\xi}^{k-1})/k, \qquad \tilde{\mathbf{h}}^k = \tilde{\mathbf{h}}^{k-1} + (\mathbf{h}^k - \tilde{\mathbf{h}}^{k-1})/k,$$

$$\widetilde{S}^k_{\tau,m} = \widetilde{S}^{k-1}_{\tau,m} + (S^k_{\tau,m} - \widetilde{S}^{k-1}_{\tau,m})/k, \quad \text{and} \quad \widetilde{\boldsymbol{\Gamma}}^k_{m'} = \widetilde{\boldsymbol{\Gamma}}^{k-1}_{m'} + (\boldsymbol{\Gamma}^k_{m'} - \widetilde{\boldsymbol{\Gamma}}^{k-1}_{m'})/k,$$

for $m = 1, 2$ and $m' = 1, 2, 3$. Theoretically, this averaging procedure automatically leads to an optimal convergence without estimating the information matrix (Polyak, 1990; Polyak and Juditski, 1992). Under some conditions, the off-line average $(\tilde{\xi}^{K_1}, \tilde{\mathbf{h}}^{K_1})$ converges to $(\hat{\xi}, s_\xi(\hat{\xi}, \mathbf{y}_o))$ almost surely, as $K_1 \to \infty$ (Zhu et al., 2005a). Finally, we can substitute $\widetilde{S}^{K_1}_{\tau,m}$ $(m = 1, 2)$ and $\widetilde{\boldsymbol{\Gamma}}^{K_1}_{m'}$ $(m' = 1, 2, 3)$ into Eq. (11) to estimate $-\partial^2_\xi \log L(\hat{\xi}; \mathbf{y}_o)$.

3.2. *Annealing stochastic approximation Monte Carlo algorithms*

The annealing stochastic approximation Monte Carlo (ASAMC) algorithm originates from the multicanonical algorithm (Berg and Neuhaus, 1991). In the past decade, the multicanonical algorithm has been studied extensively. See, for instance, $(1/k)$-ensemble sampling in Hesselbo and Stinchcombe (1995), the Wang–Landau algorithm in Wang and Landau (2001), the generalized Wang–Landau (GWL) algorithm in Liang (2004, 2005a, 2005b), and the stochastic approximation Monte Carlo (SAMC) in Liang et al. (2005) and Liang (2005c), among others. In particular, the GWL and SAMC algorithms improve the multicanonical algorithm and its variants by introducing the concept of partitioning the sample space and further extending the multicanonical algorithm from discrete system to continuum system. Computational advances, including the GWL and SAMC algorithms, led to possible solutions to many complex statistical problems, such as model selection, highest posterior density region/interval construction, and the Monte Carlo optimization, among others.

3.2.1. Multicanonical algorithm

We use the Ising model as an example to the explicate multicanonical algorithm. The Gibbs distribution of the Ising model on an $L \times L$ lattice space can be written as

$$p(f|\tau) \propto \exp\{-U(f)/\tau\}, \quad f \in S_f, \tag{14}$$

where $U(f) = -\sum_{s_i \sim s_j} \delta(s_i, s_j)$ and $S_f = \{-1, 1\}^{L^2}$. We are interested in estimating $\Omega(u) = \#\{f\colon U(f) = u\}$, called the density of states (or spectral density) of the system. We may directly use MCMC algorithms (e.g., the MH algorithm or the Gibbs sampler) to draw samples from $p(f|\tau)$ and then use the simulated samples to estimate $\Omega(u)$. However, conventional MCMC algorithms can become trapped into a local energy minimum indefinitely, rendering the simulation ineffective. The multicanonical algorithm provides an attractive solution to this difficulty. The multicanonical algorithm seeks to draw samples from a modified distribution given by

$$p_m(f) \propto \exp\{-\log \Omega(U(f))\}. \tag{15}$$

If samples can be exactly drawn from (15), then the resulting distribution for U should be uniform distribution, that is, $p_U(u) \propto 1$. Thus, the algorithm will not become trapped into a local energy minimum, because sampling from $p_m(f)$ leads to a "free" random walk in the space of energy. However, $\Omega(u)$ is unknown prior to the simulation.

The key idea of the multicanonical algorithm is to iteratively update the approximation of $\Omega(u)$, denoted as $\widehat{\Omega}(u)$, then producing Monte Carlo samples from an approximated version of $p_m(f)$. The statistical quantities related to $p(f)$ can then be estimated based on the Monte Carlo samples with the technique of importance sampling. In addition, the multicanonical algorithm is useful for optimization. For instance, a study on protein folding problems (Hansmann and Okamoto, 1997) shows that the multicanonical algorithm is much more efficient than the temperature rescaling-based algorithms, including simulated tempering and simulated annealing (Marinari and Parisi, 1992; Geyer and Thompson, 1995; Kirkpatrick et al., 1983).

3.2.2. Basic steps of the ASAMC algorithm

Let $\widetilde{U}(f)$ be the negative of the complete-data log-likelihood function of SMMs, $-\ell_c(\hat{\xi}; f, \mathbf{y}_o)$. Given $\hat{\xi}$ in Stage 1, the ASAMC algorithm is then applied to find an optimal configuration of f, denoted by $\hat{f}$, which minimizes $\widetilde{U}(f)$. The ASAMC algorithm comprises four steps as follows:

Step 1. Partition the sample space S_f into M disjoint subregions, $E_1, \ldots, E_M$, and set an arbitrary configuration $\boldsymbol{f}_0$, $\mathbf{g}_0^{\mathrm{T}} = (g_{0,1}, \ldots, g_{0,M}) = (0, \ldots, 0)$, a pre-specified parameter Δ, $U_{\min}^{(0)}$, and the search space $S_f^{(0)} = \bigcup_{i=1}^M E_i$;

Step 2. At the kth iteration, use the MH algorithm with a global proposal distribution to simulate a sample $\boldsymbol{f}_k$ from the distribution

$$p_{\mathbf{g}_k}(\boldsymbol{f}) \propto \sum_{i=1}^{I(U_{\min}^{(k-1)}+\Delta)} \frac{\psi(\boldsymbol{f})}{e^{g_{k-1,i}}} \delta(\boldsymbol{f} \in E_i),$$

where $\psi(\boldsymbol{f}) = \exp\{-\widetilde{U}(\boldsymbol{f})/t_0\}$, t_0 is a preassigned number, $\delta(\boldsymbol{f} \in E_i)$ is the Kronecker function equaling to 1 when $\boldsymbol{f} \in E_i$ and 0 otherwise, $U_{\min}^{(k-1)}$ is the minimum energy value obtained until the $(k-1)$th iteration, and $I(z)$ denotes the index of subregion where a sample $\boldsymbol{f}$ with energy $\widetilde{U}(\boldsymbol{f}) = z$ belongs to (e.g., $I(\widetilde{U}(\boldsymbol{f})) = j$ for $\boldsymbol{f} \in E_j$);

Step 3. Update the working parameter $\mathbf{g}_k$ in the following manner:

$$g_{k,i} = g_{k-1,i} + \gamma_k\big[\delta(\boldsymbol{f}_k \in E_i) - \pi_i\big], \quad i = 1, \ldots, M,$$

where $\pi_i \in (0,1)$ and $\sum_{i=1}^M \pi_i = 1$, and nonincreasing sequence γ_k $(k = 1, \ldots)$ satisfies

$$\gamma_k > 0, \quad \sum_{k=1}^{\infty} \gamma_k = \infty, \quad \text{and} \quad \sum_{k=1}^{\infty} \gamma_k^q < \infty, \tag{16}$$

where $q \in (1,2)$. Throughout the paper, we set $\gamma_k = [k_0/\max(k_0,k)]^{a_2}$ for some specified value $k_0 > 1$, where $a_2 \in (0.5, 1]$;

Step 4. Increase k to $k+1$ and update $U_{\min}^{(k)}$, M, and the sample space to $S_f^{(k)} = \bigcup_{i=1}^{I(U_{\min}^{(k)}+\Delta)} E_i$.

We have to impose several conditions on the sample space S_f in the ASAMC algorithm. In Step 1, the sample space is usually partitioned into M disjoint subregions as follows: $E_1 = \{\boldsymbol{f}\colon \widetilde{U}(\boldsymbol{f}) \leqslant u_1\}$, $E_2 = \{\boldsymbol{f}\colon u_1 < \widetilde{U}(\boldsymbol{f}) \leqslant u_2\}, \ldots, E_{M-1} = \{\boldsymbol{f}\colon u_{M-2} < \widetilde{U}(\boldsymbol{f}) \leqslant u_{M-1}\}$, and $E_M = \{\boldsymbol{f}\colon u_M \geqslant \widetilde{U}(\boldsymbol{f}) > u_{M-1}\}$, where $u_1, \ldots, u_M$ are specified real numbers such that $u_1 < \cdots < u_M$. For SMMs, $\psi(\boldsymbol{f}) = \exp\{-\widetilde{U}(\boldsymbol{f})/t_0\}$ and $w_{\psi,i} = \int_{E_i} \psi(\boldsymbol{f}) m(\mathrm{d}\boldsymbol{f})$ is the partition function of the truncated distribution of $\boldsymbol{f}$ in the subregion E_i. Furthermore, we assume that the sample space S_f is compact. This condition is trivial for some discrete systems, such as the Ising model. However, for continuous systems, we restrict S_f to a set $\{\boldsymbol{f}\colon \widetilde{U}(\boldsymbol{f}) \leqslant \widetilde{U}_{\max}\}$, where $\widetilde{U}_{\max}$ is a fixed large value so that the set $\{\boldsymbol{f}\colon \widetilde{U}(\boldsymbol{f}) > \widetilde{U}_{\max}\}$ is not of interest.

Two other important features of the ASAMC algorithm are approximately sampling from $p_m(\boldsymbol{f})$ as in the multicanonical algorithm and updating working estimates $\mathbf{g}_k$s. For simplicity, we temporarily assume that $U_{\min}^{(k-1)}$ is fixed and $I(U_{\min}^{(k-1)} + \Delta) = M$ (Liang et al., 2005). At the kth iteration, we use an MCMC algorithm to draw a sample from the distribution

$$p_{\mathbf{g}_k}(\boldsymbol{f}) \propto \sum_{i=1}^{M} \frac{\psi(\boldsymbol{f})}{e^{g_{k-1,i}}} \delta(\boldsymbol{f} \in E_i), \tag{17}$$

where $g_k = (g_{k,1}, \ldots, g_{k,M}) \in \mathcal{G}$ is an estimate of $(\log w_{\psi,1}, \ldots, \log w_{\psi,M})$ until the kth iteration. In practice, for continuum system, we set $\mathcal{G} = [-B, B]^n$ with $B = 10^{100}$. Because adding to or subtracting from $\mathbf{g}_k$ a constant will not change $p_{\mathbf{g}_k}(\boldsymbol{f})$, $\mathbf{g}_k$ can be kept in the compact set in simulations by adjusting with an additive constant. Under appropriate conditions,

$$g_{k,i} \to \begin{cases} c + \log(\int_{E_i} \psi(\boldsymbol{f}) m(\mathrm{d}\boldsymbol{f})) - \log(\pi_i + \eta), & E_i \neq \emptyset, \\ -\infty, & E_i = \emptyset, \end{cases} \tag{18}$$

where $\eta = \sum_{j \in \{i: E_i = \emptyset\}} \pi_j/(M - M_0)$ and M_0 is the number of empty subregions, as $k \to \infty$ (Liang et al., 2005). In addition, c is a constant which can be determined by imposing a constraint on $\mathbf{g}_k$. For instance, $\sum_{i=1}^m e^{g_{k,i}}$ is equal to a fixed number. Since the sample space is partitioned blindly in the ASAMC algorithm, some of the subregions may be empty, that is, $\int_{E_i} \psi(\boldsymbol{f})\,\mathrm{d}\boldsymbol{f} = 0$. The working distribution $p_{\mathbf{g}_k}(\boldsymbol{f})$ is obtained by a piecewise modification of $p(\boldsymbol{f}|\tau)$, where each subregion is associated with a different weight $e^{g_{k,i}}$ (Liang et al., 2005).

The above stochastic approximation algorithm is an annealing algorithm, because the sample space $S_f^{(k)}$ shrinks during each iteration. Theoretically, the ASAMC algorithm can find the global energy minimum if the algorithm is run long enough, but the process of locating the global energy minimum may be very slow due to the breadth of the sample space. To accelerate the process, Liang (2004, 2005c) proposed to restrict the sample space of the ASAMC algorithm to a small region during each iteration. Suppose that the subregions $E_1, \ldots, E_M$ have been arranged in ascending order by energy; that is, if $i < j$, then $U(\boldsymbol{f}) < U(\boldsymbol{f}')$ for any $\boldsymbol{f} \in E_i$ and $\boldsymbol{f}' \in E_j$. The ASAMC algorithm starts with $S_f^{(0)} = \bigcup_{i=1}^M E_i$, and then iteratively sets

$$S_f^{(t)} = \bigcup_{i=1}^{I(U_{\min}^{(k-1)} + \Delta)} E_i. \tag{19}$$

Remarkably, the ASAMC algorithm preserves the convergence (18) on the limiting sample space $\lim_{t\to\infty} S_f^{(k)}$, provided that the proposal distribution used at each iteration is global, that is, a proposal distribution $q(\boldsymbol{f}, \boldsymbol{f}')$ is global if $q(\boldsymbol{f}, \boldsymbol{f}') > 0$ for all $\boldsymbol{f}, \boldsymbol{f}' \in S_f$.

As known by many researchers, the state of the art algorithm for stochastic optimization is the simulated annealing algorithm (Kirkpatrick et al., 1983). For instance, we consider the problem of minimizing the function $\widetilde{U}(\boldsymbol{f})$. Simulated annealing works by simulating from a sequence of distributions scaled by the temperature as follows,

$$p_{t_k}(\boldsymbol{f}) \propto \exp\{-\widetilde{U}(\boldsymbol{f})/t_k\}, \quad k = 1, 2, \ldots,$$

where t_k's are called the temperatures forming a decreasing ladder $t_1 > \cdots > t_k > \cdots \geqslant 0$. Under some conditions, simulated annealing will converge to the set of global minima of $\widetilde{U}(\boldsymbol{f})$ in probability 1 when the temperature decreases sufficiently slowly, i.e., $t_k > 1/\log(L_k)$, where $L_k = N_1 + \cdots + N_k$ (Geman and Geman, 1984). In addition, N_k is the number of iterations generated from the MH algorithm in simulating from the distribution $p_{t_k}(\boldsymbol{f})$. In practice, such a slow cooling scheme is impractical.

Instead, people use a linearly or geometrically decreasing cooling scheme, but such scheme cannot guarantee that the global minima will be reached. The ASAMC algorithm does not suffer from such a pitfall. If the proposal distribution is global and the gain constants satisfy the condition (16), the ASAMC algorithm will converge to the set of global minima as the number of iterations is large. The ASAMC algorithm will result in a "free" random walk in the subspace of the subregions. Its self-adjusting ability for the acceptance of a new proposal guarantees that it will not become stuck into a local energy minimum. Hence, as a stochastic optimization algorithm, the ASAMC algorithm is potentially much more powerful than the simulated annealing algorithm (Liang, 2005c).

3.2.3. *Practical issues*

For an effective implementation of the ASAMC algorithm, several issues need to be considered.

(i) *Partitioning the sample space*: For optimization problems, the partition can be done according to the energy function. The maximum energy difference in each subregion should be bounded by a reasonable number, say, 2, to ensures that a reasonable acceptance rate is achieved for the local MH moves within the same subregion. Note that within the same subregion, sampling from the working density (17) reduces to sampling from $\psi(\boldsymbol{f})$.

(ii) *Choice of* Δ: The performance of the ASAMC algorithm depends on the value of Δ to some extent. If Δ is too large, the ASAMC algorithm may take a long time to locate the global minimum due to the breadth of the sample space. If Δ is too small, the ASAMC algorithm may also take a long time to locate the global minimum. In this case, the sample space may contain only a few isolated regions, and most of the proposed transitions will be rejected. Allowing a sampler to jump to intermediate states of high energy will increase the probability of transition from one local energy minimum to others. To compensate for the negative effect of the sample space restriction, the proposal distribution used in the ASAMC algorithm should be spread out.

(iii) *Choice of* k_0 *and the number of iterations*: The γ_k controls the moving ability of the ASAMC algorithm across subregions, and k_0 controls the speed of γ_k converging to zero. In practice, k_0 can be chosen according to the complexity of the problem. The more complex the problem, the larger value of k_0. A large value of k_0 will force the sampler to reach all subregions quickly, even in the presence of multiple local energy minima. The appropriateness of the choice of k_0 and the number of iterations can be diagnosed by examining the convergence of the run, which can be further diagnosed by examining the equality of the realized sampling frequencies of limiting subregions. As suggested by Wang and Landau (2001), a run can be regarded as converged if the sampling frequency for each of the subregions is not less than 80% of the average sampling frequency; that is,

$$\min\left\{\frac{e_i}{\bar{e}}: i = 1, \ldots, I(U_{\min} + \Delta), E_i \neq \emptyset\right\} \geqslant 80\%, \tag{20}$$

where e_i denotes the realized sampling frequency of the subregion E_i, and $\bar{e}$ is the average sampling frequency of the subregions included in the above set. If a run does not converge, the ASAMC algorithm should be re-run with more iterations or a larger value of k_0.

(iv) *Choosing the proposal function*: In the ASAMC algorithm, the global proposal distribution ensures the ergodicity of the algorithm. In practice, a global proposal distribution can be designed easily for both discrete and continuum systems. For example, in simulations from an Ising model of linear size L, a new configuration can be generated with the following steps: draw an integer T with probability ε_t $(t = 1, \ldots, L^2)$, $0 < \varepsilon_t < 1$ and $\sum_{t=1}^{L^2} \varepsilon_t = 1$; choose T spins from the set $S = \{(i, j)\colon i, j = 1, \ldots, L\}$ at random and with replacement; reset the value of each of the T spins to $+1$ or -1 with equal probability. We will call this Sampling Method (I). For a particular configuration generated with the above procedure, the transition probability is then $q(f, f') = \varepsilon_T / 2^{\mathrm{T}}$. A typical choice for the ε_t's is $\varepsilon_1 = 0.9$ and $\varepsilon_t = (1 - \varepsilon_1)/(L^2 - 1)$ for $t = 2, \ldots, L^2$. For a continuum system, $q(f, f')$ can be set to the random walk Gaussian proposal $f' \sim N(f, \sigma^2)$, with σ^2 being calibrated to have a desired acceptance rate, such as 0.25.

4. Applications

In this section, we analyzed two real data sets from imaging studies from ecology. They will be discussed to illustrate the behavior of the SAEM algorithm, the ASAMC algorithm, and their combination. All computations were done in C++ on a Dell laptop. All computer codes and executable files can be downloaded from Dr. Zhu's website:

http://www.bios.unc.edu/~hzhu/SMM/smm.tar.

4.1. Distributions of vegetation species

We consider an automulticategorical model to analyze the dataset of vegetation species in Alberta, Canada. The primary goal of this data analysis is to demonstrate the efficiency of the stochastic approximation algorithm in locating MLE in complex spatial models. In particular, through this example, we want to show the feasibility of roughly approximating the first-order and second-order derivatives of the partition function during each iteration and controlling amount of noises by using stochastic approximation (Robbins and Monro, 1951). The secondary goal is to illustrate the wide application of spatial models.

Vegetation species is in the form of an atlas map with resolution pixel equaling 0.5 °C latitude × 0.5 °C longitude; see Figure 1(a). With the aid of remote sensing and aerial photogrammetric technologies, information on four species occurrence in Alberta, Canada is documented by this format (Little, 1971; Arnold, 1993, 1995; Mitchell-Jones et al., 1999). There are total of 375 grid cells. At each site (k, l), there are a categorical response $Y(k, l)$ and 2 interesting climate covariates: $X_1(k, l)$ (absolute minimum temperature); and $X_2(k, l)$ (annual degree-days). Five major types of vegetation in Alberta are: V0 – Background, V1 – subarctic evergreen forest, V2 – boreal evergreen forest, V3 – boreal summergreen woodland, and V4 – grass prairie. Two covariates are expected to

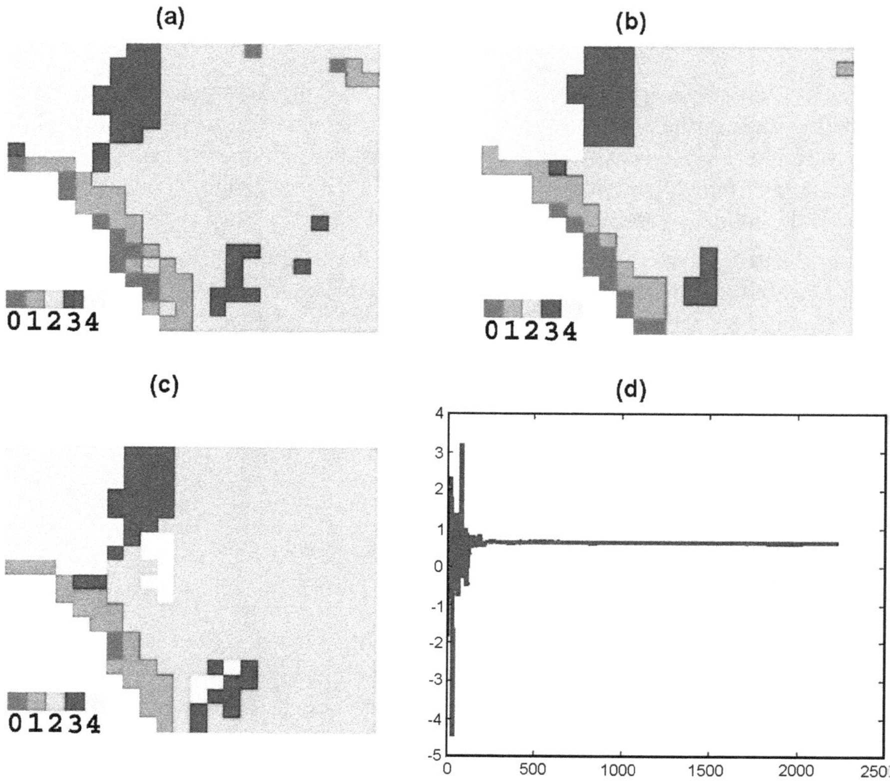

Fig. 1. Distribution of four vegetation types in Alberta, Canada: 0 = background, 1 = subarctic evergreen forest, 2 = boreal evergreen forest, 3 = boreal summergreen woodland, and 4 = grass prairie. There are four panels: (a) the observed distribution; (b) the fitted distribution; (c) the predicted distribution with annual degree-days ($X_2(k,l)$) being increased 350; and (d) $\tau(1)^k$ at each iteration of the SAEM algorithm.

be among those determining the distributions of vegetation at geographical scales and having significant changes in global warming.

Following Zhu et al. (2005b), the second-order automulticategorical regression model is assumed for $\boldsymbol{Y} = \{Y(k,l), (k,l) \in S\}$, where the conditional probability at site $(k,l) \in S$ given all the other values $Y(m,n)$ $((m,n) \neq (k,l))$ is given as follows

$$\Pr(Y_{k,l} = i \mid \text{all other sites}) = \frac{\exp\{g_{k,l}(i|\theta)\}}{\sum_{j=0}^{4} \exp\{g_{k,l}(j|\theta)\}}, \quad i = 0, \ldots, 4. \tag{21}$$

In addition, $g_{k,l}(i|\theta) = X(k,l)^{\mathrm{T}}\beta(i) + \tau(i)y^*_{k,l}(i)$ for $i = 0, \ldots, 4$, where $y^*_{k,l}(i)$ is the number of eight sites $\{(k,l-1), (k,l+1), (k-1,l), (k+1,l), (k-1,l-1), (k+1,l+1), (k-1,l+1), (k+1,l-1)\}$ colored i. To avoid redundancy, we assume that $\beta(0) = \mathbf{0}$ and $\tau(0) = 0$. The SAEM algorithm with $(a_1, b_1) = (0.8, 4)$ and $N_k = 5000$ was used to find the maximum likelihood estimates. The algorithm converged in 2231 iterations. The initial value for the vector ξ was set to be $\xi^0 = \mathbf{0}$.

Table 1
Maximum likelihood estimates of the automulticategorical model to the distribution of vegetation species data in Alberta, Canada

	Cluster 1		Cluster 2		Cluster 3		Cluster 4	
	EST	SD	EST	SD	EST	SD	EST	SD
	$\beta(1)$		$\beta(2)$		$\beta(3)$		$\beta(4)$	
Intercept	2.021	8.765	−5.269	9.003	−7.887	9.993	−18.112	15.700
$X_1(k,l)$	0.129	0.184	0.041	0.188	0.151	0.201	0.428	0.309
$X_2(k,l)$	0.003*	0.001	0.005*	0.001	0.009*	0.002	0.018*	0.005
	$\tau(1)$		$\tau(2)$		$\tau(3)$		$\tau(4)$	
	0.624*	0.161	0.526*	0.111	0.383*	0.138	0.779*	0.308

* represents that parameters are different from zero at the significance level $\alpha = 0.05$.

The obtained results are summarized in Table 1. The distribution of the vegetation species is related to the annual degree-days. The autocorrelation coefficient $\tau(i)$ $(i = 1, 2, 3, 4)$ are significantly different from zero. The fitted map $\mathbf{Y}$ of the distribution of four vegetation types in Alberta is shown in Figure 1(b). Figure 1(d) shows the estimate $\tau(1)^k$ at each iteration of the stochastic approximation algorithm. We observe that our stochastic approximation algorithm is robust to the initial value of ξ and can find MLE.

An important scientific issue is to predict the redistribution of vegetation species under various global warming scenarios. For instance, an enhanced greenhouse effect (e.g., the doubling of atmospheric concentration of CO_2) would increase the global mean temperature from 1.5 to 4.5 °C in the future 30 to 50 years. One advantage of the automulticategorical model is that the climate change effect can be quantified through odds ratio of the conditional probabilities. For example, if the annual degree-days X_2 increases by 350 (approximately equivalent to 1 °C increase in daily temperature) while other variables remain constant, then the odds ratio of the conditional probabilities of the jth vegetation presence is supposed to increase by a factor $e^{350\beta_2(j)}$. This result suggests that subarctic evergreen forest, boreal evergreen forest, boreal summergreen woodland, and grass prairie will be increased by global warming. The impact of climate change on the distributions of four vegetation types is shown in Figure 1(c).

4.2. Simulation study

Consider a degraded pixel image on a finite grid S of pixels, placing a binary random variable $Y(i,j)$ at each site (i,j) on S, a subset of a regular $M_0 \times N_0$ lattice. Let the true image be $\boldsymbol{f} = \{f(k,l)\colon (k,l) \in S\}$, where $f(k,l) = 0$ represents a white pixel and $f(k,l) = 1$ represents a black pixel. Because S is usually a irregular lattice in most applications, we consider the joint distribution of the internal site responses $\boldsymbol{f}^I = \{f(k,l)\colon (k,l) \in S^o\}$ conditional upon fixed boundary values $\boldsymbol{f}^B = \{f(k,l)\colon (k,l) \in \partial S\}$, where ∂S and S^o denote the set of all sites forming the boundary of S and the set of all internal sites of S, respectively. Following Besag (1974), the probability function of the first-order autologistic regression of $\boldsymbol{f}^I$ given $\boldsymbol{f}^B$ can be written in a Gibbsian form as follow:

$$p(\boldsymbol{f}^I|\tau, \boldsymbol{f}^B) = \exp\{\tau^{\mathrm{T}}T(\boldsymbol{f})\}/C(\tau), \tag{22}$$

where $T(\boldsymbol{f}) = \sum_{(k,l)\in S^o} f(k,l)\widetilde{X}_f(k,l)$ and $\widetilde{X}_f(k,l) = (X(k,l)^{\mathrm{T}}, \tilde{f}(k,l)/2)^{\mathrm{T}}$, in which $\tilde{f}(k,l)$ is the number of sites in $\{(k,l-1),(k,l+1),(k-1,l),(k+1,l)\}$ colored 1. Also, $X(k,l)$ is a $p\times 1$ vector of covariates at site (k,l), $\tau = (\beta^{\mathrm{T}}, \tau_1)^{\mathrm{T}} \in R^{p+1}$, $\beta \in R^p$, and $\tau_1 \in R$. Given the true image $\boldsymbol{f}$, the true observed image $\mathbf{Y} = \{Y(k,l)\colon (k,l)\in S\}$ is assumed to be conditionally mutually independent and

$$Y(k,l)|f(k,l) \sim \mathrm{Binomial}\big(1, p\big(f(k,l)\big)\big), \tag{23}$$

where $p(0) = 0$ and $p(1) = \exp(-\alpha(1)^2) \in (0,1]$. That is, if $f(k,l) = 0$, $Y(k,l) = 0$ with probability 1, while $Y(k,l) = 1$ with probability $p(1)$ and $Y(k,l) = 0$ with probability $1-p(1)$ under $f(k,l) = 1$. Thus, we obtain an SMM. In practice, scientists may consider the first-order, second-order and even higher correlation structures; see, for example, the first-order structure in Huffer and Wu (1998) and the second-order structure in Besag (1974, 1986) and He et al. (2003).

In order to check the usefulness of the proposed algorithm, we consider the following simulation study, in which the autologistic regression model is set on a 30×30 lattice and $X(k,l) = (2.5\times \sin(0.1\times (k+l)))$. In our simulation, $\beta = 1$, $\tau_1 \in \{0.2, 0.4, 0.6, 0.8\}$, and $\alpha(1) = 0.85(p(1) \approx 0.325)$. Therefore, there are three unknown parameters. To simulate the process $\boldsymbol{f} = \{f(k,l)\colon (k,l)\in S^o\}$ from (22), we use the standard Gibbs sampler. The initial state of the process is taken at random such that $X(k,l)$ is independently taken to be 1 or 0 with $1/2$ probability and the Gibbs sampler is repeated 10 000 times (10 000 Monte Carlo steps) to ensure that the equilibrium state is achieved. Afterwards, the binomial noise is added according to (23).

For each parameter vector $\xi = (\tau_1, \beta, p(1))^{\mathrm{T}}$, we generated $N = 500$ datasets. For each pseudo-observed dataset, the SAEM algorithm with $(a_1, b_1) = (0.8, 5)$ was applied to get the MLE of the unknown parameters. The initial value of ξ was set at $(0, 0, 0.5)$. In each iteration of the algorithm, the standard Gibbs sampler was used to generate $\boldsymbol{f}$ from $p(\boldsymbol{f}|\tau)$; therefore, we can estimate $\partial_\tau \log C(\tau)$ and $\partial_\tau^2 \log C(\tau)$. To simulate the process $\mathbf{Y}$ given $\boldsymbol{f}$, we used the following algorithm. If $f(k,l) = 1$, $Y(k,l)$ must be equal to 1; however, given $Y(k,l) = 0$ and other $f(u,v)$s, we have

$$\begin{aligned}&\mathrm{P}\big(f(k,l) = 1|Y(k,l) = 0,\ \text{all other values}\big)\\&\quad= \frac{(1-p(1))\exp(X(k,l)^{\mathrm{T}}\beta + \tilde{f}(k,l)\tau_1)}{1+(1-p(1))\exp(X(k,l)^{\mathrm{T}}\beta + \tilde{f}(k,l)\tau_1)}.\end{aligned}$$

The standard Gibbs sampler is also used. The number N_k was set at 30.

In this simulation study, $\beta = 1$ represents relatively strong covariate effect and τ_1 ranges in four different cases. We calculated the bias, the mean of the standard deviation estimates, and the root mean-square error obtained from the 500 estimates. The results obtained are summarized in Table 2. It can be seen that all the relative efficiencies are close to 1.0. This demonstrates that the SAEM algorithm is a useful method for optimizing SMMs.

Table 2
Bias, RMS, SD, and EFF of the maximum likelihood estimators of the noisy autologistic regression model

	True	Bias	RMS	SD	EFF		True	Bias	RMS	SD	EFF
β	1.0	−0.008	0.081	0.081	1.005	β	1.0	−0.004	0.080	0.079	1.017
τ_1	0.2	0.003	0.082	0.081	1.007	τ_1	0.4	0.004	0.073	0.077	0.945
$p(1)$	0.325	0.009	0.064	0.064	0.997	$p(1)$	0.325	0.007	0.043	0.044	0.978
β	1.0	−0.010	0.081	0.076	1.066	β	1.0	−0.002	0.083	0.083	0.991
τ_1	0.6	−0.027	0.075	0.074	1.009	τ_1	0.8	0.002	0.083	0.087	0.955
$p(1)$	0.325	0.006	0.033	0.034	0.962	$p(1)$	0.325	0.003	0.028	0.027	1.011

True denotes the true value of parameters; Bias denotes the bias of the mean of estimates; RMS denotes the root-mean-square error; SD denotes the mean of standard deviation estimates; and EFF denotes the ratio of SD and RMS.

4.3. Noisy vegetation data

We analyzed a real data on the distribution of subarctic evergreen woodland vegetation in terms of climate variables in the province of British Columbia, Canada. The subarctic evergreen woodland is in the form of an atlas map with resolution pixel equaling 0.5 °C latitude × 0.5 °C longitude. The observed map $\mathbf{Y}$ of the subarctic evergreen woodland is shown in Figure 2(a). There are total of 707 grid cells. At each site (k,l), there are a binary $Y(k,l)$ and 5 climate covariates of interest: $X_1(k,l)$ (absolute minimum temperature); $X_2(k,l)$ (annual degree-days); $X_3(k,l)$ (total actual evapotranspiration); $X_4(k,l)$ (annual soil moisture deficit); and $X_5(k,l)$ (annual snowpack). These covariates are expected to be among those determining the distribution of vegetation at geographical scales, and they are also the variables likely to change significantly because of global warming. Here, $Y(k,l)=1$ indicates that at least one subarctic evergreen woodland vegetation has been observed and $Y(k,l)=0$ indicates that subarctic evergreen woodland are either not inhabited or the subarctic evergreen woodland have not been observed. It is an obvious idea to interpret the observed map in Figure 2(a) as a degraded pixel image, in which a part of the originally black squares are white in the observed map.

We fitted the dataset by the noisy autologistic regression model (22) and (23). The stochastic approximation algorithm with $(a_1,b_1)=(0.8,4)$ and $N_k=30$ was used to find the MLE. The initial value of $\xi=(\tau_1,\beta,p(1))$ was set to be $\xi^0=(\mathbf{0},0,0.5)$. The obtained results are summarized in Table 3. The distribution of subarctic evergreen woodland vegetation is related to the absolute minimum temperature and the annual degree-days. The autocorrelation coefficient τ_1 is as high as a value at 1.52. Figure 2(c) shows the trace of $(\tau_1^k,\tilde{\tau}_1^k)$ and $(p(1)^k,\tilde{p}(1)^k)$ at each iterations of the SAEM algorithm.

We compared ICM with the simulated annealing and ASAMC algorithms, which search for the MAP $\hat{f}$ by minimizing $\widetilde{U}(f)=-\ell(f|\mathbf{Y};\hat{\xi})$ (except for a constant) given by

$$-\hat{\tau}^{\mathrm{T}}T(f)-\sum_{(k,l)\in S^o}\log\bigl[\bigl(1-\hat{p}\bigl(f(k,l)\bigr)\bigr)^{1-Y(k,l)}\hat{p}\bigl(f(k,l)\bigr)^{Y(k,l)}\bigr],$$

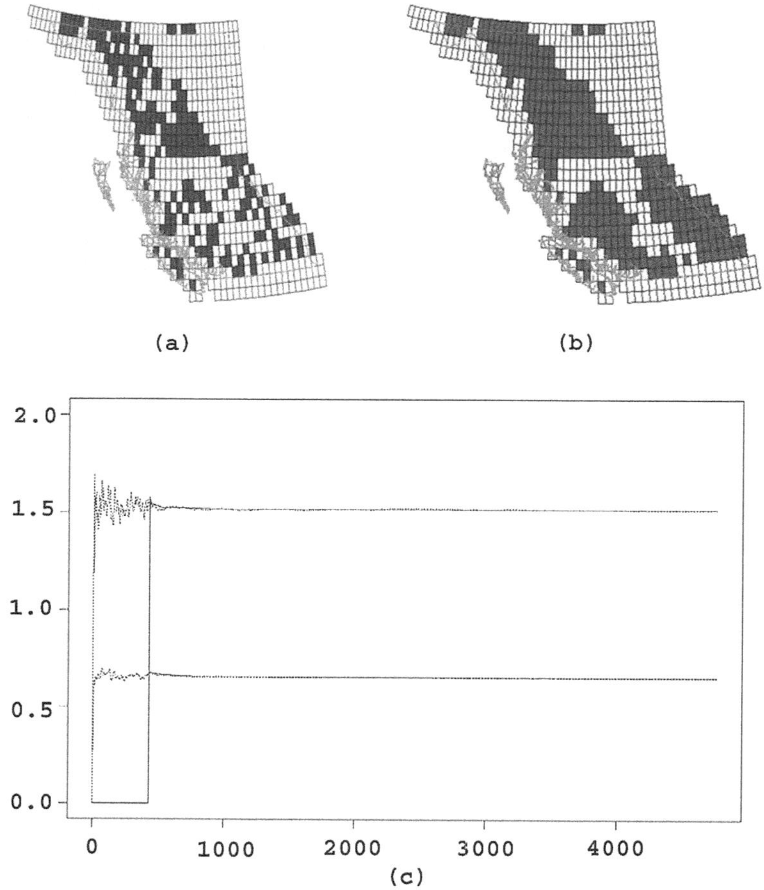

Fig. 2. Subarctic evergreen woodland data: (a) observed map; (b) restored map by the ASAMC algorithm; (c) $(\tau_1^k, \tilde{\tau}_1^k)$ and $(\alpha(1)^k, \tilde{\alpha}(1)^k)$ at each iteration of the SAEM algorithm.

where $T(\boldsymbol{f}) = \sum_{(k,l)\in S^o} f(k,l)\widetilde{X}(k,l)$. The $\widetilde{U}(\boldsymbol{f})$ contains $N_0 = 496$ variables $\{f(k,l)\colon Y(k,l) = 0\}$, because if $Y(k,l) = 1$, $f(k,l)$ must be one. Thus, the sample space has $2^{N_0} = 2^{496}$ configurations and direct searching for global minima is therefore nearly infeasible computationally. The ICM method (Besag, 1986) converged in only three iterations and leads to a local minimum 322.935. The simulated annealing and the ASAMC algorithms were run for 2×10^6 iterations. The simulated annealing located a local minimum close to 312.638, while the ASAMC algorithm located a energy minimum at 311.8. This demonstrates that the simulated annealing and ASAMC algorithms require much more computational cost, but they are able to locate the global energy minima with high probability. We include MAP $\hat{\boldsymbol{f}}$ estimated from the ASAMC algorithm in Figure 2(b).

For the simulated annealing algorithm, we considered a linear cooling scheme. We set the highest temperature $T_1 = 10$, the total number of temperature levels $K = 400$,

Table 3
Model fits to the subarctic evergreen woodland data

Iter/time		MCMC-SA algorithm ($\alpha = 0$) β						τ_1	$p(1)$
		const	X_1	X_2	X_3	X_4	X_5		
5391	EST	−0.7608	0.0371*	−0.0012*	0.0091*	0.0024	−0.0004	1.5190*	0.6534*
650s	SD	0.9639	0.0154	0.0004	0.0046	0.0016	0.0003	0.1493	0.0695

* represents that parameters are different from zero at the significance level $\alpha = 0.05$.

and the lowest temperature $T_{400} = 0.01$. At the T_t temperature level, we set $\varepsilon_1 = 0.5 + (10 - T_t)/25$ and $\varepsilon_t = (1.0 - \varepsilon_1)/(N_0 - 1)$ in Sampling Method (I) and run this procedure for 5000 iterations. The temperature decreased linearly such that $T_t = T_{t-1} - \rho$, where $\rho = (T_1 - T_{400})/(K - 1) \approx 2.504 \times 10^{-2}$. The initial configuration of $\{f(k,l)\colon Y(k,l) = 0\}$ was set as $\{f(k,l) = 0\colon Y(k,l) = 0\}$.

We presented in Figure 3(a) the index plot of the minimum values of $\widetilde{U}(\boldsymbol{f})$ at each temperature level and included in Figure 3(b) the index plots of the values of $\widetilde{U}(\boldsymbol{f})$ in the first and last 15 000 iterations. We observed that the simulated annealing algorithm led to a random walk in the sample space at high temperature. However, the simulated annealing algorithm became trapped in a local minimum 312.877 at low temperature, even though it located a minimum value at 312.638 across all temperature levels.

We applied the ASAMC algorithm to search for the minimum energy value of $\widetilde{U}(\boldsymbol{f})$ by using the following settings. The sample space was partitioned into $M = 1998$ subregions with an equal energy bandwidth: $E_1 = \{\boldsymbol{f}\colon \widetilde{U}(\boldsymbol{f}) \leqslant 310.599\}$, $E_2 = \{\boldsymbol{f}\colon 310.599 < \widetilde{U}(\boldsymbol{f}) \leqslant 310.849\}, \ldots,$ and $E_{1998} = \{\boldsymbol{f}\colon \widetilde{U}(\boldsymbol{f}) \leqslant 810.059\}$. We set $\psi(\boldsymbol{f}) = \exp(-\widetilde{U}(\boldsymbol{f})/10)$, $\pi_1 = \cdots = \pi_M = 1/M$, $\widetilde{U}_{\max} = 810.059$, $a_2 = 0.6$, $k_0 = 2500$, and $\Delta = 5$. The total number of iterations of the ASAMC algorithm was set at 2×10^6, which is the same as that of the simulated annealing algorithm. We chose the proposal distribution of Sampling Method (I), in which $\varepsilon_t = (1 - \varepsilon_1)/(N_0 - 1)$ for $t = 2, \ldots, N_0$ and ε_1 was set as 0.9 for $0 \leqslant k \leqslant 5 \times 10^5$, 0.7 for $5 \times 10^5 < k \leqslant 10^6$, and 0.5 for $k > 10^6$. The initial configuration of $\{f(k,l)\colon Y(k,l) = 0\}$ was set as $\{f(k,l) = 0\colon Y(k,l) = 0\}$.

The ASAMC algorithm outperforms the simulated annealing algorithm in this complex noisy vegetation data. In Figure 3(c), we presented the index plots of the values of $\widetilde{U}(\boldsymbol{f})$ at the 5000kth iteration and the minimum values of $\widetilde{U}(\boldsymbol{f})$ until the 5000kth iteration from the ASAMC algorithm, where $k = 0, \ldots, 400$. We observed that the ASAMC algorithm converges very quickly to a global minimum of $\widetilde{U}(\boldsymbol{f})$ at 311.8; in contrast, the simulated annealing algorithm wondered around in the sample space at the high temperature and became trapped in local minima at the low temperature (Figure 3(b)). Figure 3(d) shows the sampling frequency of each of the subregions of the ASAMC run. The 10 subregions with the lowest $\widetilde{U}(\boldsymbol{f})$ values are sampled approximately evenly. This indicates that the run has converged. Recall the diagnostic criterion given in (20) for the convergence of the ASAMC runs.

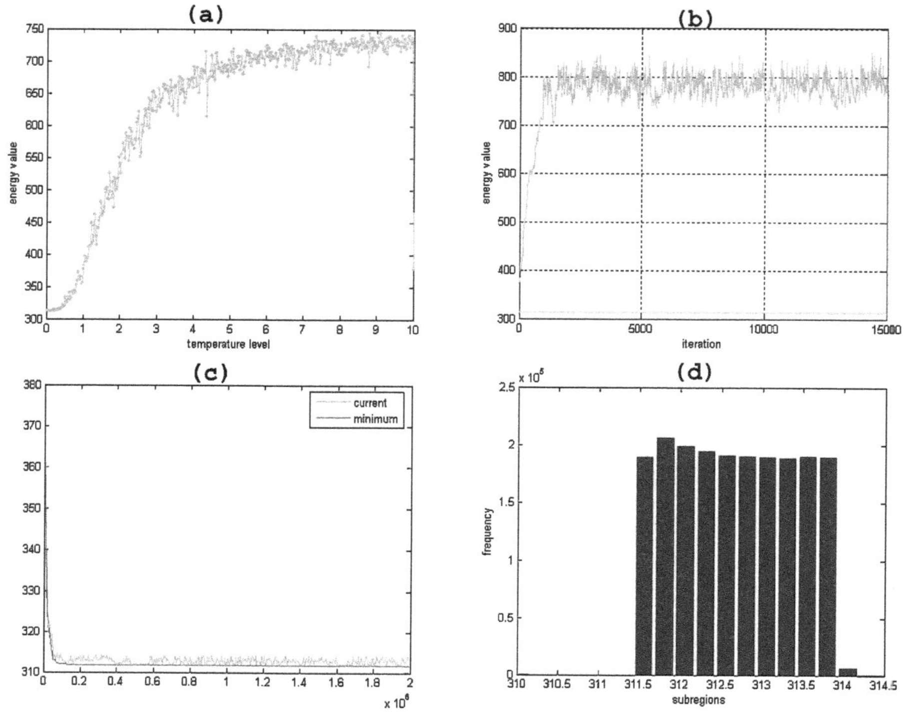

Fig. 3. Subarctic evergreen woodland data: comparison of the simulated annealing and ASAMC algorithms. $\widetilde{U}(\boldsymbol{f})$ is the negative value of the log-likelihood function of complete data, $-\ell_c(\hat{\xi}, \boldsymbol{f}; \mathbf{y}_o)$. There are four panels: (a) the index plot of the minimum energy values of $\widetilde{U}(\boldsymbol{f})$ at each temperature level from the simulated annealing algorithm; (b) the index plots of the energy values of $\widetilde{U}(\boldsymbol{f})$ in the first (red line) and last (green line) 15 000 iterations from the simulated annealing algorithm; (c) the index plots of the energy values of $\widetilde{U}(\boldsymbol{f})$ (red line) at each of the 5000kth iterations and the minimum energy values of $\widetilde{U}(\boldsymbol{f})$ (blue line) until the 5000kth iteration from the ASAMC algorithm, where $k = 0, \ldots, 400$; (d) the sampling frequency in last ten subregions from the ASAMC algorithm. (For interpretation of the references to color in this figure legend, the reader ir referred to the web version of this chapter.)

Acknowledgements

This work was supported in part by NSF grant SES-0643663 to Dr. Zhu, NSF grant DMS-0405748 and NCI grant CA104620 to Dr. Liang, by NIDA grant DA017820, NIMH grants MH068318 and K0274677 to Dr. Peterson, by the Suzanne Crosby Murphy Endowment at Columbia University College of Physicians and Surgeons, and by the Thomas D. Klingenstein and Nancy D. Perlman Family Fund. Thanks to Dr. Jason Royal for his invaluable editorial assistance and to Dr. Fangliang He for making his vegetation dataset available to us.

References

Arnold, H.R. (1993). *Atlas of Mammals in Britain.* Her Majesty's Stationery Office, London.

Arnold, H.R. (1995). *Atlas of Amphibians and Reptiles in Britain.* Her Majesty's Stationery Office, London.

Bentler, P.M., Dudgeon, P. (1996). Covariance structure analysis: statistical practice, theory, and directions. *Annals of Review Psychology* **47**, 563–592.
Berg, B.A., Neuhaus, T. (1991). Multicanonical algorithms for 1st order phase-transitions. *Physics Letters B* **267**, 249–253.
Besag, J.E. (1974). Spatial interaction and the statistical analysis of lattice systems (with discussion). *Journal of the Royal Statistical Society, Series B* **36**, 192–236.
Besag, J.E. (1986). On the statistical analysis of dirty pictures (with discussion). *Journal of the Royal Statistical Society, Series B* **48**, 259–302.
Booth, J.G., Hobert, J.P. (1999). Maximum generalized linear mixed model likelihood with an automated Monte Carlo EM algorithm. *Journal of the Royal Statistical Society, Series B* **61**, 265–285.
Bouman, C., Sauer, K. (1993). A generalized Gaussian image model for edge-preserving MAP estimation. *IEEE Transaction in Image Processing* **2**, 296–310.
Breslow, N.E., Clayton, D.G. (1993). Approximate inference in generalized linear mixed models. *Journal of the American Statistical Association* **88**, 9–25.
Delyon, B., Lavielle, E., Moulines, E. (1999). Convergence of a stochastic approximation version of the EM algorithm. *The Annals of Statistics* **27**, 94–128.
Dempster, A.P., Laird, N.M., Rubin, D.B. (1977). Maximum likelihood from incomplete data via the EM algorithm (with discussion). *Journal of the Royal Statistical Society, Series B* **39**, 1–38.
Diggle, P.J., Tawn, J.A., Moyeed, R.A. (1998). Model-based geostatistics (with discussion). *Applied Statistics* **47**, 299–350.
Geman, S., Geman, D. (1984). Stochastic relaxation, Gibbs distribution, and the Bayesian restoration of images. *IEEE Transactions on Pattern Analysis and Machine Intelligence* **6**, 721–741.
Geyer, C.J., Thompson, E.A. (1992). Constrained Monte Carlo maximum likelihood for dependent data (with discussion). *Journal of the Royal Statistics Society, Series B* **54**, 657–699.
Geyer, C.J., Thompson, E.A. (1995). Annealing Markov chain Monte Carlo with applications to ancestral inference. *Journal of the American Statistical Association* **90**, 909–920.
Gu, M.G., Kong, F.H. (1998). A stochastic approximation algorithm with Markov chain Monte Carlo method for incomplete data estimation problems. *Proceeding of National Academic Science of USA* **95**, 7270–7274.
Gu, M.G., Zhu, H.T. (2001). Maximum likelihood estimation for spatial models by Markov chain Monte Carlo stochastic approximation. *Journal of the Royal Statistical Society, Series B* **63**, 339–355.
Guyon, X. (1995). *Random Fields on a Network: Modeling, Statistics, and Applications*. Springer-Verlag, Berlin.
Hansmann, U.H.E., Okamoto, Y. (1997). Numerical comparisons of three recently proposed algorithms in the protein folding problems. *Journal of Computational Chemistry* **18**, 920–933.
Hastings, W.K. (1970). Monte Carlo sampling methods using Markov Chains and their applications. *Biometrika* **57**, 97–109.
He, F.L., Zhou, J.L., Zhu, H.T. (2003). Autologistic regression model for the distribution of vegetation. *Journal of Agricultural, Biological and Environmental Statistics* **8**, 205–222.
Hesselbo, B., Stinchcombe, R.B. (1995). Monte Carlo simulation and global optimization without parameters. *Physical Review Letters* **74**, 2151–2155.
Holland, J.H. (1975). *Adaptation in Natural and Artificial Systems*. University of Michigan Press, Ann Arbor.
Huffer, F.W., Wu, H.L. (1998). Markov chain Monte Carlo for autologistic regression models with application to the distribution of plant species. *Biometrics* **54**, 509–524.
Jalobeanu, A., Blanc-Feraud, L., Zerubia, J. (2002). Hyperparameter estimation for satellite image restoration using a MCMC maximum-likelihood method. *Pattern Recognition* **35**, 341–352.
Jiang, J.M., Jia, H.M., Chen, H. (2001). Maximum posterior estimation of random effects in generalized linear mixed models. *Statistica Sinica* **11**, 97–120.
Kirkpatrick, S., Gelatt, C.D., Vecchi, M.P. (1983). Optimization by simulated annealing. *Science* **220**, 671–680.
Lakshmanan, S., Derin, H. (1989). Simultaneous parameter estimation and segmentation of Gibbs random fields using simulated annealing. *IEEE Transactions on Pattern Analysis and Machine Intelligence* **8**, 954–963.

Lee, S.Y., Zhu, H.T. (2000). Statistical Analysis of nonlinear structural equation models with continuous and polytomous data. *British Journal of Mathematical and Statistical Psychology* **53**, 209–232.

Lee, S.Y., Zhu, H.T. (2002). Maximum likelihood estimation of nonlinear structural equation models. *Psychometrika* **67**, 189–210.

Li, S.Z. (2001). *Markov Random Field Modeling in Image Analysis*. Springer-Verlag, Tokyo.

Liang, F. (2004). Annealing contour Monte Carlo for structure optimization in an off-lattice protein model. *Journal of Chemical Physics* **120**, 6756–6763.

Liang, F. (2005a). A generalized Wang–Landau algorithm for Monte Carlo computation. *Journal of the American Statistical Association* **100**, 1311–1327.

Liang, F. (2005b). Evidence evaluation for Bayesian neural networks. *Neural Computation* **17**, 1385–1410.

Liang, F. (2005c) Annealing stochastic approximation Monte Carlo for neural network training. *Machine Learning* (revised).

Liang, F., Liu, C., Carroll, R.J. (2005). Stochastic approximation in Monte Carlo computation. *Journal of the American Statistical Association*, in press.

Little Jr., E.J. (1971). *Atlas of United States Trees*, vol. 15. U.S. Government Printing Office, Washington, DC.

Liu, J. (2001). *Monte Carlo Strategies in Scientific Computing*. Springer, New York.

Louis, T.A. (1982). Finding the observed information matrix when using the EM algorithm. *Journal of the Royal Statistical Society, Series B* **44**, 190–200.

Marinari, E., Parisi, G. (1992). Simulated tempering: a new Monte Carlo scheme. *Europhysics Letters* **19**, 451–458.

Marroquin, J.L., Santana, E.A., Botello, S. (2003). Hidden Markov measure field models for image segmentation. *IEEE Transactions on Pattern Analysis and Machine Intelligence* **25**, 1380–1387.

Metropolis, N., Rosenbluth, A.W., Rosenbluth, M.N., Teller, A.H., Teller, E. (1953). Equation of state calculations by fast computing machines. *Journal of Chemical Physics* **21**, 1087–1091.

McCulloch, C.E. (1997). Maximum likelihood algorithms for generalized linear mixed models. *Journal of the American Statistical Association* **92**, 162–170.

Mitchell-Jones, A.J., Amori, G., Bogdanowicz, W., Krystufek, B., Reijnders, P.J.H., Spitzenberger, F., Stubbe, M., Thissen, J.B.M., Vohralk, V., Zima, J. (1999). *The Atlas of European Mammals*. Poyser, London.

Möller, J. (1999). Markov chain Monte Carlo and spatial point processes. In: Kendall, W.S., Barndorff-Nielsen, O.E., van Lieshout, M.C. (Eds.), *Stochastic Geometry: Likelihood and Computation*. Chapman and Hall, London.

Neyman, J., Scott, E.L. (1948). Consistent estimates based on partially consistent observations. *Econometrica* **16**, 1–32.

Polyak, B.T. (1990). New stochastic approximation type procedures. *Automatica i Telemekh*, 98–107. English transl. in: *Automat. Remote Control* **51**.

Polyak, B.T., Juditski, A.B. (1992). Acceleration of stochastic approximation by averaging. *SIAM Journal of Control and Optimization* **30**, 838–855.

Qian, W., Titterington, D.M. (1991). Estimation of parameters in hidden Markov models. *Philosophical Transactions of the Royal Society of London, Ser. A* **337**, 407–428.

Robbins, H., Monro, S. (1951). A stochastic approximation method. *Annals of Mathematical Statistics* **22**, 400–407.

Robert, C.P., Casella, G. (1999). *Monte Carlo Statistical Methods*. Springer-Verlag, New York.

Rue, H. (2001). Fast sampling of Gaussian Markov random fields. *Journal of the Royal Statistical Society, Series B* **63**, 325–338.

Saquib, S.S., Bouman, C.A., Sauer, K. (1998). ML parameter estimation for Markov random fields with applications to Bayesian tomography. *IEEE Transactions on Image Processing* **7**, 1029–1044.

Vasconcelos, N., Lippman, A. (2001). Empirical Bayesian motion segmentation. *IEEE Transactions on Pattern Analysis and Machine Intelligence* **23**, 217–221.

Wang, F., Landau, D.P. (2001). Efficient, multiple-range random walk algorithm to calculate the density of states. *Physical Review Letters* **86**, 2050–2053.

Wei, G.C.G., Tanner, M.A. (1990). A Monte Carlo implementation of the EM algorithm and the Poor man's data augmentation algorithm. *Journal of the American Statistical Association* **85**, 699–704.

Winkler, G. (1995). *Image Analysis, Random Fields and Dynamic Monte Carlo Methods: A Mathematical Introduction*. Springer-Verlag, Berlin, Heidelberg.

Younes, L. (1989). Parameter estimation for imperfectly observed Gibbsian fields. *Probability Theory and Related Fields* **82**, 625–645.

Zeger, S.L., Liang, K.Y., Albert, P.S. (1988). Models for longitudinal data: a generalized estimating equation approach. *Biometrics* **44**, 1049–1060.

Zhang, H. (2002). On estimation and prediction for spatial generalized linear mixed models. *Biometrics* **56**, 129–136.

Zhang, H.P. (1993). Image restoration: flexible neighborhood systems and iterated conditional expectations. *Statistica Sinica* **3**, 117–139.

Zhu, H.T., Lee, S.Y. (2002). Analysis of generalized linear mixed models via a stochastic approximation algorithm with Markov chain Monte Carlo method. *Statistics and Computing* **12**, 175–183.

Zhu, H.T., Gu, M.G., Peterson, B.S. (2005a). Maximum likelihood from spatial random effect model via the stochastic approximation EM algorithm. *Statistics and Computing*, in press.

Zhu, H.T., He, F.L., Zhou, L.J. (2005b). Automulticategorical regression model for the distributions of vegetation types. *Biometrics* (revised).

Author Index

Roman numbers refer to pages on which the author (or his/her work) is mentioned. Italic numbers refer to reference pages. No distinction is made between first and co-author(s).

Subject Index

www.ingramcontent.com/pod-product-compliance
Lightning Source LLC
Chambersburg PA
CBHW071306180926
51613CB00001B/142